Self-Organization During Friction

Advanced Surface-Engineered Materials and Systems Design

MATERIALS ENGINEERING

1. Modern Ceramic Engineering: Properties, Processing, and Use in Design. Second Edition, Revised and Expanded, *David W. Richerson*
2. Introduction to Engineering Materials: Behavior, Properties, and Selection, *G. T. Murray*
3. Rapidly Solidified Alloys: Processes • Structures • Applications, *edited by Howard H. Liebermann*
4. Fiber and Whisker Reinforced Ceramics for Structural Applications, *David Belitskus*
5. Thermal Analysis of Ceramics, *Robert F. Speyer*
6. Friction and Wear of Ceramics, *edited by Said Jahanmir*
7. Mechanical Properties of Metallic Composites, *edited by Shojiro Ochiai*
8. Chemical Processing of Ceramics, *edited by Burtrand I. Lee and Edward J. A. Pope*
9. Handbook of Advanced Materials Testing, *edited by Nicholas P. Cheremisinoff and Paul N. Cheremisinoff*
10. Ceramic Processing and Sintering, *M. N. Rahaman*
11. Composites Engineering Handbook, *edited by P. K. Mallick*
12. Porosity of Ceramics, *Roy W. Rice*
13. Intermetallic and Ceramic Coatings, *edited by Narendra B. Dahotre and T. S. Sudarshan*
14. Adhesion Promotion Techniques: Technological Applications, *edited by K. L. Mittal and A. Pizzi*
15. Impurities in Engineering Materials: Impact, Reliability, and Control, *edited by Clyde L. Briant*
16. Ferroelectric Devices, *Kenji Uchino*
17. Mechanical Properties of Ceramics and Composites: Grain and Particle Effects, *Roy W. Rice*
18. Solid Lubrication Fundamentals and Applications, *Kazuhisa Miyoshi*
19. Modeling for Casting and Solidification Processing, *edited by Kuang-O (Oscar) Yu*
20. Ceramic Fabrication Technology, *Roy W. Rice*
21. Coatings of Polymers and Plastics, *edited by Rose A. Ryntz and Philip V. Yaneff*
22. MicroMechatronics, *edited by Kenji Uchino and Jayne Giniewicz*
23. Ceramic Processing and Sintering, Second Edition, *edited by M. N. Rahaman*
24. Handbook of Metallurgical Process Design, *edited by George Totten*

25. Ceramic Materials for Electronics, Third Edition, *Relva Buchanan*
26. Physical Metallurgy, *William F. Hosford*
27. Carbon Fibers and Their Composites, *Peter Morgan*
28. Chemical Processing of Ceramics: Second Edition, *Burtrand Lee and Sridhar Komarneni*
29. Modern Ceramic Engineering, Third Edition, *David Richerson*
30. Engineering Design with Polymers and Composites, *James C. Gerdeen, Harold W. Lord and Ronald A. L. Rorrer*
31. Self-Organization During Friction: Advanced Surface-Engineered Materials and Systems Design, *edited by German S. Fox-Rabinovich and George E. Totten*

Self-Organization During Friction

Advanced Surface-Engineered Materials and Systems Design

edited by

German S. Fox-Rabinovich

McMaster University

Hamilton, Ontario, Canada

George E. Totten

Portland State University

Portland, Oregon, USA

Taylor & Francis

Taylor & Francis Group

Boca Raton London New York

CRC is an imprint of the Taylor & Francis Group, an informa business

CRC Press
Taylor & Francis Group
6000 Broken Sound Parkway NW, Suite 300
Boca Raton, FL 33487-2742

Printed in the United States of America on acid-free paper
10 9 8 7 6 5 4 3 2 1

International Standard Book Number-10: 1-57444-719-X (Hardcover)
International Standard Book Number-13: 978-1-57444-719-4 (Hardcover)

Library of Congress Cataloging-in-Publication Data

Self-organization during friction : Advanced surface-engineered materials and systems design / edited by George E. Totten, German Fox-Rabinovich.
p. cm. -- (Materials engineering ; 31)
Includes bibliographical references and index.
ISBN 1-57444-719-X (alk. paper)
1. Surfaces (Technology) 2. Tribology. 3. Nanotechnology. I. Totten, George E. II. Fox-Rabinovich, German. III. Series: Materials engineering (Taylor & Francis) ; 31.

TA418.7.S45 2006
621.8'9--dc22 2006044557

Visit the Taylor & Francis Web site at
http://www.taylorandfrancis.com

and the CRC Press Web site at
http://www.crcpress.com

Preface

Friction is an extremely complex phenomenon that can be studied in depth using ideas of modern physics that deal with the problems of complexity. One of the most generic ideas considering the complexity of natural processes is the concept of self-organization.

This book is devoted to the self-organization phenomenon, physicochemical aspects of friction, and the methods of friction control using advanced materials and surface-engineering techniques.

The major topics of this book are: nonequilibrium thermodynamics, self-organization phenomena during friction and wear, tribological compatibility, and methods of friction control for heavily loaded tribosystems such as cutting and stamping tools. The key concept focuses on the issue of tribological compatibility, which is the capacity of two surfaces adapting to each other during friction, providing wear stability without surface damage for the longest period of time. This is both a generic physical and an engineering approach to the development of new wear-resistant materials and coatings.

In this context, friction control implies the existence of a stable tribosystem, which resists any instability leading to intensive wear and surface damage. From the point of view of a self-organization process, both the natural (friction-based) and synthetic processes of materials' design and engineering are outlined in this book. Therefore, we could control the synthetic processes of material engineering to encourage the evolution of natural processes occurring during friction that lead to minimal wear rate. This is the typical friction control with a positive feedback loop that results in significant tool life improvement.

The main objectives of this interdisciplinary project are: (1) to combine the fundamentals of thermodynamics and methods of material characterization including nanotribological methods, studies of tribological behavior of a wide range of materials, and (2) to present to the scientific and engineering community a new approach to the development of an emerging generation of surface-engineered self-adaptive materials. Much attention in this book is paid to the adaptive tooling tribosystems and surface-engineered materials. The main feature of these adaptive materials is that they exhibit protective properties in their structure and function that are similar to natural or biological systems. New generations of surface-engineered nanostructured materials could be considered as specific "nanomachines" that transform the tribosystems working under extreme external impact with excessive wear to those with milder friction conditions of critically decreased wear rate. It will be shown that the abnormal capacity of these materials to resist external impact is associated with their nanoscale structure, synergistical alloying, and nonequilibrium state of the surface-engineered layers. It confirms that enormous energy stored in the nanoworld has to be released in order to fully perform its protective functions.

The metallurgical design of the novel tooling materials is based on surface-engineering techniques, particularly the plasma vapor deposition (PVD) technique, as well as powder metallurgy methods. Other methods that improve the tool–workpiece tribological compatibility (such as solid lubricant application) as well as the adaptive design of cutting tools are considered as well.

This book is primarily directed at researchers in the field of material science and manufacturing engineering, demonstrating new approaches to the development of the next generation of self-adaptive materials with critically enhanced workability and wear resistance. This book could also be used as a research monograph for academic libraries.

German S. Fox-Rabinovich
George E. Totten

Acknowledgments

The preparation of this text was an enormous task, and we are indebted to various international experts for their contributions. This book is a result of long-term work of an international group of scientists from Canada, Russia, Japan, and the U.K., spanning many years of research activity. During that time, they independently worked in different fields of tribology, material science, and manufacturing. Ever since this book project started, they were working as a team. Numerous detailed discussions took place during the writing and editing of this book, which were very exciting and fruitful.

The editors would like to thank professors N. A. Bushe, L. S. Shuster, I. S. Gershman, and A.I. Kovalev for very fruitful discussions and extremely valuable comments. They also would like to thank G. Dosbaeva (McMaster University, Ontario, Canada) for metallography and the manuscript preparation, and M. Dosbaev for participation in the manuscript preparation. Many thanks to S. Koprich (McMaster University) for excellent technical support in surface-engineered materials (SEM) studies. Special thanks to the staff (especially Theresa Delforn, Allison Taub, and Gail Renard) of CRC Press, a Taylor & Francis company, for their patience and invaluable assistance.

The research was partially funded by Natural Science and Engineering Research Council of Canada and Materials and Manufacturing, Ontario.

SEM images of red abalone shell presented on the cover page are the courtesy of Professor Mehmet Sarikaya, University of Washington, Seattle.

German S. Fox-Rabinovich
George E. Totten

Acknowledgments

Editors

German S. Fox-Rabinovich, Ph.D., D.Sc., is a research scientist at McMaster Manufacturing Research Institute in Hamilton, Ontario, Canada. In 2003 he was appointed adjunct professor in the Department of Mechanical Engineering, McMaster University. Fox-Rabinovich is the author and coauthor of over 120 scientific papers, patents, and books on tribology, surface engineering, and tooling materials. He served as a committee member of International Conference of Metallurgical Coatings and Thin Films from 2003 to 2005. He received his M.S. from Technological University of Machine-Tool Engineering, Moscow, Russia, and Ph.D. and D.Sc. from Scientific Research Institute of Rail Transport, Moscow.

George E. Totten, Ph.D., FASM, is president of G. E. Totten & Associates, LLC, in Seattle, Washington, and a visiting professor of materials science at Portland State University in Oregon. Dr. Totten is coeditor of a number of books including *Steel Heat Treatment Handbook, Handbook of Aluminum, Handbook of Hydraulic Fluid Technology, Mechanical Tribology,* and *Surface Modification and Mechanisms* (all CRC Press titles), as well as the author or coauthor of over 450 technical papers, patents, and books on lubrication, hydraulics, and thermal processing. Totten is a fellow of ASM International, SAE International, and the International Federation for Heat Treatment and Surface Engineering, and a member of other professional organizations including the American Chemical Society, American Society of Mechanical Engineering (ASME), and STLE. Totten formerly served as president of IFHTSE. He received his B.S. and M.S. degrees from Fairleigh Dickinson University, Teaneck, New Jersey, and his Ph.D. degree from New York University.

Contributors

Ben D. Beake, Ph.D.
Micro Materials Ltd.
Wrexham, UK

Michael M. Bruhis, Ph.D.
McMaster University
Hamilton, Ontario, Canada

Nicolay A. Bushe, Ph.D., D.Sc.
Scientific Research Institute of Rail Transport
Moscow, Russia

Jose L. Endrino, Ph.D.
Balzers AG
Iramali, Liechtenstein

German S. Fox-Rabinovich, Ph.D., D.Sc.
McMaster University
Hamilton, Ontario, Canada

Iosif S. Gershman, Ph.D.
Scientific Research Institute of Rail Transport
Moscow, Russia

Anatoliy I. Kovalev, Ph.D.
Surface Phenomena Research Group LLC
Moscow, Russia

Lev S. Shuster, Ph.D., D.Sc.
Ufa Aviation Institute
Ufa, Russia

Stephen C. Veldhuis, Ph.D.
McMaster University
Hamilton, Ontario, Canada

Dmitry L. Wainstein, Ph.D.
Surface Phenomena Research Group LLC
Moscow, Russia

Kenji Yamamoto, Ph.D.
Kobe Steel
Kobe, Japan

Contents

SECTION I Self-Organization during Friction and Principles of Friction Control

Chapter 1 Principles of Friction Control for Surface-Engineered Materials 3

German S. Fox-Rabinovich

Chapter 2 Elements of Thermodynamics and Self-Organization during Friction 13

Iosif S. Gershman and Nicolay A. Bushe

SECTION II Tribological Compatibility and Nanotribological Characteristics to Evaluate Surface Properties during Friction

Chapter 3 Compatibility of Tribosystems 59

Nicolay A. Bushe and Iosif S. Gershman

Chapter 4 Surface Analysis Techniques for Investigations of Modified Surfaces, Nanocomposites, Chemical, and Structure Transformations 81

Anatoliy I. Kovalev and Dmitry L. Wainstein

Chapter 5 Physical and Mechanical Properties to Characterize Tribological Compatibility of Heavily Loaded Tribosystems (HLTS) 121

German S. Fox-Rabinovich, Lev S. Shuster, Ben D. Beake, and Stephen C. Veldhuis

SECTION III Self-Organization and Structural Adaptation of Heavily Loaded Tribosystems

Chapter 6 Self-Organization and Structural Adaptation during Cutting and Stamping Operations 151

German S. Fox-Rabinovich and Anatoliy I. Kovalev

Chapter 7 Tooling Materials and Some Features of Their Self-Organization: Adaptive Tooling Materials 167

German S. Fox-Rabinovich, Anatoliy I. Kovalev, Ben D. Beake, and Michael M. Bruhis

Chapter 8 Formation of Secondary Structures and the Self-Organization Process of Tribosystems during Friction with the Collection of Electric Current 197

Iosif S. Gershman

SECTION IV Adaptive Surface-Engineered Materials and Systems

Chapter 9 Surface-Engineered Tool Materials for High-Performance Machining 231

German S. Fox-Rabinovich, Anatoliy I. Kovalev, Jose L. Endrino, Stephen C. Veldhuis, Lev S. Shuster, and Iosif S. Gershman

Chapter 10 Synergistic Alloying of Self-Adaptive Wear-Resistant Coatings 297

German S. Fox-Rabinovich, Kenji Yamamoto, and Anatoliy I. Kovalev

Chapter 11 Development of the Ternary and Higher-Ordered Protective or Wear-Resistant Materials and Coatings for High-Temperature Applications and Thermodynamics-Based Principles of their Synergistic Alloying 335

German S. Fox-Rabinovich, Iosif S. Gershman, Anatoliy I. Kovalev, and Kenji Yamamoto

Chapter 12 Coolants and Lubricants to Enhance Tribological Compatibility of the "Tool–Workpiece" Tribosystem 381

Stephen C. Veldhuis, German S. Fox-Rabinovich, and Lev S. Shuster

Chapter 13 Geometrical Adaptation of Cutting Tools 411

Stephen C. Veldhuis, Michael M. Bruhis, Lev S. Shuster, and German S. Fox-Rabinovich

Index 455

Section I

Self-Organization during Friction and Principles of Friction Control

1 Principles of Friction Control for Surface-Engineered Materials

German S. Fox-Rabinovich

CONTENTS

1.1 Introduction 3
1.2 Tribological Compatibility and the Development of Advanced Materials for HLTS 5
1.3 Tribo-Films (Secondary Structures) 7
1.4 Principles of Friction Control 8
References 11

1.1 INTRODUCTION

Friction is a complex phenomenon associated with a variety of different mechanical, physical, and chemical processes [1–3]. Traditionally, this area of applied science and engineering was considered primarily a mechanical process. However, recently, friction has been viewed as a more complex phenomenon.

One of the achievements of modern physics is the development of the concept of self-organization, which deals with complex processes of nature. This concept is based on the ideas of irreversible thermodynamics [4]. Major progress in this area is associated with Nobel Prize winner I. Prigogine [4]. Irreversible thermodynamics is relevant to processes associated with the stability and degradation of structures that are formed under nonequilibrium conditions [5]. It deals with the formation of spatial, temporal, or functional structures [4,7]. The irreversible processes of such structure formations are associated with nonequilibrium phase transformations. These transformations are connected with specific bifurcation or instability points where the macroscopic behavior of the system changes qualitatively and may either leap into chaos or into greater complexity and stability [6]. In the latter case, as soon as the system passes these specific points, its properties change spontaneously because of self-organization and formation of dissipative structures.

The driving force of the self-organization process is the open system aimed to decrease entropy production during nonstationary processes. Spontaneous formation of dissipative structures is a result of symmetry perturbations that can be realized only in open systems, which exchange energy, matter, and entropy with their environments [7]. This phenomenon is a focus of attention for many researchers in different fields of science.

Heavily loaded tribosystems (HLTS) working under severe conditions associated with high temperature and stresses and exhibiting very high wear rates are considered in this book. An excellent example of HLTS are cutting and stamping tools. Effective protection of the friction surface is critically needed for severe conditions associated with modern high-performance cutting or stamping operations.

As discussed in the preceding text, interaction between frictional bodies was traditionally thought to lie entirely in the domain of mechanical response of the system to these conditions, and relatively little attention was given to the role of physical and chemical interactions between frictional bodies. Recent research has challenged that viewpoint, and it is now realized that extensive physical and chemical interactions can occur, both between the frictional bodies and the surrounding atmosphere. These considerations, together with the approach of irreversible thermodynamics and self-organization, have produced critical progress in the field of tribology.

The concepts of self-organization and irreversible thermodynamics may be successfully used for the development of advanced tribosystems and materials. These principles provide a powerful methodological tool to enhance beneficial nonequilibrium processes. These tribo-processes result in the ability of a friction surface to exhibit self-protection and self-healing properties. Utilization of these principles allows the possibility of the development of a new generation of tribosystems and materials that exhibit adaptive or smart behavior. An adaptive or smart tribosystem responds to external mechanical, thermal, and chemical forces with a positive feedback loop that leads to an improvement in the wear characteristics of a tribo-couple [1–2].

This progress is chiefly associated with the intensive research and associated activities of renowned tribologists such as B. E. Klamecki, B. I. Kostetsky, N. A. Bushe, and L. I. Bershadsky. Each has made a major contribution to the field of modern tribology. B. E. Klamecki [8] has studied wear process based on the irreversible thermodynamic approach. B. I. Kostetsky first conceived the definition of the so-called secondary structures or tribo-films and outlined some fundamentals of friction control [9–10]. L. I. Bershadsky developed the ideas proposed by B. I. Kostetsky further and introduced the concept of "dissipative heterogeneity" [11]. N. A. Bushe introduced one of the most fundamental ideas of modern applied tribology, the concept of tribological compatibility [12]. All of these ideas and concepts and their practical applications are thoroughly discussed in this book.

This book describes how to utilize these ideas for specific engineering applications. The chief theme of this book is the application of concepts of irreversible thermodynamics and self-organization to tribology and the role played by physicochemical interactions in modifying and controlling friction and wear. This topic is becoming increasingly important with the current drive for increased productivity, often under conditions in which a lubricant cannot be used (such as dry machining conditions), that characterizes modern trends for different tribological applications.

In many publications, there is much speculation related to self-organization without providing a solid scientific understanding of this phenomenon. A major feature of this book is to bring together the fundamentals of irreversible thermodynamics and the thermodynamically driven generic ideas of modern tribology while illustrating practical examples of friction control based on the presented concepts. World-renowned experts in the field have written chapters of this book, which provide the reader with a solid scientific basis of the fundamentals of adaptive tribosystems and materials.

The book is divided into four major parts. In the first part (Chapter 1 to Chapter 2), the characteristic features of friction and the role of the self-organization phenomena in helping to control wear processes are described. The second part introduces one fundamental concept of applied tribology (Chapter 3 is written by the originator of this concept — N. A. Bushe), i.e., tribological compatibility, and describes tribological, mechanical, and nanotribological characteristics to evaluate the tribological compatibility (Chapter 4 to Chapter 5).

Understanding of complex physicochemical interactions that lead to the formation of surface nanoscale tribo-films and tribo-layers has been provided by techniques such as Auger spectroscopy and electron energy loss spectroscopy. These techniques are used to study very superficial (within nano range) surface layers that are formed during friction and under certain conditions that control the durability of the tribosystem. These studies, when coupled with more conventional wear and friction experiments as well as mechanical property measurements, clearly demonstrate the positive role of surface tribo-films formed during friction in reducing total wear.

The third part of this book develops ideas of self-organization for specific applications associated with HLTS having high stresses and elevated temperatures that are encountered during the friction process (Chapter 6 to Chapter 8). Surface nanolayers play an important and sometimes critical role in wear behavior of HLTS. Although not immediately obvious, detailed experimental data have confirmed this hypothesis. This has been observed, for example, in common tool materials such as high-speed tool steels (HSS), cemented carbides, and cermets. HLTS such as current collectors are also considered in Chapter 8.

In the fourth part, a number of recent trends to enhance the performance of cutting tools by surface engineering is discussed. These include the use of monolithic or multilayered coatings, substrate modification, surface-engineered tools, and multilayered self-lubricating coatings. Throughout the discussion, the role of tribo-films is highlighted, and the concept of a "smart" coating that can respond to the cutting environment (with a positive feedback) is proposed. Finally, it is inferred that future developments of improved cutting tools will depend on a better understanding of the nature of the tribo-films formed during friction. Examples are given, illustrating how these improvements might be exploited for the improved performance of machining operations. Practical applications of the ideas of self-organization are presented for the severe conditions of high-speed or dry machining. Nanoscale crystalline, composite, or laminar coatings and a new generation of self-adaptive coatings working under extreme conditions of high-speed and dry machining are described in detail. Geometric adaptation of cutting tools is also considered in Chapter 13.

1.2 TRIBOLOGICAL COMPATIBILITY AND THE DEVELOPMENT OF ADVANCED MATERIALS FOR HLTS

The increasing demands of modern engineering have spawned the development of new advanced materials for use such as HLTS, which possesses a high level of wear resistance. The development of advanced materials can be considered to be a typical problem of engineering optimization. In this process, an integrated engineering–physical approach is used to develop novel wear-resistant materials. The key concept of this is associated with the tribological compatibility of two surfaces interacting during friction.

Tribological compatibility is the ability of a tribosystem to provide optimum friction conditions within the given range of operating conditions to meet a target criteria (see Chapter 1 to Chapter 3) [12]. Tribological compatibility is related to the capacity of two surfaces to adapt to each other during friction, providing wear stability without surface damage to the two components of the specific tribosystem for the longest (or given) period of time. The goal is to achieve a stable tool service and a predictable rate of wear with a given set of operating parameters. In this interpretation, as was outlined in the preceding text, compatibility implies an integrated optimization, both from an engineering (minimal wear rate) and physical (self-organizing) point of view.

As mentioned earlier, modern tribology, an interdisciplinary science based on mechanics, physics, chemistry, materials science, metallurgy, etc., is a very complex subject. Models that generalize this area of science and which would be acceptable for engineering applications are critically needed. As friction is the process of transformation and dissipation of mechanical energy into other kinds of energy, an energy-based approach is the most effective.

Friction is a dissipative process, and, according to the first law of thermodynamics, a portion of the mechanical energy (the work done by the external loading system, A_{fr}) is expended by the accumulation of energy into surfaces (ΔE) and the generation of heat [13]:

$$A_{fr} = Q + \Delta E$$

The energy dissipation during friction leads to the accumulation of high plastic strains in the surface layers and the formation of an ultrafine-grained structure [10]. This raises the free energy of the contact zone by creating a high density of structural imperfections in the surface layers (the activation phenomenon). The activation process transforms the surface layers into an unstable or metastable state. From the point of view of thermodynamics, transition to an equilibrium state is natural. Therefore, the activation process may be followed by passivation, i.e., the reduction in free energy of the material as a result of interaction with the environment and the generation of protective tribo-films or secondary structures (formed by friction).

A tribosystem can be considered to be an open thermodynamic system. For these systems, the second law of thermodynamics is operative. According to the principles developed by I. Prigogine, the second law does not eliminate the possibility of highly organized dissipative structures being formed in an open tribosystem. In these systems, when the excursion from equilibrium exceeds some critical value (typical for friction), the process of material ordering can proceed by the spontaneous formation of self-organizing dissipative structures [4, 14].

During friction and wear, adaptation of the materials of the tribo-couple occurs, and in many cases, leads to drastic structural changes within the surface layers. Characteristics of surface and undersurface layers (such as geometrical parameters, microstructure, and physicochemical and mechanical properties) change during the process of adaptation as a response to the external impact. The adaptation is completed in the initial stage of the life of a tribosystem, i.e., during the running-in stage. Despite evolving in a step-by-step fashion, and becoming increasingly complicated during this stage, tribo-films (secondary structures) eventually stabilize for a given tribo-pair and conditions of friction. When the characteristics of the surface layers become optimal, the running-in phase is completed and the parameters of friction (i.e., the coefficient of friction and wear rate) stabilize at a lower level [7]. The process of wear transforms to a post-running-in stationary stage.

A process of screening occurs as a result of self-organization [15]. The phenomenon of screening reflects the coordination of the rates of destructive and recovery processes in the friction zone. This is typical for a quasi-stationary state of wear, in which the processes of activation and passivation of the surface layers are in dynamic balance. External reactions usually result in the destruction of the screening phase, but these reactions and the associated process of matter exchange with the environment may provide its regeneration during friction. With the correct correlation of these processes, the state of the system is quasi-stable, corresponding to a minimum rate of entropy production. Then, according to the principle of screening [9], any kind of interaction of the frictional surfaces or destruction of the base metal should be eliminated. The contact area will be controlled by interactions of the thin layers of tribo-films, and stable friction (wear) conditions prevail as long as the dissipative structures associated with friction are self-adjustable.

During the steady stage of wear, tribological compatibility is controlled by the stability of the tribo-films formed, wear resistance of the contacting materials, their fatigue life, as well as lubricity of the surface layers (see Chapter 3). That is why various surface-engineering methodologies are considered below, which enhance the tribological compatibility of specific tribosystems (see Chapter 6 to Chapter 13). The tribological compatibility, especially for HLTS, critically depends on the intensity of surface damage during the initial (running-in) stage. Special coatings and lubricating films to prevent severe surface damage during initial stages of wear are considered in detail in Chapter 9 to Chapter 11.

It is emphasized again that tribological compatibility is a generic characteristic, and this tribosystem property is controlled by a few major processes that occur on the surface during friction, such as plastic deformation, structure transformations, and physical–chemical interaction of frictional bodies, and interaction with the environment. Different features of tribological compatibility are considered in different parts of the book. Structural aspects of tribological compatibility are described in detail in Chapter 3. To provide a bigger picture, the physical–chemical and mechanical aspect of this generic characteristic is explained in Chapter 5 and Chapter 9 by using the examples of surface-engineered cutting tools. In this case, the friction and wear rate is determined by:

- Formation of tribo-films with exceptional physical–chemical properties due to controlled changes of chemical potentials on the surface
- Post-running-in stability of the tribo-films formed that is controlled by the ability of the engineered surface to resist severe external impact

1.3 TRIBO-FILMS (SECONDARY STRUCTURES)

The self-organization (SO) phenomenon is characterized by the formation of thin (from several nanometers to a micron thick) films of the tribofilms or secondary structures (SS-s) at the friction surface. These are generated from the base material by structural modification or by interaction with the environment. It has been estimated that 90 to 98% of the work of friction can be accumulated in secondary structures, with no more than 2 to 10% in the primary structures. Thus, secondary structures represent an energy sink for the preferential dissipation of the work of friction [13, 14,16–17].

The synergistic processes of adaptation to extreme deformation and thermal,and diffusive conditions associated with friction can be concentrated in the thin layer of SS-s. Self-organizing of the tribosystem is often accompanied by a kinetic phase transition. In this case, all of the interactions are localized in a thin surface layer, the depth of which can be lower by an order of magnitude than that typically associated with damage phenomena. The rate of diffusion and chemical reactions may also increase substantially, whereas the surface layers may become ductile. In addition, the solubility of many elements may be increased and nonstoichiometric compounds may form.

There are two kinds of tribo-films (secondary structures): (1) superductile and lubricious and (2) tribo-ceramics with thermal barrier properties and possibly increased hardness and strength [17]. Tribo-films (secondary structures) of the first type (SS-I) are observed after structural activation, that is marked by an increase in the density of atomic defects at the surface. Tribo-films are supersaturated solid solutions formed by reaction with elements from the environment (most often, oxygen). Reaction between oxygen and the substrate during the self-organization process is very different from the classical case encountered in typical oxidation experiments. SS-Is are similar to Beilby layers, having a fragmented and textured structure and aligned in the shear direction [17]. In these secondary structures, the material may be superplastic (due to a nanoscale-grained or amorphous-like structure) with an elongation up to 2000%. The amorphous-like structure of SS-I may also lead to a decrease in the thermal conductivity of the surface, an important consideration in controlling friction of cutting tools [18]. Secondary structures of this type promote energy dissipation during friction. The lubricious action of these tribo-films leads to prevention of subsurface layer damage and results in a decrease of entropy production within the zone of friction.

Tribo-films (secondary structures) of the second type (SS-II or tribo-ceramics containing a higher content [percentage] of elements such as oxygen) are primarily formed by thermal activation processes. The SS-IIs are usually nonstoichiometric compounds and, as a rule, contain a deficit of the reactant. However, under heavy-loading conditions (in particular, during high-speed cutting), nonstoichiometric compounds with an excess of the reactant have been observed [8]. Secondary structures of this type (SS-II) exhibit high thermodynamic stability, thermal barrier properties, and probably high hardness. It is thought that one of the benefits they bestow is the accommodation of stress associated with friction by elastic rather than plastic deformation. The adaptation of the tribosystem, in this case, relies upon a low-intensity interaction with a frictional body in contact and the high hardness of the thin surface film formed during cutting. This results in low entropy production during friction and, obviously, leads to a wear rate decrease. On the other hand, destruction of these hard films should be prevented by proper engineering of the substrate material that should ensure effective support of the tribo-films during friction. This issue is described in further detail in Chapter 6 and Chapter 9.

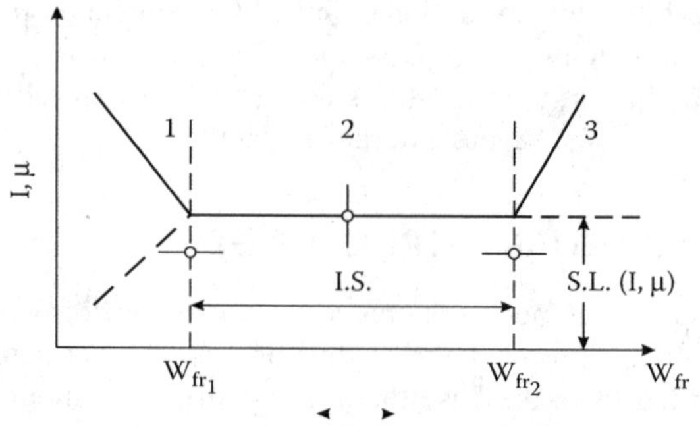

FIGURE 1.1 Diagram of wear and friction process: (1) region of nonsteady process (running-in stage); (2) region of quasi-equilibrium (steady-state) process (normal wear stage); (3) surface damage region (catastrophic or avalanche-like) wear stage. (From Ivanova, V.S., Bushe, N.A., Gershman, I.S. Structure adaptation at friction as a process of self-organization. *Journal of Friction and Wear* 1997, 18, *1*, 74–79. With permission.)

1.4 PRINCIPLES OF FRICTION CONTROL

The entire diversity of processes that occur during friction can be divided into two groups: quasi-equilibrium, steady-state processes (encountered during post-running-in stable friction and wear) and nonequilibrium, unsteady state, associated with surface damage processes as shown in Figure 1.1. Surface damage is usually observed in the initial (running-in) and final (catastrophic or avalanche-like) stages of wear. During the period of service under stable friction and wear conditions, no macroscopic damage of the contact surfaces is observed.

The results of the self-organization process may be characterized by two major criteria: interval of self-organization (IS) and level of self-organization (SL; Figure 1.1) [10]. The primary goal of friction control is to increase the IS (i.e., widen stable wear stage) and SL, i.e., decrease the wear intensity (I) and coefficient of friction (μ). This could be done by means of the tribosystem design, technological improvements, and the optimization of service conditions.

Modern technologies of surface engineering that are considered in this book provide the initial surface with a high density of lattice imperfections (see Chapter 9). These processes yield a surface with a highly nonequilibrium state that dramatically enhances beneficial passivation processes (which is associated with mass transfer) and formation of protective tribo-films (see Chapter 2 as well as "trigger" effect, considered in Chapter 9 and Chapter 11). The only limitation is the ability of this highly activated surface to accumulate additional structure imperfections that are generated during friction without deep surface damage. An initially activated, highly nonequilibrium surface, such as nanocrystalline coatings (Chapter 9), leads to explosion-like mass transfer followed by protective tribo-film formation (surface passivation process). Once self-organization and surface passivation processes are completed, the friction characteristics change critically and the wear process transforms to the after running-in, stable stage. Within this stage, friction is controlled by different mechanisms that are associated with such phenomenon as fatigue and other related phenomena (Chapter 3, Chapter 5, and Chapter 9).

Friction control in this context implies the existence of a stable tribosystem, which resists any instability leading to deep surface damage [10]. The transition from a thermodynamically nonequilibrium condition to a more stable, quasi-equilibrium condition is connected to the accelerated formation of a beneficial surface structure formed as a result of self-organization. Any tribosystem combines some features of an artificial system (engineering system, mechanism) and a natural phenomenon (the friction by itself). From the point of view of self-organizing, both natural and synthetic processes can be considered during friction. Therefore, it is necessary to control (or modify) the synthetic processes to encourage the evolution of those natural processes that lead to minimal wear rate. The problem of tribological compatibility includes developments that ensure the stabilization of the friction and wear parameters, in particular, by the development of advanced material. Chapters 9, 10, and 11 present some examples of this generic approach.

To use this generic approach for cutting conditions, a number of complex phenomena are encountered in a "tool–workpiece" tribosystem, which drastically changes the wear performance of cutting tools. The adaptation process during cutting occurs at three hierarchical levels:

1. Macrolevel — This adaptation process is associated with the chip–workpiece interface and cutting wedge geometry adaptation (see Chapter 13).
2. Microlevel — The level of asperities formed as a result of initial surface finish and during friction see Chapter 6 and Chapter 7.
3. Nanolevel — The superficial level of the workpiece–tool interface where the tribo-films are formed (see Chapter 3, Chapter 9 to Chapter 11).

At the macrolevel, the adaptation process is associated with the formation of dissipative macrostructures such as a built-up or flow zone (under different cutting conditions) and the geometrical adaptation of the cutting edge during friction. The built-up edge is considered as a dissipation structure [21] that forms under extremely unstable conditions of the attrition wear. This wear mode is associated with deep surface damage and prevents the gradual adaptation process to be developed on micro- and nanoscale levels. As the conditions at the interface are altered, for instance, by shifting the cutting conditions to a more stable area of flow-zone formation, changes in cutting wedge design, (see Chapter 13) or owing to tool material surface engineering (see Chapter 9 to Chapter 11), the adaptation is exhibited in more gradual surface transformations. These transformations occur at the microlevel and, most importantly, at the nanolevel. In the latter case, generation of protective tribo-films on the surface of the cutting tool could be enhanced, and tool life is dramatically improved [21]. Thus, the best way of friction control under severe conditions of cutting is to shift the process from macro- to micro- and, even more efficiently, to nanolevel of adaptation, when the nanoscale tribo-films start to control the wear intensity. This is accomplished by engineering solutions that significantly narrow the depth of the actual level of tool–workpiece interactions and by ensuring efficient support of the tribo-films formed during friction to improve their stability. Eventually tribological compatibility during cutting leads to two major practical improvements:

1. Tool life enhancement
2. Improved workpiece quality (e.g., a better surface finish or improved dimensional accuracy)

In subsequent chapters, specific examples will be provided.

There is a traditional methodology of advanced material development. This methodology includes some typical steps. Material designers usually start from the selection of criteria to characterize service properties of the materials. In the next step, a correlation between service properties of the material has to be developed, and its composition, structure, and engineering methods are outlined. Finally, optimization of metallurgical design of the novel material and methods of its fabrication using the selected criteria are developed.

Although there is no need to fundamentally change this classical approach, it is important to integrate concepts of modern tribology in the development of a phenomena-based algorithm for the development of advanced materials and systems. The solution can be implemented stepwise as following:

1. Assessment of the friction surface, including the analysis of self-organizing phenomena and the features of the structural adaptation, is performed. This includes investigations of structure transformations and properties of the initial materials during friction, because these surface layers possess fundamentally different characteristics that largely control wear resistance.

2. Determining the criteria to adequately characterize the advanced materials.
3. Development of an optimal engineering solution, such as the development of an adaptive material or tribosystem, ensuring the compatibility of the surfaces under severe friction.
4. Testing of the system (or material) to verify its tribological compatibility.

This phenomena-driven approach is often cited but examples of its implementation are relatively limited. In this book, examples of this concept realization (Chapter 3, Chapter 8, Chapter 9 to Chapter 11) will illustrate the beneficial application of this approach for specific applications.

The major topic of this book is associated with surface-engineered materials because:

- State-of-the-art surface-engineering technology provides exceptional possibilities for novel material development.
- New technologies of surface engineering such as plasma vapor deposition (PVD) permit the synthesis of materials under strongly nonequilibrium conditions. This results in nanoscale crystalline, composite, or laminated structures of the surface coatings providing outstanding properties, which may not be achievable using traditional methods of material fabrication.
- These novel surface-engineered materials may be designed for specific severe applications. Modern extreme applications such as high-performance machining also require strongly nonequilibrium high-temperature or high-stress conditions in which excessive wear is often encountered.

To meet these requirements, a new generation of materials is needed that may resist this external impact, such as advanced surface-engineered materials with a nanoscale structure. In Chapter 9, Chapter 10 and Chapter 11, advanced surface-engineered materials will be discussed; these enhance the resistance of the friction surface against severe external attack where traditional materials cannot be used.

Another novel direction in material science is described in this book. The final point of friction control, in general, is to ensure conditions in which elements of the tribosystem can function synergistically. This may be achieved by the development of synergistically alloyed surface-engineered materials. Details of synergistic alloying are considered in Chapter 9 to Chapter 11 with examples of nanostructured, oxidation-resistant, and wear-resistant coatings caused by self-organization that occurs for both phenomena. Thermodynamics-based principles of synergistic alloying of wear- or oxidation-resistant coatings have been discovered and are described in detail in Chapter 11.

Adaptive and smart surface-engineered materials that are overviewed in this book are an emerging generation of materials that are similar to bio-like systems. A smart material or system responds and adapts to changes in conditions or the environment by integrating functions of action and control [1–2]. They exhibit features of self-protective and self-healing behavior. To provide the development of adaptive materials and systems with a sound fundamental basis requires a clear understanding of the self-organization phenomena. Development of adaptive materials is a part of nature-mimetics, because of the ability of these materials to provide self-protection and self-healing that is typical of biomaterials. Nature mimetics is the study of nature-evolved materials, structure, and intelligence [19–20]. A key feature of the adaptive materials is their nanoscale structure. Nature also builds the most advanced bio-objects in the nanoscale world. An example of novel surface-engineered materials, described in this book, are self-adaptive nanolaminated coatings (Chapter 10 to Chapter 11). It is known that a major feature of biosystems is their ability to self-protect, self-heal, and regenerate themselves. But during self-organization, processes such as the formation of tribo-films, their destruction, and regeneration are also observed. Therefore, the surface-engineered layers that could enhance this process, for instance nanolaminates, could be considered as nature-mimicking structures. Nanolaminates are adjusted to enhance these processes of tribo-films' regeneration. They include the ordered consequence of layers that (similar to biogenerations) supply the surface with a controlled flow of matter and energy, and, if destroyed or transformed, they enhance the formation of the next generation of protective films.

It is worth noting that a design of the mollusk shell presented on the front page of the book is based on a similar concept. The architecture consists of alternating tablets of organic and nonorganic layers, which impart to the mollusk shell exceptional strength without the brittleness associated with pure inorganic phases. Mimicking this fundamental ability of nature to create the nanoscale designed materials is a challenge for the creation of new-generation materials such as nanolaminates with complex properties that are unachievable with traditional materials with larger scaled structures and may be further exploited in the next revolution in manufacturing.

REFERENCES

1. Hardwicke, C.U. Recent developments in applying smart structural materials. *JOM* 2003, *12*, 15–16.
2. Wax, S.C., Fisher, G.M., Sands, R.R. The past, present and future of DARPA's investment strategy in smart materials. *JOM* 2003, *12*, 17–23.
3. Darwin, Ch. *On the Origin of Species*. Cambridge, MA: Harvard University Press, 1859.
4. Prigogine, I. *From Being to Becoming*. WH Freeman and Company: San Francisco, CA, 1980.
5. Haken, H. Synergetics: an approach to self-organization. In *Self-Organizing Systems*, Ed., Yates, F.E. Plenum Press: New York, 1987.
6. Capra, F. *The Web of Life*. Doubleday: New York, 1996.
7. Kondepudi, D., Prigogine, I. *Modern Thermodynamics*. John Wiley & Sons: Chichester, 2000.
8. Klamecki, B.E. An entropy-based model of plastic deformation energy dissipation in sliding. *Wear* 1984, 96, 319–329.
9. Kostetsky, B.I. Structural-energetic adaptation of materials at friction. *Friction and Wear* 1985, 6, *2*, 201–212.
10. Kostetsky, B.I. An evolution of the materials' structure and phase composition and the mechanisms of the self-organizing phenomenon at external friction. *Friction and Wear* 1993, 14, *4*, 773–783.
11. Bershadsky, L.I. On self-organizing and concept of tribosystem self-organizing. *Friction and Wear* 1992, 8, *6*, 1077–1094.
12. Bushe, N.A., Kopitko, V.V. *Compatibility of Rubbing Surfaces*. Science: Moscow, 1981.
13. Kostetsky, B.I. *Friction, Lubrication and Wear in Machines*. Technika: Kiev, 1970.
14. Mansson, B.A., Lindgren, K. Thermodynamics, information and structure. In *Nonequilibrium Theory and Extremum Principles*, Eds., Sieniutycz, S., Salamon, P. Taylor and Francis: New York, 1990, pp. 95–98.
15. Ivanova, V.S., Bushe, N.A., Gershman, I.S. Structure adaptation at friction as a process of self-organization. *Friction and Wear* 1997, 18, *1*, 74–79.
16. Kostetskaya, N.B. Structure and Energetic Criteria of Materials and Mechanisms Wear-Resistance Evaluation. Ph.D. dissertation. Kiev State University: Kiev, 1985.
17. Kostetskaya, N.B. Mechanisms of deformation, fracture and wear particles forming during the mechanical-chemical friction. *Friction and Wear* 1990, 16, *1*, 108–115.
18. Gruss, W.W. Cermets, in *Metals Handbook*, 9th ed., Vol. 16, Ed., Burdes, B.P. Metals Park, OH: American Society for Metals, 1989, pp. 90–104.
19. Madou, M.J. *Fundamentals of Microfabrication: The Science of Miniaturization*. 2nd ed. CRC Press: New York, 2002.
20. Ivanova, V.S. *Fracture Synergetic and Mechanical Properties: Synergetic and Fatigue Fracture of Metals*. Moscow: Science, 1989, pp. 6–27.
21. Kabaldin, Y.G., Kojevnikov, N.V., Kravchuk, K.V. HSS cutting tool wear resistance study. *Friction and Wear* 1990, 11, *1*, 130–135.

2 Elements of Thermodynamics and Self-Organization during Friction

Iosif S. Gershman and Nicolay A. Bushe

CONTENTS

2.1 Introduction 13
2.2 Elements of Thermodynamics 15
2.2.1 Structure 17
2.2.2 Entropy and Its Fundamental Properties 17
2.2.3 The Law of Increase of Entropy (Second Law of Thermodynamics) 20
2.2.4 Entropy Production 22
2.2.5 The Stationary (Steady) State 25
2.2.6 The State with Minimal Entropy Production (Theorem by Prigogine) 26
2.2.7 Principle by Le Chatelier 28
2.2.8 Stability 29
2.2.9 Thermodynamic Stability of Equilibrium States 30
2.2.10 Stability of the Systems in a State Far from Equilibrium 32
2.2.11 Self-Organization: Dissipative Structures 33
2.3 Self-Organization during Friction 38
2.3.1 Tribosystem as an Open Thermodynamic System 38
2.3.2 A Thermodynamic Substantiation of the Existence of Surface Structures 38
2.3.3 Formation of Surface Tribo-Films (Secondary Structures) as a Process of Self-Organization 40
2.3.4 Thermodynamics of Surface Secondary Structures 41
2.3.5 Rational Selection of the Materials for Frictional Bodies 46
2.3.6 Instability of a Tribosystem 47
2.3.7 Stationary State of Surface Tribo-Films 49
2.3.8 Thermodynamics of the "Lubricating" Action of a Current 51
2.4 Conclusions 53
References 54

2.1 INTRODUCTION

This chapter presents a theoretical overview and analytical modeling of the friction phenomena based on the irreversible thermodynamics approach. Friction is a classic example of an irreversible process. The concepts of thermodynamic of irreversible processes and self-organization could be widely applied to the friction phenomena [1].

To use this approach, some generic features of friction should be outlined. It is known that wear is a generic characteristic of any tribosystem. Wear volume varies within a wide range; a large variety of different wear mechanisms exist, but the wear process is an attribute of friction. B. Klamecki suggests that wear is a fundamental characteristic of friction [2].

Some researchers [3,4] have also concluded that the other generic characteristic of friction is the formation of tribo-films or secondary (i.e., formed during friction) structures [4] on the surface that is associated with the so-called structural adaptation during friction [4]. Stabilization of the parameters of friction occurring at the running-in stage is accompanied by the formation and stabilization of tribo-films [3]. L. I. Bershadsky formulated a principle of "dissipative heterogeneity" [68]. According to this principle, the tribosystem tends to concentrate all kinds of interaction in thin surface layers of frictional bodies. Thus, tribo-films carry out protective functions, limit the depth of interaction of frictional bodies, and prevent direct contact [5].

The largest part of friction energy is accumulated within the layer of tribo-films (secondary structures). Therefore, the secondary structures are a steady zone of primary energy dissipation.

As outlined in the preceding text, wear and formation of secondary structures are fundamental processes for any tribosystem. That is why friction can be considered to be based on the fundamental laws of nature. Energy transformation occurs during friction, and therefore it is natural to regard friction based on the thermodynamic approach. Friction is a typical nonequilibrium process, and therefore this phenomenon is considered in this chapter in its relationships with nonequilibrium thermodynamics and self-organizing concepts.

Friction, in general, is a process. It is an irreversible process because energy dissipation takes place. This process changes and develops its characteristics and parameters with time, similar to that seen in evolution. Therefore, we should be interested not only in the initial and final stage of the friction process but also in the way this process has developed, as well as in the possibilities that exist to optimize this process (such as maximal speed under minimal wear rate and minimal energy dissipation for the tribosystem needed to fulfill its functionality). Traditional or equilibrium thermodynamics revolves around only the first and the last stages of spontaneously occurring processes. These are the conditions that produce a uniform distribution of energy, temperature, pressure, stresses, concentration, electric potential, etc. Friction corresponds to this state only under marginal conditions such as nullified speed of sliding, load-free conditions, and friction at rest after stress relaxation. That is why traditional thermodynamics cannot be applied to tribological applications. According to I. Prigogine [6], among the many fundamental laws of nature only the second law of thermodynamics points in the direction of time; the "arrow of time," in other words implies a direction and the possibility of evolution. That is why irreversible thermodynamics, as well as the theory of self-organization, is based on the second law of thermodynamics [7]. Under nonequilibrium conditions, some transformations could take place during irreversible processes that cannot occur with equilibrium conditions, i.e., at any temperatures, pressures, compositions, etc. For example, there are the processes that develop at entropy production decrease or at energy growth. Moreover, this nonspontaneous process development results in wear rate decrease as will be reviewed in greater detail.

The main concept of irreversible thermodynamics is that of entropy production. Entropy production characterizes the rate of entropy change when the irreversible process takes place in the system. The entropy production change characterizes the evolution of the process. Under stable conditions, the entropy production is simply related to the entropy flow, i.e., to the entropy of the system change due to interaction with the environment by matter and energy. Friction and wear could be characterized by exchange of matter and energy with the environment. That is why the first part of this chapter will be mainly focused on the concept of entropy and entropy production. The relation between entropy production and wear rate will be considered in detail in the second part of the chapter.

Wear rate develops under severe friction conditions (such as when cutting speed increases during machining or electric current growth at the sliding electric contacts). However, beginning with a definite value of the varying friction parameter (such as cutting speed or electric current), the wear rate could critically decrease. Under more severe friction conditions the wear rate could drop down even further or at least differ slightly. A connection could occur between the friction characteristics, such as the coefficient of friction decrease vs. electric current or the wear rate decrease vs. cutting speed. Meanwhile, the structures originate within the system due to intensive matter and energy flow, i.e., due to intensive irreversible processes. These structures are consistent with the energy flow and cannot exist without it. The structures that form under strongly nonequilibrium conditions are called *dissipative structures* and the process of dissipative structure formation is called *self-organization*. Dissipative structures differ from the equilibrium ones in the fact that they are the processes.

A classic example of an equilibrium structure is a crystal. However, the thermal vibrations of atoms in the crystal are chaotic because these vibrations are not related or synchronized. If, under specific conditions, there is some interrelation in the vibration of atoms, then dissipative structures could form. Once the dissipative structures have been formed, the system (of the crystal for example) is able to transfer more energy practically without any damage, as compared to the conditions when these structures are not present. This could be explained by the following: the energy that has been spent on the system damage prior to the formation of the dissipative structures is now spent on their generation. That is why wear rate critically drops in a tribosystem, owing to the initiation of the self-organization process. The self-organization during friction is usually associated with the process of the energy-rich surface structure formation. Entropy decrease and energy growth processes take place within these surface structures. If the phase and structure transformation takes place in a solid, then because the flows of matter and energy as well origins of these processes disappear, the dissipative structures disappear as well. However, the metastable structures still remain, and because the friction has been stopped, they are the secondary structures or tribo-films. The process of self-organization can start because the system has lost its stability.

The goal of this and the following chapters [3,8–12] is to develop criteria of rational selection and development of materials based on understanding the processes of friction and wear using nonequilibrium thermodynamics and the self-organizing approach. This chapter could be considered as a theoretical introduction to the entire book in which some fundamentals of wear are outlined regarding the approach of irreversible thermodynamics.

This chapter is divided into two major sections. In the first section, the fundamentals of irreversible thermodynamics are explained in detail. In the second section, the characteristic features of friction and the role of "self-organizing systems" in helping to control the wear processes are described. Throughout the discussion, the role of secondary structures and their thermodynamics are highlighted. Finally, we propose that any future development of improved tribo-materials and systems will depend on a better understanding of the nature of the secondary structures. A few examples are given as to how these improvements might be exploited for specific applications.

Using the conclusions obtained, some specific applications will be considered in greater detail in the following chapters of the book to illustrate the practical applications of generic ideas.

2.2 ELEMENTS OF THERMODYNAMICS

The concept of self-organization of systems far from equilibrium is currently widely used for fundamental research in many areas of science. For example, the instability that arises at fusion, the solidification caused by pulse laser irradiation of crystals [8], space–time ordering at the liquid-phase interface during mechanical deformation [9], structures growing at the liquid–solid interface in the course of the growth of a crystal [10], dissipative structures that are formed during the unidirectional growth of binary alloys from liquid [9], spatial heterogeneity as domains with various chemical concentration in rocks [8], and chemical reactions in biochemical systems [10].

A number of the effects described by this theory are currently considered as generic examples. First of all, this is related to the instabilities in hydrodynamic effects by Benard and Taylor [10]. A common example in chemistry is the reaction by Belousov-Zhabotinski [11]. One of the most important features of the listed effects is that they do not depend on initial conditions [12].

Schematically, the development of self-organizing processes with gradual deviation from equilibrium can be presented as follows [8]: There are no flows of matter or energy in an equilibrium system, entropy of the system has the biggest possible value under given conditions; entropy production is equal to zero; parameters of the system do not vary in due course. Equilibrium structures (for example, a crystal) could exist in the system under such conditions; energy is distributed uniformly.

Thermodynamic flows of energy, matter, charges, etc., exist within a small deviation of the system from equilibrium that is linearly related to thermodynamic forces.

If external conditions do not allow a system to reach a state of equilibrium and, at the same time, remain constant with time, the system transforms into a steady state with minimal entropy production possible under given conditions. Such a state becomes stable within an interval of external conditions. The theorem developed by I. Prigogine operates in this stationary condition [13]. According to the theorem, the flux value of disconnected thermodynamic forces in a stationary condition with the minimal entropy production goes to zero. If applied under conditions of friction, this theorem signifies that intensity of wear rate in such a stationary condition will be minimal because the thermodynamic force that is causing the fluxes of matter is not fixed.

The physical meaning of the theorem of minimal entropy production is described as follows: under conditions limited by a boundary, which do not allow a system to reach equilibrium (entropy production is equal to zero), the system is transformed to produce minimal dissipation.

I. Prigogine's theorem is validated when equilibrium exists. In a system far from equilibrium, the thermodynamic behavior is quite different and sometimes opposite to the behavior predicted by the theorem of minimum entropy production.

Steady structures can be formed in a system existing within a significant deviation from equilibrium. Benard's cells are the most well-known example of such structures [10]. These cells represent ordered convectional flows. Therefore, entropy production grows because convection represents a new mechanism of heat transfer. Matter flows are added to the heat flows. Similar structures that have formed in nonequilibrium areas have been called *dissipative structures* [7]. The given value of entropy decreases as the system transforms to a more ordered condition. Such transition has been called *self-organization* [7]. In this case, the thermodynamic force that causes the flow of matter is not fixed. However, in the system that contains Benard's cells the flow of matter grows, but this does not result in entropy decrease owing to these structures becoming ordered. That is why the generation of these flows by themselves leads to negative entropy production, whereas the total entropy production within the system itself is positive. That is why entropy production after the start of self-organization decreases as compared to the previous state (that lost its stability) being theoretically extended to the instance where self-organization is taking place [14].

It is worth noting that the Benard's cells are formed not by the particles but by the flows, i.e., the collective, synergistic (ordered) moving of particles. Instability of a thermodynamic branch of the process precedes the self-organizing phenomena [10].

The self-organization during friction consists in the formation of secondary structures on the surface. These structures represent allocated areas with increased levels of energy and modified chemical composition and structure as compared to frictional body substrates. It is shown in Reference 15, that the surface structures are dissipative. Taking into account that the formation of dissipative structures depends on the initial conditions, the phenomenon of self-organization can be considered as a fundamental characteristic of friction. According to Reference 16, formation of nonequilibrium dissipative structures improves wear resistance of the frictional bodies.

This book is devoted to the rational selection and improvement of materials for frictional bodies on the basis of studying processes of self-organizing during friction. Therefore, the methods of thermodynamics of irreversible processes and self-organization, which are used in the book, are described in this chapter in detail.

2.2.1 Structure

Structure is one of the most frequently used concepts in science. However, it does not always have a univocal definition. At the same time, there no domain of knowledge in which this concept is not used: from mathematics to biology or engineering science.

In mathematics, the concept of structure implies that the nature of the elements, in which the given set of them exists, does not play a significant role. At the same time, the relation between the elements determines the characteristics of a structure. The concept of structure in physical systems is similar. According to the definition by Kroeber [17], the system consists of elements connected by certain relations. The concept of structure includes a way of organization of elements and the characteristics of connection between them.

Spatial, temporal, and spatiotemporal structures are considered by specialists in a physical system. Atoms and molecules are usually the elements of a spatial structure. The structure is determined by their relative positioning and movement. Crystal structures serve as a vivid example of a spatial equilibrium structure. The concept of symmetry and orderliness is borrowed from crystallography [18]. Temporal structures are connected to the dynamics of a system and generally result in a periodic recurrence of some other phenomenon. The periodic changes of spatial structures are called *spatiotemporal structures*. The most known example is the reaction by Belousov-Zhabatinski [11].

It is necessary to know the difference between two kinds of structures: equilibrium and dissipative. Equilibrium structures are formed and exist during reversible transformations that are taking place at insignificant deviations from equilibrium. An example of an equilibrium structure is the aforementioned crystal. Dissipative structures [7] generate and exist due to an exchange of energy and matter with the environment under strongly nonequilibrium conditions. A known example of dissipative structures is Benard's cells. In contrast to equilibrium structures, the elements of the dissipative structures are not the atoms or other particles but the flows, i.e., collective synergistic movements of particles or other interactions between the elements of the system, including generation of new correlation between the system's parameters.

Under the term "structure formation" we will consider a generation of new properties within the system as well as relations between the elements of the system. For example, formation of a structure can be associated with the loss of uniformity in a system, reduction of a degree of symmetry, or a decrease of entropy [17].

2.2.2 Entropy and Its Fundamental Properties [18]

From the point of view of thermodynamics, entropy is the measure of the degree of organization in a system. Entropy growth is a criterion of irreversibility in a process. It is one of the major parameters in the theory of fluctuations by A. Einstein [19]. Derivatives of entropy characterize stationary conditions and the overall stability of a system. To represent the physical meaning of entropy, we are given the statistical physics definition of entropy as opposed to a phenomenological one [6].

Let us consider one of the subsystems of the equilibrium system. Its function of distribution shall be designated as "w". According to Galilee's principle of a relativity [20], it is possible to represent "w" as a function deriving exclusively from energy: $w(E)$. We can define $\Gamma(E)$ as a number of quantum conditions with energies smaller and equal to "E." In this case, the number of conditions in the interval of energy dE will be:

$$\frac{d\Gamma(E)}{dE}dE \tag{2.1}$$

Distribution of probabilities on energies will be:

$$W(E) = \frac{d\Gamma(E)}{dE} w(E) \tag{2.2}$$

if $\int W(E)\,dE = 1$, because the statistical averaging corresponds to the averaging by time, function $W(E)$ has an oscillating maximum at $E = E_{av.}$. ΔE shall represent the width of the rectangle whose height is equal to the value of function $W(E)$ in the maximum point whereas the area is equal to 1, i.e.,

$$W(E_{av.})\Delta E = 1 \tag{2.3}$$

Accounting for Equation 2.2, this expression can be written as:

$$w(E_{av.})\Delta\Gamma = 1 \tag{2.4}$$

where

$$\Delta\Gamma = \frac{d\Gamma(E_{av.})}{dE}\Delta E \tag{2.5}$$

is the number of quantum conditions covering the interval ΔE. The size of $\Delta\Gamma$, according to Reference 18, characterizes the "degree of distribution" of the macroscopic condition in a subsystem over its microscopic one. The interval ΔE coincides by the order of magnitude with the average fluctuation of energy in a subsystem; therefore, $\Delta\Gamma$ characterizes those quantum conditions in which the subsystem will spend the majority of time.

The phase volume $dq_1dp_1 \ldots dq_ndp_n = dpdq$ in which the given subsystem will spend most of the time (p, q, accordingly, are coordinates and pulses of elements of a subsystem; n is the number of degrees of freedom) corresponds to $\Delta\Gamma$ in classical statistics. In a quasi-classical case, for each quantum state there is a phase volume of $(2\pi\hbar)^n$ [21]. Therefore, the number of states in a quasi-classic case can be represented as:

$$\Delta\Gamma = \frac{\Delta p \Delta q}{(2\pi\hbar)^n} \tag{2.6}$$

The quantity $\Delta\Gamma$ is the statistical weight of the macroscopic state of the subsystem, and its logarithm is

$$S = \ln \Delta\Gamma \tag{2.7}$$

is the entropy of a system. In a classical case entropy is defined, accordingly, by the expression:

$$S = \ln \frac{\Delta p \Delta q}{(2\pi\hbar)^n} \tag{2.8}$$

The number of states $\Delta\Gamma$ cannot be less than one, and therefore entropy cannot be negative.

If each subsystem of a closed system exists in one of the $\Delta\Gamma_i$ quantum states with an entropy of S_i, the amount of various quantum states of the entire system ($\Delta\Gamma$) corresponds to

$$\Delta\Gamma = \prod_i \Delta\Gamma_i \tag{2.9}$$

$\Delta\Gamma$ from Equation 2.9 is the statistical weight of a closed system. Its logarithm subsists of the entropy (S) of a closed system. It follows from Equation 2.9 that

$$S = \sum_i S_i \tag{2.10}$$

i.e., entropy is an additive value.

A definition of entropy and some of its properties for the closed equilibrium system is given in the preceding text. However, these definitions can be generalized for the systems under any state that corresponds to the conditions of local equilibrium. The correspondence to these conditions signifies that it is possible for division of the entire nonequilibrium system into macroscopical equilibrium subsystems. This division is possible because, if the size of the subsystem is reduced, then the time of its relaxation will be reduced as well owing to the increase of the relations between the quantities of particles present at the border of a subsystem to the amount of particles in the volume.

According to Equation 2.9, the condition of a closed system as a whole could be characterized by determining the conditions in all separate subsystems ignoring the interactions within them. Then, $d\Gamma$ could be represented as a product of numbers that the quantum states of the subsystems consist of:

$$d\Gamma = \prod_i d\Gamma_i \tag{2.11}$$

(The sum of energies in a subsystem should correspond to the interval of energy values in the entire system).

It was outlined in the preceding text that the probability to ascertain the system in a certain condition depends on its energy. Energy of the closed system is constant (E_0), and therefore the probability of a presence of a system in other points of phase space should be equal to zero. To describe the function of distribution a function is introduced, such as $\delta(x) = \infty$ at $x = 0$, $\delta(x) = 0$ at $\neq 0$. The integral from the δ function for all the values of x is equal to 1. Therefore, the δ function is used to normalize the function of distribution in Equation 2.12. It is natural that the probability of discovering a system in the given conditions is proportional to the corresponding volume of the phase space or to the number of the corresponding states. The function of distribution is as follows:

$$dw = \text{const}\ \delta(E - E_0) \prod_i d\Gamma_i \tag{2.12}$$

The number of states depends on energy; therefore, it is possible to write Equation 2.12, with the condition of Equation 2.5, as:

$$dw = \text{const}\ \delta(E - E_0) \prod_i \frac{\Delta\Gamma_i}{\Delta E_i} dE_i \tag{2.13}$$

If $\Delta\Gamma_i$ is replaced by $e^{Si(Ei)}$ we will receive

$$dw = \text{const } \delta(E - E_0)e^S \prod_i \frac{dE_i}{\Delta E_i} \tag{2.14}$$

where $S = \sum_i S_i(E_i)$ is the entropy of a closed system and is a function of the exact energy values of its parts.

Equation 2.14 is the probability of all the subsystems having the energy distributed on the intervals between E_i and $E_i + dE_i$. This probability is determined by the entropy of a system. This conclusion is important because it demonstrates that the entropy of a system depends not only on energy of the system but also on the energy distribution inside it.

It was outlined earlier that average values are the most probable values of energies of the subsystems. Thus, entropy as a function of energies inside the subsystems S (E_1, E_2, ...) should have a possible maximal value within the average values of the energies in the subsystems (at a given value of the entire system's energy). According to a major principle of statistical physics, statistical averaging is completely equivalent to averaging by time [18]. Hence, the average values of subsystem energies correspond to the statistical equilibrium of a system, and entropy in this case has the greatest possible value.

2.2.3 The Law of Increase of Entropy (Second Law of Thermodynamics) [22]

With the passing of time, a closed system in a nonequilibrium state will change until it achieves a state of equilibrium. Concerning energy distribution between the subsystems, each subsequent state of the system corresponds to a more probable energy distribution. This probability is determined by exponent e^S, where the exponent signifies entropy. Hence, the state of a nonequilibrium closed system changes continuously with the gradual entropy increase up to the moment when its maximal value is achieved that corresponds to full statistical equilibrium.

If a closed system is in a nonequilibrium state, then a uniform increase of entropy within the system will be the most probable consequence of this state during the subsequent moments of time. This law of increase of entropy, or second law of thermodynamics, was discovered by R. Clausius in 1865, and L. Boltzmann did the statistical substantiation in 1870s.

Figure 2.1 illustrates the second law of thermodynamics as well as the definition of entropy as a value that is proportional to the quantity of the system's states under given energy conditions.

Figure 2.1 presents the evolution over time of the distribution of 14 molecules between two connected vessels. Each molecule has an equal probability of being in any of the vessels. A number of ways (P) in which the N molecules could be distributed within the two vessels is defined by a combinatorics formula [23]:

$$P = \frac{N!}{N_1!N_2!}$$

The P value corresponds to the number of states in Equation 2.6. The chaotic distribution of molecules is selected as an initial condition. This corresponds to 12 and 2 molecules in Figure 2.1 (state 1). Equilibrium is achieved after a significant amount of time. If the fluctuation could be ignored (state 4 in Figure 2.1, for example), then the equilibrium equals $N_1 = N_2$. If so, the P value is maximum. The quantity of possible states is related to entropy according to Equation 2.6 to

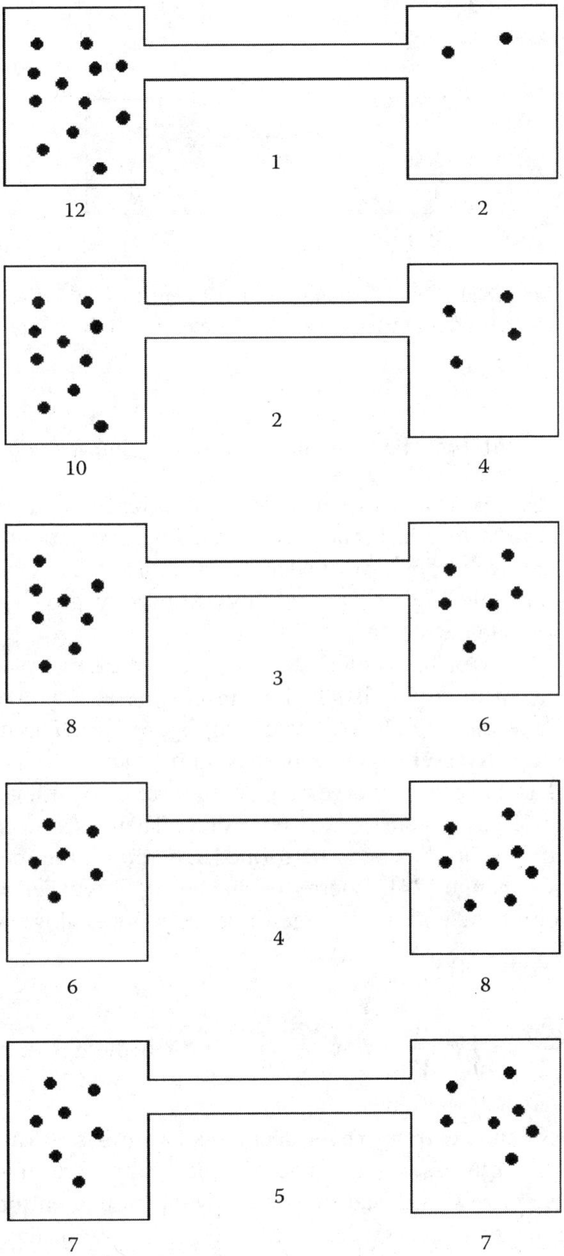

FIGURE 2.1 Evolution of the molecules distribution within two connected vessels. 1–5: sequential molecules distribution with time; 4: fluctuation.

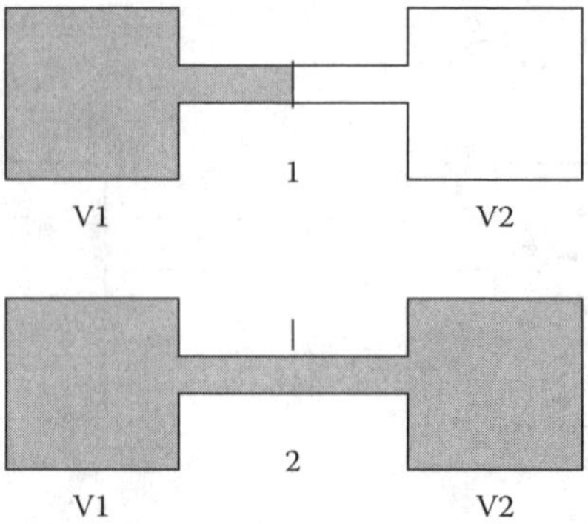

FIGURE 2.2 Nonequilibrium process of gas expansion in vacuum. 1 — initial state: an ideal gas in a vessel of volume V_1, vacuum in a vessel of volume V_2; 2 — final state: gas in the both vessels.

Equation 2.8, because $S = k \ln P$ (k is the Boltzmann constant). Entropy change is $k \ln \dfrac{\left(\frac{N}{2}\right)!}{N_1!(N-N_1)!}$ > 0. Thus, the entropy reaches its maximum under equilibrium conditions. Entropy grows with time (from the state 1 up to state 5 in Figure 2.1), excluding fluctuations. Therefore, irreversible processes are accompanied by entropy growth. This corresponds to the evolution for the state with a maximal entropy accompanied by the probability of entropy growth. It is related with the evolvement to the most probable state.

If we increase the number of molecules, we can evaluate gases by analogy, as presented in Figure 2.2. It shows a pair of thermally isolated, connected vessels with volumes V_1 and V_2. The valve between the vessels is initially closed (state 1 in Figure 2.2). The first vessel (V_1) contains 1 mol of ideal gas; the second vessel (V_2) is under vacuum (state 1 in Figure 2.2). Once the valve is opened, the gas transfers to both of the vessels and a state of equilibrium is reached. (state 2, Figure 2.2). The process of gas expansion is irreversible. There was no heat exchange with the environment ($Q = 0$) and no work has been performed. Therefore, the internal energy, as well as the temperature, does not change [24]. Energy exchange (dQ) between the vessels takes place. This permits a calculation of the entropy change to be made by analogy to the regular reversible isothermal process [24]:

$$S = \int \frac{dQ}{T} = \int \frac{pdV}{T} = \int R\frac{dV}{V} = R\ln\frac{V_1+V_2}{V_1} > 0 \text{ (} R \text{ is the universal gas constant)}$$

Thus, in all closed systems, entropy never decreases but grows or (in an ideal case) remains constant. The corresponding processes in the macroscopic systems are subdivided into irreversible and reversible. Irreversible processes are accompanied by increase of entropy of the closed system. The processes, being in reversed recurrence, cannot occur because of the subsequent decrease of entropy. During a reversible process, entropy of the closed system remains constant; hence, these processes can be reversible.

A strictly reversible process is an ideal limiting case; real processes can be considered as reversible ones only with a certain degree of accuracy.

2.2.4 Entropy Production [13]

It is known from thermodynamics that energy has a possible minimal value in an equilibrium system.

Let us consider the isolated system of two bodies that are existing in thermal balance. Entropy of such a system (S) has the probable greatest value under a given energy of a system: $E = E_1 + E_2$ (E_1 and E_2 are the energies of the two bodies). The system's entropy value will be equal to the sum of entropies of the bodies: $S(E) = S_1(E_1) + S_2(E_2)$. As is a constant, the entropy of a system is a function of one variable, and the condition of maximum entropy can be presented as:

$$\frac{dS}{dE_1} = \frac{dS_1}{dE_1} + \frac{dS_2}{dE_2}\frac{dE_2}{dE_1} = \frac{dS_1}{dE_1} - \frac{dS_2}{dE_2} = 0$$

It follows that

$$\frac{dS_1}{dE_1} = \frac{dS_2}{dE_2}$$

The inverse of the derivative of the entropy of a body with respect to its energy is its *absolute temperature:*

$$\frac{dS}{dE} = \frac{1}{T} \tag{2.15}$$

Therefore, temperatures of the bodies in equilibrium are identical. This conclusion is generalized to any number of bodies. The equality of other intensive values is also established in the equilibrium system, such as pressure, chemical and electric potential, etc. It could be interpreted as the establishment of equality of specific energy for all parts of the system.

Let us consider a system consisting of two closed subsystems if their temperatures T_1 and T_2 are varied. The quantity of energy received by each subsystem consists of two parts:

$$dE_1 = d_iE_1 + d_eE_1;\ dE_2 = d_iE_2 + d_eE_2$$

where all the terms of equation with the index i related to the energy received from each other by the subsystems and with the index e related to the energy received from the outside.

Taking into account Equation 2.15, the change of entropy within the system will be:

$$dS = \frac{dE_1}{T_1} + \frac{dE_2}{T_2} = \frac{d_eE_1}{T_1} + \frac{d_eE_2}{T_2} + d_iE_1\left(\frac{1}{T_1} - \frac{1}{T_2}\right) \tag{2.16}$$

The change of entropy consists of two parts. The first part is caused by interaction with the environment:

$$d_eS = \frac{d_eE_1}{T_1} + \frac{d_eE_2}{T_2} \tag{2.17}$$

The second part is caused by the irreversible process inside the system:

$$d_iS = d_iE_1\left(\frac{1}{T_1} - \frac{1}{T_2}\right) \tag{2.18}$$

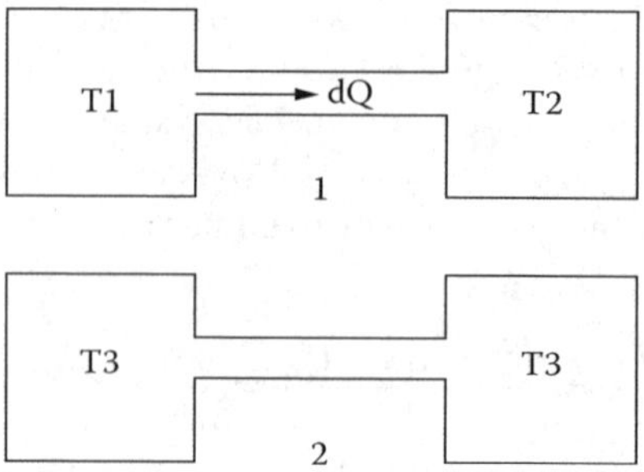

FIGURE 2.3 Nonequilibrium heat flow (dQ) between two vessels with different temperatures (T_1 and T_2) with $T_1 > T_2$.

According to the second law of thermodynamics $d_iS \geq 0$, following from Equation 2.18, energy is being transferred from a subsystem with a greater temperature into a subsystem with a lower one. Such increase of entropy because of the processes taking place inside the system (d_iS) is called *entropy production* in the thermodynamics of irreversible processes. Often, the rate of entropy growth (d_iS/dt) is called its entropy production:

$$\frac{d_iS}{dt} = \frac{d_iE_1}{dt}\left(\frac{1}{T_1} - \frac{1}{T_2}\right) \tag{2.19}$$

The right-hand side of Equation 2.19 represents the rate of an irreversible process (flow) multiplied by the difference of the appropriate intensive property. The direction of flow is determined by the negative or positive sign of this difference, which causes this flow. Figure 2.3 shows an irreversible heat flow between two parts of a system with different temperatures. This irreversible process leads to a growth in entropy. The entropy growth could be calculated in this case using Equation 2.19 if E_1 is replaced with Q in this equation.

Generally, it is possible to say that if equilibrium is achieved, the energy redistributes equally over the identical parts of a system. Thermodynamic flows such as flows of heat, particles, charges, impulse, and so forth are generated within the system. These flows tend to equalize the temperature, pressure, electric potentials, etc. The differences of these values or their gradients are thermodynamic forces. The sum of the products of thermodynamic forces and corresponding thermodynamic flows will determine entropy production. This process is accompanied by entropy growth, i.e., the process is irreversible.

It was shown in the preceding text that entropy depends on the energy of the system. Hence, irreversible processes are accompanied not only by the growth of entropy but also by the change of energy within the system (internal energy). In this case, we have to consider the reduction of internal energy because, in an equilibrium state, entropy has the possible maximum whereas energy has the possible minimum values [8]. Thus, irreversible processes should be accompanied by the dissipation of energy.

So, the change of entropy dS consists of two parts [13]. It includes a flow of entropy d_eS, caused by an interaction with the environment, and d_iS, a part of entropy change due to the processes that are taking place within the system:

$$dS = d_eS + d_iS \tag{2.20}$$

And always

$$d_iS \geq 0 \tag{2.21}$$

The entropy growth according to Equation 2.21 is a general criterion of irreversibility, as opposed to such thermodynamic potentials as free energy, entropy, etc., which are applicable in special cases (such as constants of temperature and volume, or pressure).

It is worth noting that the division of entropy change on two components (Equation 2.20) allows the establishment of a distinction between closed (isolated) and open systems. Distinction is shown in a member d_eS, which takes into account the change of entropy due to an exchange of matter in open systems.

It was shown earlier that entropy production could be expressed as the sum of products of the generalized forces and the appropriate generalized fluxes of irreversible processes:

$$\frac{d_iS}{dt} = \sum_k J_k X_k \tag{2.22}$$

where X_k and J_k are the corresponding generalized thermodynamic forces and fluxes.

It is known that under thermodynamic equilibrium: $J_k = 0$, $X_k = 0$ [13]. Therefore, it is possible to assume that there is a linear relation between forces and fluxes under a near-equilibrium condition. This assumption can be proven by the existence of empirical laws of Fick, Fourier, and Ohm [25]. Taking into consideration two irreversible processes that occur simultaneously:

$$\begin{aligned} J_1 &= L_{11}X_1 + L_{12}X_2 \\ J_2 &= L_{21}X_1 + L_{22}X_2 \end{aligned} \tag{2.23}$$

Flows and forces are connected in these equations by the phenomenological coefficients L_{jk}. The coefficients L_{jj} represent heat and electric conductivity, the coefficient of diffusion, chemical affinity, etc. The coefficients L_{jk} $(k \neq j)$ characterize interaction between two irreversible processes, for example, thermodiffusion, electrodiffusion, thermoelectric effects, etc. We have to note that the "phenomenological coefficients" are positive: $L_{jj} > 0$, but the "cross coefficients" have no determinate sign. For this case a theorem by L. Onsager [26] operates, which establishes that

$$L_{jk} = L_{kj} \tag{2.24}$$

This is L. Onsager's correlation of reciprocation [10], which shows that the mutual influence of two irreversible processes is carried out by means of the same phenomenological coefficient.

In conclusion of this section we note that any of the connected irreversible processes can flow towards the reduction of entropy, whereas the growth of entropy that is caused by the simultaneously occurring processes should be positive. For example, the diffusion of elements during thermodiffusion in the direction of the gradient of concentration results in a decrease of entropy, but this effect is compensated by entropy growth due to the flow of heat. According to Curie's principle, in a linear case, restriction is imposed on the interaction of the flows [27]. This principle states: "flows interact with each other if their tensors have identical parity" [27]. In a nonlinear area, the flows do not follow the limitations according to the aforementioned principle. Thus, in that case, any flows could correlate with each other. Dissipative structures could form from this correlation of flows in nonlinear areas that is impossible in the linear areas. The process will be shown during the analysis of the thermodynamic stability loss of the tribosystems.

2.2.5 The Stationary (Steady) State [13]

Among nonequilibrium conditions, steady states are the most interesting, because of their possible ability to exist continuously without changes.

It is known that the system is in a steady state if the variables that characterize it do not change with time [10].

An example of a steady state is a system that receives matter from the environment and then transforms through a number of intermediate reactions into the end product, which in turn dissipates into the environment. The state of such a system would be stationary if the concentration of intermediate products ceases to change through time. A stationary state should not be confused with an equilibrium state when the entropy growth is equal to zero and thermodynamic flows are nonexistent.

As proven in the preceding text, entropy, entropy production, and their density do not depend on time within a stationary state. In this case it follows from Equation 2.20 that in a steady state:

$$d_eS + d_iS = 0; \quad d_eS = -d_iS \leq 0 \tag{2.25}$$

According to Equation 2.25, for the functioning of a stationary state, the flow of entropy (equal to the value of entropy production) has to leave the system. Therefore, in a stationary state, it is impossible to start a flow of entropy because it depends on the production of entropy, i.e., the condition of a system. This is important from the perspective of the relation between wear rate and entropy production. We have to take into consideration that the flow of entropy includes the entropy of wear particles. Then, as it will be shown in the following text, under similar conditions, the decrease in entropy production leads to the reduction of wear rate.

Based on experimental observation, it is reported in Reference 28 that a system operating under the influence of time-independent factors will in time achieve a steady state.

Lastly, we have to outline an important property of a stationary state, its capability to remain stable in relation to small deviations [29].

2.2.6 The State with Minimal Entropy Production (Theorem by Prigogine) [6]

As we already know, the system's entropy in a stationary state remains constant Equation 2.25. However, the production of entropy behaves in a definitive fashion if linear relations Equation 2.23 and conditions Equation 2.24 are fulfilled.

The boundary conditions often do not allow a system to reach equilibrium. An example of such case is a system that is similar to the one considered in Subsection 2.2.4, but it consisting of two open subsystems that operate under the constant differentiation of temperatures. Exchange of matter and heat could also take place. In this case, the phenomenological equations of flows of matter and heat will be:

$$\begin{aligned} J_B &= L_{11}X_B + L_{12}X_T \\ J_T &= L_{21}X_B + L_{22}X_T \end{aligned} \tag{2.26}$$

and entropy production will be:

$$\frac{d_iS}{dt} = J_TX_T + J_BX_B \tag{2.27}$$

Under condition of

$$X_T = \text{const.} \tag{2.28}$$

Entropy production does not vary within a stationary state. Taking into account Equation 2.26 to Equation 2.28:

$$\frac{\partial}{\partial t}\left(\frac{d_iS}{dt}\right) = \frac{\partial J_T}{\partial t}X_T + \frac{\partial J_B}{\partial t}X_B + \frac{\partial X_B}{\partial t}J_B =$$

$$L_{21}\frac{\partial X_B}{\partial t}X_T + L_{11}\frac{\partial X_B}{\partial t}X_T + \frac{\partial X_B}{\partial t}(L_{11}X_B + L_{12}X_T)$$

and considering Equation 2.24:

$$\frac{\partial}{\partial t}\left(\frac{d_iS}{dt}\right) = 2\frac{\partial X_B}{\partial t}J_B = 0 \tag{2.29}$$

From Equation 2.29 two decisions are following:

$$X_B = \text{const.}; \quad J_B = 0 \tag{2.30}$$

The condition (Equation 2.30) is equivalent to the condition of minimum entropy production. Using Equation 2.26 to Equation 2.28 it is possible to receive:

$$\frac{d_iS}{dt} = L_{11}X_T^2 + 2L_{21}X_TX_B + L_{22}X_B^2 \tag{2.31}$$

Having taken a derivative from Equation 2.31 on X_B at constant X_T, we get:

$$\frac{\partial}{\partial X_B}\left(\frac{d_iS}{dt}\right) = 2(L_{21}X_T + L_{22}X_B) = 2J_B = 0 \tag{2.32}$$

It is necessary to consider that as entropy production is essentially positive, the conditions of extremum (Equation 2.30 and Equation 2.32) are related to the minimum.

Conditions Equation 2.30 and Equation 2.32 are equivalent if the linearity of the ratio in Equation 2.26 is fulfilled. This conclusion is relevant to any amount of thermodynamic forces and flows. Therefore, the theorem of minimal entropy production (Prigogine's theorem) [6] has two basic formulations.

The first shows that if there is a condition of linear correlation between the flows and the appropriate thermodynamic forces (Equation 2.26), then the steady state is characterized by the minimal entropy production.

According to the second formulation, if there are "n" thermodynamic forces and flows within the system and "k" forces are fixed in a stationary state of minimal entropy production, then the flows of unfixed forces disappear [27].

The physical meaning of Prigogine's theorem is as follows: a system that cannot come to an equilibrium state because of external conditions attempts to enter a condition with the lowest energy dissipation. According to this theorem, flows of nonfixed forces disappear within a stationary state. The thermodynamic forces that cause the flows of matter in tribosystems are not fixed. That is why we can expect that the wear rate under stable conditions will be minimal with an extremely low contribution to entropy production. These speculations will be used during thermodynamic analysis and the development of analytic expression for the lubricating action of an electric current (Chapter 8).

The minimum entropy production in a stationary state allows the assumption that the internal nonequilibrium processes always operate towards the reduction of entropy production [13]. Hence, the system in a stationary state with minimal entropy production can not change from this state spontaneously in accord with inequality

$$\frac{\partial}{\partial t}\left(\frac{d_i S}{dt}\right) < 0 \tag{2.33}$$

If a system slightly deviates from this state owing to fluctuations, internal processes occur that return it to its initial condition. Therefore, this state of the system is stable. Thus, the principle by Le Chatelier [18] is relevant to these stationary states.

2.2.7 Principle by Le Chatelier

In equilibrium thermodynamics, the principle by Le Chatelier is formulated as follows: "the external influence that deviates a body from equilibrium enhances the processes in this body, which, in turn, strive to weaken the results of this influence."

Let us consider S as the entire entropy of an isolated system of a body with an environment, and "x" and "y" are the values related to the body. Thus, the condition $\partial S/\partial x = 0$ means that the body itself is in equilibrium, and the condition $\partial S/\partial y = 0$ means that the body is in equilibrium with its environment.

Let us designate:

$$X = -\frac{\partial S}{\partial x}; \quad Y = -\frac{\partial S}{\partial y}$$

In the condition of full thermodynamic equilibrium entropy is maximum, i.e., $X = 0$, $Y = 0$, and also $(\partial X/\partial x)_y > 0$, $(\partial Y/\partial y)_x > 0$.

If the equilibrium with the environment is broken as a result of an insignificant external influence and if the value x varies on Δx, then the condition: $X = 0$ is broken. A change in the value X as a result of external influence will be $(\Delta x)_y = (\partial X/\partial X)_y \Delta x$. Change of x at the constant y results in infringement of the condition $Y = 0$, i.e., the infringement of the body's internal equilibrium. After the restoration of this equilibrium the value ΔX will be equal to $(\Delta X)_{Y=0} = (\partial X/\partial x)_{Y=0}\Delta x$.

At comparison of both values ΔX, according to Le Chatelier's principle:

$$\left(\frac{\partial X}{\partial x}\right)_y > \left(\frac{\partial X}{\partial x}\right)_{Y=0} > 0 \tag{2.34}$$

or

$$\left|(\Delta X)_y\right| > \left|(\Delta X)_{Y=0}\right| \tag{2.35}$$

If we evaluate the change in the value x of Δx, as the degree of external influence on a body and ΔX as the degree of the change in properties of a body as a result of this influence, then the inequality in Equation 2.35 shows that, upon the restoration of the body's internal equilibrium after external influence, the value ΔX decreases. Thus, inequalities Equation 2.34 and Equation 2.35 comprise the principle formulated above by Le Chatelier.

Following the results of the previous section, Le Chatelier's principle can be applied to stationary states with minimal entropy production.

The principle by Le Chatelier explains that the system tends to reduce the external impact. As an example, friction is the external impact on a tribosystem and wear is the result of this impact. According to the principle by Le Chatelier, we can expect that the tribosystem will respond to more severe friction conditions in the course of reducing wear intensity. This concept will be speculated upon in the thermodynamic analysis of the tribo-film formation during friction.

Le Chatelier's principle is closely connected to the instability of some state or other of a thermodynamic system.

2.2.8 Stability

According to Reference 7, the description of the newly ordered conditions that are far from equilibrium (for example, outside of a thermodynamic branch's stability) should be based on the fluctuation theory. We shall consider, as an example, the stability of a laminar flow of liquid. When viewed on a line graph, if a small fluctuation in kinetic energy $\delta E\kappa_{in}$ occurs, then a small deviation in speed can be observed. If $\delta E\kappa_{in}$ disappears at $t \rightarrow \infty$, then the current is stable. If $\delta E\kappa_{in}$ grows with time, then a new type of current forms. Under a value in excess of the critical value of the Reynolds number, the current becomes turbulent [28]. It is possible to assert that the new structure is the result of instability and originates from fluctuations.

Under equilibrium and near-equilibrium conditions, the system reacts to the fluctuations in a way that returns it back to the undisturbed condition. This is a constituent of Le Chatelier's principle [18] and as a consequence of the theorem of minimal entropy production [7]. However, existence of hydrodynamic instability shows that this statement is not applied to conditions far from equilibrium. In this sense, the Benard's cell is the huge fluctuation that is supported by flows of energy and matter which, in turn, are determined by boundary conditions.

To draw a conclusion on the general conditions of stability it is necessary to define stability by itself [7].

Let x be a set of independent variables, characterizing the state of a system at any point. Change of a system state at this point would be:

$$x = \varphi(t, t_0, x_0) \tag{2.36}$$

where x_0 corresponds to a state of the system at the moment of time t_0. Under the condition of validity Equation 2.36, in the vicinity of x_0 that is determined by the deviation "δ," the change "φ" during initial disturbance δ at the moment of time "t" is characterized by:

$$y(t) = \varphi(t, x_0 + \delta) - \varphi(t, x_0) \tag{2.37}$$

From the continuity of φ, we can assess that the value of $|y(t)|$ is small if the values of $|\delta|$ and "t" are also small. It is necessary to note that $|y(t)|$ is the distance $\sqrt{(\Sigma y_i^2)}$ in the space within states.

Based on these considerations, the definition of stability (Equation 2.36) should be as follows: if, for any value of $\varepsilon > 0$, there is a value of $k(\varepsilon) > 0$ such that

$$|\varphi(t, x_0 + \delta) - \varphi(t, x_0)| < \varepsilon \tag{2.38}$$

under all the values of t and at $|\delta| < k(\varepsilon)$, then Equation 2.36 is stable according to Lapunov [30]. In these conditions, the disturbance movement (Equation 2.36) strives to return to its initial movement at $t \rightarrow \infty$. Thus, if the positive sum of y^2 does not grow, i.e.,

$$\frac{d(y^2)}{dt} \leq 0 \tag{2.39}$$

at all the values of t, then Equation 2.36 will be stable. However, Equation 2.39 is only a sufficient condition of stability. A function with a defined sign such as y^2, resulting in a condition of stability Equation 2.39, could be classified as a Lapunov function. Such functions play an important role in the theory of stability, particularly in thermodynamic stability; therefore, we shall consider their application from that perspective [31].

Let us consider a system in which the evolution is defined by the variables X_i (for example, by concentration of the intermediate products of chemical reactions):

$$\frac{dX_i}{dt} = F_i(X_i) \tag{2.40}$$

i.e., F is the resulting speed of production of the component X_i. If at $X_i = 0$ and $F = 0$, then the state, at this point, is in equilibrium. We will use the concept of Lapunov's function to understand whether an equilibrium state is an attracter or not (i.e., the established state into which the given system evolves [32]) for the initial values (that are not equal to 0) of concentration considering that the function of concentration $G(X_i)$ is positive in the observed range of concentrations if $G = 0$ at $X_i = 0$. Taking into account Equation 2.40, the function G will vary with time at a change of concentration as following:

$$\frac{dG}{dt} = \sum_i \frac{\partial G}{\partial X_i}\frac{dX_i}{dt} = \sum_i \frac{\partial G}{\partial X_i} F_i(X_i) \tag{2.41}$$

According to the theorem by Lapunov, the equilibrium state of a given system can be an attracter if the sign of the derivative dG/dt is opposite to the sign of the G function (so in this case a derivative should be negative).

In physics, the stable movements (Equation 2.36) gradually change the function φ at a constant change of the x parameters. Accordingly, under unstable movement, spasmodic changes arise as a sudden reaction of the system to a slight change of external parameters. Such a change could be defined as a catastrophe [32]. Similar catastrophes (as will be further described in Chapter 9) occurring with respect to dependence of friction parameters on friction conditions are often related to the formation of dissipative structures.

2.2.9 Thermodynamic Stability of Equilibrium States

The theory of thermodynamic stability of an equilibrium state was developed by J. Gibbs and improved by F. Duhem [33].

The first law of thermodynamics is usually formulated as follows:

$$dQ = dE + pdV \tag{2.42}$$

i.e., the energy received by the system from its environment (dQ) is expended on the change of internal energy (dE) and on performing work (pdV). Taking into account that irreversible processes are proceeding within the system, the second law of thermodynamics is viewed as an inequality by Carnot–Clausius [34]:

$$dS_i = dS - dQ/T \geq 0 \tag{2.43}$$

Having substituted Equation 2.42 in Equation 2.43, we observe the second law of thermodynamics for the isolated systems under constant temperature and pressure in motion:

$$TdS_i = TdS - dE - pdV > 0 \tag{2.44}$$

In fact the Equation 2.44 determines the criterion of thermodynamic equilibrium's stability. This state is stable if the disturbances do not correspond to inequality as presented in Equation 2.44.

Let us designate δ as the symbol for small but random increments. Then the criterion of stability will be:

$$\delta E + p\delta V - T\delta S \geq 0 \tag{2.45}$$

The inequality in Equation 2.45 shows that when fluctuations satisfy inequality of Equation 2.44, the state of the system is stable.

Stability of equilibrium systems under specific conditions follow from Equation 2.44 and Equation 2.45:

$$\delta E \geq 0, \text{ at } S,\ V = \text{const.} \tag{2.46}$$

i.e., in a stable equilibrium the internal energy is minimal; therefore, in equilibrium:

$$(\delta E)_{eq.} = 0 \tag{2.47}$$

Under infinitesimal deviations, Equation 2.46 and Equation 2.47 are

$$(\delta^2 E)_{eq.} > 0 \tag{2.48}$$

It follows from Equation 1.45 that:

$$(\delta S)_{eq.} \leq 0 \text{ at } E \text{ and } V = \text{const.} \tag{2.49}$$

Analogically, under infinitesimal disturbances, Equation 2.48 can be written as:

$$(\delta^2 S)_{eq.} < 0 \tag{2.50}$$

For receiving specific stability conditions the J. Gibbs's formula will be applied [7]:

$$T\delta s = \delta e + p\delta v - \sum_{\gamma} \mu_\gamma \delta N_\gamma \tag{2.51}$$

μ_γ and N_γ are the chemical potential and the concentration of matter γ, respectively. Subscripted letters signify the local values of corresponding extensive values.

The second differential of entropy can be received from differentiating the formula by J. Gibbs by appropriate independent variables:

$$\delta^2 s = \delta T^{-1}\delta e + \delta(pT^{-1})\delta v - \sum_{\gamma} \delta(\mu_\gamma T^{-1})\delta N_\gamma \tag{2.52}$$

Having substituted known thermodynamic ratios and identities [7,18,35] in Equation 2.52, we shall receive a concrete type of second differential of entropy:

$$\delta^2 s = -\frac{1}{T}\left[\frac{c_v}{T}(\delta T)^2 + \frac{\rho}{\chi}(\delta v)_N^2 + \sum_{\gamma'\gamma} \mu_{\gamma\gamma'}\delta N_\gamma \delta N_{\gamma'}\right] \tag{2.53}$$

where c_v is the thermal capacity at a constant volume, χ is the isothermal compressibility, $\mu_{\gamma'\gamma} = \left(\frac{\partial \mu_\gamma}{\partial N_{\gamma'}}\right)_{T,p,N\gamma}$. Equation 2.53 represents a quadratic equation with a negative sign.

Substituting Equation 2.53 in Equation 2.50, we get:

$$(\delta^2 S)_{eq.} = \int \rho \delta^2 s dV = -\int \frac{\rho}{T}\left[\frac{c_v}{T}(\delta T)^2 + \frac{\rho}{\chi}(\delta V)^2 N_\gamma + \sum_{\gamma\gamma'} \mu_{\gamma\gamma'}\delta N_\gamma \delta N_{\gamma'}\right] dV \tag{2.54}$$

For the observation of stability conditions, the integral in Equation 2.54 should be negative. The expression under the integral is a quadratic equation. To make this expression positive, the following conditions should be applied:

$$c_v > 0; \quad \chi > 0; \quad \sum_{\gamma\gamma'} \mu_{\gamma\gamma'} x_\gamma x_{\gamma'} > 0 \tag{2.55}$$

The inequalities Equation 2.55 are, respectively, conditions of thermal, mechanical, and diffusion stability. The physics interpretation of Equation 2.55 is the following: the system responds to the disturbances resulting in its heterogeneity, by striving to restore the uniformity that corresponds to Le Chatelier's principle [18].

2.2.10 Stability of the Systems in a State Far from Equilibrium

By distributing the entropy near its equilibrium state into series, we obtain (eq. stands for equilibrium):

$$S = S_{eq.} + \delta S_{eq.} + \frac{1}{2}\delta^2 S_{eq.} + \ldots \tag{2.56}$$

Since

$$S_{eq.} = \max,\ \delta S_{eq.} = 0,\ \text{and}\ \frac{1}{2}\delta^2 S_{eq.} < 0 \tag{2.57}$$

The condition in Equation 2.57 corresponds to a condition of stability in an equilibrium state Equation 2.50. We cannot apply it to states that are far from equilibrium as the distribution into the series of Equation 2.56 is received in a state of near-equilibrium. However, in compliance with the conditions of local equilibrium, assuming it is stable, we get the following:

$$\delta^2 s < 0 \tag{2.58}$$

With the account of Equation 2.54 and Equation 2.55 this value is a defined, negative function of the independent variables' increase, which are included in J. Gibbs's formulas (Equation 2.51 and Equation 2.52) for the local state of a dissipative system. Hence, the stability of a system can be characterized based on the function $\delta^2 s$ as the Lapunov's function. Thus, the local condition of stability will be:

$$\frac{\partial}{\partial t}(\delta^2 s) \geq 0 \tag{2.59}$$

It is necessary to note that the application of the condition in Equation 2.59 for the variables included, usually, in J. Gibbs's equation (energy, volume, and concentration) results (at equilibrium state) in the conditions witnessed in Equation 2.55, i.e., in the principle by Le Chatelier. Therefore, it is concluded in Reference 7, that the inequality in Equation 2.59 allows the generalization of Le Chatelier's principle. Thus, for a stable system with $\delta^2 s$ as a function by Lapunov, the principle is carried out automatically.

The principle of local equilibrium assumes validity of Equation 2.59 at every point of the volume and at any moment in time. Owing to this, and also in the course of continuity of the function $\delta^2 s$, integration of Equation 2.59 based on volume of the system will give a sufficient condition of stability for the given volume:

$$\frac{\partial}{\partial t}(\delta^2 S) \geq 0 \text{ at}: \delta^2 S < 0 \tag{2.60}$$

The derivative by time from $\delta^2 S$ in Equation 2.60 is connected, as shown in Reference 7, with entropy production caused by disturbance, i.e.,

$$\frac{1}{2}\frac{\partial}{\partial t}(\delta^2 S) = \sum_n \delta X_n \delta J_n \geq 0 \tag{2.61}$$

The sum in the right-hand side of the equation could be defined as the excess of entropy production. Values δX_n and δJ_n are the deviations of appropriate fluxes and forces in a stationary state. If the inequality in Equation 2.61 is observed from the beginning of a disturbance, then the given state is stable. However, during some processes or the interaction of various processes, it is possible to receive negative contribution to the excess of entropy production, which grows with the increase of disturbance. In this case, the given state can become unstable. (A positive excess in entropy production is a sufficient but not a necessary condition of stability.) Only after passing through instability can the process of self-organizing begin. This will be used during the analysis of the formation of dissipative structures under different friction conditions (see examples in Chapter 9 to Chapter 12).

2.2.11 Self-Organization: Dissipative Structures

Following from the definition and physical interpretation of entropy given in Subsection 2.2.2, the equilibrium state should be characterized by the absence of orderliness. Such a disorder, according to the conclusions of the previous section, will continue until deviations from equilibrium are caused by damped fluctuations or stochastic disturbances. According to the theorem by Prigogine, during regular deviation from equilibrium, i.e., when an increase occurs in a parameter that describes a system's condition or its external influence, the steady (and close-to-equilibrium) state is stable. The slope in phase space that describes this condition and depends on a varied parameter could be called a *thermodynamic branch*. However, if a certain critical value of the parameter is exceeded, owing to a change of an inequality's negative or positive sign Equation 2.61, the state of the system could become unstable. Then, a small disturbance will withdraw the system from the thermodynamic branch.

The new order that has been established within the system could correspond to an ordered state, i.e., a new state with a smaller entropy in comparison to a chaotic condition. In this case, a bifurcation (the emergence of a new solution to the equation under a critical value of a parameter) takes place at a critical point, resulting in a new branch of possible solutions.

I. Prigogine has categorized similarly ordered configurations, which appear outside a thermodynamic branch's stability area, as dissipative structures [7]. The selection by the system of solutions that arise from the point of bifurcation, i.e., process of the formation of dissipative structures, is called self-organization [7].

Self-organization is the choice that the system makes between a few possible solutions at the point of bifurcation (i.e., the point of a possible occurrence of a second solution for the equation that describes the process).

This choice is controlled by the laws of probability. Self-organization under conditions far from equilibrium leads to the growth of the system's complexity. The second solution could only occur in a nonlinear area. During a self-organization process, stable structures form spontaneously in nonlinear open dissipative systems. Such structures could only occur during collective (synergistic) interactions. The common conditions necessary for nonequilibrium phase transformations that lead to the formation of new dissipative structures are:

1. Dissipative structures form only in open systems because an energy flow that compensates for the losses due to the dissipation is only possible in these systems, ensuring the existence of higher ordered states.
2. Dissipative structures form in macro systems. Collective or synergistic interactions are possible in these systems that are necessary for their reconstruction.
3. Dissipative structures form only in the systems that follow nonlinear equations for macro functions.
4. Nonlinear equations have to permit the change of a solution's symmetry for the formation of dissipative structures. Such change, for instance, can be witnessed in the transformation of a molecular into a convective mass transfer, as in Benard's cells.

It was outlined earlier that the system's stability loss, followed by the formation of dissipative structures, is a probabilistic process. Even if the system lost its stability (i.e., the entropy production is negative) dissipative structures could either form or not form under the same conditions. This is the major difference between dissipative structures and equilibrium phase transformation, which always take place under specific conditions (for instance, melting when a particular temperature is reached). This happens because self-organization depends on stochastic processes including fluctuations.

An extensive amount of literature is devoted to experimentation, theory, and the description of similar structures [36–43]. Therefore, we will not describe specific structures here but, instead, present general results.

Only nonlinear equations can contain more than one solution. Therefore, bifurcations and, hence, the occurrence of ordered states can take place within a system in which the processes follow nonlinear laws.

The first similar results were achieved in hydrodynamics as shown by Benard (1900) and Taylor (1924) [44]. Benard's cells are described in the majority of works on self-organization [10]. Therefore, we shall only briefly consider one aspect of this effect. A spatial cellular structure arises in a thin layer of liquid when the gradient of temperature exceeds critical value. However, heat conductivity is a linear effect, which cannot cause a bifurcation. Therefore, Benard's cells appear as the result of interaction between heat conductivity and gravitation. It is possible to neglect gravitation's influence on the thin layer of liquid, which is in equilibrium. However, for conditions far from equilibrium it is necessary to take gravitation into account. Thus, in this case, formation of a spatial periodic structure occurs as a result of an interaction between two or more processes.

The same conclusion can be made about the instability of chemical processes. As a rule, oscillatory chemical reactions correspond to an ordered state. Plenty of such reactions (a most detailed review can be found in Reference 45) are widely known. However, the reaction by Belousov–Zhabatinski [11] is the most well known. I. Prigogine considers this reaction historical, because it has shown that, in a condition far from equilibrium, matter acquires new properties [9] when macroscopical areas in a reactionary mixture simultaneously and strictly periodically change color. It evidences the occurrence of long-distance correlations in conditions far from equilibrium, which are absent in equilibrium states. In the chemistry of nonequilibrium processes, nonequilibrium spatial and temporal structures [46–47] were observed.

A classic example of self-organization in material science is the process of martensite formation. The speed of the formation of martensite is very high (in the range of the speed of sound) that indicates a nondiffusion mechanism of the process [48]. Martensite is a nonequilibrium phase, i.e., an oversaturated solid solution of carbon in iron. It is known that the equilibrium phase is an eutectoid (perlite): a mixture of two phases — ferrite and cementite. The martensite formation is the result of an interaction of two processes: thermal conductivity and eutectic phase transformation. According to the principle by Curie [27], these processes cannot interact in a linear area because they are characterized by the tensors in different ranges: heat flow is a vector (a tensor of first range) and the speed of phase transformation is a scalar (a tensor of zero range). The conditions of martensite's formation could not be characterized as an equilibrium phase transformation by specific temperature but by the cooling rate. The rate is characterized by the temperature gradient and heat flow.

Any structure formed under nonequilibrium conditions will have a smaller entropy in comparison to those formed in equilibrium ones. If a system that is removed from equilibrium is isolated instantly, it will attain an equilibrium condition having increased entropy, according to the second law of thermodynamics. However, it does not always result in the appearance of dissipative structures. The thermodiffusion phenomenon, i.e., the separation of gases in a mixture of gases influenced by a gradient of the temperatures has been known since the 19th century [10]. In this case, diffusion follows the gradient of concentration, i.e., the entropy decreases. A general increase in entropy of the given system occurs because of the flow of heat. Thermodiffusion is the result of linear processes: the difference in concentration is proportional to the difference of temperatures. The difference in concentration forms gradually within a gas mixture, instead of changing abruptly as it should during self-organization. Aside from that, the molecules move independently during diffusion, i.e., the process of diffusion is chaotic in this sense. Therefore, thermodiffusion is not a process of self-organization. However, in this case, the thermodiffusion process leads to negative entropy production. Thus, thermodiffusion reduces the total entropy production in a system. It is worth noting that entropy production reduces not only because, for instance, of the increase of thermal conductivity but also as a result of another nonequilibrium process. We can assume that self-organization begins only because of the occurrence of a new process (which takes place during negative entropy production) as a result of the interaction between two or more processes. As will be shown in the second part of the chapter, a tribosystem could lose its stability as a result of the interaction between two or more independent processes. That is why strictly nonequilibrium nanostructures have improved wear resistance (see Chapter 9 to Chapter 11 for details).

An essential property in all effects of self-organization is that their stationary modes practically do not depend on the initial conditions. L. Bertalanffy has named this property as equifinality [29]. Behavior of the system near equilibrium is determined by the initial and boundary conditions. In contrast, the behavior of the system far from equilibrium, i.e., during self-organizing, depends on the concrete mechanism and small deviations, and can result in large-scale consequences. A certain autonomy or self-organization can be attributed to such systems. After passing through instability, the system "forgets" its initial conditions. Therefore, the properties and composition of the dissipative structures do not depend on initial conditions.

Despite the fact that self-organization is usually preceded by instability, the opposite is not always fulfilled. Therefore, an analysis of specific processes is necessary to verify that self-organization occurs. That is why there is no precise general definition of this concept, and it is believed that such definitions are not necessary [36]. To draw a conclusion about the self-organization process, it is necessary to investigate which of the considered phenomena is more highly ordered during their evolution. As shown in Reference 49 and Reference 50, it is possible to estimate a relative degree of orderliness in open systems on experimental data using the S theorem by Klimantovich. According to this theorem, if the entropy-to-energy ratio decreases, it is possible to conclude that the process of self-organization is taking place. In fact, the decrease of entropy under a definite level of energy within the system indicates its ordering [49].

Generally the theorem by Prigogine does not cover nonlinear irreversible processes including the process of self-organization. However, the principle of "minimum entropy production" during self-organizing is formulated in Reference 51 and Reference 38. This principle asserts that entropy production in "new" and higher-ordered conditions, arising as a result of self-organization, has lower entropy production compared to the "old" condition, which is theoretically expanded over the unstable area. The processes that occur in dissipative structures, formed during self-organizing, have higher orderliness in comparison to the processes before self-organization. Usually this takes place during the so-called "cooperative processes," when the particles that did not previously interact with each other new interact intensively after self-organization begins.

Let us illustrate the process of self-organization with the formation of dissipative structures in the following method. A typical break point in the amount of energy (q) that transfers during the self-organization process is shown in Figure 2.4. Different parameters such as temperature difference between hot and cold surfaces, as for the case of Benard's cells [10], could be used as a factor of equilibrium. For friction phenomena, which are the subject of this book, it could be speed, loading, current value, coefficient of friction, etc. To the left of the break point (Figure 2.4), the system transfers energy using thermal conductivity, diffusion, and other linear mechanisms. In this case a part of the energy dissipates and entropy production grows as the factor of equilibrium increases. The rest of the energy accumulated by the system is termed stored energy in tribology [1]. The frictional body undergoes intensive surface damage if the critical amount of energy is reached. This critical amount is usually assumed to be equal to the melting heat [1]. However, because the break point is achieved, as shown in Figure 2.4, some processes that need more energy are initiated in the frictional body. Therefore, more energy is removed from the friction surface and less energy is transferred toward surface damage or intensive wear. Usually a break point can

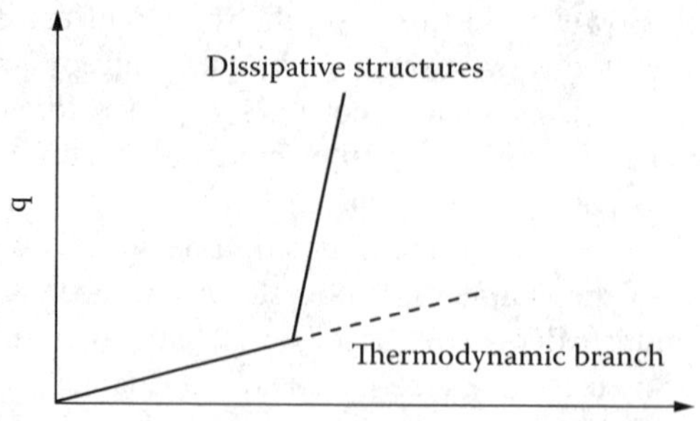

FIGURE 2.4 Amount of energy (q) that is transferring within the system vs. nonequilibrium factor (intensification of the external conditions).

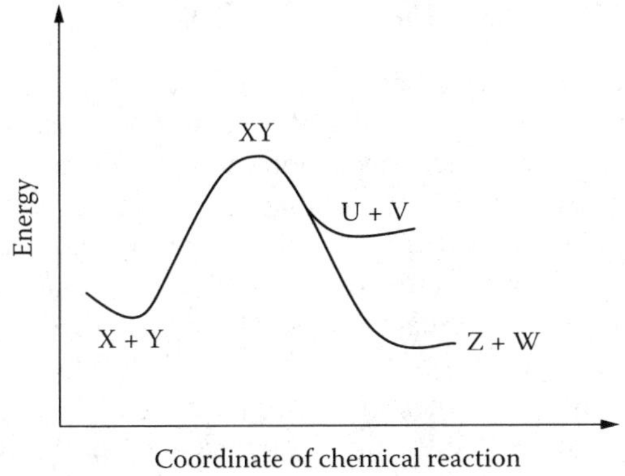

FIGURE 2.5 Energy of chemical reaction vs. coordinate of chemical reaction, which indicates on a stage of its processing, where $X + Y$ are the reagents involved in the reaction, XY – the activated molecules, $Z + W$ are the products of the reaction at usual (equilibrium) chemical reaction, and $V + U$ are the products of the reaction at nonequilibrium chemical reaction.

be observed at the curve of wear intensity versus external conditions, and thus the wear rate critically drops.

The processes that develop with energy growth and, correspondingly, with entropy reduction are nonequilibrium processes. These processes cannot occur spontaneously. An example of such a process is shown in Figure 2.5. It represents the dependence of chemical reaction's energy on the coordinate of chemical reaction, which is indicated on the stage of its processing. Energy of reagents involved in the reaction (X + Y) is higher than energy of the products of reaction (Z + W) during a usual (equilibrium) chemical reaction. In contrast, the energy of reagents involved in reaction (X + Y) is lower than the energy of products of the reaction (V + U) during a nonequilibrium (nonspontaneous) chemical reaction. The latter processes allow the removal of energy from the friction surface. Specific mechanisms of these processes will be discussed in further detail in the corresponding chapters (Chapter 3 and Chapter 8). Such nonequilibrium processes take place alongside entropy reduction (i.e., with negative entropy production) that corresponds to the structure's definition (see Section 2.1.1). Therefore, the initiation of these processes corresponds to the formation of dissipative structures. However, according to the second law of thermodynamics, general entropy production has to be positive. That is why some processes with positive entropy production have to occur simultaneously with nonequilibrium processes to enable the general entropy production to be positive.

The interaction of different processes with each other under the intensification of external conditions (nonequilibrium factor in Figure 2.4) could result in the decrease of wear rate as a result of enhanced intensity of nonequilibrium processes. After the break point is passed, almost all the energy supplied to the system (Figure 2.4) is used up on nonequilibrium processes. In this case, entropy production does not change. The specific examples of wear rate decrease (or at least stabilization) for different tribosystems under analysis will be shown later in the corresponding chapters of the book (Chapter 8, Chapter 10, and Chapter 11). Finally, we have to note that self-organization and the formation of dissipative structures could begin (a break point in Figure 2.4) only after stability of the system is lost. The loss of stability, as was outlined in the preceding text, is a stochastic process. A system could either lose its stability or not lose it. In the latter case, the system develops along the thermodynamic branch (the dotted line in Figure 2.4). If the system develops alongside the thermodynamic branch, wear intensity will grow under the intensification of external conditions (nonequilibrium factor in Figure 2.4). That is why the challenge faced by tribo-material science and engineering is to enhance nonequilibrium processes and shift the initiation of these processes to milder service conditions, as will be shown in Chapter 8 to Chapter 11.

2.3 SELF-ORGANIZATION DURING FRICTION

Since the late 1970s several works have been published that considered the friction phenomenon based on the theory of self-organization [52]. Unfortunately, the majority of these works have a declarative character, often with incorrect conclusions and premises. Moreover, some of the works that mention the use of nonequilibrium thermodynamics investigate a tribosystem using the principles of equilibrium thermodynamics that is associated with the concept of "stored" energy [53]. Much literature exists that does not directly concern self-organization and thermodynamics, but is nevertheless indirectly associated with these issues. Usually it is related to the studies of surface structures generated during friction [54–58].

However, it is worth noting that there are a limited number of publications in which the process of friction is correctly described (from the physical point of view) based on the concepts of irreversible thermodynamics and the theory of self-organization. First of all, the works of B.E. Klamecki [3,59–62], L.I. Bershadsky and B.I. Kostetsky [4,68] should be mentioned. These researchers pioneered the application of the nonequilibrium thermodynamics concept in tribology. They have also tried to prove the applicability of this concept in analyzing friction phenomena and in analytical modeling of the friction process using the principles of nonequilibrium thermodynamics.

Investigations of these scientists and further activity in this field [63] showed that the main effects that occur during friction are concentrated within a thin, surface layer. A thermodynamic analysis of this layer's state and the relation of this state to a wear process is considered in details in the following sections.

2.3.1 Tribosystem as an Open Thermodynamic System [64]

Assuming a tribosystem is an isolated equilibrium one, its entropy will depend on internal energy. Taking into account that internal energy of a system will be equal to the difference between its full energy (E) and the kinetic energy of a sliding body ($mv^2/2$) with a mass "m" and with a speed of sliding "v," the entropy of a tribosystem (S) will be written as [18]:

$$S = S(E - mv^2/2) \tag{2.62}$$

In an isolated system full energy is constant, and therefore, under the differentiation of entropy by speed, we get:

$$\frac{\partial}{\partial v} S(E - mv^2/2) = -mv/T \tag{2.63}$$

According to the second law of thermodynamics, entropy of an isolated system can only increase. From Equation 2.63 it follows that the speed of sliding in such a system should be equal to zero. Hence, a tribosystem cannot be considered as an equilibrium isolated system because in such a system macroscopical movement of its parts is impossible. The relative speed of sliding indicates a nonequilibrium situation in an open system. Therefore, we shall further consider a tribosystem as an open nonequilibrium thermodynamic system.

2.3.2 A Thermodynamic Substantiation of the Existence of Surface Structures

One of the major principles in tribology, i.e., the principle of secondary dissipative heterogeneity has been formulated by Kostetsky and Bershadsky [68]. According to this principle a phenomenon of the materials' structural adaptability during friction occurs when all kinds of interaction between frictional bodies are localized within a thin layer of surface tribo-films (or secondary structures).

The balance of energy received by a frictional body is as follows:

$$E_0 = \Delta U + A \tag{2.64}$$

where E_0 is the part of energy generated at friction and transformed into a frictional body; ΔU, the change of a body's internal energy; and A, the dispersion of energy during its transformation from a contact zone into a frictional body.

To prevent energy transforming into a frictional body, a major part of the energy generated during friction has to dissipate. Following from Equation 2.64: $A = E_0 - \Delta U(S)$, where S is the entropy of a frictional body.

Let us differentiate A by entropy of a final state (considering energy E_0 as a constant):

$$\frac{\partial A}{\partial S} = -\left(\frac{\partial \Delta U}{\partial S}\right) = -\Delta T \tag{2.65}$$

where ΔT is the increase of temperature in a body.

Following from Equation 2.65, A will diminish with the increase of S as ΔT is positive. Entropy grows as the irreversible process transfers energy from a heated zone to a cooler one. Therefore, the biggest value of A corresponds to the lowest increase of entropy of a frictional body and, under a constant time of interaction, to the lowest entropy growth rate.

During direct contact of frictional bodies entropy growth rate due to external impact (dS_e/dt in Equation 2.20) is equal under conditions of low energy exchange:

$$\frac{dS_e}{dt} = q\left(\frac{1}{T_s} - \frac{1}{T_c}\right) \tag{2.66}$$

where T_S and T_C are the temperatures of the frictional body surface and contact zone, respectively.

Therefore, energy dissipation hardly occurs, and the process of friction transforms to a catastrophic mode of surface damage according to the classification by Kostetsky [4]. To prevent entropy growth it is necessary to introduce an intermediate body into the system. In this intermediate body a dispersion of energy takes place. Temperatures of the friction zone and the corresponding surface of an intermediate body should be equal (T_c), as well as the temperatures of the frictional body and the adjacent surface of an intermediate body (T_s); then isothermal transition of energy takes place. In the ideal case the process of temperature decrease of an intermediate body from T_c down to T_s should be adiabatic. However, a gradient of temperature is established because the process is taking place on a permanent basis. Such a process of energy transfer is similar to the cycle by Carnot, which occurs without change in entropy [65]. The surface tribo-films (secondary structures) serve as an intermediate body during friction.

Surface tribo-films (or secondary structures) are necessary for energy dispersion during its transformation from a friction zone into a frictional body. To obtain minimal wear rate, a dissipation of energy should occur under the lowest rate of entropy growth. Surface secondary structures perform protective functions, limiting the interactions (occurring during friction) into the depth of frictional bodies and reducing the intensity of such interaction. Therefore, their formation is relevant to Le Chatelier's principle [66]. In fact we can assume that the formation of secondary structures is a tribosystem's response to an external impact, i.e., friction. This reaction is aimed at reducing the external impact by decreasing wear rate. If the tribosystem cannot form stable secondary structures, it tries to stop the friction by means of strong seizure or jamming.

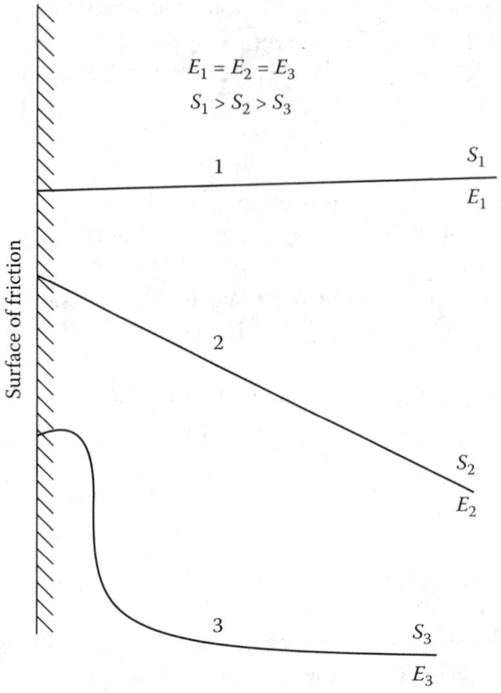

FIGURE 2.6 Schematic distribution of energy (E) in a frictional body under various conditions.1 — equilibrium conditions; 2 — a steady state in absence of the surface tribo-films (secondary structures); 3 — a steady state in presence of the surface tribo-films (secondary structures). S_1, S_2, and S_3 are the entropies of corresponding states. E_1, E_2, and E_3 are energies of corresponding states. At $E_1 = E_2 = E_3$, $S_1 > S_2 > S_3$.

2.3.3 Formation of Surface Tribo-Films (Secondary Structures) as a Process of Self-Organization

The surface tribo-films (secondary structures) store more than 90% of total friction energy [67]. From a thermodynamics perspective, they are the stable zone of a dominating dissipation of energy.

There is no accurate criterion for the self-organization process [14]. Therefore, it is necessary to make a comparative estimation of entropy in possible states for complex processes under consideration. Such an estimation will allow conclusions to be made about the occurrence of self-organizing processes.

Once two bodies, initially under different conditions, achieve equilibrium, the parameters (thermodynamic forces) of the system become equal, and energy as well as matter flows disappear. It might happen only during the discontinuance of friction. The external impact does not allow the system to reach an equilibrium state during friction. Under constant external conditions the system strives to reach a stationary state. In that state the parameters of a system remain constant, but in a frictional body energy and potentials are distributed nonuniformly. This results in the flows of matter, heat, etc.

If surface tribo-films (secondary structures) do not form, the energy and potentials are distributed as a gradually decreasing function into the depth of a frictional body (Figure 2.6).

Energy supplied to the system during friction could be bigger than the dissipation of energy through linear channels (heat and electrical conductivity and diffusion). It can either lead to catastrophic wear mode or to the formation of surface structures [3]. In the latter case, an allocated zone forming with enhanced energy dissipation, i.e., the formation of secondary surface structures in a frictional body is accompanied by energy redistribution.

Let us consider three possible states of a frictional body, with a similar amount of energy supplied to the system. Schematic distribution of energy (E) in a frictional body under various conditions is given in Figure 2.6.

In Figure 2.6, state 1 corresponds to equilibrium conditions, state 2 corresponds to a stationary state with the absence of surface tribo-films (secondary structures) and state 3 is a steady state with the presence of the surface tribo-films (secondary structures; S_1, S_2, and S_3 are the entropies of corresponding states).

$$\text{At } E_1 = E_2 = E_3,\ S_1 > S_2 > S_3$$

State 1: This is the equilibrium state. The energy is distributed uniformly; entropy of a body has maximum value.

State 2: This is a stationary state without surface tribo-films (secondary structures). Energy distributes as a gradually decreasing function with the change of the distance from the friction zone. Accordingly, entropy decreases as compared to an equilibrium state owing to nonuniform distribution of energy.

State 3: This is a stationary state with the presence of tribo-films (surface secondary structures). Most of the energy is concentrated within a narrow zone of tribo-films (surface secondary structures). Entropy decreases in an even larger degree as compared to the stationary state without surface tribo-films.

Therefore, the entropy of a sliding body decreases because of formation of tribo-films (surface secondary structures). According to the S theorem by Klimontowich [48], in a self-organizing system the relation of entropy to energy decreases. We can conclude that the formation of these tribo-films (surface secondary structures) during friction results in decrease of coefficient of friction and wear rate and corresponds to the initiation of the self-organization process and formation of dissipative structures. Self-organization of a tribosystem leads to the accumulation of a major part of the energy within a thin surface layer of the secondary structures. Nonuniform energy distribution at the depth of a frictional body leads to the decrease in entropy production as compared to uniform energy distribution (see later text). Enhanced energy dissipation within a layer of secondary structures could be achieved in two ways: first, for example, is the application of refractory compound-based coatings that are usually termed "hard" coatings (see Chapter 9). In this case intensive energy dissipation leads to intensive generation of heat within surface layers. A hard coating can resist this heat generation. However, these coatings usually have low thermal conductivity (see Chapter 9), and this enhances the heat generation on the surface. We believe that further improvement of hardness for this type of coating is a dead end way for the development of novel coatings. The second way is the application of materials for coating fabrication that enhances nonequilibrium processes under operation. Thus, the energy formed during friction will expand to nonequilibrium processes, which are accompanied by accumulation of energy and entropy reduction instead of heat generation and surface damage. It will result in wear rate decrease. Once self-organization has started, a sudden change in energy accumulation and correspondingly a decrease of wear rate occurs.

2.3.4 Thermodynamics of Surface Secondary Structures

Let us consider a system that consists of a single frictional body and an energy source within the friction zone. The source has no mass; hence, it does not possess entropy. Change of entropy, relative to this single body, will be taken into account. Entropy of a frictional body may vary because of: (1) heat transfer from the source into the body and (2) accumulation of different chemical compounds with their own entropy, i.e., by external impact, which could be called *entropy flow* (dS_e). Entropy

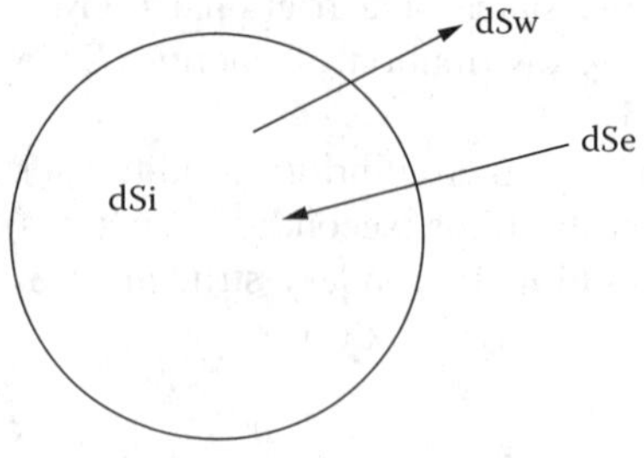

FIGURE 2.7 Entropy change for the frictional body. Here, dS_i is the entropy production ($dS_i > 0$); dS_w, the entropy of the wear particles ($dS_w < 0$); and dS_e, the flow of entropy without entropy of the wear particles.

of the body (dS_i) may vary because of entropy production when heat and other flows come inside it; at the same time, some physical and chemical processes take place within the body. The entropy flow due to heat exchange with the source results in an increase of entropy in the body because heat transfers from a source with a higher temperature into a body with a lower temperature. The chemical compounds that are introduced during the process of friction have their own entropy (dS_m), and because they accumulate on the surface, they tend to decrease the difference between the chemical potentials of the environment and the body, thus increasing its overall entropy. From the second law of thermodynamics, entropy production is positive, ($dS_i > 0$), leading to the increase in entropy of the body. Thus, all the effects mentioned earlier result in entropy growth. However, if the self-organizing process occurs, the entropy has to decrease (at an identical level of energy). The decrease of entropy during friction may be caused by wear. The entropy decrease as a result of a wear process (dS_w) compensates and exceeds entropy growth due to the production of entropy and the flow of entropy into the body. The diagram of a frictional body's entropy change is presented in Figure 2.7. It is known from literature that stable tribo-films (secondary structures) form during the running-in stage of a wear process [68]. Owing to the reasons outlined in the earlier text, it is quite natural that during the running-in stage high wear rate takes place [68].

However, entropy decrease during friction could be a result of: (1) intensive wear (worst case scenario, which is observed quite often) and (2) beneficial structure and phase transformations that take place on the surface (best-case scenario, which is considered in detail in this book). A few practical examples of this situation are thoroughly studied, for instance, in Chapter 9, Figure 9.30. After the surface tribo-films (secondary structures) have formed, two variants are possible: (1) increase of entropy in secondary structures (if the tribo-films are equilibrium, Figure 2.6b) or (2) decrease of entropy (in the case of nonequilibrium tribo-films, Figure 2.6c) as compared to the entropy of reacting substances (i.e., materials of the frictional bodies, environment, lubricants, etc.).

Thus, surface secondary structures can have an equilibrium or nonequilibrium nature. In the first case, the entropy of their formation (dS_f) is positive and, in the second, negative. Let us consider both variants. At the initial stage of the formation of tribo-films, (secondary structures) entropy increase (dS) of the frictional body consists of the following components:

$$dS = dS_e + dS_i + dS_m + dS_f - |dS_w| \tag{2.67}$$

where dS_i is the entropy production regardless of the entropy of physical and chemical transformations within the layer of surface secondary structures; dS_e, the entropy flow regardless of the entropy of compounds that have been formed during friction on the surface; dS_m, the change of entropy due to the introduction during friction of compounds with their own entropy; dS_f, the change of entropy due to the formation of tribo-films (surface secondary structures); and dS_w, the change of entropy as a result of a wear process.

As was mentioned earlier, dS_e is positive, dS_i is positive according to the second law of thermodynamics, dS_m is positive because the compounds formed during friction possess their own entropy, and entropy of the formation of surface tribo-films (secondary structures) dS_f may be either positive or negative, for example, $dS_f > 0$ during formation of non saturated solid solutions or stoichiometric chemical compounds. In contrast, $dS_f < 0$ if the structures and phases that are formed during friction do not correspond to the equilibrium constitutional diagrams or if the chemical reactions with negative chemical affinity take place.

Change of entropy as a result of a wear process is always negative because the substances that are removed as a result of the process take away their own entropy. Therefore, $|dS_w|$ is negative in Equation 2.67.

During the running-in stage of wear, a relative area covered by tribo-films grows continuously [68]. Considering this growth, we may present the wear process as follows. The components of environment, lubricant, and counterbody deposit on a friction surface during sliding. Interacting with each other and with a material of the frictional body, they form the surface tribo-films (secondary structures), which are subject to wear. However, because experiments show that a relative area covered by tribo-films continuously grows [68], a conclusion can be made that only a part of them, $h(h < 1)$ is worn. Then:

$$dS_w = h|dS_f| + hdS_m + h|dS_s| \tag{2.68}$$

where dS_s is the change of entropy of a frictional body because wearing of the substrate material. At $dS_f > 0$, Equation 2.68 with regard to Equation 2.62 transforms to:

$$dS = dS_e + dS_i + (1 - h)|dS_f| + (1 - h)dS_m - h|dS_s| \tag{2.69}$$

The only negative term is the last one ($-hdS_s$) on the right-hand side of Equation 2.69. This term largely controls the wear rate of the frictional body.

There is supposed to be a certain amount of entropy decrease to start a self-organization process (in our case, to form the tribo-films, or surface secondary structures). Following from Equation 2.69, increased wear rate is required for self-organization to begin.

At $dS_f < 0$, Equation 2.68 with regard to Equation 2.70 transforms to:

$$dS = dS_e + dS_i - (1 - h)|dS_f| + (1 - h)dS_m - h|dS_s| \tag{2.70}$$

The right-hand side of the Equation 2.70 has two negative terms. One is related to the wear (hdS_s) and another to the chemical–physical transformations within the surface layers of a frictional body $(1 - h)|dS_f|$. Therefore, the same amount of decrease in entropy, as in the case of Equation 2.70, will be attained with a lower wear rate of the frictional body.

Thus, a very important conclusion could be made — the wear intensity of the frictional body is much lower if the surface secondary structures are in a nonequilibrium state as compared to those in that of equilibrium. As a consequence of this conclusion, it is possible to assume that nonequilibrium processes occurring in the tribo-film layers (surface secondary structures) of a frictional body can result in significant wear rate decrease. In contrast, equilibrium processes occurring in a frictional body could result in an increase of wear rate.

Yet, considering that the sum of dS_f and dS_i is the total entropy production, their sum should be positive, and hence, if $dS_f < 0$,

$$dS_i - |dS_f| > 0 \tag{2.71}$$

It means that the compensation of entropy production exclusively as a result of its decrease at nonequilibrium transformations within the layer of surface secondary structures is impossible. This is illustrated in the example of structure transformation during plastic deformation [69]. A formation of fragmented structures takes place during the large plastic deformation during friction [70]. During the process of fragmentation, the dislocations, stress, and subsequently the energy of deformation localize on the boundaries of fragments. There is a practically defect-free structure inside the fragments. Such nonuniformity of distribution of deformation energy promotes entropy decrease in a frictional body. However, this decrease of entropy is a part of entropy production, which as a whole should be positive according to Reference 64. Therefore, the wear of a frictional body is necessary during the formation and growth of the tribofilms (surface secondary structures).

In the stationary state the surface tribo-films (secondary structures) complete their evolution and the processes are stabilized in the frictional body. Therefore, the quantity of matter coming from the environment and the counterbody should be equal to the amount of matter leaving the body during the wear process, and therefore, in the stationary state, $h = 1$.

Having differentiated the entropy increment in Equation 2.68 and Equation 2.69, we obtain the entropy rate:

For $dS_f > 0$:

$$dS/dt = dS_e/dt + dS_i/dt + (1 - h)|dS_f|/dt + (1 - h)dS_m/dt - h|dS_s|/dt \tag{2.72}$$

For $dS_f < 0$:

$$dS = dS_e/dt + dS_i/dt - (1 - h)|dS_f|/dt + (1 - h)dS_m/dt - h|dS_s|/dt \tag{2.73}$$

In a stationary state, Equation 2.72 and Equation 2.73 will be equal:

$$dS/dt = dS_e/dt + dS_i/dt - |dS_s|/dt = 0 \tag{2.74}$$

The first two terms in the right-hand side are positive. The last term in the right-hand side appears with the minus sign. This term represents the wear of substrate material in a frictional body. Thus, the wear of the substrate material has to take place in order to maintain a stationary state of the tribosystem. Following from Equation 2.74, the lower the entropy production at a fixed entropy flow, the smaller the wear rate. The scheme of entropy balance for the frictional body is presented in Figure 2.8.

J.L. Klimontovich [48,51] applied Prigogine's theorem of minimal entropy production for the processes of self-organization. His formulation is as follows: Entropy production in a new stationary state that occurs after a current nonequilibrium phase transformation is lower than entropy production of an old state of system expanded to an unstable area (Figure 2.4). The given formulation confirms a conclusion that in a stationary state the intensity of the wear process for nonequilibrium

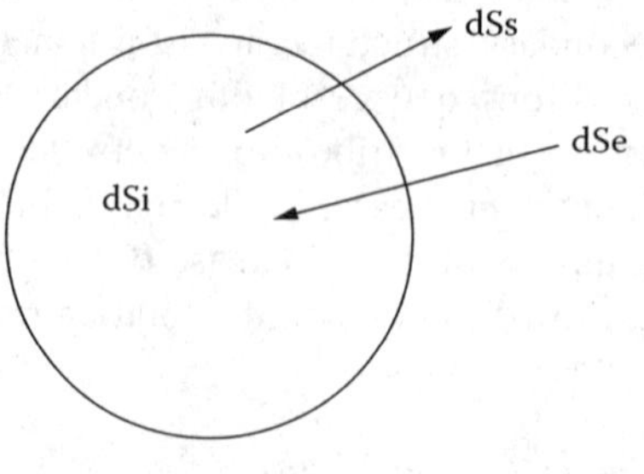

FIGURE 2.8 Stationary state of the frictional body. Here, dS is the change of entropy of a frictional body; dS_i, the entropy production; dS_e, the flow of entropy; dS_s, the change of entropy of a frictional body due to the wear of the substrate material; and $dS = dS_i + dS_e - dS_s = 0$ and $dS_s = dS_i + dS_e$. The lower is entropy production (dS_i) at a fixed flow of entropy (dS_e) the smaller is a change of entropy due to the wear of the substrate material.

surface tribo-films (secondary structures) is lower compared to ones in equilibrium. It happens because self-organization leads to the formation of dissipative structures. This formation is followed by a decrease in entropy. That is why after self-organization starts entropy production decreases, i.e., $|dS_s|/dt$ decreases in Equation 2.74, where S_s is the entropy of a frictional body's substrate material. Taking into account that entropy is an additive value, the wear rate of the frictional body will decrease once the $|dS_s|/dt$ drops down under other equal conditions. One of the most important practical issues is the intensification of service conditions for different tribo-pairs. It relates, for instance, to the intensification of the machining process (see Chapter 11 for details) and other applications. That is why the effect of service conditions' intensification on the friction phenomena should be studied in further detail.

Evolution of a tribosystem is accompanied by the formation of the tribo-films (surface secondary structures) because a tribosystem without the surface secondary structures is unstable to external disturbance. Thus, formation of surface tribo-films (secondary structures) corresponds to Le Chatelier's principle [68]. A tribosystem responds to the external disturbance by forming surface secondary structures, which reduce the intensity of the wear process. If a system cannot form the surface secondary structures, a catastrophic wear mode such as seizure develops. The seizure can also be considered as a specific adaptability of the system because there is no relative sliding of the bodies and, therefore, there is no friction within the zones of seizure. Thus, owing to seizure, the system tends to stop friction, whereas the formation of surface secondary structures facilitates friction.

The stationary state generates stability for the surface secondary structures, and its analysis is therefore of great interest.

Entropy exchange between surface secondary structures and the environment occurs during friction because of the flow of matter and energy from the contact zone. This results in entropy increase (dS_{t1}/dt) due to the transfer of energy into the frictional body. At the same time, owing to wear, it leads to a decrease in entropy (dS_{t2}/dt).

The rate of entropy change for surface secondary structures in a stationary state (dS_t/dt) can be represented as:

$$dS_t/dt = dS_{t1}/dt - dS_{t2}/dt + dS_{ti}/dt = 0 \tag{2.75}$$

where dS_{ti}/dt is the entropy production within surface secondary structures.

It follows from Equation 2.72 that entropy production will be:

$$dS_{ti}/dt = dS_{t2}/dt - dS_{t1}/dt \tag{2.76}$$

The term dS_{t2}/dt is a part of entropy flow, which is expanded to increase the entropy of the frictional body and corresponds to the term dS_e/dt in Equation 2.74. Owing to the additivity of entropy, the smaller the value of dS_{ti}/dt, the lower the wear rate. Under constant external conditions, dS_{t2}/dt remains constant as well. Following from Equation 2.76, under conditions of minimal wear rate, when the surface tribo-films (secondary structures) occur:

$$dS_{ti}/dt \rightarrow \min \tag{2.77}$$

Thus, under conditions of minimal wear rate in the presence of surface secondary structures, their stationary state is characterized by minimal entropy production. This was shown experimentally in Reference 71. In the presence of the dissipative structures a state with minimal entropy production could be stable with change of various friction parameters. That is why (as will be shown in the following text) in the example of sliding electric contacts at a stable state with minimal entropy production, the coefficient of friction will decrease correspondingly with the electric current values (Chapter 8). A similar situation will take place during cutting when the wear rate will drop with cutting speed (Chapter 10 and Chapter 11). Therefore, a correlation between the parameters of friction could occur in this state, but until self-organization has started this correlation cannot occur. This conclusion is widely used for various practical applications (see Chapter 8 to Chapter 11).

2.3.5 Rational Selection of the Materials for Frictional Bodies

The theorem by Klimontovich connects two basic conclusions of the present chapter:

- Intensity of wear decreases after nonequilibrium processes that developed during entropy decrease have begun
- Intensity of wear drops with the decrease of entropy production

After the self-organization process has started, dissipative structures begin to form. The formation of dissipative structures is associated with the initiation of nonequilibrium processes. According to the Klimontovich theorem [50], entropy production in a stationary state of dissipative structures should be less than the entropy production in a stationary state that existed prior to the beginning of the self-organization process. Hence, the intensity of a wear process in the presence of dissipative structures will be lower than the intensity of a wear process without dissipative structures under the same external conditions.

The physical explanation of decrease in wear intensity after the beginning of nonequilibrium processes is that a significant amount of energy has to be spent on the formation of dissipative structures. Therefore, after nonequilibrium processes have been initiated, there is less energy available within the tribosystem for the frictional bodies' surface damaging processes that are associated with intensive wear modes.

It was shown earlier that, according to the Klimontovich S theorem [50], formation of surface secondary structures during friction corresponds to the initiation of the self-organization process and the generation of dissipative structures. The structures are stable because their state does not depend on entry conditions, and their formation could be characterized by sufficient repeatability. Such states of a system are known as attractors [72]. Hence, the composition of surface tribo-films (secondary structures) depends only slightly on the entry conditions. Therefore, the selection of materials for the frictional bodies as well as lubricants can be performed based on the results of detailed investigations of the composition and structure of tribo-films (secondary structures) as shown in detail in Chapter 6 to Chapter 13. During these studies it is absolutely necessary to identify the nonequilibrium physical and chemical processes of the formation of surface tribo-films (secondary structures), because the nonequilibrium processes are responsible for the wear rate decrease. After that it is necessary to change the compositions of the frictional bodies or other substances participating in friction to increase the intensity of nonequilibrium processes and to shift the initiation of these processes to milder friction conditions. It is possible, for instance, to introduce some specific catalysts of appropriate nonequilibrium processes into the substances participating in friction [10]. The rational selection of the materials for frictional bodies based on the results of studies of surface tribo-films' (secondary structures') composition and structure for some specific applications will be shown later (Chapter 3 and Chapter 8).

Although the nature of dissipative structures somewhat depends on the entry conditions, it is reliant even more strongly on boundary conditions. In particular, the compositions of frictional bodies can be used as a boundary condition during friction. Hence, the compositions of the tribo-films (secondary structures) will depend on the compositions of frictional bodies. It appears that the friction of bodies with compositions that strongly differ from each other (at the same counter-body) could not be recommended. Each frictional body will form the typical tribo-films that belong to this specific body. The tribo-films structures formed by another body with totally different characteristics could intensively wear out in this case. Thus, the process of friction of such bodies can be accompanied by undesirable intensive wear processes. Some specific examples of this complex phenomena will be shown later in Chapter 3 and Chapter 8.

2.3.6 Instability of a Tribosystem

The independence of the nature of dissipative structures and surface tribo-films (secondary structures) from the entry conditions is related to the fact that a process of self-organizing begins after the system loses the stability of its previous state [10].

The tribosystem can pass through this instability several times because of evolution. [73]. An increasing external effect, e.g., the increase in sliding speed or frictional force, causes a systematic deviation of a given system from the equilibrium state, and therefore dissipative structures are able to form within the system. In the first instance, it loses its stable equilibrium state with the occurrence of a relative nonzero sliding speed. This can be attributed to the fact that, in the equilibrium state, the macroscopic particles of the system cannot move against each other. Further losses in stability take place owing to the evolution of secondary structures on the friction surfaces.

As was shown in the beginning of the chapter, the possibility of thermodynamic stability loss is defined by the plus or minus sign for the value of excess entropy production. According to Equation 2.22, the entropy production is defined as the sum of the products of generalized forces on the appropriate generalized flows of irreversible processes.

Let us assume that friction is the only independent source of energy dissipation in the system. In this case the entropy production has the following form:

$$dS_i/dt = J_t X_t \tag{2.78}$$

Accounting $X_t = -BgradT$, $J_t = -qB\text{grad}T = kpv$, we obtain:

$$dS_i/dt = (kpv)^2/BqT^2 \tag{2.79}$$

where k is the coefficient of friction; p, the loading; v, the sliding speed; q, the heat conductivity; T, the temperature; and B, the area of the contact.

Excessive entropy production has the following forms:

As the load varies:

$$\partial(\delta^2 S)/2\partial t = v^2(p\partial k/\partial p + k)^2(\delta p)^2/(BqT^2) \tag{2.80}$$

As the speed varies:

$$\partial(\delta^2 S)/2\partial t = p^2(v\partial k/\partial v + k)^2(\delta v)^2/(BqT^2) \tag{2.81}$$

The right-hand side of expressions Equation 2.80 and Equation 2.81 are quadratic forms, i.e., they can be only positive (considering that q is always greater than 0). Hence, we have to emphasize that when friction is the only source of energy dissipation the system does not lose its stability and consequently self-organization does not occur.

In real systems some other sources of energy dissipation can be available along with friction, e.g., superposition of electromagnetic fields including passage of current, radiation, various oscillation processes, etc. In addition to friction, some physical–chemical interaction always takes place between frictional bodies, the environment, and the lubricant. With significant deviations from an equilibrium state, the specific chemical kinetics will also be important in providing the stability of the given state of a system.

Physical and chemical interactions of a frictional body with the environment or a counterbody results in formation of surface tribo-films. According to B. Kostetsky, the properties of superficial secondary structures considerably differ from the properties of a frictional body [4]. Equation 2.80 and Equation 2.81 include heat conductivity. Heat conductivity will change during friction because of the formation of surface secondary structures, i.e., the change of heat conductivity will characterize physical and chemical processes occurring on the friction surface. Then, in the expressions signifying excessive entropy production, not only the coefficient of friction but also heat conductivity values will vary. In this case, Equation 2.80 and Equation 2.81 will be as follows:

As the load varies:

$$\frac{\partial(\delta^2 S)}{2\partial t} = (v^2 / BqT^2)\left(p\frac{\partial k}{\partial p} + k\right)\left(\left(p\frac{\partial k}{\partial p} + k\right) - (kp/q)\frac{\partial q}{\partial p}\right)(\delta p)^2 =$$
$$(v^2 / BqT^2)\left(\left(p\frac{\partial k}{\partial p} + k\right)^2 - (kp/q)\left(p\frac{\partial k}{\partial p}\frac{\partial q}{\partial p} + k\left(\frac{\partial q}{\partial p}\right)^2\right)\right)(\delta p)^2 \quad (2.82)$$

As the velocity varies:

$$\frac{\partial(\delta^2 S)}{2\partial t} = (p^2 / BqT^2)\left(v\frac{\partial k}{\partial v} + k\right)\left(\left(v\frac{\partial k}{\partial v} + k\right) - (kv/q)\frac{\partial q}{\partial v}\right)(\delta v)^2 =$$
$$(p^2 / BqT^2)\left(\left(v\frac{\partial k}{\partial v} + k\right)^2 - (kv/q)\left(v\frac{\partial k}{\partial v}\frac{\partial q}{\partial p} + k\left(\frac{\partial q}{\partial v}\right)^2\right)\right)(\delta v)^2 \quad (2.83)$$

Excessive entropy production in Equation 2.82 and Equation 2.83 can become negative because of a negative sign "–" before the second term in the brackets. Whereas excessive entropy production could become negative, the sum $\left(p\frac{\partial k}{\partial p}\frac{\partial q}{\partial p} + k\left(\frac{\partial q}{\partial p}\right)^2\right)$ or $\left(v\frac{\partial k}{\partial v}\frac{\partial q}{\partial v} + k\left(\frac{\partial q}{\partial v}\right)^2\right)$ should remain positive. It is defined by the sign of the products $\frac{\partial k}{\partial p}\frac{\partial q}{\partial p}$ and $\frac{\partial k}{\partial v}\frac{\partial q}{\partial v}$. These products should be positive. Thus, the system of friction can lose its stability if the coefficient of friction and heat conductivity simultaneously increase or decrease with loading or speed. The increase of the coefficient of friction with loading or speed can result in a seizure. The increase of heat conductivity will be observed in this case. It can be called the degenerative case of self-organization when the system tries to stop the friction and wear process with the help of seizure (there is no friction within a zone of seizure). Reduction of the coefficient of friction with the loading or sliding speed is associated, as a rule, with the formation of surface secondary structures and corresponds to the process of self-organization.

Generally, it is possible to make a very important conclusion that the process of self-organization during friction is possible if one more or more independent processes, except friction itself, are affecting a body. The analysis of known self-organizing systems shows that dissipative structures form as a result of the interaction of two or more processes. For example, the formation of the Benard cells is the result of interaction of thermal conductivity and gravity [28]. Turing structures [47] are formed as a result of interaction between chemical reaction and diffusion. Martensite formation is the result of the interaction of thermal conductivity and phase transformation [48]. On the other hand, thermodiffusion and similar phenomena [9] are not the result of self-organization, because they are caused by a single process. They are triggered by this specific process and develop gradually without sudden change. Therefore, the system does not lose its thermodynamic stability.

Under varying conditions, self-organization could accelerate or slow different processes. To be initiated, self-organization needs two or more independent external impacts on the system. Moreover, this conclusion could be expanded for other physical processes, for instance, high-temperature oxidation as observed in Chapter 11. As will be shown in that chapter, the oxidation rate could be controlled by self-organization if a second nonequilibrium process exists in addition to the isothermal oxidation. This is a relaxation process within strongly nonequilibrium nanostructures.

2.3.7 Stationary State of Surface Tribo-Films

Self-organization and formation of dissipative structures take place in a nonlinear area. Therefore, Curie's principle is not operative in this area [13]. Therefore, the processes described by tensors of ranks of different parity can cooperate inside dissipative structures. For example, the speed of chemical processes is a scalar, i.e., it is a tensor of a zero rank; the thermal flow is a vector, i.e., it is a tensor of the first rank. These processes do not interact with each other on a linear area, but are able to interact within dissipative structures. In a stationary state of the dissipative structures, minimal entropy production keeps the value within a wide range of changes of external conditions. Strictly speaking, Prigogine's theorem does not operate under the conditions of the existence of the dissipative structures. However, according to the Klimontovich theorem [50], entropy production after self-organization will be lower than at previous conditions. Therefore, it is possible to assume that, for the new stationary condition Prigogine's theorem can be applied. In this case, it is possible to neglect the flows of unfixed thermodynamic forces. Thermodynamic forces that cause a flow of matter are not fixed. Therefore, the terms in the equation for entropy production connected to flows of matter, i.e., the wear process, can be neglected. Entropy production will be defined by expression Equation 2.79. Let us assume that entropy production does not vary within the certain range of loading and speed of sliding change. Then, it will follow from Equation 2.79 that:

$$k^2 = CBqT^2/(pv)^2 \tag{2.84}$$

where C is the constant value of entropy production.

As described in Equation 2.84, in a stationary state of dissipative structures, at minimal wear rate, relations could also appear between a coefficient of friction and loading and sliding speed. In this case, the coefficient of friction can decrease with values of loading and speed of sliding.

Mechanisms of these relations could be different, and they are usually connected to nonequilibrium processes. These processes can be nonequilibrium chemical reactions in which the material of a frictional body cannot participate as a reagent. It can be partition of phase components, such as allocation of a soft phase in a surface layer. Such effects are observed in bronze–plastic bearings and also in bearings from aluminum or copper alloys [74]. In these materials, the soft phase had a uniform distribution in the structure of the alloy. Owing to the occurrence of the aforementioned relations the soft phase concentrates within a surface layer [74]. After that the distribution ceases to be equal and so an ordering of structure occurs. It should be accompanied by a decrease in entropy (Chapter 3).

We can assume that, at the very moment when the dependence of the coefficient of friction on other friction parameters occurs, it can be used as a criterion of self-organizing process initiation and formation of dissipative structures. The examples of this approach are the following: lubricating action of a current, i.e., reduction of the coefficient of friction with an electric current in a sliding electric contact [5]; effect of selective carry, i.e., the reduction of the coefficient of friction in a specific environment in which the nonequilibrium processes of ion separation occur [75]; and

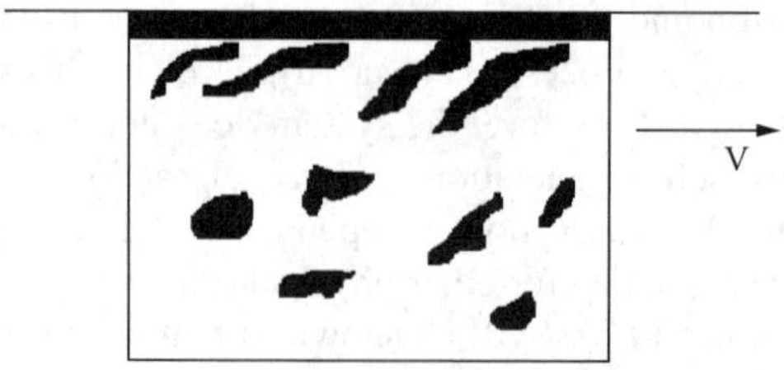

FIGURE 2.9 Self-organization during friction of the frictional body with a hard matrix (light field) and inclusions of a soft phase (dark field). The self-organization is developing by the accumulation of the soft phase on the friction surface. Before friction the soft phase was evenly distributed in the structure. This is typical for example for the Al–Sn-based bearing alloys.

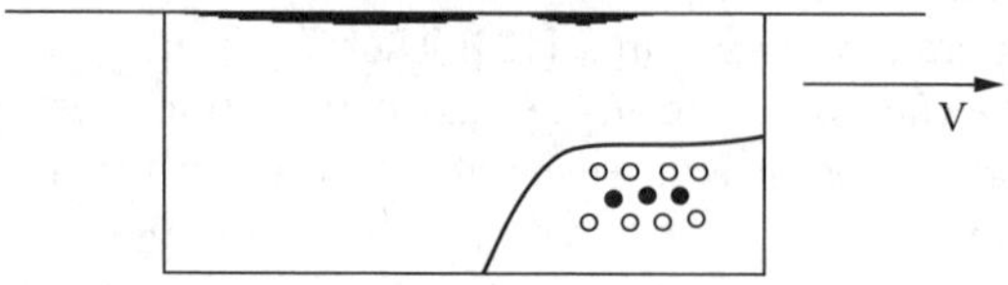

FIGURE 2.10 Self-organization during friction of the frictional body with a structure that consists of solid solution of a soft metal in a hard matrix. This is typical for Cu–Sn-based alloys. This structure corresponds to the binary constitutional diagram. — the atoms of the hard material (a solvent); — the atoms of the solved soft metal. Self-organization develops in the segregation of the solved metal from solid solution with a formation of island-like films of the soft metal (dark spots) on the friction surface.

effect of abnormal low friction, i.e., reduction of the coefficient of friction under irradiation on frictional surfaces [75].

A few examples of self-organization are shown in Figure 2.9 to Figure 2.12. Figure 2.9 and Figure 2.10 show a structural self-organization. In the first case (Figure 2.9), this is self-organization of a two-phase alloy with a hard matrix and evenly distributed soft inclusions. Such structure is typical for a bearing Al–Sn-based alloy. Self-organization of this alloy consists in soft-phase accumulation on the friction surface. The process will be described in detail in Chapter 3. In the second case (Figure 2.10), this is self-organization of a material with a structure that consists of a single-phase solid solution of a soft material in a hard one. Such structure is typical for Cu–Sn alloys. This structure corresponds to the binary constitution diagram. Self-organization of this alloy consists in the soft-phase segregation from a solid solution on the friction surface. The process will be also considered in detail in Chapter 3.

Figure 2.11 presents one of the examples of self-organization due to friction. The example presented is a result of interaction with the environment when nonequilibrium chemical reaction takes place that results in a decrease of entropy. The process will be considered further in Chapter 8 for the sliding electric contacts.

Figure 2.12 presents a second example of self-organization due to friction. Figure 2.12 shows the self-organization process of the frictional body with a complex composition. Different modes

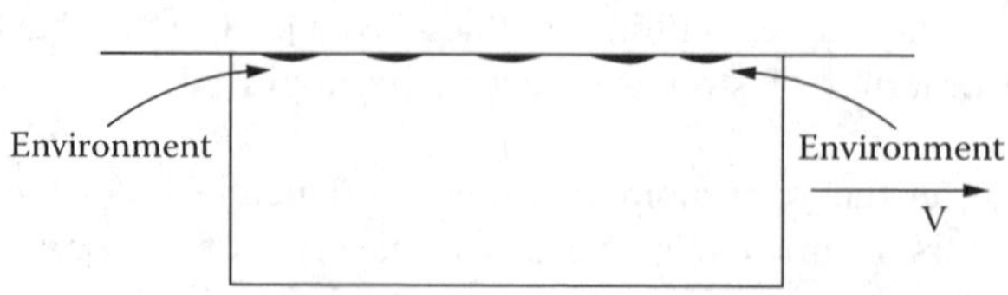

FIGURE 2.11 Self-organization during friction owing to interaction of the frictional body with the environment. Self-organization develops in nonequilibrium reactions on the friction surface with entropy decrease. As a result, the products of these nonequilibrium reactions are forming on the surface (dark spots).

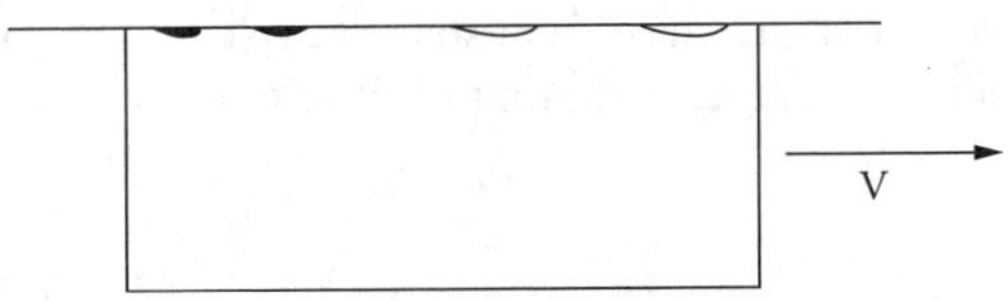

FIGURE 2.12 Self-organization of the frictional body with a complex composition when a few different mechanisms of friction takes place on the surface. Self-organization develops in a selective mass transfer of the specific elements to the zones of surface that are working under various friction conditions. For example, dominating mass transfer of heavy elements to the zone of seizure is taking place. Zones of the friction surface enriched by heavy elements are forming (light spots) and the wear rate drops. Mass transfer of light elements to the zone of dry sliding is also taking place. Zones of friction surface enriched by light elements are also forming (dark spots), which also results in wear rate reduction.

of friction (dry sliding and seizure) take place simultaneously. Self-organization consists in selective mass transfer of the specific elements to the areas with varying wear modes. Within the zone of regular sliding, a mass transfer of light element dominates, whereas within a seizure zone, the heavy elements dominate similarly. This is typical for higher-ordered quaternary wear-resistant coatings, which are further described in Chapter 9 to Chapter 11. It is necessary to note that all of the effects outlined occur with entropy decrease and are accompanied by a significant drop of wear rate.

2.3.8 Thermodynamics of the "Lubricating" Action of a Current

In the present chapter, the system of friction with current collection is used as an example of the thermodynamic conclusions outlined earlier. Other practical applications of these principles are shown in details in the corresponding chapters (Chapter 3 to Chapter 11). Application of Prigogine's theorem [6] and the Le Chatelier–Brown generalized principle [66] is demonstrated as an example of strong-current sliding electrical contacts. An analytical expression describes the "lubricating" action of an electric current and corresponds to the experimental data obtained. The sliding current collection tribosystem can be considered according to the self-organization approach because two independent processes occur under operation, i.e., current collection and friction. It is interesting to find out what dissipative structures could form during the interaction of these processes.

In thermodynamics of irreversible processes, entropy production is defined by Equation 2.22. There are three basic thermodynamic flows: flow of heat from friction, electric current, and flow of matter within the surface secondary structures of a sliding electric contact. Thermodynamic forces and flows (X_e, J_e) in this contact may be considered as fixed ones because of the continuous passage of the electric current. A similar conclusion can be drawn for the coefficient of friction, forces, and flows (X_t, J_t). This, however, does not apply to the thermodynamic forces associated with the flow of matter. According to the Prigogine theorem [6], the flows of unfixed forces vanish in a stationary state.

The flow is not determined by operation conditions. Ignoring the matter flows, we get:

$$\frac{dS_i^*}{dt} = J_t X_t + J_e X_e \tag{2.85}$$

where J_e and J_t are the flow of electrical current and of heat and X_e is the voltage.

Using Equation 2.79 it is possible to represent Equation 2.85 as:

$$\frac{dS_i^*}{dt} = \frac{(kpv)^2}{qBT^2} + \frac{J_e X_e}{T} \tag{2.86}$$

Entropy production is minimal and stable in a stationary state under minimum wear rate within certain limits of the parameters' change, including electric current. Therefore, at the change of an electric current value:

$$\left(\frac{d}{dJ_e}\right)\frac{dS_i^*}{dt}=0 \tag{2.87}$$

$$\left(\frac{d}{dJ_e}\right)\frac{dS_i^*}{dt}=\frac{2k\left(\frac{dk}{dJ_e}\right)(pv)^2}{qBT^2}+\frac{X_e}{T}=0;\quad \frac{dk^2}{dJ_e}=-\frac{qBTX_e}{(pv)^2};$$
$$k=\sqrt{k_0^2-\frac{J_eTBqX_e}{(pv)^2}} \tag{2.88}$$

where k_0 is the coefficient of friction without a current.

Equation 2.88 is obtained if we assume that the heat conductivity and voltage in the surface secondary structure do not vary with an electric current. This expression is an analytical expression for the effect of electric current "lubricating" action. Following from Equation 2.87:

$$C=(1-k^2/k_0^2)/J_e=(qBTX_e)/(pvk_0)^2 \tag{2.89}$$

According to Equation 2.89, C should remain constant with current.

It follows from Equation 2.89 that the coefficient of friction is reduced with current passing through the sliding contact. Moreover, the influence of current becomes more significant with the increase of voltage in surface secondary structures (X_e). It corresponds to the results received in Reference 74 on the effect of lubricating action of a current during current collection of metal–graphite and metal–polymer electric brushes. It was outlined in Reference 74 that an increase of copper content in composite materials reduces or (at the copper content of 90 wt%) suppresses the effect of lubricating action of a current. The increase of the copper contents in nonmetallic materials results in the decrease of contact voltage between a brush and a copper counterbody (X_e in Equation 2.89). According to Equation 2.88, the effect of lubricating action of a current is reduced with the decrease of contact voltage.

Equation 2.88 and Equation 2.89 were tested on a service bed of the contact system with current variation from 0 to 720 A. A direct current was passed through the sliding contact. Data on measured currents ratio k^2/k_0^2 and calculated values of C depending on the current are presented in Table 2.1.

From the data given in Table 2.1 it follows that the lubricating action of a current could be demonstrated at the values of a current above 300 A. Because the current value is 400 A, it is possible to get the constant value C from Equation 2.89 practically independent of the current. By increasing the current 2.4 times, from 300 to 720 A, the coefficient of friction is reduced by 5.5 times.

It is necessary to note that the effect of the lubricating action of a current was not observed on a copper counterbody with removed surface secondary structures.

The factor of correlation of the conformity of experimental values k/k_0 and the current (Equation 2.89) is equal to 0.91.

Existence of the lubricating action of a current effect and its conformity to the analytical expression (Equation 2.89) can be used as proof of the conclusion that the state of the surface secondary structures with minimal entropy production is stable in relation to the change of an electric current at a low wear rate. The generalized principle of Le Chatelier–Brown [66] could be used in this case. In this state, change of entropy production with the variation of a current is

TABLE 2.1
Data on Measured Currents, Ratio k^2/k_0^2 and Calculated Values of C Depending on the Direct Current

Current (A)	k/k_{01}[a]	k/k_{02}[a]	k/k_{03}[a]	C (2.25)
0	1	1	1	—
80	1.29	1.21	1.15	—
100	1.80	1.63	1.48	—
300	0.90	0.87	0.88	0.00064
400	0.69	0.67	0.70	0.0013
460	0.47	0.44	0.46	0.0017
540	0.41	0.39	0.42	0.0015
560	0.27	0.26	0.28	0.0014
660	0.26	0.25	0.27	0.0014
700	0.20	0.20	0.22	0.0014
720	0.18	0.18	0.19	0.0014

[a] k_{01}, k_{02}, k_{03} are various values of factor of friction without a current.

compensated by the change of the coefficient of friction but not by the occurrence or change of wear rate (i.e., flow of matter). Thus, the minimal wear rate takes place even if the current values within the system are changing.

Equation 2.88 was created considering the absence of a flow of matter (condition of the insignificant wear rate) and according to the conditions of Prigogine's theorem [6]. Significant decrease in wear rate under the lubricating action of a current was observed many years ago [75]. The conditions of the Prigogine theorem were observed in this case. In conclusion, the hypothesis about the insignificant wear intensity in a steady state is confirmed. The occurrence of new connections between elements of a system (in this case, between a coefficient of friction and a current) indicates new structure formation or, in other words, ordering of the system.

It should be noted that $k_{02} = 0.8\ k_{01}$ and $k_{03} = 0.5k_{01}$. Following from the data shown in Table 2.1, despite various values of the initial coefficient of friction without a current, the relation k/k_0 depends only on the value of a current and does not depend on the value of k_0. Such independence of the entry conditions can be used to confirm that the processes of self-organization have been initiated in the tribosystem and that the structures formed are stable dissipative structures.

Basic changes in the tribosystem occur within a thin surface layer in a zone of friction, and they are related to the formation of surface tribo-films. It is possible to assume that, if the effect of a lubricating action of a current occurs and wear rate decreases, then the structures that are formed on the surface are dissipative structures. The mechanism of this phenomenon and the details of development of wear resistant materials for current collection materials based on self-organization concept are described in further details in Chapter 8.

2.4 CONCLUSIONS

The fundamentals of friction phenomena based on the concept of irreversible thermodynamics were described in detail.

It was shown in the present chapter that wear rate of nonequilibrium surface tribo-films (secondary structures) is lower than that compared to the wear rate of equilibrium surface tribo-films (secondary structures).

Taking into account the independence of dissipative structures from entry conditions, i.e., their "constancy" under the given conditions of friction, it is obviously possible to perform the development and rational selection of different tribological materials based on the investigations of the characteristics of surface tribo-films (secondary structures). During development of the wear-resistant materials, major attention should be paid not only to their volumetric properties but also (and mainly) to their ability to enhance nonequilibrium processes (the processes that are taking place with entropy decrease) on the friction surface. That is why the development of wear-resistant materials should be mainly focused on the enhancement of strong nonequilibrium processes under milder friction conditions. This direction of research could be considered as a trend for future developments of novel generation of tribo-materials. The traditional ways of improvement of wear-resistant materials are their alloying or surface engineering (such as special heat treatment, hardening, coating deposition etc.). As a result of this engineering, the characteristics of the wear-resistant materials, such as mechanical properties (hardness, strength, fatigue strength, fracture toughness, and so on), conductivity (thermal and/or electrical) heat or thermal resistance, and other volumetric properties, improve. However, it is necessary to emphasize that volumetric properties are not directly related to friction and wear. Surface layers that form during friction have a composition and properties that significantly differ as compared to the volumetric properties of the frictional body. That is why the traditional ways of development of wear-resistant material using volumetric properties do not always guarantee achievement of the highest possible wear resistance. Therefore, the examples of the development of wear-resistant materials that is based on the concepts of self-organization and irreversible thermodynamics are presented in Chapter 8 to Chapter 11.

It is shown in this chapter that wear rate drops because of the initiation of nonequilibrium processes (processes that develop with entropy decrease and energy accumulation) in the tribo-system. When the wear-resistant material is developing, it is necessary to (1) select their composition, structure, and thermodynamic state in a way so as to enhance the nonequilibrium processes on the frictional surface and (2) shift the initiation of processes to milder friction conditions. To understand in depth the features of nonequilibrium processes that have been taking place during friction, it is necessary to investigate the formed secondary structures. Various examples of this methodology are presented in Chapter 8 to Chapter 11. For enhancement of nonequilibrium processes and to shift them to milder friction conditions, some catalyst could be used. An example of such alloying is given in Chapter 8. In order to initiate the self-organization process that leads to the decrease of wear rate, an interaction of two or more independent processes in the system is mandatory. A higher number of independent irreversible processes in the system increase the probability of self-organization. If the materials of the frictional bodies have a nonequilibrium structure, the process of relaxation is enhanced. In this case, the relaxation process interacts with the process of friction. Interaction of these processes could lead to a significantly higher probability of initiation of the self-organization process of a frictional body with a nonequilibrium structure than that of an equilibrium one. Some specific applications of this principle as well as other thermodynamic principles outlined in this chapter are widely presented in the corresponding chapters of the book. Examples of self-organization and wear rate decrease using surface-engineered materials with nonequilibrium structures are presented in Chapter 9 to Chapter 11.

REFERENCES

1. Tabor, E.D. Friction as dissipative process. In *Fundamentals of Friction: Macroscopic and Microscopic Processes*, Eds., Singer, I.L., Pollock, H.M., Vol. 220, NATO ASI Series E: Applied Sciences. Kluwer Academic Publishers: Dordrecht, 1992, pp. 3–24.
2. Bushe, N.A., Kopitko, V.V. *Compatibility of Rubbing Surfaces*. Moscow: Science, 1981.
3. Klamecki, B.E. A thermodynamic model of friction. *Wear* 1980, 63, 1, 113–120.
4. Kostetsky, B.I. *Surface Strength of the Materials at Friction*. Technica: Kiev, 1976.

5. Gershman, I.S., Bushe, N.A. Thermodynamic aspects of the existence of stable secondary structures on surfaces of sliding contacts at high current. *Journal of Friction and Wear* 1989, 10, *2*, 24–29.
6. Prigogine, I. *Etude thermodynamique des processus irreversible*. Desoer: Liege, 1947.
7. Glansdorff, P., Prigogine, I. *Thermodynamic Theory of Structure, Stability and Fluctuations*. Wiley-Interscience: London, 1970.
8. Ebeling, W., Engel, A., Feistel. R. *Physik der evolutionsprozesse*. Springer-Verlag: Berlin, 1990.
9. Prigogine, I. *The End of Certainty*. Free Press: New York, 1997.
10. Prigogine, I., Kondepudi, D. *Modern Thermodynamics*. John Wiley & Sons: New York, 1999.
11. Zhabotinski, A.M. *Oscillations of Concentrations*. Science: Moscow, 1974.
12. Kudriavtsev, I.K. *Chemical Instabilities*. Moscow State University: Moscow, 1987.
13. Prigogine, I. *Introduction to Thermodynamics of Irreversible Processes*. Charles C. Thomas: Springfield, IL, 1955.
14. Klimontovich, Yu.L. Is turbulent motion chaos or order? Is hydrodynamic or kinetic description of turbulent motion really natural? *Physica* 1996, B228, 51.
15. Gershman, I.S., Bushe, N.A. Self-organization of structures of rubbing bodies. Proceedings of All Union Conference on Problems of Synergetica. Ufa 1989, 86–87.
16. Gershman, I.S., Bushe, N.A. Realization of dissipative self-organization on friction surface of tribosystems. *Journal of Friction and Wear* 1995, 16, *1*, 61–70.
17. Ebeling, W. *Strukturbildung Bei Irreversiblen Prozessen*. BSB B.Q. Teubnern: Verlagsgesellshaft, 1976.
18. Landau, L.D., Lifshits, E.M. *Statistical Physics*. Science: Moscow, 1976.
19. Prigogine, I. *Nonequilibrium Statistical Mechanics*. John Wiley & Sons: New York, 1962.
20. Landau, L.D., Lifshits, E.M. *Mechanics*. Science: Moscow, 1973.
21. Landau, L.D., Lifshits, E.M. *Quantum Mechanics*. Science: Moscow, 1974.
22. Prigogine, I. *Non-Equilibrium Thermodynamics: Variation Techniques and Stability*. University of Chicago Press, Chicago, 1965.
23. Feynman, R., Leighton, R., Sands, M. *The Feynman Lectures on Physics*. Addison-Wesley: Reading, MA, 1963.
24. Onsager, L. Reciprocal relations in irreversible processes. *Phys. Rev.* 1931, 37, 405–426, 38, 2265–2279.
25. De Groot, S.R., Mazur, P. *Nonequilibrium thermodynamics*. North-Holland: Amsterdam, Netherlands, 1962.
26. Nicolis, G., Prigogine, I. *Self-Organization in Nonequilibrium Systems*. John Wiley & Sons: New York, 1977.
27. Bertalanffy, L. *Biophysik des Fliessgleichgewichtes*. Braunschweig: Vieweg, 1953.
28. Pars, L.A. *A Treatise on Analytical Dynamics*. Oxbow: Woodridge, CT, 1979.
29. Prigogine, I. *From Being to Becoming*. Freeman and Company: San Francisco, CA, 1980.
30. Arnold, V.I. *Theory of Catastrophes*. Science: Moscow, 1990.
31. Prigogine, I., Defay, R. *Chemical Thermodynamics*. Longman: London, 1967.
32. Planck, M. *Treatise on Thermodynamics,* 3rd ed. Dover: New York, 1945.
33. Zhuhovitskii, A.A., Shvartsman, L.A. *Physical Chemistry*. Metallurgy: Moscow, 1987.
34. Prigogine, I., Stengers, I. *Order Out of Chaos*. Bantam: New York, 1984.
35. Prigogine, I., Stengers. I. *Entre le Temps et L'eternite*. Fayard: Paris, 1988.
36. Klimontovich, Yu.L. *Turbulent Motion and Structure of Chaos*. Science: Moscow, 1990.
37. Ivanova, V.S., Balankin, A.S., Bunin, I.G., Oxogoev, A.A. *Synergetic and Fractals in Physical Metallurgy*. Science: Moscow, 1994.
38. Gyarmati, I. *Non-Equilibrium Thermodynamics: Field Theory and Variational Principles*. Springer-Verlag: New York, 1970.
39. Basarov, I.P., Gevorkian, E.V., Nickolaev, P.N. *Nonequilibrium Thermodynamics and Physical Kinetics*. Moscow State University: Moscow, 1989.
40. Zigler, G. *Extreme Principles of Thermodynamics of Irreversible Processes and Mechanics of Continuum*. Mir: Moscow, 1966.
41. Stratonovich, R.L. *Nonlinear Nonequilibrium Thermodynamics*. Science: Moscow, 1985.
42. Chandrasekhar, S. *Hydrodynamic and Hydro Magnetic Stability*. Clarendon: Oxford, 1961.
43. Kapral, R., Showalter, K. *Chemical Waves and Patterns*. Kluwer: Dordrecht, 1995.

44. Chanau, J., Lefever, R. *Inhomogeneous Phases and Pattern Formation. Physica*, 1995, A 213, *1–2*.
45. Turing, A.M. The chemical basis of morphogenesis. *Philosophical Transactions of the Royal Society of London*, 1952, B 237, 37.
46. Klimontovich, Yu.L. Problems in statistic theory of open systems: criteria of relative degree of order of states during self-organization. *Physics-Uspehi* 1989, 158, 59.
47. Klimontovich, Yu.L. Determination of relative degree of order in open systems basing on S-theorem and experimental data. *Technical Physics Letters* 1988, 14, 631.
48. Klimontovich, Yu.L. Introduction. In *Physics of Open Systems*. Yanus: Moscow, 2002.
49. Prigogine, I., Nicolis, G., Babloyante, A. Thermodynamics and Evolution. *Physics Today* 1972, *11–12*, pp. 23, 38.
50. Ludema, K.C A review of scuffing and running-in of lubricated surfaces, with asperities and oxides in perspective. *Wear* 1984, 100, *3*, 315–331.
51. Krishna, K.V., Pramila, B.N. Wear mechanism in hypereutectic aluminum silicon alloy slidings against steel. *Scripta Metallurgica* 1990, 24, *2*, 267–271.
52. Maurice, D.R., Courtney, T.H. Modeling of mechanical alloys. *Metallurgical and Material Transactions A,* 1995, 26A, 2431–2435.
53. Ackerson, B.J., Clark, N.A. Shear-induced melting. *Physical Review Letters* 1981, 46, 242–251.
54. Eberhart, M.E., Latanision, R.M., Johnson, K.H. The chemistry of fracture: A basis for analysis. *Acta Metallurgica* 1985, 33, *10*, 1769–1783.
55. Buckley, D.Y. *Surface Effects in Adhesion, Friction, Wear and Lubrication*. Elsevier: Amsterdam, 1981.
56. Klamecki, B.E. An entropy-based model of plastic deformation energy dissipation in sliding. *Wear* 1984, 96, *3*, 319–329.
57. Klamecki, B.E. Energy dissipation in sliding. *Wear* 1982, 77, *3*, 115–128.
58. Klamecki, B.E. Wear — entropy production model. *Wear* 1980, 58, *2*, 325–330.
59. Bershadsky, L.I., Iosebidse, D.S., Kutelia, E.R. Tribosynthesis of graphite — diamond films and its employment for obtaining structurally. *Thin Solid Films* 1991, 204, 275–293.
60. Tabor, E.D. The role of surface and intermolecular forces in thin film. In *Microscopic Aspects of Adhesion and Lubrication*, Ed., Georges, J.M. Elsevier: Amsterdam, 1982, 651–679.
61. Heilmann, P., Rigney, D.A. An energy-based model of friction and its application to coated systems. *Wear* 1981, 72, 195–217.
62. Clapeyron, E., Clausius, E., Sadi Carnot. Reflection on the motive force of fire. In *The Second Law of Thermodynamics*, Ed., Mendosa, E. Peter Smith: Gloucester, MA, 1977.
63. Laidler, K.J. *The World of Physical Chemistry.* Oxford University Press: Oxford, 1993.
64. Rabinowicz, E. Friction fluctuation. In *Fundamentals of Friction: Macroscopic and Microscopic Processes*, Eds., Singer, I.L., Pollock, H.M., Vol. 220, NATO ASI Series E. Applied Sciences. Kluwer Academic Publishers, Norwell, MA, 1992, pp. 25–34.
65. Karasik, I.I. Running-in as reflection of the fundamental property of tribosystems structural adaptability. *Journal of Friction and Wear* 1993, 14, *1*, 121–128.
66. Rubin, V.V. *Large Plastic Deformations and Destruction of Metals*. Metallurgy: Moscow, 1986.
67. Garbar, I.I., Skorinin, J.V. Metal surface layer structure formation under sliding friction. *Wear* 1978, 51, 327–336.
68. Bershadsky, L.I. Self-organization of tribosystems and conception of wear resistant. *Journal of Friction and Wear* 1992, 13, *6*, 1077–1094.
69. Lasota, A., Mackey, M. *Probabilistic Properties of Deterministic Systems*. Cambridge University Press: Cambridge, 1994.
70. Ebeling, W., Engel, A. Models of evolutionary systems and their application to optimization problems. *Systems Analysis Modelling Simulation* 1986, 3, 377.
71. Chichinadze, A.V. *Foundations of Tribology*. Mashinostroenie: Moscow, 2001.
72. Garkunov, D.N. *Tribotechnics*. Mashinostroenie: Moscow, 1999.
73. Silin, A.A. Behavior and stability of the externally activated tribosystems. *Journal of Friction and Wear* 1980, 5, 791–796.
74. Konchits, V.V. Friction interaction and current collection in sliding electric contact of composite with metal. *Journal of Friction and Wear* 1984, 5, *1*, 59–67.
75. Van Brunt, C., Svage, R.H. Carbon-brush contact films. *General Electric Review* 1944, 47, *8*, 17–38.

Section II

Tribological Compatibility and Nanotribological Characteristics to Evaluate Surface Properties during Friction

3 Compatibility of Tribosystems

Nicolay A. Bushe and Iosif S. Gershman

CONTENTS

3.1 Introduction: Concept of Compatibility of Tribosystems 59
3.2 Selection of Criteria for Tribological Compatibility 60
3.3 Compatibility of Tribosystems during Different Stages of Wear 60
3.3.1 Compatibility during the Running-In Stage 60
3.3.2 Compatibility during Post Running-In Period 64
3.3.3 Tribological Compatibility under the Domination of Liquid Lubricant 65
3.3.4 Compatibility in a Mode of Mixed Friction 66
3.3.5 Compatibility in Dry (Lubricant-Free) Friction Conditions 69
3.4 Physical Concepts of Tribological Compatibility 72
3.5 The Role of the Self-Organizing Process in Tribological Compatibility 73
3.6 Conclusions 75
References 77

3.1 INTRODUCTION: CONCEPT OF COMPATIBILITY OF TRIBOSYSTEMS

Tribological compatibility could be defined as the ability to provide optimum conditions within a given range of operating parameters by the chosen criteria [1]. Once tribological compatibility is achieved, the operation of a specific tribosystem is optimized depending on its functions and operating conditions. If elements of the tribosystem are compatible, they quickly adapt to each other. However, the desirable optimum state of a system strongly depends on the specific applications and given service conditions. Besides, the features of compatibility depend both on operational and functional requirements.

The concept of tribological compatibility is associated with the concept of self-organization of tribosystems.

The ability of frictional materials to adapt to each other and to the conditions of friction, provided the given durability, is defined by the proper selection of structural materials, lubricants, and the parameters of the system's design. Rational selection provides acceptable dynamic characteristics, including temperature regimes (states) of the tribosystem's operation. As a result, the compatibility of a tribosystem affects its reliability. Different aspects of compatibility of tribosystems are studied in Reference 2 to Reference 6.

The goal of this chapter is to overview the main features of tribological compatibility under different friction conditions and to associate the concept of tribological compatibility with the self-organization phenomenon.

3.2 SELECTION OF CRITERIA FOR TRIBOLOGICAL COMPATIBILITY

Criteria of a tribosystem's compatibility can be established based on a set of parameters that fully characterize the work of the tribo-couple. These parameters are usually determined during laboratory testing. It is mandatory to examine the physical, chemical, and mechanical aspects of friction and wear processes, including impact of the environment and lubrication.

A desirable optimum state of the tribosystem can be determined depending on its operating conditions and specific applications.

There are a few practical examples of tribological compatibility in various tribosystems [1]:

1. Brake systems: Tribological compatibility means that the materials of frictional bodies provide a high coefficient of friction values at low wear rates.
2. Plain bearings: Long fatigue life at a low wear rate. Galling resistance is also very important in this case.
3. Sliding electric contacts: A stable current collection at a low wear rate.
4. Cutting and stamping tools: Low wear rate and high surface finish of the machined part.
5. Under overloading or an emergency situation during service of a tribosystem: Maximal life of the tribosystem prior to the deep surface damage or failure.

Engineering optimization should be completed based on the definite wear characteristics of specific tribosystems. Features of tribological compatibility vary for the specific modes of friction and are controlled by operating conditions, such as loading, speed, and temperature. The process of compatibility is directed by response of the system to variations in modes of friction that could arise due to unstable operating conditions. A beneficial response to still more severe friction conditions can result in the formation of specific surface structures at the actual areas of tribo-contact. These surface structures vary greatly for specific tribosystems. They could be thin, easily mobile films, oxide layers with high antigalling properties, or soft (lead-based alloys and polymers) or hard (usually transitional metal-based) compounds [7].

The response of the tribosystem to increase in the applied load is shown in Figure 3.1 for two bearing alloys containing soft structural components: aluminum alloy, containing 20% tin, and copper alloy, containing 22% lead. From Figure 3.1, it can be inferred that the tribological characteristics of an aluminum-based alloy are better than the characteristics of a copper-based alloy. This can be explained by the features of the alloy's frictional behavior after a continuous layer of oil has been disturbed. Constant monitoring of the lubricating layer's electric resistance R shows that a liquid lubricating mode prevails in an aluminum-based alloy and a mode of boundary lubrication dominates in a copper-based alloy (Figure 3.2).

Materials with good tribological compatibility are capable of reducing quickly to a level of contact pressure. Such reduction is more easily performed for a soft alloy, but this is also possible for harder elements of the tribosystem. It is desirable that the deformations of local highly loaded sites are mainly concentrated within surface layers. In this case, the rule of the positive gradient of shear strength (by Kragelski) [8–10] is realized.

3.3 COMPATIBILITY OF TRIBOSYSTEMS DURING DIFFERENT STAGES OF WEAR

3.3.1 Compatibility during the Running-In Stage

It is necessary to differ the compatibility of a tribosystem during the running-in stage and the post-running-in period. The running-in process can be conditionally divided into two separate phases:

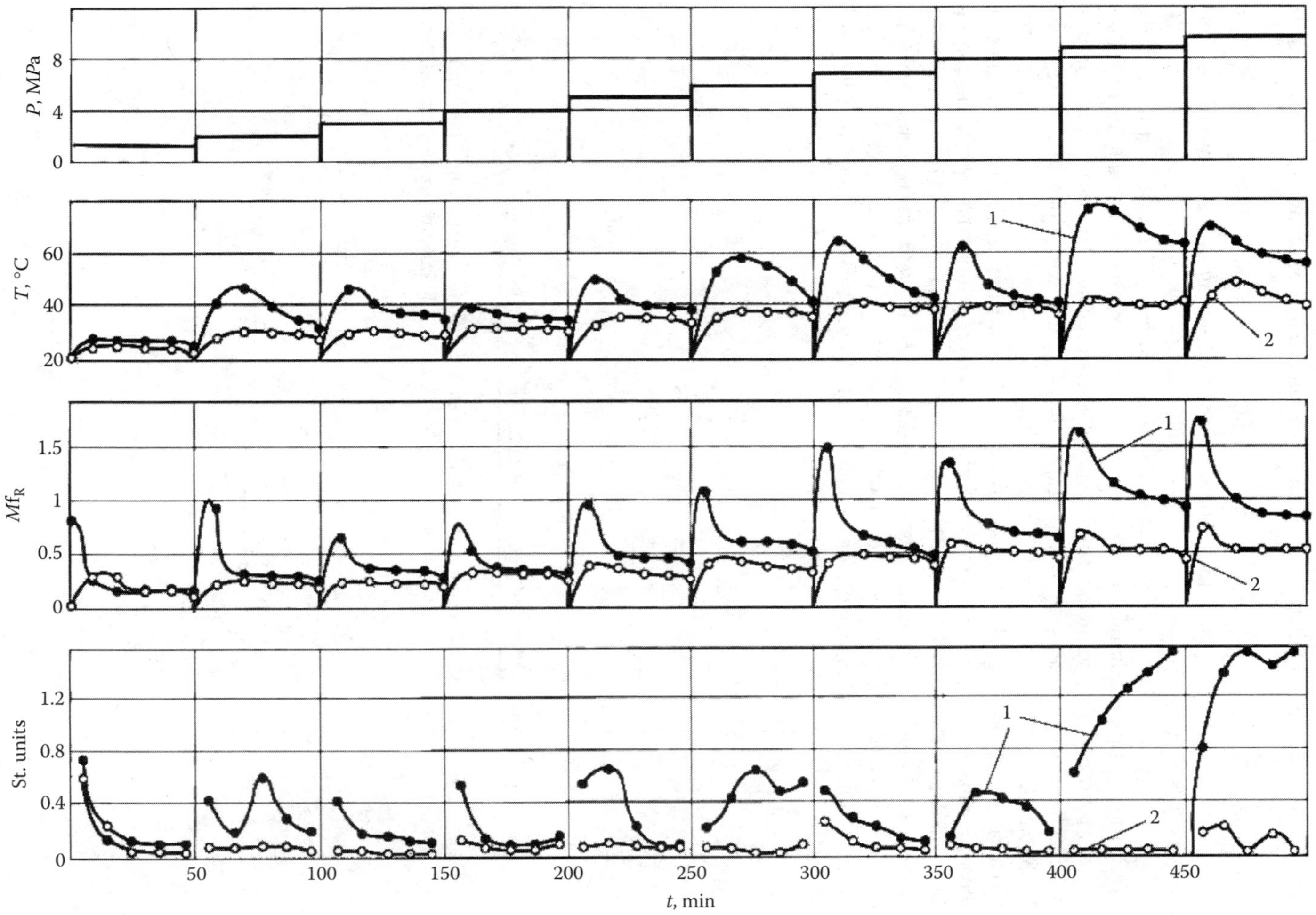

FIGURE 3.1 Change of temperature (T°C), the moment of friction (M_{fr} in N m) and wear rate (standard units) vs. step growing load applied (P) for (1) the lead bronze with 22% Pb and (2) the aluminum–tin alloy with 20% Sn.

1. Intensive running-in stage, when the frictional bodies' interactions occur at a macrolevel. Planimetric and actual area of contact are increased as a result of plastic deformation and wear during this period.
2. After a sufficient area of contact is achieved, interaction of frictional bodies continues at the microlevel because of increase in the amount and actual area of individual contacts.

After the running-in stage is completed, surface equilibrium roughness begins to form (regardless of the sizes and character of the initial asperities) [8]. Self-organization of the tribosystem simultaneously occurs as a result of structural adaptability during friction.

During the running-in stage, intensive wear rate is observed under the near to critical (at the threshold of seizure) friction conditions [11]. Therefore, the ranges of the loading and speed follow the decrease of Sommerfeld's parameters ($\mu\omega/p_a$) and the coefficient of friction f_T according to the known dependence by Hersey–Stribeck (Figure 3.3). Friction conditions in this case are semiliquid lubricated.

Bearing capacity significantly differs depending on the type of tribo-pair. After the running-in stage, the bearing capacity of the aluminum-based alloys containing more than 9% tin is considerably higher than that of lead-based bronzes (Figure 3.4). It is important that the maximal bearing capacity during the running-in stage does not decrease during further work of the tribosystem.

The running-in process of the frictional component is of critical importance for a wide class of internal combustion engines, especially for a "rear side of the bearing or crankshaft" tribo-pair.

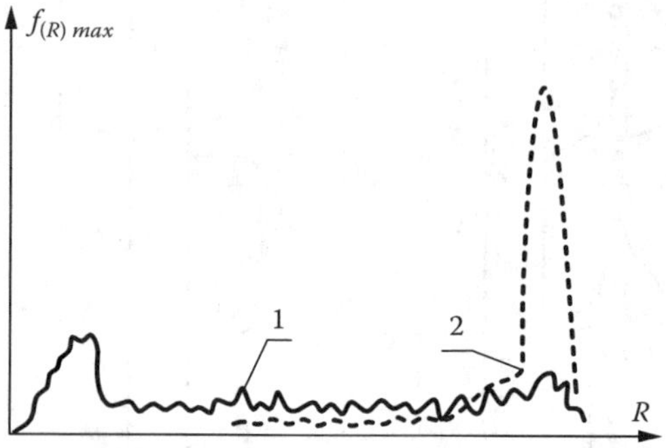

FIGURE 3.2 Density of distribution $f(R)_{\max}$ of the electric resistance (R) of a lubricating layer under the "mixed" friction conditions for (1) the lead-based bronze with 30% Pb and (2) aluminum–tin alloy with 20% Sn.

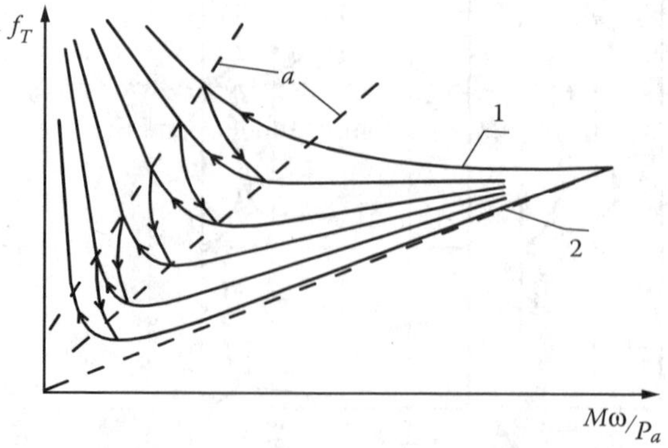

FIGURE 3.3 Change of coefficient of friction (f) and a parameter by Sommerfeld ($\mu\omega/p_a$) during the running-in stage of friction: (1) before running-in stage; (2) after running-in stage. a — Range of regulation in a process of running-in.

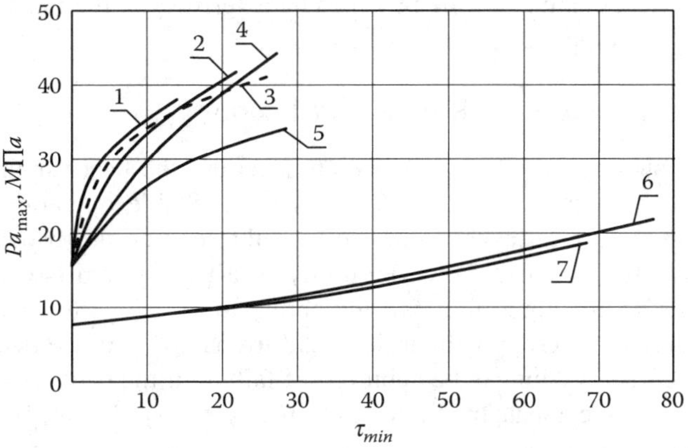

FIGURE 3.4 Change of the maximal normal pressure ($P_{a\ max}$) with time (τ) during friction of aluminum–tin alloys, copper–lead alloys, and Babbitt in contact with steel. 1 — Al–50% Sn; 2 — Al–10% Sn; 3 — Babbitt; 4 — Al–9% Sn; 5 — Al–3% Sn; 6 — Cu–30% Pb; 7 — Cu–1% Sn–22% Pb.

The running-in process depends on the lubricant used, the metal of a crankshaft, and, especially, the antifrictional alloy of the bearing. The rate of the running in depends on the intensity of the wear process and plastic deformation. The latter is characterized by the degree of distribution and the type of process (laminar or turbulent metal flow).

Individual-contact interactions of friction surfaces with subsequent creations of separate centers of galling; build-ups and tearing-off zones precede the formation of seizure during the running-in stage. The degree of hardening and the depth of plastic deformation control the intensity of surface damage. If plastic deformation takes place within the thin and mobile surface layers, the tribo-pair will operate without the further formation of seizure zones. This process can be enhanced by the deposition of special soft metal, solid lubricant coatings, or by the application of self-lubricating alloys and composite materials [12].

A major feature of self-lubricating materials is the interaction of soft phases and a hard matrix during elastic and plastic deformation [12]. Intensive squeezing of the soft phase occurs during the running-in stage. Soft-phase squeezing occurs because of the difference in the yield point and range of plastic deformation of the corresponding phase of material under stresses above a matrix yield point. Thin protective films form on a friction surface as a result of the soft-phase transfer. These films are capable of forming a softened zone on the conjugated surfaces during friction.

In some cases, the formation of the softened zone is associated with friction under conditions of hot plastic deformation. Sometimes, this is the result of lower hardening of the surface layers rather than the ones on the subsurface at an equal density of dislocations. Softening of the surface layers can result from lubricant impact, for example, under conditions of selective transfer [13–14].

Adaptability of surface layers during the running-in stage can approach a stationary state. In a stationary state, loss of matter as a result of wear is compensated by the formation of new surface layers. Apparently, structural conformity of the two surfaces is generated at a reduced speed of this process. This state, as explained in Part II, corresponds to a stationary state with minimal entropy production and, hence, minimal wear.

Plastic deformation of the surface layers is of great importance during the running-in process. Thermomechanical effects enhance the plastic deformation on the friction surface. Surface plastic deformation is accompanied by an intensive wear process during the running-in stage. Finally, the planimetric and actual areas of contact increase [15] and the optimum surface roughness forms under the given conditions, which is independent of the initial surface roughness [8]. The independence of roughness on the entry conditions indirectly confirms the conclusion of Chapter 2 about the formation of dissipative structures during the running-in process. The rate of the running-in process and its completeness are controlled by the characteristics of the materials of a tribosystem and their tendency to be compatible.

3.3.2 Compatibility during Post Running-In Period

The running-in stage should provide stable operation during the post running-in period without surface damage throughout the lifetime of a tribosystem. Stable operation of the tribo-pair is controlled by the state of surface layers formed during the running-in period.

During the post running-in period, serviceability of a tribosystem is controlled by the wear resistance of materials, their fatigue life, and the ability to resist against overloading, as well as infringements of a lubricating layer without deep and irreparable surface damage. Prolonged and stable operation of the tribosystem can be achieved, if failure of the tribosystems is determined by natural wear. It is possible to evaluate the lifetime of a tribosystem if the wear rate and the acceptable wear range are known [8,16–17].

Increase of wear rate and intensive surface damage during the post running-in period depends on the micro- and macrodamages that formed during the previous running-in stage. These surface damages consist of the following: (1) damages of separate structural components, (2) damages of separate sites of the structural components, and (3) severe damage to the majority of the frictional surface. Local destructions of the structural components under overloading fall under the first category. In metals, these damages develop as the result of low-cycle fatigue. Thus, the weaker or brittle structural components are the first to be damaged. For instance, if babbits are used in bearings, microcracks will occur in cubic crystals of the hard Sn–Sb phase [18]. They become the centers of the cracks' initiation and propagation within the depth of the surface layer. Destruction of the relatively large conglomerates of eutectoid is observed at the surface of frictional components made from tin-based bronze.

It is shown in Reference 19 that the contact strength of rail steel is significantly reduced in the presence of nonmetallic aluminum oxide inclusions. Sometimes, prior to the end of the running-in stage, destructions arise because of contact fatigue, resulting in the disintegration of metal on the surface of the rails.

The second kind of surface damage occurs in practice more frequently. This happens because the processes that take place on the friction surface develop on periodically varying planimetric sites of friction. These sites have a small area of actual contact. On these rather small areas, separate sites become excessively overheated because of overloading. If the temperature reaches recrystallization, softening of the actual site is unavoidable.

The intensity of surface damage depends on the materials of the tribo-pair. For example, the damage rate of steel and pig-iron pins increases with a rise in the melting temperature of the bearing material. Local damage of steel pins that are working in contact with lead bronze is higher than that of the pins working in contact with an aluminum–tin alloy. The sites of damage are not observed for pins working in contact with soft babbits. Damages such as cracks are observed on the steel or cast-iron pins if heat treatment (such as quenching, nitriding, or cementation) has been performed. For the surface without heat treatment, notches and deep sites of surface damage are typical.

Damages that occur on the major part of a frictional surface could be caused by severe running-in conditions accompanied by direct metal contact of the frictional bodies. Increased wear can result in the removal of a layer of soft coating. This critically reduces the workability of the bearings made from tin, bronze, or aluminum–tin alloys with low tin content [20].

Heat generation under severe friction conditions can result in the softening of the surface layer. The depth of the softening for bearings made from aluminum–tin and aluminum–lead alloys is around 6 to 8 μm if the tribosystem operates under conditions close to jamming. Softening of the surface layer can be beneficial because it fulfills the rule of "positive gradient of properties" [1]. At the same time, however, fatigue resistance reduces. Therefore, it is desirable that a thinner layer be involved in the softening process. In this case, the fatigue strength will not be reduced.

The state of surface layers of metals after the running-in stage is completely dependent on the distribution of internal stresses within the depth of an antifrictional layer. The value and the sign of residual stresses depend on the modes of friction and the type of lubricant used. To improve the wear resistance, it is necessary to form compression stresses within the surface layers of the frictional bodies [21]. During operation without a lubricant, internal compressive stresses within the surface layers form. Differences in the character of the distribution of residual stresses depends on the surface-active additives contained in the lubricants [22] (see Chapter 13). The distribution of residual stresses within the surface layers also depends on the initial condition of the metal.

3.3.3 Tribological Compatibility under the Domination of Liquid Lubricant

Hydrodynamic friction is fulfilled if the thickness of an oil layer is 0.1 μm and above [23]. The height of the asperities should be less than the thickness of the lubricating layer. Under conditions of liquid friction domination, the surfaces of the tribo-pair are separated by a thick layer of lubricant. After a short period, the lubricant layer experiences damage that shifts the friction conditions to a boundary friction mode. Under this mode of friction, a layer of liquid forms on the surface having nonvolumetric properties [24].

Active interaction of boundary layers results in corrosion and cavity damages of the frictional metal surfaces, even in the liquid friction mode [25]. Types of surface damage are affected by the structural and lubricating materials used. Corrosion of surface layers enhances the wear rate or surface damage of a tribosystem. In special cases, a selective corrosion of anodized sites on the surface causes selective transfer. Selective transfer is a nonequilibrium process and therefore, results in the decrease of wear rate. The surfaces of bearings are most exposed to a lubricant corrosion attack.

It is shown in Reference 25 that the bearing alloys are ranked depending on their corrosion resistance in the following order: tin babbitt, babbitt with alkaline metals, and lead bronze. The antifrictional properties of lead bronze are controlled by a major structural component, which is subjected to corrosion. As a result of significant corrosion, the friction taking place on the copper surface causes the failure of the accelerated bearings.

Corrosion of bearing alloys in a lubricative environment occurs because of the electrochemical mechanism. According to Reference 24, the following reactions develop corrosion:

$$M + AO_2 \rightarrow AO + MO$$

$$MO + 2HA \rightarrow MA_2 + H_2O$$

where M is a metal, AO_2 a peroxide, MO a metal oxide, ON an organic acid, and AO a ketone.

Cavitation damages to the bearings' frictional surfaces are the result of a continuous distraction of the lubricating layer. Bubbles or cavities arise because of the pressure decrease in a liquid below the pressure of vaporization. Such conditions arise during the sudden change of a liquid speed, during its flow around obstacles, which is followed by the formation of turbulences; and at the separation of liquid flow from the surface. Bubbles slam within the increased pressure zone, producing a pulsating impact on the surface during cavitation. The metal surface gets destroyed because of corrosion and surface fatigue damage. Cavitation damages were observed on babbitt layers of the bearings. The damaged area expands over a major part of the friction surface of these bearings. Such surface damage forms as the result of a drop in oil speed or the presence of water in the oil. The cavitation resistance of various antifrictional materials was evaluated by measuring the depth of the damaged area [26]. Resistance to cavitation of the aluminum–tin alloy was two times higher than the resistance of lead bronze and eight times higher than the resistance of a babbitt. Cavitation damages (numerous cavities that result in the generation of a sponge-like surface morphology) can form on the surface and also propagate in the depth of a frictional body. Surface damages are accompanied by the formation of fatigue cracks; fatigue failure prevails if heavy damage occurs.

Cavitation resistance is not always directly connected to the strength of the materials. Therefore, it is incorrect to assume the nature of this process solely based on the fatigue phenomenon. This kind of surface damage should be considered as corrosion fatigue. It is related to bearings that function in mineral oils. During long-term service, these oils are oxidized and form aggressive surface-active substances. Surface layer damage depends on the microstructure of an alloy. Destruction of the hard structural components occurs in the heterogeneous structure of a material that is typical of babbitts bronze and aluminum alloys. Failure of the antifrictional alloys containing the soft structural components is caused by the loss of soft-phase particles. If cavitation happens, localization of the impact loading is very important. Therefore, it is desirable to have fine crystals of a second phase.

Cavitation resistance depends on the ability of a material to absorb the energy introduced to the surface. If the energy of hydraulic impacts is spent not only on deformation and surface damage but also on phase transformations, the incubatory period is extended and the cavitation resistance of the materials increases. The unstable austenite steel belongs to this type of alloy. During cavitation, phase transformations occur in these steels with the formation of the martensite of deformation. It is necessary to note that nonequilibrium phase transitions occur because they require plenty of energy. The formation of the nonequilibrium martensite phase confirms this hypothesis. In contrast, equilibrium phase transitions occur spontaneously.

3.3.4 Compatibility in a Mode of Mixed Friction

Mixed-lubrication conditions are part hydrodynamic and part boundary lubricative. In this case, the behavior of the tribosystem is controlled by properties of both the surface and lubricant. Properties of the lubricating layer differ from the volumetric characteristics. To achieve tribological compatibility, it is necessary to ensure the reliable operation of the tribosystem in the absence of a permanent liquid lubrication. The formation of the mixed-lubrication mode for various antifrictional materials is defined by the so-called "transitive temperature."

Figure 3.5 presents the coefficient of friction vs. the temperature of lubrication for the aluminum–tin alloys and lead bronzes. Tests were performed using a steel sample as a counterbody. There is a significant difference in values of the transitive temperatures once liquid lubrication conditions transfer to the mixed conditions. Low transitive temperatures of the lead bronze enhance the probability of seizure formation depending on the increasing temperatures of oil.

The transitive temperature of aluminum–tin alloys grows with the soft-phase content in the alloy (Figure 3.6). The transitive temperatures depend not only on antifrictional alloys used but on the material of the counterbody as well. During testing in contact with a high-strength cast iron, low transitive temperatures at a high coefficient of friction values are observed. Low antifrictional properties of the high-strength cast iron containing spherical graphite can be explained by the plastic deformation of the soft ferrite phase located around graphite globules during the polishing. It results in the formation of beadings on the strain-hardened metal. These beadings are not eliminated during subsequent polishing.

The general change of characteristics under mixed-friction conditions can be evaluated using a parameter by Sommerfeld ($\mu\omega/p_a$), where μ is the viscosity, ω the speed, and p_a the pressure, according to the curve by Hersey-Stribeck (Figure 3.7) [27]. According to Reference 27, Sommerfeld's parameter depends on the load applied to the bearing, the characteristic of elasticity, roughness of both surfaces in contact, radial backlash, radius of the shaft, Poisson's ratio, material of the shaft and the bearing, etc.

Antifrictional alloys can be divided into two groups based on their microstructure: (1) a hard phase (usually an intermetallic compound) interspersed in a soft matrix and (2) a soft phase interspersed in a hard matrix.

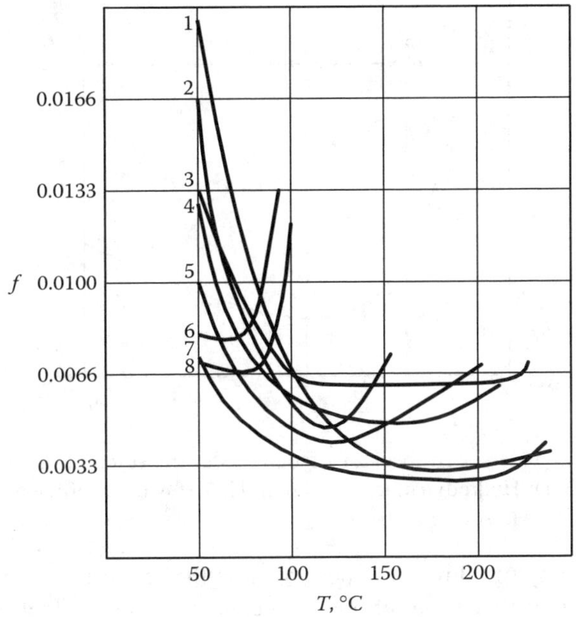

FIGURE 3.5 Coefficient of friction (*f*) vs. the temperature (*T*) of oil for aluminum alloys and lead bronzes. 1 — Al–20% Sn (powder metallurgy); 2 — Al–9% Sn; 3 — Al–15% Pb; 4 — Al–9% Pb–1% Si–1% Ni; 5 — Al–20% Sn; 6 — Cu–22% Pb–1% Sn; 7 — Al–15% Pb–3% Sn; 8 — Cu–30% Pb.

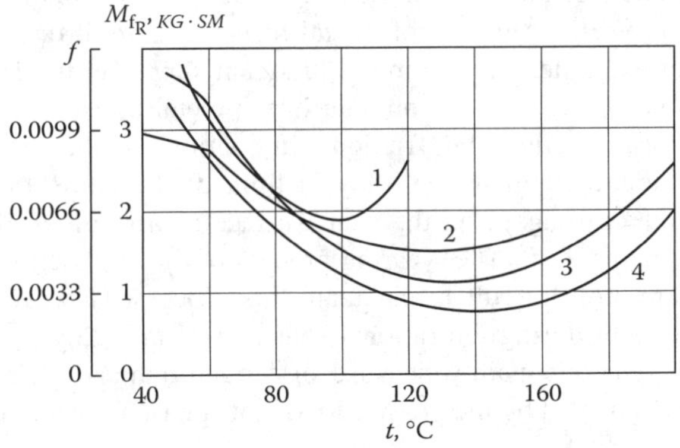

FIGURE 3.6 Coefficient of friction (*f*) and moment of friction (M_{fR}) of aluminum alloys vs. the temperature of oil during friction in contact with soft steel. 1 — Al–1% Cu–1% Ni–2,5% Si; 2 — Al–1% Cu–1% Ni–2% Si–3% Sn; 3 — Al–1% Cu–1% Ni–2% Si–9% Sn; 4 — Al–1% Cu–1% Ni–2.5% Si–12% Sn.

For the alloys of the first group (the soft babbitts), hard crystals intrude in the soft matrix during this phase wearing, and gradually the amount of the hard phase grows on the surfaces of friction [28]. It results in the decrease of wear rate. However, an excessive amount of hard phase could lead to fatigue failure. For alloys of this group, which are harder than babbitts (copper alloys and aluminum), higher stresses are required for the plastic deformation a soft matrix. It results in nonuniformity of the plastic deformation on the surface of friction. Hence, plastic deformation will expand on the deeper surface layers. Thus, the babbitt will have lower deformation and adhesive components of friction that improves their antifrictional properties.

For the second group of alloys, interaction of the surfaces is controlled by the behavior of soft structural component friction [29]. Aluminum–lead, aluminum–tin, aluminum–cadmium, and lead

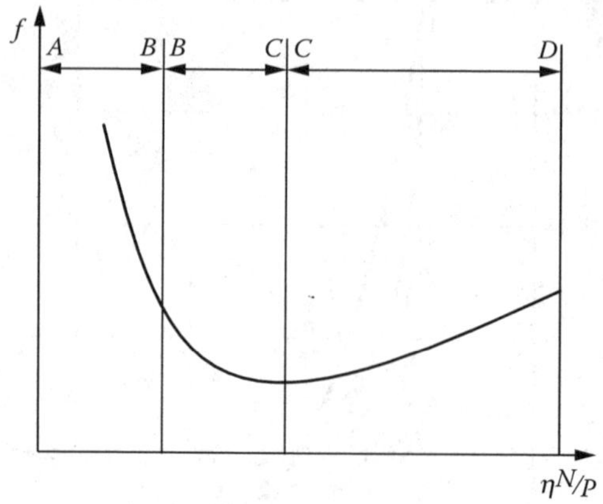

FIGURE 3.7 The diagram by Hersey-Stribeck of friction modes transition [21]. A–B: Boundary lubrication; B–C: Mixed lubrication; C–D: Hydrodynamic lubrication. Here, f is the coefficient of friction and $\eta N/p$, the nondimensional parameter, or Hersey's number.

bronze belong to this group of alloys. It is possible to allocate the following stages of interaction during the mixed friction of this group of alloys in contact with steel (e.g., aluminum–tin alloys). During the metal-to-metal contact, the soft structural component is exposed to plastic deformation and the hard matrix to elastic deformation. This is possible if a certain ratio of the hardnesses of these phases and a significant difference of their modules of elasticity are achieved. As a result of plastic deformation and destruction of the soft metal, "comet-tail"-like beadings are formed. Simultaneously, seizure occurs between the soft phase and the surface of a shaft made from steel or cast iron. As a result, a thin protective film of soft metal forms (Figure 3.8). If the friction conditions become more severe (for example, a pause in the lubricant supply to the friction surface), a thick strip of tin appears on the surface, i.e., the amount of the metal transferred grows.

After the formation of protective film, friction will continue on the shaft surface covered with the layer of tin. The amount of tin in the alloy is limited; therefore, the protective layer will periodically wear out and form again on the shaft surface as a result of the wear process of an antifrictional alloy hard matrix [30]. The wearing process of the antifrictional alloy hard matrix is stochastic, and it is, therefore, difficult to calculate this process. However, some considerations have to be taken into account during the rational selection of the alloys' compositions. The hard matrix particles should wear out mainly because of fatigue damage. The minimal value of the matrix strength must be controlled by the possibility of soft-phase transfer, and the maximal values

FIGURE 3.8 Formation of the secondary structure from tin during friction of the aluminum–tin alloy. (White spots are Sn, black spots are Al).

have to prevent fatigue failure. For aluminum–tin alloys, the ratio of the microhardness of a matrix to the microhardness of a soft phase is 3:6. Thus, considering the severe operating conditions of internal combustion engines, it is necessary to provide the high resistibility of an antifrictional layer to prevent fatigue failure. To do this, it is necessary to increase the matrix strength. This can be done by means of alloying the copper matrix with tin and zinc and using heat treatment and cold hardening. Reduction of the matrix wear rate, due to its increase in hardness, should be accompanied by simultaneous reduction of the intensity of soft-phase wear. For this purpose, it is necessary to increase the hardness of the soft phase. Such hardening can be achieved by means of soft-phase alloying using components that form a solid solution with this phase. For example, owing to the addition of 0.2% of bismuth to an aluminum–tin alloy, the hardness increases by 20 to 30%. Hardness can be also increased by better soft-phase dispersion. If a soft phase is dispersed, the compatibility of the tribosystem is improved. It is necessary to consider the influence of working temperature on hard-to-soft phases hardness ratio. Owing to different degrees of these phase softenings, the hard-to-soft phase hardness ratio at working temperatures of the bearing could be around 6:12.

Under conditions of boundary lubrication, the temperature at the contact has a significant impact on the change of friction mode. The lubricating layer loses its ability to separate the surfaces of friction once a critical temperature is achieved, resulting in the increase of the coefficient of friction and wear. Further rise in temperature can result in a seizure. Sometimes, additives to oil promote beneficial interactions with the friction surface that leads to the decrease of wear intensity [27]. Usually, these interactions have a nonequilibrium character, and as shown in Chapter 2, result in the decrease of wear rate and coefficient of friction.

3.3.5 Compatibility in Dry (Lubricant-Free) Friction Conditions

Friction without a lubricant is a very severe condition that should only be allowed in emergency cases in regular tribosystems. Such situations are typical for tribosystems working in vacuum, brake engineering, current collection, etc. This mode of friction is also common in dry machining conditions. Satisfactory work under dry friction conditions occurs once the thin surface layers participate in the process of friction. Failure of the thin layers does not lead to deep surface damage. In this case, the tribosystem's service conditions correspond to the principle of the positive gradient of mechanical properties [1].

Stable friction of the bearing alloys containing a soft phase occurs because of the formation of a film of soft material [31]. This film serves as a solid lubricant, enhancing the operation of the friction surfaces. Once seizure and corresponding metal transfer occur, some severe types of surface damage develop, such as galling and jamming. These damages can result in the termination of the relative movement of frictional components and eventually in their failure.

It is extremely important to consider the physical mechanism of seizure for the severe friction conditions outlined earlier. The physical substance of the seizure process could be evaluated based on the energy hypothesis [32], the film hypothesis [33], the diffusion hypotheses [34], and the dislocation movements during these processes [35].

Physical contact between two bodies precedes the formation of the metal junction of two metals. During this contact, conjunction of atoms as a result of surface plastic deformation leads to their interaction due to the van der Waals forces. Chemical interaction of surfaces is also possible. Formation of a physical contact with subsequent development of a zone of seizure is controlled by the plastic deformation.

Three stages of the seizure process can be considered (Figure 3.9). During the first stage (Figure 3.9a), plastic deformation at the working temperature leads to the damage of surface oxides and absorbed films. Deformation develops independently on each surface of the tribo-couple. During the second stage (Figure 3.9b), the centers of seizure nucleate. Plastic deformation develops during the interaction of the two metals' crystals. However, significant deformation of separated surface

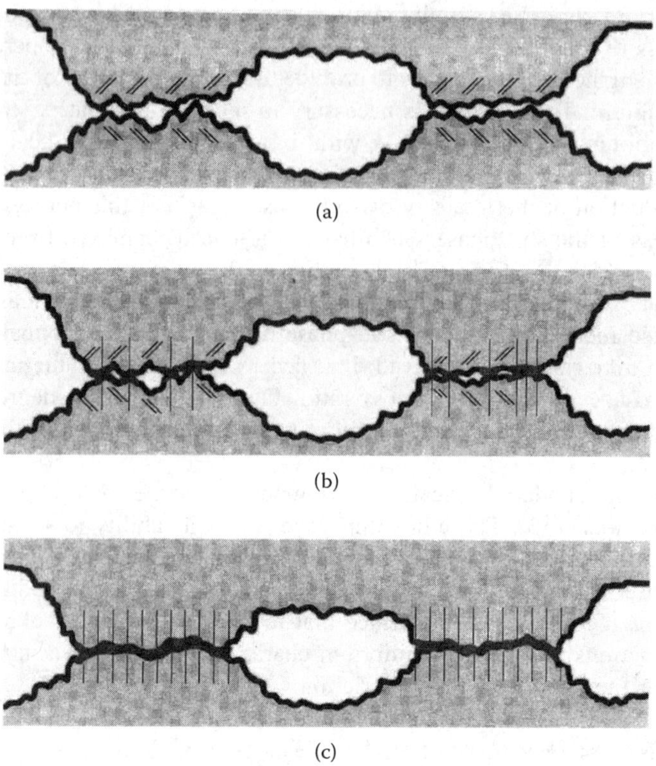

FIGURE 3.9 A scheme of three stages of seizure: (a) Mechanical interaction: deformation within frictional bodies takes place independently (skew hatch); (b) Initial stage of seizure: deformation occurs conjointly (vertical hatch). The sites with independent deformation are also depicted (skew hatch); (c) Final stage of seizure: conjoint deformation of the frictional bodies occurs at all sites of seizure.

layers still takes place. During the third stage (Figure 3.9c), the seizure of microscopic volumes occurs. Plastic deformation of conjoined crystals takes place. The physical contact usually has an elastic–plastic character [36]. Plastic-to-elastic deformation ratio depends on the operational conditions and mainly on the temperature.

The process of the formation of physical contact can be divided into three stages. Intensive deformation and metal hardening take place during the first stage under growing pressure. The area of physical contact ($S_{p.c.}$) depends on the pressure:

$$S_{p.c.} = A + B\lg p_a \tag{3.1}$$

where A and B are the constants that depend on the crystal structure of a metal and p_a is the pressure.

In some cases, such as titanium-based alloys, plastic deformation of the surface layers can be accompanied by change in their chemical composition [37].

During the first stage, surface layers prepare to form an adequate number of active centers. The active center is defined as the area around the dislocation that outcrop from the surface, including the center of interaction and the zone with radius 15b (b is the module of Burger's vector) around a nucleus of dislocation. The number of excited atoms in the active centre depends on the energy of indignation and the potential energy barrier.

During the second stage, the seizure process develops under constant external pressure. Development of physical contact occurs because of the creep of metal. The level of stress does not allow dislocations to separate from the atoms of impurity. The movement speed of dislocations is controlled by the diffusion of atoms. The energy of a physical contact activation during the second stage coincides with the energy of activation of atomic diffusion, such as for impurity atoms of the Armco iron and titanium [37]. Development of physical contact intensifies with temperature. It is shown experimentally that physical contact during the second stage is a process of the internal stresses' relaxation within a metal surface layer [38].

The third stage is observed at high temperatures in excess of $0.59T_L$ (T_L is the melting temperature). At this stage, physical contact develops because of a high-temperature creep. The surface is activated once the oxide and absorbed films are destroyed, and physical contact of the two metals is achieved. Increase of the number of the sites of contact leads to the reduction of plastic deformation resistance. The activated surface enhances dislocation fields outcropping on the surface. Stress fields in a crystal are not counterbalanced by matter on the other side of the surface, causing the appearance of forces that push the dislocations to the surface. Dislocations outcropping on the surface result in the formation of a step equal to the Burgers vector. Nucleating of active centers controls the formation of strong conjunctions between the crystals of the two surfaces. The intensity of dislocation outcropping on the surface (within the zone of conjunction) is controlled by plastic deformation.

Under the action of normal and tangential stresses, conditions of electron shell interaction and diffusion initiation are created. For pure metals or single-phase alloys, these processes lead to the collectivization of electrons; therefore, metallic bonds are formed between the atoms of crystal lattice. During the interaction of metals with metal-base compounds, oxides, semiconductors, and intermetallics, a covalent bonding forms. Formation of a strong bond between two surfaces during a seizure is a result of a gradual increase of the seizure nucleus area, which arises on the sites of the activated centers. At the same time, deformation of the already-bonded crystals will take place. On the bond-free area, intensity of the activated center initiation and new centers of seizure formation will arise. Increase of energy of the active centers intensifies the atoms of the counterbody surface. This results in the acceleration of a new bond formation.

The mechanism of the conjunction of two different metals can be presented as follows: Atoms outcropping on the interface are caused by the functioning forces between them. A positive linear dislocation is formed if the lattice parameters are different. The atoms strive to converge in a crystal with the bigger lattice parameter, and also strive to draw apart in a crystal with a smaller lattice parameter. The energy of such dislocation formation depends on the difference in the values of lattice parameters, shear modules, and Poisson's ratio.

The formation of such dislocations is enhanced by the excitation of atoms. As a result, during the conjunction of two crystalline bodies, the density of dislocations grows at the interface. It leads to hardening of the zone of seizure if temperature at the interface is lower than the temperature of recrystallization. The seizure of homogeneous metals results in the formation of an interface that is similar to the intercrystallite boundaries. Seizure of different metals leads to the formation of boundaries which are similar to interphase boundaries.

Volumetric conjunction develops after the formation of strong bonding and the disappearance of seizure-free zones. Deformation of the already-bonded grains develops further. The deformation is expanded towards the layers below and above the interface of the metals because the strength of the metal within the zone of seizure is higher than in the metal core. This is applicative to a majority of metals except to those with a low temperature of recrystallization (such as tin and lead) and metals with low-intensity hardening during cold plastic deformation (such as zinc and cadmium).

The seizure process during friction is not static. During the relative sliding of surfaces, points of contact are permanently regenerating. Compression and shear strains develop. The development of volumetric plastic deformation during friction is less intensive.

Less energy is required to form welded asperity junctions (5 to 10 times) [39] compared to contact welding because of the localization of deformation and heating within a thin contact layer. Accumulation of shear plastic deformation during friction leads to the intensification of adhesive bonding. Once the limiting amount of the adhesive bonds is achieved, surface damage begins.

Damage nucleation is caused by the diffusion of vacancies and their coalescence. Accumulation of the limiting energy results in the occurrence of "the centers of melting" [40]. According to Reference 40, the crystal lattice in this case is almost a liquid phase.

Localization of deformation at friction within centers of actual contact enhances nucleation of the surface damage centers [41]. Concentration of deformation results in seizure and, afterward, in the damage of welded asperity junctions.

A microcrack develops during the constant plastic flow of metal. Thus, the cracks initiate and propagate on the formed centers of contact simultaneously with the development of plastic deformation. Crack formation leads to fracturing and wearing of metal volumes near the surface. Deeper surface damage and increased wear under low-sliding-speed conditions will occur if the surface layers are hardened. Hot deformation will take place under higher sliding speeds and increased heat generation. In this case, the wear rate will be lower and the polishing of friction surfaces will occur. Once the physical contact has occurred, optimal surface roughness will form at an any speed for the given friction conditions.

The formation of adhesive bonds during friction is accompanied by increase of the area and amount of microcontact sites as a result of shear deformation. Damage to welded asperity junctions develops simultaneously with heat generation due to friction. It can lead to intensive seizure or even welding of elements in the tribosystem.

Development of the seizure process is accompanied by metal transfer onto the counter body surface. Layers transferred can prevent deep surface damage and improve antifrictional properties of the tribosystem. There is a critical thickness of the transferred layer for the given friction condition. Once this thickness is exceeded, seizure with further surface damage arises.

Processes occurring in metals during dry friction lead to either seizure or friction under boundary lubrication conditions. The first process is the result of seizure with subsequent deep surface damage to the frictional body. The second process is controlled by the localization of deformation within thin surface layers, which perform a protective function. Under high sliding speeds [42], such a protective layer is formed as the result of the melting of thin surface films formed during friction. Thus, the liquid lubricating layer that forms follows the laws of hydrodynamics. Low values of the coefficient of friction are typical in this case. If the temperature of the surface during friction does not exceed the temperature of recrystallization, deep surface damage of the frictional body can be prevented if the alloys containing a phase with a low temperature of melting have been used. During seizure of such metals, this phase acts as a solid lubricant because of concentration within the surface layers. If the amount of this phase is sufficient, severe dry-friction conditions transform to a milder form of boundary lubricating friction.

3.4 PHYSICAL CONCEPTS OF TRIBOLOGICAL COMPATIBILITY

Special states of the surface layers form during friction. An active layer and a specific subsurface layer are generated. The thickness of the active layer varies from a fraction of a micron to tens of microns, depending on the friction conditions. The thickness of the subsurface layer can reach a few millimeters. The surface layer is exposed to plastic deformation and is involved in physical and chemical tribo-reactions. Several absorption layers cover the surface layer. This can provide a mode of boundary lubrication friction.

Surface plastic deformation has its specific features. After a dislocation outcrops on the absolutely clear surface, its energy of deformation decreases because the surface enhances this process. This results in a force that pushes out a dislocation onto the surface. This force is approximately equal to the force of interaction of two dislocations of an opposite sign (the second dislocation is a mirror image of the first dislocation on the other side of the surface). Reduction of energy once the dislocation outcrops on a surface is compensated by the increase in surface energy during the formation of a step on the surface. Work of the formation of the step is proportional [43] and $b^2\sigma$

TABLE 3.1
The Ratio of the Force of Mirror Display to the Force Interfering with the Dislocation Outcropping on the Surface for Various Metals

Metal	Pb	Au	Al	Cu	Sn	Zn	Fe	Cd	Ag
g	0.89	1.43	1.48	1.62	1.8	2.25	2.36	2.55	1.7

(b is the module of the Burgers vector, σ is the surface energy). Generally, the ratio of the force of the mirror image to the force interfering with the dislocation outcropping from the surface is equal to:

$$g = Gb/c\sigma \tag{3.2}$$

where G is the shear module and c is a numerical factor, $c = 4$. This ratio for different metals is presented in Table 3.1.

As inferred from Table 3.1, the forces interfering with the dislocation outcropping on the surface are higher than the forces of the mirror image for all pure metals excluding Pb. According to Reference 1, to initiate the outcropping of dislocations on the surface, it is necessary to accumulate 100 dislocations at 0 K under a stress of about 10^6 P. Generally, the creation of a new surface occurs under the accumulation of n dislocations under external stress τ_1, where:

$$n = \frac{Gb}{4\pi(1-\mu)\tau\gamma} \tag{3.3}$$

where μ is Poisson's ratio and γ the width of a dislocation, $\gamma < 10b$.

The surface prevents dislocations to outcrop because there are some areas with increased density of dislocation within a surface layer. The surface is covered by films of various types. The force that prevents outcrop of dislocations from the surface depends on the difference in the modules of elasticity of film and substrate metal, difference in lattice parameter, grain size values, etc.

It is necessary to note that the speculations outlined here are true for small plastic deformations. However, an active surface layer at friction exists under conditions of major plastic deformations. It is shown in Reference 44 for volumetric deformations and in Reference 45 for surface deformations at friction that a fragmented structure is the only stable structure under large plastic deformation conditions. Fragmented structures are the sites of a metal with practically no dislocations (i.e., fragments) surrounded by the boundaries, which consist of a complex interlacing of dislocations. The density of dislocations at the boundaries of the fragments grows with increase in the degree of deformation. The fragments remain practically dislocation free. Under minor deformations, the dislocations do not interact with each other and distribute chaotically in the volume of metal. During major deformations, the dislocations are ordered. Finally, the sites with the highest energy (of the fragments boundaries) are located around the sites with the lowest energy (the fragments) [7].

3.5 THE ROLE OF THE SELF-ORGANIZING PROCESS IN TRIBOLOGICAL COMPATIBILITY

Practically, all the mechanisms of tribological compatibility correspond to the conditions of self-organization presented in Chapter 2. Surface structures that form during friction provide compatibility for the tribosystems. These surface layers are the allocated zones with increased internal energy. Besides, the structure and properties of these layers critically differ from those of the substrate material. Studies of secondary structures have shown that these surface tribofilms deviate even more from equilibrium under increasingly severe friction conditions. This deviation from equilibrium occurs because of the separation of elements. Before friction, elements and structural components are distributed uniformly within the entire volume of a material. The material can either be single-phase (solid solution) or multiphase. Surface films that form during friction usually consist of a soft phase of substrate material [31]. With other conditions being equal, entropy of a material with nonuniformly distributed structural components is lower than that of a material with uniformly distributed ones. Under any temperature and pressure, the process of separation of structural components, with the formation of macroscopic nonuniformity, can begin spontaneously. Therefore, formation of this nonuniformity is a nonequilibrium process. The coefficient of friction and the wear rate are reduced once the nonequilibrium processes begin, corresponding to the conclusions of Chapter 2.

Secondary structures formed during friction were investigated in Reference 46 for the anti-frictional copper-based alloys with tin and lead additions. Under liquid friction conditions within a layer of secondary structures, the content of tin in a solid solution was increased compared to the initial structure of the cast material. After the conditions of friction become more severe, the amount of tin on the surfaces is increased by a greater degree and the areas without tin disappear completely. The formation of these structures corresponds to the rule by Charpy [47]. Once the metals of the tribo-pair component start to intertact with each other, noncontinuous films of lead and carbon appear on the surface of friction. Under near-seizure conditions, inclusions of almost pure tin are found on the surface of friction. These inclusions transform to tin-based films after conditions of friction become more severe [31].

Under dry-friction conditions, the formation of the almost pure copper zone takes place on the surface of the sliding bearings made from tin bronze (in particular, on the rear side surface of the loose leaf of the bearings).

The segregation of pure tin from a solid solution under near-seizure conditions and the occurrence of the sites of pure copper in the structure of bronze under dry-friction conditions are extreme cases of element separation with the formation of a macroscopic nonuniformity of elements. It is necessary to note that, prior to friction, the structure of tin bronze was homogeneous (i.e., a solid solution). The processes of such nonequilibrium (dissipative) structure formation are accompanied by decrease in wear rate.

By themselves, tin or copper are not nonequilibrium formations in contrast to the supersaturated solid solutions or nonstoichiometric chemical compounds. According to the equilibrium binary constitutional diagram of Cu–Sn, within the given range of compositions, temperatures, and pressures, tin cannot be segregated from the copper-based solid solution [48]. Therefore, the process of the tin segregation from the copper-based solid solution can only be nonequilibrium. Such a process is accompanied by the reduction of entropy. The cause of this process is not increase in temperature or pressure but increase in the gradients of these characteristics as well as their chemical potentials.

As with any nonequilibrium process, segregation of tin from the nonsupersaturated solid solution needs plenty of energy. Thus, once the process is initiated, a major portion of friction energy is spent on the separation of tin from the solid solution instead of on surface damage. Therefore, a significant decrease in wear intensity is related to the self-organization process.

It is necessary to note that self-organization, according to the generalized principle by Le Chatelier–Brown [31], should reduce the impact of external changes on the system. Considering the tribosystem, self-organization should reduce wear rate under intensifying friction conditions (see Chapter 2). Wear rate decrease can be achieved by two methods: (1) termination of friction or (2) decrease of the coefficient of friction [49]. In both cases, reduction of entropy production occurs and, therefore, the wear rate is reduced. Upon the termination of friction, it drops down to zero. Decrease in the coefficient of friction could be a result of the formation of secondary structures, which are enriched either by tin or due to the effect of the lubricating action of a current [49].

The process of self-organization during friction of antifrictional tin-containing copper alloys is related to element separation. This separation during friction consists of the segregation of tin from a solid solution and formation of tribo-films (secondary structures) that are enriched by tin with sites consisting of pure tin films. The mechanism of such site formation could be related to the outward diffusion of tin. Lead has a similar behavior during the friction of lead-containing bronzes. However, lead does not dissolve in copper; therefore, the basic mechanism of the formation of lead-enriched secondary structures is most probably lead extrusion [46].

Based on this research, antifrictional alloys have been developed in which the mechanism of self-organization caused by separation of the elements was enhanced compared to tin and lead bronzes. The alloys have been designed as materials with a heterogeneous structure, which have a great ability to extrude the soft phase under a lower applied load compared to bronze. Aluminum–tin alloys and aluminum–tin–lead alloys are the materials of this type. These alloys consist of two phases: an aluminum-based matrix and tin and/or tin–lead inclusions. The tin–lead eutectic is formed at lower melting temperature compared to tin or lead. Therefore, the eutectic has a significantly higher mobility than its components under similar loads and temperatures. Thus, the formation of secondary structures with a low melting temperature occurs faster and under lower loads and temperatures. This results in a higher load-bearing capacity and a more intensive running-in stage of these alloys compared to the lead-based bronzes. Figure 3.8 presents the formation of secondary structures from tin during friction of the aluminum–tin alloy.

As outlined previously, to form welded asperity junctions the formation of dissipative structures occurs during the running-in stage of wear with plenty of energy required for these nonequilibrium processes. Thus, less energy can be spent by the system for surface-damaging mechanisms of wear. However, plenty of energy is also required to initiate these nonequilibrium processes. A solution to this problem is to intensify the friction conditions. It is shown in Reference 14 that the running-in stage under conditions that begin from increased loadings with a subsequent load decrease lead to lower wear rates, and the running-in stages are shorter than the conditions that start under lower loading conditions with subsequent load growth (a typical loading regime). This is illustrated in Figure 3.10a and Figure 3.10b. The figures present the change of wear debris concentration in oil (I) for the bearing (Al) and the shaft (Fe) and values of brake power (ΔN) steps of growth with time of the running-in stage (diesel engine, under a typical and novel loading regime).

Secondary structures that formed on the surface of an aluminum-based composite material during friction of a brake shoe were investigated in Reference 50. Secondary structures, in general, have a strongly nonequilibrium state. They are the supersaturated solid solutions of silicon in aluminum. The silicon content in aluminum was around 30% in contradiction with the constitutional diagram. Significant wear rate decrease has been observed during formation of nonequilibrium secondary structures. Similar secondary structures have been observed on the friction surface of an antifrictional aluminum–tin alloy under near-seizure threshold conditions. The nonequilibrium Al–Si–Fe–Mn phase was formed. Once this phase occurred, seizure was not observed in contrast to the sites of the surface free of this phase. Similar secondary structures were formed on the surface within zones of maximal heating of the piston aluminum alloys. In all these cases, the presence of silicon in these alloys was mandatory. The friction surface structure investigation proved that the nonequilibrium secondary structures were the most wear resistant.

3.6 CONCLUSIONS

Tribological compatibility is the ability of a tribosystem to provide optimum friction conditions within the given range of operating conditions to meet a target criterion. Tribological compatibility is related to the capacity of two surfaces to adapt to each other during friction and provide wear stability without surface damage to the two components of the specific tribosystem for the longest

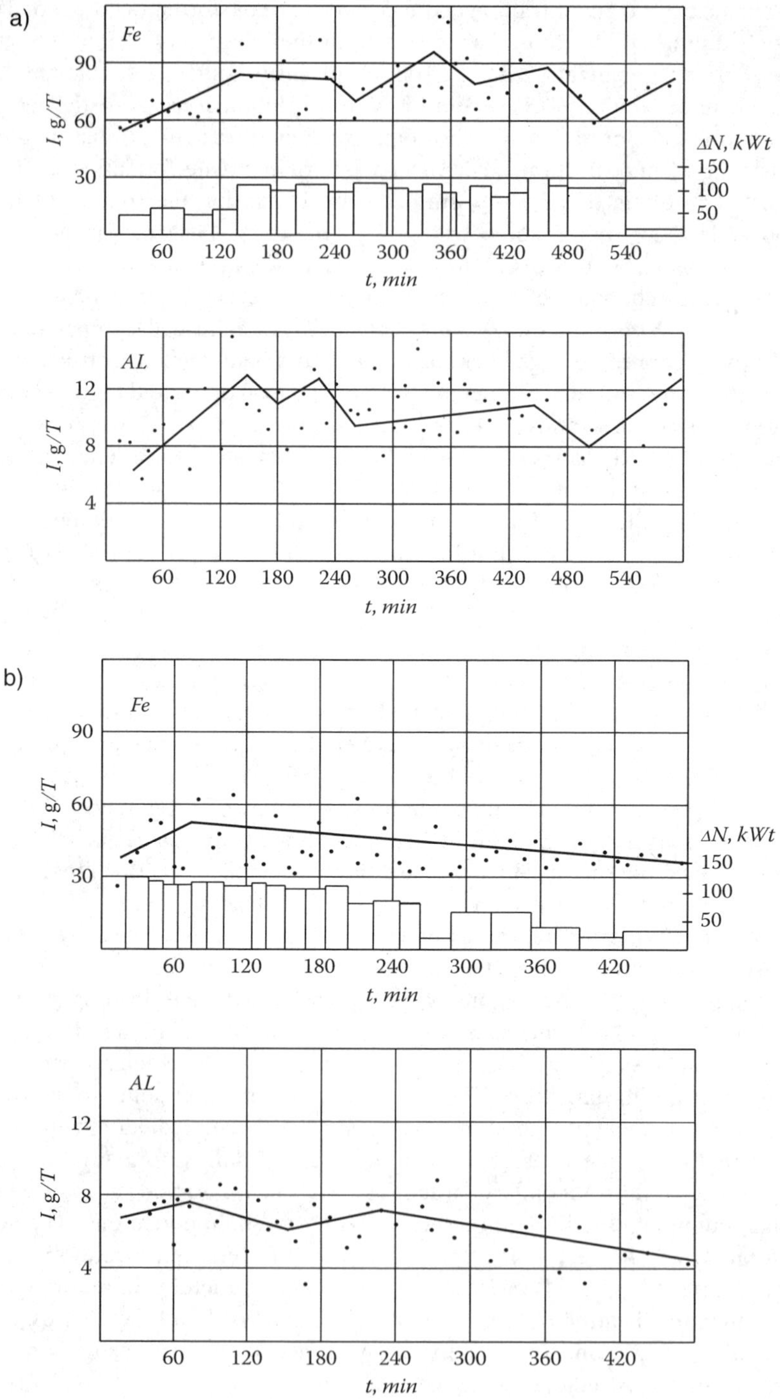

FIGURE 3.10 (a) Change of the wear debris concentration in oil (I) for the bearing (Al) and of the shaft (Fe) and the values of the of brake power (ΔN) steps of growth vs. the time of the running-in (diesel engine, under a typical loading regime); (b) Change of the wear debris concentration in oil (I) for the bearing (Al) and of the shaft (Fe) and the values of the of brake power (ΔN) steps of growth vs. the time of the running-in (diesel engine, under a novel loading regime).

(or given) period of time. The compatibility of tribosystems during different stages of wear and under different modes of wear is considered, and corresponding practical examples are presented.

The examples presented confirm the conclusion stated in Chapter 2 that nonequilibrium (dissipative) secondary structures possess a lower wear rate than those in equilibrium.

According to the theorem by Curie [51], processes with tensors of ranks of different evenness do not interact with each other in a linear area. Nonlinear processes can cooperate without restrictions. The interaction of such processes is of interest with reference to the formation of dissipative structures. A textbook example of dissipative structures is the Benard cells [52], which appear as the result of interaction between a thermal flow and gravitation. Examples of this effect in tribology could be: (1) the lubricated action of a current, whose thermodynamic analysis is presented in Chapter 2, and (2) the occurrence of a positive gradient of mechanical stresses during friction (a well-known rule by Kragelsky) [1]. Under tribological compatibility conditions within the surface layer of antifrictional materials, a positive gradient of mechanical stresses is formed as a result of friction, i.e., the mechanical stresses are increased with the depth of secondary structures [1]. Let us consider the conditions of equilibrium when relative movement of the two bodies is absent but shear stresses between them are still present (friction at rest). Then the negative gradient of the mechanical stresses will exist, i.e., the mechanical stresses will decrease within the depth of a surface layer. Hence, the occurrence of a positive gradient of mechanical stresses during friction is a nonequilibrium process. This effect is associated with the formation of secondary structures on the basis of a soft structural component. Here, on one hand, interaction of the mechanical stresses and strains (a tensor of the second rank) occurs, and on the other hand, the directed flow of matter, such as diffusions (a tensor of the first rank), are taking place. It is necessary to note that, if the soft phase is concentrated within a surface layer, the entropy of a frictional body will be less than the entropy of a body with a uniform distribution of structural components. Reduction of entropy is one of the major features of the self-organization process.

Thus, practically all the processes that enhance tribological compatibility are nonequilibrium, and they also enhance self-organization.

REFERENCES

1. Bushe, N.A., Kopitko V.V. *Compatibility of Rubbing Surfaces*. Science: Moscow, 1981.
2. Hruchov, M.M., Babichev, M.A. *Abrasive Wear*. Science: Moscow, 1970.
3. Kostetsky B.I. *Friction, Lubrication and Wear in Machines*. Technica: Kiev, 1970.
4. Bershadsky, L.I. Self-organization of tribosystems and conception of wear resistant. *Friction and Wear* 1992, 13, *6*, 1077–1094.
5. Matveevski, R.M., Buyanovski, I.A., Lazovskaya, O.V. *Resistance of Lubricant Environments to Seizure at Friction in a Mode of Boundary Greasing*. Science: Moscow, 1978.
6. Evdokimov, V.D. *Reversivity of Friction and Quality of Machines*. Technica: Kiev, 1977.
7. Bushe, N.A., Gershman, J.S., Mironov, A.E. Processes of self-organizing tribosystem at friction with current collection and without current collection. 2nd World Tribology Congress 2001, Vienna, pp. 68–76.
8. Kragelski, I.V., Dobychin, N.M., Kombalov, V.S. *Fundamentals in Calculations of Friction and Wear*. Mashinostroenie: Moscow: 1977.
9. Starostin, N.P. *Calculation of Tribotechnical Parameters during Sliding*. Science: Moscow. 1999.
10. Shapovalov, V.V. Relation of friction and dynamic characteristics of mechanic systems. *Friction and Wear* 1991, 6, 732–738.
11. Karasic, I.I. *Ability of Bearing Materials of Sliding to Running-In Stage*. Science: Moscow, 1978.
12. Gershman, J.S., Bushe, N.A., Mironov, A.E. Proceedings of World Tribology Congress III, WTC 63889 2005. Washington D.C.
13. Lukiants, V.A. *Physical Effects in Mechanical Engineering*. Mashinostroenie: Moscow, 1993.

14. Bushe, N.A., Volchenkov, A.V., Sokolov, B.N., Zelinsky, V.V. Running-in modes improvement for diesel engines. *Railway Transportation* 1989, 7, 17–21.
15. Goryacheva, I.G. *Mechanics of Interaction at Friction*. Science: Moscow, 2001.
16. Chichinadze, A.V. *Foundations of Tribology*. Mashinostroenie: Moscow, 2001.
17. Pronnicov, A.S. *Reliability of Machines*. Mashinostroenie: Moscow, 1978.
18. Nakajima, K., Isogai, A. Electron microprobe study of the effect of abrasion of the surface of alloy crystals. *Wear* 1967, 10, 2, 151.
19. Nickolaev, R.S. *Causes of Parts Failure on Rolling Stock and Rails*. Transzheldorizdat: Moscow, 1954.
20. De Gee, A.W.J., Vaessen, G.H.G., Begelinger, A. The influence of composition and structure on the sliding wear of Cu-Sn-Pb alloys. *Transactions of ASLE* 1969, 12, 1, 44.
21. Gane, N., Skinner, J. The generation of dislocations in metals under a sliding contact and the dissipation of frictional energy. *Wear* 1973, 25, 3, 381.
22. Deryagin, B.V., Churaev, N.V., Muller, V.M. *Surface Forces*. Science: Moscow, 1987.
23. Petrusevich, A.I. General revue of contact-hydrodynamic theory of greasing. *Izvestia Akademii nauk USSR* 1951, *2*, 47–56.
24. Diachkov, A.K. *Sliding Bearings under the Liquid Friction Conditions*. Mashinostroenie: Moscow, 1955.
25. Zundema, G.G. *Service Characteristics of Lubricants*. Gostechizdat: Moscow, 1957.
26. Gee, A.W. Material research and tribology. T.N.O. News 1971, 6, *8*, 445–451.
27. Tabor, E.D. Friction as dissipative process. In *Fundamentals of Friction: Macroscopic and Microscopic Processes*. Eds., Singer, I.L., Pollock, H.M., Vol. 220, NATO ASI Series E: Applied Sciences. Kluwer Academic Publishers: Norwell, MA, 1992, pp. 3–24.
28. Hruchov, M.M., Kuritsyna, A.D. Research of structure changes of working surface of babbits during friction and wear process. *Friction and Wear in Machines*. Science: Moscow, 1950.
29. Ferrante, J., Buckley, D.H. A review of surface segregation, adhesion and friction studies performed on copper-aluminum, copper-tin and iron-aluminum alloys. *Trans. ASLE* 1972, 15, *1*, 18.
30. Bushe, N.A., Goryacheva, I.G., Makhowskaya, Yu.Yu. The effect of the phase composition of anti-friction aluminum alloys on their self-lubrication in friction. *Friction and Wear* 2002, 23, *3*, 89–96.
31. Gershman, J.S., Bushe, N.A. Thin films and self-organization during friction under the current collection conditions. *Surf. Coat. Technol.* 2004, 186, 405–411.
32. Semenov, A.P., Savitskii, M.E. *Metal-Fluorine-Plastic Bearings*. Mashinostroenie: Moscow, 1976.
33. Ainbinder, S.B. *Cold Welding of Metals*. Science: Moscow, 1957.
34. Kazakov, N.F. *Diffusion Welding in Vacuum*. Mashinostroenie: Moscow, 1968.
35. Astrov, E.I. *Plated Multilayered Metals*. Metallurgy: Moscow, 1965.
36. Shorshorov, M.H. Physical and chemical bases of ways of connection of diverse materials. Metallurgy: Moscow, 1966.
37. Heinicke, G. *Tribochemistry*. Akademie-Verlag: Berlin, 1984.
38. Andarelli, G., Maugis, D., Courtel, R. Observation of dislocations crested by friction on aluminum thin foils. *Wear* 1973, 23, *1*, 21.
39. Ruff, A.W. Measurements of plastic strain in copper due to sliding wear. *Wear* 1978, 46, *1*, 25.
40. Ivanova, V.S. *Synergetic and Fractals in Physical Metallurgy*. Science: Moscow, 1994.
41. Makushok, E.M. *Mechanics of Friction*. Nauka i Technica: Minsk, 1974.
42. Chichinadze, A. V., Goriunov, V.M., Piskunov, Ju.M. Study of friction at unsteady high-speed conditions. In *News of Theory of Friction*, Ed., Chichnadze, A.V. Science: Moscow, 1966, pp. 91–100.
43. Lichtman, V.I., Schukin, E.D., Rebinder, P.A. *Physical and Chemical Mechanics of Metals*. Science: Moscow, 1962.
44. Rybin, V.V. *Large Plastic Deformations and Destruction of Metals*. Metallurgy: Moscow, 1986.
45. Grabar, I.I., Skorinin, J.V. Metal surface layer structure formation under sliding friction. *Wear* 1978, 51, 327–336.
46. Gershman, I.S., Bushe, N.A., Mironov, A.E., Nikiforov, V.A. Self-organization of secondary structures at friction. *Friction and Wear* 2003, 24, *3*, 329–334.
47. Charpy, G. Les alliages blancs dit antifriction. *Bull. Soc. d'Encouragement pour I'1nd Nationale*, 1898.
48. Hansen, M., Anderko, K. *Constitution of Binary Alloys*. McGraw-Hill: New York, 1958.

49. Gershman, I.S., Bushe, N.A. Dissipative structure formation and self-organization on friction surface of tribosystems. *Friction and Wear* 1995, 16, *1*, 61–70.
50. Semenov, B.I., Semenov, A.B., Ignatova, E.V. Nucleation and evolution of self-ordering in dry friction pair: aluminum matrix-based composite — brake lining. *Materialovedenie (Material Sciences Transactions)* 2000, 3, 27–34.
51. De Groot, S.R., Mazur, P. *Nonequilibrium Thermodynamics*. North Holland: Amsterdam, 1962.
52. Prigogine, I. *From Being to Becoming*. W.H. Freeman and Company: San Francisco, CA, 1980.

4 Surface Analysis Techniques for Investigations of Modified Surfaces, Nanocomposites, Chemical, and Structure Transformations

Anatoliy I. Kovalev and Dmitry L. Wainstein

CONTENTS

4.1 Introduction 82
4.2 X-Ray Photoelectron Spectroscopy (XPS, ESCA) 83
 4.2.1 Quantitative Analysis of the Chemical Composition of the Surface 85
 4.2.2 Investigation of Electronic Structure and Chemical Bonds 85
 4.2.3 Multicomponent Structure of the Spectra 86
 4.2.4 Chemical Environment — Chemical Shift 87
 4.2.5 Extra-Atomic Relaxation 88
4.3 Auger Electron Spectroscopy (AES) 89
 4.3.1 Physical Basis 89
 4.3.2 AES Quantitative Analysis 90
 4.3.3 High-Resolution AES 92
4.4 Determination of Interatomic Distances on Surfaces Using Electron Spectroscopy — Electron Energy Losses Fine Structure (EELFS) Technique 92
 4.4.1 Physical Basis of the EELFS Method: Experimental Procedure and Peculiarities of Data Processing 92
 4.4.2 Quality and Reliability of EELFS Results 94
4.5 High-Resolution Electron Energy Loss Spectroscopy (HREELS) 97
 4.5.1 Vibrational Spectroscopy 97
  4.5.1.1 Normal Modes of Vibration of Polyatomic Molecules 99
  4.5.1.2 The HREELS Experimental Results 101
 4.5.2 Bulk and Surface Plasmons 103
 4.5.3 Interband Transitions 104
4.6 Secondary Ions Mass Spectroscopy 107
 4.6.1 Physical Basis of the Technique 107
 4.6.2 Relative Sensitivity Factors 109
 4.6.3 Sensitivity and Detection Limits 109
 4.6.4 Mass Spectra 110
 4.6.5 Depth Resolution 112
  4.6.5.1 Standards for RSF Measurement 112

4.6.5.2 Bulk Analysis 112
4.6.5.3 Sample Charging and Charge Neutralization by Electron Bombardment 113
4.6.6 Isotope Ratio Measurements 113
4.7 The Modeling and Electron Spectroscopy of Modified Surfaces 113
4.8 Electron Spectroscopy Experimental Conditions 118
References 119

4.1 INTRODUCTION

The need for new physical methods of surface analysis is caused by recent advances in many fields of science and engineering, such as solid-state physics, chemistry, metallurgy, and others. These methods have a special significance in connection with the development of new technologies.

The surface state of these materials is crucial under fabrication and exploitation because the surface structure, chemistry, and reactivity determine the materials' intended functions.

The development of nanotechnologies for surface modification and the study of physical phenomena of the surface and interface at the nanoscale level impose special analytical technique requirements. The methods of surface analysis satisfy these requirements. They make it possible to determine the concentration of trace elements and the atomic and electronic structure of the finest and most complex coatings and layers with a thickness close to several interatomic distances. Naturally, various methods of surface analysis can satisfy these conditions by different amounts. But their combination allows us to investigate the properties of substances in nanovolumes. The only drawback is the inability to study the processes (*in situ*) that are developing rapidly in nanoscale time. "Freezing" the high-speed processes makes it possible to investigate them stage by stage. Such methods and approaches are widely used in this work.

All the methods of surface analysis in this study are based on different aspects of interaction of radiation and matter.

At sufficient beam energy, the excitation of electrons from the external and internal orbitals of atoms occurs on the surface. The electron spectroscopy equipment analyzes the number of electrons as a function of the kinetic energy. Similar to other spectroscopic techniques, the qualitative and quantitative analysis of chemical composition of substances is possible on the basis of the obtained electron spectra. The x-ray characteristic emission has a depth of output of about 500 nm, the mean free path of the electrons is 10 nm, and the depth for the elastically scattered ions is within a single atomic monolayer.

Specifically, these basic conditions predetermine the high surface sensitivity of the methods of electron and ion spectroscopy by comparing it to that of x-ray spectral analysis. When studying the microheterogeneities in layers with thickness of about 1 μm, the most acceptable technique is x-ray spectral analysis. For investigations of thin surface layers (tens and hundreds of angstroms), it is advisable to use x-ray photoelectron and Auger spectroscopy. The spectroscopy of low-energy ion scattering allows investigation of the structure and composition of single-layered surfaces. Emission lines observed on the electron spectra are the result of electron transitions from different electron levels in the atoms. The aim of high-resolution photoelectron spectroscopy is to study the fine structure of lines, which can then be used to determine the electronic structure and nature of the chemical bonds. Chemical bonds are the most important characteristics of surface chemistry. The methods of electron and ion spectroscopy possess unique depth resolution and make it possible to analyze the finest surface layers. This advantage differentiates them from other methods of spectral analysis. The limitation of electron emission methods, such as models of heterogeneous composition, has led to the development of methods with high spatial resolution.

All methods of electron spectroscopy can be divided into two groups. The first group includes those in which the focused electron beam is directed onto the surface of samples and electrons are either partially elastically scattered (back-scattered electrons) or excited, causing the emission of secondary electrons. In this case, the spectra of the elastically scattered or secondary electrons are investigated. The most common methods used in this group are Auger electron spectroscopy (AES) and the spectroscopy of the characteristic electron energy losses, including plasmon losses spectroscopy and vibration spectroscopy. The developing method of this group is the analysis of the extensive energy loss fine structure (EELFS) spectra, which allows researchers to determine the atomic structure on the surface.

The second group unites the methods in which the emission of electrons from the sample surface model is excited by a photon beam (for example, by the radiation of x-ray tubes or by synchrotron sources). In this case, the secondary electrons, which left the atoms on the surface as a result of the complete quantum absorption, are investigated (one quantum absorbed one photoelectron left).

X-ray photoelectron spectroscopy of the outer and inner shells of atoms is based on this effect. Despite the fact that electron emission is widely used for investigating surface features, which has resulted in the further development of surface science, all methods based on this effect possess many serious limitations. The ultrahigh vacuum and special surface preparation of samples to prevent random pollution are necessary for investigations.

Electron spectroscopy relates to nondestructive test methods. However, fine changes in the surface state in the course of research should always be considered when surface phenomena are under investigation. The durable irradiation of the surface by electrons or x-rays can cause the ionization of atoms on the surface, the phenomena of adsorption and desorption, and the dissociation of molecules or chemical reactions on the surface. All these negative surface transformations should be considered when obtaining data, many of which can be neglected by special experimental approaches.

The success of fine-surface studies is based on the use of very specific equipment combined with the skills of researchers and their profound knowledge of the physics of the interaction of radiation with the substance. Till now, electron spectroscopy relates to nonstandard research methods in which the craftsmanship of the experimenter determines the success of the study.

The equipment for electronic spectroscopy consists of a complex high-vacuum system. This system is outfitted with controlling and analyzing electronic equipment and is supplied with units for computer control and processing of experimental data. Typical vacuum pressures in the work chamber of a spectrometer can reach 10^{-10} Pa.

The present work was carried out using ESCALAB-MK2, an electron spectroscope manufactured by Vacuum Generators (U.K.). The spectrometer is equipped with an electron energy analyzer of the "hemispherical capacitor" type, a scanning electron gun LEG200, a monochromatized x-ray Al K_α source, a gun of monochromatized low-energy electrons EMU50, scanning ion gun AG61, mass-spectrometer of the quadrupole type SQ300, a cooling–heating device providing sample temperatures ranging from –196°C to 650°C, and finally, units for mechanical sample cleaning and fracture at high vacuum.

4.2 X-RAY PHOTOELECTRON SPECTROSCOPY (XPS, ESCA)

X-ray photoelectron spectroscopy (XPS) is used for electron structure investigations of the surface, composition, chemical bonds, oxidation degree, and details of the valence band. This information relates to a depth of 1 to 10 nm from the surface. All elements, except H and He, can be detected at concentrations as low as approximately 1 — 0.01 atomic percent. The best lateral resolution is about 10 μm. The most common applications of surface analysis are for organic and inorganic compounds, semiconductors, alloys, etc.

Usually, XPS is based on the energy spectrum of electrons knocked out from the atoms of the surface layer of the material (the photoeffect) by primary x-radiation. The most widely used x-ray sources are Al K_α or Mg K_α, with quantum energies of 1486.6 and 1253.6 eV, respectively.

Under the irradiation of the x-ray beam with quantum energy, $h\nu = E_0$, the surface of the sample emits low-energy electrons ranging from 0 and E_0, called *photoelectrons*.

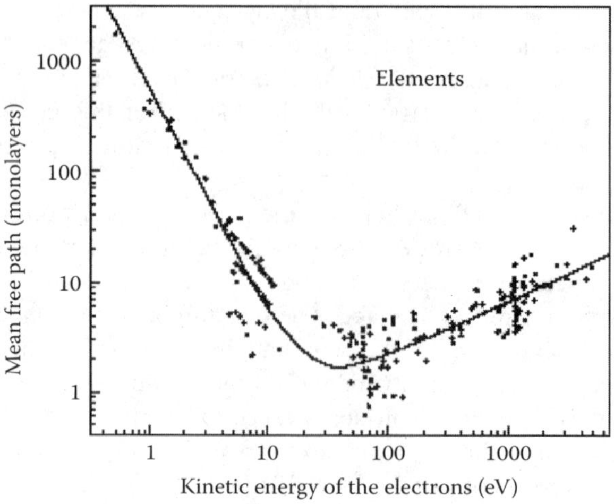

FIGURE 4.1 Variation of the mean free path in relation to kinetic energy of the electrons. (From Seah, M.P. *Surface Analysis by Auger and X-Ray Photoelectron Spectroscopy.* SurfaceSpectra Ltd/I M Publications, Eds., Briggs, D., Grant, J.T., National Physical Laboratory: Teddington, U.K., 2003, pp. 167–189. With permission.)

By measuring the kinetic energy of the emitted electrons (E_{kin}), the binding energy (E_b) can be determined for the photoelectrons detected ($E_b = h\nu - E_{kin}$, in a first approximation). The chemical analysis is possible by knowing the binding energies associated with the various atoms and with their electron levels.

Therefore, XPS analysis allows us to obtain a spectrum representing the number of electrons detected (the ordinate axis) vs. the binding energy values (the abscissa axis). This spectrum comprises two sets of information:

1. The position of the peaks, i.e., the binding energy value of the maximum point of the peak, which determines the chemical elements present on the sample surface.
2. The shape of each of the peaks determines the concentration of the chemical elements.

The presence of peaks at particular energies indicates the presence of a specific element in the sample under study. Furthermore, the intensity of the peaks is related to the concentration of the element within the sample region. Thus, the technique provides a quantitative analysis of the surface composition, alternatively known as *electron spectroscopy for chemical analysis* (ESCA).

XPS is indeed a surface-analysis technique because the photoelectrons emitted can only travel a few nanometers in the sample. The average distance that an electron can travel without losing part of its energy is called the *mean free path*, a value that is detailed in Figure 4.1 and which varies from 1 to 10 nm for the energies considered here.

Figure 4.1 clearly shows the far-reaching scattering of the points measured around the function

$$1 = k \cdot Ec^{0.5} \tag{4.1}$$

where k is a constant that limits the accuracy of the XPS quantitative analysis.

Each line on the spectrum corresponds to the occupied atomic orbital of the studied elements and reflects the ionization potential of the orbital. All lines on the spectrum are characterized by the half-width, intensity, and position on the energy scale. The intensity of the line is connected with the ionization cross section and the quantity of atoms of a specific element. Its width is determined by the time of life of a vacancy at this electron level and by instrument widening. The asymmetric form of the line is because of the numerous processes by which electrons can lose energy prior to their exit from the sample. Miscellaneous electron processes accompanying photoionization weaken the intensity of the lines, which leads to the appearance of satellites and extensive background noise.

4.2.1 Quantitative Analysis of the Chemical Composition of the Surface

The primary information directly accessible from an XPS spectrum is the chemical nature of the atoms making up the surface. The presence of photoelectron peaks at cleanly set binding energies directly reveals the chemical nature of the emitting atoms. Other available information is the position of the binding energy peaks. It is important to remember that the chemical environment of the emitting atom will alter the binding energy of the electrons.

Regarding surface heterogeneity of the samples, the electron emission is not isotropic. However, it is possible to investigate the chemical composition of heterogeneous materials with a tolerance less than 1% (only in special occurrences). There are special methods that make it possible to successfully use XPS as a nondestructive method for quantitative analysis of the chemical composition of homogeneous samples. Measuring the intensities of the characteristic lines of each element and comparing them with the reference spectra from clean substances and calibration graphs with corrections for elastic scattering and absorption, one can obtain data about the chemical composition. In the case of thin microheterogeneities on the surface where the layer thickness is comparable with the mean free path value, one should consider the influence of the matrix on the final quantitative results.

The practice of building depth profiles is widely used in quantitative analysis of surface layers with a thickness of approximately 10^{-8} to 10^{-6} m. For this purpose, ion etching of various elements combined with the acquisition of the electron spectra can be used, which represents the surface chemical composition for different depths. The information recorded by the computer is then converted into graphs representing the change of concentration of different elements as a function of distance from the initial surface.

Figure 4.2 presents the XPS spectra of an Al_2O_3 (001) single crystal after P implantation (ion doping) with a dose of 10^{16} ions/cm^2 and accelerating voltage $E = 100$ keV. The depth distribution of phosphorus can be estimated by using Ar^+ ion etching and recording the XPS spectra. The etching speed was approximately 1 monolayer/sec. The intensities of Al 2s and P 2p photoelectron lines are changed with etching time. This effect is related to the increase of the phosphorus concentration in the Al_2O_3 matrix at depths of 36 to 40 nm.

One should take into account that depth profiling is a surface-destructive technique. Ion etching produces the effects of selective sputtering, ion implantation, and ion-stimulated desorption, which influences the authenticity of the information obtained.

For this reason, in the case of XPS, ion etching is used with caution. If possible, other methods such as techniques of angular dependences or angle-resolved XPS (ARXPS) are preferred. The ARXPS method allows us to calculate the depth of the bedding of segregation under the surface of the sample on the basis of one electron spectrum by determining the ratio of the integral intensity of the characteristic line and background close to it and comparing this value with the one obtained for an alloy of homogeneous composition.

4.2.2 Investigation of Electronic Structure and Chemical Bonds

These are the main tasks for XPS, and the so-called chemical shifts and multicomponent structures of spectra are used to obtain the solution.

The basic equation used for the calculation of electron binding energy follows the law of conservation of energy:

$$h\nu = E_{kin} + e\varphi + E_b \tag{4.2}$$

where $h\nu$ is the quantum energy of primary radiation, φ is the work function of an electron, and E_{kin} and E_b are the kinetic and binding energies of an electron, respectively.

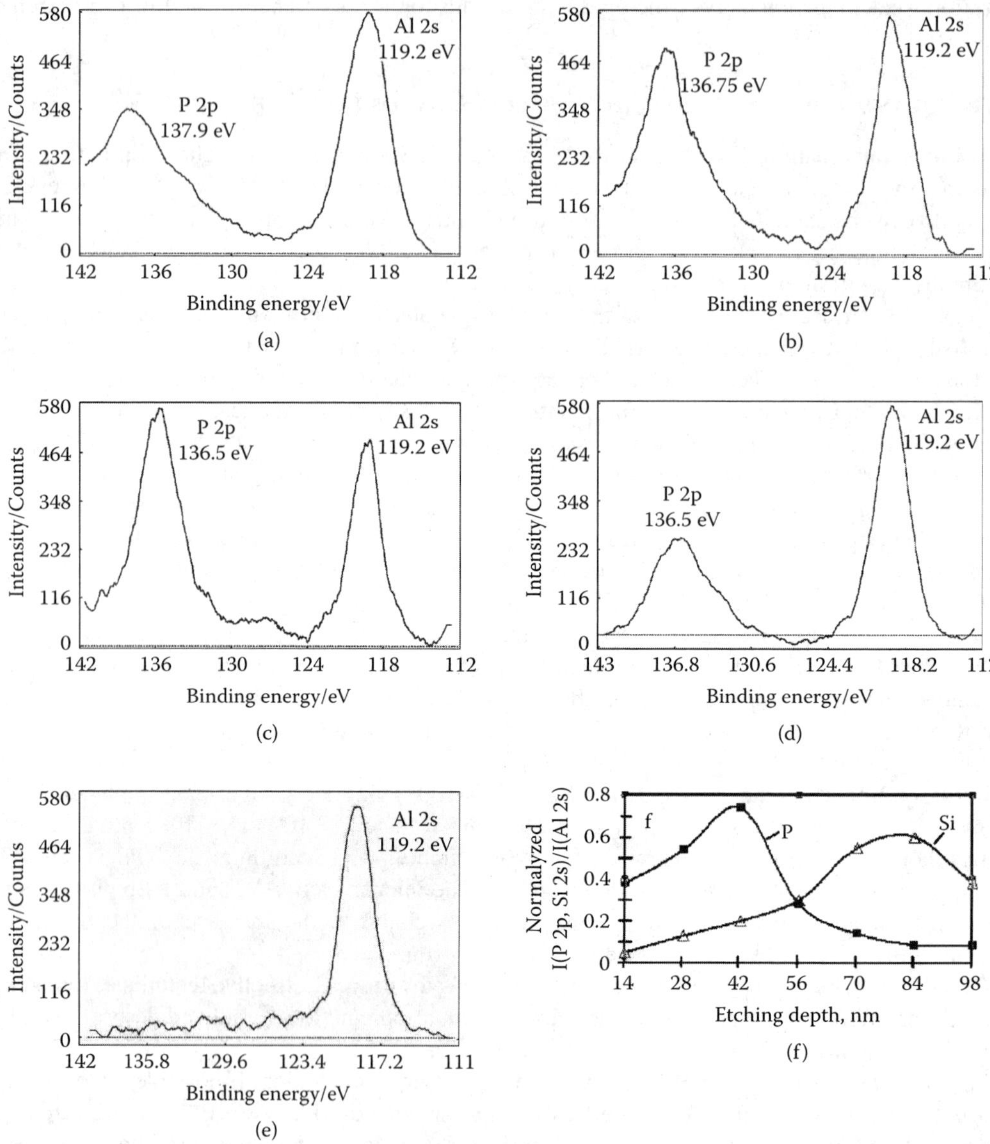

FIGURE 4.2 Photoelectron spectra of P 2p and Al 2s (a–e) for Al_2O_3 single crystal after Si and P ion implantation, annealing at 700°C during 2 h and Ar^+ ion etching during 1, 3, 4, 5, and 7 min, consequently, and the dependence of normalized I(P 2p), I(Si 2s)/ I(Al 2s) from etching depth (f). The intensity of the photoelectron lines were divided by the photoelectron cross section (sensitivity factor) of each element.

4.2.3 Multicomponent Structure of the Spectra

As a result of photoionization, the ion can be found in one of a limited number of final states. Thus, the removal of one electron from the p-shell of an atom filled with even numbers of electrons can lead to the formation of ions characterized by the azimuthal quantum number *I*, which possesses one unpaired electron with a spin of ± 1/2. As a result, a spin-orbital interaction can arise with the formation of two ionized states with I = 3/2 and I = 1/2. These states possess different energies, and the line on the electron spectrum reflecting this ionization is the doublet.

Excitation of electrons from an atom that has an unpaired electron in the initial state (transition metals with unfilled d-shell, rare-earth elements, actinides, etc.) can lead to the formation of an ion with two unpaired electrons. This results in a variety of different final states and is reflected on the spectrum as multiple splitting of the lines.

4.2.4 Chemical Environment — Chemical Shift

The exact binding energy of an electron depends not only on the level from which photoemission occurs, but also on:

1. The formal oxidation state of the atom
2. The local chemical and physical environment

Changes in either give rise to small shifts in the peak positions in the spectrum — so-called chemical shifts.

Such shifts are readily observed and interpreted in XP spectra (unlike in Auger spectra) because the technique:

- Is of high intrinsic resolution (because core levels are discrete and generally of a well-defined energy)
- Is a one-electron process (thus simplifying the interpretation)

In practice, the ability to resolve between atoms exhibiting slightly different chemical shifts is limited by the peak widths, which are governed by a combination of factors, particularly:

- The intrinsic width of the initial level and the lifetime of the final state
- The line-width of the incident radiation, which for traditional x-ray sources can only be improved by using x-ray monochromators
- The resolving power of the electron-energy analyzer

In most cases, the second factor is the major contributor to the overall line width.

As a rule, the binding energy increases with the degree of oxidation. Atoms of a higher positive oxidation state exhibit a higher binding energy owing to the extra coulombic interaction between the photo-emitted electron and the ion core. The ability to discriminate between different oxidation states and the chemical environments is one of the major strengths of the XPS technique.

A complex structure of XPS Cr $2p_{1/2}$ and Cr $2p_{3/2}$ spin-orbital doublet is shown in Figure 4.3. The XPS spectrum was recorded from a worn crater on a (Ti–Cr–Al)N-coated HSS cutting tool after its exploitation. Chemical shifts of Cr 2p lines are related to existing Cr–N and Cr–O bonds and different oxidation states of this element. Chromium suboxides are formed during wear of hard coatings, which can influence the wear-resistance properties.

In chemical compounds, the relative changes in binding energy of the inner shell are small (less than 5%) in comparison to the value for the valence bands (up to 20%). However, the absolute values of these changes for the inner shells are frequently similar and even higher when compared to valence electrons and can be measured on the photoelectron spectra.

The *chemical shift* is defined as the change in the electron-binding energy in a chemical compound in comparison with the binding energy of the same energy level in the free atom or the pure elemental substance:

$$\Delta E_B = E(M) - E(I) \tag{4.3}$$

where $E(M)$ and $E(I)$ are the binding energies of the electron in a molecule and free atom.

The chemical shifts are determined experimentally for all elements in the compounds. It can be positive or negative, depending on the presence of the donor–acceptor bonds. The high absolute

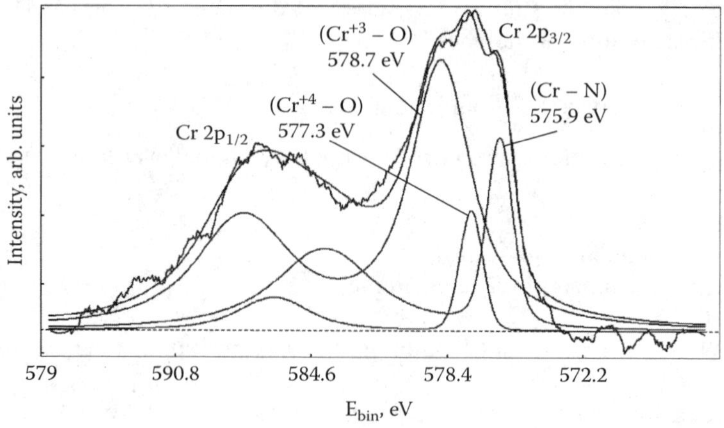

FIGURE 4.3 The chemical shifts of Cr 2p photoelectron line, which are tied with tribo-oxidation of (Ti–Cr–Al)N hard coating on HSS cutting tool during exploitation.

values of the chemical shifts for the inner electrons in the atoms of many organic and inorganic compounds ensured the success of the XPS application. For this reason, the XPS technique is also known as *electron spectroscopy for chemical analysis* (ESCA). The table of values for the chemical shifts measured by XPS for these substances is listed in Reference 1.

A change in the binding energy of the electrons attributable to the chemical environment is small for steels and alloys, and it does not produce large effects on the photoelectron spectra. In addition, it can also be registered and investigated.

The data that can be obtained from the chemical shift of photoelectron lines allow researchers to determine the change of the effective charge of atoms in the chemical compounds by comparing it to the pure element. A good correlation between the chemical shifts and the specific heat of the chemical compounds' formation is observed. The concepts behind the chemical shift are widely used for the study of chemical adsorption and oxidation processes on the surface of miscellaneous metals and alloys.

4.2.5 Extra-Atomic Relaxation

The photoelectron line on the spectrum reflects the binding energy and occupation of the electronic states at the electron level in the atom. The presence of vacancies at this level affects the position of the spectrum lines. This influence is more significant in the Auger transition, when the twofold charged ion in the final state is created. The existence of the "hole" at the electron level leads to the polarization of the atom, which introduces energy in the form of extra-atomic relaxation, $R_{ea.}$ The calculation of the binding energy of an electron is as follows:

$$E_b = h\nu - E_{kin} - e\varphi - R_{ea} \tag{4.4}$$

The energy of extra-atomic relaxation affects the value of the observed chemical shift. Regarding the Auger process, in contrast to photoemission that causes a double-charged ion, it is clear that the chemical shifts of Auger lines are larger than those of the photoelectron ones. This difference is characterized by the so-called Auger parameter α:

$$\alpha = 2R_{ea} \tag{4.5}$$

The Auger parameter is widely used in XPS for more precise interpretations of chemical shifts and for the determination of the nature of chemical surroundings of atoms. The determination of this value for clean substances and chemical compounds makes it possible to investigate the state of the electrons not only in valence bands, but also in inner shells at the formation of the chemical bonds.

4.3 AUGER ELECTRON SPECTROSCOPY (AES)

Auger electron spectroscopy is used for investigation of the microheterogeneities of the chemical composition and distribution of elements on the surface, as well as, for determination of the chemical state of these elements.

Studies of chemical effects on the Auger spectra and the procedures of quantitative analysis of thin surface films were developed. The design of Auger spectrometers and Auger microprobes with a high spatial resolution on the order of hundreds of nanometers contributes to its wide application in analytical situations.

In turn, the success of high-spatial-resolution AES has stimulated the development of photoelectron spectroscopy techniques with a high locality of analysis. Now, AES is the most popular method for surface analysis in the fields of materials science, metallurgy, chemistry, and semiconductor technology. In combination with other methods of spectroscopy, this technique makes it possible to perform layer-by-layer analysis together with ion etching and high locality. This ensures that the acquisitions of the distribution of chemical elements in the microvolumes are of unquestionable value for the AES method.

In AES, an incident electron beam strikes the surface of the sample and electrons of different and lower energies are ejected. The electrons are then analyzed. Similar to XPS, only electrons from the top 10 nm can escape without scattering or losing energy. All elements except H, He, and Li can be detected at amounts as low as 0.01 to 1 atomic percent. The best lateral resolution for small objects is almost as low as 10 nm, which makes AES a dominant method of analysis in the semiconductor field. AES also has applications in many areas of corrosion and fracture analysis, thin-film analysis, and many other surface investigations.

4.3.1 PHYSICAL BASIS

In the AES method, the emission of secondary electrons is the result of the excitation of atoms on the surface under the primary electron beam with energy E. A small portion of the electrons (order 10%), the so-called Auger electrons, leave the specimen after interorbital transitions. During ionization of the atom, the electron can pass into an unexcited state without the emission of a quantum of x-radiation. This results in the transition of an electron from a higher energy level to the formed vacancy with the excitation of an Auger electron.

Figure 4.4 presents the diagram of output of photoelectrons from the K level and Auger electrons from the L level (KLL electron) under the action of an external radiation beam. The emission of Auger electrons is often observed with the participation of the valence band: KLV, KVV transitions. Despite the fact that Auger electrons can be excited by x-rays, the excitation by primary electrons is usually used in AES owing to the higher output of Auger electrons. Obviously, the energy positions of Auger lines are independent of the primary electron's energy. The ionization cross section of atoms under the primary beam is significantly distinguished, depending on its nature. The maximal ionization cross section of the inner shells of atoms under the electron's excitation is observed when the primary electron's energy exceeds the binding energy of the corresponding level by four times. Several series of Auger lines are observed on the electron spectra. The KLL lines are observed for the elements starting from boron. The Auger lines are often weak and blurred on the $N(E)$ spectra, and Auger spectra are thus represented typically in the form of first-order derivatives $dN(E)/d(E)$. The LMM Auger lines of Os and RbOH in the $dN(E)/d(E)$ form are present in Figure 4.5 and Figure 4.6. The multicomponent structure of the spectra is linked to the complex Auger electron transitions and plasmon excitations.

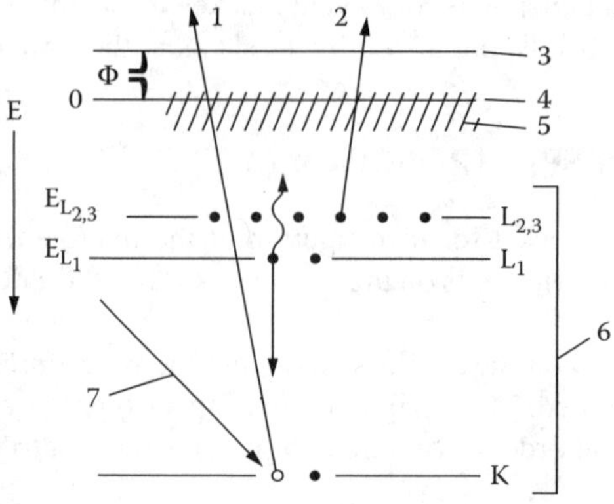

FIGURE 4.4 Energy-level diagram showing the filling of core holes on levels K and L1, giving Auger electron emission from level L2,3. 1 — Photoelectron; 2 — Auger electron KL1L2,3; 3 — vacuum level; 4 — Fermi level; 5 — valence band; 6 — core levels; 7 — initial energy beam; and Φ — work function. (From Kovalev, A.I., Scherbedinsky, G.V. *Modern Techniques of Metals and Alloys Surface Study*. Metallurgy: Moscow, 1989, p. 191. With permission.)

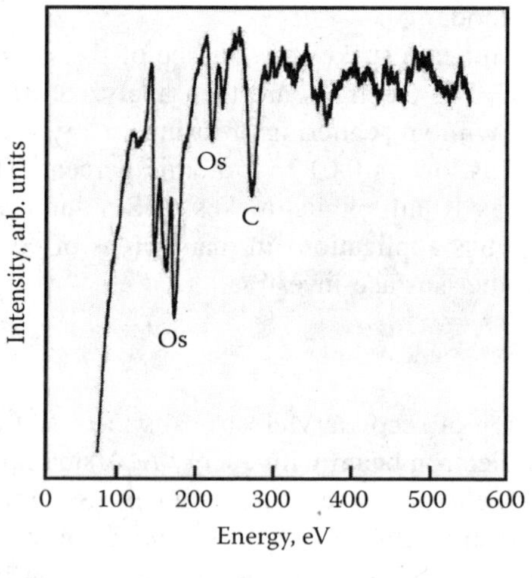

FIGURE 4.5 Auger spectrum of osmium.

4.3.2 AES Quantitative Analysis

Evaluating the possibilities of quantitative AES analysis and other methods of electron spectroscopy, it is necessary to note that there are many factors that reduce its accuracy and the reproducibility of results. This method is nonstandard, and it is difficult to compare its results with the standard methods of chemical analysis. A fundamental limitation is associated with the investigation of heterogeneous samples. For such objects, the emission of electrons from the surface is nonuniform and very difficult to estimate in practice. The difference between the experimental (real) and computed values of the depth of the output of electrons is another source of errors, which can reach as high as 30 to 36%.

Surface roughness and diffraction effects can also influence the results of quantitative analysis, especially in the study of single crystals. The individual calculation procedures of AES are developed in connection with different practical tasks. This hampers the comparison of data obtained in different works. For these reasons, AES cannot compete with other methods of chemical composition analysis.

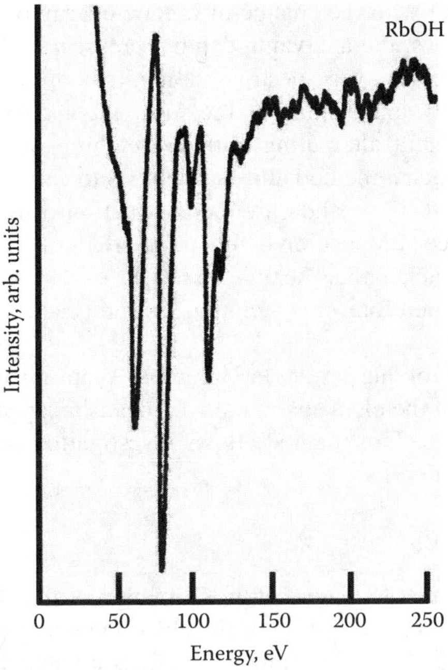

FIGURE 4.6 MNN Auger spectrum of rubidium.

A basic and unquestionable advantage of the method over other standard surface chemical analysis techniques, such as x-ray fluorescent microanalysis, is its high surface sensitivity. AES, as with many methods of electronic spectroscopy, makes it possible to analyze the finest surface layers with the thickness of several atoms, which is impossible for other methods of chemical, radioisotope, or spectral analysis.

For samples with homogeneous compositions, the concentration of elements on the surface is determined in view of the factor of relative element sensitivity S_j using the standard Auger spectra for pure components, neglecting the influence of matrix effects:

$$C_j = \frac{I_j / S_j}{\sum_i (I_i / S_i)} \tag{4.6}$$

where C_j is the concentration of the jth element, at. % and I_j is the intensity of the characteristic line on the spectrum. The values of S for various elements are listed in standard tables or determined from the reference spectra for pure substances. This relation allows us to define the contents of many elements in multicomponent alloys and compounds. The accuracy of such a method is higher than 30%. Special algorithms should be carried out to determine the chemical composition of the finest surface films with thicknesses comparable or less than the depth of the Auger electrons' output.

The important task during surface investigation is determination of the chemical composition of the sample at various depths (or layer-by-layer analysis). The methods used for this purpose depend on the depth of interest; the depth scale is comparable or bigger than the length of the electrons' mean free path, i.e., from tens to hundreds of angstroms. In the first case, it is necessary to take into account the exponential deterioration of the signal at depths that raise the accuracy of definition of concentration of the elements in thin films. If the depth under investigation is less than the electron free path (5 nm), scientists use the procedure that takes into account the change of intensity of absorption edge with the change of kinetic energy of the primary electrons. As the primary electrons' energy grows, the analyzing depth increases.

In the case of thicker-layer analysis, destructive methods are applied, such as ion etching or angle laps. Depth profiling by ion etching is the most labor-consuming technique. It requires recording of the electron spectra alternating with ion etching of the sample surface. This can sometimes last many hours. Such a method allows scientists to carry out fine research of multilayer coatings. Serious limitations of the method are the destruction of the surface under the ion beam and the alteration of the surface composition owing to an effect of selective sputtering when lighter elements leave the sample quicker than heavy ones. The ion beam also destroys the crystalline structure of the surface and, therefore, this technique is rarely used for investigations of coatings that are several microns thick.

In this case, a method involving an angle laps cross section together with line scan AES is used. The depth distribution of the elements in the coatings is analyzed using angle metallographic laps with an angle of 5 degrees. This method allows investigation of the chemical composition to depths up to several tens of microns.

4.3.3 High-Resolution AES

The change of the charging state of an atom at the formation of the chemical bonds leads to a shift in the electron spectrum lines. Redistribution of the electron density depending on donor–acceptor properties of the elements in a chemical compound defines a value and a sign for these shifts that increase together with the strengthening of the interatomic bonds. These shifts of Auger lines in chemical compounds can reach from 10 to 20 eV when compared with the pure components.

The spectroscopy of Auger transitions with participation of a valence band is widely used. This technique is applied for precise research of the chemical interatomic bonds arising from adsorption and segregation. For interpretation of the Auger spectra with transitions in a valence band, the energy calculations for multiple electron transitions is used, taking into account the various matrix effects. The computer calculations, particularly the deconvolution method, allow us to find the electron-distribution valence band.

4.4 DETERMINATION OF INTERATOMIC DISTANCES ON SURFACES USING ELECTRON SPECTROSCOPY — ELECTRON ENERGY LOSSES FINE STRUCTURE (EELFS) TECHNIQUE

4.4.1 Physical Basis of the EELFS Method: Experimental Procedure and Peculiarities of Data Processing

Close to the lines of back-scattered, photo, and Auger electrons on the low-kinetic energy side of the electron spectra, there is always a background gradually decreasing in intensity. It is determined by the flow of electrons subjected to nonelastic scattering by electron–electron interactions. In contrast to the photo-effect initiated by x-rays or ultraviolet radiation with the energy of quantum being completely absorbed, the electron excitation at the atomic level causes the emission of electrons in a continuous spectrum of kinetic energy. On passing the thinnest surface layers, the intensity flow of such electrons undergoes nonelastic and elastic scattering as well. Owing to all of these processes, emission of electrons that have a continuous spectrum of energy (and the intensity of background on electron spectra) considerably increases when the sample is irradiated by electrons. Because of the wave nature of electrons, in the process of their scattering in the nearest atomic surrounding and under certain conditions of their waves' existence in the crystal lattice, there is a possibility for the phenomena of resonance and antiresonance. This leads to modulation of the electron wave function amplitude in a given point of space and, in particular, in the detector. These conditions are determined both by the electron wavelength (or energy) and by the distance from the emission center to the scattering center (or interatomic distance). As a result, there are modulations in the intensity of the detected signal from electrons with different kinetic energy on the emission spectrum.

Emission of electrons and their nonelastic scattering in the nearest atomic surrounding are probabilistic by nature. Because of this, the amplitude modulation of the signal intensity gets diffused over the energy scale and creates an extended fine structure of background close to the lines of elastic scattering and Auger lines. This fine structure can range between 250 and 300 eV. Using the line of back-scattered electrons, we can obtain the average picture of interatomic distances on the surface (between all atoms). Using Auger lines, the distances from selected atoms to their neighbors can be calculated. Depending on the wave vector of electron k, the oscillation on such a spectrum has the following equation [3,4]:

$$\chi(k) = (-1)^l \cdot \sum_j \frac{N_j \cdot e^{-2\sigma_j^2 \cdot k^2}}{kR_j^2} \cdot e^{-2R_j/\lambda(k)} \cdot A_j(k,\pi) \cdot \sin[2kR_j + \varphi_j(k)]$$

where R_j is the distance between the centers of emission and scattering, N_j is the coordination number, $A_j(k, \pi)$ represents the amplitude of back-scattering, $\varphi(k)$ is the phase shift of an electron at its scattering, σ_j is the correcting factor for the heat oscillation of the atoms in the lattice, $\lambda(k)$ is the depth of electron exit, and $l = 1$ or 2. To determine the length of the atomic bonds in the nearest coordination spheres (averaged and partial functions of radial distribution), the Fourier analysis of this amplitude modulation is applied. In the present work, we used the Fourier transformation of the type:

$$F(R) = \int_{k_{min}}^{k_{max}} \chi(k) \cdot k^n \cdot W(k) \cdot \sin(kR)dk \tag{4.7}$$

where $k \sim \sqrt{E_0 - E}$, $\chi(k)$ corresponds to the modulated fine structure, k_{min} and k_{max} are the onset and end of the spectrum processing range, k^n is the weighting factor ($1 \leq n \leq 3$), and $W(k)$ is the window function. A theoretical basis for such an analysis is analogous to that presented in Reference 3.

The EELFS spectra were registered using the electron spectrometer ESCALAB MK2. It is equipped with an electron gun LEG200 at vacuum pressure of 2×10^{-8} Pa, an accelerating voltage between 1 to 1.5 kV in CAE mode (pass energy = 2 eV), and a slit C3 of 1.5 mm. The spectra-recording parameters were the following: range 250 eV, 1200 channels, 10 min/scan, 10 scans.

The following procedure is used to distill the structure-dependent oscillations from electron spectra.

1. The spectrum is recorded in integral mode ($N(E)$) below the back-scattered electron line (to get the averaged inter-atomic distances) or Auger line (to get distances from definite element). To make the accurate peak synthesis, some points at the peak foot are required.
2. The following sequence of numerical filtering is applied:
   a. Removal of the slow background (appearing because of analyzer settings) by parabola approximation.
   b. Peak and inelastic "tail" synthesis using the algorithm described in Reference 4; subtraction of the synthesized peak and normalization of the losses spectrum.
   c. Recalculation of the losses spectrum to K-space.
   d. Fourier transformation.

The random interference (noise) influences the quality of the background removal and the Fourier transformation. Decreasing the noise using only prolonged spectrum acquisition can increase the experiment duration and occasional contamination of the sample surface, even in ultrahigh vacuum. Therefore, in addition to the selection of a reasonable signal acquisition time, the sliding-band smooth algorithm does not produce a harmonic distortion of spectrum:

$$F'(x_i) = \sum_{k=-n}^{n} \varphi_k \cdot F(x_{i+k}) \tag{4.8}$$

The smoothing factors φ_k in Equation 4.8 are selected in the way that the processed function $F'(x)$ is equal to initial function $F(x)$ when $F(x) = ax^2 + bx + c$ or $F(x) = ax^3 + bx^2 + cx + d$:

$$\varphi_k = \frac{9n^2 + 9n - 3}{(4n^2 + 4n - 3)\cdot(2n+1)} - 15 \cdot k^2 \cdot (4n^2 + 4n - 3)\cdot(2n+1) \tag{4.9}$$

To interpret the experimental data, the information on interatomic distances obtained by other techniques can be used, where possible — information on bulk crystalline structure from x-rays and electron diffraction experiments, Raman spectroscopy, etc. In other cases, when the crystallographic data is not available, we can compare the averaged (below the line of back-scattered electrons) and partial (below the set of Auger lines) pictures of interatomic distances to determine the distances between definite elements. In this case, we can use the data on atomic radii to interpret the system of interatomic distances.

4.4.2 Quality and Reliability of EELFS Results

The atomic structures of surfaces were investigated by EELFS in the following systems: grain boundary segregations [5], hard diamondlike films [6], and adapting tool materials [7]. The correlation between theoretical (from crystallographic data) and experimentally determined interatomic distances is shown in Figure 4.7. One can see a good correlation between theory and experiment. The error does not exceed 0.2E. Moreover, the trend line crosses the X axis at –0.1E, which corresponds to a narrowing of interatomic distances on the surface because of surface tension.

As an example, we investigated the TiO_2 (rutile) reference sample with purity 99.9%. Interatomic distances in this system, obtained by EELFS, are shown in Figure 4.8a. A good resolution at distances of up to 6 Å is observed. The distances on the Fourier transformant are in good correspondence with crystallographic data for the nearest dense coordination spheres.

Figure 4.9a and Figure 4.9b present the EELFS data from a diamondlike amorphous hydrogenated hard coating (α:C–H) applied to silicon and glass substrates. The films were prepared by deposition on the substrates in the reactor with parallel plates. The RF-fed (13.56 MHz) electrode diameter (16 cm) and grounded electrode diameter (20 cm) were placed at a distance of 4.4 cm. After cleaning the substrate, electrodes, and a chamber with oxygen and argon ions, the system was filled with methane used as a carbon source and hydrogen. The flow of hydrogen was 60 to 80 l/min and, for methane, 1 to 4 l/min. The voltage between electrodes was 60 to 120 V, and the pressure in the chamber was 2×10^{-2} Pa.

The interatomic distances in the systems are listed in Table 4.1 and Table 4.2. One can see that carbon bonds with different lengths (sp, sp^2, sp^3) are well resolved. Moreover, the C–H and H–H distances can be measured, which is extremely difficult for other techniques of surface atomic structure investigations.

The EELFS method, therefore, has shown good reliability and reproducibility for the determination of interatomic distances on the surface, especially in systems containing three or more kinds of atoms. In our case, we received good resolution owing to the recording spectra in $N(E)$ mode

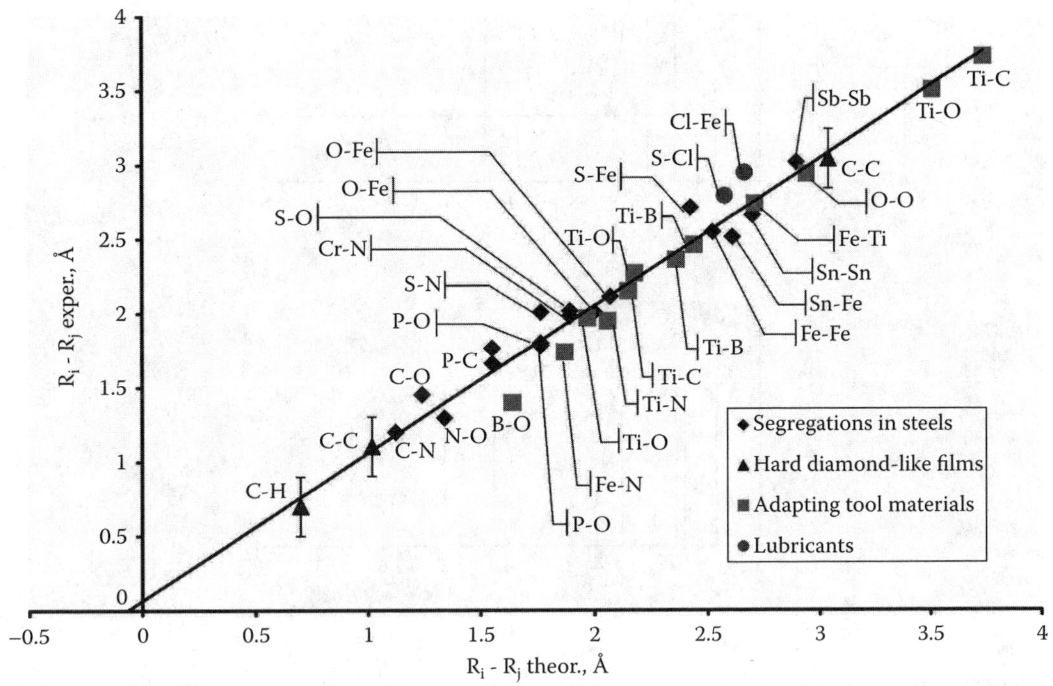

FIGURE 4.7 Comparison of theoretical (from crystallographic data) and experimental (from EELFS experiments) interatomic distances on surfaces.

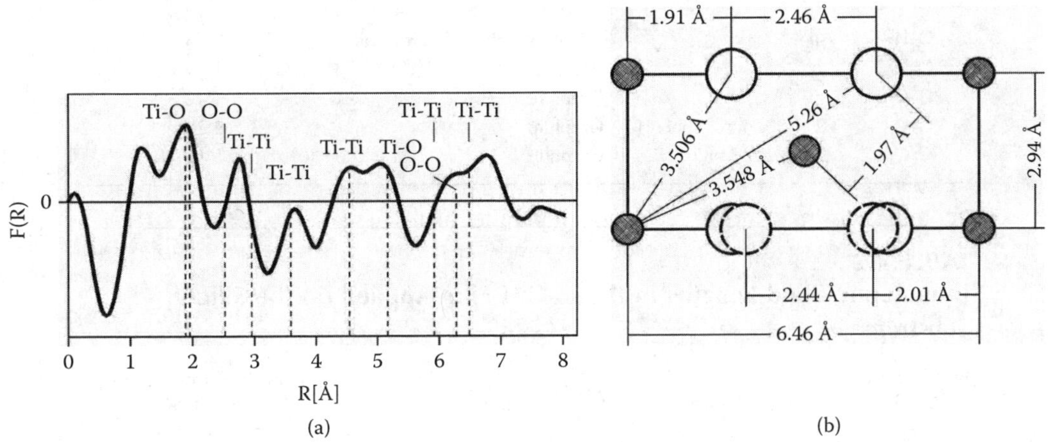

FIGURE 4.8 Radial distribution function from (a) TiO_2 (rutile) reference sample and (b) rutile crystalline lattice.

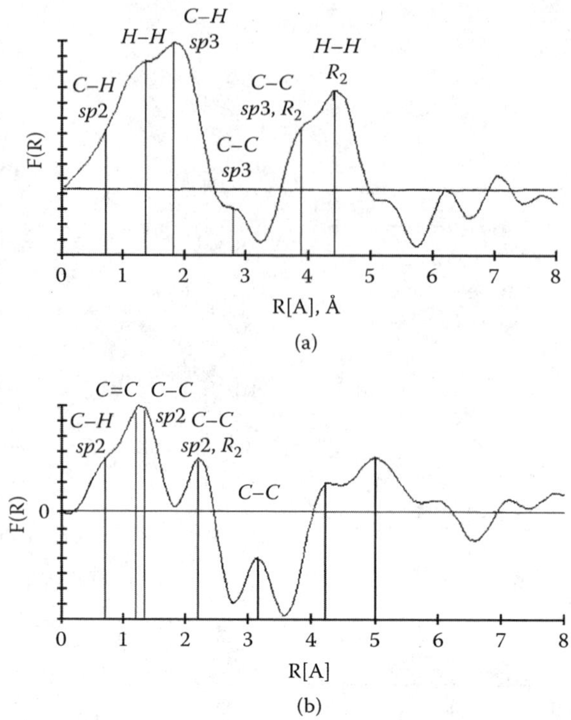

FIGURE 4.9 Radial distribution function from α:C–H film applied on (a) the glass and (b) silicon substrate.

TABLE 4.1
Interatomic Bond Lengths in the α–C:H Film Applied on a Glass Substrate

Interatomic Bond	Hybridization		Length of Interatomic Bond (*E*)
C–H	sp^2	Arom., graphite, free radicals	0.73
C=C			1.39
C–C	sp^2	Graphite	1.42
C–C	R_2 (2nd coord. sphere)	Graphite	2.20
C–C	Hydrogen bonds	Graphite	3.16

TABLE 4.2
Interatomic Bond Lengths in the α–C:H Film Applied on the Silicon Substrate

Interatomic Bond	Hybridization		Length of Interatomic Bond (*E*)
C–H	sp^2	Arom., graphite, free radicals	0.70
C–C	sp^2	Graphite	1.39
C–H	sp^3	Diamond	1.80
H–H	R_1 (1st coord. sphere)		2.80
C–C	sp^3	Diamond	3.75
H–H	R_2 (2nd coord. sphere)		4.40

with further numeric filtering, in comparison to other authors who have used dN/dE or d^2N/dE^2 spectrometer modes.

4.5 HIGH-RESOLUTION ELECTRON ENERGY LOSS SPECTROSCOPY (HREELS)

This section of electronic spectroscopy includes various methods to research the composition and atomic structure of surface films based on the energy losses of electrons scattered near the surface or at small depths. The fundamentals, spectroscopy equipment, requirements examples of practical application for methods such as vibrational spectroscopy, and plasmon losses spectroscopy are considered in Reference 8 to Reference 10. These methods give unique data on the structure of the adsorbed films, which can be extremely useful for the research of thin-film coatings, oxidation, implantation, and catalysis.

HREELS can investigate the energy losses of low-energy electrons and, because of this, it has become one of the most useful tools for basic research of the physical and chemical phenomena of surface science. One of the most important features of the method is its high surface sensitivity, allowing it to define vibrating energies of adsorbed atoms and molecules at concentrations less than 0.001 monolayer (1 monolayer = 10^{15} atoms/cm^2). It can also identify the centers of adsorption and the coordination of the adsorbed molecules on the substrate. In comparison with AES, spontaneous contamination of the surface limits usage of this method despite its high sensitivity. The HREELS method allows us to detect traces of residual gases (CO and H) in the chamber of a spectrometer, whereas AES is unable to define hydrogen directly. In most cases, the complex groups, including C, H, O, and F on the surface of various materials, are the subject of attention. Thus, characteristic vibration modes on the electrons' energy-loss spectra can be investigated in the range of several eV. Until now, opportunities for this technique's application were limited to qualitative information. Quantitative characteristics can be obtained based on the dynamic modeling of the vibrational spectra with full accounts of multiple scattering. This approach requires the combination of calculation and experiment. Reliable information can be acquired on the spectrometers, with a monochromatic source of slow electrons and hemispherical capacitor analyzers, with high-energy resolution of up to several meV.

4.5.1 Vibrational Spectroscopy

Vibrational spectroscopy provides the most definitive means of identifying the surface species generated upon molecular adsorption and the species generated by surface reactions. Any technique that can be used to obtain vibrational data from solid-state or gas-phase samples (IR, Raman, etc.) can be applied to the study of surfaces. In addition, there are a number of techniques that have been specifically developed to study the vibrations of molecules at interfaces.

There are, however, only two techniques that are routinely used for vibrational studies of molecules on surfaces:

1. Infrared reflection–absorption spectroscopy (of various forms, e.g., IRAS and MIR)
2. Electron energy loss spectroscopy (EELS)

HREELS is a technique utilizing the inelastic scattering of low-energy electrons to measure the vibrational spectra of the surface species. Superficially, it can be considered the electron analog of Raman spectroscopy.

The high sensitivity (< 0.1% of a monolayer) and broad spectral range (0 to 1000 meV, or 0 to 8000 cm^{-1}) makes EELS an ideal tool for exploring the properties of a wide variety of surfaces. These include (but are not limited to) single-crystal metals, semiconductors and insulators,

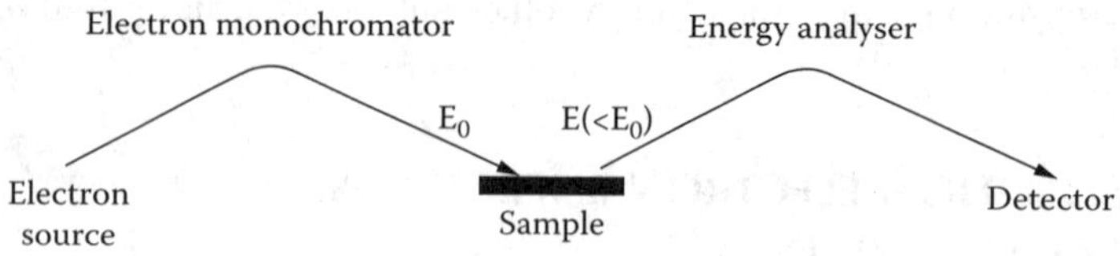

FIGURE 4.10 Schematic representation of the EELS experiment.

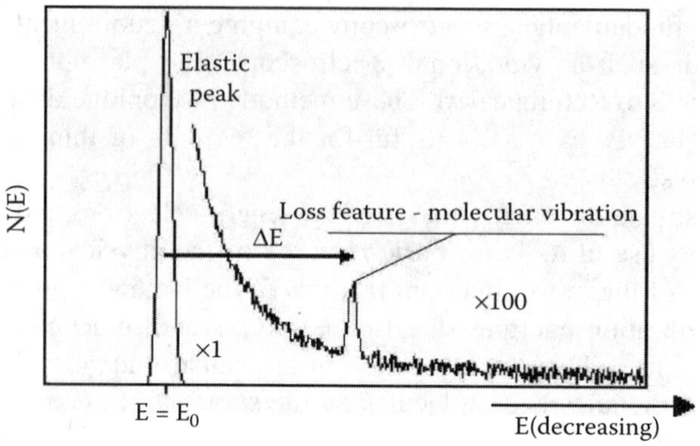

FIGURE 4.11 A schematic electron energy-loss spectrum.

evaporated films, polycrystalline foils, and polymers. At present, the resolution is limited to 4 meV (32 cm^{-1}), depending on the scattering properties of the surface under study.

Because the technique employs low-energy electrons, it is necessarily restricted to use in high-vacuum (HV) and UHV environments. However, the use of such low-energy electrons ensures that it is a surface-specific technique and is the vibrational technique of choice for the study of most adsorbates on single-crystal substrates.

The basic experimental geometry is fairly simple as illustrated schematically in Figure 4.10. It involves using an electron monochromator to produce a well-defined beam of electrons of fixed incident energy and then analyzing the scattered electrons using an appropriate electron energy analyzer.

A substantial number of primary electrons are elastically scattered ($E = E_0$). This gives rise to a strong elastic peak in the spectrum. On the low-kinetic energy side of this main peak ($E < E_0$), additional weak peaks are superimposed on a mildly sloping background. These peaks correspond to electrons that have undergone discrete energy losses during the scattering from the surface (Figure 4.11).

The magnitude of the energy loss, which is given by:

$$\Delta E = (E_0 - E) \quad (4.10)$$

is equal to the vibrational quantum (i.e., the energy) of the vibrational mode of the adsorbate excited in the inelastic scattering process. In practice, the incident energy E_0 is usually in the range of 5 to 10 eV (occasionally up to 200 eV), and the data is normally plotted against the energy loss (frequently measured in meV).

The incoming electrons take note of the oscillating dipoles that are present on the surface. Often, these arise from the vibrational modes of the molecular adsorbates that are present. They are also sensitive to the chemisorption bonds on the surface. Furthermore, phonons on the surface of semiconducting substrates (not metallic) also interact with the electron beam.

The scattering of primary low-energy electrons is linked to their electrostatic interaction with a surface. Elastic scattering occurs when the change in momentum vector k of the electron upon reflection is minimal. These electrons electrostatically interact with the surface dipoles such that coulombic energy transfer occurs while the electron is still 100 to 200 Å above the surface. The electron in vacuum feels not only the surface dipole (of adsorbed film) but also the response of the conduction electrons to this dipole. This response is known as the image dipole.

Such an electron is scattered specularly with an energy loss characteristic of the energy it deposited in the vibrational mode. Hence, such information is exactly like an infrared spectrum.

The elastic dipole scattering of low-energy electrons has a narrow angular spread about the specular position and the long-range nature of the interaction as a direct consequence of an electrostatic interaction of the surface dipole with the incident and outgoing electrons.

Inelastic electron scattering is related to the short-range interaction of the electron with the surface atomic potential, which is modulated at vibrational frequencies. These electrons may lose energy into the corresponding vibrational mode of the surface-adsorbate complex. The scattering is more isotropic (not in the specular direction, but everywhere) but the energy losses still reflect vibrational excitations in the adsorbate. The angular distribution of peaks around the specular direction can distinguish between peaks that result from different scattering modes (dipole scattering is dominant).

For metallic substrates and a specular geometry, scattering is principally by a long-range dipole mechanism. In this case, the loss features are relatively intense, but only those vibrations giving rise to a dipole change normal to the surface can be observed.

By contrast, in an off-specular geometry, electrons lose energy to surface species by a short-range impact-scattering mechanism. In this case, the loss features are relatively weak, but all vibrations are allowed and may be observed.

4.5.1.1 Normal Modes of Vibration of Polyatomic Molecules

Even at absolute zero, molecular vibration occurs because the molecule cannot have a point of energy less than zero. Polyatomic molecules undergo complex vibrations, which may be resolved into a limited number of normal modes of vibration for the diatomic molecule. Any motion of this molecule can be obtained by superimposing six kinds of motions (Figure 4.12).

Vibrations pass through the center of mass of molecules about normal modes of vibration. In this case, both atomic displacements occur at the same frequency and in phase. These modes are

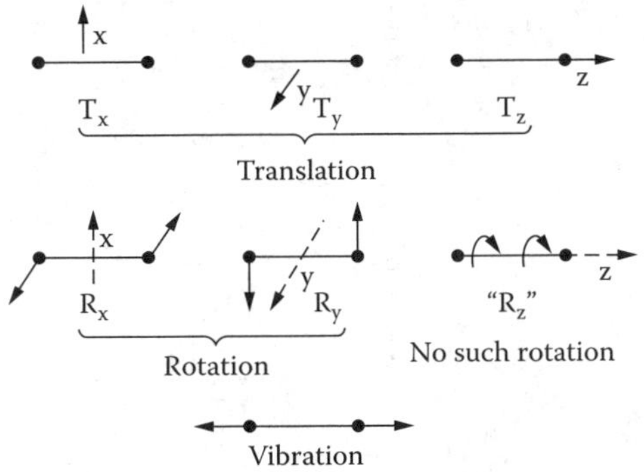

FIGURE 4.12 The six degrees of freedom of diatomic molecule. (From Harris, D.C., Bertolucci, M.D. *Symmetry and Spectroscopy: An Introduction to Vibrational and Electronic Spectroscopy*. Dover Publications: New York, 1989. With permission.)

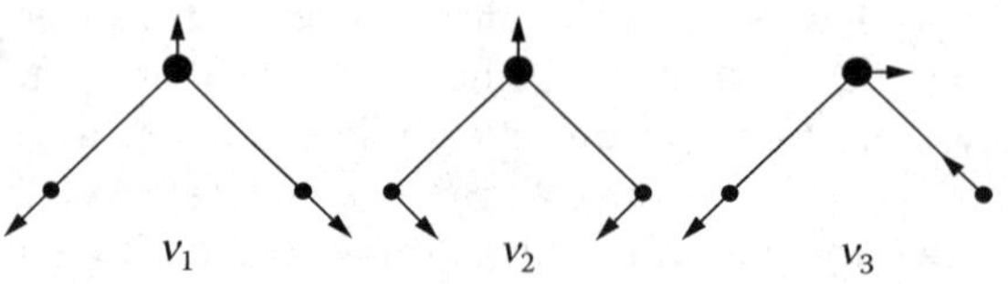

FIGURE 4.13 The three normal modes of vibration of H_2O. (From Harris, D.C., Bertolucci, M.D. *Symmetry and Spectroscopy: An Introduction to Vibrational and Electronic Spectroscopy.* Dover Publications: New York, 1989. With permission.)

defined such that the potential energy of the diatomic molecule can be expressed by Equation 4.11, where atoms A_1 and A_2 move a distance Δr:

$$V = (1/2)\sum_i \lambda_i q_i^2 \tag{4.11}$$

where λ is a constant, and

$$q = r - r_0 = \Delta r(A_1) + \Delta r(A_2) \tag{4.12}$$

Each normal mode of vibration will form a basis for an irreducible representation of the point group of the molecule [12]. For example, the three normal vibrations of water are shown in Figure 4.13.

The vibrations are described by the following abbreviations: ν, stretching; δ, deformation (bending); ρ_w, wagging; ρ_r, rocking; ρ_t, twisting; π, out of plane; as asymmetric; *s*, symmetric; *d*, degenerate.

The combination of different vibrations for the XeF_4 molecule is shown in Figure 4.14. There are seven different vibrational energies for the nine normal modes of XeF_4.

In HREELS (dipole scattering mode), the incident electrons interact primarily with the long-range fields set up in the vacuum by the oscillating dipoles. The fields created by the oscillation of a dipole are screened by a factor of 1/ε if the dipole is orientated parallel to the surface, or not at all if the dipole is orientated perpendicular to the surface. The only vibrations that can be excited are those that create dipoles with a component perpendicular to the surface.

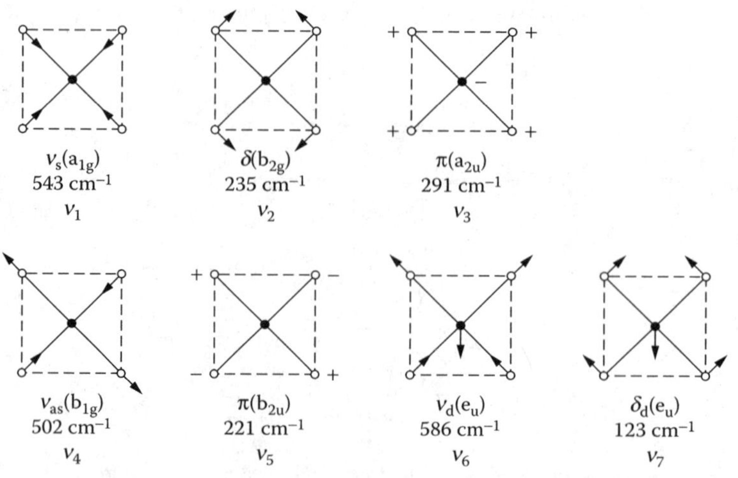

FIGURE 4.14 The normal modes of vibration for a planar XeF_4 molecule. (From Harris, D.C., Bertolucci, M.D. *Symmetry and Spectroscopy: An Introduction to Vibrational and Electronic Spectroscopy.* Dover Publications: New York, 1989. With permission.)

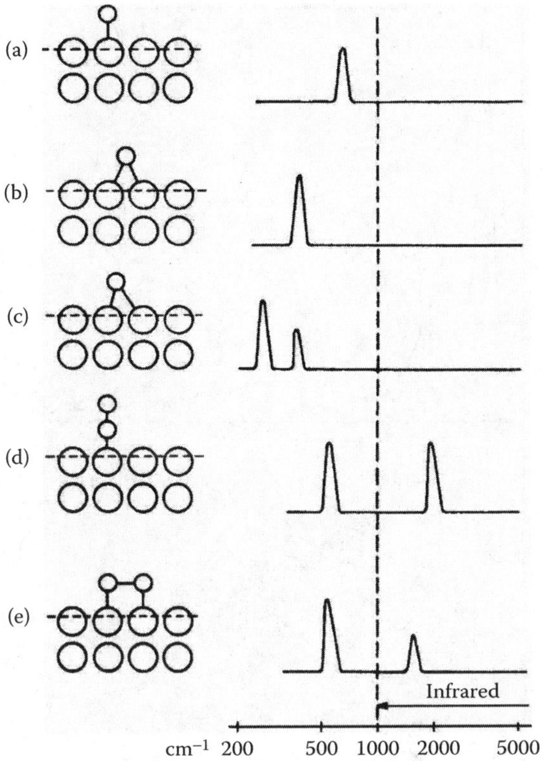

FIGURE 4.15 The imaginary adsorption states showing the expected loss peaks associated with each structure. (From Woodruff, D.P., Delchar, T.A. *Modern Techniques of Surfaces Science*, 2nd ed. Cambridge University Press: London, 1994. With permission.)

We can see in Figure 4.15 that the low-frequency mode derives from the vibration of isolated atoms perpendicular to the surface. The high-frequency mode originates from the stretching vibration of the quasi-molecule, whereas the frequency split is proportional to the square root of the mass ratio.

4.5.1.2 The HREELS Experimental Results

The high-energy resolution of the incident beam is achieved by monochromatizing an electron beam with a hemispherical sector capacitor. The gun, which has a hemispherical monochromator, operates similar to the spectrometer energy analyzer EMU-50 (VG, U.K.) used as a source of monochromatized low-energy electrons. This system produces a high beam current. The alignment of this device is very important to achieve the ultranarrow primary electrons line. The beam energy width for EMU-50 was ~8.0 meV using the negatively charged graphite standard. The interpretation of the HREELS spectra follows that of the reflection-absorption infrared spectroscopy (RAIRS). The RAIRS spectra are usually cast in wave numbers, whereas HREELS are in meV (1 meV = 8.065 cm^{-1}). The resolution in RAIRS is typically 0.25 meV for single crystals and 10 to 20 meV for polycrystals.

The interpretation of EELS is based on experiments using reference samples, literature data, and calculations. The methods of vibrational spectra modeling are offered in Reference 12. But the online calculations can be performed at http://www2.chemie.uni-erlangen.de/services/telespec/.

Figure 4.16a and Figure 4.16b presents the vibrational spectra of the PVD nonstoichiometric Ti–C coating with 30% of excessive carbon that forms diamondlike and amorphous phases, before

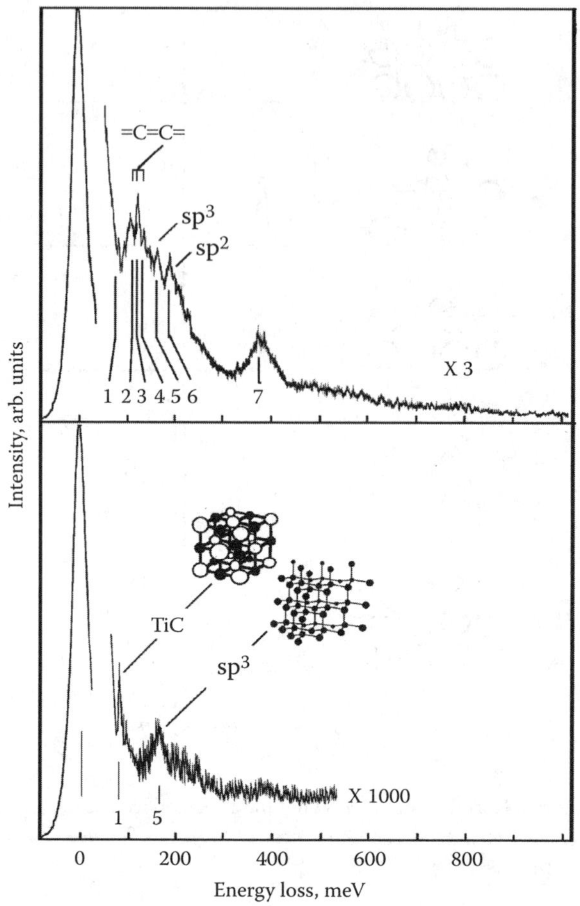

FIGURE 4.16 Vibrational spectra of Ti–C PVD biocompatible coating; (top) before and (bottom) after Ar^+ ion etching. (From Kovalev, A.I., Wainstein, D.L., Karpman, M.G., Sidakhmedov, R.Kh. Experimental verification of PCT diagrams for TiC and ZrC PVD coatings and determination of free carbon state by AEA, XPS and HREELS methods. *Surface and Interface Analysis* 2004, 36, *8*, 1174–1177. With permission.)

ion etching, and after 900 sec of Ar^+ etching. The interpretation of these spectra is presented in Table 4.3. The ion etching allowed us to investigate the peculiarities of the phonon spectra in coatings at different depths. The upper spectrum includes a high number of peaks tied with carbon but not bound in carbides. The peaks (2–4) at 107, 127, and 130 meV correspond to stretching vibrational modes for =C=C= bonds. Thus, the carbyne clusters we observed are probably polycumulene, i.e., consistent with the chemical bonding of beta-carbyne reported in the literature [13]. The peak at 164 meV corresponds to diamond phase. The other (broader) peak at 189 meV originates from the amorphous carbon. The analogous Raman spectrum of the a-C magnetron sputtered films was presented in Reference 13. The intense and broad peak (7) at 375 meV corresponds to the stretching frequency of C_mH_n hydrocarbons.

Ion etching at 900 sec has changed the spectrum significantly. The argon etching allowed us to reach deeper layers of the coating. The peaks (1) and (5) at 78 and 165 meV are clearly determined. The first one corresponds to TiC, and the second one relates to the diamondlike carbon with sp^3 hybridization. Evidently, these features relate to the decomposition of carbyne and decreasing of carbon sp^2 fraction, and, therefore, increase in the number of fourfold coordinated carbon atoms (sp^3) and TiC concentration in PVD coating. This data is in agreement with investigations of other PVD nanocomposite TiC–α-C:H coatings.

TABLE 4.3
Interpretation of HREEL Spectra

Number of Peaks	Experimental Position (meV)	Character of Bonds	Raman Spectroscopy (cm^{-1}/meV)
1	**78**	**C in TiC**	**475/78**
2	107	Carbon, A2u	867/107
3	127	Carbine =C=C=	970/127
4	131	Carbine =C=C=	1070/130
5	**165**	**Diamond, sp^3**	**1330/164**
6	189	Carbon, sp^2	1530/189
7	375	C_mH_n	2900–3080/359–381

Source: Kovalev, A.I., Wainstein, D.L., Karpman, M.G., Sidakhmedov, R.Kh. Experimental verification of PCT diagrams for TiC and ZrC PVD coatings and determination of free carbon state by AES, XPS and HREELS methods. *Surface and Interface Analysis* 2004, 36, *8*, 1174–1177. With permission.

4.5.2 Bulk and Surface Plasmons

The dominant mechanism for the electron energy loss in solids is the phenomenon of dielectric energy loss or *plasmon loss*.

As an electron moves through matter, it may lose energy to the surrounding free electrons in the material through collective electron–electron interactions. These collective losses are called *plasmon interactions*. Plasmons are the quanta of longitudinal plasma oscillations. For example, the electron gas in a metal displays oscillations against the positively charged background made up by the ion cores. This specific type of behavior is owing to the long-range part of the Coulomb interaction making the electrons move in a coherent and organized manner. Loss interactions can occur in bulk states or surface states of free electron "gas." Because the electrons lose their kinetic energy, they appear to gain binding energy. This leads to additional peaks at higher binding energy values. The energy of the characteristic energy losses depends on the material. Thus, the study of plasmon excitations provides a good probe for multiple-particle theories of interacting electrons. It is also capable of providing insight into fundamental questions regarding the electronic properties of solids.

The intensity depends on both the nature of the material and the kinetic energy of the electron passing through the material. The cross-section for plasmon loss decreases monotonically with increasing energy according to

$$\sigma = \frac{\hbar\varpi_p / 2r_0}{E_0 \ln\{[(1+y_p)^{1/2} - 1]/[x - (x^2 - y_p)^{1/2}]\}} \tag{4.13}$$

where E_0 is the kinetic energy of primary electrons, $x = (E_0/E_F)^{1/2}$, $y_p = \hbar\varpi_p/E_F$ is the kinetic energy of the electron, E_F is the Fermi energy, r_0 is the Bohr radius, ϖ_p is the plasmon frequency, and $\hbar$ is the Planck constant.

The theoretical spectral intensity of each plasmon peak is related to its energy, the energy of primary electrons E_p. In accordance with the dielectric theory of losses, the intensity is determined by the expression:

$$I_m\left[-\frac{1}{\varepsilon}\right] = \frac{E\Gamma E_p^2}{E^2 - E_p^2 + Eh\Gamma} \tag{4.14}$$

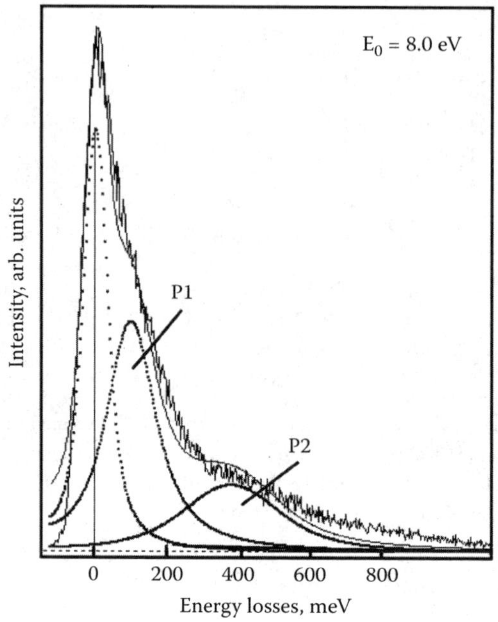

FIGURE 4.17 HREEL spectrum of nanocrystal PVD wear-resistant (Ti35–Al65)N coating on HSS end mill. The spectra are recorded with a primary electron energy of 8 eV (FWHM = 0.065 eV).

The plasmon peak intensity in the spectrum of electron energy loss is connected with the excitation of collective oscillations among conduction electrons and is determined by their concentration. This permits us to determine the effective concentration of conduction electrons on the basis of principles of the dielectric theory of energy losses, using the integral intensity of plasmon peaks:

$$n_{eff} = \frac{2\varepsilon_0 M}{\pi h^2 e^2 n_a} \int I_m \left[-\frac{1}{\varepsilon} \right] E' dE' \tag{4.15}$$

The surface plasmons' structure of ELS cannot be resolved in the case of nanocrystallic or amorphous coatings, as is shown in Figure 4.17. The curve-fitting procedure was used for analysis of the fine structure of these spectra. P1 and P2 are surface plasmon peaks.

4.5.3 Interband Transitions

Another loss mechanism is the creation of electron-hole pairs. The electron-hole pairs in a metal can be created with infinitely small energies by lifting an electron from an energy level just below the Fermi energy to a level just above. An electron-hole creation makes a contribution to the dielectric function at all energies. For a semiconductor, the smallest energy for an electron-hole pair creation is the energy of the fundamental gap. In semiconductors, a structure in the dielectric function can be found that corresponds to excitations over the gap. At slightly lower energies, excitons are found. For both metals and semiconductors, so-called critical points in the band structure give rise to strong features in the dielectric function. A critical point, for example, is a situation in which the occupied and unoccupied bands are parallel in a larger region of *k*-space. Then the optical transitions from the region all have the same energy and provide a light emission in a narrow-wavelength range.

The system of nanoinclusions of Si in the SiO_2 matrix (SiO_2:Si) attracts a great deal of attention owing to its luminescence ability in the visible and near-IR range spectrum. The doping of P, in this case, increases these luminescent properties. The following investigations explain features of the analysis of occupied and nonoccupied levels in such semiconductor nanocomposites.

Single-crystal samples of thinned plate Si (100) with 10 × 10 × 2 mm dimension were investigated. The samples were ion-doped by P or, after oxidation, were simultaneously doped by P and Si. The thermally grown SiO_2 films on silicon were used as the target for Si implantation. The oxidized Si samples, with an oxide film thickness of about 0.6 μm, were subjected to ion implantation by Si^+ ions (10^{17} cm^{-2}) at an accelerating voltage of 150 kV. After ion implantation, the samples were annealed at 1000°C for 2 h and doped by P ions with a total dose of 3×10^{16} cm^{-2} and annealed at 1000°C for 0.5 h. The annealing was done in an N_2 atmosphere. After the implantation of Si in SiO_2 and the annealing of the samples, nanoinclusions of Si (quantum dots) were formed in surface layers.

Figure 4.18, from top to bottom, shows a high-resolution energy losses spectra (HREELS) for a Si (100) single-crystal surface in initial state (1), after P doping (2), and differential spectrum

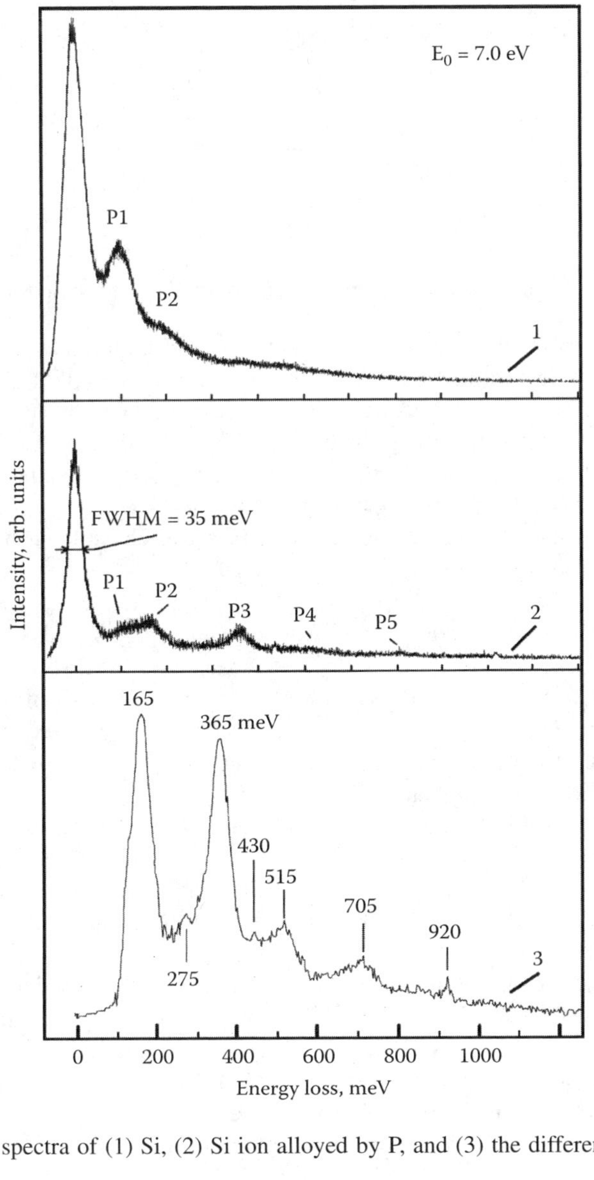

FIGURE 4.18 HREEL spectra of (1) Si, (2) Si ion alloyed by P, and (3) the differential spectrum.

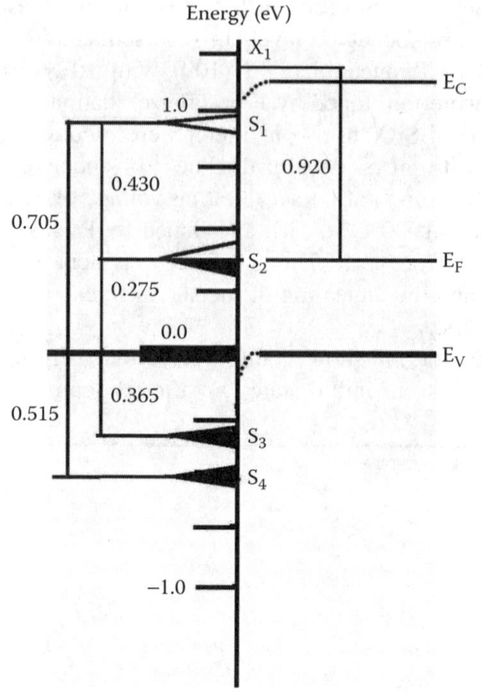

FIGURE 4.19 Scheme of intra- and interband transitions in Si:SiO_2 nanocomposite.

obtained by subtracting spectrum 1 from spectrum 2. The subtraction of spectra allowed us to remove the structure of spectra related to inelastic losses and to amplify the input of P doping in the picture of characteristic losses. According to these results, the intensities of some characteristic losses' peaks are increased at P doping. The excitation of electrons in this low-energy range is provided by surface states and intrazone transitions. The types of excited electron transitions were determined by accounting the energy positions of the points with maximal integral DOS in the Brillouin zone (in valence and conductivity bands) that were calculated by LMTO method. The scheme of these transitions is presented on Figure 4.19.

There is agreement between the experimental and calculated data, which allows us to reach the following conclusions. In silicon, at Γ_{25}' valence band splitting takes place in the light hole band Γ_{25}'(l) and heavy hole band Γ_{25}'(h). Second electron states (SES) for the Si (100) face include of two groups. The first SES group in the forbidden band has density maxima at 0.92, 0.705, 0.430, and 0.275 eV energies. The second group of the surface states is located in the valence band and consists of the surface resonance near the valence band top of –0.165, surface state band with $E =$ –0.365 energy, and a local surface state of –0.515 eV.

As one can see from Table 4.4 and Figure 4.19, addition of P increases the electron transitions:

1. Between the integral density-of-state maxima in the valence band and conduction band
2. From the surface states in the valence band to unfilled surface states and to the conduction band
3. From the valence band top to surface states located in the forbidden band
4. Between the surface states in the forbidden band

The interband transition analysis is very important for investigation of band gap features in semiconductors and density of states above Fermi level. Figure 4.20 presents the EELS spectra of P doped Si:Al_2O_3 nanocomposite after Ar^+ ion etching during (a) 1 min, (b) 3 min, and (c) 6 min.

TABLE 4.4
Interpretation of Electrons Transitions on Energy Losses Spectra from Si (100) Surface after P Ion Doping

Excited Electron Transition	Transition Energy (eV)	Mean energy Loss Values (eV) (Figure 4.18)
S_r–E_v	~0.0	0.165
S_3–Γ_{25}'(h) (p-type)	0.35–0.40	0.365
Γ_{25}'(h)–S_2	0.34–0.36	0.275
S_2–S_1	0.49	0.430
S_4–Γ_{25}'(h)	0.56	0.515
Γ_{25}'(I) – S_1	0.74	0.705
S_2–$X_1(\Sigma_1)$	0.98	0.920

Source: Kovalev, A.I., Wainstein, D.L., Tetelbaum, D.I., Hornig, W., Kucherenko, Yu. N. Investigation of the electronic structure of the phosphorus-doped Si and SiO_2: Si quantum dots by XPS and HREELS methods. *Surface and Interface Analysis* 2004, 36, *8*, 958–962. With permission.

The speed of ion sputtering was approximately 1 monolayer/sec. The nanocomposite structure was formed after sequential Si and P ion implantation in sapphire matrix and annealing at 700°C during 2 h. The peaks at 6.95, 6.55, and 6.50 eV are related to excitation of electrons' transitions through the band gap. Its intensity is related to concentration of P in nanocomposite and its influence on DOS over the band gap. The depth distribution of P and Si in this nanocomposite are shown in Figure 4.2f. The P concentration has a maximum depth of 40 nm. For this reason, the intensity of HOMO–LUMO transitions (intensity of common losses pike) increases (Figure 4.20b).

4.6 SECONDARY IONS MASS SPECTROSCOPY

4.6.1 Physical Basis of the Technique

Secondary ions mass spectroscopy (SIMS) is a destructive analytical technique used for the compositional elemental and molecular analysis of surfaces and interfaces. The excellent and wide elemental sensitivity, along with good depth resolution, make SIMS an extremely powerful technique for surface analysis.

Depending on the technique used, the surface being probed may constitute just the top monolayer of atoms (which some consider be the only true surface) or it may extend several microns beneath the top monolayer. The experiments are carried out under ultrahigh vacuum. SIMS operates on the principle that bombardment of a material with a beam of ions with high energy (1 to 30 keV) results in the ejection or sputtering of atoms from the material. A small percentage of these ejected atoms leave as either positively or negatively charged ions, which are referred to as *secondary ions*. The secondary ions emitted from the sample surface are detected in a mass spectrometer that uses electrostatic and magnetic fields to separate the ions according to their mass-to-charge ratio. Ions of different mass-to-charge ratios are measured by changing the strength of the magnetic field. The off-axis location of the detector prevents neutral particles (such as photons) from adding to the signal. Generally, SIMS is considered a surface-analysis method with a spatial resolution laterally on the order of 1 μm in certain cases.

The primary ion dose can be carefully controlled so that the damage to the surface is negligible, and the technique is essentially considered nondestructive (so-called static SIMS). It gives good molecular information content and, as with Auger, can be used for surface microanalysis at high

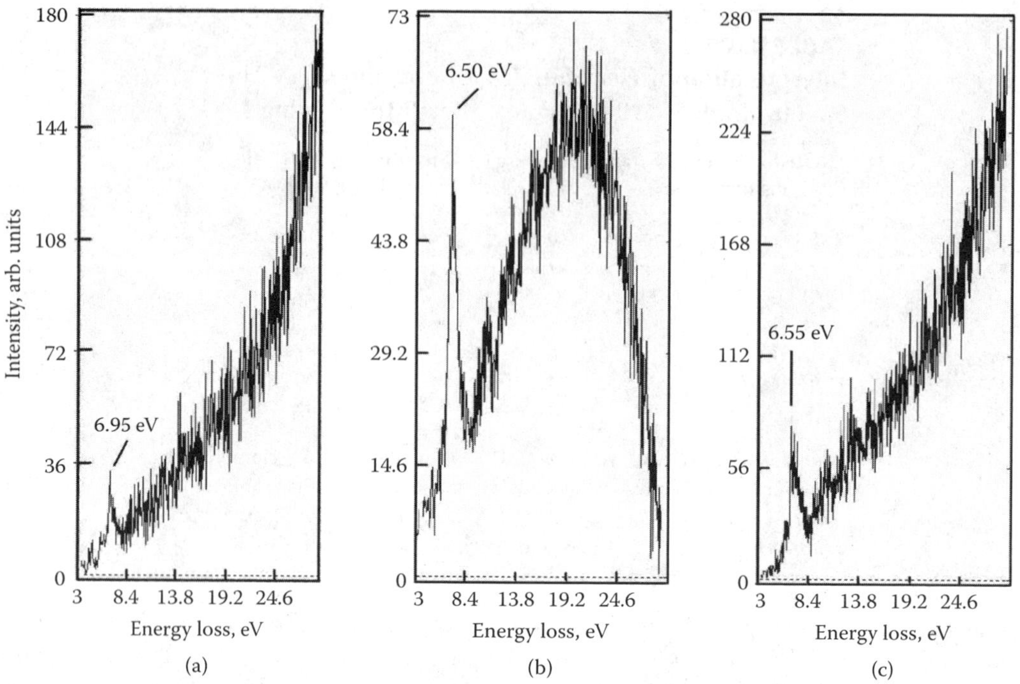

FIGURE 4.20 Electron energy loss spectra for P doped nanocomposite, containing Si nanocrystals in sapphire matrix in initial state (a) after annealing at 700°C during 2 h, (b) after Ar^+ ion etching during 3 min (depth of 42 nm), and (c) 5 min (depth of 70 nm). The spectra are recorded with a primary electron energy of 39.05 eV (FWHM = 0.05 eV).

spatial resolution. At low primary ion currents, a surface monolayer may exist for hours before being completely removed.

A larger number of ions per unit area can also be used to sputter subsurface layers of the sample (so-called dynamic SIMS). Dynamic SIMS determines the elemental composition and trace levels of impurities and dopants in solid materials. This mode of operation allows chemical depth profiles to be obtained. As the sputtered ions escape from shallow depths, the sputtering of the sample has to be prolonged to extend the analytical zone of the sample into deeper regions of the bulk material. Monitoring secondary ion emission in relation to the sputtering time, therefore, allows depth profiling of the sample's composition. Layers up to 10,000 ≈ thick can be depth-profiled using SIMS.

These techniques are complementary to the electron spectroscopies because they provide higher sensitivity and give precise molecular information. On unknown samples, it is common to use a combination of electron spectroscopy technique and a mass spectrometry technique for surface characterization.

SIMS is a very sensitive surface analytical tool. The method offers several advantages over other composition-analysis techniques, namely:

1. The ability to identify all elements, including H and He
2. The ability to identify elements present in very low concentration levels, such as dopants in semiconductors

The yield of secondary ion sputtering, which affects the SIMS sensitivity, depends on the specimen's material, the specimen's crystallographic orientation, and the nature, energy, and incidence angle of the primary beam of ions.

4.6.2 Relative Sensitivity Factors

SIMS has one major drawback that gives rise to very significant problems with quantification. The ionization yield of a given sputtered element may vary by several orders of magnitude, depending on the composition of the matrix in which it is located. For example, the ionization probability of Ti will be 10 to 10,000 times greater in an oxide sample compared to a metallic sample. This phenomenon prevents SIMS from becoming an easy tool for quantitative analysis, except when dealing with ideal samples such as a single crystal of silicon.

Quantitative analysis by SIMS uses relative sensitivity factors defined according to the following equation:

$$\frac{I_R}{C_R} = RCF_E \frac{I_E}{C_E} \tag{4.16}$$

where RCF_E is the relative sensitivity factor of the element, I_E is the secondary ion intensity for element E, I_R is the secondary ion intensity for reference element R, C_E is the concentration of E, and C_R is the concentration of R.

The major (or matrix) element is usually chosen as the reference. Substituting M (matrix) for R (reference) and rearranging gives the following equation.

In the trace element analysis, we can assume that the matrix elemental concentration remains constant. The matrix concentration can be combined with the elemental *RSF* to give a more convenient constant, *RSF*.

$$C_E = RCF_E \frac{I_E C_M}{I_M} \tag{4.17}$$

The *RSF* is a function of the element of interest and the sample matrix.

$$C_E = RCF \frac{I_E}{I_M} \tag{4.18}$$

Note that *RSF* and *CE* have the same concentration units. This is the most common form of the *RSF* equation.

4.6.3 Sensitivity and Detection Limits

The SIMS detection limits for most trace elements is between 1×10^{12} and 1×10^{16} atoms/cc. In addition to the ionization efficiencies (RSFs), two other factors can limit sensitivity. The output of an electron multiplier is referred to as *dark counts* or *dark current* if no secondary ions are striking it. This dark current arises from stray ions and electrons in instrument vacuum systems and from cosmic rays. Count-rate limited sensitivity occurs when sputtering produces less secondary ion signal than the detector dark current. When the SIMS detector analyzes elements, the introduced level constitutes sensitivity limited by a background. Oxygen present as a residual gas in the vacuum systems is an example of an element with background-limited sensitivity. The analyte atoms sputtered from mass spectrometer parts back onto the sample by secondary ions constitute another source of background. Mass interferences also cause background-limited sensitivity.

For samples with homogeneously dispersed analyte, bulk analysis provides better detection limits than depth profiling, usually by more than an order of magnitude. Faster sputter rates increase the secondary ion signal in bulk analysis. The fastest possible sputtering requires intense primary ion beams that sacrifice depth resolution because they cannot be focused as required for flat-bottom (rastered) craters. Otherwise, bulk analyses are similar to depth profiles. Ion intensity data are displayed as a function of time. This provides a means for verifying that the sample is indeed homogenous. In a typical heterogeneous sample, the analyte is concentrated in small inclusions that produce spikes in the data stream.

The secondary ion current I_s for a selected ion of mass m (or, more correctly, of mass/charge ratio m/z, where z is almost always unity in SIMS) is given by

$$I_s(m) = I_p y \alpha T C(m) \tag{4.19}$$

where I_p is the primary ion current, y is the sputter yield, α is the ionization probability, T is the overall transmission of the energy and mass filters (i.e., the fraction of the sputtered ions of a given mass which are actually detected), and $C(m)$ is the concentration of the detected species in the sputtered volume.

For a quadrupole mass analyzer, T is on the order of 0.1% and is approximately inversely proportional to m. Therefore, the sensitivity of the instrument decreases as higher-mass ions are selected by the mass filter. The total yield of sputtered particles y of neutral and charged atoms of mass m is typically equal to between 1 and 20 per incident primary ion. It is also a function of the mass and energy of the primary ion and its angle of incidence (peaking strongly at ~60° to the surface normal). The respective probabilities, α+ and α-, are the probabilities that the sputtered particle will be either a positive or negative ion. Ionization probabilities vary dramatically across the elements and are additionally very sensitive to the electronic state of the surface.

The secondary ion yield $y\alpha$ can therefore vary by over four orders of magnitude for different elements and may be matrix-sensitive. For example, the yield of Mg+ from clean Mg vs. MgO is 0.01/0.9 (≈ 0.01), whereas the yield of W+ from clean W vs. WO_3 is 0.00009/0.035 (≈ 0.003).

SIMS is therefore inherently nonquantitative, i.e., there is no simple given relationship between concentration of a given mass, $C(m)$, and peak intensity, $Is(m)$. However, for a given elemental species in a fixed matrix, and provided the concentration range is not too great, the intensity ratio of the elemental peak to that of a matrix-related peak will usually follow a linear relationship with concentration. This allows composition to be determined if a suitable standard calibration material is available.

The high sensitivity of SIMS requires a high sputter rate (high primary beam energy, high primary flux, and low angle of incidence), a large analytical area, and high useful yield. Approximately 100 secondary ions per sec are required to give good analytical statistics. A useful yield of 10^{-3} is rather good, and a sputter rate of 10 nm/sec is high enough for an analytical area of 1000 μm^2. Under these conditions, a concentration of 10^{13} atoms/cm^3 could be detected.

Figure 4.21 shows a high sensitivity of SIMS for characteristics of trace elements in pure metal. The spectrum clearly shows H, Li, C, N, O, Na, Al, Si, K, Ca, Sc, Ti, V, Mn, Fe, Ni, Cu, and well-resolved isotopes of Rb. The peak intensities show the isotopic and chemical compositions of rubidium (Table 4.5).

4.6.4 Mass Spectra

Mass spectra of the secondary ions are recorded in a selected mass range by continuous monitoring of the ion signal while scanning a range of mass-to-charge (m/z) ratios. The mass spectrum detects both atomic and molecular ions. Secondary ions containing more than one atom are called molecular ions in SIMS. Note that the term "molecular ion" also finds use in organic mass spectrometry, where it refers to the parent ion before any fragmentation. In the case of computer control, the mass analyzer must scan in small steps to ensure that all mass-to-charge (m/z) ratios are sampled. Ten steps per mass unit are common. At higher mass resolution, ten mass increments per peak

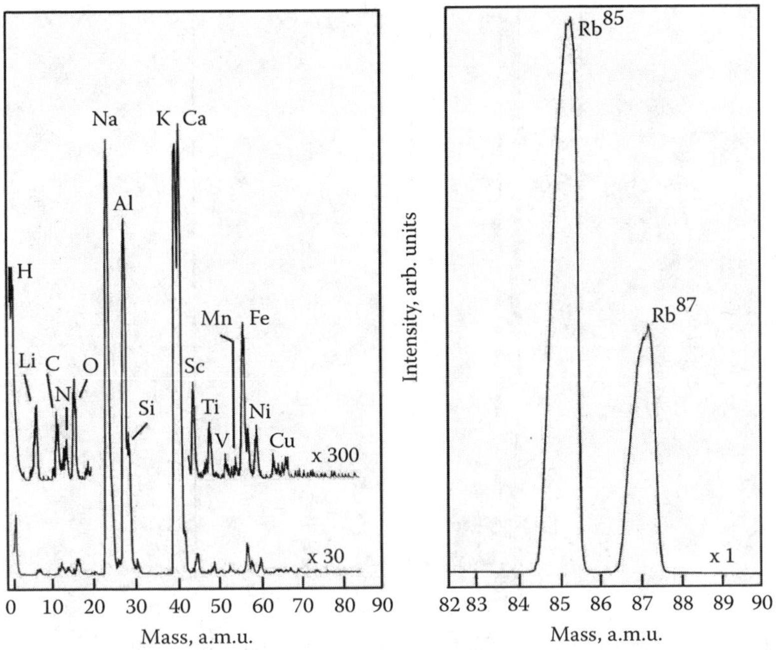

FIGURE 4.21 Positive SIMS spectra from Rb sample.

TABLE 4.5
Interpretation of Secondary Ions Mass Spectra (Figure 4.21)

Atomic Number	Element	Concentration wt%	at. %	Intensity 3×10^3 (Imp/sec)	Relative Intensity (I_x/I_{Rb})	Relative Sputtering Coefficient
37	Rb	99.6269	98.8243	59,700	1	—
2	Li	0.0001	0.0012	18	3.01E–4	0.250
12	Mg	0.015	0.0523	220	3.68E–3	0.0703
11	Na	0.15	0.5531	1,070	1.79E–2	0.0323
13	Al	0.15	0.4713	870	1.45E–2	0.0307
19	K	0.001	0.0022	1,060	1.77E–2	8.04
20	Ca	0.01	0.0212	1,110	1.85E–2	0.872
22	Ti	0.01	0.0177	14	2.34E–4	0.0132
23	V	0.015	0.0250	5	8.37E–5	0.00335
25	Mn	0.01	0.0154	37	6.19E–4	0.0402
26	Fe	0.001	0.0015	44	7.37E–4	0.491
28	Ni	0.001	0.0014	12	2.01E–4	0.143
29	Cu	0.01	0.0133	10	1.67E–4	0.0125

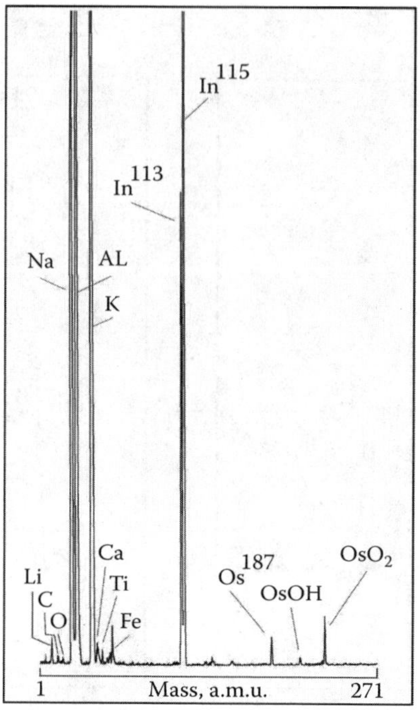

FIGURE 4.22 Positive SIMS spectrum of the Os^{187} isotope on indium substrate.

width adequately define the peak shape. A mass spectrum with a mass range of 100 has at least 1000 data channels for which a reasonable analysis time is 0.1 sec per channel.

The spectrum on Figure 4.22 was obtained using precise tuning of Quadrupole SQ-300 analyzer for higher mass resolution in wide mass range. The low intensity of Osmium ion peaks is because of the low sputtering yield of these heavy ions at 3000-eV Ar+ bombardment.

This figure demonstrates a unique combination of high sensitivity and high elemental discrimination, which enables multielement isotopic analysis of surfaces.

4.6.5 Depth Resolution

The depth resolution depends on flat-bottom craters. ESCALAB-MK2 provides uniform sputter currents by sweeping a focused primary beam in a raster pattern over a square area. To obtain accurate depth resolution, the data acquisition system ignores all secondary ions produced when the primary sputter beam is at the end of its raster pattern.

4.6.5.1 Standards for RSF Measurement

Quantitative SIMS analysis requires standard materials to measure *RSF* values. Because ion yields depend on the analyzing element, the sputtering species and the sample matrix, separate *RSF*s must be measured for each. The reference samples produced by ion implantation are the best ones.

4.6.5.2 Bulk Analysis

For samples with homogeneous composition, bulk analysis provides better detection limits than depth profiling, usually by more than an order of magnitude. Faster sputter rates increase the secondary ion signal in bulk analysis. The fastest possible sputtering requires intense primary ion beams that sacrifice depth resolution because they cannot be focused as required for flat-bottom (rastered) craters. Otherwise, bulk analysis is similar to depth profiles. Ion intensity data are displayed as a function of time. This provides a means for verifying that the sample is indeed homogenous. In a typical heterogeneous sample, the analyzed substance is concentrated in small inclusions that produce spikes in the data stream.

4.6.5.3 Sample Charging and Charge Neutralization by Electron Bombardment

The primary ion beam, secondary ions, and secondary electrons produce a net electric current at the sample surface. If the sample material conducts, the current flows through the sample into the instrument. However, insulating samples undergo charge build-up. Sample charging diffuses the primary beam and diverts it from the analytical area, often eliminating the secondary ion signal entirely. Sample charging also changes the energy distribution of the secondary ions, which affects their transmission and detection by the mass spectrometer. When the sample is a thin dielectric on a conducting substrate, a strong electric field develops. Mobile ions such as carbon, nitrogen, sodium, and lithium migrate in the electric field, and depth profiles or mass spectra no longer reflect the original compositions of the layers.

Electrons compensate for positive charge build-up, which results from positive primary ions or negative secondary ions and electrons. In our investigation, we used the bombardment by electrons with the energy of 25 eV. Low-energy electron beams work better because higher energies produce more than one secondary electron for every incoming electron. Low-energy electron beams are more easily implemented in quadrupole SIMS instruments, making quadrupoles the system of choice for insulating materials.

4.6.6 Isotope Ratio Measurements

Isotope ratio measurements are operationally similar to depth profiles except that the precision and accuracy requirements are higher. Because all isotopes of the same element have similar chemical properties, ionization and detection efficiencies remain nearly constant for the different isotopes. A precision of 0.1% is common. Error analysis indicates that precision is limited mainly by Poisson-counting statistics. To attain these accuracies, Quadrupole SQ-300 mass spectrometer of ESCALAB-MK2 must be carefully tuned, and interferences eliminated. Figure 4.23 and Figure 4.24 show the high mass resolution of the Ag and Pd isotopes used for interpretation of the chemical composition of the outer layer and sublayers of Pd coating on Ag matrix. The spectrum was recorded before and after ion etching of Ag with the Pd coating.

Occurrence of Ag lines on a complex spectrum is observed after short-term Ar ion etching of Pd coating sample. The intensity of the lines for the isotopes Ag and Pd will be well coordinated to the natural isotopic composition of these elements. The ratio of the isotope signal intensities must be corrected for slight variations in detection efficiencies at different masses and for slight variations that depend on signal intensity. These corrections are usually larger than the range of expected isotope ratios.

Figure 4.25a and Figure 4.25b show high mass resolution of Osmium isotopes in an isotope-enriched metal and a natural one. The spectra were recorded after tuning the SQ-300 quadrupole mass spectrometer in the heavy mass region.

4.7 THE MODELING AND ELECTRON SPECTROSCOPY OF MODIFIED SURFACES

The explanation of complex phenomena on surfaces is not a simple task, even if it is based on experimental results. The computer modeling of electron structures and electron and vibrational spectra is widely used in electron spectroscopy to resolve many uncertainties. Some examples are

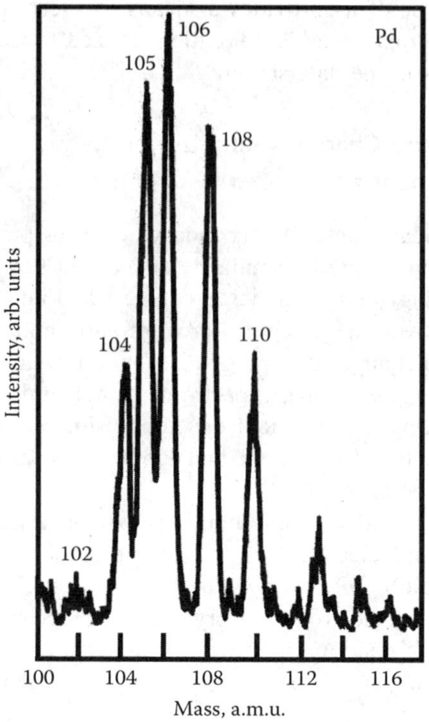

FIGURE 4.23 The positive SIMS spectrum of Pd coating on Ag catalyst before ion etching.

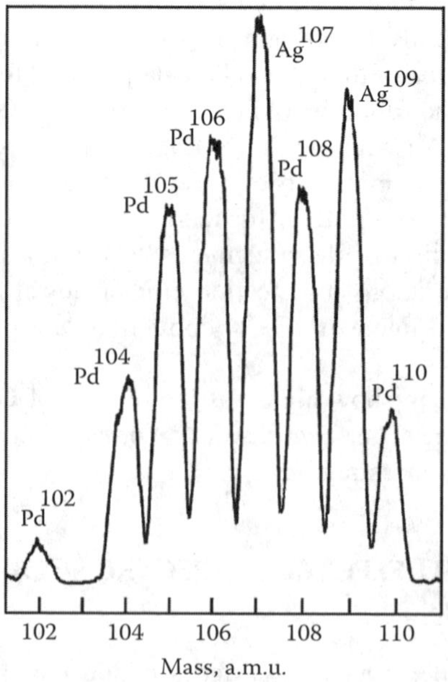

FIGURE 4.24 The positive SIMS spectrum of Pd coating on Ag catalyst after ion etching.

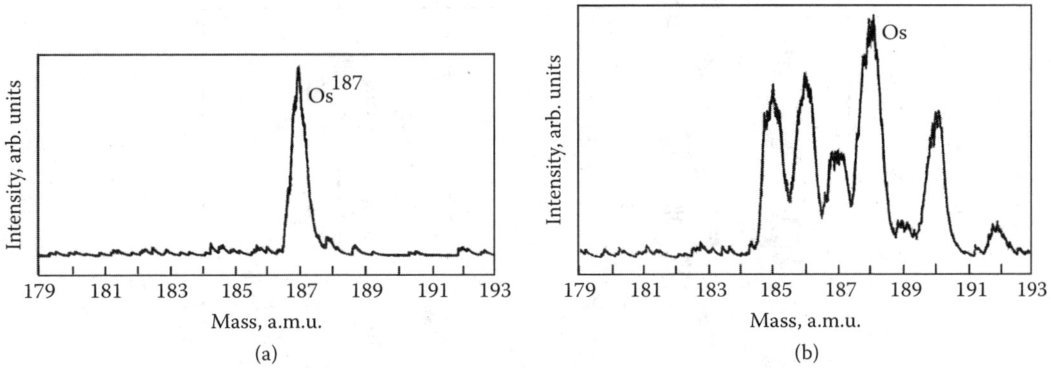

FIGURE 4.25 The positive SIMS spectra of (a) Os^{187} and (b) the natural osmium.

TABLE 4.6
Phase Composition of PVD Coatings Based on Ti and Zr Nitrides

Nitrogen Pressure (Pa)		Substrate Temperature (°C)		Phase Composition	
Calculated	Experiment	Calculated	Experiment	Calculated	Experiment
1.06	1.06	200–300	250 ± 25	$TiN_{1,20}$	$TiN_{1,17}$
0.04	—	200–300	—	$TiN_{0,96}$	—
0.008	0.008	200–300	250 ± 25	$TiN_{0,84}$	$TiN_{0,83}$
1.06–0.4	0.7–0.8	200–900	250 ± 25	$ZrN_{0,90}$	$ZrN_{0,88}$
0.04–0.004	0.008	200-800	250 ± 25	$ZrN_{0,84}$	$ZrN_{0,82}$

cited that will demonstrate the basic computational methods that are used during investigations of nanostructured material and coating surfaces (Table 4.6).

The carbides and nitrides of transition metals are widely used for hard, wear-resistant, biocompatible, and corrosion-resistant coatings. All these compounds have a wide homogeneity region on phase-equilibrium diagrams and, correspondingly, various compositions.

Alteration of chemical composition of these equilibrium phases is connected to changes in the vacancies' concentration in the crystalline lattice. Nanostructured coatings are nonequilibrium phases oversaturated by point defects when compared to equilibrium ones. The coating model explains the role of chemical composition and crystalline structure in relation to the mechanical and corrosion strengths.

The TiN and ZrN crystals have a simple cubic lattice (space symmetry group O_h^5). The cubic lattice constants are equal to $a = 0.424$ nm and $a = 0.432$ nm, and the interatomic distances are $d = 0.211$ and $d = 0.216$ nm. The nearest surrounding Ti or Zr atoms form a tetrahedron of 4N atoms. The cubic unit cell contains 27 atoms in the initial state and 26 atoms in the lattice with vacancy. The central metallic atom is substituted by vacancy in these nitride lattices.

The energy band structure of the system considered has been calculated by means of the self-consistent density functional methods (ZIND). The calculations associated with this study were performed using the ÄrgusLab and HyperChem 7.5 software (www.planaria-software.com, www.hyper.com). The ZINDO method is the most suitable semiempirical method to determine the structure and energy of molecules or crystals for the first and second transition row metals. These metals have a wide range of valences, oxidation states, spin multiplicities, and possess unusual bonding situations. In addition, the nondirectional nature of the metallic bonding is less agreeable to a ball-and-spring interpretation [15]. The ZINDO method is based on an Intermediate Neglect of Diatomic Differential Overlap Hamiltonian (INDO). Specifically, it is the INDO1/s parameter-

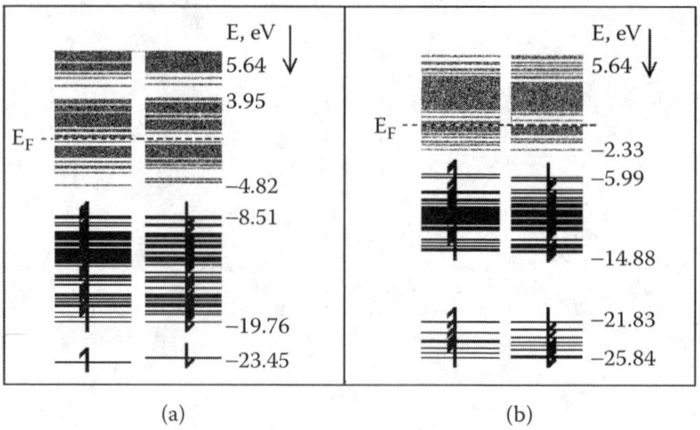

FIGURE 4.26 Calculated molecular orbitals of ZrN cluster. a) Without vacancy; b) with vacancy.

ization firstly published by Zerner in the 1970s. This method has enjoyed tremendous success in predicting excited-state properties of a wide variety of compounds. The electronic structure of the following elements can be calculated by using this method: H; Li–F; Na–Cl; K–Zn; Y–Cd.

The molecular orbital calculations on the metal yield less reliable results as compared to the results obtained from organic compounds. Nevertheless, these quantum mechanical calculations are very useful to predict the wide spectrum of interatomic bonding and the interpretation of the physical properties of the metals and their compounds. The properties include dipole moment, polarizabilites, total electron density, total spin density, electrostatic potential, heat of formation, orbital energy levels, vibrational normal modes and frequencies, strength, and plasticity.

This method is used mainly for calculations of electron density (DOS), molecular orbitals (MO), and the electrostatic potential (ESP) of crystals and molecules.

Figure 4.26 and Figure 4.27 demonstrate the energy diagrams of the calculated molecular orbitals (MO) of ZrN and TiN (a) without vacancies, and (b) with vacancies. The resulting set of levels includes a set of valence and conduction bands. The electronic band structure of these nitrides is well known and described in detail elsewhere. Therefore, we do not present a detailed discussion of our calculated results (Figure 4.1); however, they are in agreement with known results that are obtained by means of other calculation methods [16]. Our calculations show (Figure 4.27a) that the total density of the states of TiN comprises roughly three fine structures. The lowest energy

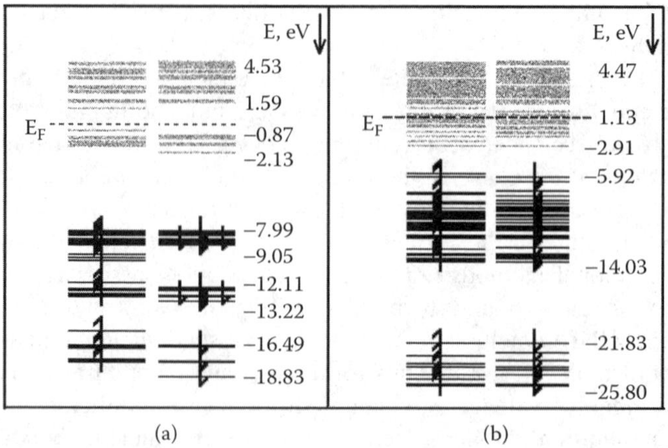

FIGURE 4.27 The calculating P-C-T diagram of phase equilibrium for ion-plasma deposited ZrN coatings.

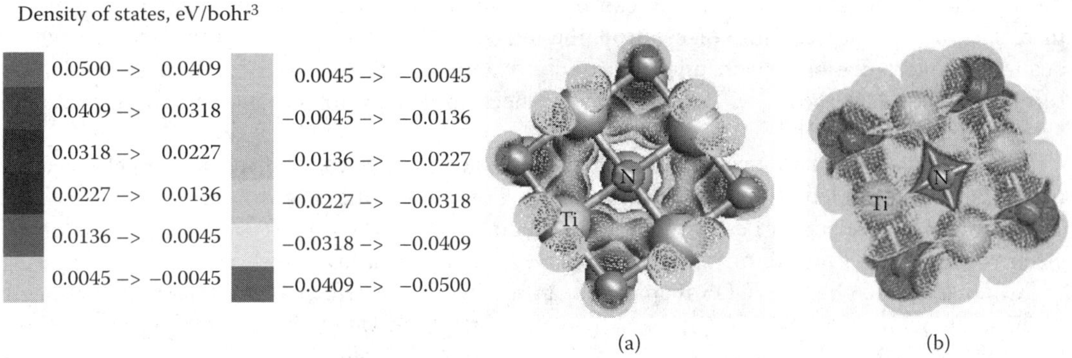

FIGURE 4.28 Calculated molecular orbitals of TiN cluster. a) Without vacancy; b) with vacancy.

structure is centered on –16.6 … –7.1 eV and is composed mainly of electrons from the 2s bands of N atoms and the 3p and 3d bands of Ti. The middle energy structure is centered on –2.7 eV. It is primarily composed of electrons from the 2p (valence) bands of the N atoms and electrons from the 3d (valence) and 3p bands of the Ti atoms. The narrowest, highest energy structure is centered on 1.2…4.2 eV and is primarily composed of electrons from the 3d bands of the Ti atoms. The valence band is full, and the conduction band has to accommodate 1N electrons. Three of the four valence electrons from Ti complete the filling of the valence band. The one remaining electron from the Ti atom populates the conduction band, thus making TiN a metallic conductor.

Comparing Figure 4.27a and Figure 4.27b, one can see that enrichment of ZrN by vacancies decreases the electron density localization in the crystalline lattice nodes, decreases the band gap width, and increases the electron density near the Fermi level. This data confirms a decrease of the covalent part and an increase of the metallic part in the interatomic bonds in the crystalline lattice of nonstoichiometric ZrN.

In an ESP-mapped density surface, the electron density surface gives the shape of the surface, and the value of the ESP on that surface gives the colors, or shadings in Figure 2.28 and Figure 4.29. The electrostatic potential is the potential energy felt by a positive "test" charge at a particular point in space. If the ESP is negative, this is a region of stability for the positive test charge. Conversely, if the ESP is positive, this is a region of relative instability for the positive test charge.

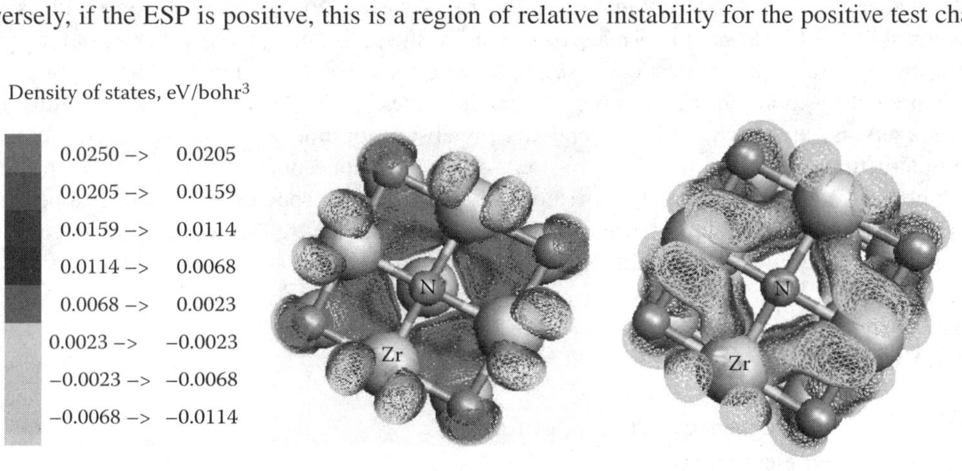

FIGURE 4.29 Molecular orbitals of (a) 27 atomic cluster of ZrN and (b) 26 atomic cluster of ZrN with central positioned vacancy.

Thus, an ESP-mapped density surface can be used to show regions of a molecule that might be more favorable to nucleophilic or electrophilic attack, making these types of surfaces useful for qualitative interpretations of chemical reactivity. Another way to think of ESP-mapped density surfaces is that they show "where" the frontier electron density for the molecule is greatest (or least) relative to the nuclei.

The shadings are the values of the ESP at the points on the electron density surface. The shading map is given on the left. Note the large shaded region around the nitrogen end of the molecule. There is enhanced electron density here. The shading indicates the most negative regions of the electrostatic potential where a positive test charge would have a favorable interaction energy.

Comparing the obtained DOS maps, one can see that point defects increase the bond strength of metal atoms more in TiN than in ZrN. The electrostatic potential on the DOS surface is equal to –0.0045 … –0.0227 eV/Bohr^3 and 0.0068 … –0.0023 eV/Bohr^3 for Ti and Zr nitrides, correspondingly. For this reason, the chemical activity of the TiN nanoparticle surface is significantly lower than ZrN.

4.8 ELECTRON SPECTROSCOPY EXPERIMENTAL CONDITIONS

The chemical and phase composition of the surface of the modified layers, as well as the "cutting tool–workpiece" interface, were studied by means of Auger electron spectroscopy (AES). The analysis of the extended fine structure of the electron energy losses spectra (EELFAS) and secondary ion mass spectroscopy (SIMS) was performed using the VG ESCALAB-MK2 spectrometer. Scanning Auger spectroscopy was used to analyze the composition on the surface of the wear crater on the rake face of the cutting tool at different stages of wear. In each of these cases, several sections of size 1.5 × 1.5 mm were chosen for analysis. A TV sweep-speed for the primary electrons was used at a 2000 Å beam diameter. The Auger signal was recorded in the mode with *CRR* set equal to 2 V, at a speed of 2.1 eV/sec, and with primary electron energy at E = 2500 eV. The phase composition of the surface of the crater wear was analyzed with the aid of scanning mass spectroscopy of secondary positive ions. For this purpose, an argon ion beam 0.5 m in diameter was scanned in a synchronized mode with the television sweep rate at an accelerating voltage of 5 keV and an Ar pressure of 2×10^{-5} Pa. Under these conditions, the ion etching speed did not exceed 0.5 monolayers/min. The analysis was made in a nearly static mode. Amorphization and fine structural changes in the nearest atomic neighbors of the surface-modified samples were investigated by means of the EELFAS method. This involved analyzing the extended fine structure of the electron spectra close to the line of elastically scattered electrons. Recently, this method, applied in its reflection electron mode, has been used for precise investigations of the atomic structure in thin surface layers. By analogy with the extended x-ray absorption fine structure (EXAFS) method [3], the fine structure of the electron spectra contains information about the structure of the nearest atomic neighbors at the surface. The methods of mathematical spectra processing adopted in the EXAFS spectroscopy [17] can be used to analyze the fine structure of the electron spectra, which enables one to determine the lengths of the atomic bonds.

ABBREVIATIONS

ADAS Angular dependent Auger spectroscopy
AES Auger electron spectroscopy
ARAES Angle resolved Auger electron spectroscopy
ARUPS Angle resolved ultraviolet photoelectron spectroscopy
ARXPS Angle resolved x-ray photoelectron spectroscopy

CLS	Characteristic loss spectroscopy
EDS	Energy distribution spectroscopy
EELFS	Extended energy-loss fine structure (spectroscopy)
EELS	Electron energy-loss spectroscopy
ELS	Electron-loss spectrometry
EPMA	Electron probe microanalysis
EPXMA	Electron probe x-ray microanalysis
ESCA	Electron spectroscopy for chemical analysis
EXAPS	Electron-excited x-ray appearance potential spectroscopy
EXELFS	Extended electron-energy-loss fine structure (spectroscopy)
HREELS	High-resolution electron-energy loss spectroscopy
IAES	Ion-excited Auger electron spectroscopy
IETS	Inelastic electron tunneling spectroscopy
IPMA	Ion probe microanalysis
IR	Infrared spectroscopy
RS	Raman spectroscopy
ISS	Ion-scattering spectrometry
PESIS	Photoelectron spectroscopy of the inner shell
RAIRS	Reflection absorption infrared spectroscopy
SAM	Scanning Auger microscopy
SEM	Scanning electron microscopy
SESCA	Scanning electron spectroscopy for chemical applications
SEXAFS	Surface extended x-ray absorption fine structure (spectroscopy)
SIMS	Secondary ion mass spectrometry
STM	Scanning tunneling microscopy
UPS	Ultraviolet photoelectron spectroscopy
XAES	X-ray-excited Auger electron spectroscopy
XPS	X-ray photoelectron (or photoemission) spectroscopy

REFERENCES

1. Seah, M.P. *Surface Analysis by Auger and X-Ray Photoelectron Spectroscopy.* SurfaceSpectra Ltd/I M Publications, Eds., Briggs, D., Grant, J.T., National Physical Laboratory: Teddington, U.K., 2003, pp. 167–189.
2. Kovalev, A.I., Scherbedinsky, G.V. *Modern Techniques of Metals and Alloys Surface Study.* Metallurgy: Moscow, 1989, p. 191.
3. Wainstein, D.L., Kovalev, A.I. Fine determination of inter-atomic distances on surface using extended energy-loss fine structure (EELFS) data: peculiarities of the technique. *Surface and Interface Analysis* 2002, 34, 230–233.
4. Sherwood, P.M.A. Auger and x-ray photoelectron spectroscopy. In *Practical Surface Analysis, Auger and X-Ray Photoelectron Spectroscopy,* 2nd ed., Vol. 1, Eds., Briggs, D., Seah, M.P. John Wiley and Sons: Chichester, 1990, p. 555.
5. Kovalev, A.I., Mishina, V.P., Stsherbedinsky, G.V., Wainstein, D.L. EELFS method for investigation of equilibrium segregations on surfaces in steel and alloys. *Vacuum* 1990, 41, *7–9*, 1794–1795.
6. Wainstein, D.L. Analysis of Extended Fine Structure of Electron Energy Losses Spectra for Determination of Atomic Structure of the Solids Surface, Ph.D. dissertation. MISIS: Moscow, 2000.
7. Kovalev, A.I., Wainstein, D.L., Mishina, V.P., Fox-Rabinovich, G.S. Investigation of atomic and electronic structure of films generated on a cutting tool surface. *Journal of Electron Spectroscopy and Related Phenomena* 1999, 105, 63–75.

8. Vickerman, J., Swift, A. Secondary ion mass spectroscopy: the surface mass spectrometry. In *Surface Analysis: The Principal Techniques*, Ed., Vickerman, J.C. John Wiley & Sons: New York, 1997.
9. Woodruff, D.P., Delchar, T.A. *Modern Techniques of Surfaces Science*, 2nd ed. Cambridge University Press: London, 1994.
10. Wilson, E.B., Decius, J.C., Cross, P.C. *Molecular Vibrations: The Theory of Infrared and Raman Vibrational Spectra*. Dover Publications: New York, 1955.
11. Pemble, M. Vibrational spectroscopy from surfaces. In *Surface Analysis: The Principal Techniques*, Ed., Vickerman, J.C. John Wiley & Sons, New York, 1997, 267–307.
12. Harris, D.C., Bertolucci, M.D. *Symmetry and Spectroscopy: An Introduction to Vibrational and Electronic Spectroscopy*. Dover Publications: New York, 1989.
13. Kovalev, A.I., Wainstein, D.L., Karpman, M.G., Sidakhmedov, R.Kh. Experimental verification of PCT diagrams for TiC and ZrC PVD coatings and determination of free carbon state by AES, XPS and HREELS methods. *Surface and Interface Analysis* 2004, 36, *8*, 1174–1177.
14. Kovalev, A.I., Wainstein, D.L., Tetelbaum, D.I., Hornig, W., Kucherenko, Yu. N. Investigation of the electronic structure of the phosphorus-doped Si and SiO_2: Si quantum dots by XPS and HREELS methods. *Surface and Interface Analysis* 2004, 36, *8*, 958–962.
15. Zerner, M.C. Semi-empirical molecular orbital methods, reviews. In *Computational Chemistry II*, Eds., Lipkowitz, K.B., Boyd, D.B. VCH: New York, 1991, p. 313.
16. Nemoshkalenko, V.V., Kucherenko, Yu.N. *The Computing Methods of Physics in Solid State Theory.* Naukova Dumka Publisher: Kiev, 1986.
17. De Crescenzi, M., Chainet, E., Derrien, J. Evidence of extended fine structures in the Auger spectra: a new approach for surface structural studies. *Solid State Commun.* 1986, 57, 487.

5 Physical and Mechanical Properties to Characterize Tribological Compatibility of Heavily Loaded Tribosystems (HLTS)

German S. Fox-Rabinovich, Lev S. Shuster, Ben D. Beake, and Stephen C. Veldhuis

CONTENTS

5.1 Introduction..121
5.2 The Characteristics of Friction as a Criterion of Tribological Compatibility 127
5.3 Micro- or Nanomechanical Properties to Characterize the Tribological Compatibility of Surface-Engineered Tools 129
References 146

5.1 INTRODUCTION

Friction is a complex phenomenon and proper selection of the properties responsible for the wear resistance and the wear behavior of the heavily loaded tribosystems (HLTS) is one of the challenges to enhance tribological compatibility.

In this chapter, we will focus on the "workpiece–cutting tool" tribosystem. Cutting and stamping tools are examples of HLTS working under high-stress and high-temperature conditions. Hardness is one of the major properties of surface-engineered tooling materials. During the last few decades major progress in coating developments was associated with hardness improvements. Traditional Ti–N coatings with a microhardness of 23 to 25 GPa were first used for cutting systems. Recently, some state-of-the-art coatings have a microhardness that doubles and sometimes even triples this value [1–2]. However, this significant progress does not always result in direct tool life improvements. For instance, the highest tool life improvement, when compared to the Ti–N coating, is associated with the Ti–Al–N coating with hardness around 30 to 35 GPa. It is obvious that a number of parameters are needed to characterize service properties of HLTS. A few preliminary conclusions can be made based on data published elsewhere [3]. The graphs shown in Figure 5.1 to Figure 5.5 are based on the selected data presented in [3–4]. They show some properties, as well as tool life data, for cutting tools with and without coatings. The tool life during turning of major types of machined materials such as structural steel (1040, Figure 5.1), Inconel (Figure 5.2), and titanium alloy (Figure 5.3) were analyzed. A number of different parameters were measured such as frictional characteristics (friction force vs. temperature, Figure 5.1 to Figure 5.3b), and the oxidation resistance

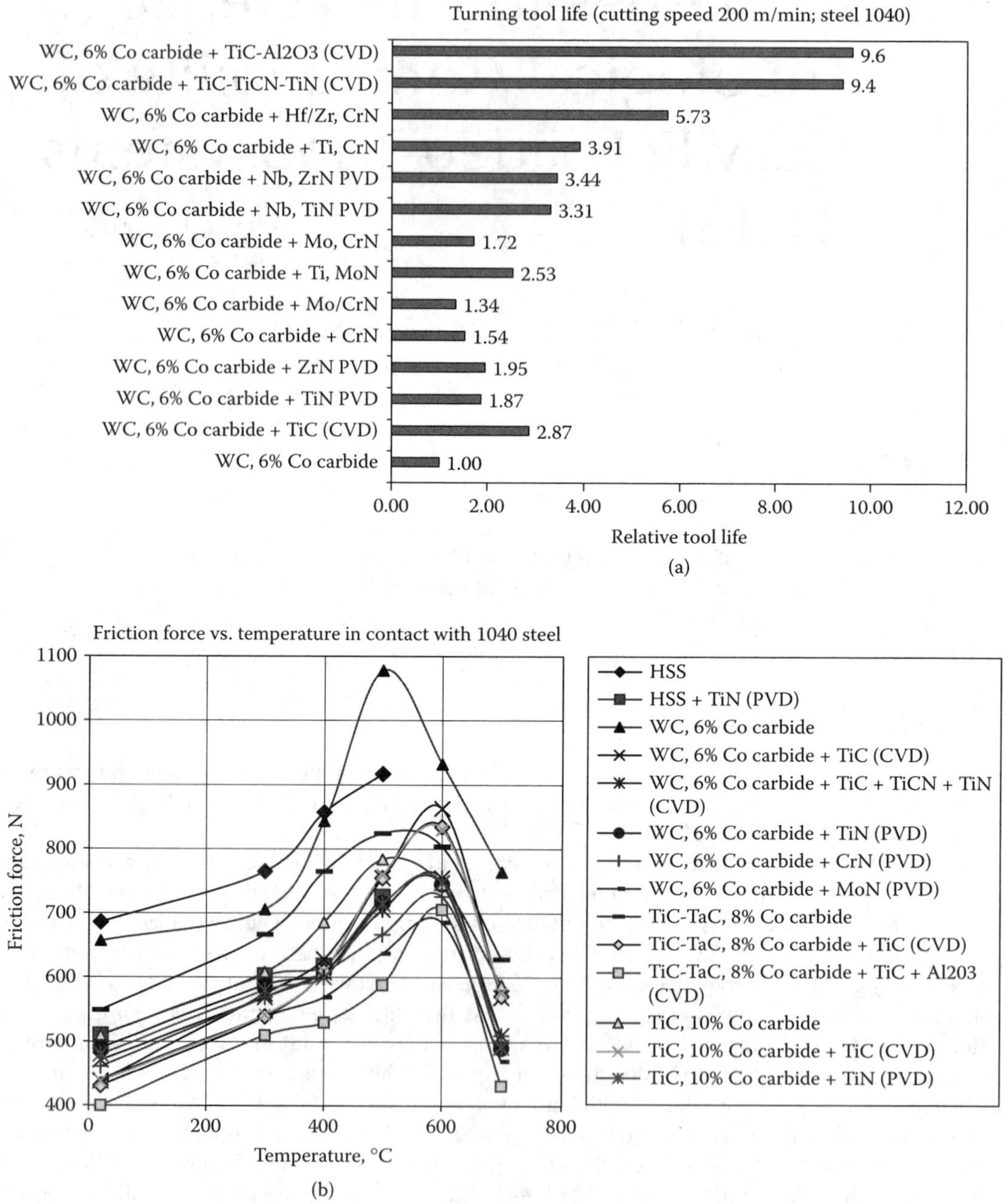

FIGURE 5.1 Tool life for the cutting tools with coatings and friction forces in contact with 1040 steel.

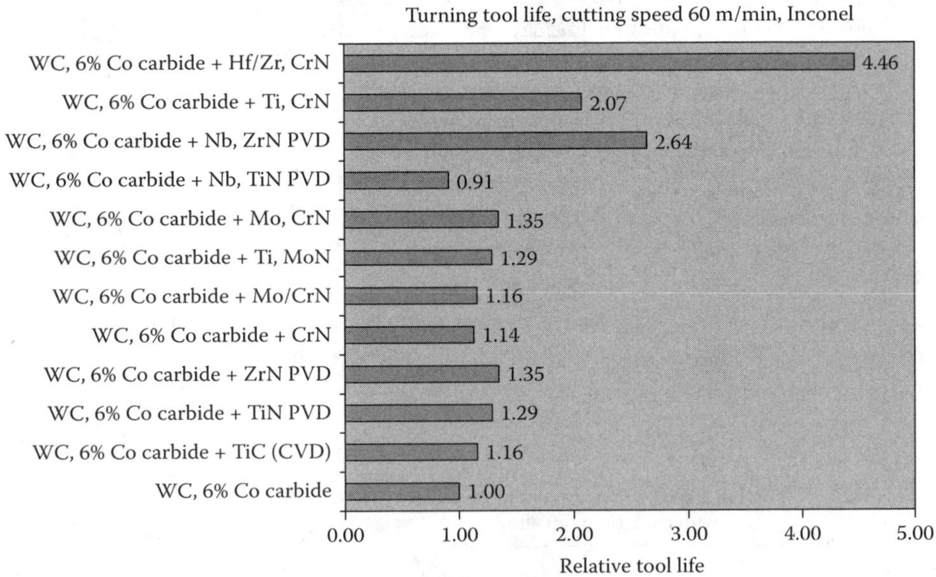

(a)

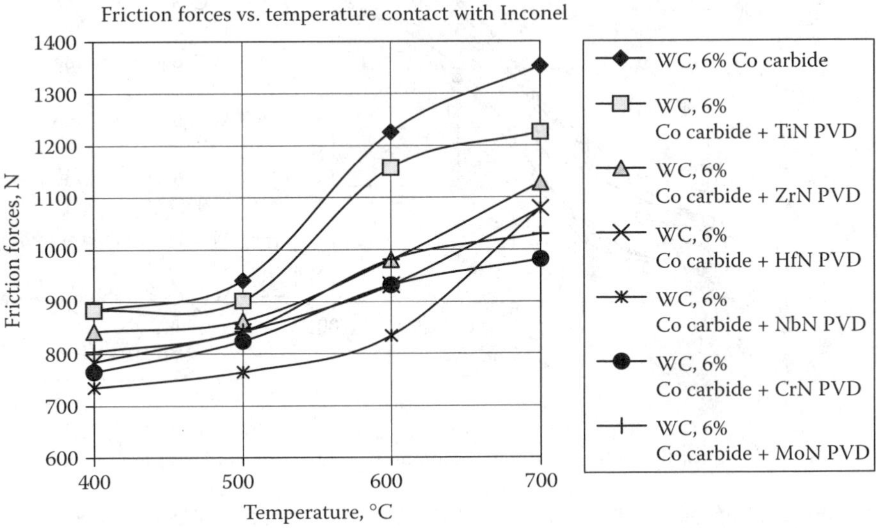

(b)

FIGURE 5.2 Tool life for the cutting tools with coatings and friction forces in contact with Inconel.

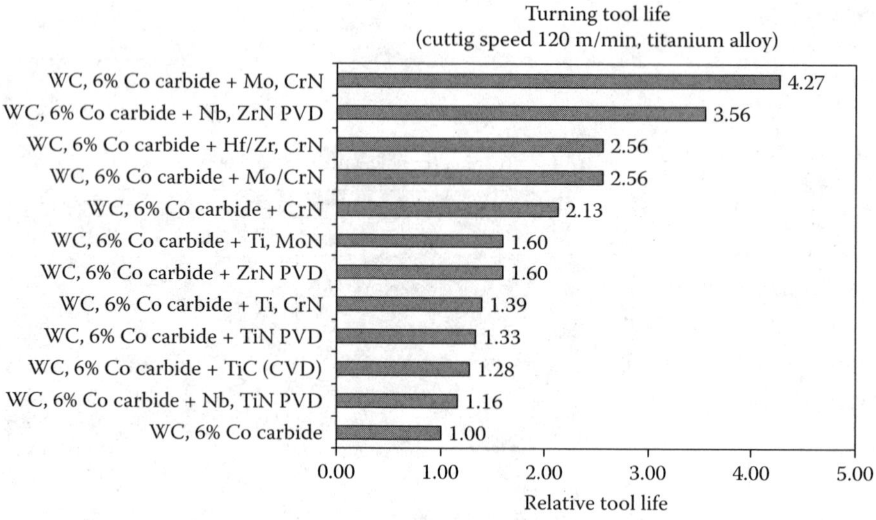

(a)

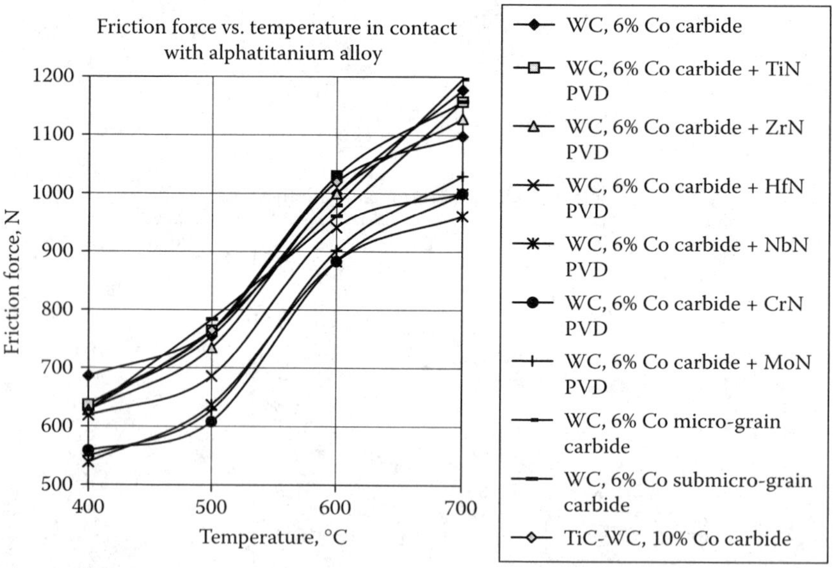

(b)

FIGURE 5.3 Tool life for the cutting tools with coatings and friction forces in contact with titanium alloy.

of the coated tooling materials (Figure 5.4). These parameters have been related to the data on the thermal conductivity of the different refractory compounds (Figure 5.5) [4]. (See also Table 5.1.)

The data presented show that the service properties of the cutting tools critically depend on the workpiece materials to be machined. The tools with multilayered CVD coatings with an Al_2O_3 outer layer have the highest tool life during machining of structural 1040 steel (Figure 5.1a). The coatings of this type have the lowest friction force (Figure 5.1b) and the highest oxidation stability (Figure 5.4), as well as low thermal conductivity (Figure 5.5). It is well known that the performance of CVD coatings exceeds that of PVD coatings during turning. The best tool life among cutting tools with PVD coatings has an (Hf–Zr–Cr) N composition. This result is associated with the elevated temperature stability of the deposited compounds because of formation of strong inter-

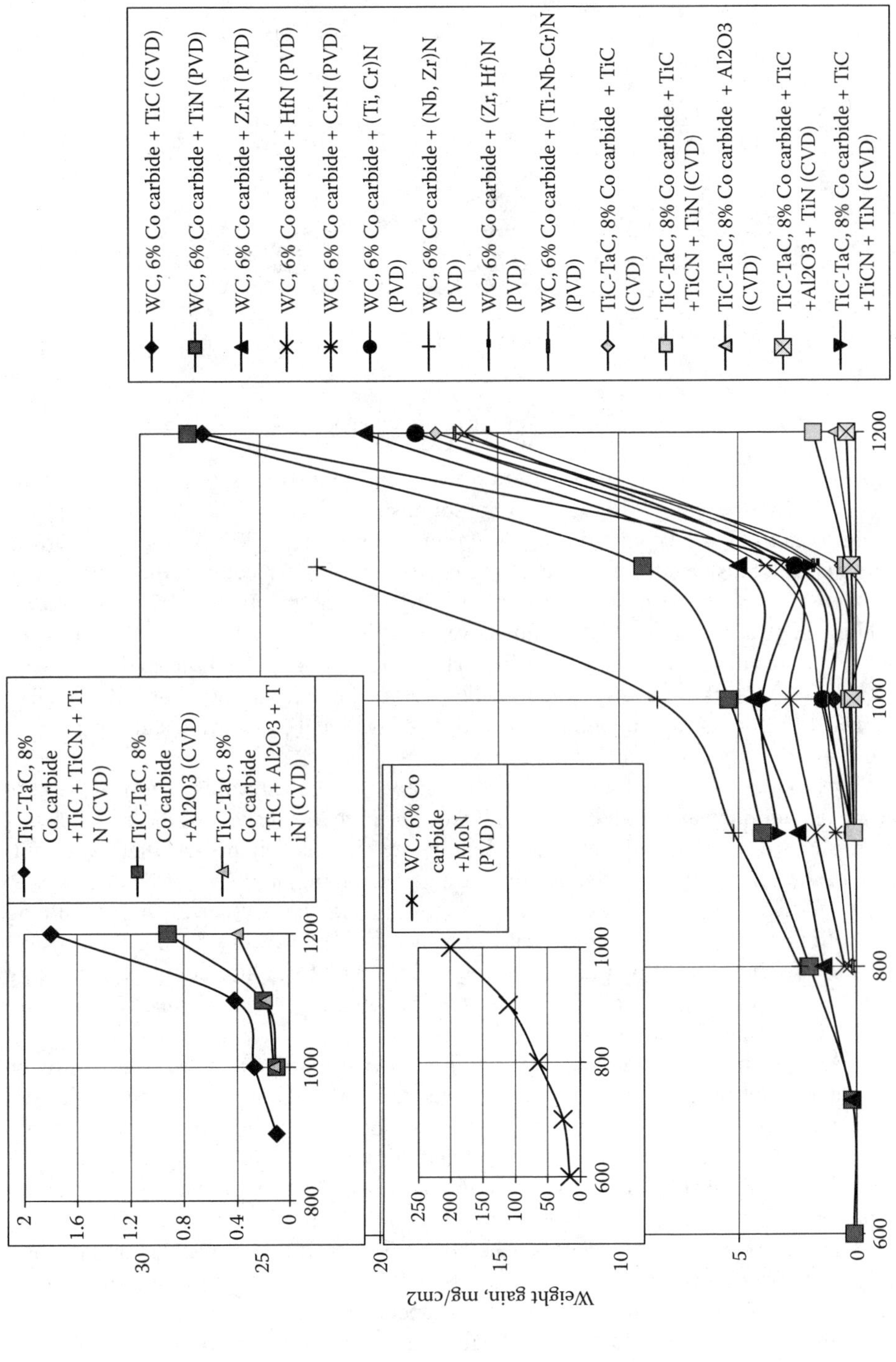

FIGURE 5.4 Oxidation resistance of different coatings at elevated temperatures.

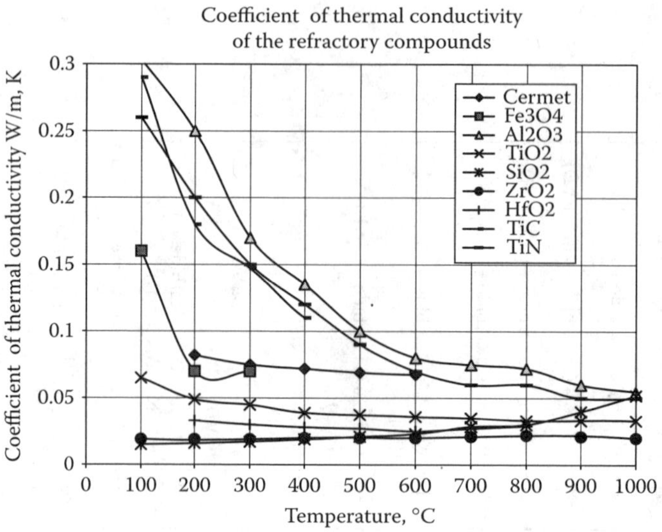

FIGURE 5.5 Coefficient of the thermal conductivity of refractory compounds.

atomic bonds [5]. At the same time, the low friction force and low oxidation resistance that are typical for Cr–N and Mo–N coatings (Figure 5.4) do not ensure a high tool life (Figure 5.1a).

Severe cutting conditions are typical during machining of Inconel [6]. The major feature of Inconel machining is an intensive adhesive interaction between the workpiece and the cutting tool. This results in intensive heat generation during cutting [6]. The relatively low friction force of Mo–N and Cr–N coatings are because of their intensive oxidation at elevated temperature (Figure 5.2b), but this does not result in an elevated tool life. The highest tool life and thermal stability is attained with a (Hf–Zr–Cr)N PVD coating. The low thermal conductivity of the Hf–O; Zr–O tribo-films that are formed on the surface of tools with (Hf–Zr–Cr)N coatings during cutting play an important role in tool life improvement (Figure 5.5).

The features of the cutting-tool wear behavior are different during the turning of alpha-titanium alloys. It is known that an intensive rutile formation takes place on the surface of the chips during cutting of titanium alloys [6–7]. The rutile is an excellent thermal barrier (Figure 5.5) that transforms the major portion of the heat flow generated during cutting into a tool. On the other hand, it is also known that a very intensive stress concentration near the cutting edge takes place during machining of titanium [6]. Both these factors result in a very intensive wear rate of the cutting tools. Tools with the highest tool life show a coating that intensively oxidize and form lubricious oxides during cutting (Figure 5.3a). These are MoN and Cr–N coatings. The highest wear resistance in this case is the (Mo,Cr)N PVD coating (Figure 5.3b).

Based on the data obtained we can conclude that a number of parameters are needed for the comprehensive characterization of cutting tools' wear behavior for specific applications. It is very difficult to decide which set of parameters is needed for specific applications. However, a generic concept of compatibility exists in tribology (Chapter 3), and this approach could help to characterize the wear behavior of any tribosystem. This concept combines the scientific and engineering understanding of the problem of wear resistance improvement. As was outlined in Chapter 3, the concept of tribological compatibility covers the aspects of the physical–chemical and mechanical phenomena that are taking place during friction. During the initial and most important running-in stage of wear, tribological compatibility implies generation of the specific conditions that prevent intensive surface damage. This is possible when a proper design of a tribosystem is developed and optimization of its operating parameters is achieved. This is directly associated with tribological parameter improvements. The tribological parameters could combine the other physical properties outlined in the preceding text such as oxidation stability and thermal conductivity in a generic manner (Figure 5.1 to Figure 5.4). But during the post running-in stage of wear when the wear rate is stabilized, the mechanical characteristics of the surface layers such as hardness or hot hardness (and corresponding structural characteristics) start to play a decisive role. Therefore, in this chapter we are focusing on the development of the tribological and mechanical parameters to characterize the tribological compatibility of heavily loaded tribosystems (HLTS).

5.2 THE CHARACTERISTICS OF FRICTION AS A CRITERION OF TRIBOLOGICAL COMPATIBILITY

Tribological compatibility as outlined in Chapter 1 and Chapter 3 is the ability to provide an optimal state of tribosystem within a given range of operating conditions using the chosen criteria. In other words, tribological compatibility is related to the capacity of the two surfaces to adapt to each other during friction, providing wear stability without damage to the two frictional bodies for the longest (or given) period of time.

We have to emphasize once again that the concept of tribological compatibility as is shown in Chapter 3 has a different meaning for the different stages of wear such as the running-in and stable wear stages.

During the running-in stage of wear the process of seizure determines the tribological compatibility of the friction system. The method of tribological compatibility evaluation is known [8–10] and includes the measurement of the maximum load that initiates seizure (the seizure threshold). This load value is a criterion of tribological compatibility. The tribo-couple with a higher seizure threshold has a better tribological compatibility. However, this method does not consider heat generation during friction under seizure conditions and has a relatively low accuracy because it is difficult to fully imitate the specific working conditions of different tribo-pairs.

Another method of tribological compatibility evaluation includes the application of normal loads and sliding speeds to imitate a real working condition of the specific tribo-pair [11]. Tribological compatibility can be evaluated as a specific criterion of the frictional bodies' interaction using the temperature of the seizure initiation. The seizure threshold is normally observed visually, but this method is time consuming and its accuracy is low. A more accurate method of evaluation of tribological compatibility should be developed. The development of a specific criterion of tribological compatibility for the HLTS is needed. This criterion has to be more accurate when compared to the visual observations [11] and should correspond to the specific working conditions. To realize this idea, the phenomenon of seizure (galling) formation has to be considered. The seizure initiation during friction indicates that an external friction transforms to an internal one [12,13]. This is the so-called threshold of external friction. Under conditions of plastic strains at the frictional bodies interface, the external friction is impossible if the following inequality takes place:

$$\frac{h}{r_1} + \frac{\tau}{cG} > \frac{1}{2} \tag{5.1}$$

where h is the depth of the asperities' indentation; τ, the shear strength of the adhesion bonds under condition of plastic strains at the frictional bodies interface; r_1, the actual asperities radius; cG, the ultimate load corrected for a hardening as a result of the plastic deformation of the frictional bodies in contact.

As a first approximation [13]:

$$cG = P_{rn} \tag{5.2}$$

where P_{rn} is the maximal normal stress that corresponds to the plastic contact conditions.

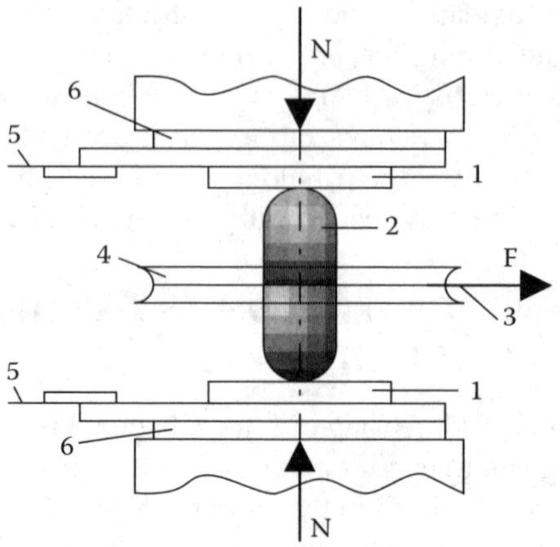

FIGURE 5.6 Schematic diagram of friction test apparatus. 1 — specimen made of workpiece material; 2 — specimen made of tooling material; 3 — driving rope; 4 — driving disk; 5 — electrical contact wires; 6 — isolation system.

The actual measurements show the relative indenter's indentation (h/r_1) corresponds to the stresses P_{rn} for a wide range of engineering materials. This parameter lies within the range of 0.12 to 0.15. This allows us to evaluate the tribological compatibility based on the equation:

$$\tau/cG \cong \tau/P_{rn} \tag{5.3}$$

The latest ratio could be obtained on the measurements using a special tribometer under the normal loads equal to P_{rn} (see details later in the text).

Thus, the ratio τ/P_{rn} could be considered as a criterion of tribological compatibility for HLTS. Using Equation 5.1 and the actual value of h/r_1 = 0.12 – 0.15, we can consider the following inequality as a condition of an external friction threshold and seizure initiation:

$$\tau/P_{rn} \geq 0.38 - 0.35 \tag{5.4}$$

The bigger the difference between the actual $\frac{\tau}{p_{rn}}$ ratio and Equation 5.4, the more favorable are the conditions of external friction taking place, as a result of which the wear rate decreases. The lower the difference, the higher the possibility of seizure initiation. If this ratio is equal to Equation 5.4, the external friction is not possible because the seizure forms immediately. Based on these considerations, the method of the tribological compatibility evaluation was developed. [14].

The method to evaluate cutting or stamping tools tribological compatibility includes the following: tests were performed using a specially designed apparatus (Figure 5.6) described in Reference 14. A rotating sample (pins with spherical heads) of the coated substrate or any tooling material was placed between two polished specimens made of workpiece material. To simulate tool friction conditions, the specimens were heated by resistive means through a temperature range from 20 to 1000°C. A standard force was applied to provide plastic strain in the contact zone. The moment the plastic strain initiation could be easily observed on the surface of the polished samples, the corresponding rotating pin made a spherical print. The tribological compatibility criterion (coefficient of friction value) was determined as the ratio of the shear strength induced by the adhesive bonds between the tool and the workpiece and the normal contact stress developed on the contact surface under the test temperatures $\frac{\tau}{p_{rn}}$. No fewer than three tests were usually performed for each kind of material. The scatter of the friction parameter measurements was usually found to be close to 5%.

The tribo-couple that ensures a lower $\frac{\tau}{p_{rn}}$ value under the given friction condition has a lower possibility of seizure at the contacting frictional bodies interface. The tribological compatibility of this tribo-couple is believed to be better. The temperature of the tribological compatibility threshold is the temperature at which seizure occurs on the surface of the frictional body with a lower hardness, which corresponds to the maximum $\frac{\tau}{p_{rh}}$ value.

We have to emphasize that this test adequately imitates the friction conditions at the tool–workpiece interface of a majority of cutting and stamping operations, because the tool–workpiece adhesive interactions mainly control the wear behavior of these heavy loaded tribosystems.

The method outlined in the preceding text can also be used for seizure-free conditions. If seizure does not occur, the τ stress can be a characteristic of energy dissipation due to the breakage of the bonds formed at the frictional bodies interface. This could be used, for instance, as a criterion of tribological compatibility for the lubricants such as PFPE (see Chapter 9, Figure 9.3).

5.3 MICRO- OR NANOMECHANICAL PROPERTIES TO CHARACTERIZE THE TRIBOLOGICAL COMPATIBILITY OF SURFACE-ENGINEERED TOOLS

The characteristics of friction described earlier can obviously be used as one of the major parameters to characterize the tribological compatibility of HLTS during the running-in stage of a wear process when the tribological characteristics play a decisive role (Chapter 3). But, for the post running-in stable stage of wear, some micromechanical properties of surface layers should also be considered to evaluate the tribological compatibility. In this case, tribological compatibility means the ability of the surface to provide steady work of the tribosystem without surface damage during its lifetime (see Chapter 3). In the post running-in period, tribological compatibility of the system is associated with wear resistance of the friction surface, which is mainly controlled by the hardness and the hardness changes with temperature, as well as with its resistance to fatigue-damage accumulations. A number of the different micromechanical properties can be associated with the ability of the friction surface to dissipate and accumulate energy of deformation during friction.

Micro- and nanoindentation has widely been applied for the practice of characterization of thin coatings [15]. Microhardness is a well-known basic mechanical property, and classic theories of wear [16] emphasize hardness as a major property that determines the wear resistance of a surface. This is true for stable wear conditions [13], but given the wide variety of wear modes experienced in practice, some additional characteristics should be considered. The majority of hard PVD coatings are currently used for tool life improvement. With tooling applications, a combination of high coating hardness with some toughness is critically needed [17–18]. It is topical to develop one or more parameters that can meaningfully capture a combination of these properties. One of these parameters is the hardness-to-elastic-modulus (H/E) ratio. It has been recognized that the ranking of materials according to their hardness and elastic modulus, the H/E ratio, provides extremely close agreement to the materials ranking in terms of wear behavior [19,20]. The H/E ratio can then be used to characterize the wear behavior of different cutting tools. The primary data presented in Reference 21 for Al–Ti–N PVD coatings was transformed to H/E ratios vs. the drill's tool life. The results are presented in Figure 5.7, and they show that a higher H/E ratio corresponds to a lower tool life during drilling operations when an adhesive wear mechanism dominates. This implies that a coating with a combination of high hardness and elastic modulus, corresponding to a high H/E ratio, has a lower wear resistance when compared to coatings that have a combination of relative

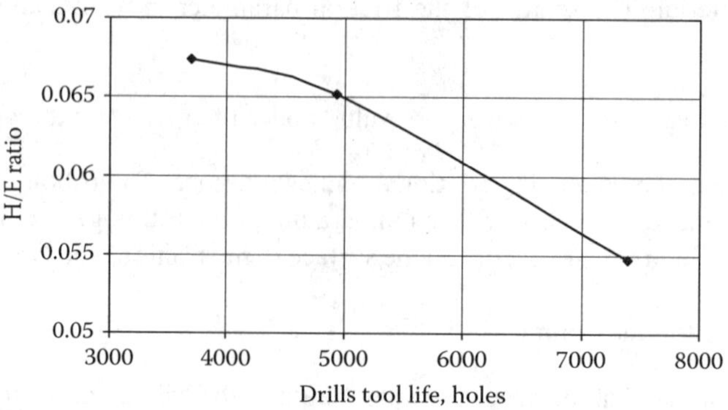

FIGURE 5.7 H/E ratio vs. drills tool life with Al–Ti–N PVD coatings. (From Sato, K., Ichmiya, N., Kondo, A., Tanaka, Y. Microstructure and mechanical properties of cathodic arc ion-plated (Al,Ti)N coatings. *Surf. Coat. Technol.* 2003, 163–164, 135. With permission.)

low hardness and elastic modulus, which corresponds to a low H/E ratio. This result shows that for heavily loaded tooling applications, in which intensive deformation of the surface layers is taking place during friction, a so-called elastic strain to failure (H/E) [20] as well as a reserve of plastic properties [22] are critically needed to achieve an improvement in tool life for a coated tool.

Another parameter that could be used to characterize the service properties of surface-engineered cutting tools is related to plasticity of the coating [23–26]. In this study we are trying to work out a new interpretation of the plasticity index as a parameter related to the coating's wear resistance under specific working conditions [50]. This parameter is similar in physical meaning and complimentary to H/E and is based on an energetic approach. Friction accompanies the processes of plastic deformation, surface damage, and wear. All these processes are dissipative in nature [27]. The entire diversity of the processes that take place during friction can be divided into two groups: quasi-equilibrium steady state processes, which are encountered during steady friction and wear and nonequilibrium unsteady state processes, which are associated with surface damage (see Chapter 1, Figure 1.1) [18,27]. Wear resistance of any material in general and coatings in particular depend to a great extent on the ability of a surface to (1) accumulate and (2) dissipate the energy of deformation during friction [28]. The ability of the surface to accumulate energy during elastic deformation, as well as to dissipate energy during irreversible plastic deformation, can be estimated by the nanoindentation method [29–33]. The total work of elastic and plastic deformation can be estimated based on the respective areas associated with the loading curve, whereas the work of elastic deformation can be estimated using the area between the unloading curve and the Y axes; the work of plastic deformation corresponds to the area between the loading and unloading curves [31–36]. The remaining challenge is to determine the connection between the data of nanoindentation tests and the wear resistance of coatings under different wear conditions [37].

The goal of this study is to investigate the relationship between the measured characteristics of energy dissipation or accumulation during nanoindentation and the other micromechanical and tribological characteristics, as well as the wear resistance of hard coatings under different wear conditions such as those experienced by cutting and stamping tools.

The experimental studies were performed on hard PVD Ti–N coatings. Hard Ti–N coatings were deposited by arc evaporation PVD technique [38–39]. The deposition parameters were as follows: (1) arc current from 50 to 160 A; (2) bias voltage from 150 to 350 V; (3) temperature of deposition, 500°C.

The studies were performed in two stages. During the first stage the Ti–N coatings were deposited within a very wide range of nitrogen pressure from 6.5×10^{-3} Pa to 2.5 Pa. These pressures are rarely used in practice but were needed for research purposes. At pressures lower then 10^{-2} Pa the coating has a very high amount of "droplet" phase present, and at pressures above 1.0 Pa the

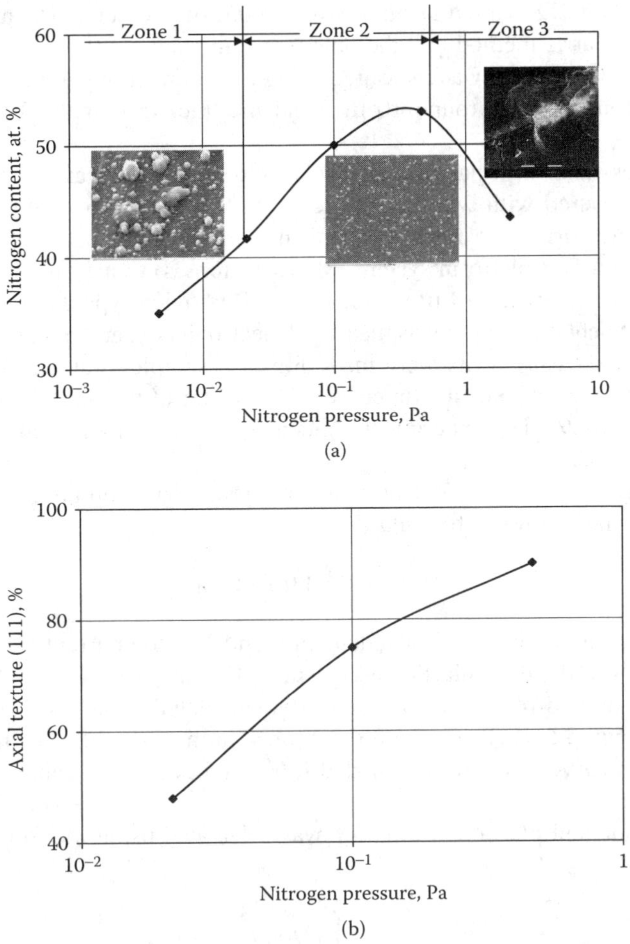

FIGURE 5.8 Chemical composition (a) and axial texture (b) of the Ti–N coatings vs. nitrogen pressure in the chamber during the coating deposition. Zone 1: a very low nitrogen pressure, below 10^{-2} Pa, when PVD coatings with poor morphology are forming (excessive amount of the "droplet" phase); Zone 2: the working range of nitrogen pressure used in research and industrial practice; Zone 3: very high nitrogen pressure, where the coatings with poor adhesion to the substrate are forming. (From Fox-Rabinovich, G.S., Veldhuis, S.C., Scvortsov, V.N., Shuster, L.Sh., Dosbaeva, G.K., Migranov, M.S. Elastic and plastic work of indentation as a characteristic of wear behavior for cutting tools with nitride PVD coatings. *Thin Solid Films* 2004, 469–470, 505–512. With permission.)

adhesion of the coating to the substrate is poor [38,39]. This strongly affects the characteristics of the coatings (see Figure 5.8a). During the second stage of research, a narrower range of nitrogen pressure was used. The range of nitrogen pressure during the arc deposition of the Ti–N coating in the chamber lay between 10^{-2} to 1.0 Pa. Nitrogen pressures of 1.0 Pa and below were used for the deposition of coatings for cutting-tools applications, whereas a nitrogen pressure of 10^{-2} Pa was used for coatings in which a high hardness was needed [39].

Filtered arc coatings were deposited in the PVD unit using a magnetic filtering module [40–41] with the same deposition conditions. Two types of substrate materials were used. The first one was an M2 high-speed steel (hardness HRC 64–66), and the second one was made from cemented carbide substrates.

The chemical composition of the coatings was studied by scanning Auger electron spectroscopy (AES) on an ESCALAB-MK2 (VG) spectrometer. Phase analysis and axial texture measurements were determined by an XRD method using a DRON 3.0 diffractometer. The axial TiN texture was determined using the Harris method and the pole figure methods [42].

The thickness of the coatings was measured using a ball crater apparatus. The thickness of the Ti–N coatings was found to be around 10 μm, and the thickness of the Ti–Al–N coatings was around 2.5 μm.

The microhardness and the work of elastic–plastic deformation were evaluated using load vs. displacement data measured with two types of testers: (1) a computer-controlled nanoindentation tester MTI-3M for measuring the micromechanical characteristics of the Ti–N coating (10-μm thickness) and (2) Nano Test platform system (Micro Materials) to measure mechanical properties of Ti–Al–N and Al–Cr–N coatings (3-μm thickness). A Berkovich-type diamond indenter was used in these testers. The indentation depth was measured electronically, and the indentation curves were then evaluated. Thirty measurements were made for each sample. Standard methods of statistics were used to calculate average indentation curves. The scatter of the microhardness measurements was found to be around 2%. The experimental indentation curve closest to the calculated average was used for the analysis.

A typical loading and unloading indentation curve is shown in Figure 5.9. The microhardness was calculated using the following formula:

$$HV = F/26.43(h^2)\ \text{GPa} \tag{5.5}$$

where F is a stepwise increasing load of indentation and h is the current depth of indentation at the loading curve in μm (depth of elastic and plastic deformation).

The work of elastic and plastic deformation was calculated using the loading and unloading indentation curves (Figure 5.9). To measure these work terms for each specimen the loading and unloading indentation curves were approximated by a third-degree polynomial. The parameter of determination was $R^2 = 1$.

The work of elastic and plastic deformation was calculated using the following formula [43]:

$$A = \int_0^h f(h)dh \tag{5.6}$$

The total work of elastic and plastic deformation, A, was estimated by integrating the loading curve of the F function in the range from 0 up to the current depth of indentation, h. The integration step was chosen to be equal to the load increase step (in our case 200 mN). The range of loads was from 0 to 2500 mN (Figure 5.9).

The work of elastic deformation, A_{eo}, was determined by integrating the unloading curve after the curve was normalized to the origin of coordinates (Figure 5.9).

The work of plastic deformation or irreversible work of indentation [24], A_p, was determined by $A_p = A - A_{eo}$.

Using the data obtained, the coefficient K_e was calculated as the ratio of elastic to the total work of deformation during indentation. This is given by:

$$K_e = A_{eo}/A$$

This parameter was then taken to be an elastic recovery parameter [17] or a parameter of energy accumulation.

The plasticity index [16] or parameter of dissipation (energy losses during plastic deformation) can be estimated using the following formula:

$$K_d = 1 - K_e$$

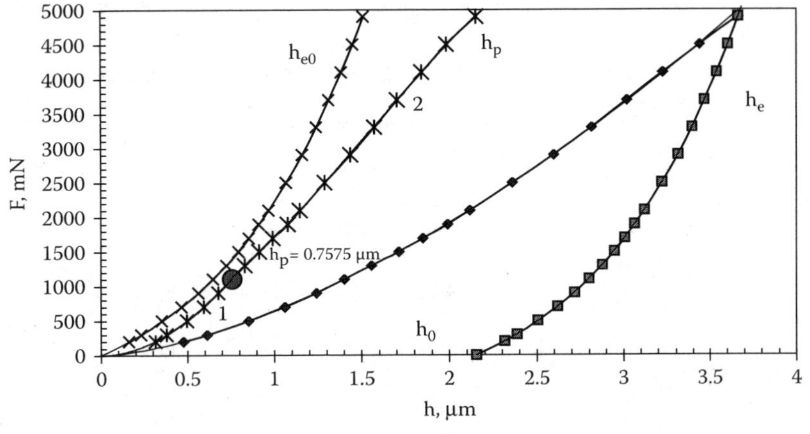

(a)

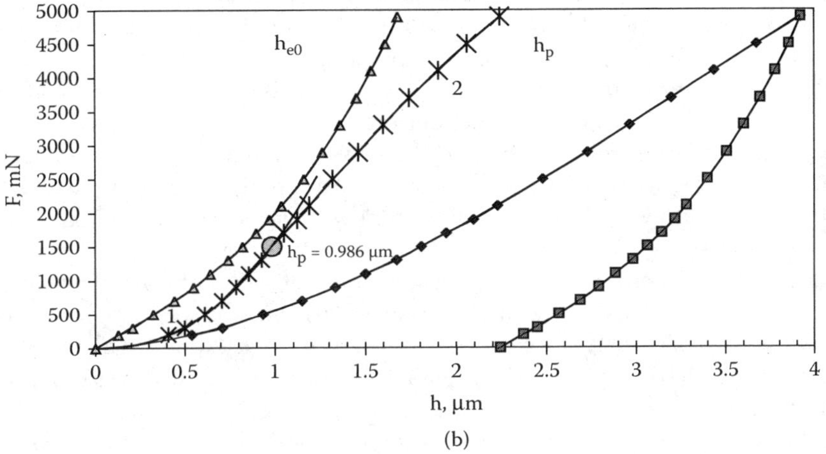

(b)

FIGURE 5.9 Loading and unloading indentation curves for Ti–N coatings: (a) arc PVD, (b) filtered arc PVD coating. Here, F is a stepwise increasing load of indentation; h is the current depth of indentation at the loading curve, μm (depth of elastic and plastic deformation); h_e is the current depth of indentation at the unloading curve, μm (elastic deformation); h_{eo} is the depth of elastic indentation normalized to origin of coordinates, μm (Figure 5.1); and h_p is the depth of plastic deformation after unloading completing, calculated as $h_p = h - h_{eo}$, μm. (From Fox-Rabinovich, G.S., Veldhuis, S.C., Scvortsov, V.N., Shuster, L.Sh., Dosbaeva, G.K., Migranov, M.S. Elastic and plastic work of indentation as a characteristic of wear behavior for cutting tools with nitride PVD coatings. *Thin Solid Films* 2004, 469–470, 505–512. With permission.)

To emphasize the physical meaning of the parameter based on the energetic approach, we suggest naming this characteristic the "microhardness dissipation parameter" (MDP) of a coating.

The curve of plastic deformation after unloading, calculated as $h_p = h - h_{eo}$ (μm), was plotted, where h_{eo} is the depth of elastic indentation normalized to the origin of coordinates (see Figure 5.9). The curve of plastic deformation, h_p, was plotted. The curve was a result of the line approximation obtained using the mathematical proceeding of the experimental graphs. Every curve had its own equation of approximation (Table 5.1). Two segments (1 and 2) were observed for the curve. Every segment had a coefficient of determination R^2 in the range of 0.9997 to 0.9999. When the equation of the curve changes a cusp on the curve occurs (Figure 5.9). The first segment (1) characterizes the superficial layer with a thickness of around 10% of the total thickness of the coating. This is well known from the literature as the coating's "intrinsic" hardness [44–46]. The

TABLE 5.1
The Equations for the Curve of Plastic Deformation h_p in Figure 5.1

	Regular Ti–N Arc PVD Coating (Figure 5.1a)	
Segment 1	$0<h<0.7575$ μm	$F = 190.19h_p^{1.9315}$; $R^2 = 0.9997$
Segment 2	$0.7575<h<2.1535$ μm	$F = 38.037h_p^3 + 175.95h_p^2 + 19.866h + 12.482$; $R^2 = 0.9999$
	Ti–N Filtered Arc PVD Coating (Figure 5.1b)	
Segment 1	$0<h<0.986$ μm	$F = 155.36h_p^{2.2918}$; $R^2 = 0.9999$
Segment 2	$0.986<h<2.242$ μm	$F = 24.633h_p^3 + 85.325h_p^2 + 197.574h_p$; $R^2 = 1$

Source: Fox-Rabinovich, G.S., Veldhuis, S.C., Scvortsov, V.N., Shuster, L.Sh., Dosbaeva, G.K., Migranov, M.S. Elastic and plastic work of indentation as a characteristic of wear behavior for cutting tools with nitride PVD coatings. *Thin Solid Films* 2004, 469–470, 505–512. With permission.

second segment (2) characterizes a deeper layer in which the substrate material starts to be involved in the plastic deformation of the surface zone. The cusp in the line of plastic deformation is related to this well-known phenomenon. The position of the cusp could be a source of important information for the characterization of a coating's mechanical properties, as described in detail later in the text. Based on the data obtained, the value of MDP was calculated for the superficial layer with a thickness of around 10% of the total thickness of the coating.

The wear resistance of the coating under heavily loaded sliding conditions was determined by the wear rate value using a special friction apparatus. The friction tests were made using a shaft surface method under heavy loads applied (400 MPa) with dry sliding conditions. For these tests, the sliding speed was set at 0.5 m/sec. The time of one friction test was half an hour. Friction tests were performed using a ring made of W1 hardened tool steel (HRC 55) and a flat sample made of D2 cold-worked tool steel (HRC 60) with Ti–N coatings used in this study. The wear intensity was determined as the ratio of the crater wear to the length of friction. No less than five measurements were performed for each tribo-couple. The accuracy of the wear intensity measurement was found to be approximately 9%.

The coefficient of friction was determined using two methods. The first method was aimed at imitating the domain of cutting associated with relatively low cutting speeds. High-speed steel (HSS) cutting tools are usually used for this domain of application [6,18]. For this test, contact with a continuously new surface of the workpiece material was performed to imitate typical machining conditions. The cross-sliding cylinder method was used under heavy applied loads (1000 N). The friction parameter was measured at room temperature. The cylindrical specimen was made of W1 steel (hardness HRC 25–30) and was slid across the two short samples made of M2-cold-worked tool steel (HRC 62) with Ti–N PVD coatings. No less than three measurements were made for each tribo-couple. The second method, which was described earlier in the text (Figure 5.6), was to evaluate the tribological compatibility of the coatings. This method aimed at imitating the adhesive interaction between the carbide cutting tool and the workpiece for conditions in which the cutting speeds, as well as the temperature at the interface, is high.

The tests of the coated tools were performed under varying conditions simulating the different mechanisms of wear experienced during cutting. Moderate cutting speeds were chosen to simulate the adhesive wear mechanism predominantly. The cutting parameters are shown in Table 5.2. The arc PVD Ti–N coatings were evaluated under turning tests using square indexable inserts made of M2 HSS to simulate adhesive and adhesive-fatigue wear mechanisms [38,42]. The filtered arc PVD Ti–N coatings were tested under conditions of face milling to simulate adhesive wear [42].

The microhardness of Ti–N coatings (Figure 5.10a), as well as the microhardness dissipation parameter (MDP), (Figure 5.10b) was measured vs. the nitrogen pressure in the chamber during

TABLE 5.2
Cutting Parameters Data

Workpiece Material		Cutting Parameters		
Steel	Hardness, HB	Speed (m/min)	Depth of Cut (mm)	Feed (mm/rev)
		Turning Tests		
Indexable Turning HSS Inserts with Arc PVD Ti–N Coatings				
1040	200	45	1	0.33
		Face-Milling Tests		
Indexable Carbide Inserts with Arc PVD Coatings				
1040	200	50	3	10
		End-Milling Test		
Carbide End Mills with Ti–Al–N PVD Coatings				
Stainless steel grade 304	200	63	3	10

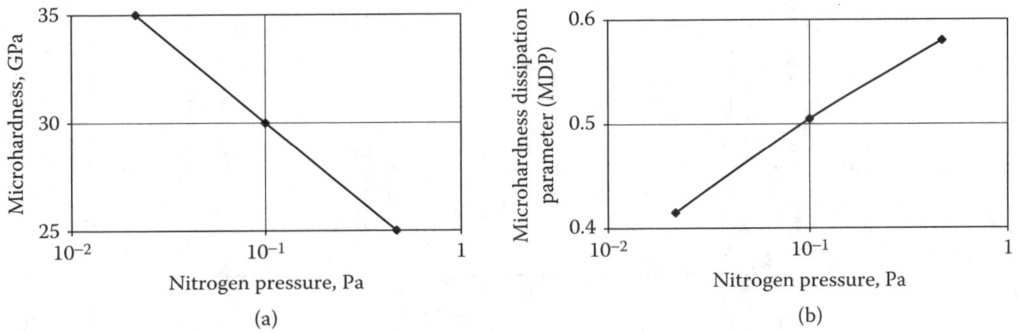

FIGURE 5.10 (a) Microhardness and (b) microhardness dissipation parameter of the Ti–N coatings vs. nitrogen pressure in the chamber during the coating deposition. (From Fox-Rabinovich, G.S., Veldhuis, S.C., Scvortsov, V.N., Shuster, L.Sh., Dosbaeva, G.K., Migranov, M.S. Elastic and plastic work of indentation as a characteristic of wear behavior for cutting tools with nitride PVD coatings. *Thin Solid Films* 2004, 469–470, 505–512. With permission.)

the coating's deposition. It was shown that these parameters are strongly influenced by the coating's phase composition. The high microhardness is achieved when the two-phase α-Ti+TiN composition changes into three phase α-Ti+Ti_2N+TiN composition [39,42].

The best coating wear resistance during dry sliding-friction conditions corresponds to the high hardness of the coatings (Figure 5.11a).

The friction parameter under conditions imitating adhesive interaction of the coated tool with the workpiece material gradually decreases versus the nitrogen pressure growth (Figure 5.11c). Based on the literature it is assumed that the low friction parameter is caused by the high (111) axial texture of the Ti–N coating deposited under high nitrogen pressure [39].

The best wear resistance for HSS turning inserts with Ti–N coating in the domain of adhesive or adhesive-fatigue wear mechanisms (Figure 5.11b) corresponds to the lowest coefficient of friction value (Figure 5.11c). A similar result was achieved under conditions of face milling using carbide

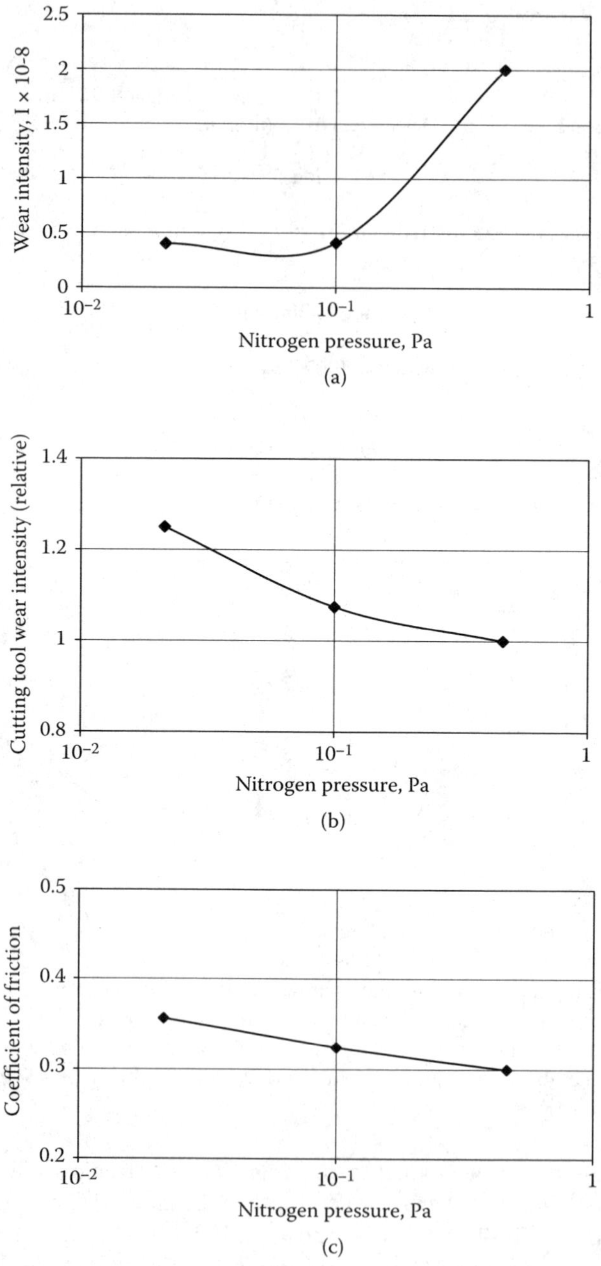

FIGURE 5.11 Wear resistance of the Ti–N coatings under (a) sliding conditions, (b) during turning under adhesive wear mechanism, and (c) coefficient of friction under conditions imitating the contact of the coated tool with permanently juvenile surface of the workpiece material. (From Fox-Rabinovich, G.S., Veldhuis, S.C., Scvortsov, V.N., Shuster, L.Sh., Dosbaeva, G.K., Migranov, M.S. Elastic and plastic work of indentation as a characteristic of wear behavior for cutting tools with nitride PVD coatings. *Thin Solid Films* 2004, 469–470, 505–512. With permission.)

TABLE 5.3
Properties of Ti–N Coatings

	Mechanical Properties		
Coatings	Microhardness (GPa)	MDP Value	Position (Depth) of the Cusp, μm, on the Curve of Plastic Deformation (Figure 5.9)
Regular arc PVD TiN	29 to 30	0.56	0.76
Filtered arc PVD TiN	26 to 28	0.66	0.98

Source: Fox-Rabinovich, G.S., Veldhuis, S.C., Scvortsov, V.N., Shuster, L.Sh., Dosbaeva, G.K., Migranov, M.S. Elastic and plastic work of indentation as a characteristic of wear behavior for cutting tools with nitride PVD coatings. *Thin Solid Films* 2004, 469–470, 505–512. With permission.

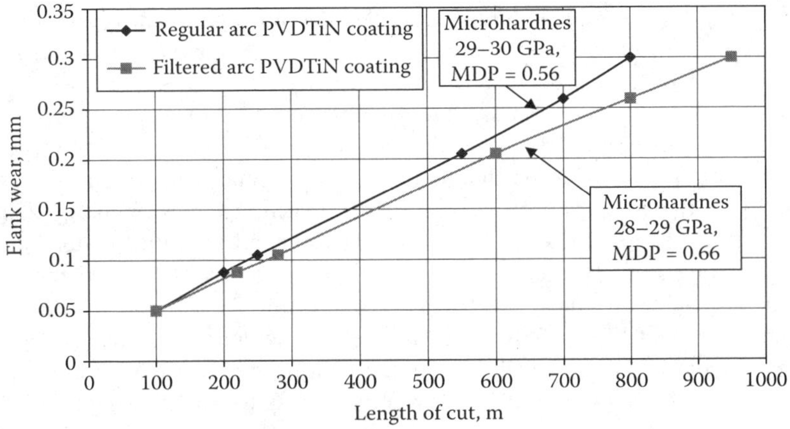

FIGURE 5.12 Flank wear vs. length of cut for Arc PVD and Filtered Ti–N PVD coatings.

tools with Ti–N coatings (Table 5.2 and Table 5.3). In this case the adhesive-wear mechanism [42] also dominates (Figure 5.12). The different types of coatings with a high value of MDP demonstrate improved tool life. Two types of Ti–N coatings were studied. The first one was a regular arc deposited coating, and the second one was a filtered arc deposited coating [40–41]. The mechanical characteristics of the coatings are shown in Table 5.3. The coated cutting tool's life data is presented in Figure 5.12.

The correlation between the microhardness dissipation parameter (MDP) and the wear resistance of the Ti–N coatings was observed. Under stable conditions of sliding wear, microhardness controls the wear resistance and the value of MDP should be relatively low within the range 0.4 to 0.5 (see Figure 5.11). But the requirements of the surface properties change when the adhesive-wear mechanism starts to dominate. This is typical for cutting operations. Adhesive wear is caused by the formation of welded asperity junctions between the chip and the tool face. The subsequent fracture of the junctions by shear leads to microscopic fragments of the tool material being torn out and subsequently adhering to the chip or the workpiece. This can lead to intensive wear of the cutting tool. Under adhesive-wear conditions, the ability of the surface to dissipate a significant part of the energy of mechanical deformation becomes beneficial [18]. That is why the best wear resistance of cutting tools with Ti–N coatings working in the domain of adhesive wear corresponds to higher values of MDP, i.e., the ratio of energy dissipated during friction to the total energy of elastic–plastic deformation should be higher, within the range above 0.5. For turning with Ti–N-coated HSS cutting tools, the optimal value of MDP is 0.58; see Figure 5.4. The ability of the coatings to dissipate energy during friction is exhibited by its beneficial friction characteristics (Figure 5.10 and Figure 5.11). This results in turning tool life enhancement for low and moderate cutting speed conditions.

The same conclusion can be made for the face-milling operation when the conditions of adhesive wear also dominate. A further increase in the MDP value for the Ti–N coatings from 0.56 to 0.66 results in an increase in tool life (see Table 5.3). This benefit was achieved through the application of advanced filtered arc deposited Ti–N PVD coating techniques [40–42]. The "droplet" phase elimination, as well as the grain size refinement in this case, results in a coating with better fracture toughness [40]. This is a necessary property for tool life improvement in milling with interrupted cutting activity.

It should be emphasized that MDP is not the only parameter that can be used to characterize the ability of a coating to dissipate energy. The line of plastic deformation, h_p, obtained mathematically by studying the loading and unloading curves (see the experimental details in the preceding text and Figure 5.9) during nanoindentation could also be used for these purposes. This curve has a cusp and the position of this cusp is important. The depth of this point is directly related to the value of MDP (K_d). This implies that the higher the value of K_d, the deeper the cusp is at the point at the line of plastic deformation occurs (Table 5.3).

The cause of this phenomenon could be explained based on a generic energetic approach. The position of the cusp is related to the moment when the substrate starts to get involved in the plastic deformation process during indentation. This depends on the hardness of the coating and on the ability of the coating to resist plastic deformation as well as energy dissipation. The harder the coating, the deeper the cusp of the curve associated with plastic deformation will be. Owing to the substrate's involvement at this point, the layer beneath the indenter of the harder coating should be more resistant to plastic deformation at the surface. In our case, one can observe an opposite effect (Table 5.3). The hardness of the coatings under analysis is approximately between 28 and 30 GPa. The slightly harder coating (regular TiN with the microhardness 29 to 30 GPa) has a cusp position at 0.75-μm depth. The K_d for this coating is 0.56. The filtered Ti–N coating with a slightly lower microhardness (26 to 28 GPa) has a cusp position at 0.98-μm depth. But the K_d value for this coating is more than 0.66. This means that during the penetration of the indenter into the coating's layer with the higher K_d value, the substrates' involvement in the plastic deformation process of the surface layer is delayed. This point will occur later for the higher depth of the indenter's penetration because of higher energy dissipation during indentation. More energy is being dissipated, and thus less energy is spent on the deformation of the layer beyond the indenter. At a lower depth of penetration, the amount of energy is just not sufficient to start the deformation of the substrate material. We can conclude that the position (depth) of the cusp can be used to characterize the ability of the coating to dissipate energy during indentation.

A similar result was achieved for the condition of end milling of 1040 steels using carbide tools with Ti–Al–N and Al–Cr–N coatings, in which adhesive or adhesive-fatigue wear mechanisms took place (Figure 5.13). Cutting parameters were as following: spindle speed = 560 rev per min, depth of cut: 3 mm, width of cut = 10 mm, feed = 63 mm/min, tool: solid carbide end mill (12-mm diameter, 4 flutes). The thickness of the coatings was within the range of 2.5 to 3 μm.

Nanoindentation of the thin coatings deposited on the mirror-polished cemented carbide substrates was performed using the Micro Materials Nano Test system. Indentation was performed in load control mode to 25 mN, at 1 mN/sec, with a 5-sec hold period at peak load for creep and a 60-sec hold period at 90% unload for the thermal drift correction. The choice of this load resulted in a contact depth around 5% of the total coating thickness to minimize the substrate contribution and provide a reasonable estimate of elastic modulus. A Berkovich-type diamond indenter was used. The parameters measured were as follows: microhardness, reduced elastic modulus, elastic

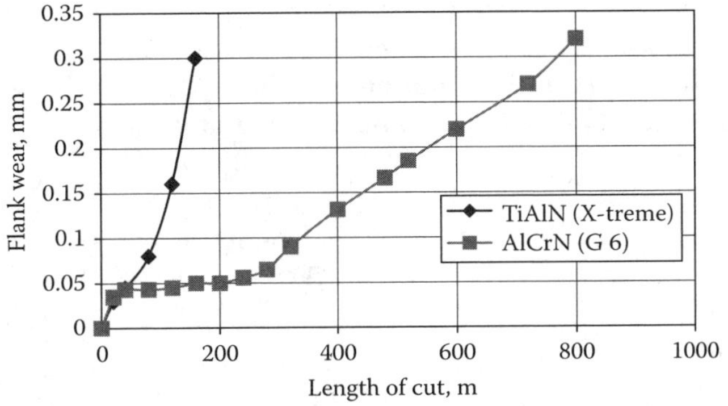

FIGURE 5.13 Flank wear vs. length of cut for Ti–Al–N, Al–Cr–N coatings. (From Fox-Rabinovich, G.S., Beake, B.D., Veldhuis, S.C., Endrino, J.L., Parkinson, R., Shuster, L.S. Effect of mechanical properties measured at room and elevated temperatures on wear resistance of cutting tools with TiAlN and AlCrN coatings, *Surf. Coat. Technol.* 2006, 20–21, 5738–5742. With permission.)

TABLE 5.4
Characteristics of Ti–Al–N Coatings

	Mechanical Properties	
Ti–Al–N Coatings	**Microhardness (GPa)**	**MDP Value**
Monolayered	25	0.76
Nanolayered	22	0.66

recovery parameter, and microhardness dissipation parameter (Table 5.4). The parameters outlined were evaluated using the load vs. displacement data measured with a computer-controlled nanoindentation. The indentation depth was measured electronically, and the indentation curves were then evaluated [43–46].

Measurements of microhardness were performed vs. temperatures at 25, 250, and 500°C correspondingly under an applied load of 25 mN, using the high-temperature system. Conditions for high-temperature tests were as follows: 36 indents to 25 mN at 1 mN/sec, repeated at each temperature (25, 250, and 500°C, correspondingly).

Scratch tests have been performed using Nano Test Scratching Module with 25-μm radius diamond probe at room and elevated temperatures (up to 500°C). No less than three tests were performed for each sample. The details of this technique are published elsewhere [47–50].

As was outlined in Chapter 3, the tribological compatibility under stable wear conditions is also associated with the ability of the surface to resist fatigue-damage accumulation during friction. In Chapter 6, we will show that the fatigue-damage accumulation is one of the characteristics that is responsible for tool wear behavior. That is why we have measured fatigue life of surface-engineered layers using nanoimpact fatigue test and related this parameter with other micromechanical characteristics.

The pendulum impulse impact configuration of the NanoTest system was used for the impact testing [51]. A solenoid connected to a timed relay was used to produce the repetitive probe impacts on the surface. A cube-corner diamond indenter test probe was accelerated from a distance of 11 μm from the surface to produce each impact at applied loads of 15 mN. The experiments were computer controlled so that repetitive impact occurred (at the same position). The frequency was 1 impact every 4 sec, while the test time was 600 sec. Every test was repeated 5 times at different locations

TABLE 5.5
Summary of Measurement Performed during Nanoindentation and Nanoscratch Tests for Ti–Al–N as well as Al–Cr–N Coatings at Room Temperature

Sample	Microhardness (GPa)	Reduced Elastic Modulus (GPa)	Elastic Recovery Parameter	Microhardness Dissipation Parameter (MDP)	Final Impact Depth/μm	Scratch Crack Propagation Resistance $L_{c1}(L_{c2} - L_{c1})$
TiAlN	30 ± 8.4	383.7 ± 81.8	0.32	0.466	1.91 ± 0.31	4.4
AlCrN	24.7 ± 10	383.6 ± 87.7	0.25	0.597	1.66 ± 0.13	8.1

Source: Fox-Rabinovich, G.S., Beake, B.D., Veldhuis, S.C., Endrino, J.L., Parkinson, R., Shuster, L.S. Effect of mechanical properties measured at room and elevated temperatures on wear resistance of cutting tools with TiAlN and AlCrN coatings, *Surf. Coat. Technol.* 2006, 20–21, 5738–5742. With permission.

on each sample. Repetitive impacts were performed at the same position of the sample. A sharp probe was used to induce the fracture quickly. The depth of failures was monitored vs. time. A longer time to fracture of the coating under analysis is related to a more tough and durable coating.

A summary of measurements performed for Ti–Al–N as well as Al–Cr–N coatings under room temperatures are shown in Table 5.5. Standard deviations are shown in the table.

The measured values of microhardness together with reduced elastic modulus sorted into ascending order are shown in Figure 5.14. The data presented show that Ti–Al–N coatings are harder when compared to Al–Cr–N coatings, but the stiffness of the coatings is similar.

The micromechanical characteristics measured vs. temperature are shown in Figure 5.15. The microhardness decreases more intensively with temperature for Ti–Al–N coatings. The Al–Cr–N coatings have a higher microhardness at 500°C when compared to Ti–Al–N coatings (Figure 5.15a). The elastic modulus reduces with temperature for Ti–Al–N coatings but increases for Al–Cr–N coatings (Figure 5.15b). The mechanical properties related to the coatings plasticity (H/E ratio as well as microhardness dissipation parameter) increase vs. temperature. They show some difference at room temperature but become very close at elevated temperatures (Figure 5.15c and Figure 5.15d).

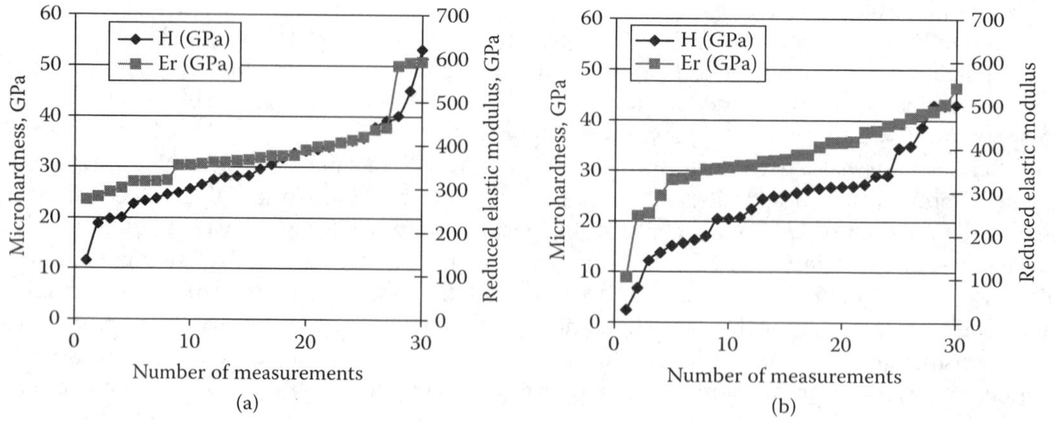

FIGURE 5.14 The measured values of microhardness together with reduced elastic modulus for: (a) Ti–Al–N and (b) Al–Cr–N coatings sorted into ascending order. (From Fox-Rabinovich, G.S., Beake, B.D., Veldhuis, S.C., Endrino, J.L., Parkinson, R., Shuster, L.S. Effect of mechanical properties measured at room and elevated temperatures on wear resistance of cutting tools with TiAlN and AlCrN coatings, *Surf. Coat. Technol.* 2006, 20–21, 5738–5742. With permission.)

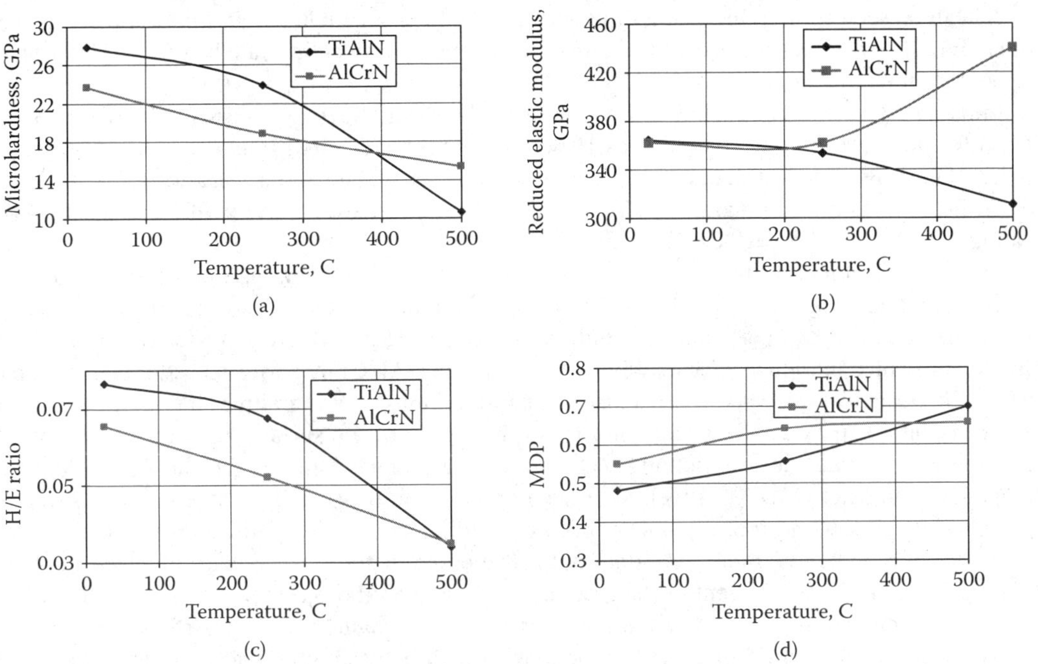

FIGURE 5.15 Micromechanical characteristics for Ti–Al–N and Al–Cr–N coatings vs. temperatures (25, 250, and 500°C): (a) microhardness, (b) reduced elastic modulus, (c) H/E ratio, (d) plasticity index (MDP). (From Fox-Rabinovich, G.S., Beake, B.D., Veldhuis, S.C., Endrino, J.L., Parkinson, R., Shuster, L.S. Effect of mechanical properties measured at room and elevated temperatures on wear resistance of cutting tools with TiAlN and AlCrN coatings, *Surf. Coat. Technol.* 2006, 20–21, 5738–5742. With permission.)

Scratch testing data vs. temperature is presented in Table 5.6. The L_{c1} load has been identified as the "lower critical load," which corresponds to a first crack event during low-load scan over the scratch track and in optical microscopy. The "higher crucial load" L_{c2} load corresponds to a dramatic failure of the coatings. The parameter $CPR_s = L_{c1}(L_{c2} - L_{c1})$ was calculated based on the data obtained (Table 5.5 and Table 5.6). This characteristic was suggested as a "scratch crack propagation resistance" and could be a measure of the fracture toughness of the coatings [47–50]. It can be seen from Table 5.6 that the values of L_{c2}, as well as CPR_s, increases slightly with temperature on Al–Cr–N coatings but drops on Ti–Al–N coatings.

TABLE 5.6
Critical Loads in Scratch Test — Variation with Temperature

Coating	Temperature (°C)	First Crack Event Load, $L_{c1}(N)$	Dramatic Failure Load, L_{c2} (N)	CPR_s Parameter $L_{c1}(L_{c2} - L_{c1})$
Al–Cr–N	25	2.8 ± 0.2	5.7 ± 0.2	8.1
	500	2.3 ± 0.4	6.0 ± 0.4	8.5
Ti–Al–N	25	2.1 ± 0.4	4.2 ± 0.5	4.4
	500	1.5 ± 0.2	3.7 ± 0.2	3.3

Source: Fox-Rabinovich, G.S., Beake, B.D., Veldhuis, S.C., Endrino, J.L., Parkinson, R., Shuster, L.S. Effect of mechanical properties measured at room and elevated temperatures on wear resistance of cutting tools with TiAlN and AlCrN coatings, *Surf. Coat. Technol.* 2006, 20–21, 5738–5742. With permission.

As can be seen from Table 5.5, a direct correlation of the scratch crack propagation resistance with H/E ratio and MDP values could be found. All of these parameters could be used to characterize the fracture toughness of the coating.

Impact fatigue test data is shown in Figure 5.16. Dramatic fracture occurs for Ti–Al–N coating after a few impacts; time to fracture was 10 sec (Figure 5.16a). Small fractures were observed for Al–Cr–N coating; time to fracture was 45 sec (Figure 5.16b). The impact behavior of Ti–Al–N and Al–Cr–N coatings are summarized in Table 5.5. The probe final depth was 1.91 μm for Ti–Al–N coatings and 1.66 μm for Al–Cr–N coatings.

The tool life of end mills with Ti–Al–N coatings is much lower compared to cutting tools with Al–Cr–N coatings (Figure 5.13). It is known from the literature that the tool working under conditions of interrupted cutting, in particular end milling, undergoes intensive adhesive-fatigue wear [6,42]. The data presented in Figure 5.13 and Figure 5.16 show that Al–Cr–N performs better in nanoimpact and milling tests. As was shown previously under conditions of end milling, the high hardness of wear-resistant coatings has to be combined with the ability of the surface layer to dissipate a part of the energy generated during friction [52]. Such coating characteristics as E/H ratio, microhardness dissipation parameter (MDP), as well as scratch propagation resistance (CPR_s), may be predictors of fracture toughness and tool life under specific conditions of the experiments performed.

We have to make one more important note. The depth of the surface layer involved in plastic deformation and the subsequent surface damage determines the optimal value of MDP. All the previous examples were related to adhesive-fatigue wear mechanism with superficial surface damage associated with fracture of asperity junctions formed between chip and tool face. The deeper surface layers are not involved in the process. But more unstable modes of wear exist [6] when deep surface damage around tens of microns occur. This is related to the idea outlined in Chapter 1 about different hierarchal levels (nano-, micro-, and macrolevels) of tool–workpiece interaction during cutting. For instance, an intensive build-up edge formation takes place during machining a ductile SS 304 stainless steel. This is a very unstable attrition wear mode because the build-up edge eventually tears off the surface, resulting in deep surface damage to the cutting tool (Figure 5.17) [6,7]. In this case, the adhesive interaction between a tool and workpiece, within a layer only a few microns thick, further develops in a deeper surface damage (around a few tens of microns; see Figure 5.17), and a chipping of the cutting edge occurs. The microstructure and properties of the coating strongly affects the chipping intensity. Under these conditions, the layer of surface coating has to have the capacity to accumulate (absorb) the energy produced from deformation [52–54]. This characteristic probably controls the intensity of surface damage under unstable wear conditions such as the attrition wear, or the catastrophic stage of the wear (see Chapter 9 [53–54]). In this specific case when a level of tool–workpiece interaction shifts from the micro- to the macrolevel, a coating that has a higher value of MDP shows a lower tool life and more intensive chipping (i.e., deeper surface damage, Figure 5.18).

To characterize surface response effect within the surface layer associated with deep surface damage (chipping), the depth of Berkovich tip penetration indentation during nanoindentation was high. The nano-indentation data shown in Figure 5.18 was obtained using a depth of indentation close to 6 μm because the actual depth of surface damage was around 30 μm (Figure 5.18). The difference in MDP values was observed only within a 2- to 3-μm thick surface layer that corresponds to the thickness of the coatings deposited. The properties of this relatively thin surface layer control the chipping behavior of end-mill cutters under unstable attrition wear mode.

That is why the carbide end mills with two different Ti–Al–N coatings (nanolayered and monolayered) showed different tool lives depending on the value of MDP (Figure 5.19). The higher tool life is associated with the nanolayered coatings having a lower MDP value due to the improved capacity of this coating to accumulate the energy of deformation (Figure 5.19). This results in a significantly lower intensity of surface damage to the tool (Figure 5.17a). The value of the MDP in this case probably plays no less important a role than the coefficient of friction values. A lower coefficient of friction (Figure 5.20) obviously decreases the sticking of the workpiece material to

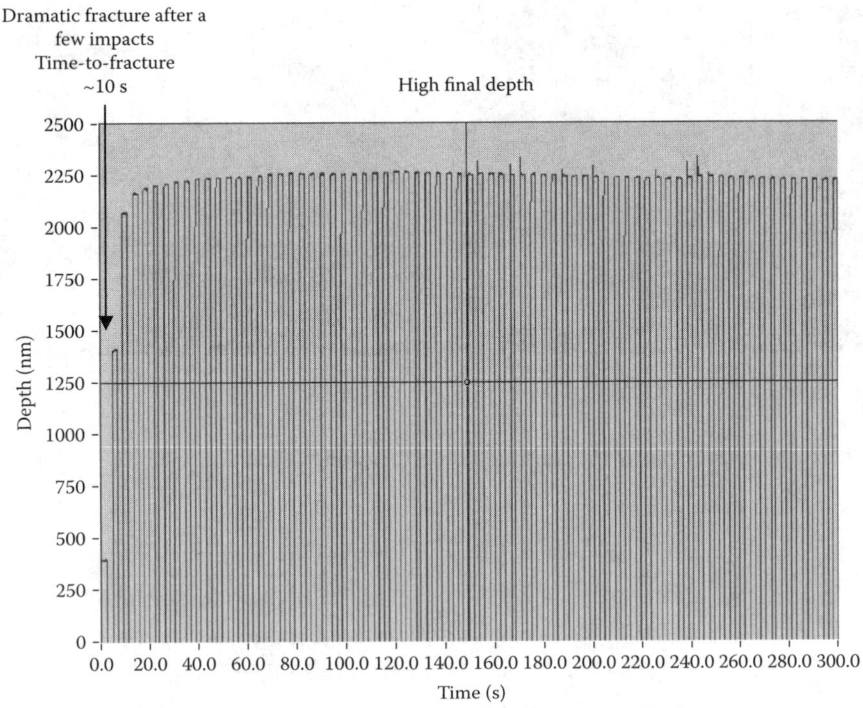

(a)

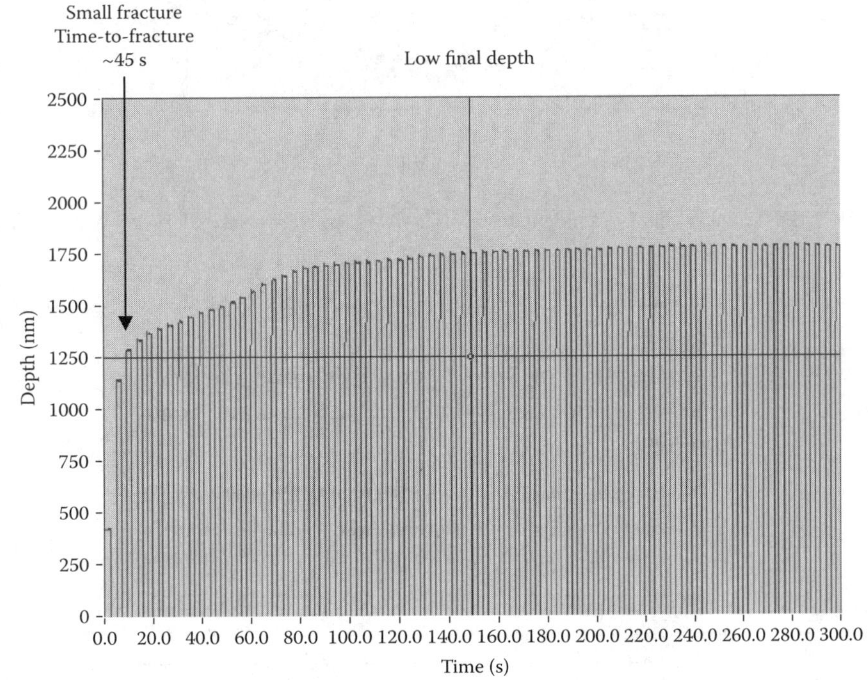

(b)

FIGURE 5.16 Typical results of impact fatigue tests for (a) Ti–Al–N and (b) Al–Cr–N coatings. (From Fox-Rabinovich, G.S., Beake, B.D., Veldhuis, S.C., Endrino, J.L., Parkinson, R., Shuster, L.S. Effect of mechanical properties measured at room and elevated temperatures on wear resistance of cutting tools with TiAlN and AlCrN coatings, *Surf. Coat. Technol.* 2006, 20–21, 5738–5742. With permission.)

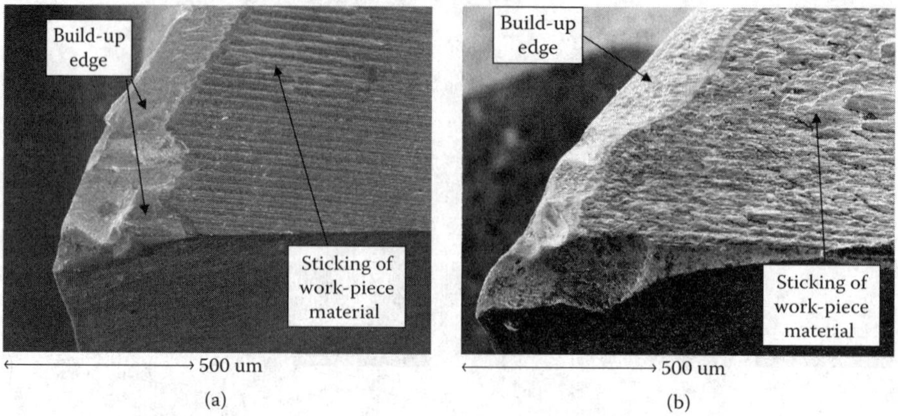

FIGURE 5.17 SEM micrographs of the flank surface of the carbide end mills with (a) Ti–Al–N nanolayered and (b) Ti–Al–N monolayered PVD coatings.

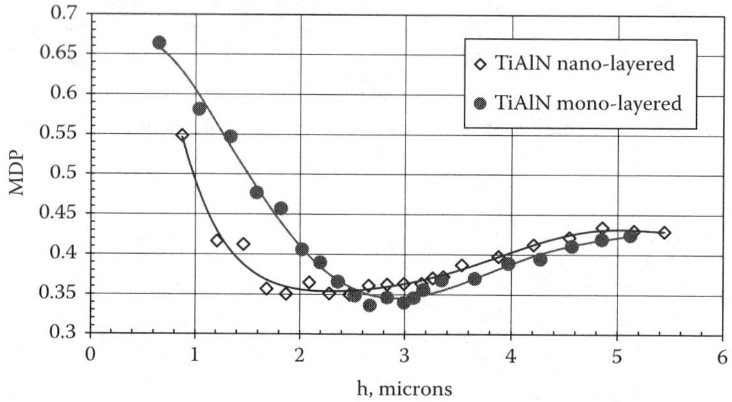

FIGURE 5.18 MDP value vs. depth of indentation for Ti–Al–N nanolayered and Ti–Al–N monolayered PVD coatings.

TABLE 5.7
Characteristics of the TiN (arc-PVD) and Duplex (TiN/Ion Nitriding) Coatings on D2 Substrate (Load Applied = 2.5 N)

	Mechanical Properties	
Coatings	Microhardness (GPa)	Elastic Recovery Parameter Value
Monolayered TiN	9.3	0.163
Duplex (TiN/ion nitriding)	11.9	0.224

the nanolayered coating (Figure 5.17a). However, on the other hand, the values of the coefficient of friction are close for both types of coatings considered under the cutting tests performed in this study (within the temperature range 500 to 600°C; Figure 5.20). A similar result was obtained earlier for a catastrophic stage of wear when the threshold of seizure was passed [53–54] (see Chapter 9, Figure 9.20).

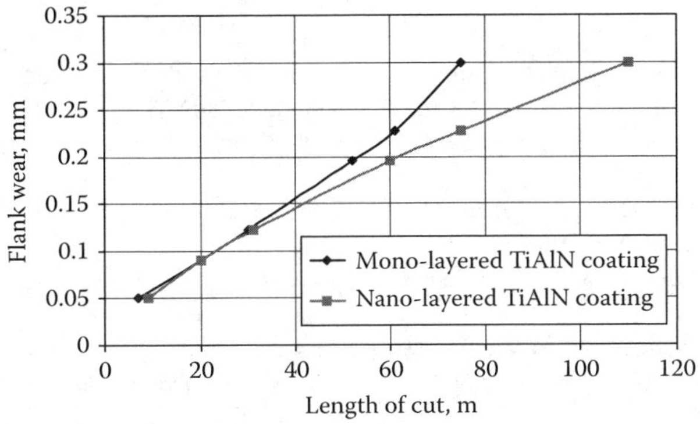

FIGURE 5.19 Tool life for end-mill cutters with Ti–Al–N nanolayered and Ti–Al–N monolayered PVD coatings.

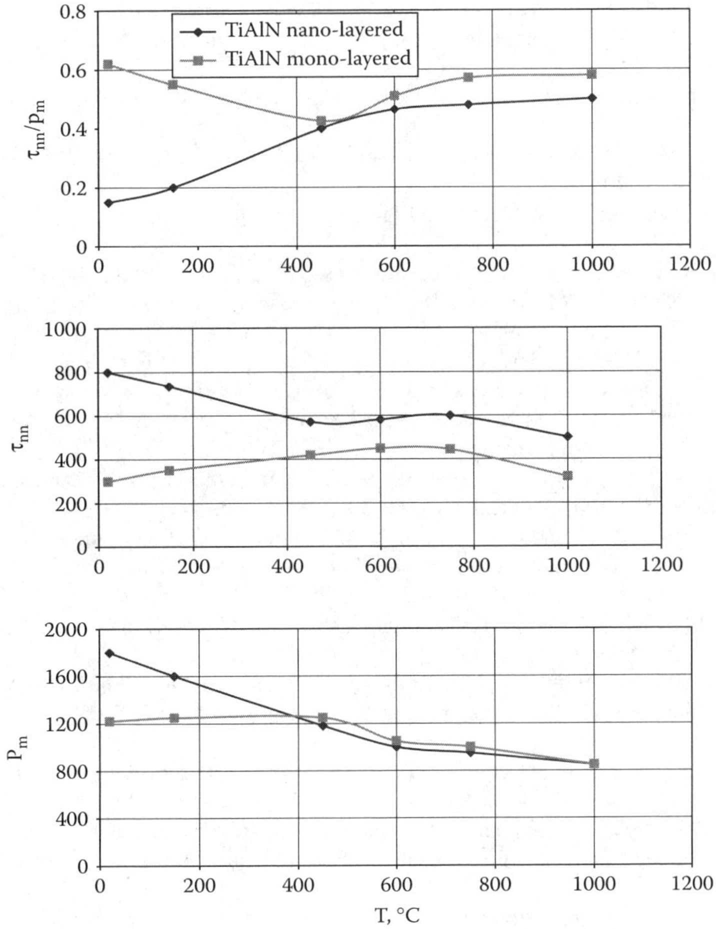

FIGURE 5.20 Coefficient of friction of the Ti–Al–N monolayered and nanolayered coatings vs. temperature in contact with 304 stainless steel.

REFERENCES

1. Neiderhofer, A., Nesladek, P., Mannling, H.-D., Moto, K., Veprek, S., Jilek, M. Structural properties, internal stress and thermal stability of nc-TiN/a-S_3N_4, nc-TiN/$TiSi_x$ and nc-(Ti_{1-y} Al_y Si_x)N superhard nanocomposite coatings reaching the hardness of diamond. *Surf. Coat. Technol.* 1999, 120–121, 173–178.
2. Musil, J., Zeman, H., Hrubý, P., Mayrhofer, H. ZrN/Cu nanocomposite films-novel superhard material. *Surf. Coat. Technol.* 1999, 120–121, 179–183.
3. Tretiakov, I.P., Vereshchaka, A.S. *Cutting Tools with Coating*. Mashinostroenie: Moscow, 1994, 17–297.
4. Samsonov, G.V., Vinnitsky I.M. *Heavy Melting Compounds*. Mashinostroenie: Moscow, 1976, 44–56.
5. Holmberg, K., Matthews, A. *Coating Tribology: Principles, Techniques, and Application in Surface Engineering*. Elsevier Science B.V.: Amsterdam, 1994, 257–309.
6. Trent, E.M., Wright, P.K. *Metal Cutting,* 4th ed. Butterworth: Woburn, MA, 2000, p. 211.
7. Trent, E.M., Suh, N.P. *Tribophysics*. Prentice-Hall: Englewood Cliffs, NJ, 1986, 125–489.
8. Bushe, N.A., Kopitco, V.V. *Tribological Compatibility.* Nauka: Moscow, 1981, p. 128.
9. Bushe, N.A. Tribo-engineering materials. *International Engineering Encyclopedia: Practical Tribology — World Experience,* Vol. 1. Science and Technique Centre: Moscow, 1994, pp. 21–29.
10. Beliy, V.A., Ludema, K., Mishkin, N.K. *Tribology: Studies and Applications — USA and USSR Experience*. Mashinostroenie: Moscow, Allerton Press: New York, 1993, pp. 202–452.
11. Shuster, L.S. *Adhesive Interaction in Metal Solids*. Gilem: Ufa, 1999, pp. 4–196.
12. Shuster, L.S. Tribological compatibility during forming of titanium alloys. *J. Friction Wear* 1993, 18, *5*, 914–921.
13. Kragelsky, I.V., Alisin, V.V. *Friction, Wear and Lubrication*. Mashinostroenie: Moscow, 1978, pp. 160–230.
14. Shuster, L.S. Method of Tribological Compatibility Evaluation. Patent of USSR (Russia) # 1335851 MKI G01 3/56, 1979.
15. Pharr, G.M. Measurement of Mechanical Properties by Ultra-low Load Indentation. *Mater. Sci. Eng.* 1998, A 253, 151–159.
16. Archard, J.F. Contact and rubbing of flat surfaces. *J. Appl. Phys.* 1953, 24, 981.
17. Cheng, Y.-T., Cheng, C.-M. Analysis of indentation loading curves obtained using conical indenters. *Philos. Mag. Lett.* 1998, 77, 39.
18. Fox-Rabinovich, G.S., Kovalev, A.I., Weatherly, G.C. Tribology and the design of surface engineered materials for cutting tool applications. In *Modeling and Simulation for Material Selection and Mechanical Design*, Eds., Totten, G., Xie, L., Funatani K. Marcel Dekker: New York, 2004, 301–382.
19. Orbele, T.L. Properties influencing wear of metals. *J. Met.* 1951, 3, 438.
20. Leyland, A., Matews, A. On the significance of the H/E ratio in wear control: a nanocomposite coating approach to optimized tribological behavior. *Wear* 2000, 246, *1–2*, 1–11.
21. Sato, K., Ichmiya, N., Kondo, A., Tanaka, Y. Microstructure and mechanical properties of cathodic arc ion-plated (Al,Ti)N coatings. *Surf. Coat. Technol.* 2003, 163–164, 135.
22. Halling, J. Surface films in tribology. *Tribologia* 1982, 1, *2*, 15.
23. Brisccoe, B.J., Fiori, L., Pelillo, E. Nano-indentation of polymeric surfaces. *J. Phys. D: Appl. Phys.* 1998, 31, 2395–2405.
24. Andrievskii, R.A., Kalinnikov, G.V., Hellgren, N., Sandstorm, P., Stanskii, D.V. Nanoindentation and strain characteristics on nano-structured boride/nitride films. *Phys. Solid State* 2000, 42, *9*, 1624–1627.
25. Catledge, S.A., Vohra, Y. Structural and mechanical properties of nano-structured metalloceramic coatings on cobalt chrome alloys. *Appl. Phys. Lett.* 2003, 82, *10*, 1625–1627.
26. Catledge, S.A., Vohra, Y., Woodard, S., Venugopalan, R. Structure and mechanical properties of functionally-graded nanostructured metalloceramic coatings. *Mater. Res. Soc. Symp. Proc.* 2003, 781–786.
27. Kostetsky, B.I. Evolution of structure, phase conditions and mechanism of self-organization of materials in external friction. *J. Friction Wear* 1993, 14, *4*, 773.
28. Kabaldin, J.G., Kojevnikov, N.V., Kravchuk, K.V. Study of wear of HSS cutting tools. *J. Friction Wear* 1990, 11, *1*, 130.

29. Cheng, Y.-T., Li, Z., Cheng, C.-M. Scaling relationship for indentation measurements. *Philos. Mag. A* 2002, 82, *10*, 1821–1829.
30. Sakai, M. Energy principle of the indentation-induced inelastic surface deformation and hardness of brittle materials. *Acta Metall. Mater.* 1993, 41, 1751.
31. Cheng, Y.-T., Cheng, C.-M. Relationships between hardness, elastic modulus and the work of indentation. *Int. J. Solid Struct.* 1999, 39, 1231.
32. Haisworth, S.V., Candler, H.W., Page, T.F. Analysis of nanoindentation load-displacement loading curves. *J. Mater. Res.* 1996, 11, 1987.
33. Rother, B. Depth-sensing indentation measurements as mechanical probes for tribological coatings. *Surf. Coat. Technol.* 1996, 86–87, 535.
34. Sjostrom, H., Hultman, L., Sundgren, J.-E., Hainsworth, S.V., Page, T.F., Theunssen, G.S.A.M. Structural and mechanical properties of carbon nitride CN_x (0.2 × 0.35) films. *J. Vac. Sci. Technol. A* 1996, 14, *1*, 56.
35. Page, T.F., Hainsworth, S.V. Using nanoindentation techniques for the characterization of coated systems: a critique. *Surf. Coat. Technol.* 1993, 61, *3*, 201–208.
36. Veprek, S. The search for novel, superhard materials. *J. Vac. Sci. Technol. A* 1999, 17, *5*, 2401–2420.
37. Fox-Rabinovich, G.S., Beake, B.D., Veldhuis, S.C., Endrino, J.L., Parkinson, R., Shuster, L.S. Effect of mechanical properties measured at room and elevated temperatures on wear resistance of cutting tools with TiAlN and AlCrN coatings, *Surf. Coat. Technol.* 2006, 20–21, 5738–5742.
38. Fox-Rabinovich, G.S. Structure of complex coatings. *Wear* 1993, 160, *1*, 67–76.
39. Dosbaeva, G.K. The Development of Multi-Layered TiN PVD Coating for Metal Working Tools Applications. Ph.D. thesis, Technological University of Machine-Tool Engineering, Moscow, 1989.
40. Konyashin, I., Fox-Rabinovich, G.S. Nanograined titanium nitride thin films. *Adv. Mater.* 1998, 10, *12*, 952.
41. Fox-Rabinovich, G.S., Weatherly, G.C., Dodonov, A.I., Kovalev, A.I., Veldhuis, S.C., Shuster, L.S., Dosbaeva, G.K., Wainstein, D.L., Migranov, M.S. Nano-crystalline FAD (filtered arc deposited) TiAlN PVD coatings for high-speed machining application, *Surf. Coat. Technol.* 2004, 177–178, *1*, 800–811.
42. Fox-Rabinovich, G.S. Scientific Principles of Material Selection for Wear-Resistant Cutting/Stamping Tools basing on Surface Structure Optimization, D.Sc. thesis, Scientific Research Institute of Rail Transport, Moscow, Russia, 1983.
43. Haasen, P. *Physical Metallurgy.* Cambridge University Press: London, 1978, p. 521.
44. Tian, J., Han, Z., Yu, X., Li, G., Gu, M. Two-step penetration: a reliable method for the measurement of mechanical properties of hard coatings. *Surf. Coat. Technol.* 2004, 176, *3*, 267–273.
45. Ohmura, T., Matsuoka, S. Evaluation of mechanical properties of ceramic coatings on a metal substrate. *Surf. Coat. Technol.* 2003, 169–170, 728.
46. Lee, K.W., Chung, Y-W., Chan, C.Y., Bello, I., Lee, S.T., Karimi, A., Patscheider, J. An international round-robin experiment to evaluate the consistency of nanoindentation hardness measurements of thin films. *Surf. Coat. Technol.* 2003, 168, *1*, 57.
47. Zhang, S., Sun, D., Fu, Y., Du, H. Effect of sputtering target power on microstructure and mechanical properties of nanocomposite nc-TiN/a-SiN_x thin films. *Thin Solid Films* 2004, 447–448, 462–467.
48. Voevodin, A., O'Neill, J.P., Zabinsky, J. Nanocomposite tribological coatings for aerospace applications. *Surf. Coat. Technol.* 1999, 116–119, 36–45.
49. Ichimura, H., Ishii, Y. Effects of indenter radius on the critical load in scratch testing. *Surf. Coat. Technol.* 2003, 165, *1*, 1–7.
50. Zhang, S., Sun, D., Fu, Y., Du, H. Toughening of hard nano-structural thin films: a critical review. *Surf. Coat. Technol.* 2005, 198, 2–8.
51. Beake, B.D., Smith, J.F. Evaluating the fracture properties and fatigue wear of tetrahedral amorphous carbon films on silicon by nano-impact testing. *Surf. Coat. Technol.* 2004, 177–178, 611–615.
52. Fox-Rabinovich, G.S., Veldhuis, S.C., Scvortsov, V.N., Shuster, L. Sh., Dosbaeva, G.K., Migranov, M.S. Elastic and plastic work of indentation as a characteristic of wear behavior for cutting tools with nitride PVD coatings. *Thin Solid Films* 2004, 469–470, 505–512.
53. Fox-Rabinovich, G.S., Weatherly, G.C., Kovalev, A.I., Korshunov, S.N., Scvortsov, V.N., Veldhuis, S.C., Shuster, L. Sh., Dosbaeva, G.K., Wainstein, D.L., *Surf. Coat. Technol.* 2004, 187, *2–3*, 230–237.
54. Fox-Rabinovich, G.S., Kovalev, A.I., Wainstein, D.L. Smart Multi-layer Coating. German Patent 201 16 404.3, 2002.

Section III

Self-Organization and Structural Adaptation of Heavily Loaded Tribosystems

6 Self-Organization and Structural Adaptation during Cutting and Stamping Operations

German S. Fox-Rabinovich and Anatoliy I. Kovalev

CONTENTS

6.1 Introduction 151
6.2 The Features of Wear Behavior and Self-Organization during Stamping 153
6.3 The Features of the Self-Organizing Process during Cutting 163
6.4 Conclusion 165
References 165

6.1 INTRODUCTION

Metalworking, including metal cutting and stamping operations, is associated with mechanical and thermal processes that involve intensive plastic deformation of the workpiece ahead of the tool tip and severe frictional conditions at the interfaces of the tool, chip, and the workpiece. Most of the work of plastic deformation and friction is converted into heat. During cutting, about 80% of the heat leaves with the chip, but the other 20% remains, increasing the temperature of the tool. The pressure at the nominal contact area during machining is approximately 10^3 N/m^2, i.e., high pressure dominates and extreme local temperatures (up to 1300°C) can be encountered, especially during the high-speed cutting of steels and other materials that are hard to machine [1]. The surface of the tool continuously comes into contact with untouched workpiece material that is unaffected by the environment, e.g., by oxidation. Freshly produced chips and the surface of stamped material interact chemically with the tool material. The contact friction behavior at the tool–workpiece interface has been shown to be adhesion related. The current understanding of this aspect of machining is focused on the friction of a clean (in a physicochemical sense) and oxidized surface of a tool and workpiece through adhesive interaction [2]. At high cutting and stamping speeds (typical of modern manufacturing operations) or stamping at elevated temperatures, the mechanical and physical conditions at the tool–metal interface are far removed from the "classical friction" situation in textbooks, where mechanical interaction of the asperities prevails.

These severe service conditions can lead to intensive surface damage of the cutting and stamping tools. Several mechanisms of tool wear have been identified for machining tools: (1) adhesive wear, (2) cratering (or chemical) wear on the rake face of the cutting tool because of chemical instability, including diffusion and dissolution, (3) abrasive and (4) fatigue wear, most often low-cycle fatigue (Figure 6.1) [1–3]. Some authors have also discussed the electrochemical and delamination wear of HSS cutting tools and, especially, stamping tools. Delamination is related to fatigue wear [4].

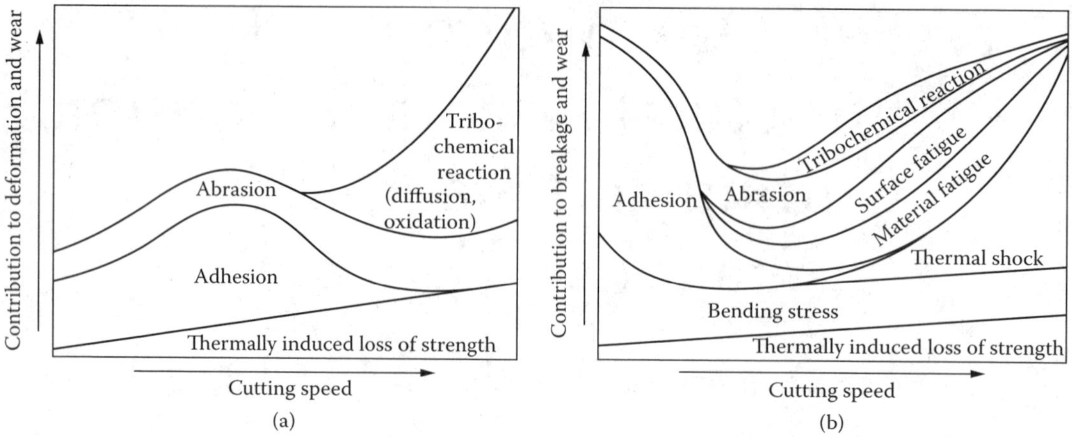

FIGURE 6.1 Failure mechanisms in (a) continuous cutting and (b) interrupted cutting and stamping. (From Loffler, J.H.W. Systematic approach to improve the performance of PVD coatings for tool applications. *Surf. Coat. Technol.* 1994, 68–69, 729–740. With permission.)

Adhesive wear is caused by the formation of welded asperity junctions between the chip or workpiece and the tool face. The subsequent fracture of the junctions by shear leads to microscopic fragments of the tool material being torn out and adhering to the chip or workpiece (see Chapter 3, Figure 3.9). This kind of wear occurs on the flank face in low-speed cutting when the contact temperatures are relatively low, as well as under stamping conditions such as blanking, drawing, and bending [3]. Abrasive wear is caused by hard particles of carbides or oxides in the work material or by highly strain-hardened fragments detached from the unstable built-up edge of the tool. Wear due to chemical instability is very important in high-speed metal cutting and stamping because of the high temperatures prevailing at the contacting surfaces; this includes both diffusion wear and solution wear. *Diffusion wear* is characterized by material loss due to the diffusion of atoms of the tool material into the workpiece. *Solution wear* describes the wear mechanism that takes place when the wear rate is controlled by the dissolution rate rather than by convective transport. An oxidation reaction with the environment can produce a scaling of the cutting edges [1]. As shown in Figure 6.1a, for a given workpiece material, adhesive wear is found mainly at relatively low cutting temperatures, corresponding to low cutting speeds. Wear due to chemical instability, including effects such as diffusion and oxidation, appears at high cutting speeds. Abrasive wear occurs under all cutting conditions. This type of wear is important for low-speed machining of cast iron and steels, but is less important for cutting conditions in which wear due to chemical instability prevails. However, this mechanism of wear is very important for high-speed machining when oxidation of the workpiece surface leads to the formation of oxide scales, which can be detached from the surface of the tool.

It must be noted that during actual machining, the external loads applied affect both surfaces and subsurface layers, leading to deep surface damage and chipping of the cutting edge. This is especially important for interrupted machining processes (Figure 6.1b), e.g., milling and stamping, because cyclic thermal and mechanical stress can lead to crack initiation and propagation [5–6].

For cold-work stamping tools, the wear mechanisms outlined in the preceding text are most common, but for hot-work stamping tools, the tribo-chemical wear mode (high-temperature corrosion) becomes significant [7]. The major cold-work stamping operation is blanking, and blanking dies are most widely used in industry. This is why we will consider the wear behavior of blanking dies in detail. Low-cycle friction fatigue is a primary phenomenon affecting the tool life and the wear behavior of blanking dies. We can consider the characteristic features of D2 cold-work tool steel, which are most widely used for blanking dies manufacturing. The structural transformations within the surface layer and tribo-film formation have been studied in detail in relation to the impact these phenomena have on the wear behavior of blanking dies.

Blanking dies have working surfaces, which can be considered as an alternating load concentrator. Working temperatures are relatively low and usually do not exceed 200°C, but the actual temperatures can be much higher for high-performance stamping conditions, especially for high-speed stamping, which is typical of modern machinery [3–11]. The magnitude and area of load distribution over the top and flank surfaces of a die are different. Top surface loads are high and not evenly distributed. Maximal loads act along the edge and decrease farther away. Workpiece–die contact does not occur over the total face surface of the die but over a narrow ring zone next to the edge. The contact zone width is small and comparable to the thickness of a blanked workpiece. Adhesive interaction with a workpiece at the top surface is minimal because of the formation of oxides and lubricious films covering the surface of the sheet to be blanked [9]. The actual loads on the flank surface are much lower, though. However, owing to reverse punctuated contact with a juvenile workpiece surface being cut off, intensive seizure occurs, resulting in build-up formations (Figure 6.2) [9,11]. For the majority of the workpiece materials to be blanked, the flank surface wear rate is higher compared to that of the top surface [10,11]. To enhance the stamping tools' life, a clear understanding of the wear phenomenon is necessary. To realize this goal, we will investigate the wear behavior of blanking dies in detail and outline the principles of friction control based on the studies performed [12].

6.2 THE FEATURES OF WEAR BEHAVIOR AND SELF-ORGANIZATION DURING STAMPING

The investigation of blanking die wear behavior was carried out with punches used for the blanking of the electric motors components. Blanks were punched out of a magnetic steel sheet containing 2.8% silicon [3]. The thickness of the sheet was 0.5 mm. The wear values were measured based on the clearance between punches and die changes during the operation process.

It is known that D2-type, cold-worked stamping-tool steels, containing 12% chromium, are most commonly used for blanking dies. The composition of these steel types is given in Table 6.1. The structure and properties of these steels vary within a wide range, depending on the parameters of the heat treatment (Table 6.2).

We characterized the die surface using various metallography methods. In doing so, we studied integral structural features within the layer undergoing plastic deformation during operation. The thickness of this surface layer is around 10 μm. Together with the integral features, we studied the structural characteristics distribution over the width of the transformed surface layer. X-ray diffraction (XRD) analysis was performed using the Roentgen diffractometer DRON 3.0 with Co K_α emission. The parameters to be studied are as follows: the amount of retained austenite determined from the austenite to martensite line intensity ratio, the carbon concentration in martensite determined by the method of the third moment of the line $(200)_\alpha$, the martensite line width $(200)_\alpha$, and the residual stresses by Roentgen strain measurements (sin ψ^2 method [11,12]). To assess the distribution of the structural characteristics within the depth of the superficial surface layer, a Roentgen sliding-beam method [13] was used. Changes in the chemical composition of the surface contact layer of the punches during friction have been studied by Auger spectroscopy of angle lap cross sections [14] by means of an ESCALAB electron spectrometer manufactured by the VG Company, U.K.

Features distinguishing atomic crystalline structures, as well as electron structures of the punch surface at different stages of wear, were studied by means of electron loss spectroscopy (EELS). Specifically, two modifications of the method were used: extended energy-loss fine-structure spectroscopy (EELFS) and plasmon-loss spectroscopy. All the details of the methods applied are described in detail in Chapter 4.

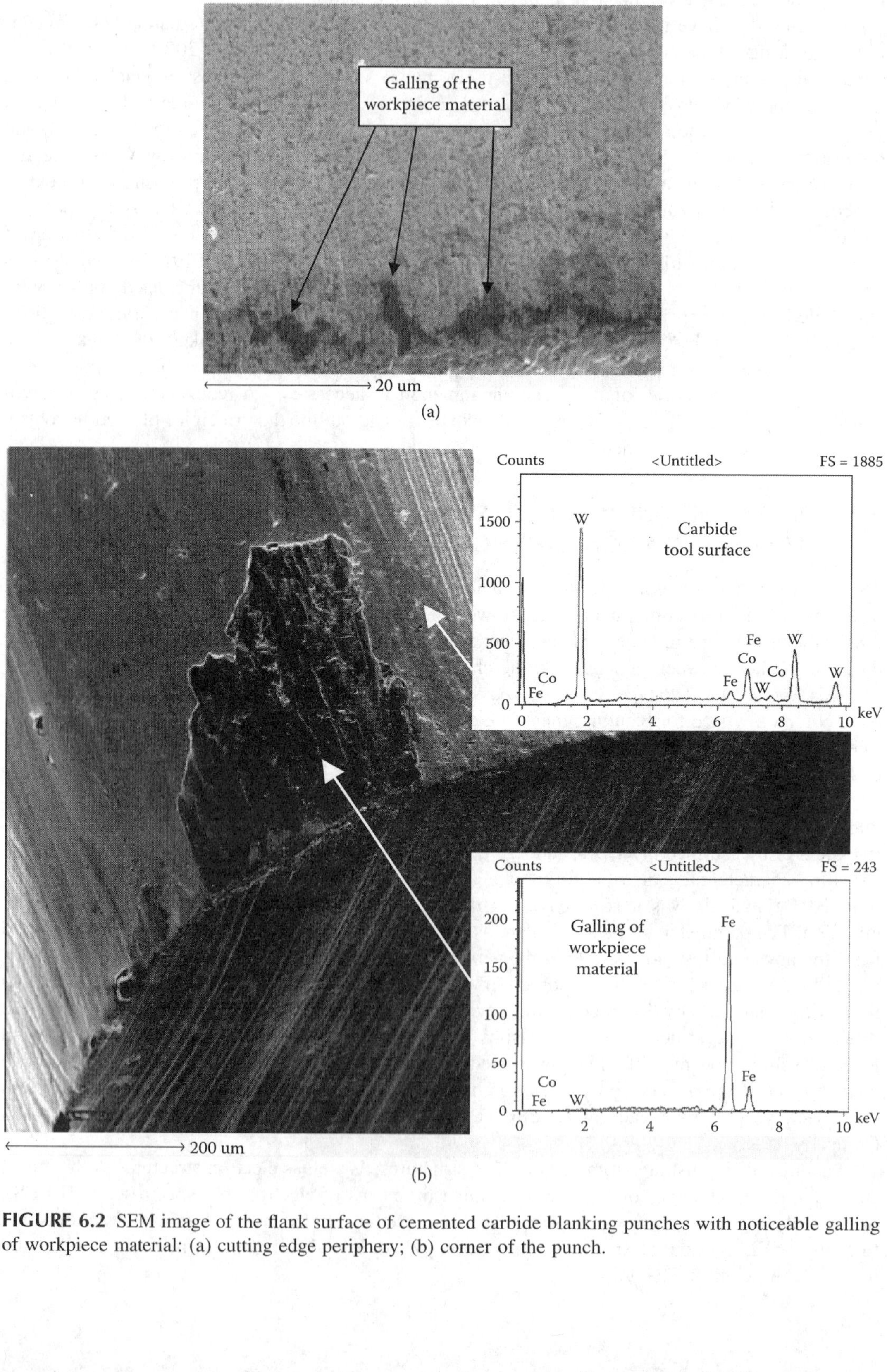

FIGURE 6.2 SEM image of the flank surface of cemented carbide blanking punches with noticeable galling of workpiece material: (a) cutting edge periphery; (b) corner of the punch.

TABLE 6.1
Composition of Die Steels with 12% Cr

Steel Grade	C	Si	Cr	Mo	V
D3	2.0–2.2	0.15–0.75	11.5–13.0		
D2	1.45–1.65	0.15–0.75	11.0–12.5	0.4–0.6	0.15–0.3

TABLE 6.2
Structure of D2 Steel after Heat Treatment

	Structure									
	Carbide Phase						Retained Austenite			
				Size (μm)						
Steel Grade	Content (Vol%)	Type	Hardness (GPa)	Eutectic	Secondary	Martensite Hardness	Hardness (GPa)	Content (vol%)	Grain Size, (μm)
D2	15–17	M_7C_3	16.0	5–10	0.5–1.0	7.5–8.0	3.0–3.5	Up to 20	12–17

In the study, we investigated the change in structure and properties of the stamping-tool surface layers vs. the number of strokes. The initial structure and properties of the die surface are critically important to the wear process. It is known that after grinding the surface of blanking dies made of D2 steel usually forms a secondary hardened layer approximately 7- to 10-μm deep. The structure of such a layer features nontempered martensite and some amount of retained austenite (up to 25%) [14]. In addition, pronounced residual tensile stresses up to 300 to 400 MPa are induced in this layer (Table 6.3).

It was found that tool wear follows the fatigue mechanism [3]. Adhesion of the workpiece material to the flank surface of a die intensifies the wear intensity. Fatigue and adhesion wear types are accompanied by the abrasive wear caused by the carbide particles. These carbide particles are torn off during friction as a result of relatively intensive wear of the surrounding martensite matrix of D2 steel [14]. The stamping-tool life is directly related to the duration of the specific stages of wear such as the running-in phase, stable stage of wear, intensive wear stage, as well as to the surface damage intensity during friction (Figure 6.3). Figure 6.3 exhibits the stamping-tool wear curve (relative die wear, ΔL) vs. the number of strokes. It can be seen that the most intensive surface damage occurs during the initial running-in stage of tool wear that corresponds to the ideas outlined in Chapter 2. This is caused not only by intensive surface plastic deformation but also by an unfavorable initial microstructure and stress distribution as a result of secondary hardening during grinding of the stamping tool (Table 6.3) [14].

Structural transformations occur during the initial stages of the die operation within the contact zone. Two main processes during friction, i.e., plastic deformation and thermal heating, competitively affect the surface structure and properties of the stamping tool. It was found that the phenomena associated with plastic deformation, which intensify the $\gamma \rightarrow \alpha$ transformation, prevail during the running-in stage. For example, an amount of retained austenite within the surface layer decreases from 17 to 10% (Table 6.3). The retained austenite transforms to martensite as a result of the plastic deformation, but the hardness and amount of carbon in the martensite within the surface layer remain practically unchanged (see Table 6.3). XRD structural analysis performed using the sliding-beam method has shown that the most intensive $\gamma \rightarrow \alpha$ transformation is found within the surface, in an approximately 2-μm deep layer of the die (Figure 6.4). The amount of retained austenite decreases to 4.0 % within this layer. As the distance from the surface increases, the intensity of the $\gamma \rightarrow \alpha$ transformation decreases at a depth of 7 μm, and the amount of retained

TABLE 6.3
Structure and Properties of the Friction Surface of Blanking Punches made of D2 Steel after Heat-Treatment, Grinding, and Service

	Structure Characteristics			Properties			
Flank Surface of the Die after the Process of:	Amount of Carbon in the Martensite (%)	Martensite $(200)_\alpha$ Line Width $B\times10^{-3}$ Radian	Retained Austenite (%)	Microhardness $H_{0.5}$ (GPa)	Roughness (Rz) μm	Residual Stress (MPa)	The Depth of the Area Under Analysis (μm)
Heat treatment	0.55	32.9	22.0	7.2		+300	10
						+300	300
Grinding and lapping	0.53	26.6	17.0	7.2	0.16	+400	
Service, number of strokes							
20,000			10.0		0.6	+480	
30,000	0.56		10.0	7.0	0.6	+460	
70,000	0.47		10.0	6.2	1.0	+870	10
						+440	300

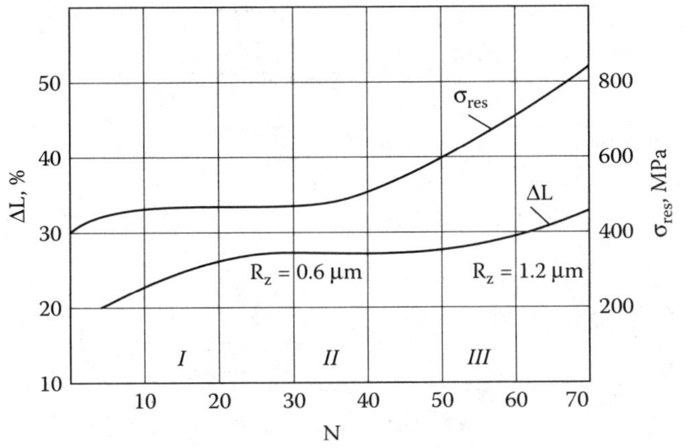

FIGURE 6.3 Wear curve (relative stamping-tool wear, ΔL) and residual stress vs. number of strokes. I — running-in stage; II — post running-in stage of stable wear; III — intensive wear. (From Fox-Rabinovich, G.S., Kovalev, A.I. Characteristic features of blanking die wear with consideration for the change in composition, structure and property of contact surfaces. *Wear* 1995, 189, 25–31. With permission.)

austenite is much higher around 10 to 12% (Figure 6.4a). This result demonstrates that the depth of the zone of plastic deformation during friction is around 7 to 10 μm, having a higher intensity of deformation on the surface and decreases further with depth. It is worth mentioning that the depth of the worn layer (around 10 to 15 μm) is practically equal to the depth of the zone of surface plastic deformation.

In the die operation process, secondary tensile residual stresses develop on the surface. Their values are seen to be associated with the intensive $\gamma \rightarrow \alpha$ transformation (Table 6.3).

When the running-in stage is over, the wear process achieves a steady-state phase. In accordance with current knowledge (see Chapter 1 and Chapter 2), this is caused by the structural adaptation of the "tool–workpiece" tribosystem. As a result of the process of self-organization and structural adaptation, the tribo-films formed during friction are aimed at reducing and even preventing surface

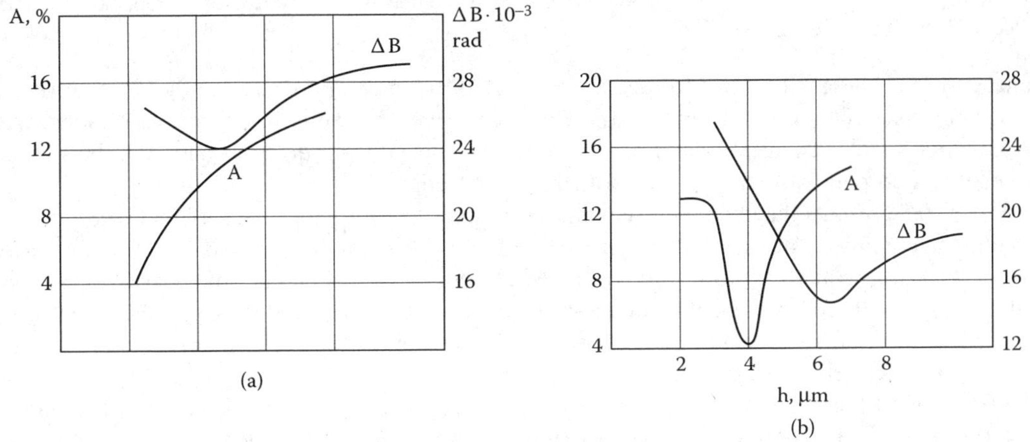

FIGURE 6.4 Structural changes on the surface of punches after: (a) 30,000 strokes, (b) above 70,000 strokes. X-ray Roentgen sliding-beam method. (From Fox-Rabinovich, G.S., Kovalev, A.I. Characteristic features of blanking die wear with consideration for the change in composition, structure and property of contact surfaces. *Wear* 1995, 189, 25–31. With permission.)

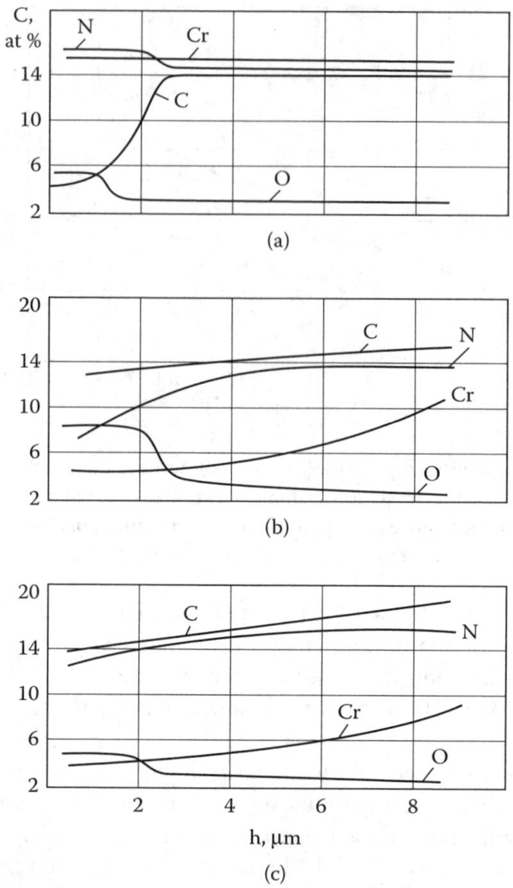

FIGURE 6.5 Chemical composition of the surface of punches made on D2 steel after: (a) initial stage; (b) 30,000 strokes; (c) above 70,000 strokes. (From Fox-Rabinovich, G.S., Kovalev, A.I. Characteristic features of blanking die wear with consideration for the change in composition, structure and property of contact surfaces. *Wear* 1995, 189, 25–31. With permission.)

damage [4,15]. These tribo-films formed are a result of the interaction between the surface friction and the environment (oxygen from the air). The development of these thin films is associated with the oxidation of the base metal that is accompanied by intensive plastic deformation of the surface. Eventually, amorphous-like layers are formed (of the Beilby layers type) [15–16].

Figure 6.5 presents the chemical elements' depth profile within the tool surface layer. The data was obtained by means of scanning Auger spectroscopy of the angle laps cross section. The results indicate that thin oxide tribo-films are developed at both the initial and steady wear stages. It was found that, after 30,000 strokes, the thickness of the oxide tribo-film and the oxygen concentration on the surface are about 2.5 times higher compared to the initial stage. Considering the low oxygen solubility in steels, these differences are substantial.

Oxidation of the tool surface occurs during the steady-state wear stage because of heat generation at stamping. This leads to the redistribution of the main alloying elements of D2 steel as a result of the diffusion phenomena. Growth in the thickness of the iron oxide film results in the surface being depleted of chromium within a depth of 8 to 10 μm at the tool surface layer. At the same time, the nitrogen and carbon distribution along the layer of this depth becomes even more uniform than that initially after grinding (Figure 6.5a and Figure 6.5b). Redistribution of the chemical elements within the surface layer at later stages of the wear (after 70,000 strokes) is

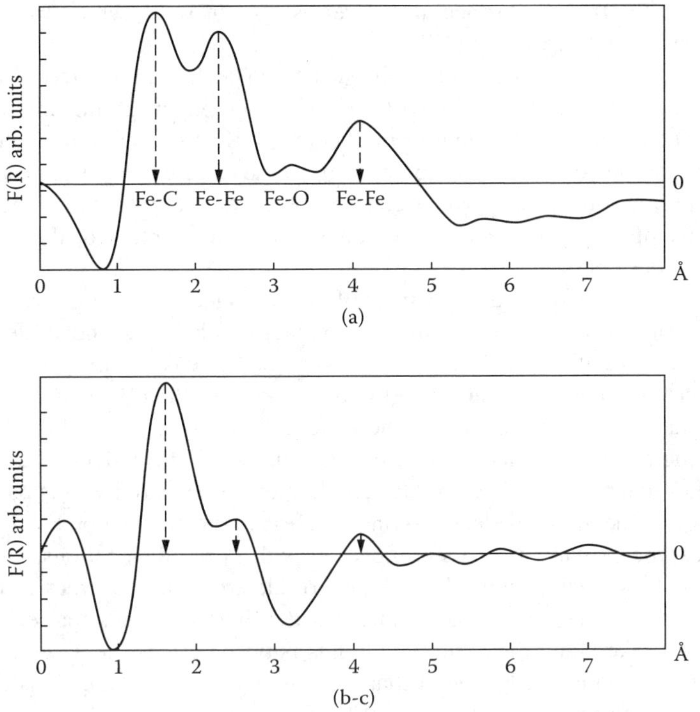

FIGURE 6.6 Fourier transforms of the EELFFS close to the line of elastic scattered electrons for the superficial layers forming on the surface of the stamping tools: (a) initial stage; (b) after 30,000 strokes; (c) after 70,000 strokes. (From Fox-Rabinovich, G.S., Kovalev, A.I. Characteristic features of blanking die wear with consideration for the change in composition, structure and property of contact surfaces. *Wear* 1995, 189, 25–31. With permission.)

associated with the destruction of the iron oxide tribo-film and the damage of the unprotected surface of the substrate material.

Changes in the atomic crystalline structure of the surface were observed during different stages of the stamping tool's wear using the EELFS method (See Chapter 4). Figure 6.6 presents the Fourier transforms of the nearest atomic surroundings obtained after processing the spectra of back-scattered electrons from the surface of the punches at the initial stage and after their service. A distance from the origin of coordinates to the maximum points corresponds to the length of interatomic bonds in the nearest coordination spheres. A detailed analysis of this picture based on the analysis of the fine structure of spectra close to the KLL C line made it possible to identify the separate peaks on Figure 6.6a. Pronounced peaks at interatomic distances of 1.5, 2.4, and 4.3 Å (Figure 6.6a) can be observed on the Fourier transforms $F(R)$ at the initial state. This is an indication that the presence of order in the atoms' arrangement is in at least three coordination spheres surrounding each atom of iron. Determined values of the interatomic distances are in agreement with crystallographic characteristics of α-Fe. On the worn surface, there is an appreciable attenuation of the maximum intensity on the Fourier transform $F(R)$ at interatomic distances around 2.4 and 4.3 Å. This indicates on the amorphization of the surface structure layer and disappearance of order in atoms' arrangement in the second and third coordination spheres. The amorphization of the surface structure can be caused by intensive accumulation of dislocations, as well as the other crystalline structure defects within the surface layer during wear.

As the number of die strokes increases, a change in the surface energy occurs, which is associated with accumulating defects within the thin surface layer. This is proved by studies of the electron structure in the friction surface at different stages of wear, which was made using high-resolution plasmon-loss spectroscopy [17].

Figure 6.7a and Figure 6.7b present the fine structure spectra in the proximity of backscattered electron lines obtained from the surface of the blanking punch at the initial stage and after subsequent stages of wear. An excitation beam of primary electrons with monochromatized energy approximately 16 eV and semiwidth of the order of 30 meV were used. The pronounced peaks, P_1 and P_2, are volume plasmon peaks, which can be observed in the spectra (Figure 6.7a to Figure 6.7c). The intensity of the peaks decreases gradually with the number of the strokes. The details of the method are considered in Chapter 4.

The plasmon peak intensity in the spectrum of primary electrons energy loss is associated with the excitation of collective oscillations among conductive electrons and is determined by their concentration. After normalizing an integral intensity for P_2 peaks to the respective intensities of back-scattered electron lines, we could determine a change in the effective density of conductive electrons on the surface of the punches at the different stages of wear (Table 6.4). Initially, the concentration of the conduction electrons within the punch's surface layers was n_{eff} = 100%, then, after 30,000 strokes, the relative n_{eff} was 60.8%, and, after 70,000 strokes, this parameter decreased to 31.4%. A significant decrease in electron concentration in the conduction band of the tool surface being worn out can be explained by a significant growth in the dislocations density and electron polarization around these dislocations. This change of electron surface structure causes substantial changes in the atomic interactions. This means that the metallic bonds that existed on the initial surface of the tool weaken, and the ionic-type bonds between atoms increases with the number of strokes of the blanking punches. It is vital that the outcropping of dislocations causes the surface energy to grow, resulting in surface damage [18].

The aforementioned transformation of electron structure suggests that at the end of the stable stage of wear the tribo-films tend to get destroyed by detaching from the surface and aggregate in the form of spheres 1 to 3 mm in size [19,20]. This phenomenon is seen in a number of tribosystems [17,18]. Similar debris was observed at the friction surface of blanking dies using the SEM technique [4].

As mentioned earlier, dislocations outcropping on the friction surface as a result of plastic deformation (10-μm deep) near the surface layer is the main cause for the intensive surface damage to the blanking tools during service. This is a typical low-cycle fatigue wear mode [4,21]. As soon as the layer of the protective tribo-film is destroyed, the friction process starts to undergo substantial and unfavorable changes. Surface roughness increases appreciably from 0.6 μm to 1.2 μm (Table 6.3). Thermal phenomena are enhanced on the surface, and a secondary hardened zone is formed. This is illustrated by the Roentgen structural analysis data obtained using the sliding-beam method (Figure 6.4b). As we can see in the Figure 6.5b, the layers nearest to the surface (3.5-μm deep) contain a significant amount of retained austenite (around 13%). At the same time, we found a considerable widening of the martensite line up to 26×10^{-3} radian. This indicates that the surface structure contains a mixture of nontempered martensite and retained austenite that is typical of the secondary hardened layers [2]. The amount of retained austenite and the widening of the martensite line decreases significantly in the subsurface layers, which indicates the tempering that corresponds to the heating temperatures, approximately 723 K. This conclusion is based on the systematic investigations of the structure vs. regimes of D2 steel heat treatments [4]. The microstructural changes shown earlier correspond to the reduction in the microhardness of the tool surface (Table 6.3). At the same time, surface layers are depleted of chromium (Figure 6.6). This indicates the intensification of the diffusion processes on the surface owing to increased heat generation during the catastrophic stages of wear.

The residual stress values and sign confirm the results of the microstructure studies and the microhardness measurements. Tensile residual stresses are induced on the surface, and their values gradually approach 875 MPa (Table 6.3). It was found that both the wear curve of the blanking punches and the curve of residual stress vs. the number of strokes look similar (Figure 6.3). One can assume that the residual stress is a property that controls the wear resistance of cold-worked stamping tools.

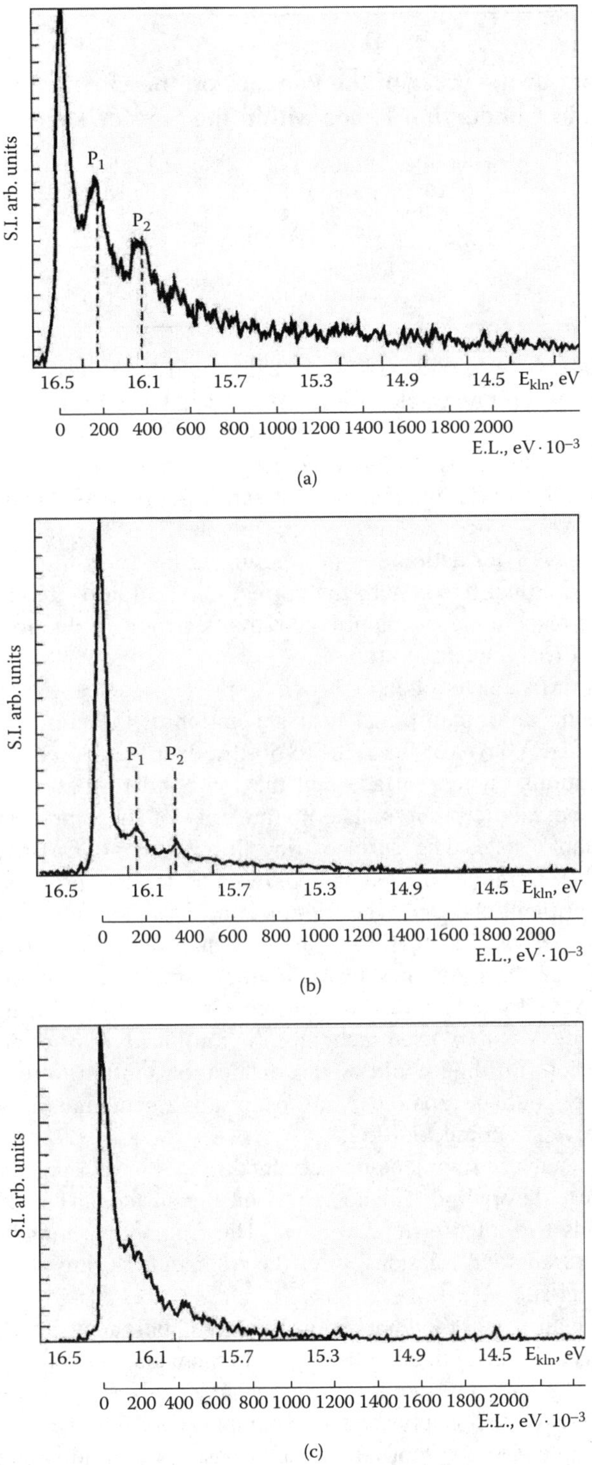

FIGURE 6.7 Spectrum of electron energy losses for the punches made of D2 steel: (a) initial state; (b) after 30,000 strokes; (c) after 70,000 strokes. (From Fox-Rabinovich, G.S., Kovalev, A.I. Characteristic features of blanking die wear with consideration for the change in composition, structure and property of contact surfaces. *Wear* 1995, 189, 25–31. With permission.)

TABLE 6.4
The Impact of the Wear of the Punches on the Effective Number of Electrons in Conduction Bands within the Surface Layer

Sample	Plasmon Integral Intensity P_2 ($\times 10^{-3}$ imp sec^{-1})	Normalization Intensity (relative units)	N_{eff} (%)
Initial stage	300	0.337	100
30,000 strokes	112	0.205	60.8
70,000 strokes	77	0.105	31.4

Thus, the heat resistance of cold-work tool steel should be no less than 723 to 772 K during blanking. Simultaneously, it is preferable to develop the residual compressive stress on the stamping-tool surface.

It is known that the tribo-films' formation during the stable stage of wear is associated with the process of screening [15]. The phenomenon of screening reflects the coordination of the rates of destructive and recovery processes in the friction zone. This is typical of a pseudo-stationary state where the processes of formation and destruction of the tribo-films are in dynamic balance on the surface of the frictional body. With the correct correlation of these processes; the state of the system is stable, corresponding to a minimum rate of entropy production. We can conclude that this is a pseudo-linear thermodynamic process.

The role of the subsurface layer, though, is of extreme importance. Intensive thermomechanical processes such as plastic deformation and heat generation at the contact area occur during the service of stamping tools. A layer at least 30- to 50-μm deep is involved in these processes. The outcropping of dislocations on the surface and the ion-bond formation indicates the nonlinear character of the thermodynamic process. The nonlinearity of these processes obviously deviates the system from a stable state. The surface tribo-films undergo fragmentation and lose their protective ability [19,20]. To reduce the intensity of these destructive processes, surface engineering (surface hardening) is critically needed. The friction control of stamping tools by means of surface engineering could be based on the creation of the structure of the surface layer that enhances the linear processes and prevents or postpones the accumulation of defects within the subsequent surface damage. This can be done by means of different techniques of surface engineering described in Chapter 9. One of the most widely used techniques is ion nitriding or duplex coatings deposition [4]. It was shown that ion nitriding could hinder dislocation motion to the surface [22]. Even an initially damaged surface could be repaired by means of subsequent ion nitriding [22]. This approach could be used to improve stamping-tool life.

The phenomenon of surface fragmentation occurs during the wear process [18]. The fragments' boundaries prevent the outcropping of dislocations on the surface. The stabilization of the fragments' boundaries leads to tool life improvement. The fragments' boundaries stabilization, i.e., enhancement of close orientation relationship of the fragments during wear, can also be done by means of ion nitriding [18].

As mentioned, the major wear mode of stamping tools is fatigue wear. That is why it is obvious that a similar approach be recognized for fatigue life enhancements [22]. The specimen was subjected to fatigue loading, then nitrided, and finally subjected to fatigue testing again. This leads to fatigue life improvement. More sophisticated techniques based on the understanding of the wear behavior are also available. One type of technique is based on the idea of friction control with a positive feedback loop. In this case, the surface engineering is aimed at preventing or postponing the transformation of the stable wear stage to the catastrophic one. This can be done by bringing the tribosystem back to the pseudo-linear thermodynamic state, particularly by the formation of a surface structure (in particular, by means of ion mixing) [23] that accelerates the generation of protective tribo-films and elongates the stable wear stage (see Chapter 9).

The approach presented is related to the ideas discussed in Chapter 1. It is associated with in-depth investigations of wear behavior of specific tribosystems, particularly stamping tools, and identification of the methods that promote beneficial processes and the inhibition of processes that lead to surface damage and wear. We are convinced that this approach to friction control is the best way to achieve the highest efficiency of surface engineering.

The trend in stamping in the last decade is high performance machining. This means that high-speed stamping has started to dominate, and the thermal impact is also becoming more significant. More heat-resistant tooling materials should be used for this specific domain of application, such as heat-resistant powder tool steels and cemented carbides.

6.3 THE FEATURES OF THE SELF-ORGANIZING PROCESS DURING CUTTING

The features of the self-organization phenomenon during cutting are quite different. In contrast to the cold-work stamping tools, in which the majority of the tools are made from tool steels, cutting tools are mainly produced from hard metals such as cemented carbides. To understand the features of the self-organizing phenomenon during cutting, one should understand the processes occurring at the surface of the "cutting tool–workpiece" tribosystem over a range of cutting speeds [25,26]). Studies of the cutting of structural steel (~1040) have shown that tool life can vary dramatically with speed (Figure 6.8). At low speeds of cutting (up to 50 m/min), lying in the domain of cutting with HSS tools, intensive build-up formation takes place. The transformation of the machined material at the tool surface is frequently observed during metalworking and is common in structural steels. At the same time, oxygen (from the air) penetrates into the cutting zone. Fe-containing chip fragments will react with oxygen and carbon at the tool surface and form a boundary layer of both iron carbides and oxides. This leads to a built-up edge. The formation of a built-up layer is considered to be the result of self-organization of the tool–workpiece tribosystem and dissipative structure formation at low cutting speeds. The built-up layer is a dissipative structure or composite "third body," which consists of heavily deformed and refined machining material, as well as oxides, nitrides, and other compounds generated during cutting. The built-up layer is similar in many ways to a composite material. The "ceramic-like" built-up layer offers significant protection to the tool surface. However, the stability of a built-up layer as a dissipative structure is very low, especially when cemented carbide tools are used under attrition wear conditions. For example, the adhesive interaction of tungsten carbide grains with a machined part can lead to microcrack formations. These cracks are generated at the interface of the phases because of the cyclical stress action at the points of adhesion with the workpiece, and this lead to separation of carbide grains from the tool surface. The principal mechanism of wear (and hence the dissipation of energy) of cemented carbide tool surface layers is the formation of microcracks and breaking off of the carbide grains. The failure of the built-up layer results in significant tool surface damage, as well as cutting edge breakage and chipping (Figure 6.1b) [5–6]. The formation of a built-up layer demonstrates that the formation of a dissipative structure with low stability can also lead to increased wear if the friction surface does not ensure stable wear behavior of the dissipative structure formed. This is typical for cemented carbide cutting tools working under attrition wear conditions.

At cutting speeds higher than 50 m/min (used without a coolant), the process of seizure intensifies. On the surface of the tool face close to the cutting edge, a zone of plastic contact with a high friction coefficient will be formed [26]. It results in the formation of a thin layer of heavily worked material at the tool–chip interface. The wear rate of the carbide tool is reduced with an increase in the cutting speed of up to 50 to 80 m/min. At optimum cutting speeds for carbide tools (50 to 80 m/min), the contact processes at the tool surface become nearly constant, whereas the formation of the built-up layer decreases and a flow zone forms [27]. The formation of the flow zone with an increase in the cutting speed is the outcome of self-organization of the tribosystem, leading to a stabilization of friction. In this situation, the tungsten carbide grains undergo

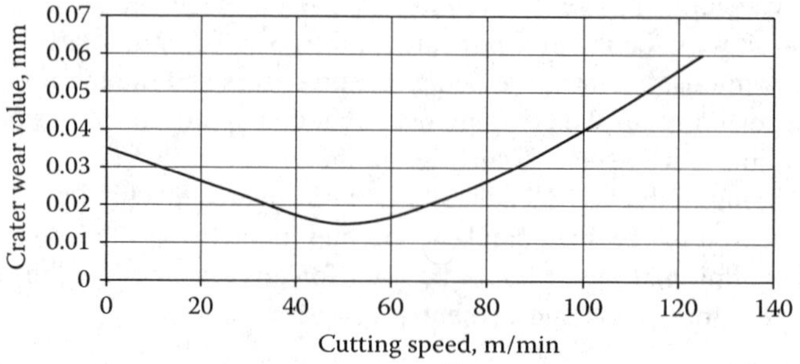

FIGURE 6.8 Wear intensity during turning of 1040 steel using CC cutting tool.

considerable fragmentation. This reduces the wear rate because of a decrease in the volume of the spalled fragments (Figure 6.8).

With a further increase in cutting speeds (more than 100 to 150 m/min), the wear rate again increases owing to an intensification of the diffusion processes and the separation of carbide grains from the tool (Figure 6.8). These high cutting speeds lie in the domain of application of cermets and ceramic cutting tools, as well as surface-engineered cutting tools.

One of the principal features associated with cutting is a rapid increase in the dislocation density near the surface with the deformation being localized in a thin layer. Under extreme cutting conditions, this layer can undergo dynamic recrystallization, leading to a very fine-grained structure. This structure appears to be very stable during wear because it has the ability to both effectively accumulate and dissipate energy [25–26]. The increase in dislocation density in this local volume is accompanied by activation of the surface layers of the tool, leading to further interactions with the environment. This can result in the formation of passivating surface structures (i.e., solid solutions of oxygen in the metal or tribo-oxides), which can control any damaging adhesive interactions at the cutting tool–workpiece interface. Fragmentation of the carbides (or metal matrix) followed by subsequent formation of energy-absorbing, oxygen-containing surface film (at optimal cutting speeds) can also be considered to be an example of a cutting tool–workpiece self-organizing phenomenon, which occurs during the running-in stage of wear.

Table 6.5 displays the values of the limiting energy accumulation capacity for refractory compounds that are widely used for cermets and cemented carbides. These compounds can also be formed on the friction surface as tribo-films.

TABLE 6.5
Limiting Energy Accumulation Capacity for Refractory Compounds

Compound/Material	Limiting Energy Accumulation Capacity (kJ/mol)		Compound/Material	Limiting Energy Accumulation Capacity (kJ/mol)	
	$\Delta E_{773\ K}$	$\Delta E_{1273\ K}$		$\Delta E_{773\ K}$	$\Delta E_{1273\ K}$
HSS T1	77.1		Fe_2O_3	130.1	
HSS M2	68.2		ZrO_2	180.2	
Mo_2C	112.1		TiO	82	52
WC	123	107	TiO_2	103	66
TiC	155	129	Ti_2O_3	247	173
TiN	140.2		WO_3	115	54
ZrN	143.3		MoO_3	31	
HfN	149.6		CoO	79	51

The oxygen-containing tribo-films that are formed on the surface during friction have a high-energy accumulation capacity, resulting in tool life improvement. Under low-cutting-speed conditions, oxygen-containing tribo-films such as Fe–O, Mo–O, and W–O are formed for the cutting tools made of HSS, whereas W–O, Co–O, and Ti–O tribo-films are formed for cutting tools made of cemented carbides. Under optimal cutting conditions for cemented carbide tools, high-oxygen-content compounds such as WO_3, TiO_2, and Ti_2O_3 can occur. The most energy-accumulating compound is Ti_2O_3 (Table 6.5). The elevated values of the energy accumulation capacity of Ti_2O_3 oxides is possibly one of the reasons for the improved tool life of TiC-containing carbides compared to WC ones [28–29].

6.4 CONCLUSION

The self-organization process during cutting and stamping can possibly be used in the proper selection of tooling materials, as well as for the development of surface-engineering techniques, as shown in subsequent chapters (Chapter 7 and Chapter 9). More attention will be given to specific tooling materials with enhanced adaptation ability. These adaptive materials constitute a promising generation of novel materials. The fabrication of these materials can be accomplished through surface engineering and powder metallurgy treatments (see Chapter 7 and Chapter 9).

REFERENCES

1. Van der Heide, E., Schipper, D.J. Tribology in metal forming. In *Mechanical Tribology: Materials, Characterization, and Applications.* Eds., Liang, H., Totten, G. Marcel Dekker: New York, p. 307.
2. Geller, Yu. A. *Tool Steels.* Progress: Moscow, 1986, p. 427.
3. Poznak, L.A., Skrinchenko I.A., Tishaev, S.N. *Die Steels.* Metallurgia: Moscow, 1980.
4. Fox-Rabinovich, G.S., Scientific Principles of Material Selection for Wear-Resistant Cutting/Stamping Tools basing on Surface Structure Optimization. D.Sc. thesis, Russian Scientific Research Institute of Rail Transport: Moscow, 1993, p. 475.
5. Viereggs, G., *Zerspaunung der Eisenwekstoffe.* Stahleisen: Dusseldorf, 1970.
6. Kramer, G., Fortschritt-Berichte, V. *Arc PVD Beschichtung von Hartmetallen fur den unterbrochenen Schnitt.* VDI Verlag: Dusseldorf, 1993.
7. Michailenko, F.P. *Blanking Dies Wear Resistance.* Mashinostroenie: Moscow, 1986.
8. Maeda, T., Isamu, A. Electron microscopic observation on worn surface of blanking tools: wear resistance of die tools. *Proc. Int. Conf. on Machining*, Bratislava, 1974, 378–383.
9. Olsson, D.D., Bay, N. Analysis of pitch-up development in punching, CIRP annals 2002 manufacturing technology. *Annals of the International Institution for Production Engineering Research*, Vol. 51, *1*, 185–191. Colibri Ltd.: Uetendorf, Switzerland, 2002.
10. Fox-Rabinovich, G.S., Kovalev, A.I. Characteristic features of blanking die wear with consideration for the change in composition, structure and property of contact surfaces. *Wear* 1995, 189, 25–31.
11. Umansky, Ya.S. *Crystallography Roentgenography and Electron Microscopy.* Metallurgia: Moscow, 1982.
12. Rusakov, A.A. *The Metal Roentgenography.* Atomizdat: Moscow, 1977.
13. Ribakova, L.M., Kucsenova, L.I., Basov, S.V. Roentgenography method of this surface layers structure changes study. *Zavodskaya Laboratory* 1973, 3, 293.
14. Kovalev, A.I., Scherbedinsky, G.V. *Modern Method of Metals and Alloys Surface Analysis.* Metallurgia: Moscow, 1989.
15. Kostetsky, B.I. Structural-energetic adaptation of materials in friction. *J. Friction Wear* 1985, 6, *2*, 201.
16. Bergmann, E., Vogel, J., Simmen, L. Failure mode analysis of coated tools. *Thin Solid Films* 1987, 153, 219–231.
17. Yosizawa, N., Yomada, Y., Shrishi, M. Structure of amorphous hydrogenated carbon films prepared from plasma deposition. *Carbon* 1993, 31, 1019–1055.

18. Kostetskaya, N.B. Mechanism of deformation, failure and debris forming in mechanical and chemical friction. *J. Friction Wear* 1990, 11, *1*, 108–154.
19. Kabaldin, Y.G., Kojevnikov, N.V., Kravchuk, K.V. HSS cutting tool wear resistance study. *J. Friction Wear* 1990, 11, *1*, 130–135.
20. Scott, D., Mills, G.H. Spherical debris — its occurrence, formation and significance in rolling contact fatigue. *Wear* 1973, 24, *2*, 235–242.
21. Suh, N.P. New theories of wear and their application for tool material, *Wear* 1980, 62, 1–20.
22. Alsaran, A., Kaymaz, I., Celik, A., Yetim, F., Karakan, M. A repair process for fatigue damage using plasma nitriding. *Surf. Coat. Technol.* 2004, 186, *3*, 333–338.
23. Fox-Rabinovich, G.S., Veldhuis, S.C., Kovalev, A.I., Korshunov, S.N., Shuster, L.Sh., Weatherly, G.C., Dosbaeva, G.K., Scvortsov, V.N. Improvement of surface engineered coatings for HSS cutting tools by ion mixing. *Surf. Coat. Technol.* 2004, 187, *2–3*, 230–237.
24. Loffler, J.H.W. Systematic approach to improve the performance of PVD coatings for tool applications. *Surf. Coat. Technol.* 1994, 68–69, 729–740.
25. Kabaldin, Y.G., Kojevnikov, N.V., Kravchuk, K.V. HSS cutting tool wear resistance study. *J. Friction Wear* 1990, 11, *1*, 130–135.
26. Kabaldin, Y.G. The structure-energetic approach to the friction wear and lubricating phenomenon at cutting. *J. Friction Wear* 1989, 10, *5*, 800–801.
27. Trent, E.M., Suh, N.P. *Tribophysics*. Prentice-Hall: Englewood Cliffs, NJ, 1986, 125–489.
28. Brookes, K.J.A. Hardmetals and other hard materials. In *International Carbide Data 2003*. Hertfordshire, U.K., p. 220.
29. Gruss, W.W. Cermets. In *Metals Handbook,* 9th ed., Vol. 16, Ed., Burdes, B.P. American Society for Metals: Metals Park, OH, 1989, pp. 90–104.

7 Tooling Materials and Some Features of Their Self-Organization: Adaptive Tooling Materials

German S. Fox-Rabinovich, Anatoliy I. Kovalev, Ben D. Beake, and Michael M. Bruhis

CONTENTS

7.1 Major Tooling Materials....167
7.1.1 Universal Cutting-Tool Materials....167
7.1.1.1 Cemented Carbides....168
7.1.2 Specific Cutting-Tool Materials....176
7.1.3 Adaptive Cutting-Tool Materials....177
7.1.3.1 Frictional and Wear Behavior and Self-Organization of Adaptive Cutting-Tool Materials....178
7.1.4 Functional Graded Materials for Cutting Tools....194
References....195

7.1 MAJOR TOOLING MATERIALS

Tooling materials have traditionally been chosen for their excellent hardness and wear resistance under severe conditions (high stresses and temperatures) associated with machining and stamping operations. It is possible to classify tooling materials according to their different characteristics and domains of application.

7.1.1 UNIVERSAL CUTTING-TOOL MATERIALS

The principal focus of this chapter is on the so-called "universal" cutting-tool materials and their modifications (HSS and HSS-based materials and cemented tungsten carbides). These materials are used in a broad range of applications, not only for the machining of steels but also for many other materials. A major trend in the metallurgical design of cutting-tool materials is the application of universal tool materials using different methods of surface engineering.

There are several ways to improve the wear resistance of regular tool materials. One method is refining the structural components of traditional tool materials using powder metallurgy methods (fine-grained powder HSS and, recently, nanograined cemented carbides). The application of powder metallurgical HSS tools results in less surface damage under conditions of adhesive and attrition wear [1–2]. This is most important for the relatively brittle, high-cobalt HSS tool steels [3]. Other methods are also available, for example, the refining of carbide particles (and martensite grains in the case of high-speed steels) as a result of a surface laser treatment [2].

7.1.1.1 Cemented Carbides

7.1.1.1.1 Carbides for Cutting Tools

Owing to high competition in the market and the trend to improve productivity while reducing costs of metalworking, cemented carbides have became more popular for a variety of tooling applications such as cutting and stamping tools.

In comparison to tool steels, cemented carbides [3,4] are harder and more wear resistant but they exhibit a lower fracture resistance. On the other hand, they have a lower thermal conductivity than HSS. The major trend in advanced cemented carbide grades is refinement of the grain sizes. In recent years, WC–Co alloys with submicron carbide grain sizes have been developed for applications requiring more edge strength and minimal surface damage. This can be done by additions of small (within the range 0.25 to 3%) amounts of tantalum, niobium, chromium, vanadium, titanium, hafnium, and other carbides of transitional metals [5]. In most instances, their basic purpose is grain growth inhibition, maintaining a consistently fine structure without perceptible recrystallization. Vanadium carbide (VC), the most effective additive for grain growth inhibition, leads to unacceptable embrittlement if used alone. Usually VC is used in conjunction with tantalum carbide, which significantly improves the toughness of cemented carbides when in solid solution. Tantalum carbide alone is the most frequently used grain-growth-inhibiting additive. However, chromium carbide is probably the most advantageous of all additives. Besides a general improvement in properties at room and elevated temperatures, chromium carbide produces lubricious oxide-like tribo-films during friction. This film resists build-up formation and diffusion attack when cutting materials such as titanium (see Chapter 5, Figure 5.3) and steel. This carbide addition is also beneficial for the cemented carbide grades used for high-speed stamping tools.

It is worth mentioning that nanograined WC–Co composites have the potential to replace in prospective the standard materials for tools and dies because the fracture toughness and wear resistance of WC–Co can be significantly increased [6]. The properties are improved because nanoscale particles (1 to 100 nm) usually have physical properties different from those of large particles (10 to 100 μm). It has been found that nanoparticles exhibit a variety of previously unavailable properties depending on particle sizes, including a set of mechanical and physical properties and surface reactivity [7]. Recent experiments have shown that nanomaterials have improved mechanical properties, such as hardness and ductility, which are critically important for brittle materials [8].

The unique properties of nanoscale particles and nanograin bulk materials can be attributed to two basic phenomena. The first phenomena is that the number of atoms at the surface or grain boundaries in these materials is comparable to that of the atoms located in the crystal lattice. Thus, the chemical and physical properties are increasingly dominated by the atoms at these locations. The second phenomenon is the "quantum size effect." When particles approach the nanometer size range, their electronic and photonic properties can be significantly modified as a result of the absence of a few atoms in the lattice, thus resulting in relaxation of the lattice structure. These phenomena can be considered nonequilibrium processes (see Chapter 2).

The major engineering problem with nanoscale WC–Co materials is their consolidation technique. During the consolidation of nanoscale powders, the particles tend to coarsen rapidly from the nanoscale up to the microscale level (a fracture of a micron, i.e., a few hundred nanometers) and lose a majority of their potential properties. We have to emphasize that traditional techniques of materials fabrication are based on quasi-equilibrium physicochemical processes and ensure the formation of an equilibrium structure of the materials. This is why the application of traditional methods for the production of nanoscale materials are usually not successful. Success can be achieved if the techniques applied ensured the process be far from equilibrium conditions. These nonequilibrium techniques are the result of the complex of thermal, chemical, and stress gradients

TABLE 7.1
Dissolution Rate of Refractory Compounds vs. Temperature Relative to TiC

Material	Dissolution rate (°C)		
	100	500	1100
WC	1.1×10^{10}	5.4×10^{4}	3.2×10^{2}
TiC	1.0	1.0	1.0
TaC	2.3	1.2	8.0×10^{-1}
TiB_2	9.9×10^{1}	8.5	2.8
TiN	1.0×10^{-8}	1.8×10^{-3}	2.2×10^{-1}
HfN	2.5×10^{-12}	3.8×10^{-5}	2.5×10^{-2}
Al_2O_3	1.1×10^{-24}	8.9×10^{-11}	4.1×10^{-5}

Source: From Santthanam, A.T., Tierney, P. Cemented Carbides. In *Metals Handbook,* 9th ed., Vol. 16, Ed., Burdes, B.P. American Society for Metals: Metals Park, OH, pp. 71–89. With permission.

within the system when its state is close to the bifurcation point, i.e., the point when the system loses stability (see Chapter 2 for details). Under these conditions, the parameters of diffusion increase drastically and intensive mass transfer occurs. Eventually, self-organization and dissipative structure formation takes place [9].

There are a few emerging techniques to synthesize nanoscale cemented carbides that are strongly nonequilibrium, such as plasma-activated sintering, quick hot isostatic pressing (HIP) [6], or cold (gas dynamic) spray deposition of WC–Co nanosized coatings [10]. Another nonequilibrium technique is spark plasma sintering, the direct competitor of hot-pressing processes (single-axis or isostatic).

The principle leading to powder-material densification combines a single axis-like compaction and a high-intensity electrical discharge (from 2,000 to 20,000 A in a few milliseconds). It thus allows obtaining parts to achieve 100% material density in a short time (a few minutes) [11]. This technology can be applied in various application fields: wear materials, cutting tools, cermets, and functionally graded materials (FGM). This process retains the structural state of the initial powders. Particularly, it allows fabrication of the nanostructured materials (dense nanomaterials), notably ceramics, at ambient temperatures. Electric field-assisted sintering (FAST) is also an emerging technology for the fabrication of metals, ceramics, and their composites starting from powders. It has the potential to densify nanosized or nanostructured powders but avoids the coarsening, which accompanies standard densification routes [11].

Typical machining applications of cemented carbides include a wide variety of solid carbide milling and drilling tools. Mixed tungsten–titanium–tantalum carbides are used for steel machining to resist chemical (diffusion) wear. Tungsten carbide diffuses rapidly into the chip surface (Table 7.1), but a solid solution of tungsten carbide and titanium carbide resists this type of chemical wear. Titanium carbide is more brittle and less abrasive resistant than tungsten carbide (Table 7.2). For this reason, tungsten carbide alloys have a better wear resistance for machining of cast iron when abrasive wear is significant and some plasticity of the hardmetal is needed. The amount of titanium carbide added to the tungsten carbide–cobalt alloys is limited. On the other hand, the addition of titanium carbide has a noticeable effect on crater (diffusion) wear rate. The content of TiC should be substantially increased for high-speed machining applications when the "cratering" tendency becomes more severe. Generally, the TiC content is no more than 18%, but in some cases it can be as high as 30%. During the high-speed cutting of steels, a surface phase transformation resulting in the formation of oxygen-containing stable secondary structures of the Ti–O type (see

TABLE 7.2
Relative Resistance of Different Chemical Compounds against Abrasive Wear and to Dissolution in Iron at 700°C

Material	Relative Wear Resistance against Abrasive Wear	Material	Relative Wear Resistance against Dissolution In Iron
SiC	0.004	Al_2O_3	0.0000
WC	0.008	TiO_2	0.0000
Si_3N_4	0.030	TiO	0.0000
Al_2O_3	0.075	HfN	0.0009
HfN	0.28	TiN	0.018
HfC	0.34	HfC	0.035
ZrC	0.79	$TiCo_{0.75}O_{0.25}$	0.32
TiC	1.0	ZrC	0.36
TaC	1.0	TiC	1.0
$TiCo_{0.75}O_{0.25}$	1.3	TaC	1.1
HfB_2	1.6	NbC	1.9
NbC	2.2	TiB_2	5.3
TiO_2	2.2	Si_3N_4	250
TiO	2.8	WC	5200
Mo_2C	110	Mo_2C	12000
TiN	170	SiC	24000

Source: Kramer, B.M., Judd, P.K. Computation design of wear coating. *J. Vac. Sci. Technol.* 1985, 3, *6*, 2439–2444. With permission.

Table 7.2) might occur. The mechanism for the Ti–O formation is described in some detail in the following pages for adaptive materials. There is a lack of information in the literature regarding the self-organization of these alloys during cutting, but it is clear that the formation of protective tribo-film structures can increase the tool life significantly at elevated cutting speeds.

Although WC–TiC–Co hard materials are widely used, they were surpassed long ago by WC–TiC–Ta(Nb)C–Co cemented carbides in applications in which higher strength combined with crater wear resistance was required. These grades today are the most popular class of hard metals used mainly for high-speed steel cutting. The most significant contribution of tantalum carbide is that it increases the hot hardness of the tool, which, in turn, reduces thermal deformation (see Chapter 13) [12]. Competing directly with carbonitride cermets and silicon nitride ceramics (see the following text), the carbides of this class can take very heavy cuts at high speeds on all types of steels, nickel-based superalloys, and cast iron, in which large amounts of heat and high pressures are generated at the cutting edge. They do not, however, possess either the abrasion resistance of fine-grained WC–Co carbides or the extreme chemical and crater resistance of titanium carbide-based cemented carbides.

7.1.1.1.2 Carbides for Stamping Tools [13]

Although initial costs of tungsten carbide stamping tools are likely to exceed those of conventional tools made of tool steel, the tungsten carbide upgrade can often provide significant end-use cost reductions.

The majority of cemented carbides used for stamping-tool applications are tungsten carbide usually with a cobalt binder. The key to determining the properties of a tungsten carbide material for stamping-tool applications is the relationship between the carbide grain size and the binder percentage. The smaller the carbide grain size, the greater the hardness and abrasive resistance, but lower the impact toughness. Larger carbide grain size increases toughness but reduces hardness.

When the binder percentage is lowered, impact toughness declines. Conversely, raising the binder percentage improves shock resistance.

The carbide grades most widely used for stamping-tool application are so-called conventional grades with a 1.0-μm to 6.0-μm tungsten carbide grain size. A typical application of cemented carbides for stamping tools is electrical fractional motor lamination dies. When a fine, keen cutting edge is also needed, submicron carbide grades have been used with 0.7-μm being the average grain size for specific applications, such as caulking punches. Ultrafine submicron carbide grades with 0.5-μm grain size have been used where a strong and exceptionally keen cutting edge is required. Ultrafine submicron tungsten carbide provides high fracture toughness for a given hardness. This combination of properties is especially advantageous where thin, sharp sections may be subjected to moderate shock loads. However, there is an important relationship between the tungsten carbide grain size and the binder composition. As suggested earlier, shock loads can be absorbed by a larger grain size and higher binder percentage. However, if the carbide binder percentage is too high (above 20%), the grade may become too ductile for the application. A condition known as swaging or peening could occur, causing the binder to plastically deform and the die to fail prematurely. When the carbide grain size is too small and the binder percentage does not give the wear part enough ductility, the carbide grade may be too hard and lack enough impact toughness for the specific application. Fracturing, cracking, or chipping may cause premature tool failure. Shock, wear, and toughness properties can be balanced for optimum performance by carefully adjusting grain size. Larger grain sizes and higher binder percentages can absorb shock. To sustain severe impact shock load, therefore, a coarser grain size should be used with a medium-to-high range of binder percentage. For a less severe bending load, a finer grain size would be more appropriate with a low-to-medium binder level. After the optimization of a carbide grain size and binder content, the specific type of binder has to be determined. Binder systems directly affect the life of the tungsten carbide die or wear part. If straight abrasive wear is a concern, a cobalt binder system is recommended. If corrosion poses a threat, a nickel binder system is generally more suitable.

Adding tantulum–niobium and titanium carbides can remedy conditions such as galling or hot adhesion of the workpiece material. Owing to these carbide additions, the hot hardness can be improved but brittleness occurs. At high temperatures, these additions form a lubricious oxide tribo-film on the working surfaces. The oxide layer formation also reduces heat production on the surface. For the greatest galling resistance, tantalum carbide of up to 25% can significantly improve tool performance, particularly at temperatures above 500°C but only for applications when high-impact toughness is not needed [5].

For a longer stamping-tool life, it is necessary to determine the optimization of the cemented carbide alloying (Table 7.3). For instance, a cemented carbide (Figure 7.1) containing 12% Co binder and medium-coarse (around 2 to 3 μm) WC grains significantly outperforms cemented carbide with 14% Co and smaller (approximately 0.5 μm) grains during high-performance stamping (300 strokes/min) of 1019 magnetic steels. SEM images of the worn surface of the cemented carbide punch with 14% Co binder shows numerous cracks forming on the surface as a result of intensive adhesive-fatigue wear. Co binder content decreased to 12% enhances the stamping-tool life but does not significantly improve the surface morphology of the worn punch (Figure 7.1). The Co binder intensively squeezes off the surface during friction, which weakens the coarser-grained carbide phase (Table 7.3). The fatigue testing was performed to confirm this hypothesis.

As outlined in Chapter 6, the tool life of blanking tools is directly associated with the low-cycle-fatigue phenomenon. A special fixture was fabricated to study surface failure mechanisms that are related to fatigue, primarily regarding the chipping resistance of blanking punches (Figure 7.2). Square blanking punches (6 × 6 × 65 mm) were made of cemented carbides as shown in Table 7.3. The anvil was made from O1 tool steel with hardness HRC 20. Fatigue testing was performed using MTS servo-hydraulic mechanical test systems. Load and punch strokes were continuously recorded during the test. The fatigue test procedure consisted of applying 200,000 load cycles at

TABLE 7.3
Characteristics of Cemented Carbide Punches for Blanking-Tools Applications

		Grain Size									
Grade	Manufacturer	(mm)	Category	Cobalt (%)	Other Alloys (i.e., TiC, TaC, NbC) (%)	Hardness (HRA)	Density (g/cm^3)	Transverse Rupture Strength (ksi)	Compressive Strength (ksi)	Modulus of Elasticity (ksi) $\times 10^3$	Fracture Toughness (MPa $m^{1/2}$)
CD-650	CARBIDIE	0.4–1.0	Submicron	15	0.3-VC	89.0–90.5	13.9–14.1	550	650	80	12
CF-H40S	PLANSEE	1.0–2.0	Medium	12	1.15 Cr_3C_2, 0.2-TiC, 0.4-TaC/NbC	90	14.3	372	715	87	11.91
K3109	CARBIDIE	3.0–6.0	Medium-coarse	12	0.3-TiC, 0.3-TaC, 0.3-NbC	88.0–89.0	14.25–14.5	470	635	82.2	14

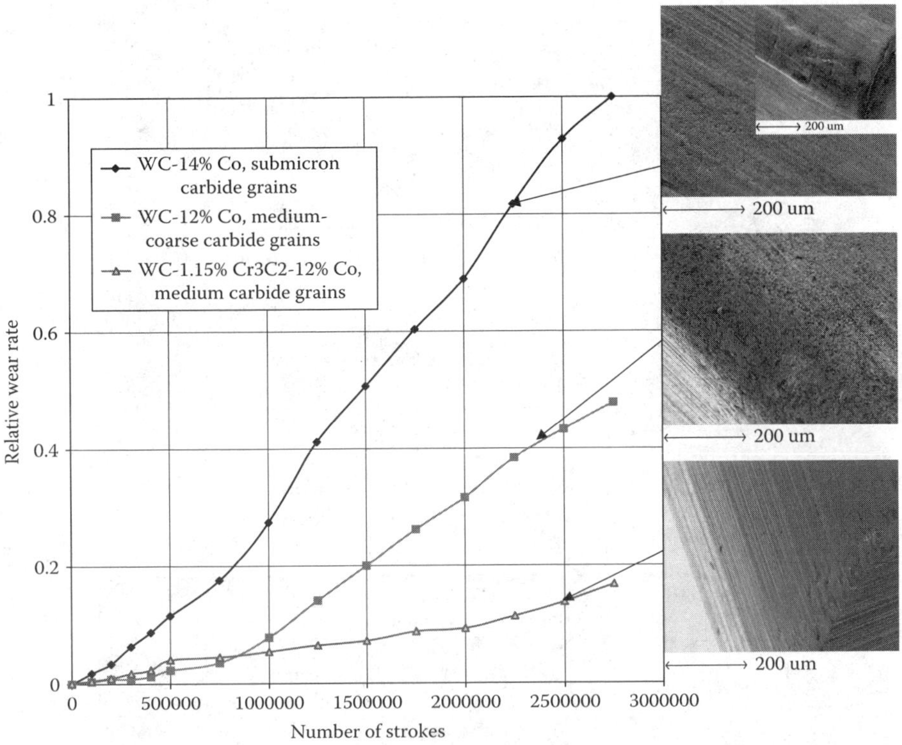

FIGURE 7.1 Relative wear rate vs. number of strokes for cemented carbides with different Co binder amount (12 to 14%) and different metallurgical design (composition and grain size) and SEM images of the cemented carbide punches' worn surface. High-performance stamping (300 strokes/min) of the 1019 magnetic steels.

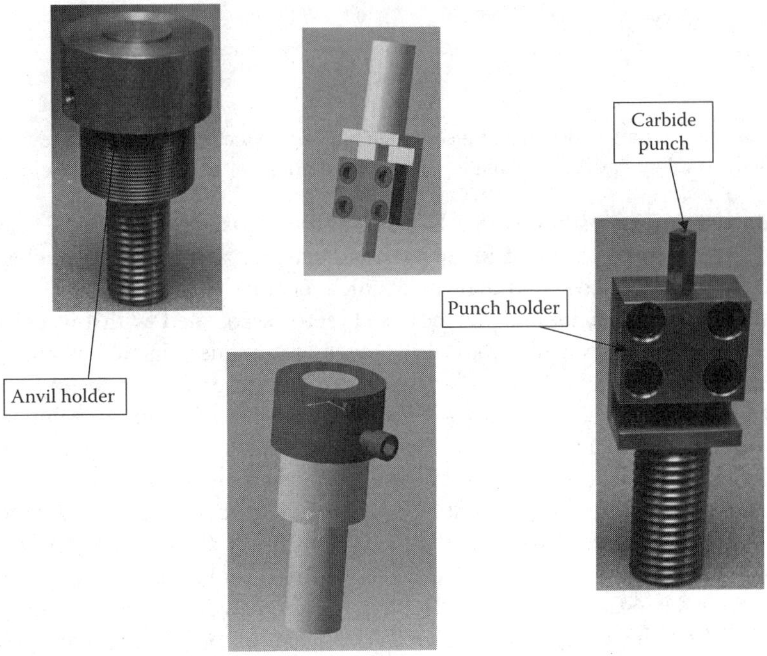

FIGURE 7.2 Design of experimental fixture for testing the chipping resistance of the cemented carbide punches.

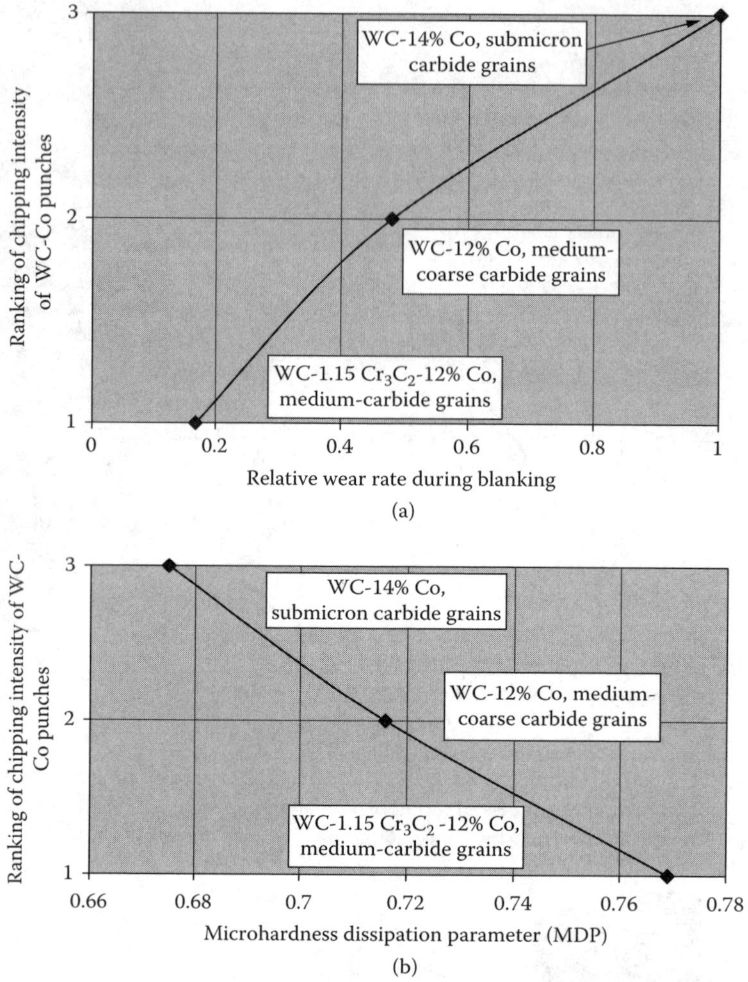

FIGURE 7.3 Relative chipping resistance of the carbide blanking punches after 200,000 cycles vs. (a) relative wear intensity and (b) microhardness dissipation parameter (MDP) of the cemented carbides studied.

room temperature. A cycle loading of 8,500 N was applied using a sinusoidal waveform with a frequency of 5 Hz. Relative chipping intensity was measured after 200,000 cycles and based on the microstructure of the punches and their mechanical characteristics (Figure 7.3 and Figure 7.4). The wear resistance of the blanking punches is directly associated with the chipping intensity (Figure 7.1 and Figure 7.3a) because both phenomena are fatigue related (see Chapter 6).

The surface area cross sections of the carbide punches after fatigue testing are presented in Figure 7.4. The data presented in Figure 7.4 show intensive carbide grain fragmentation and Co binder removal within the contact zone of the blanking punches after fatigue testing. The voids originate from selective removal of the cobalt phase. This is typical for WC–14% Co carbides that have the highest chipping and wear intensity (Figure 7.1, Figure 7.3a, Figure 7.4a to Figure 7.4c). Void nucleation is still quite intensive for carbides with lower Co content (around 12%) having coarser carbide grains (Figure 7.4d to Figure 7.4f). Minimal void accumulation is observed within the contact zone for the WC–12% Co alloy with medium (1 to 2 μm) grain sizes (Figure 7.4g to Figure 7.4i). Based on the Energy Dispersive X-ray Spectroscopy (EDS) data presented for Co binders (Figure 7.4a, Figure 7.4d, Figure 7.4g), it can be inferred that void nucleation depends on the amount of Co and the thickness of cobalt binder interlayer. Medium-grained cemented carbides

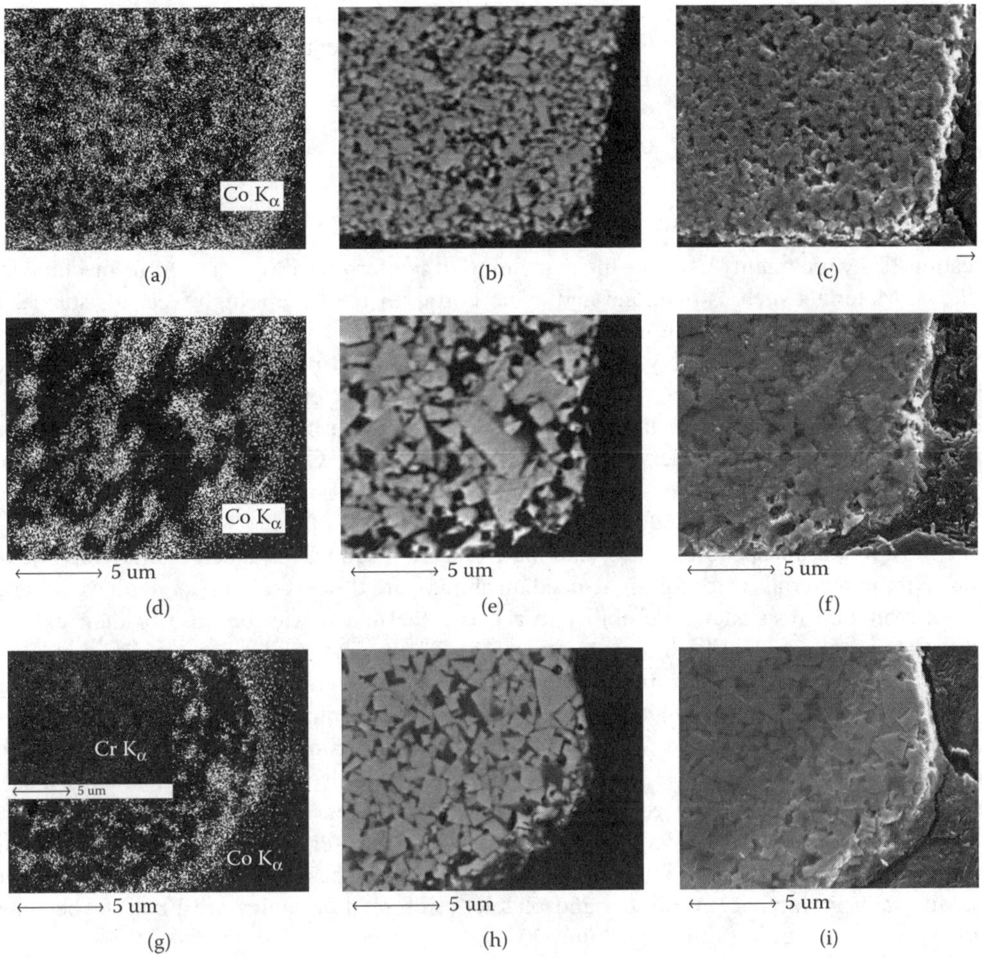

FIGURE 7.4 Secondary electron microscopy (c–i), back-scattering images (b–h), and elemental maps (Co–Kα, Cr–Kα, a–g) of the cemented carbides studied: (a–c) WC–14% Co, submicron grained carbide; (d–f) WC–12% Co, medium-coarse grained carbide; (g–i) WC-1, 15 Cr_3C_2–12% Co, medium grained carbide. Fatigue tests; load applied 8500 N, 200,000 cycles.

containing 12% of cobalt binder with thinner Co binder interlayer have better chipping resistance than the coarser grained ones with a thicker Co layer.

Some mechanical properties of cemented carbides, such as the microhardness dissipation parameter (see Chapter 5), are found to be responsible for the chipping resistance of the materials. Cemented carbides with higher elastic properties have better chipping resistance. This is based on the ideas presented in Chapter 5. It means that chipping resistance, as well as the intensity of adhesive-fatigue wear of the blanking punches, is related to the ability of surface and subsurface layers to accumulate energy of deformation during fatigue cycling (Figure 7.3).

The addition of Cr carbide to cemented carbides with optimized Co content (12%) and grain sizes further improves the tool life of blanking punches. This is related to the formation of Cr oxide as a result of the phase transformation on the surface of the stamping tool during tribo-oxidation. This oxide improves lubricating properties and critically reduces the galling intensity of the workpiece material on the surface of the stamping tool (Figure 7.1). It is worth noting that lubricious Cr tribo-oxides also play a very important role in the tool life improvement of cutting tools with Ti–Al–Cr–N coatings (see Chapter 10 for details). Owing to the formation of this oxide, the tool wear intensity decreases and the wear rate stabilizes (Figure 7.1). This is a typical feature of tribological adaptive materials. Thus, the Cr_3C_2-containing cemented carbide can be considered an adaptive tooling material.

7.1.2 Specific Cutting-Tool Materials [5]

Specific cutting-tool materials (e.g., ceramics, diamonds, in particular polycrystalline diamonds [PCD], or cubic boron nitride [CBN]) have a unique serviceability but a limited domain of application. They are mainly used for the machining of nonferrous alloys or hard-to-machine steels and alloys. Materials such as diamond and cubic boron nitride have a high wear resistance. This is due to their ability to dissipate the energy generated during friction into thin surface layers of atomic dimensions. CBN has very high hardness and abrasion resistance, coupled with extreme chemical stability when in contact with ferrous alloys at high temperatures. It has the ability to machine both steel and cast irons at high speeds. Currently, their high costs compared to cemented carbides have prevented their wider use in high-speed machining. CBN should mainly be considered as a finishing-tool material because of its extreme hardness and brittleness. Machine tool and setup rigidity for CBN, as with diamond, is critical.

The most important specific cutting-tool materials are ceramics [14] such as alumina. The main advantage these materials offer for high-speed machining are their excellent hot hardness, chemical inertness, oxidation resistance, and ability to act as a thermal barrier because of their extremely low thermal conductivity at high temperatures [15]. The main drawback of ceramics though, is their low fracture toughness. The typical application of these ceramics is the finish machining of steels and cast iron but newer high-strength alumina–zirconia ceramics in high-speed machine tools are becoming popular. The latest fine-grained alumina-based composites have sufficient strength for milling cast iron at speeds of up to 1000 m/min. They also have exceptional resistance to cratering when machining steel (see Table 7.2). An important innovation has been the introduction of alumina-based ceramics reinforced with silicon carbide single-crystal "whiskers." The immensely strong whiskers toughen the aluminum oxide. This development has enabled dramatic improvements in tool life and productivity. Whisker- and platelet-reinforced ceramics are likely to become an important cutting material in the near future [5].

Recently, one of the most effective ceramic cutting-tool materials has been developed based on silicon nitrides (Si_3N_4 of trademark SiAlON). SiAlON combines high strength with high hardness and shock resistance. The material is designed to machine high-nickel alloys and cast irons. Ceramics such as SiAlON tend to form intensive adhesive interaction with certain grades of steel.

Other ceramic compositions also look promising for future applications, such as the TiB_2-based ones. They combine superior hardness with higher strength than most other ceramics. This material also has excellent abrasion resistance as well as resistance to chemical attack.

Surface-engineered ceramics and functionally graded materials (FGM) are the most advanced materials in this class for future application. There are clear technical advantages in applying different methods of surface engineering for ceramic tools, such as deposition of PVD and CVD coatings. An increasing number of coated ceramics are expected to be in service in future years. Potential improvements include enhanced hardness, fatigue strength, and resistance to propagation of surface cracks [5]. The excellent stability of ceramics suggests the composition of the surface layer of functionally graded tool material. The material should generate alumina on the friction surface during interaction with the environment. This changes the cutting conditions by minimizing interaction with the workpiece and increasing service stability and heat-flow redistribution from the surface of the tool to the chip and the surrounding environment.

Detailed research shows that tribo-films that are formed on the surface of ceramics at high temperature have exceptional friction properties and strongly improved wear behavior [15].

7.1.3 Adaptive Cutting-Tool Materials

The main characteristic of these materials is the formation of a stable protective tribo-film during cutting. The generation of these films leads to the concentration of an interaction between the cutting tool and the workpiece within a thin surface layer that prevents intensive surface damage. Cermets are typical representatives of this class of material.

Cermets usually contain titanium carbide or titanium carbonitrides (and, recently, more complex titanium–molybdenum–carbon–nitrogen and titanium–tungsten–carbon–nitrogen compounds) as the hard refractory phase, comprising approximately 30 to 85% by volume the tool [16]. The metallic binder phase can consist of a variety of elements such as nickel, cobalt, iron, chromium, molybdenum, and tungsten.

The crater and the flank wear resistance of titanium carbide and titanium carbonitride cermet tool materials are superior to those of conventional cemented carbide (WC) tools. Nitrogen additions to the hard phase lead to a higher wear resistance [16]. Titanium carbonitrides are the primary materials used for cutting-tool applications. Titanium nitride and cubic boron nitride are excellent cermets when they are combined with a hard binder metal. Cermets are more wear resistant and allow for higher cutting speeds than tungsten carbides. The main drawback of traditional cermets is a lack of toughness and thermal shock resistance, but additions of molybdenum carbide and tantalum–niobium carbides have broadened the scope of their applications [17]. Cermets possess many of the characteristics of a tool material that is capable of filling the gap between conventional cemented carbides and ceramics. The mechanical properties of cermets were recently improved because of significant grain refinement. This has widened the area of future application of these materials.

As previously mentioned, one of the characteristic features of cermets is the formation of a thin layer of a protective tribo-film providing some lubricity at the tool surface. The formation of a stable tribo-film results in excellent surface finish and better tolerance on longer production runs [16], which is because of enhanced adaptability of these materials during cutting and stamping [14]. These attractive properties can be illustrated by comparing the shape of wear curves of cemented carbide and cermet tools. Cermets have better adaptability, as shown by the lower wear value, during the running-in phase and more stable cutting behavior during the normal wear stage because of the formation of protective and lubricious tribo-films (Figure 7.5) [16]. The stability of tribo-films is controlled by the thermodynamic stability of the compounds formed at the tool–

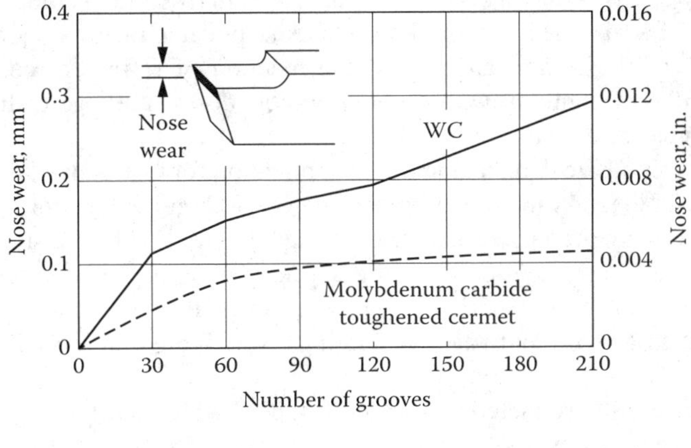

FIGURE 7.5 Wear comparison between cemented carbide and cermet cutting tools with 4135 alloy steel.(From Gruss, W.W. Cermets. In *Metals Handbook*, 9th ed., Vol. 16, Ed., Burdes, B.P. American Society for Metals: Metals Park, OH, 1989, pp. 90–104. With permission.)

workpiece interface. The thermodynamic stability of the compounds can be assessed by the enthalpy of formation of various cutting-tool materials [16].

Two types of cermets are used for cutting or stamping tools:

- Those with a ductile metal matrix in which the refractory phase content is more than 50%. The typical application of these materials is as turning inserts for high-speed cutting.
- Those with a hard steel matrix in which the matrix content is more than 50%. Typical application of these materials includes end mills for moderate cutting speeds and different types of stamping tools.

7.1.3.1 Frictional and Wear Behavior and Self-Organization of Adaptive Cutting-Tool Materials

As outlined in Chapter 6, one of the problems of traditional cutting-tool materials such as cemented carbides, especially at low and moderate cutting speeds when adhesive wear predominates, is connected with the low relaxation properties of the materials. Wear resistance is strongly correlated to the relaxation properties of the material, especially at the unstable wear modes during the running-in or catastrophic stages. If the tool material has a structure that is unable to effectively absorb or dissipate the energy generated by friction (a concern with the majority of traditional tool materials), damaging surface-relaxation processes (e.g., adhesion to the machined part or crack formation on the tool surface) dominates during cutting. Under these conditions the formation and stability of thin, surface tribo-films becomes questionable, and the hard materials that are used for cutting cannot ensure excessive wear during unstable wear stages. To avoid this situation, the hard materials must possess favorable relaxation properties [18].

Powder HSS, modified by the addition of titanium compounds, can adapt to the conditions of cutting and have better relaxation properties (compared to WC–Co alloys), giving tribological compatibility under low-speed cutting conditions. The metallurgical design of these materials is based on the application of the principles of screening by self-organization, as discussed earlier in Chapter 1. The screening effect prevents direct interaction of the cutting tool and workpiece, thus avoiding the consequent destruction of the tool metal. The localization of tool–workpiece interactions in the thin surface layers of tribo-films prolongs tool life.

Currently, tool materials based on HSS (or the other high-alloyed steel) made by sintering and hot extrusion of powders also contain between 10 to 50% of high-melting compounds (e.g., Sandvik Coronites, Ferro-Tic, Ferro-Titanit, deformed composite powder materials [DCPM]) [19–21]. A characteristic feature of these tool materials is the reaction of refractory compounds (carbides, carbonitrides, or nitrides of titanium) with the environment during cutting, leading to the formation of oxygen-containing surface layers (tribo-films).

It is important to emphasize that the domain of application for materials of this type lies between HSS and cemented carbides. As mentioned, the trend of modern manufacturing is the wide application of universal tooling materials, primarily cemented carbides with wear-resistant coatings. But adaptive materials have clear potential for future applications. The features of wear behavior and the structural adaptation of these materials can be used to understand similar phenomena in other adaptive materials, such as hard and adaptive coatings, which are considered in detail in Chapter 9 to Chapter 11.

The wear resistance of these materials has been studied with regard to changes in composition and structure to the surface of the tool during operation. The materials studied include M2 and T15 types of HSS, as well as deformed composite powder materials (DCPM) with an addition of 20% TiC. The wear resistance was assessed during the turning of 1040 carbon steel by using cutting indexable tetragonal inserts (with a side length of 12 mm). The cutting parameters used in the tests had speeds of 55 to 70 m/min, a depth of cut of 0.5 mm, and a feed rate of 0.28 mm/rev.

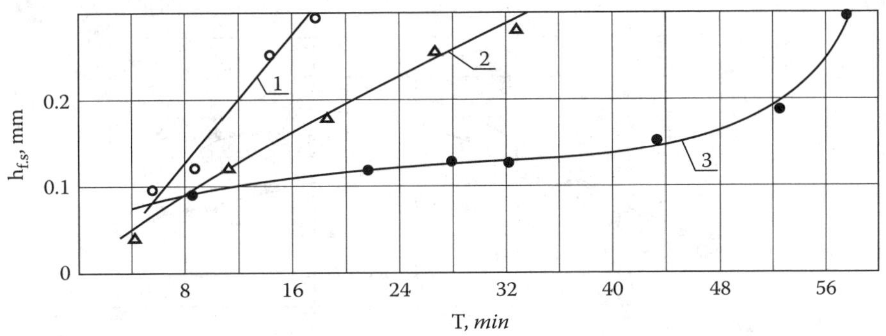

FIGURE 7.6 The dependence of the flank wear value of cutting tools on the cutting time: (1) HSS M2; (2) HSS T15; (3) deformed composite powder material (DCPM). Turning test data acquired with 1040 steel. Parameters of cutting: speed (m/min): 55; depth (mm): 0.5; feed (mm/rev): 0.28. (From Fox-Rabinovich, G.S., Kovalev, A.I., Shuster, L.S., Bokiy, Y.F., Dosbaeva, G.K., Wainstein, D.L., Mishina V.P. Characteristic features of wear in HSS-based compound powder materials with consideration for tool self-organization at cutting. Part 1. Characteristic features of wear in HSS-based deformed compound powder materials at cutting. *Wear* 1997, 206, 214–220. With permission.)

The frictional properties of the tribo-pair under analysis were determined with the aid of a tribo-meter, the design of which is described in Chapter 5 [22]. The adhesion component of the friction coefficient responsible for wear at low and medium cutting speeds (typical for HSS tools) [23] was used as a measure of the friction. This parameter was defined as the ratio of the resistance to shear of the adhesion bonds (formed between the sample made of the tool material and the workpiece under test) to the short-time tensile yield strength of the softer contact body at the test temperature. The value of this parameter is simply a measure of the resistance of the joint to shear. The friction condition at the surface of a cutting tool will be similar to that for which the value of the parameter was measured.

The results of the wear resistance tests are given in Figure 7.6 [19]. As can be seen, the wear resistance of HSS tools is 2.0 to 3.5 times lower than that of DCPM tools. This reduction was associated with a significantly lower coefficient of friction of DCPM compared to HSS (Figure 7.7) [19], and with a broadening of the range of stable wear (Figure 7.6). This result lies within the trends outlined previously for friction control (see Chapter 1). Within the stable wear stage, the rate of wear for DCPM is much lower than that of HSS (Figure 7.6, curves 1–3). Although the hardness and heat-resistance values of the HSS grade T15 and DCPM are similar, the wear resistance of the latter is significantly higher (Table 7.4). In our opinion, the lower wear intensity of the DCPM tool material is related to the presence of titanium carbides in the structure and their subsequent transformation to oxygen-rich compounds during cutting [2].

When studied by secondary ion mass spectroscopy (SIMS), the analysis of typical wear craters revealed the formation of oxygen-containing phases. The data in Figure 7.8 demonstrate that the transformation of titanium carbide into an oxygen-containing phase starts in the initial stage of wear (during the running-in process; Figure 7.8a). With further operation, there is increased surface oxide formation at the bottom of the wear crater. This process is accompanied by stabilization of the wear processes (Figure 7.6, Figure 7.8b to Figure 7.8c) and an expansion of the stable wear stage. Evidently, this is determined by the phenomenon of self-organization that is connected with the emergence of tribo-films (titanium–oxygen compounds), which function as stable solid lubricants [19].

The cross-section made across the cutting-tool face in the wear zone is shown in Figure 7.9a. The corresponding distributions of Ti, O, and C along the direction I–I, as obtained by electron spectroscopy, are given in Figure 7.9b. In the left part of the micrograph, an iron-based build-up can be seen. The right part of the micrograph shows the distribution of dispersed hardening phases

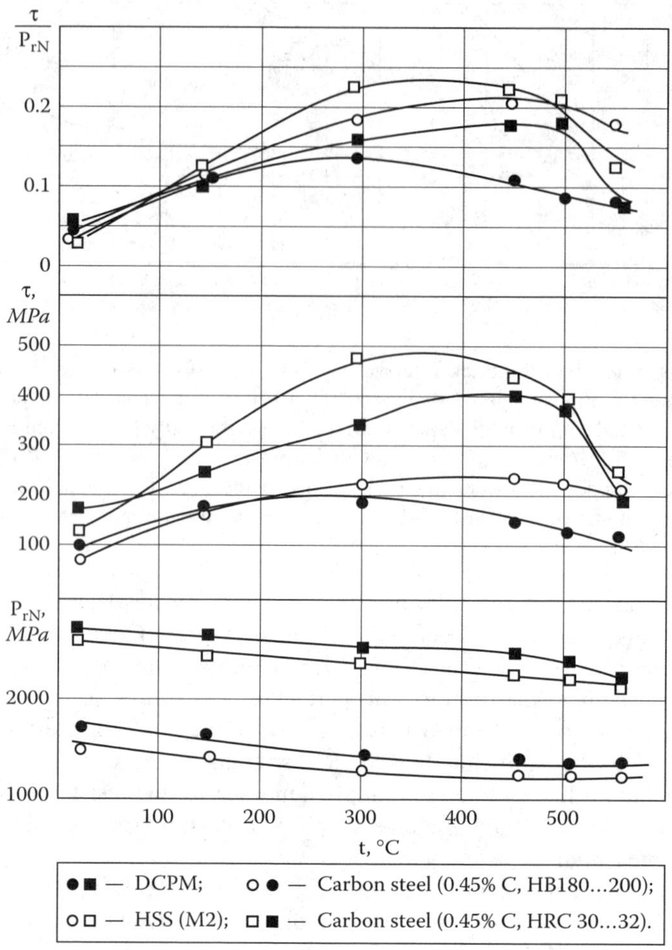

FIGURE 7.7 Impact of the test temperature on the frictional and wear characteristics as determined from wear contact tests for the DCPM with a 20% TiC addition. (From Fox-Rabinovich, G.S., Kovalev, A.I., Shuster, L.S., Bokiy, Y.F., Dosbaeva, G.K., Wainstein, D.L., Mishina V.P. Characteristic features of wear in HSS-based compound powder materials with consideration for tool self-organization at cutting. Part 1. Characteristic features of wear in HSS-based deformed compound powder materials at cutting. *Wear* 1997, 206, 214–220. With permission.)

in the HSS-based DCPM. There are angular (dark) particles of titanium carbide (less than 8 μm in crosssection) as well as dispersed tungsten and molybdenum carbides (less than 0.2 to 1.5 μm in diameter) uniformly distributed in the HSS matrix. In the surface layers of the tool material, there is a zone of intense plastic deformation less than 5 μm in depth. There, dispersed particles of a titanium-containing phase have been drawn out parallel to the wear surface, forming a discontinuous film. The titanium carbides in the wear zone have been transformed into oxides (Figure 7.8 and Reference 21). Titanium oxides are known to be much more plastic than titanium carbides, accounting for the plastic deformation of particles in the surface layers of the HSS-DCPM on cutting.

These results are confirmed by Auger spectroscopy. Figure 7.9b represents the distribution of the intensity of the characteristic Auger KLL lines for O, C, and the LMM (418-eV) line of Ti along the I–I direction in Figure 7.9a. The analysis volume includes the built-up layer (of 1040 steel), the built-up layer–wear crater boundary, and the DCPM volumes beneath the wear crater. At the interface, the titanium compounds show an increased concentration of oxygen and decreased

TABLE 7.4
Properties of HSS and HSS-based DCPM Materials

	Heat Treatment		Physico-Mechanical Properties			
Material	Hardening Temperature (°C)	Temperature of Tempering (°C)	Hardness after Heat Treatment (HRC)	Bending Strength (MPa)	Impact Toughness (kJ/m²)	Thermal Stability (°C)
M2	1220	Triple treatment at 560°C	63–65	3200	400	610
T15	1240	Triple treatment at 560°C	67–68	2400	220	645
HSS-based DCPM	1210	Triple treatment at 560°C	69–70	2000	80	655

Source: Fox-Rabinovich, G.S., Kovalev, A.I., Weatherly, G.C. Tribology and the design of surface engineered materials for cutting-tool applications. In *Modeling and Simulation for Material Selection and Mechanical Design*, Eds., Totten, G., Xie, L., Funatani K. Marcel Dekker: New York, 2004, pp. 301–382. With permission.

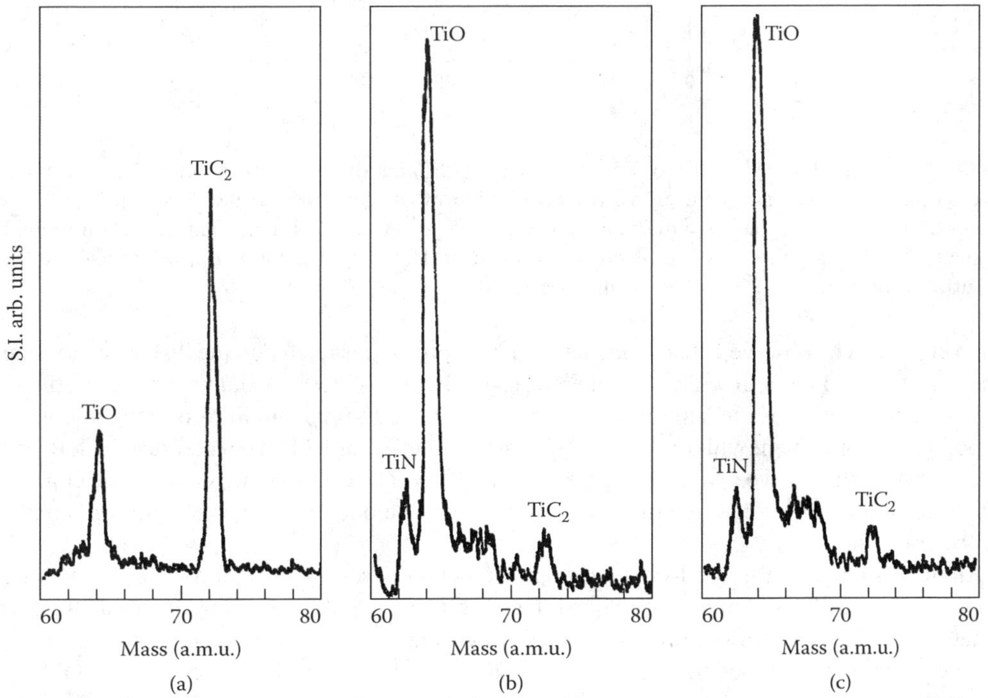

FIGURE 7.8 Mass spectra of a wear crater in a cutting tool made of HSS-based DCPM with a 20% TiC addition, determined as a function of the cutting time: (a) 4 min, (b) 20 min, (c) 24 min. (From Fox-Rabinovich, G.S., Kovalev, A.I., Shuster, L.S., Bokiy, Y.F., Dosbaeva, G.K., Wainstein, D.L., Mishina V.P. Characteristic features of wear in HSS-based compound powder materials with consideration for tool self-organization at cutting. Part 1. Characteristic features of wear in HSS-based deformed compound powder materials at cutting. *Wear* 1997, 206, 214–220. With permission.)

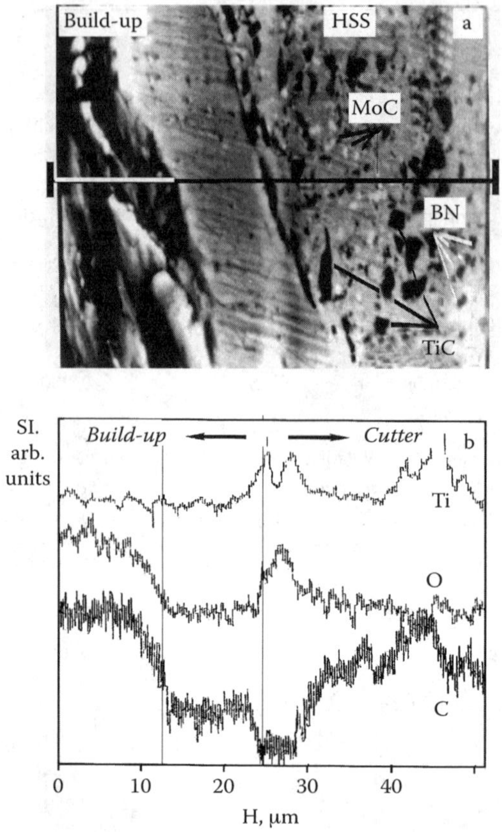

FIGURE 7.9 Microphotograph of tool friction surface with films of secondary structures: (a) general view of the surface using secondary electrons, (b) distribution of oxygen close to the "built-up-crater" contact surface (SI, intensity of signals, arbitrary units). (From Kovalev, A.I., Fox–Rabinovich, G.S., Wainstein, D.L., Mishina, V.P., Studying the structure of films generated on HSS–based deformed powder materials cutting tool surface. *Wear*, 238, 2, 2000. With permission.)

carbon content. The observed change in chemical composition is related to the instability of titanium carbides. Owing to the high cutting temperatures (in excess of 450°C) and pronounced affinity for oxygen, titanium adsorbs the latter from the environment and forms thin films of oxygen-containing compounds, in agreement with the SIMS data presented in Figure 7.8. The total plastic deformation of these particles at the wear surface is greater than 600%. The crystal structure of these compounds is believed to differ from the titanium oxides that would be obtained under equilibrium conditions (see the following text).

An understanding of the self-organizing phenomenon is very important for the development of advanced tool materials. An area of major interest in these studies (regarding materials science) is the nature of the tribo-films formed under severe cutting conditions. According to the principles of modern tribology, one of the main methods to control friction is the creation of stable tribo-films at the tool surface. The more stable the tribo-films, the greater the tool life. The development of protective tribo-films can be manipulated by alloying or surface engineering. The type of tribo-films formed during cutting strongly depends on the conditions of cutting and the type of the material under analysis.

A detailed study of the physicochemical parameters of the tribo-films formed during cutting using a tool made of DCPM was done using AES, ELS, and EELFS methods. To interpret the atomic structure of the tool wear surface, data obtained in this work by the EELFS method was

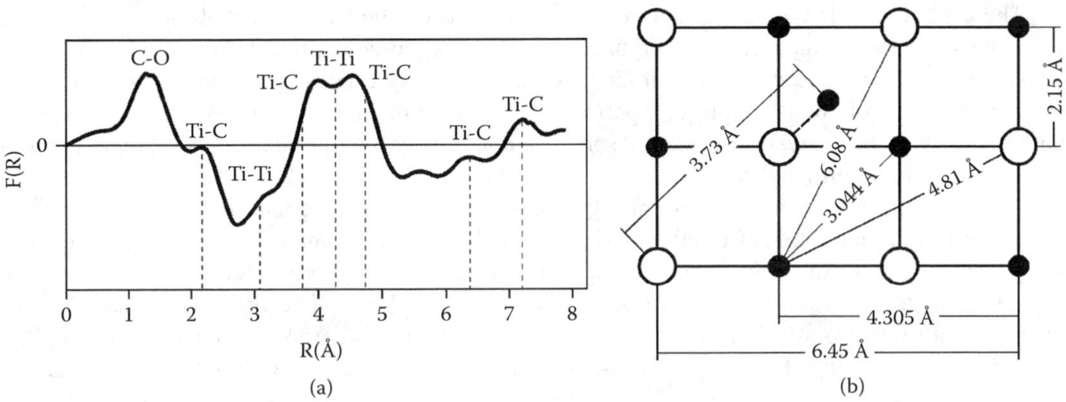

FIGURE 7.10 (a) Fourier transform of EELFS close to the line of back-scattered electrons for TiC specimen, Ep=1500 eV, (b) cubic lattice of titanium carbide (○ = Ti, ● = C). (From Kovalev, A.I., Fox–Rabinovich, G.S., Wainstein, D.L., Mishina, V.P., Studying the structure of films generated on HSS–based deformed powder materials cutting tool surface. *Wear*, 238, 2, 2000. With permission.)

compared to TiC and TiO_2 standards. Figure 7.10 presents the Fourier transform data obtained by analyzing the extended electron energy loss fine structure (EXELFS) for titanium carbide (TiC) with a cubic (B1) structure. The positions of the main peaks (Figure 7.10a) are consistent with the interatomic distances for a (100) plane in the cubic lattice of titanium carbide (see Figure 7.10b).

Figure 7.11a shows data for TiO_2 with a rutile (C4) structure. We can identify the type of bonds by using partial functions F(R) obtained from the analysis of the fine structure of spectra close to the characteristic Auger lines of oxygen and titanium. By comparing this data with those given in Figure 7.10a, we can see that TiO_2 has a more complex crystalline structure than TiC. This explains the greater number of F(R) function peaks. The positions of the main peaks are again in good agreement with the inter atomic distances for a (100) plane in the TiO_2 lattice. The complete analysis of all the peak positions by the Fourier transform method shows that the interatomic distances, O–O and Ti–O, in the secondary structures are different from those discussed in the literature (see Figure 7.11b). This may be related to a deviation from stoichiometry, because data on the interatomic distances were obtained during the measurement of this characteristic within a few angstrom thick surface layer. This data could differ from the equilibrium values.

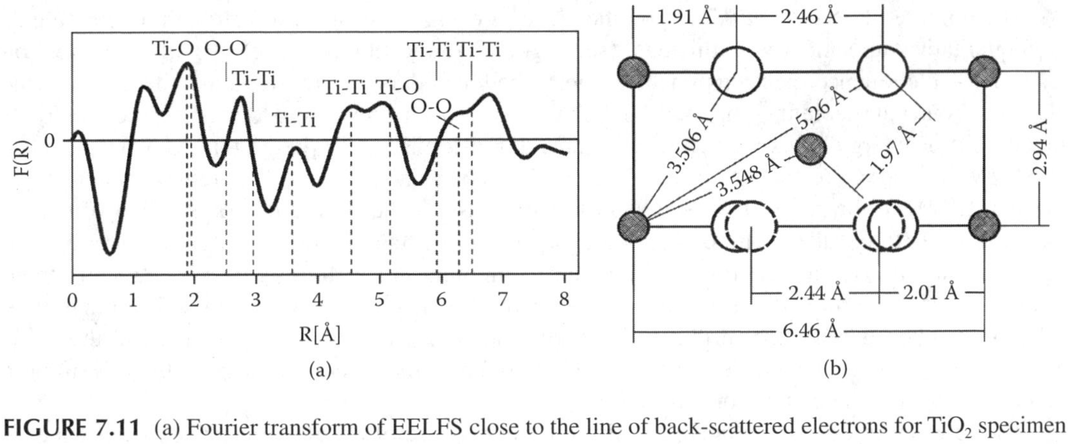

FIGURE 7.11 (a) Fourier transform of EELFS close to the line of back-scattered electrons for TiO_2 specimen with rutile structure, Ep=1500 eV, (b) cubic lattice of titanium oxide (o = Ti, = = O). (From Kovalev, A.I., Fox–Rabinovich, G.S., Wainstein, D.L., Mishina, V.P., Studying the structure of films generated on HSS–based deformed powder materials cutting tool surface. *Wear*, 238, 2, 2000. With permission.)

The evolution of the atomic structure in surface films on the wear crater of the cutting tool is well illustrated by the data given in Figure 7.12. The oxygen-containing films in the wear crater are significantly enriched with titanium and oxygen after only 5 min of cutting (see Figure 7.12a to Figure 7.12d). As this takes place, a periodicity in the arrangement of atoms of various types is observed both in the nearest coordination sphere and at greater interatomic distances, up to approximately 7 Å (see Figure 7.12b).

As noted previously, the interatomic distances in these oxygen-containing films differ from those observed in equilibrium titanium oxides, including rutile (compare with Figure 7.11). The very thinnest films form, whose atomic structure is close to the supersaturated α-solid solutions of oxygen in titanium. After 15 min of cutting, the degree of long-range order is reduced but the intensities of peaks from higher-order coordination spheres are less pronounced (see Figure 7.12c). After 30 min of cutting (Figure 7.12d), the translational symmetry at large interatomic distances disappears and peaks at R greater than 4 Å are lost.

The adaptability of the surface layer to external thermomechanical effects is the physical basis of such evolution. The surface layer is gradually converted to an amorphous state during the wear process (after a cutting times of about 15 min). When a steady state condition is reached, i.e., after the development of the tribo-films is completed, the surface generates amorphous-like films having an effective protective function. The lattice instability of the solid solution of oxygen in titanium finally leads to complete amorphisation of the wear surface. A similar effect was observed earlier from EELFS data for Ti–N coatings on worn cutting tools [24].

Typical EELS spectra of TiC, TiN, and TiO_2 obtained with a 30.0-eV primary electron beam are shown in Figure 7.13. The elastic peak has a 30-meV FWHM (full width at half maximum). The high-resolution structures of the spectra are represented at 1000× magnification after normalizing. The experimental curves are approximated by Gaussian peaks in each spectrum in the range of 1- to 9-eV energy losses.

In the series of titanium compounds (such as carbide–nitride–oxide) TiC–TiN–TiO, the number of 4s electrons in the atomic sphere of the metal decreases correspondingly. These electrons are transferred to the 2p orbital of the nonmetal atom, and the band energy is lowered because of the increasing Ti–X attraction in this series. Accordingly, it is more likely that, on being excited, electrons would pass from the 2p orbital to the 3d orbital. The peak observed at 6.8 eV corresponds to the X 2p→Ti 3d transition (Figure 7.13). The intensity of lines at about 6.8 eV is increased in the series TiC–TiN–TiO_2. The Ti 3d orbital is hybridized with X 2p orbital, but the energy of this p–d orbital interaction is shifted downward along the energy scale as one moves from TiC to TiO, i.e., E_b(Ti–C) = 4.5 eV, E_b(Ti–N) = 4.8 eV, and E_b(Ti–O) = 6.89 eV [15].

The observed line shift at 6.7 eV on the electron energy loss spectra for titanium compounds is in qualitative agreement with this data (see Figure 7.13a to Figure 7.13c). A partial distribution analysis of the valence electrons enables one to better understand the particular features of the chemical bonds in titanium compounds and on the wear surface of cutting tools. The lines at 1.6 eV in the electron energy loss spectra reflect intraband transitions $t_{1g}\rightarrow t_{2g}$ in the π band, whereas those at 3.1 eV correspond to transitions $2t_{2g}\rightarrow 3e_g$ in the σ band of Ti atoms. The reduced intensity of lines at 3.1 eV in the series TiO_2→TiN→TiC is because of the decreased contribution of the dσ electrons of Ti to covalent ion bonds. This arises when the 4s electrons of Ti are transferred into the 2p orbital of a nonmetal atom in a titanium compound: this orbital is more completely filled in the case of oxygen than in either nitrogen or carbon. For this reason, when the 3d Ti and 2p X orbitals are hybridized, the contribution of the dσ electrons of Ti is less pronounced in the oxide and more expressed in the nitride and carbide. Consequently, the oxide has considerably less strength and hardness than the carbide or nitride [15].

The interaction of Ti–Ti atoms is realized at the expense of dσ electrons. The metallic nature of this compound is related to the high density of dσ electrons. As seen in Figure 7.13, the intensity of $t_{1g}\rightarrow t_{2g}$ transitions in the π band is relatively insignificant in TiC but is much higher in TiO_2. This implies that the density of conduction electrons is low in TiO_2 but is higher for TiC and TiN.

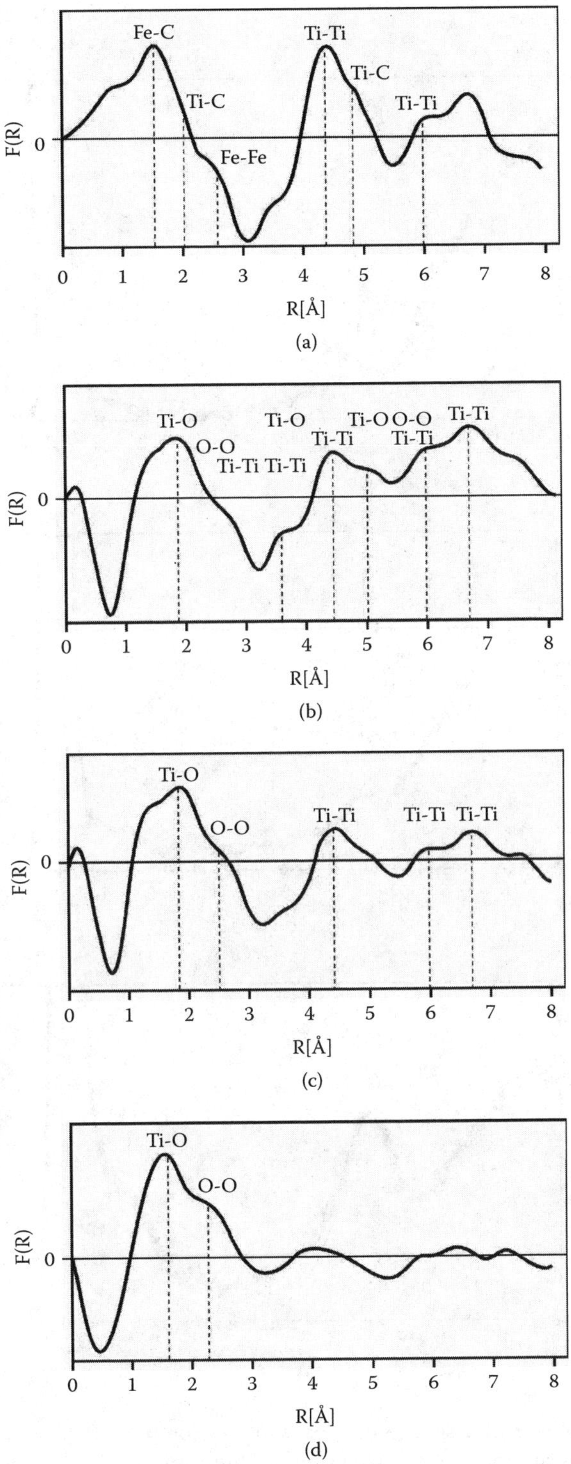

FIGURE 7.12 Fourier transform of EELFS close to the line of back-scattered electrons for a cutting tool made of DCPM after cutting times of: (a) initial stage, (b) 5 min, (c) 15 min, (d) 30 min. (From Kovalev, A.I., Fox–Rabinovich, G.S., Wainstein, D.L., Mishina, V.P., Studying the structure of films generated on HSS–based deformed powder materials cutting tool surface. *Wear*, 238, 2, 2000. With permission.)

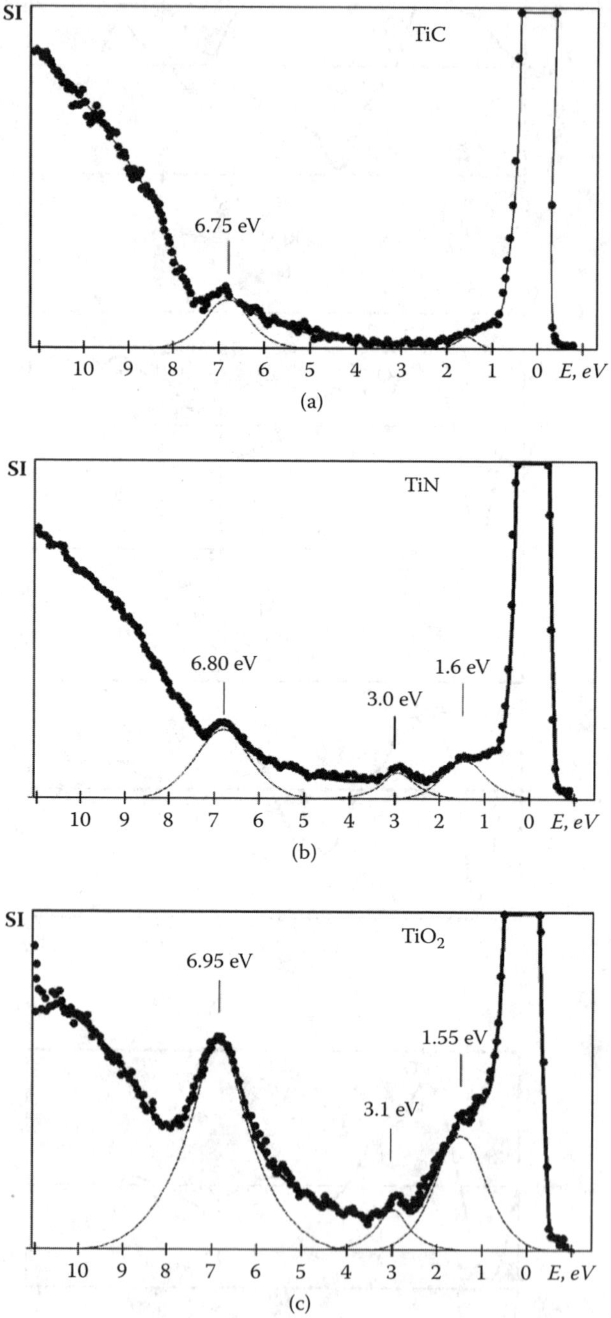

FIGURE 7.13 Representative EELS spectra of energy loss region 0 to 10 eV below the elastic peak of (a) TiC, (b) TiN, (c) TiO_2. The spectra were obtained using a 30-eV primary electron beam. (From Kovalev, A.I., Fox–Rabinovich, G.S., Wainstein, D.L., Mishina, V.P., Studying the structure of films generated on HSS–based deformed powder materials cutting tool surface. *Wear*, 238, 2, 2000. With permission.)

This is consistent with the electrical conductivity data of these compounds, which is extremely low for the dielectric TiO_2 but is 16,400 $(m)^{-1}$ for TiC and TiN, respectively [15].

The replacement of carbon with oxygen in titanium compounds was shown to change their properties significantly. Thus, the oxidation of TiC at 823 K for 30 min influences the electronic structure of the material. The electron spectrum acquires some features that are specific to TiO_2 (see Figure 7.14a and Figure 7.13c). After oxidation of TiC, the intensity of the lines at 6.7 eV and 1.6 eV is substantially enhanced. On oxidation, the titanium carbide loses its metallic properties and acquires those of a dielectric. In this case, we observe a reduced concentration of conduction electrons and a localization of the electron density, both in metal and nonmetal atoms. This is shown by the increased intensity of the peak at 1.6 eV corresponding to π states in the 3d band of titanium (see Figure 7.13a).

It was noted earlier that an intense oxidation of TiC could be observed during the operation of a DCPM tool [18,20]. In this case, the nature of the phase transformation differs significantly from that found on heating a TiC standard up to 823 K for 30 min. Figure 7.14b and Figure 7.14c present the electron energy loss spectra from the wear crater after 5 and 30 min for DCPM tool sample operations. As the wear time increases, the spectra display a somewhat increased intensity of peaks at 6.8 and 3.1 eV. Peaks corresponding to plasmon losses (p1 and p2) appear, and the peak at 1.6 eV is significantly attenuated.

The thin tribo-films in the wear crater of the cutter are associated with the formation of a supersaturated solid solution of oxygen in titanium owing to the oxidation of titanium carbide. In this case, we observe an increase in the electron density in the 2p orbital of the nonmetal (peak 6.8 eV) as well as an enhanced filling of the $d\sigma$ electron band of titanium atoms (peak 3.1 eV). These effects are similar to those encountered in the model oxidation of TiC (Figure 7.14). There are, however, substantial differences. As the cutting time increases, the effects brought about by the crystalline structure of phases become significantly weakened in the electron spectrum. The splitting of the 3d orbital into π and σ states degenerates the intensity of $t_{1g} \rightarrow t_{2g}$ transitions. This characteristic is reduced and so is the density of π electrons, which are related to the long-range Ti–Ti bonds in the lattice [along the diagonals in the (100) planes]. These distinctive features of the electron structure are related to amorphization and to the increasing role of short-range interatomic bonds. Of considerable interest is the appearance of plasmon loss peaks p1 and p2 in the spectra of Figure 7.14b and Figure 7.14c because of the growing concentration of conduction electrons. The delocalization of π electrons close to the titanium atoms enhances the metallic nature of the bonds in the amorphous films developed on the friction surface. These specific traits of electronic and atomic structural changes might help to explain the unique mechanical properties of the tribo-films or secondary structures of the first type (see Chapter 1). The high wear resistance and good frictional properties of DCPM tools are associated with complex structural and phase transformations on the surface, among them being TiC oxidation and the development of thin protective amorphous films. The secondary structures are saturated or supersaturated (amorphous) solid solutions of oxygen in titanium, whose electron structure is characterized by a high density of conduction electrons giving metallic characteristics.

These results show that the tribo-films formed during cutting not only increase the DCPM tool life but also change the friction characteristics (Figure 7.7). The amorphous-like tribo-films of the first type behave as a solid lubricant with enhanced tribological properties [25].

Additional alloying of the DCPM might be beneficial. For example, the partial substitution of titanium carbide by aluminum oxide, which is stable under cutting (see Table 7.2), leads to a decrease in the friction coefficient (Figure 7.15) and an increase in the wear resistance of the tool (Figure 7.16). The decrease of the coefficient of friction when Al_2O_3 is added is important because it increases the wear resistance and also lowers the cutting temperature at the tool surface [26,27]. Alloying often cannot be implemented by traditional metallurgical methods because this may induce an undesirable change in the properties of the cutting material. We took a different approach by making small additions of low-density compounds, which are relatively unstable at the operational

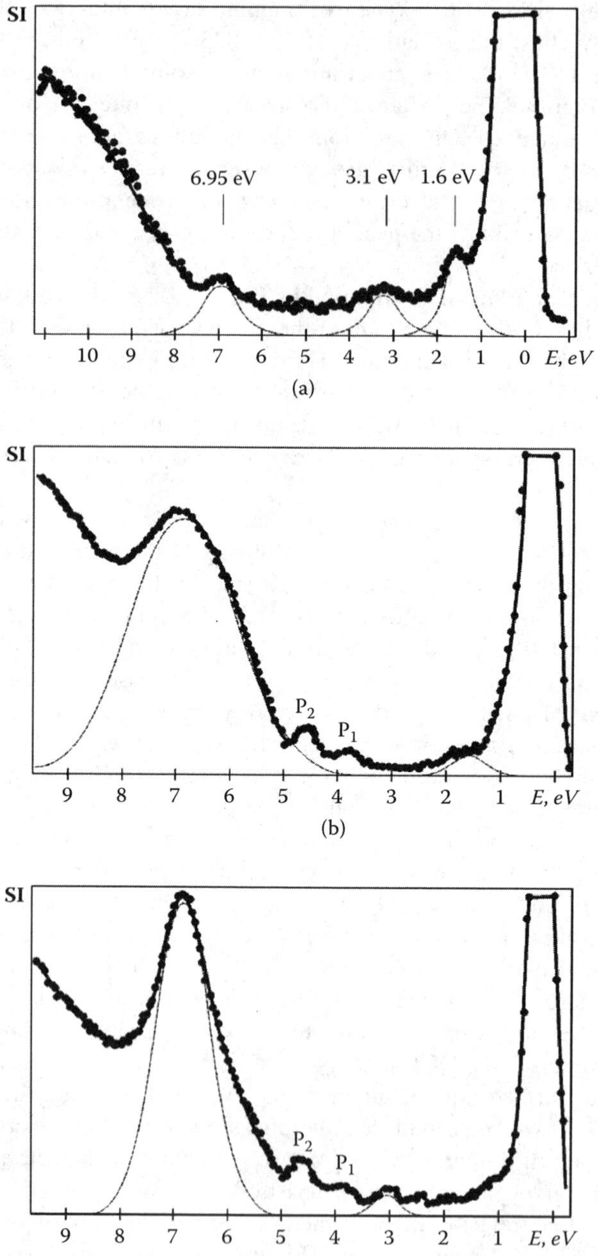

FIGURE 7.14 Representative ELS spectra: (a) after oxidation of TiC by heating up to 823 K for 30 min in air, (b) wear surface of DCPM cutting tool after 5-min operation, (c) wear surface of DCPM cutting tool after 30 min operation. The spectra were obtained using a 30-eV primary electron beam. (From Kovalev, A.I., Fox–Rabinovich, G.S., Wainstein, D.L., Mishina, V.P., Studying the structure of films generated on HSS–based deformed powder materials cutting tool surface. *Wear*, 238, 2, 2000. With permission.)

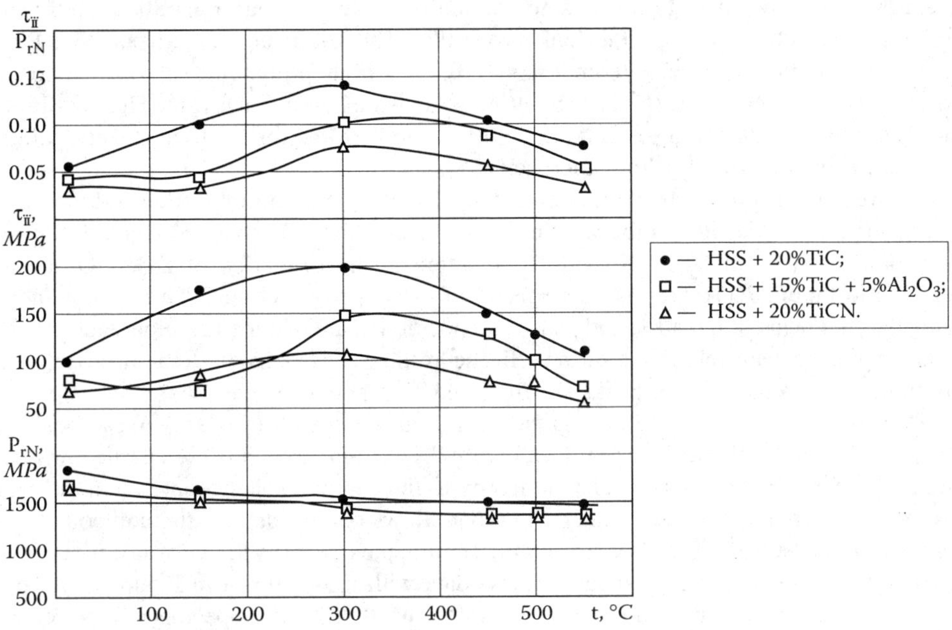

FIGURE 7.15 Impact of the test temperature on the coefficient of friction for the DCPM with 15% TiC + 5% Al_2O_3; 20% TiC + 2% BN and 20% TiCN additions. (From Fox-Rabinovich, G.S., Kovalev, A.I., Shuster, L.S., Bokiy, Y.F., Dosbaeva, G.K., Wainstein, D.L., Mishina V.P. On characteristic features of alloying HSS-based deformed compound powder materials with consideration for tool self-organization at cutting. Part 2. Cutting tool friction control due to the alloying of the HSS–based deformed compound powder material. *Wear* 1998, 214, 2. With permission.)

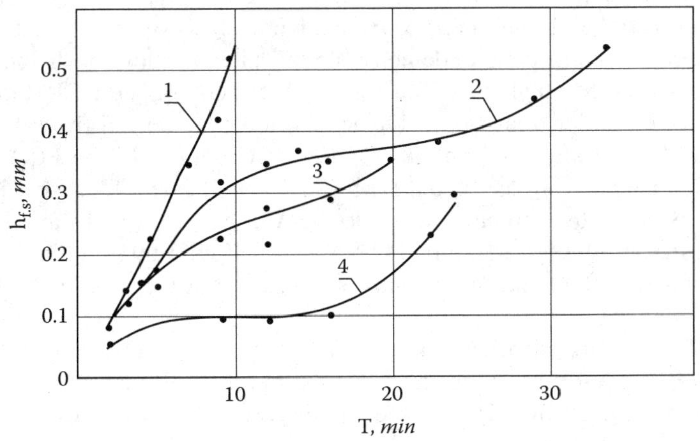

FIGURE 7.16 Wear curves of the adaptive cutting-tool materials: (1) DCPM with 20% TiC, (2) DCPM with 20% TiC and 2% BN, (3) DCPM with 15% TiC and 5% Al_2O_3, (4) DCPM with 20% TiCN. Turning test data acquired with 1040 steel. Parameters of cutting: speed (m/min): 90; depth (mm): 0.5; feed (mm/rev): 0.28. (From Fox-Rabinovich, G.S., Kovalev, A.I., Shuster, L.S., Bokiy, Y.F., Dosbaeva, G.K., Wainstein, D.L., Mishina V.P. On characteristic features of alloying HSS-based deformed compound powder materials with consideration for tool self-organization at cutting. Part 2. Cutting tool friction control due to the alloying of the HSS–based deformed compound powder material. *Wear* 1998, 214, 2. With permission.)

temperatures. This allowed us to use this compound in relatively small quantities (up to 2 wt %) with minimal possible impact on the bulk properties. The solid lubricant (hexagonal BN) was chosen as the additional alloying compound [26]. The high probability of oxygen-containing secondary phases formed from BN during cutting was also taken into account. The possibility that TiC and BN might oxidize and generate thin surface oxide films for exploitation in cutting tools can be assessed by a thermodynamic approach [15].

SIMS investigations have shown that on cutting DCPM with a boron nitride addition, oxygen-containing compounds develop at the wear crater surface, which is associated with a set of parallel disassociation reactions of BC, BN, and TiC, leading to the formation of BO, TiO, and TiBN. Figure 7.17a to Figure 7.17c presents the spectra of the positive and negative ions obtained upon analyzing the chemical composition and phase composition of a BN-doped carbide steel investigated at various depths beneath the crater surface. In the volume closest to the tool surface, (0.15 μm) there is an increase in intensity of peaks O^-, BO^+, and TiO^+ and a decrease in the intensity of peaks BN^-, BC^+, and TiC_2^+ compared to data gathered at a greater depth (0.6 μm). Weak peaks corresponding to TiBN and TiBO also appear (see Figure 7.17a to Figure 7.17c). As a rule, the observed ions in the SIMS method cannot be related directly to the compounds encountered in the analyzed regions, but they provide a useful "fingerprint" that allows one to identify the compound.

Figure 7.18 presents the SIMS data from different depths beneath the wear crater surface for a specimen of the same material. Comparing these data with those shown in Figure 7.17c, one can see an appreciable weakening of the intensity of the peak BC^+, the disappearance of peak TiC_2^+, as well as pronounced enhancement of peaks BO^+, $TiBN^+$, and $TiBO^+$, characteristic of the formation of thin surface tribofilms. These are compounds of titanium and boron with oxygen and nitrogen, formed as a result of the carbide and nitride reaction with the atmosphere. Based on the information presented in this figure, one can estimate a thickness of the tribo-films that are formed during cutting. It is very important to emphasize that the thickness of the stable tribo-film layer does not exceed 0.1 to 0.15 μm (Figure 7.18). This means that the thickness of these films does not exceed 100 nm, i.e., lie in a nanoscale area. That is why we can claim that all the studies presented in this book on the physicochemical transformation of the surface are studies directly related to nanotribology.

The degree to which particular crystalline structures can develop secondary structures depends on the composition of the tool material. A comparative analysis of the nearest atomic neighbors using EELFS spectroscopy was done to demonstrate this phenomenon for the phases that form on the wear crater surfaces of M2 high-speed steel, DCPM, and BN-doped DCPM tools. Figure 7.19a to Figure 7.19c shows Fourier transforms obtained from data collected from the surface of wear craters in HSS and DCPM alloyed with either TiC or TiC with BN. The F(R) functions feature pronounced peaks in the range 1–2 Å, 4–5 Å, and 7–8 Å for all cases. Using partial F(R) functions obtained by analysis of the fine structures close to the Auger lines of C, B, and Ti, it was possible to interpret the nearest-neighbor interatomic bonds. It was found that Fe–O bonds are typical of the HSS sample, whereas B–O and Ti–B bonds are observed at the wear crater of the BN-doped DCPM sample.

The results of these investigations have shown that the composition of the tool material determines the composition of the phases developed at the friction surface in the cutting zone, but it also exerts an influence on the perfection of the crystalline structure of the new phases. The thinnest films of iron oxide formed on the HSS tool friction surface are crystalline, as shown by the pronounced F(R) function maxima at $R \approx 4.3$ Å and $R \approx 7.4$ Å (see Figure 7.19a). These are tribo-films or secondary structures of the second type, i.e., oxides whose composition is close to being stoichiometric [4,6] (see Chapter 1).

As seen in Figure 7.19b and Figure 7.19c, tools made of a TiC-containing composition of DCPM and of a [TiC, BN]-doped DCPM possess very different secondary structures in the cutting zone, which are quite distinct from the surface films formed at the HSS tool surface (Figure 7.19a). The surface of the DCPM tool reveals the development of tribo-film or secondary structures of the first type, i.e., supersaturated solid solutions having an amorphous-like structure [4]. This is shown

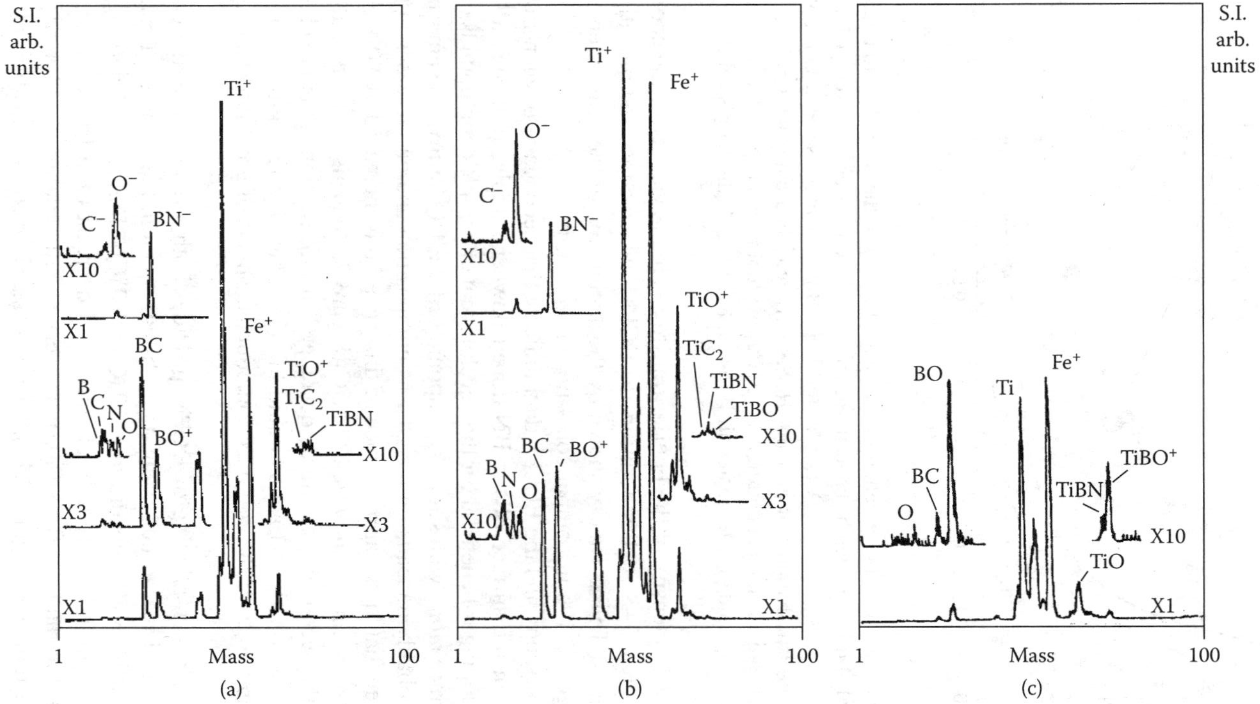

FIGURE 7.17 Mass spectra of secondary ions of BN-doped DCPM specimen after 4 min cutting determined at different depths beneath the surface of a wear crater: (a) 0.5 μm, (b) 0.15 μm, (c) at the surface. (From Fox-Rabinovich, G.S., Kovalev, A.I., Shuster, L.S., Bokiy, Y.F., Dosbaeva, G.K., Wainstein, D.L., Mishina V.P. On characteristic features of alloying HSS-based deformed compound powder materials with consideration for tool self-organization at cutting. Part 2. Cutting tool friction control due to the alloying of the HSS–based deformed compound powder material. *Wear* 1998, 214, 2. With permission.)

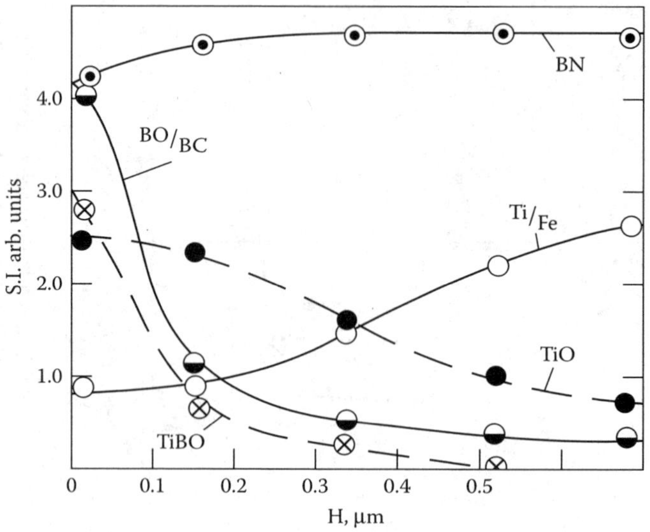

FIGURE 7.18 Change in absolute and relative values of the peak intensity of the secondary ion mass spectra for the BN-doped DCPM tool at different depths beneath the surface of the wear crater developed after 4-min cutting. (From Fox-Rabinovich, G.S., Kovalev, A.I., Shuster, L.S., Bokiy, Y.F., Dosbaeva, G.K., Wainstein, D.L., Mishina V.P. On characteristic features of alloying HSS-based deformed compound powder materials with consideration for tool self-organization at cutting. Part 2. Cutting tool friction control due to the alloying of the HSS–based deformed compound powder material. *Wear* 1998, 214, 2. With permission.)

by the attenuation of the peak intensity of the Fourier transforms in the vicinity of the coordination spheres at R ≈ 4–5 Å and R ≈ 7–8 Å. BN additions to DCPM enhance the amorphization effect for oxygen-containing phases. This is clearly shown by the attenuation of the peak intensity in the vicinity of 4–5 Å (see Figure 7.19c in comparison with Figure 7.19b).

Thus, both mass spectroscopy and the EELFS data indicate that secondary, oxygen-rich, amorphous structures develop on the surface of the BN-doped powder sample. Judicious alloying of carbide steel through BN doping is beneficial for the development of complex compounds of TiB_xO_y that appear on the tool surface along with simpler compounds of the TiO family. The amorphization of secondary structures probably depends on the DCPM composition and on increases in the level of BN alloying. One can see that the wear resistance of this material is increased by 80% compared to the DCPM having a base composition with 20% TiC (Figure 7.16). This suggests that alloying enhances the stability of the secondary structures developed during friction. This is enhanced by the presence of BO-type compounds that act as liquid lubricants (at elevated cutting temperatures [26]) and promote the adaptability of the complex compounds. This is of great importance for tool wear resistance.

Finally, it is possible that such alloying of DCPM provides both a reduction in the coefficient of friction and a broadening of the normal wear stage. In our opinion, the same goal can also be achieved in HSS-based DCPM by the substitution of TiC with TiCN. Figures 7.15 and Figure 7.16 demonstrate that this substitution is extremely effective and decreases the friction coefficient to abnormally low values (in the range of 0.03–0.05) at a service temperature of 500 to 550°C and significantly increases the tool life (Figure 7.16). In this case, the self-organization mechanism differs somewhat from the process used for materials alloyed with BN. With TiCN, the diffusional transfer of nitrogen into the chip arises from dissociation of TiCN during cutting [24]. The increased nitrogen concentration on the contact surface of the chip is a direct consequence of the mass transfer of nitrogen formed by dissociation of nitrides and carbonitrides. Such mass transfer takes place under extreme temperature and stress conditions encountered in the friction zone. This mass transfer

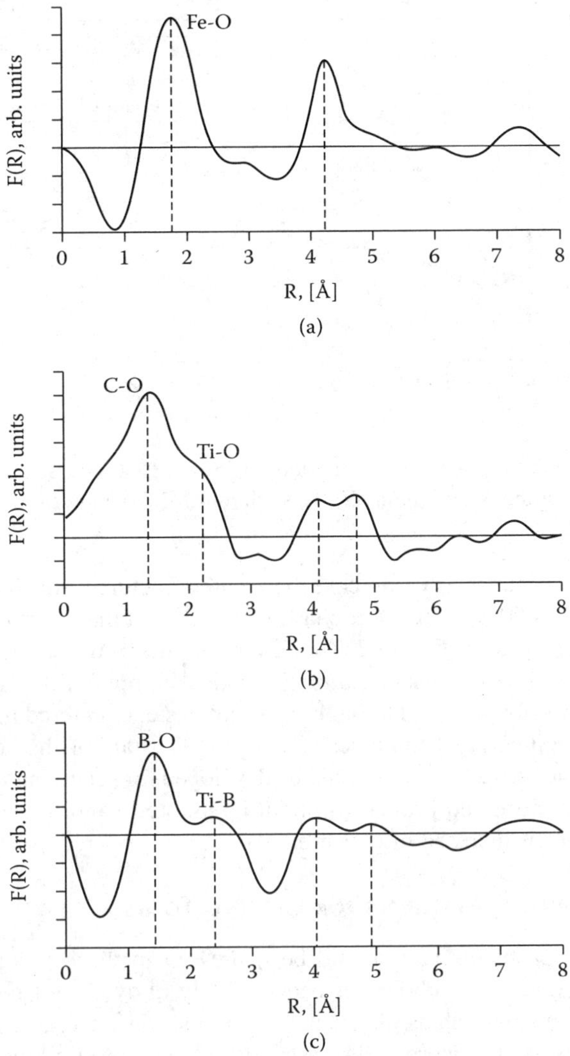

FIGURE 7.19 Fourier transforms of extended fine structure within the range of 250eV close to the line of elastic scattering of electrons with primary energy E = 1500 eV from the wear crater surface in a cutting tool made of (a) M2 HSS, (b) DCPM, and (c) BN-doped DCPM. (From Fox-Rabinovich, G.S., Kovalev, A.I., Shuster, L.S., Bokiy, Y.F., Dosbaeva, G.K., Wainstein, D.L., Mishina V.P. On characteristic features of alloying HSS-based deformed compound powder materials with consideration for tool self-organization at cutting. Part 2. Cutting tool friction control due to the alloying of the HSS–based deformed compound powder material. *Wear* 1998, 214, 2. With permission.)

could be beneficial for tool life enhancement [1]. It is worth noting that state-of-the-art cermets are also based on TiCN.

Therefore, the selection of titanium carbonitride for cermets and titanium nitride as the hard phase in Sandvik Coronites seems to be completely reasonable from the standpoint of tribological compatibility. The behavior of titanium nitride is comparable to a carbonitride [20,28] and this has been confirmed experimentally by the analysis of the self-organizing phenomenon of PVD Ti–N coatings (see Chapter 9).

Another important factor is the thermal stability of DCPM compared to cemented carbides. The importance of the thermal stability of this type of material is evident at elevated cutting speeds.

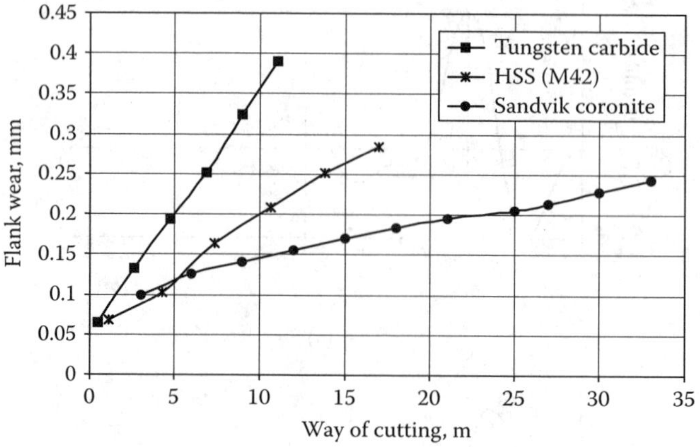

FIGURE 7.20 Comparative tool life of the cutting-tool materials. End-mill test data. Machined material — 1040 steel. Parameters of cutting: speed (m/min): 21; depth (mm): 3.0; width (mm): 5; feed (mm/flute): 0.028; cutting with coolant.

One way to improve this property is to increase the volume fracture of the hard phase in the DCPM; e.g., Sandvik Coronite has 50% of the hard phase (TiN) [28] and its wear resistance is higher than either HSS or cemented carbides (Figure 7.20). So the benefits of an adaptive material are obvious from the standpoint of both wear resistance and tribological compatibility. However, the application of these materials is currently limited, although the problems encountered might be partially solved by improved surface-engineering techniques. For example, state-of-the-art coating technologies together with the use of adaptive materials could be developed for functionally graded tool materials, i.e., materials whose structure and properties might be tailored from the core to the surface. That might result in a significant increase in tool life.

7.1.4 Functional Graded Materials for Cutting Tools

There are two pressing problems that should be addressed in the development of materials for cutting tools. The first one is the choice of economically alloyed tool materials because of the severe shortage of core metals such as tungsten. This problem is becoming more relevant although manufacturers cautiously try to ignore it. The second problem is to find the optimum combination of different, and even contradictory, properties in the same cutting material. New innovative materials are needed. The material of the interest will have to combine such diverse properties as high hardness and ductility with excellent tribological properties at elevated temperatures. A recent development is the production of a functionally graded material, whose properties (wear resistance and strength) vary from the core to the surface. This class of functionally graded materials (FGM) combines a high wear resistance at the surface with high strength and toughness of the core. The intermediate layer has graded properties that lie between those of the core and the surface. Promising developments in this field have been reported by the Laboratory of Materials Processing and Powder Metallurgy of the Helsinki University of Technology. They have developed a novel functionally graded material having a surface ceramic layer, a graded WC–cermet composition with high crack resistance, and a cemented carbide core with excellent toughness. A high crack-resistance parameter value (K_{1C} = 25 MPa m$^{1/2}$) at a hardness of 1500 HV (typical for tool steels with half the hardness) was found (http://www.hut.fi/Units/LMP/).

There are two methods used for FGM processing. The first one, noted in the preceding text, is a surface-engineering method. This method has unique possibilities and versatility. Other methods, principally those based on powder metallurgy, are also widely used. Combinations of these two methods have recently been put into practice. Thus, tools manufactured by ordinary sintering processes, having high-toughness cemented carbide substrates with high wear-resistance ceramic coatings and functionally graded interlayers, show excellent wear resistance [29]. FGM can have superior wear resistance, resistance to fracture, and good thermal shock resistance in comparison to conventional cermets, with a beneficial compressive residual stress distribution [30]. Ceramics have also been recently developed with functionally graded structures. In order to combine high hardness and high toughness, graded ceramics of Al_2O_3+TiC (surface)–Al_2O_3 + Ti (inner core) and SiAlON (inner core) have been successfully developed [31]. Functionally graded ceramic tools exhibit better cutting abilities than regular ceramic tools [32]. FGM have also been successfully used for milling applications [33]. Functionally graded powder materials are normally used for high-speed roughing operations. For this specific application, a functional gradient material with outer layer containing WC–Co cemented carbide, which improves fracture toughness on the TiC–TaC(Nb)C–Co substrate, was developed. The functional gradient layer forms an excellent base for subsequent CVD or PVD coatings, such as multilayered CVD (TiC+TiCN+TiN+Al_2O_3) + PVD TiN and even more sophisticated coatings [5]. Functionally graded powder materials can also be successfully employed in the domain of HSS tools' application, particularly with functionally graded cement carbide and hard PVD coatings [30].

REFERENCES

1. Trent, E.M., Wright, P.K. *Metal Cutting*, 4th ed. Butterworth-Heinemann: Woburn, MA, 2000.
2. Fox-Rabinovich, G.S., Kovalev, A.I., Weatherly, G.C. Tribology and the design of surface engineered materials for cutting tool applications. In *Modeling and Simulation for Material Selection and Mechanical Design*, Eds., Totten, G., Xie, L., Funatani K. Marcel Dekker: New York, 2004, pp. 301–382.
3. Geller, Y. *Tool Steels*. Mir: Moscow, 1978, p. 657.
4. Santhanam, A.T., Tierney, P. Cemented carbides. In *Metals Handbook*, 9th ed., Vol. 16, Ed., Burdes, B.P. American Society for Metals, Metals Park: OH, pp. 71–89.
5. Brookes, K.J.A. *Hardmetals and Other Hardmaterials*, 3rd ed. International Carbide Data: Hertfordshire, U.K., 1998, p. 220.
6. Shao, G.-Q., Duan, X.-L., Xie, J.-R., Yu, X.-I., Zhang, W.-F., Yuan, R.-Z. Sintering of nanocrystalline WC-Co composite powder. *Rev. Adv. Mater. Sci.* 2003, 5, 281–286.
7. Ashley, S. Small scale structure yields big property payoffs. *Mech. Eng.* 1994, 116, 2, 52–57.
8. Hadjipanyayis, G.C., Siegel, R.W. *Nanophase Materials: Synthesis, Properties and Applications*. Kluwer: Dordrecht, Netherlands, 1994, 227–246.
9. Ivanova, V.S. *Fracture Synergetic and Mechanical Properties: Synergetic and Fatigue Fracture of Metals*. Moscow: Science, 1989, pp. 6–27.
10. Kim, H.-J., Lee, C.-H., Hwang, S.-Y. Fabrication of WC-Co coatings by cold spray deposition, *Surf. Coat. Technol.* 2005, 191, *188*, 335–340.
11. Shen, Z. Spark plasma sintering of alumina. *J. Am. Ceram. Soc.* 2002, 85, 1921–1927.
12. Schneider, G. Cutting-tool materials. *Tooling and Production*, 2001, 2, 87.
13. Powell, R. Cemented tungsten carbides for tools, dies and wear parts. *Tooling and Production*, 1999, 1, 58.
14. Kostetsky, B.I. *Surface Strength of the Materials at Friction*. Technica: Kiev, 1976, 76–154.
15. Yang, Q., Senda, T., Kotani, N., Hirose, A. Sliding wear behavior and tribofilm formation of ceramics at high temperatures, *Surf. Coat. Technol.* 2004, 184, *2–3*, 270–277.
16. Gruss, W.W. Cermets. In *Metals Handbook*, 9th ed., Vol. 16, Ed., Burdes, B.P. American Society for Metals: Metals Park, OH, 1989, pp. 90–104.
17. Fujisawa, T. Cermet cutting tool consisting of titanium carbonitride with high resistance to thermal shocks. Japanese Patent Kokai Tokkyo Koho JP 2000 54,055 (Cl. C22C29/04) 220, February 20, 2000.
18. Shevela, V.V. Internal friction as a factor of wear resistance of the tribosystems. *J. Friction Wear* 1990, 11, *6*, 979–986.

19. Fox-Rabinovich, G.S., Kovalev, A.I., Shuster, L.S., Bokiy, Y.F., Dosbaeva, G.K., Wainstein, D.L., Mishina V.P. Characteristic features of wear in HSS-based compound powder materials with consideration for tool self-organization at cutting. Part 1. Characteristic features of wear in HSS-based deformed compound powder materials at cutting. *Wear* 1997, 206, 214–220.
20. Uchida, N., Nakamura, H. Influence of chemical composition of matrix powders on some properties of TiN dispersed and carbide enriched HSS. *Proc. 12th Int. Plansee Seminar*, Vienna 1989, 2, 541–555.
21. Fox-Rabinovich, G.S., Kovalev, A.I., Shuster, L.S., Bokiy, Y.F., Dosbaeva, G.K., Wainstein, D.L., Mishina V.P. Characteristic features of wear in HSS-based compound powder materials with consideration for tool self-organization at cutting. Part 2. Cutting tool friction control due to the alloying of the HSS-based deformed compound powder material. *Wear* 1998, 214, 279–286.
22. Shuster, L.S. *Adhesive Interaction in Metal Solids*. Gilem: Ufa, 1999, pp. 4–196.
23. Trent, E.M., Suh, N.P. *Tribophysics*. Prentice-Hall: Englewood Cliffs, NJ, 1986, 125–489.
24. Fox-Rabinovich, G.S., Kovalev, A.I., Afanasyev, S.N. Characteristic features of wear in tools made of HSS with surface engineered coatings. Part 2. Study of surface engineered HSS cutting tools by AES, SIMS and EELFAS methods. *Wear* 1996, 198, 280–286.
25. Kostetsky, B.I. An evolution of the materials' structure and phase composition and the mechanisms of the self-organizing phenomenon at external friction. *J. Friction Wear* 1993, 14, *4*, 773–783.
26. Beliy, V.A., Ludema, K., Mishkin, N.K. *Tribology: Studies and Applications — USA and USSR Experience*. Allerton Press: New York, 1993, pp. 202–452.
27. Tretiakov, I.P., Vereshchaka, A.S. *Cutting Tools with Coatings*. Mashinostroenie: Moscow, 1994, pp. 17–297.
28. Oskarsson, R., Von Holst. P. Sandwick Coronite — A New Compound Material for End Mills. Metal Powder Report 1989, 44, *12*, 44–56.
29. Zhao, C., Vandeperre, L., Vleugeks, J., Van Der Biest, O. Innovative processing/synthesis ceramic, glasses, composites III. *Ceram. Trans.* 2000, 108, 193–201.
30. Nomura, T., Ikegaya, A. Functionally graded cemented carbide tools. *Mater. Integr.* 1999, 12, *10*, 13–19.
31. Moriguchi, H., Nomura, T., Tsuda, K., Isobe, K., Ikegaya, A., Moriyama. K. Design of functionally graded cemented carbide tools. *Funtai Funmatsu yakin Kyokai* 1998, 45, *3*, 231–236.
32. Van Der Biest, O., Vleugels, J. Perspectives on the development of ceramic composites for cutting tool application. *Proc. Int. Conf. on Cutting Tools and Machining Systems*, Atlanta, 2001, pp. 156–161.
33. Schmauder, S., Melander, A., McHugh, P.E., Rohde. J. New tool material with structure gradient for milling application. *J. Phys.* 1999, 4, *9*, 147–156.

8 Formation of Secondary Structures and the Self-Organization Process of Tribosystems during Friction with the Collection of Electric Current

Iosif S. Gershman

CONTENTS

8.1 Introduction....197
8.2 Instability of the Sliding Electric Contacts....198
8.3 Experimental Study of Instability of the Tribosystem with Current Collection....200
8.4 Mechanisms of the Stability Loss at Change of Current....203
8.5 Studies of Contact Resistance Change with Current during Sliding....206
8.6 Studies of the Secondary Structures on the Friction Surface....209
8.6.1 Properties of Current-Collecting Materials and Their Wear Resistance....210
8.6.2 Composition of the Secondary Structures Formed on the Surface of the Copper Contact Wires....212
8.7 Natural Graphite-Based Current-Collecting Material....216
8.8 Compatibility of the Current-Collecting Materials Working in Contact with a Wire Simultaneously....218
8.9 Impact of the Catalyst on the Wear Resistance of Sliding Electric Contacts....222
8.9.1 Effect of CO_2 Supply to the Sliding Electric Contact Area on the Wear Rate of the Current-Collecting Materials....222
8.9.2 Impact of the Catalyst on the Wear Resistance of the Sliding Electric Contacts....223
8.10 Conclusions....226
References....226

8.1 INTRODUCTION

Self-organization of the sliding electric contacts during friction will be considered in the present chapter. The occurrence of self-organization will be shown using an example of friction with current collection (sliding electric contacts). This concept will be used for the explanation and for the rational selection of frictional materials as well.

Friction processes with current collection were chosen as an example because this type of process is widely used and is of importance for practical applications (such as electric motors, generators, etc.). The self-organization of friction with current collection is of interest because there are two independently controlled processes that are operating within the system: friction and the passage of an electric current through the contact. In the other tribosystems, as well as in tribosystems with current collection, a physical and chemical interaction with the environment, counterbody, and lubrication always take place in addition to the friction itself. However, the physical and the chemical interactions are not controlled processes. The system "chooses" the types of physical and chemical interactions with the environment by its dependence on the state of the system and the operating conditions. The characteristics of friction (a load applied, sliding speed, and a coefficient of friction *in situ*) and the parameters of the current collection (current and contact voltage *in situ*) in the tribosystem with current collection are controlled independently. The interaction of these processes is of great interest. The processes of current collection, heat conductivity, chemical interaction, and mechanical stresses are described by the tensors with ranks of various evenness. Therefore, these processes can interact under conditions that are far from equilibrium, i.e., in the field of nonlinear relations between the thermodynamic forces and flows. In contrast, there can be no interaction between them in a linear area.

The interaction of friction and current collection is shown in the example in Chapter 2 (Equation 2.88 and Equation 2.89) for the lubricating action of a current and in which the coefficient of friction is reduced with current under low-wear-rate conditions. Analytical expressions for the lubricating action of the current are obtained based on the assumption that the two processes (friction and current collection) interact with each other in a steady state with minimal entropy production under wear-free conditions.

The coefficient of friction starts to depend on the current values once a certain value of the current is achieved. This fact and the independence of the k/k_0 relation (see Chapter 2, Table 2.1) vs. k_0, i.e., from the entry conditions, can be used as an indirect validation of the formation of dissipative structures.

Experimental validation of the analytical expression for the lubricating action of a current and formation of the dissipative structures shows that the concepts of self-organization and nonequilibrium thermodynamics could be used to explain the friction phenomena, in particular the sliding electric contacts.

Self-organization begins because the system loses stability. Thus, the self-organizing of the sliding electric contacts will be considered, starting with the condition of its stability loss.

8.2 INSTABILITY OF THE SLIDING ELECTRIC CONTACTS

At least two independent sources of energy dissipation are presented in the tribosystem with current collection. They are: (1) friction and (2) the passage of current through the sliding contact. Thus, there are two flows of fixed forces within the system: the flow of heat towards the frictional bodies and an electric current. Entropy production for this case will be written down in a way similar to Equation 2.86:

$$\frac{dS_i^*}{dt} = \frac{(kpv)^2}{\lambda BT^2} + \frac{J_e^2 R}{T} \qquad (8.1)$$

where R is the electric resistance of the contact.

A friction system with current collection differs from the usual friction system because of the passage of an electric current through the sliding contacts. Therefore, it is of interest to study the influence of an electric current on the tribosystem's stability.

For this purpose, we shall consider an excess of entropy production at a change of current. It follows from Equation 2.61 and Equation 8.1 that:

$$\frac{\partial}{2\partial t}\delta^2 S = \frac{(pv)^2}{T^2 B\lambda}\left(\frac{\partial k}{\partial J_e}\right)^2 (\delta J_e)^2 + \frac{1}{T}\left(R + J_e \frac{\partial R}{\partial J_e}\right)(\delta J_e)^2 \tag{8.2}$$

It can be seen from this equation that the fluctuation of electric current is related to friction (the first summand of the right-hand side) and can only be positive as it represents the quadratic form.

Part of the excess of entropy production is related to the passage of an electric current (the second summation in Equation 8.2). This contribution appears to promote the change of sign of this value. This contribution is connected to the derivative $\frac{\partial R}{\partial J_e}$. If this derivative is negative, then the product is also negative, i.e. $J_e \frac{\partial R}{\partial J_e} < 0$. If the absolute values of positive contributions to the excess of entropy production are exceeded, the last one can be negative. Hence, the given state of the system can lose stability.

Thus, to achieve a probable loss of the disturbed stable condition, it is necessary for the tribosystem with current collection to follow the inequality:

$$\frac{\partial R}{\partial J_e} < 0 \tag{8.3}$$

This condition is reached when the contact resistance drops vs. the electric current through the contact. The contact resistance can decrease because of the growth of the area of contact as a result of the frictional body softening and because of the formation of secondary structures on the frictional surface. These secondary structures are semiconductors.

Equation 8.2 is obtained based on the assumption that the heat conductivity (λ) and current (j_e) are independent characteristics. If it is assumed that the heat conductivity depends on the current, then the expression for the excess of entropy production Equation 8.2 will be as follows:

$$\frac{\partial}{2\partial t}\delta^2 S = \frac{(pv)^2}{T^2 B\lambda}\left(\left(\frac{\partial k}{\partial J_e}\right)^2 - \frac{k}{\lambda}\frac{\partial \lambda}{\partial J_e}\frac{\partial k}{\partial J_e}\right)(\delta J_e)^2 + \frac{1}{T}\left(R + J_e \frac{\partial R}{\partial J_e}\right)(\delta J_e)^2 \tag{8.4}$$

The first term on the right-hand side of this equation in contrast to Equation 8.2 can become negative if the derivatives of heat conductivity and coefficient of friction with respect to current have identical signs. If the condition in Equation 8.3 is fulfilled, it is possible to expect that $\frac{\partial \lambda}{\partial J_e}$ > 0. Hence, the first term of the right-hand side of Equation 8.4 can become negative if:

$$\frac{\partial k}{\partial J_e} > 0 \tag{8.5}$$

It follows from Table 2.1 that before the effect of the lubricating action of the current starts to develop, the coefficient of friction increases with the current.

If the conditions in Equation 8.3 and Equation 8.5 are fulfilled, the tribosystem with current collection can lose stability. This can happen because the positive production of entropy excess is a sufficient but not necessary condition of stability.

If one of these conditions is fulfilled, then the tribosystem with current collection can lose its stability. After this occurs, a dissipative structure can appear.

If the conditions of Equation 8.3 and Equation 8.5 are fulfilled simultaneously, the probability of the tribosystem losing stability and the formation of dissipative structures increase.

To conclude this part of the chapter, we have to note that the contact resistance, the heat conductivity, and the coefficient of friction can vary with the electric current as a result of the frictional bodies interaction with the environment, i.e., owing to physical and chemical changes within the surface layers (i.e., secondary structures).

8.3 EXPERIMENTAL STUDY OF INSTABILITY OF THE TRIBOSYSTEM WITH CURRENT COLLECTION

A testing apparatus was made to simulate the operation conditions of a "contact wire–head current-collection" system and to study the relation between the wear rate of the materials investigated and the electric current passing through the contact [1] or (Figure 8.1). A ring (200-mm diameter and 3-mm thickness) made of cold-worked copper wire was soldered to a copper disc made of a 0.5-mm-thick foil. The disc was rotated around the ring axis with a velocity of 60 r/min. Two specimens were fabricated from the current-collecting materials and were pressed on the wire at diametrically opposite points. The pressing load was 10 N. The length of the linear contact with a current-collecting specimen was 8 mm. An alternating current was passed through the sliding contacts. To ensure the transverse motion of the wire, the centre of rotation was displaced by 10 mm from the centre of the wire ring. The wear rate of the wire was determined by means of measuring its diameter using a micrometer. The wear of the specimens were determined by the mass loss using an analytical balance. Current-collection samples were made of standard artificial graphite-based contact inserts.

Sliding electric contacts differ from other tribosystems by the passage of an electric current. Therefore, we have discovered the effect of currents on tribosystems. The data on wear rate of the current-collecting material vs. current is presented in Figure 8.1. A similar dependence was observed for the copper wire. Curve 1 (a wire without initial tribofilms) was obtained from the testing data of the copper wire with the surface tribo-films removed before each test. Curve 2 (a wire with the initial tribo-films) was obtained from the testing data of the wire with the surface tribo-films obtained using a current of 80 A.

It is possible to indicate five segments on the curve in Figure 8.1. The first segment indicates that the wear rate of the current-collecting material does not vary with the current while it increases

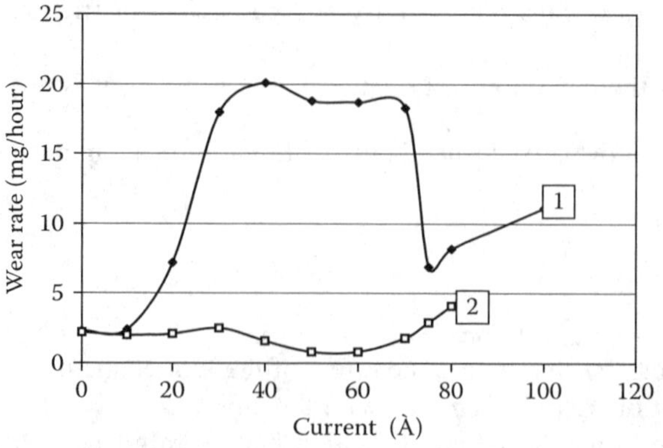

FIGURE 8.1 The wear rate of the artificial graphite-based current-collecting material vs. current: (1) wire without the initial films and (2) wire with the initial films formed under current value of 80 A.

from 0 to 10 A. The wear rate sharply grows almost by an order of magnitude at the change of a current from 10 to 30 A. From 30 to 70 A, the wear rate remains stable again. As the current increases from 70 to 75 A, the wear rate sharply decreases by approximately three times. Finally, during the last segment, further increases in current to 100 A results in the wear rate to growing gradually again. At current values of 100 A the wear rate is almost twice as large when compared to the wear intensity at a current of 30 A.

The first stability loss on the "wear rate vs. current" curve (Figure 8.1, curve 1) occurs at the current values above 10 A. This is related to the softening of the surface layers of the copper wire. This can lead to the growth of the area of contact and corresponds to the decrease of the contact resistance. This is associated with the condition of instability indicated in Equation 8.3. The softening of the surface layers of the wire can occur because of the increase in temperature as a result of a passage of electric current. However, significant irreversible hardness reduction is a result of the recrystallization of the surface layers of the wire. Intensive growth of the wear rate of the contact wire and current-collecting material at recrystallization is confirmed elsewhere [2–3]. It is shown in Reference 4 that up to 150°C, the temperature impact on the wear rate within a contact zone is insignificant. It is also shown in Reference 3 that at 200°C, local recrystallization of surface layers of the contact wire occurs. This leads to an intensive wear rate increase of the contact wire and the current-collecting material. Recrystallization occurs once the temperature threshold is achieved. Thus, it can be expected that once an electric current achieves a certain value, the result of a stability loss will occur as followed from Equation 8.2. According to this equation, the excess of entropy production can change signs when a negative contribution (that is proportional to the electric current value) exceeds the positive contribution by its absolute value.

The metallographic studies performed prove indirectly that stability loss takes place owing to the recrystallization of the surface layers of a contact wire. Subsurface layers of the copper contact wires that are working in contact with the coal current-collecting materials have a recrystallized microstructure [4].

Secondary structures are formed on the copper wire at small values of current (0 to 30) owing to the transfer of graphite from the current-collecting material. The particles of graphite cover part of the copper surface without any physical and chemical interaction. At the values of current above 30 A, recrystallization and oxidation of the copper surface begins simultaneously and a CuO oxide film forms. This oxide is stable within a range of temperatures between 300 and 800°C [5]. Intensive wear rate growth of the current-collecting material during recrystallization can be related to the increase in contact area that corresponds to the increased consumption of graphite from the current-collecting material. Within a range of current values between 30 and 70 A, these processes become stationary; a new condition of the system becomes steady and the wear rate becomes constant.

It is worth noting that, within a given range of electric current, during the processes described in the preceding text the secondary structures formed have not a dissipative but an equilibrium character. They consist of copper, copper oxides that are stable under given temperature conditions, and graphite. The recrystallization of copper is an equilibrium process for the given temperature. The wear rate grows with passage of the equilibrium processes.

The second loss of stability occurs at current values above 70 A. This leads to a significant wear rate decrease of the current-collecting material, by approximately three times, as the current increases from 70 to 75 A (Figure 8.1, curve 1). The sharp decrease of the wear intensity occurs within a narrower range of current variation; therefore, we can call the process behavior that follows that of curve 1 in Figure 8.1 a catastrophe (i.e., a sudden change of the function with a gradual change of the argument) [6]. The catastrophe occurs during a stability loss by the system [6]. It is related to the transformations that are taking place in the secondary structures.

The condition in Equation 8.3 is fulfilled within this range owing to the semiconductor nature of the new secondary structures because their electric resistance reduces with the electric current. Study results of the secondary structures' electric properties are presented in Reference 7. A typical curve of the secondary structures resistance for the copper contact wire that works in contact with

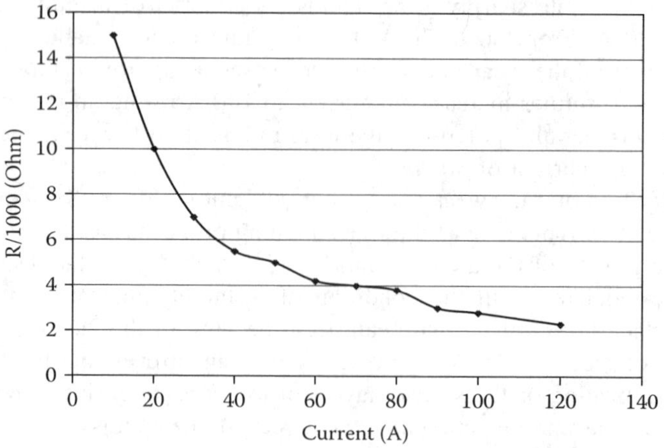

FIGURE 8.2 The resistance of secondary structures formed on the surface of a copper contact wire working with an artificial graphite-based current-collecting material vs. electric current.

the artificial graphite-based current-collecting material vs. electric current is shown in Figure 8.2, where the resistance of the secondary structures is reduced with the growth of an electric current. The resistance of the secondary structures was determined under static conditions (without sliding).

The intensive wear rate decreases for the current-collecting material at a current value of 70 A. This can be related to the adsorption of carbon on the surface of the secondary structures from the atmosphere but not from the current-collecting material.

It was outlined in Chapter 2 that the structure, behavior, properties, and state of the dissipative structures do not depend on the entry conditions. This means that the composition, structure, and properties of the dissipative secondary structures are controlled by the composition of the frictional bodies and the environment. Hence, after identical modes of operation they should have a certain composition. Studies of the secondary structures composition of the copper contact wires show [8] that they always contain copper, a protoxide of copper and carbon (as graphite). Under relatively mild modes of friction, they also contain a copper oxide. Because the dissipative secondary structures have a certain composition, a carbon supply from the atmosphere results in the reduction of its supply from the coal counterbody. This leads to wear rate decrease.

The second curve (curve 2, Figure 8.1) is obtained under conditions of sliding for a wire with secondary structures present on the surface. These films have been retained on the surface after testing at a current value of 80 A. Under conditions of the sliding for the wire with secondary structures present on the surface, the wear rate slightly varies with the current (Figure 8.1). The wear rate for this condition is lowered by an order of magnitude at a current value of 20 to 70 A. For a current of 75 to 80 A, the wear rate is decreased by two times when compared to the wear rate for the wire without secondary structures on the surface under the corresponding current conditions.

During testing a slight dependence of the wear rate of the current-collecting material vs. the current values with previously formed secondary structures shows that the processes of their formation have already been completed. In this case for each specific current value, the stationary processes that are taking place within the system are stable. This is proved by the effect of the lubricating action of the current. This effect takes place under steady state conditions with minimal entropy production. Under such conditions, the growth of entropy production due to the electric current increase is compensated by the decrease in the coefficient of friction with hardly any change in wear rate. As a result, the system returns to the initial steady state with minimal entropy production and minimal wear rate. Thus, stability of the steady state with minimal wear rate is provided.

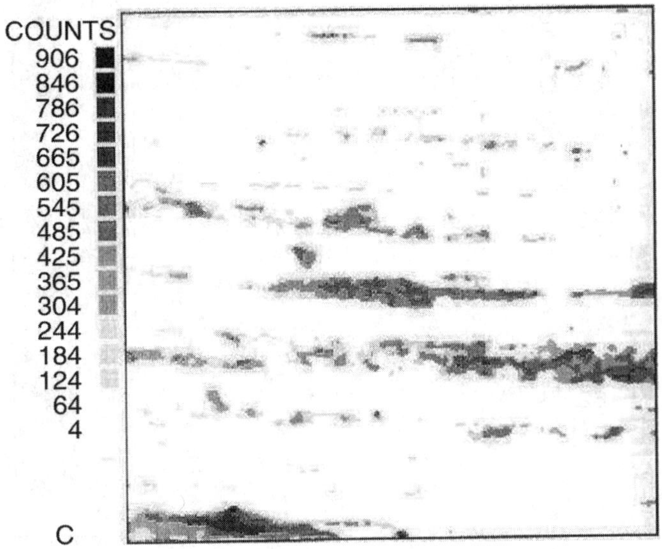

FIGURE 8.3 The carbon spread on the friction surface of a copper contact wire under the current value of 0 A. 512 × 512 μm.

8.4 MECHANISMS OF THE STABILITY LOSS AT CHANGE OF CURRENT

The wear rate increase with current-collection intensity is quite natural. In contrast, wear rate decrease with further growth of current-collection intensity is quite unexpected. Therefore, the mechanism of the second instability (curve 1, Figure 8.1) is of considerable interest. If the mechanism of instability is established, it will provide proof for the rational selection of the current-collecting material. Such material passes through instability under milder conditions and reduces the wear rate during earlier stages of the process.

To understand the mechanism of instability, a surface of copper wire after friction with current collection was studied in contact to an artificial graphite-based material. Microstructural studies and the investigation of the chemical elements' distribution on the working surfaces of a wire have been made using SEM and EDX. The images obtained were investigated using metallographic analyzer. During the investigation of the copper wire, attention was mainly focused on the carbon presence. In contrast, during the studies of the carbon-based current-collection material, attention was mainly on the copper presence. Two structural modifications of the carbon were found on the working surface of the copper wire, the fist one being the film that covers the surface of the wire. This carbon was transferred from the current-collecting material and "spread" on the working surface of the wire, with the islandlike carbon films extended in the direction of friction.

Typical x-ray images of the carbon films are shown in Figure 8.3 to Figure 8.6. As follows from Figure 8.3 to Figure 8.6, the area of the carbon films that are transferred from the counterbody grow significantly at a current value of 55 A as compared to the current values of 0 and 20 A. The area of the transferred carbon films at a current value of 80 A is lower than at the value of 55 A. This is supported by metallography data. The total area of the transferred carbon films has been determined. The relative areas of the working surfaces of the wires that are covered by the transferred carbon are shown in Table 8.1. This data correlates with the results of metallographic studies presented in Figure 8.3 to Figure 8.6.

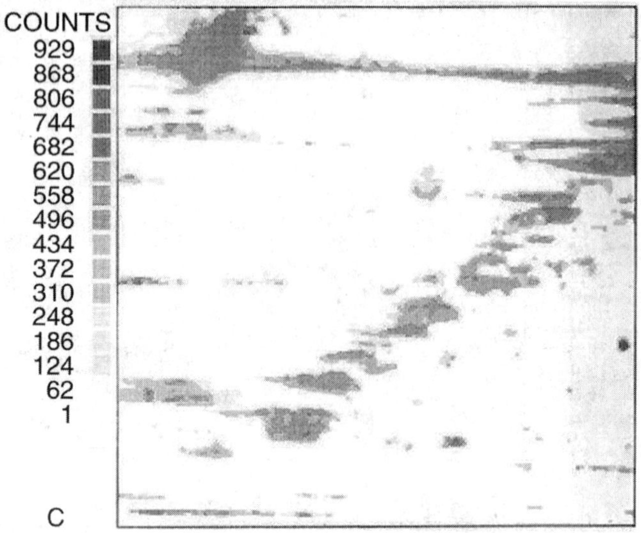

FIGURE 8.4 The carbon spread on the friction surface of a copper contact wire under the current value of 20 A. 512 × 512 μm.

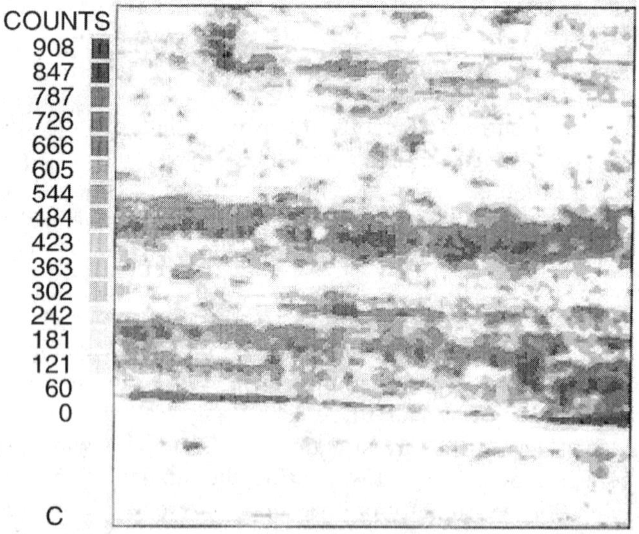

FIGURE 8.5 The carbon spread on the friction surface of a copper contact wire under the current value of 55 A. 512 × 512 μm.

Table 8.1 presents the data on the wear rates of the current-collecting material vs. the current values. As follows from the data shown in Table 8.1, the larger the area of the working surface of the wire covered by the transferred carbon, the higher the wear rate of the wire and current-collecting material. This data contradicts traditional beliefs that the formation of graphite on a surface of a contact wire enhances the wear rate decrease for both contacts [9].

To clarify the situation, the transferred carbon-free sites of the friction surface of the wire were investigated. However, carbon is also present on these sites, but the source of this carbon is different. This "introduced" carbon was evenly distributed on a surface of the copper matrix as a fine, no-bigger-than-1-μm-diameter inclusions. The amount of the introduced carbon in the copper matrix (mass percentage) is shown in Table 8.1.

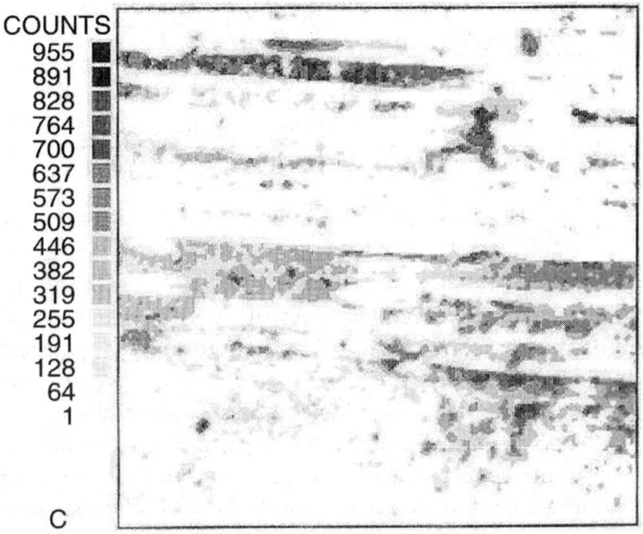

FIGURE 8.6 The carbon spread on the friction surface of a copper contact wire under the current value of 80 A. 512 × 512 μm.

TABLE 8.1
Carbon Contents on a Surface of a Copper Wire and Wear Rate of Sliding Contacts vs. Current (Artificial Graphite-Based Current-Collecting Materials)

Current Value (A)	0 (without friction)	0	20	55	80
Transferred carbon (% of the area of a surface of a copper wire)	—	45	50	90	60
Carbon content in a copper matrix (% mass)	7	7	8	2.5	5.5
Wear rate of a current-collecting mater. (mg/hr)	—	2.5	6	18	6
Wear rate of a copper wire (μ/hr)	—	0.2	0.4	1.1	0.3

The wear rate increases with the amount of introduced carbon in the copper matrix. Such a dependence of wear rate and the amount of the introduced carbon in the copper matrix vs. current values allows us to assume that the mechanism of the event that has occurred varies under different current conditions. At current values lower than 20 A, carbon is introduced in the copper matrix as a result of adsorption. It is introduced into the copper matrix by the adsorption of carbon dioxide on the copper surface because the carbon contains in a copper matrix prior to the beginning of friction (Table 8.1). If the current values are below 20 A, the amount of the introduced carbon does not depend on the current values whether or not the friction has started. This happens prior to friction and during friction. At current values above 20 A, recrystallization of the cooper surface layers occurs [4], which results in irreversible growth of the contact area and decrease of the adsorbed carbon concentration in the copper.

The increase of the carbon content in the copper matrix occurs as a result of nonequilibrium reactions of the carbon reduction by copper at the current values above 70 A:

$$4Cu + CO_2 = C + 2Cu_2O \tag{8.6}$$

$$2Cu + CO_2 = C + 2CuO \tag{8.7}$$

TABLE 8.2
Affinity of Reactions 8.6, 8.7

	Affinity (kJ/mole)		
Reactions	**T = 298 K**	**T = 1000 K**	**T = 2000 K**
(8.6)	–102	–200	–342
(8.7)	–269	–271	–477

The second instability is dependent on the values of wear rate vs. current (curve 1, Figure 8.1) and is related to the aforementioned reactions. According to the data presented in Table 8.2 [10], the affinity of reactions in Equation 8.6 and Equation 8.7 are negative at various temperatures and therefore they can only take place forcibly at nonequilibrium conditions.

Following from the data shown in Table 8.1, the wear rate of the contacts decreases compared to the amount of carbon in the copper matrix. The beneficial role of CO_2 is outlined in Reference 11 to Reference 12. In these works, the composition of the secondary structures on the surface of copper sliding contacts at the passage of a current of 100 μA was investigated using Auger spectroscopy. During the work of the contacts in vacuum seizure took place, whereas in a wet CO_2 environment seizure did not occur. It is believed that the formation of monomolecular films of CO_2–H_2 on the contacting surfaces takes place. Beneficial impacts of CO_2 on the intensity of wear process and its influence on the average coefficient of friction values for copper on a copper tribocouple are outlined in Reference 13. It was shown in Reference 14 that gaseous addition to the atmosphere, in particular H_2S, enhances wear rate, and the coefficient of friction decreases for noble metals such as Au, Pt, and Pd.

Thus, passage of the reactions in Equation 8.6 and Equation 8.7 has a favorable effect on the wear resistance of sliding contacts. Hence, to reduce wear rate of the sliding electric contacts it is desirable to initiate the nonequilibrium reactions in Equation 8.6 and Equation 8.7 under milder conditions, i.e., before the recrystallization occurs.

8.5 STUDIES OF CONTACT RESISTANCE CHANGE WITH CURRENT DURING SLIDING

The sign of the contact resistance derivative controls the possible occurrence of instability for the tribosystem with current collection (Equation 8.3). Therefore, the dependence of the contact resistance with respect to current is investigated in this section.

The dependence of the contact resistance with current was determined *in situ*, during the process of sliding. The measurements were made using the apparatus described in Section 8.3.

Figure 8.7 presents the contact resistance vs. current. The contact resistance, as well as the resistance of the secondary structures (Figure 8.2), is reduced with current. The general trend is the gradual decrease of the resistance with current. However, there are two areas of intensive resistance decrease: at current values of 40 and 70 A. This is proved by the dependence of the derivative dR/dI on the current (Figure 8.8) for the graph shown in Figure 8.7.

The current values (40 and 70 A) with the minimal values of derivative dR/dI correspond to the points at which the given state stability loss have the highest possibility. At these values of current, the wear rate changes significantly (Figure 8.1), especially at the current value of 70 A. This confirms two conclusions: (1) the possibility of the loss of the system's stability if the ratio Equation 8.3 is fulfilled and (2) the system losses its stability at sudden changes in the wear rate (Figure 8.1).

The contact resistance (Figure 8.7) was determined in the presence of the secondary structures that are typical for the given current values. Hence, this dependence can be related to curve 1 in Figure 8.1. Curve 2 in Figure 8.1, represents the sliding conditions with the secondary structures,

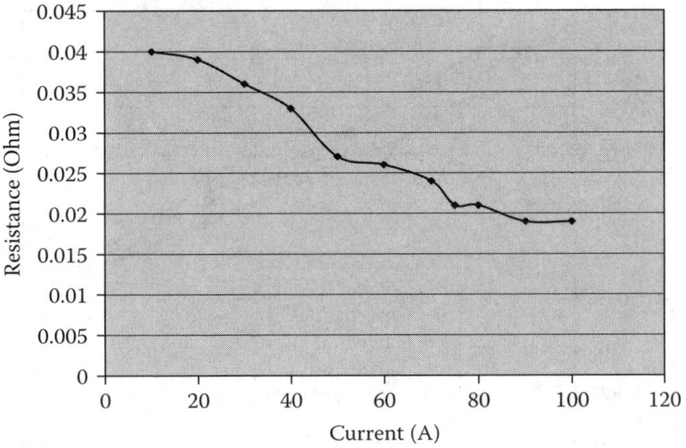

FIGURE 8.7 Resistance vs. current for the contact between a copper contact wire and an artificial graphite-based current-collecting material.

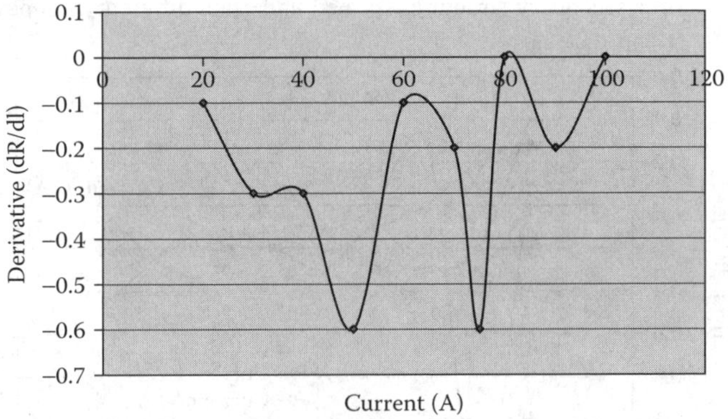

FIGURE 8.8 The derivative *dR/dI* vs. current for the graph presented in Figure 8.7.

which are typical for current values of 80 A. The wear rate is lower on curve 2 as compared to curve 1 (Figure 8.1). Therefore, the dependences of the contact resistance with current under the conditions of sliding on the surface of the secondary structures formed under various currents values are of great interest. For this purpose, sliding of the samples made of artificial graphite-based materials on the surface of the copper wire were performed for 3 h under current values of 0, 30, 55, and 80 A. Then, for each contact, the dynamic volt–ampere characteristics were determined for current values from 0 to 100 A. The contact resistance vs. current is shown in Figure 8.9. The derivatives d*R*/d*I* vs. current for the curves shown in Figure 8.9 is presented in Figure 8.10.

As follows from Figure 8.9, the contact resistance reduces with current for all the secondary structures formed on the surface. However, a gradual decrease of the contact resistance is observed only for contacts with secondary structures that are formed under current-free conditions (except for the area from 10 to 20 A), as well as under current values of 30 A. This is proved by the graphs shown in Figure 8.10. The derivatives *dR/dI* for these contacts are negative and vary slightly from –0.2 to –0.3 within all the ranges of current change. The resistance of the contact with secondary structures (formed at the current values of 55 A) grows if the current increases to 40 A. At current values above 40 A the contact resistance starts to reduce. It follows from the corresponding curve in Figure 8.10 that the derivative *dR/dI* gradually reduces from 0.3 to 0 with a current change from 20 to 40 A. The derivative become negative at a current value above 40 A; it remains constant

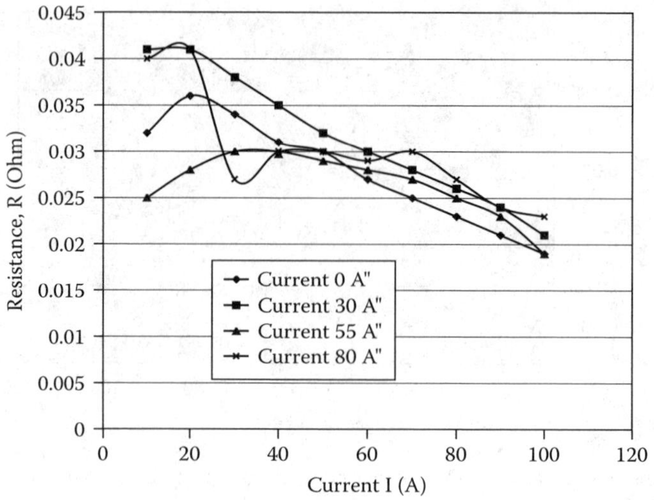

FIGURE 8.9 Resistance vs. current for the contact between a copper wire and an artificial graphite-based current-collecting material for secondary structures formed under the different current conditions.

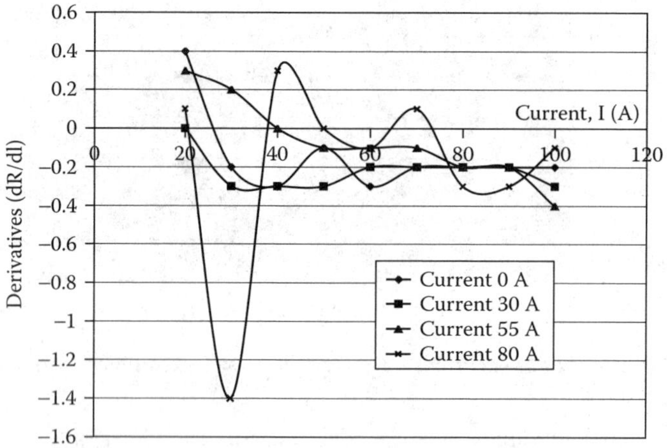

FIGURE 8.10 The derivatives *dR/dI* vs. current for the dependences presented in Figure 8.9.

within a range of –0.1 at current values from 50 to 70 A. The derivative *dR/dI* decreases to –0.2 at a current change from 70 to 90 A and reaches –0.4 at a current value of 100 A.

The dependence of the resistance of the contact with the secondary structures formed with a current value of 80 A is not as unequivocal as the dependences considered previously. A sharp decrease in the resistance is most noticeable as the current increases from 20 to 30 A. The contact resistance does not change as the current increases up to 70 A. The resistance is reduced gradually with the current growth from 70 A to 100 A. The derivative dR/dI (Figure 8.10) sharply decreases correspondingly from 0.1 to –1.4 within a range of current values from 20 to 30 A. It grows up to 0.3 with a current increase up to 40 A. The derivative fluctuates in the neighborhood of 0 up to a current value of 80 A. It becomes negative after 70 A, and reaches the value –0.1 at 100 A.

The following observations can be made if we compared the graphs presented in Figure 8.9 and Figure 8.10 with curve 1 in Figure 8.1. The reduction of resistance for the contacts with secondary structures (formed at current values of 0 and 30 A) vs. current occurs at approximately constant values of *dR*/d*I*. The negative value of *dR/dI* indicates that the stability loss is possible for the given state. However, the constant value of the derivative shows that the probability of the stability loss does not grow with the current. In fact, according to the curve 1 of Figure 8.1 the system does not lose its stability at current values equal to 0 and is quite far from the next instability. At current values of 30 A, the system exists after it has passed through current instability but is also still quite far from the next stability loss.

The d*R*/d*I* value is positive up to a current of 40 A for the contact with the secondary structures previously formed at a current of 55 A (Figure 8.10). At larger currents, the d*R*/d*I* value becomes negative and reaches –0.4 at currents above 80 A, with a trend for further decrease. This means that the probability for stability loss increases for larger current values. It follows from Figure 8.1 (curve 1), that a current of 55 A in the system exists in a steady condition with high wear rate; however, it is closer to instability (of 70 A) than under the previous conditions.

The dependence of *R* and *dR*/*dI* on current (Figure 8.9 and Figure 8.10) is of a great interest for the contact with secondary structures formed at current values of 80 A. The dependence of *R* on current is practically identical to all the contacts at currents above 30 A. A sharp decrease in the resistance within the current values of 20 to 30 A (Figure 8.9) corresponds to the minimum in the derivative *dR*/*dI* in Figure 8.10. This reaches a value of –1.4, which is lower than for other areas. This tells us that the probability for instability to occur at this point is maximal. The derivative *dR*/*dI* fluctuates in the neighborhood of 0 (from –0.1 to 0.1) at currents above 30 A, i.e., the probability of stability loss within this range of current is not significant. Let us compare the graphs *R* and *dR*/*dI* vs. current with the graphs shown in Figure 8.1. The instability (with a high probability due to a minimum of the derivative at 20 to 30 A) is related to the beginning of the intensive nonequilibrium reactions Equation 8.6 and Equation 8.7. Thus, if the secondary structures are formed at current values of 80 A, these reactions start earlier than during sliding on a juvenile surface (Figure 8.1, curve 1). In this case, the nonequilibrium secondary structures start to form prior to the beginning of intensive recrystallization of the surface copper layers. That is why significant wear rate growth (Figure 8.1, a curve 1), which is caused by the recrystallization, does not occur under the conditions of sliding on the surface of the previously formed secondary structures (Figure 8.1, curve 2).

The early start of the dissipative secondary structures formation takes place because these secondary structures contain natural graphite because the reactions in Equation 8.6 and Equation 8.7 have already occurred within these layers. The presence of natural graphite, which acts as a catalyst, shifts the onset of these reactions to the area of lower temperature [15].

Instability can occur in the tribosystem with current collection if the contact resistance decreases with current. Thus, the conclusion of Chapter 2 that dissipative secondary structures formation after the tribosystem passes through instability results in a decrease of the wear rate is confirmed experimentally, including the conditions when friction becomes more severe. Investigations of the mechanism of self-organization have shown that the carbon, which is formed from the atmosphere as a result of nonequilibrium processes of carbon dioxide decomposition, plays the most important role.

8.6 STUDIES OF THE SECONDARY STRUCTURES ON THE FRICTION SURFACE

The secondary structures formed on friction surfaces wear out during friction. It is known that the nonequilibrium (dissipative) secondary structures, other factors being equal, have a lower wear rate than equilibrium ones (see Chapter 2). The secondary structures formation follows the Le Chatelier–Brown's principle, which says that a tribosystem tries to reduce the result of an external impact (in our case it is friction), i.e., tries to reduce the wear rate. Thus, the secondary structures fulfill protective functions. The composition and properties of the secondary structures as dissipative structures do not depend on the entry conditions but on the boundary conditions. The boundary

TABLE 8.3
Current-Collecting Materials and Their Properties

Material (Basis)	Density (g/cm³)	Specific Electric Resistance (μOhm.m)	Hardness (HS)	Content of Metal (%)
Artificial graphite-based material	1.75	15	50	0
Powder metal material (iron)	7.8	0.28	120HB	75Fe-10Cu-14Pb-1Sn
MY7D (coke)	2.5	8	90	30Cu-8Pb
Natural graphite-based material	1.68	5	40	0

conditions are controlled by the composition of the frictional bodies. Formation of secondary structures is a natural process.

Therefore, it is reasonable to select the compositions of frictional bodies' materials based on the results of investigations on secondary structures. The frictional bodies should be selected so as to "help" the tribosystem to form nonequilibrium secondary structures as soon as possible and with smaller losses.

In this section the studies of the compositions of various secondary structures are described. But first, we will present experimental data showing that there is no direct relation between volumetric properties of the current-collecting materials and their wear resistance.

8.6.1 Properties of Current-Collecting Materials and Their Wear Resistance

Three major types of current-collecting materials were selected for the studies. They are as follows: (1) artificial graphite-based materials, (2) carbon–copper coke-based material, and (3) iron-based metal powder material. The composition and the properties of the materials are shown in Table.8.3.

A high-temperature pitch was used as a binder for the artificial graphite-based material. The material is thermochemically processed at high temperatures (above 900°C) to polymerase the binder.

The material grade MY7D is fabricated by the impregnation of a porous coke blank by liquid copper under high pressure and heating in a protective atmosphere.

The powder metal material is a sintered porous blank made of a mixture of iron (85 to 90%) and copper (10 to 15%) powders impregnated with a Pb–5% Sn alloy.

Intensive heating of the materials takes place during friction with electric current collection. Therefore, a change in the properties of these materials as well as decreases in their weight and volume, with temperature, have been investigated. The data is presented in Table 8.3. The samples were heated in air for 1 h at the given temperature. Major attention was paid to the threshold temperature when other characteristics were significantly changed.

Significant changes of the properties are observed after heating to 500°C for the artificial graphite-based material and MY7D. The hardness of these materials drops by 2 to 5 times. The MY7D material lost about half its volume and more than half its weight. The specific electric resistance for the artificial graphite-based material grew by more than 1.7 times. Therefore, the critical temperature for these materials is approximately 500°C.

Significant changes in properties are observed for a powder metal material after heating to 700°C. The specific electric resistance was increased by 11 times. Therefore, the critical temperature of this material is about 700°C.

Figure 8.11 and Figure 8.12 show the graphs of the wear rate of the current-collecting materials and of the copper wire vs. current. It can be seen from the graphs that there is no direct correlation between the wear rate and the properties of the materials. The properties of materials can differ by an order of magnitude; for example, the specific electric resistance of the metal-carbon-based

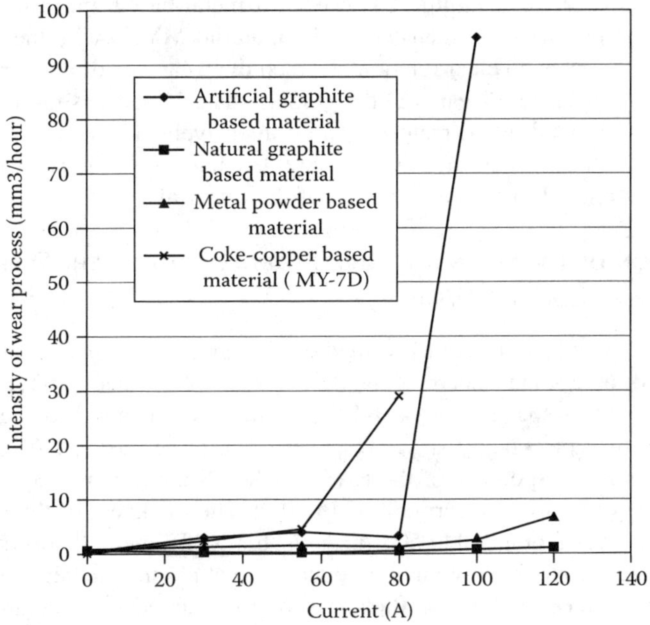

FIGURE 8.11 Wear rate of the different current-collecting materials vs. current.

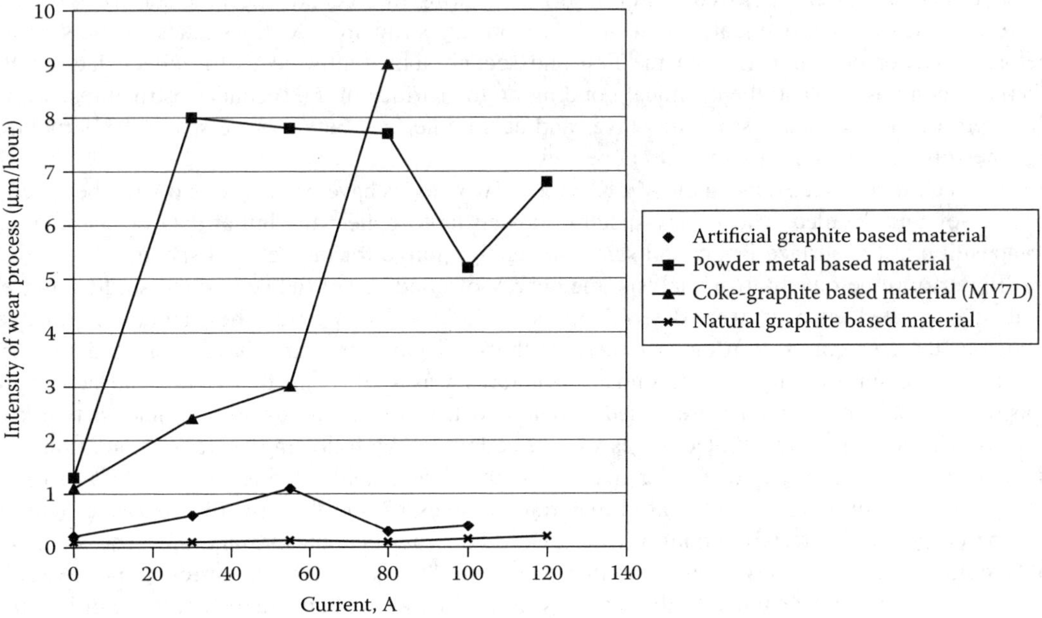

FIGURE 8.12 Wear rate of the contact copper wire vs. current for different current-collecting materials.

material differs by an order of magnitude. However, a metal-based material, which has a higher wear resistance, wears out the wire intensively. The material MY7D with the highest strength has the largest wear rate. There is no clear relation between the wear rate of the wire and the properties of the materials. The copper wire wears out the graphite-based materials more slowly, whereas the wire wears out the coal-metal-based materials more intensively. There is also no influence of the threshold temperature on the wear rate. Thus, if any dependence of the wear rate vs. the material properties exists, it is related to the composition of the materials.

8.6.2 Composition of the Secondary Structures Formed on the Surface of the Copper Contact Wires

Investigations of the secondary structures formed on the surface of copper contact wires have been performed using electronic Auger spectroscopy (EAS), x-ray photo electronic spectroscopy (XPES), and secondary neutral mass spectroscopy (SNMS) methods. Elemental and chemical compositions of the surface tribo-films (secondary structures) were determined using EAS and XPES methods correspondingly. The actual spot size of the EAS and XPES analysis was about 1 cm^2. The depth of the surface layer investigated was around 30 to 50 Å. The thickness of the secondary structures formed on the surface was around 10 to 50 μm. Thus, the XPES method allows us investigate only a small fraction of the secondary structures' thickness. To study the chemical inhomogeneities and the chemical bonds formed within the thickness of the secondary structures, layer-after-layer material removal methods were used. Two major methods of material removal are most widely used to obtain XPES depth profiles, i.e., ionic etching and mechanical polishing [16].

The major advantage of ionic etching is that the object is bombarded by the ions in the chamber of analysis. However, the low speed of ionic etching (0.1 to 0.2 μm/h), the change of elemental composition, and the chemical bonds during the etching [17] make this method inappropriate for secondary structures studies. This is why mechanical polishing was the preferred method of analysis. The samples were mechanically polished twice. During the first stage of polishing, 3- to 7-μm-thick surface layer were removed from the surface. During the second stage, the surface layer was decreased by a thickness equal to that of the secondary structures. XPES studies of the samples before and after polishing in both the first and second stages allowed us to collect data on the chemical composition and the chemical bonding on the surface of the secondary structures, within the depth of the secondary structures layer and at the interface between the secondary structures and the substrate material (a contact cooper wire).

The data was collected using the Mg K_α mode. Two modes have been used to obtain the spectra, i.e., survey and detailed. The survey spectra allowed us to collect the initial data on the element composition in the surface layers and select the energy range for the detailed spectra.

To perform the quantitative analysis, the energy of bonding for the compound studied as well as their integrated intensity were determined using the detailed spectra. The chemical inhomogeneities of the elements within the thickness of the secondary structures have been studied using SNMS. The composition of the secondary structures was studied on the friction surface of the copper contact wires, which were tested under field test conditions using a contact network of railway. The corresponding contact wires were tested under the following operating conditions: (1) in contact with an iron-based powder material with a dry graphite lubricant (sample #1), (2) in contact with an artificial graphite-based material (samples #2 and #4), and (3) in contact with an iron-based powder material without a dry graphite lubricant (samples #3 and #5). The wear rate of sample #3 was two times higher than that of sample #5 under identical current conditions.

The elemental and chemical analysis data is shown in Table 8.4 and Table 8.5. It is worth noting that the content of copper after the second polishing is excessive because the polishing continued until the surface of the contact wire substrate material was developed.

As follows from the data obtained, the bases of all the secondary structures formed are copper. On the other hand, the composition of the secondary structures is controlled by the composition

TABLE 8.4
Elemental Composition of Secondary Structures of Contact Wires

Number of Sample and Polishing	Concentration of Elements (% mass)							
	Cu	Fe	O	C	Cl	S	Sn	Pb
1	63.8	11.6	2.3	20.8	0.8	0.7		
1-1	45.6	30.9	6.4	15.3	0.4	0.8		0.6
1-2	93.5	2.7	0.5	2.9		0.1	0.1	0.2
2	36.3	14.2	3	46.5				
2-1	63.7	9.8	1.2	21.7				
2-2	92.3	4	0.8	2.9				
3	53.2	20.3	3.9	5	0.3	0.3	7.6	9.4
3-1	44.4	33.4	4.9	3.2	0.2	0.2	2.8	10.9
3-2	80.4	10.5	2.3	1.8			1.3	3.7
4	62		1.1	36.5		0.4		
4-1	52.9		2.4	44.5	0.5			
4-2	94.9		0.5	4.5	0.1			
5	29.2	30.4	5	16.1	0.1	1.2	5.3	12.7
5-1	33.6	35.3	5.3	11.6		0.9	3.6	10
5-2	72	10.9	2.2	8.3	0.1	0.1	2.3	4.1

of the current-collecting materials that have been used. However, all the secondary structures contain copper and carbon irrespective of the current-collecting material used. Free carbon is represented in the composition as natural graphite. The secondary structures of the contact wires working in contact with the iron-based current-collecting materials contain 35 mass% of iron. The secondary structures of the contact wires working in contact with the artificial graphite-based current-collecting materials can contain up to 46 mass% of carbon. Copper transfers to the secondary structures from the contact wires, whereas iron, lead, and tin transfer from the current-collecting materials, sulfur and carbon from the lubricant, atmosphere, and current-collecting materials, and oxygen and chlorine from the atmosphere.

The composition of the secondary structures for sample #1 differs from the composition of samples #3 and #5, whereas they have identical counterbodies, i.e., current-collecting iron-based material. The main difference is that samples #3 and #5 contain significant amounts of lead and tin (mainly in the oxide films), and there is practically no lead and tin in sample #1. The difference of the compositions is caused by the dry graphite lubricant presence on the surface of sample #1 during friction.

It is believed that the self-lubricating properties of the iron-based current-collecting material are provided by the Pb–5% Sn impregnation of the alloy [8]. The content of the impregnation material in the current-collecting material is around 12 to 20 mass%, whereas the contents of iron is approximately 65 to 75 mass% and the rest is copper. If the lead alloy provides the lubricating properties to be spread on the copper surface, then the relative iron concentration within the layer of the secondary structures should be higher than its amount in the current-collecting substrate material. The amount of iron in the current-collecting material is approximately –3 to 3.5 times higher as compared to the lead alloy. The data presented in Table 8.4 show that the ratio of iron to lead and tin concentration within the layer of secondary structures of sample #3 is around 1.2 to 2.3, whereas sample #5 it is around 1.7 to 2.6. There is practically no lead and tin within the layer of secondary structures of sample #1. Thus, if the dry graphite lubricant is supplied to the

TABLE 8.5
Chemical Content of Secondary Structures of Contact Wires

No. of Sample and Polishing	Concentration of Elements and Chemical Compounds (% mass)															
	Cu	Cu_2O	Cu_2S	C	Fe_3C	Fe_2O_3	Cl	Fe	SnO_2	PbO_2	PbO	CuO	S	Pb	SnO	Pb_3O_4
1	56.4	5.1	3.5	20.2	7.9	6.1	0.8									
1-1	39.2	3.3	3.7	14.2	14.3	24.4	0.4				0.5					
1-2	91.8	1.5	0.3	2.9	2.1	1.1			0.3							
2	28	8.9		45.8	9.5	7.8										
2-1	65.2	3		21.2	2.6	2.8		5.2								
2-2	86.9	3.4		2.8	0.6	3.8		2.5								
3	50.2	2.6	1.6	1.4	14	3.1		9.6	4.6		10.2					
3-1	43	0.9	0.8	2.1	16.2	10.9	0.2	10.7	3.5		11.7					
3-2	81.2			1.6	3.3	5.6		3.6	1.6							3.1
4	51.6	10.2	1.9	36.3												
4-1	33.9	21.4		44.2			0.5									
4-2	90.9	4.5		4.5			0.1									
5	19.1	6.9	4	15.3	17.7	3.7	0.1	11.5	6.6	4.5	9.3	1.3				
5-1	26.4	5.2	3.2	9.3	30.6	9.7			4.7	5.7	5		0.2			
5-2	71.7		0.3	8.2		5.1	0.1	7.2		2.4	1.8			2.1	1.1	

contact area, it performs the lubricating action instead of the lead alloy. This is proved by the presence of the significant amount (20.2 mass%) of carbon (graphite) (Table 8.5).

Without dry graphite lubricant supplied to the contact area (samples #3 and #5), the iron to tin concentration ratios within the layer of the secondary structures is reduced by 1.5 to 2 times compared to the ratio in the current-collecting material. Therefore, it can be assumed, based on the elemental composition data of the secondary structures (Table 8.4), that the lead alloy fulfilled the lubricating functions. However, the analysis of the chemical composition (Table 8.5) shows that all the tin and almost all the lead transform into oxides. The lead can be observed in a free state only at the interface between the layers of secondary structures and the copper contact wire (sample #5). The free iron to free tin ratio in this case is around 3.5 in the current-collecting material. Thus, the lead alloy does not have a significant lubricating effect under the electric sliding contact conditions.

A significant amount of carbon within the layer of the secondary structures in samples #3 and #5 is of interest. The dry graphite lubricant has not been supplied to the contact area. The iron-based current-collecting material practically has no carbon. The maximum amount of carbon within the layer of the secondary structures in samples #3 and #5 is 5 and 16.1 mass%, respectively. The maximum amount of the free carbon (graphite) in the samples is 4.1 and 15.3 mass%, respectively. The source of the carbon in the layer of secondary structures can only be supplied from the atmosphere. Carbon presented within the layer of secondary structures of samples #3 and #5 is a result of nonequilibrium chemical reactions Equation 8.6 and Equation 8.7.

Samples #5 and #3 were made of contact wires. The wear rate of sample #5 is twice as low as that of sample #3, but the friction conditions are identical. It follows from Table 8.4 and Table 8.5 that the amount of graphite within the layer of secondary structures on the surface of sample #3 is almost four times lower than that of sample #5. This carbon should be formed as a result of nonequilibrium processes.

Figure 8.13 and Figure 8.14 exhibit the depth profiles of the element composition of the secondary structures on the surface of wires working in contact with the metal and carbon materials correspondingly. The depth profiles are obtained using the SNMS method. They show that the composition of the secondary structures does not vary qualitatively within their depth.

Thus, the rational selection of the wear-resistant current-collecting materials providing low wear rate of the contact wires should be performed based on their composition and not on their volumetric properties. The composition of the current-collecting materials should also enhance the nonequilibrium processes on the frictional surfaces. These nonequilibrium processes should be

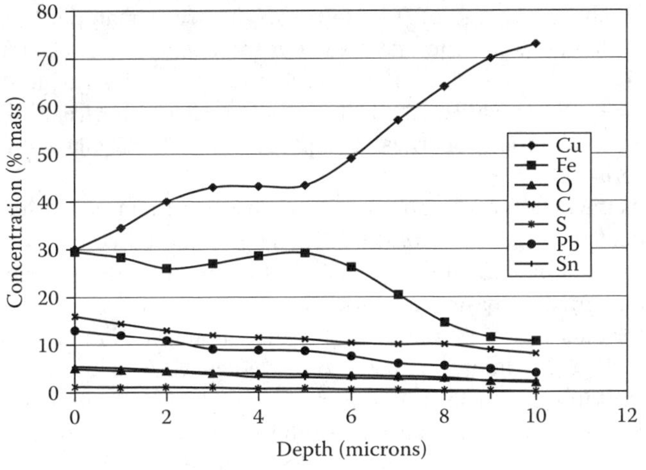

FIGURE 8.13 Elemental depth profiles of the secondary structures of the contact wires working in contact with the iron-based metal powder material.

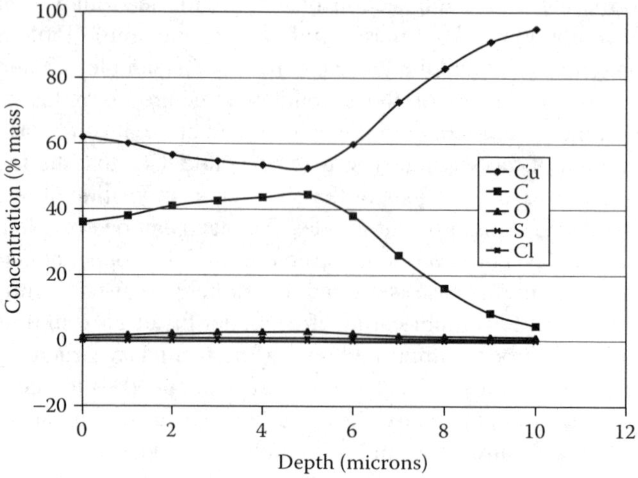

FIGURE 8.14 Elemental depth profiles of the secondary structures of the contact wires working in contact pair with artificial graphite-based material.

initiated and intensified by both the friction and current collection. It is known [15] that natural graphite reduces the temperature at the beginning of these reactions and plays the role of a catalyst for these processes. It is worth noting that the search of wear-resistant materials based on the results of secondary structure studies is possible because of the stability of the dissipative structures and their independence of the initial conditions.

8.7 NATURAL GRAPHITE-BASED CURRENT-COLLECTING MATERIAL

The material was made from particles of natural graphite and a pyrolitic carbon was used as the binder. The properties of the material are shown in Table 8.3. The specific electric resistance of the natural graphite-based material is higher by an order of magnitude than the specific electric resistance of the metal-based material. The material has the lowest hardness. The critical temperature of the material is around 600°C. The coefficient of friction for the carbon-based materials in contact with the copper lies within the range of 0.13 to 0.15. The set of material properties are far from optimum but have the lowest wear rate and wear out the copper wire less intensively than other materials (Figure 8.11 and Figure 8.12). The wear rate of the natural graphite-based material depends slightly on the current value. Thus, the tribological characteristics of the material also depend slightly on its volumetric properties. The processes that take place on the friction surface control these characteristics.

Table 8.6 presents the data for the natural graphite-based current-collecting material. This data is similar to that presented in Table 8.1 for the artificial graphite-based current-collecting material. As follows From Table 8.6, the amount of transferred carbon does not depend on the current, in contrast to the artificial graphite-based current-collecting material.

It follows from the data shown in Table 8.6 that the amount of carbon in the copper matrix, as well as the wear rate, depends slightly on the current values. Carbon appears in the copper matrix as a result of nonequilibrium chemical reactions Equation 8.6 and Equation 8.7. A recrystallization of the surface layers of copper begins during sliding of the natural and artificial graphite-based current-collecting materials at current values of 15 to 30 A. It is possible to assume that the nonequilibrium reactions Equation 8.6 and Equation 8.7 begin at smaller current values for the natural graphite than for the artificial graphite. For the artificial graphite, they begin at current values of 70 A (Figure 8.1). Taking into account that the wear rate of the natural graphite-based

TABLE 8.6
Carbon Content on a Surface of a Copper Wire and Wear Rate of Sliding Contacts vs. Current (Natural Graphite-Based Current-Collecting Material)

Current (A)	0 (Without Friction)	0	20	55	80
Transferred carbon (% of the area of a surface of a copper wire)	—	32	30	31	35
Carbon in a copper matrix (% mass)	4.8	6.1	5.7	4.8	4.9
Intensity of wear process of current-collecting material (mg/hr)	—	1.8	1.1	0.9	1.1
Intensity of wear process of a copper wire (μ/hr)	—	0.1	0.1	0.13	0.1

material practically does not vary during the recrystallization of copper surface layers, it is possible to assume that the reactions Equation 8.6 and Equation 8.7 start at current values lower than 30 A.

Nonequilibrium reactions Equation 8.6 and Equation 8.7 start once the system passes through instability. A tribosystem with current collection can lose its stability if the condition of Equation 8.3 is observed. Figure 8.15 presents a graph of the contact resistance vs. current for the natural graphite-based current-collecting material. It follows from Figure 8.15 that a significant decrease of the contact resistance occurs at current values of 10 to 20 A. This can be proved by the derivative dR/dI vs. current graph (Figure 8.16) for the dependence presented in Figure 8.15. The derivative reaches a minimum (–0.8) at the current values of 10 to 20 A, and it remains close to 0 under different current conditions. An obvious minimum at the curve shown in Figure 8.16 allows us to assume that nonequilibrium reactions begin before the recrystallization of copper or simultaneously with the recrystallization. As a result, the wear rate does not increase during recrystallization.

Thus, the selection of wear-resistant current-collecting materials cannot be made based on the volumetric properties only. During the development of novel wear-resistant current-collecting materials, significant attention should be paid to their ability to enhance nonequilibrium physical and chemical processes on the friction surfaces under milder operation conditions. This conclusion can also be expanded to other materials and tribosystems as will be shown in the corresponding chapters of this book.

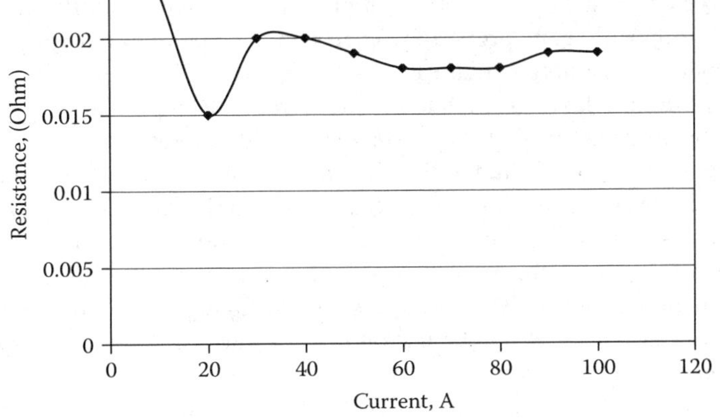

FIGURE 8.15 The resistance vs. current for the copper contact wire and the natural graphite-based current-collecting material.

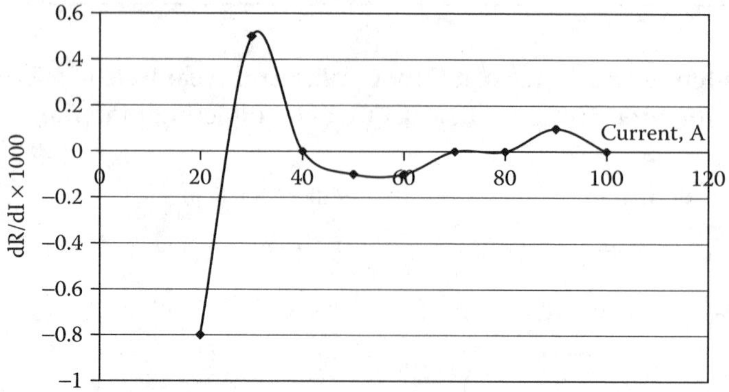

FIGURE 8.16 The derivative *dR/dI* vs. current for the dependence presented in Figure 8.15.

8.8 COMPATIBILITY OF THE CURRENT-COLLECTING MATERIALS WORKING IN CONTACT WITH A WIRE SIMULTANEOUSLY

In this section, we shall consider the characteristics of the dissipative structures, in particular their stability and independence from entry conditions.

Under severe friction conditions, a tribosystem generates nonequilibrium secondary structures, which are dissipative structures. Once they have been formed, the wear rate drops as compared to the equilibrium secondary structures (see Chapter 2). The composition and properties of the nonequilibrium secondary structures, with other things being equal, are controlled by the composition and properties of the frictional materials, i.e., their boundary conditions and does not depend on the initial conditions. Thus, during friction of various materials in contact with the same counterbody, each of the materials tries to form tribo-films with specific composition, structure, and properties.

Formation of secondary structures occurs during the running-in stage of wear in which the wear rate is increased. Thus, the tribological compatibility of the various materials working in contact with the same counterbody is of interest from the point of view of wear intensity.

Let us consider the friction of two current-collecting materials working simultaneously with a copper contact wire. The materials consist of an iron-based metal powder (Table 8.3) and an artificial and natural graphite-based carbon (Table 8.3).

Because the composition, properties, and state of dissipative structures are independent of the entry conditions, we can assume that every current-collecting material will form secondary structures with specific composition and properties. Their composition does not vary significantly within the layer of secondary structures (Figure 8.13 and Figure 8.14). Hence, if a current-collecting material is changed, a new layer of secondary structures will appear after a full removal of the previous layer. This is accompanied by an increase in the wear rate of the frictional bodies. Simultaneous operation of two current-collecting materials can lead to a situation in which the frictional bodies will permanently work in the running-in mode. This corresponds to the wear rate growth of a contact wire. Tests have been performed using the apparatus described in Section 8.3. The graphs of the wear rate of the copper wire with a current for the different combinations of current-collecting materials are shown in Figure 8.17 and Figure 8.18. The graphs of the wear rate of the current-collecting materials vs. the current for the various combinations of these materials are presented in Figure 8.19 to Figure 8.22.

As follows from the graphs shown in Figure 8.17 and Figure 8.18, the wear rate of a copper wire in contact with a metal material or with graphite materials is much lower than the wear rate of both current-collecting materials if they are working at the same time. This difference becomes significant once the current reaches a values of 20 A, and it grows considerably after the current

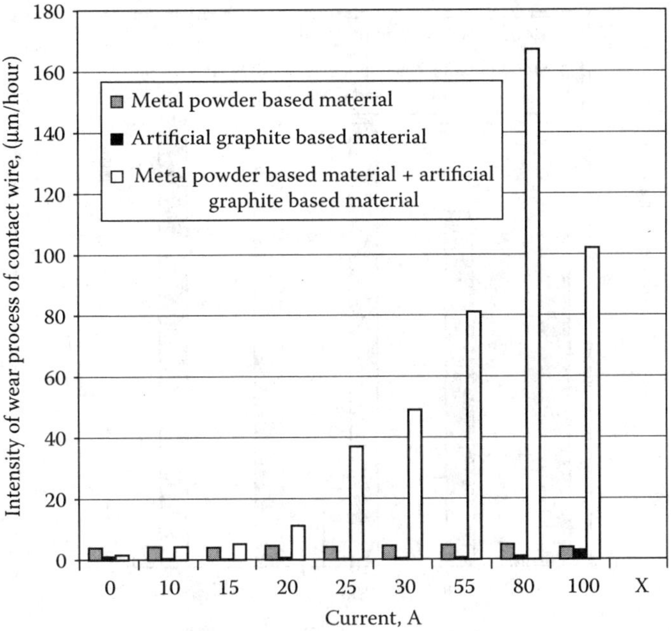

FIGURE 8.17 The wear rate of a copper wire vs. current for the combination of an artificial graphite-based and an iron-based metal powder current-collecting materials.

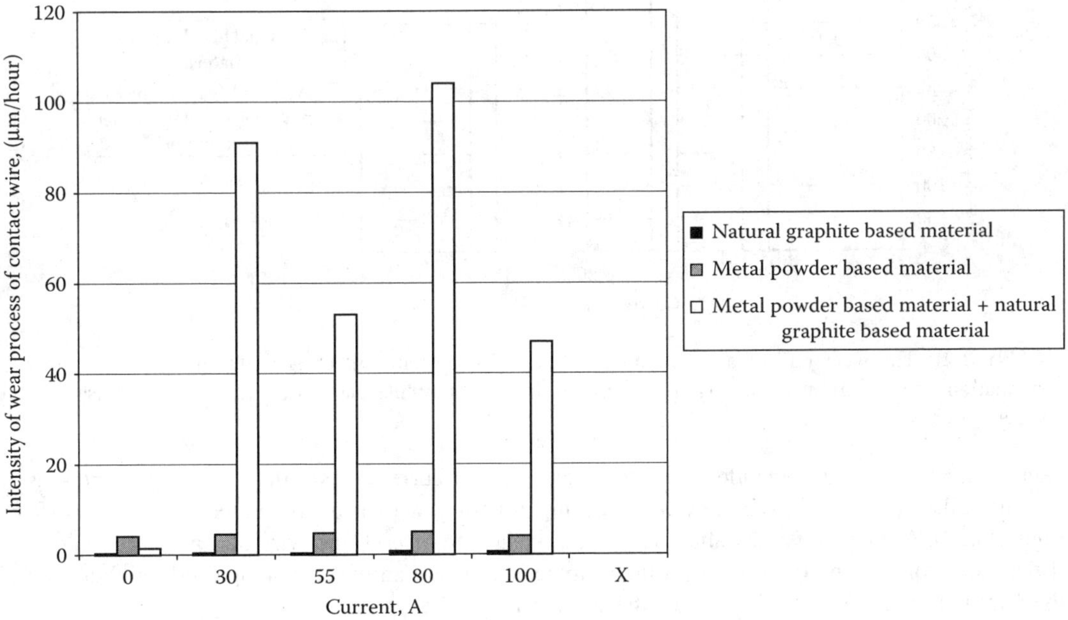

FIGURE 8.18 The wear rate of a copper wire vs. current for the combination of a natural graphite-based and an iron-based metal powder current-collecting materials.

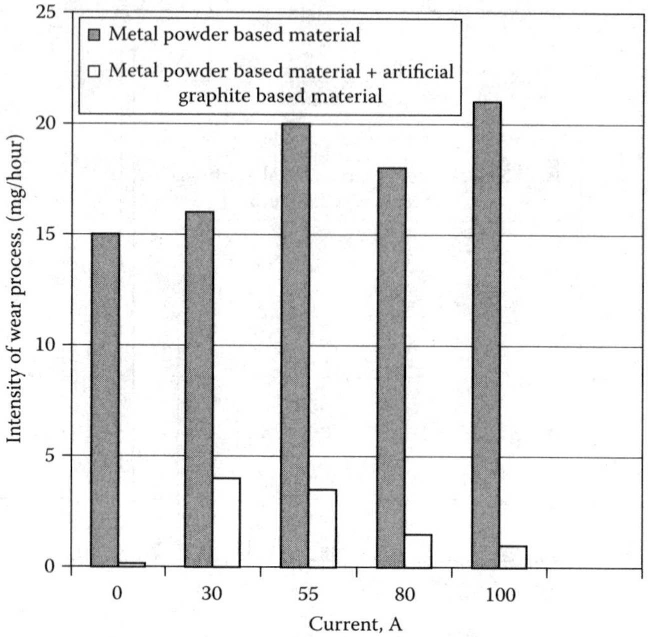

FIGURE 8.19 The wear rate of an iron-based metal powder current-collecting material vs. current for the combination of the current-collecting materials: iron-based metal powder material an artificial graphite-based material.

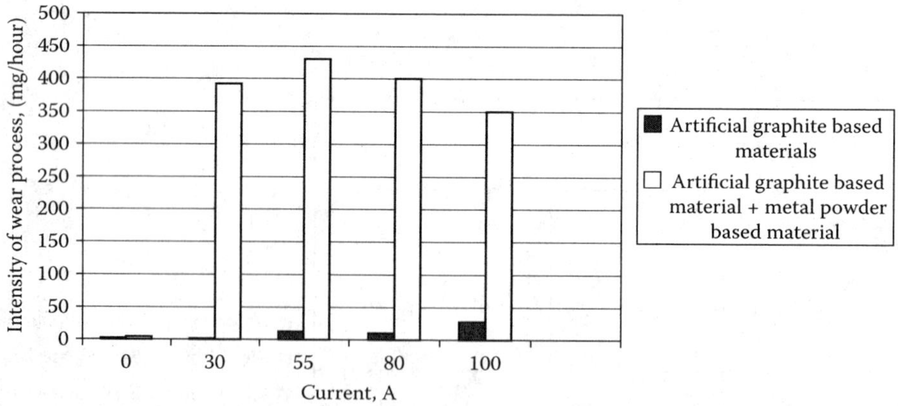

FIGURE 8.20 The wear rate of an artificial graphite-based current-collecting material vs. current for the combination of the current-collecting materials: an artificial graphite-based material and iron-based metal powder material.

value is above 25. The dependence of the wear rate on current is similar for the graphite-based current-collecting materials if they are working together with a metal material (Figure 8.19 to Figure 8.22). At current values above 20 A, the wear rate of both the wire and graphite materials, if they are working in contact with a metal material, are by an order of magnitude higher than if the wire works in contact with one of these materials.

The wear rate of the metal material that is working together with the graphite material is approximately 1.5 to 15 times lower than during work without the graphite material (Figure 8.19 and Figure 8.20). Thus, if by working together the graphite materials could be a source of lubrication for the metal powder material but not for the wire.

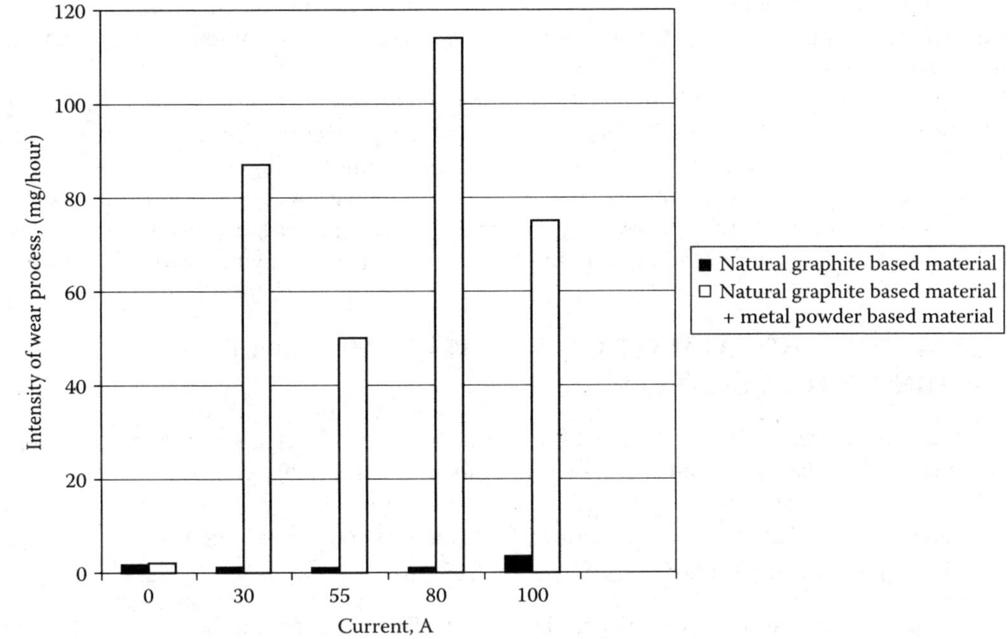

FIGURE 8.21 The wear rate of a natural graphite-based current-collecting material vs. current for the combination of the current-collecting materials: a natural graphite-based material and iron-based metal powder material.

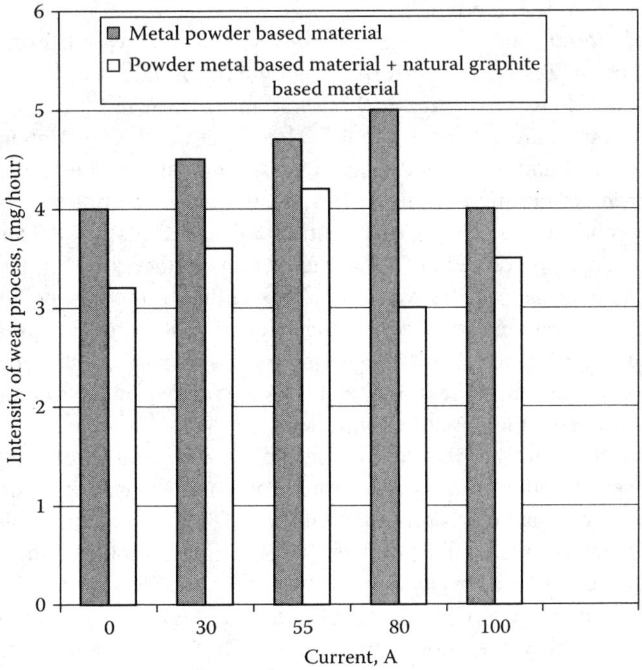

FIGURE 8.22 The wear rate of an iron-based metal powder current-collecting material vs. current for the combination of the current-collecting materials: iron-based metal powder and a natural graphite-based material.

This is related to the independence of the dissipative structures from the initial conditions. Each material tries to form its specific secondary structures and makes the tribosystem work permanently in the running-in mode.

During simultaneous work of both metal and graphite current-collecting materials, an intensive growth of the wire and the coal material wear rates starts from the current values of 20 to 25. This value of current corresponds to the occurrence of the first instability during the work of the artificial graphite-based material. This occurrence of instability is related to the recrystallization of the surface layers of the copper contact wire. Therefore, an intensive wear rate of the contact wire is caused by the increased wearing out of the recrystallized "soft" copper layers by the hard metal material.

8.9 IMPACT OF THE CATALYST ON THE WEAR RESISTANCE OF SLIDING ELECTRIC CONTACTS

The catalyst in the current-collecting materials, as well as increase of CO_2 content in the atmosphere, should enhance the nonequilibrium reactions of Equation 8.6 and Equation 8.7.

8.9.1 Effect of CO_2 Supply to the Sliding Electric Contact Area on the Wear Rate of the Current-Collecting Materials

From the start, it is necessary to understand the impact of CO_2 content on the wear resistance of sliding electric contacts. Later on we can understand the direct impact of the catalyst on intensity of the nonequilibrium reaction Equation 8.6 and Equation 8.7.

Experiments have been performed using the apparatus described in Section 8.3. The CO_2 gas was permanently supplied to the zones of contact of the current-collecting material with a copper wire. The impact of the carbon dioxide was investigated for: (1) iron-based current-collecting materials impregnated with a tin–lead alloy and (2) artificial graphite-based material. The properties of the materials are shown in Table 8.3.

The wear rate of the current-collecting materials vs. current was determined during friction with the current collection. Tests were performed using current values of 0, 30, 55, and 80 A. This corresponds to the linear density of current: 0, 4, 8.5, and 11.5 A/mm.

The CO_2 content in the atmosphere is about 0.01%. That is why the amount of carbon dioxide needed for the reactions Equation 8.6 and Equation 8.7 to occur can be formed as a result of the oxidation of the carbon-containing current-collecting material. The pressured feed of CO_2 to the sliding contact zone enhances the reactions Equation 8.6 and Equation 8.7 under relatively mild friction conditions. The pressured feed of CO_2 to the sliding contact zone for the iron-based current-collecting material has not changed the wear rate. This is due to the absence of graphite from the surface of the copper, which enhances the reactions Equation 8.6 and Equation 8.7 at a relatively low temperature. It is worth noting that the major mechanism to reduce the friction and wear intensity of the iron-based current-collecting material is the formation of structures that are enriched by tin on a friction surface of the copper counterbody.

To investigate the impact of carbon dioxide on the wear resistance of the graphite material, a special material was selected in which the wear rate grows suddenly at a certain current value. The wear rate vs. current for this material is presented in Figure 8.23. A sudden growth of this material's wear rate starts with a current value of 40 A. The carbon dioxide supply to the sliding contact zone of the material did not change the wear rate of the current-collecting material at current values below 40 A. The corresponding curve is shown in Figure 8.23. As follows from the graphs presented in Figure 8.23, the CO_2 supply does not change the wear rate for current conditions below 40 A. No significant wear rate increase was observed under current conditions above 40 A if the CO_2 was supplied to the sliding contact area. It is possible to assume that an increase in the partial pressure of CO_2 in the atmosphere leads to the enhancement of the reactions in Equation 8.6 and

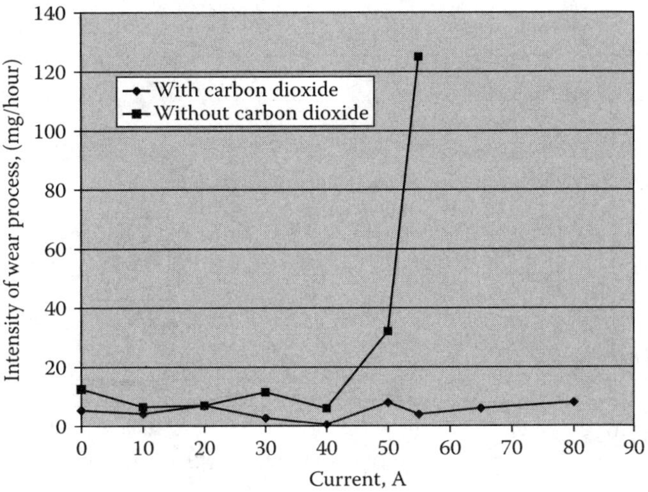

FIGURE 8.23 The wear rate of a graphite-based current-collecting material vs. current with submission and without the carbon dioxide supply to the contact zone.

Equation 8.7. As a result, more energy is generated during friction and the passage of the electric current could be spent on the nonequilibrium reactions of Equation 8.6 and Equation 8.7. Correspondingly, the reduced energy generated during friction can be spent on the advancement of surface damage from the sliding contact.

8.9.2 Impact of the Catalyst on the Wear Resistance of the Sliding Electric Contacts

As it was outlined earlier, nonequilibrium processes lowers the wear rate (see Chapter 2), in particular the wear rate of the sliding electric contacts Equation 8.6 and Equation 8.7. The increase in the CO_2 partial pressure in the atmosphere enhances the reactions Equation 8.6 and Equation 8.7. The higher the amount of CO_2 involved in the reactions, the greater the energy spent on the nonequilibrium processes and less the energy spent on friction surface damage. In practice, the CO_2 supply to the sliding contact area is very difficult or even impossible to realize. Therefore, to enhance the reactions in Equation 8.6 and Equation 8.7 a catalyst can be used. There is a list of data on the use of catalysts for the decomposition of CO_2 [17–23]. The industrial Cr–Ni catalyst was used to find out the impact of the catalyst on the wear rate of current-collecting materials. The powdered catalyst was added to natural graphite and pitch-based current-collecting material. The amount of the catalyst in the current-collecting material was the following: 0, 3, 6, and 9%, correspondingly. Properties of the materials are shown in Table 8.7.

TABLE 8.7
Properties of Current-Collecting Materials with Catalysts

Content of Catalysts (%)	Density (g/cm³)	Hardness (HS)	Specific Electric Resistance (μOhm.m)
0	1.64	31	9.7
3	1.65	33	10.1
6	1.67	34	13.4
9	1.68	28	14.4

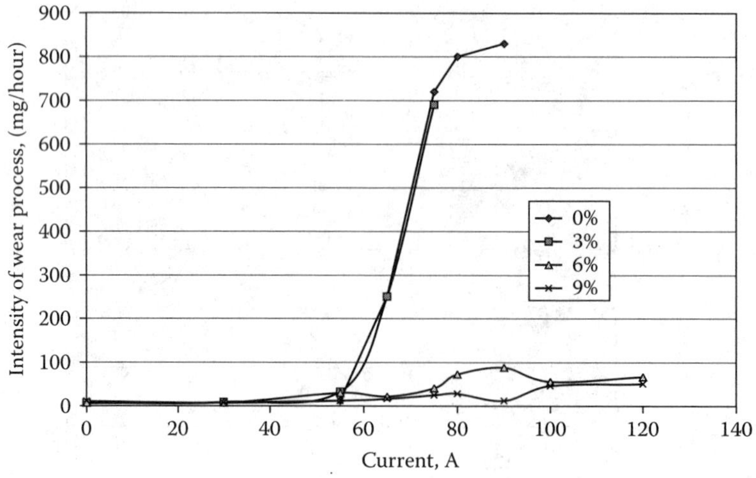

FIGURE 8.24 The wear rate of a graphite-based current-collecting materials with different amounts of the catalysts (0, 3, 6, and 9 mass%) vs. current.

The test has been performed using the apparatus described in Section 8.3. The wear rate of the current-collecting materials vs. current was determined. The current values varied from 0 to 120 A. This corresponds to the linear density of current from 0 to 17 A/mm. The data obtained is presented in Figure 8.24.

The graphs in Figure 8.24 show that, up to current values of 55 A, the wear rate of the materials is approximately identical and does not depend on the current. At a current value of 65 A, the wear rate of the current-collecting materials that contain no catalyst or contains about 3% of the catalyst grows intensively by 40 times as compared to the wear rate of these materials during friction without any current. At the current value of 75 A, the wear rate of these materials is increased approximately by 100 times as compared to the wear rate of the materials during friction without any current. The wear rate of the current collection materials that contains 6 to 9% of the catalyst is increased approximately by 2 times at a current value of 65 A as compared to the wear rate of these materials during friction without any current. Further, as the current increases up to 120 A, the wear rate of these materials changes slightly, fluctuating within a range that is approximately by 2 to 4 times higher as compared to the friction without any current.

The surface friction of the copper counterbody was investigated using EDX. The surface of the copper samples were analyzed after friction under the current value of 80 A. The test was performed using the following materials: (1) containing 9% of the catalyst and (2) without the catalyst. The wear rate intensity of the current-collecting material with 9% of the catalyst is 10 times lower than that of the current-collecting material without the catalyst.

Two structural versions of carbon were observed on a working surface of the copper counterbody (see a similar result in Section 8.4). The first one is the island-like film that covers the surface of the counterbody. This carbon was transferred from the current-collecting material and "spread" on the working surface of the copper. These carbon films were extended in the direction of the applied friction. The carbon x-ray images of these films are shown in Figure 8.25 and Figure 8.26. Metallography analysis has shown that the area of the copper counterbody's friction surface, covered by the graphite that is transferred from the surface of the catalyst-containing material (Figure 8.25), is approximately 1.5 times lower than the area of the surface formed in contact with the material without the catalyst (Figure 8.26).

The analysis of the carbon content was performed within the areas of the frictional surface of the copper, which are not covered by the transferred graphite. These areas are marked on Figure 8.25 and Figure 8.26. The carbon content within these areas, which were working in contact with

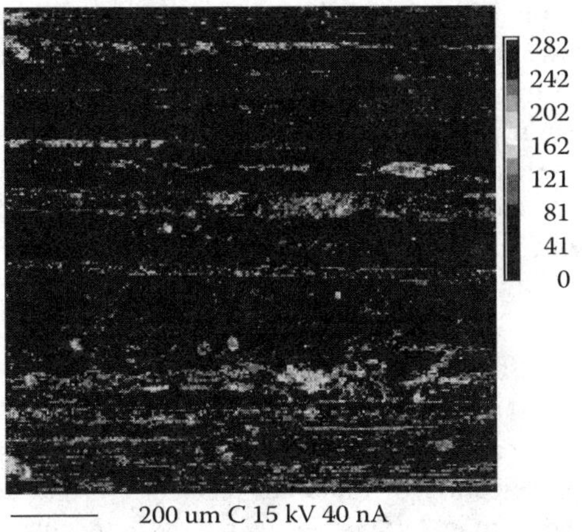

FIGURE 8.25 Carbon spread on the surface of a copper counterbody during friction of a current-collecting material with the catalyst under the current value of 80 A.

FIGURE 8.26 Carbon spread on the surface of a copper counterbody during friction of a current-collecting material without the catalyst under the current value of 80 A.

no catalyst, contained material that is approximately 2 times lower compared to a friction surface of copper that was working in contact with a material that contains 9% of the catalyst.

Thus, because the wear rate is increased, the amount of carbon introduced into the copper matrix grows. In contrast, the amount of the graphite transferred mechanically onto the surface of the copper is reduced. The carbon is introduced into the copper matrix as a result of nonequilibrium reactions Equation 8.6 and Equation 8.7. The catalyst increases the speed of these reactions. The energy that is spent on these reactions is increased. Therefore, the energy that can be spent on surface damage is reduced.

It is necessary to emphasize that the properties of the materials containing 6 to 9% of the catalyst are much worse than the properties of the materials without the catalyst or with only 3% of the catalyst. Their specific electric resistance is 2 times higher and the hardness is lower than 20%. Nevertheless, their wear rate is much lower. The only difference between the materials is the catalyst content. It confirms once again that the main characteristics, which control the wear rate, are not the material's volumetric properties but the ability to perform nonequilibrium processes during friction.

Application of the concepts of self-organization and nonequilibrium thermodynamics for the rational selection of wear-resistant materials for tribosystem with current collection was discussed in this chapter. The positive impact of the nonequilibrium processes on the wear rate reduction was presented. The equilibrium processes can lead to a wear rate increase. The materials for the frictional bodies should be selected in a way to enhance the nonequilibrium processes and shift the process to milder friction conditions. It is possible to do this by adding catalysts to the frictional materials. The nonequilibrium processes have been identified based on the results of the frictional surfaces investigations.

8.10 CONCLUSIONS

In the present chapter tribosystems with current collection are considered in detail based on the concept of self-organization. Selection of the wear-resistant current-collecting materials was performed using volumetric and contact properties using the concepts of the self-organization and nonequilibrium thermodynamics. The methodology of the wear-resistant material selection based on the concepts of self-organization and nonequilibrium thermodynamics can be presented as follows:

1. Investigations of the composition and properties of the secondary structures that are formed on the friction surfaces when the tribosystem has minimal wear rate under the given conditions.
2. Understanding of the general characteristics of the secondary structures that are typical for the frictional bodies in contact with the specific tribosystems.
3. Studies of the processes that control the generic feature of the secondary structures. It is desirable to understand which nonequilibrium processes are responsible for the secondary structures formation.
4. A proper selection of the materials or lubricants for frictional bodies can be done to enhance these nonequilibrium processes.

REFERENCES

1. Lancaster, J.K. Instabilities in the frictional behavior of carbons and graphites. *Wear* 1975, 34, 275–290.
2. Kubo, S., Tsuchiya, H., Ikeuchi, J. Wear properties of metal/carbon composite pantograph sliders for conventional electric vehicles. *Q. Rep. Railway Tech. Res. Inst.* 1997, 38, *1*, 25–30.
3. Berent, V.Y., Gershman, I.S. Secondary structures on surfaces of sliding contacts at high current. 3. Mechanisms of formation, growth and destruction of secondary structures. *J. Friction and Wear* 1990, 11, *1*, 85–92.
4. Oshe, E.K., Zimina,.U. Investigation of processes of defects formation on the surface oxides under high temperature copper oxidizing. *Met. Prot.* 1983, 1, *5*, 745–750.
5. Arnold, V.I. *Theory of Catastrophes*. Science: Moscow, 1990.
6. Berent, V.Y., Gershman, I.S., Zaichikov, A.V. Secondary structures on surfaces of sliding contacts at high current. 2. Electric properties of secondary structures. *J. Friction and Wear* 1989, 10, *6*, 1019–1025.
7. Gershman, I.S., Penskii, N.V. Investigations of secondary structures formation in friction conditions with current pick-off. *J. Friction and Wear* 1995, 16, *1*, 126–131.
8. Cuptsov, Yu.E. *Increasing of Contact Wire Service Life*. Transport: Moscow, 1972.
9. Heinicke, G. *Tribochemistry*. Akademie-Verlag: Berlin, 1984.
10. Hwang, B.H., Singh, B., Vook, R.W., Zhang, I. In situ auger electron spectroscopy characterization of wet CO_2-lubricated sliding copper electrical contacts. *Wear* 1982, 78, 7–17.
11. Singh, B., Zhang, I., Hwang B.H., Vook, R.W. Microstructural characterisation of rotating Cu-Cu electrical contacts in vacuum and wet CO_2 environments. *Wear* 1982, 78, 17–28.

12. Chang, Y.J., Kyhlmann-Wilsdorf, D. Effects of ambient gases on friction and interfacial resistance. *Trans. ASME J. Tribology* 1988, 110, *3*, 508–515.
13. Pope, L.E., Peebles, D.E. Gaseous contaminants modify the friction and wear response of precious metal electrical contact alloys. *IEEE Trans. Compon., Hybrids, Manuf. Technol.* 1988, 11, *1*, 124–133.
14. Fialkov, A.S. *Carbon: Interlayer Compounds and Composites Based on It.* Aspect Press: Moscow, 1997.
15. Prutton, M. *Introduction to Surface Physics.* Clarendon Press: Oxford, 1994.
16. Chuang, T.I., Brundle, C.R., Wandelt, K. An x-ray photoelectron spectroscopy study of the chemical changes in oxide and hydroxide surfaces induced by Ar^+ ion bombardment. *Thin Solid Films* 1978, 53, 19–27.
17. Lu,Y., Xue, J.Z., Shen, S.K. Activation of CH_4, CO_2 and their reactions over Co catalyst studied using a pulsed-flow micro-reactor. *React. Kinet. Catal. Lett.* 1998, 64, *2*, 365–371.
18. Eremin, A.V., Ziborov, V.S., Shumova, V.V., Voiki, D., Roth, P. Formation of O (D-1) atoms in thermal-decomposition of CO_2. *Kinet. Catal.* 1997, 38, *1*, 1–7.
19. Sakakini, B.H., Tabatabaei, J., Watson, M.J., Waugh, K.C. Structural changes of the Cu surface of a $Cu/ZnO/Al_2O_3$ catalyst, resulting from oxidation and reduction, probed by Co infrared-spectroscopy. *J. Mol. Catal. A-Chemical* 2000, 162, *1–2*, 297–306.
20. Ehrensberger, K., Palumbo, R., Larson, C., Steinfeld, A. Production of carbon from carbon-dioxide with iron-oxides and high-temperature. *Ind. Eng. Chem. Res.* 1997, 36, *3*, 645–648.
21. Hadden, R.A., Sakakini, B.H., Tabatabaei, J., Waugh, K.C. Adsorption and reaction induced morphological changes of the copper surface of a methanol synthesis catalyst. *Catal. Lett.* 1997, 44, *3–4*, 145–151.
22. Wang, H.Y., Au, C.T. Carbon dioxide reforming of methanol to signals over SiO_2-supported rhodium catalysts. *Appl. Catal. A-General* 1997, 155, *2*, 239–252.
23. Tsuji, M., Yamamoto, T., Tamaura, Y., Kodama, T., Kitayama, Y. Catalytic acceleration for CO_2 decomposition into carbon by Rh, Pt or Ce impregnation onto Ni(II)-bearing ferrite. *Applied Catal. A: General* 1996, 142, *1*, 31–45.

Section IV

Adaptive Surface-Engineered Materials and Systems

9 Surface-Engineered Tool Materials for High-Performance Machining

German S. Fox-Rabinovich, Anatoliy I. Kovalev, Jose L. Endrino, Stephen C. Veldhuis, Lev S. Shuster, and Iosif S. Gershman

CONTENTS

9.1 CVD Coatings 232
9.2 PVD Coatings 232
9.3 PVD Coatings for General-Purpose Machining 233
9.3.1 Monolithic and Multilayered PVD Titanium Nitride Coatings 233
9.3.1.1 Wear Behavior and Self-Organization of Ti–N PVD Coatings 233
9.3.1.2 Frictional Wear Behavior and Self-Organization of Ti–N PVD Coatings 237
9.3.2 Duplex Coatings 238
9.3.2.1 Friction and Wear Behavior and the Features of Self-Organizing of Duplex Coatings 243
9.3.3 Self-Lubricating Coatings 244
9.3.4 "Smart," Multilayered, Wear-Resistant Coatings 245
9.3.5 Wear-Resistant Coatings for High-Performance Machining Applications 258
9.3.5.1 Hard Nanoscale Crystalline/Composite/Multilayered Coatings 259
9.3.5.2 Adaptive Coatings for Cutting-Tools Applications 266
References 291

Surface engineering has recently become one of the most effective ways to improve the wear resistance of tool materials. The favorable effects that are associated with surface engineering for tooling application are shown in Table 9.1.

Some methods of surface engineering most widely used in practice for tooling application are chemical vapor deposition (CVD), physical vapor deposition (PVD), and thermal diffusion methods (especially for stamping applications).

TABLE 9.1
Improved Performance of Surface-Engineered Cutting Tools

Favorable Effect	Improved Performance of Cutting Tools	Practical Advantages
Increased hardness, especially at elevated temperatures	Reduced surface deformation of the tools at the actual temperatures of cutting	Increased tool life
Improved diffusion barrier and chemical stability properties	Reduced diffusion and dissolution of the tooling material during machining	Increased productivity Increased tool life
Reduced adhesion at the workpiece–tool interface	Less material transfer from the tool surface	Increased productivity of the machining operation
Enhanced adaptability	Lower friction forces and heat generation, beneficial heat distribution during cutting	Improved workpiece quality (better surface finish, improved dimensional accuracy), increased tool life

Source: Holleck, H. Properties of titanium based hard coatings. *J. Vac. Sci. Technol.* 1986, A4, 2661–2676. With permission.

9.1 CVD COATINGS

CVD is a high-temperature process (1000°C) used to deposit refractory layers on hardmetal substrates. Most often, carbides, carbonitrides (mainly Ti), and oxides (mainly Al) are typically deposited by the CVD technique. An example of an advanced CVD coating is the multilayered TiC + TiCN + Al_2O_3. This type of multilayered coating is superb without a rival wear resistance during turning operations. When compared to coatings deposited by the PVD process, the adhesion of CVD coatings to cemented carbide substrates is superior. The major problem with carbide tools during high-speed machining is plastic deformation of the cutting edge [2] owing to high temperature creep. It is known that TiC has better creep strength at the actual temperatures of high-speed machining when compared to tungsten-carbide-based hard metals [2]. In addition, the outer Al_2O_3 layer of the multilayered CVD coating has high hot hardness and low thermal conductivity that critically enhance tool life of the coated tools. The latest generation of CVD multilayered coating for high-speed-machining applications combine CVD and PVD methods. For instance, a multilayered TiC + TiCN + Al_2O_3 coating additionally includes the outer Ti–N PVD layer. This layer protects the brittle Al_2O_3 layer against damage during running-in stage of wear and improves adaptability of the CVD multilayered coating.

Quite recently, a plasma-activated CVD coating (PACVD) was developed. The plasma application results in significant reduction in the temperature of deposition, down to 450 to 600°C. Different types of coatings, including nanoscale as well as nanocomposite [3], could be deposited using this technique on versatile substrate materials.

9.2 PVD COATINGS

PVD coatings are most widely used for a variety of cutting and stamping-tool applications. This coating type has the following advantages when compared to CVD ones. First of all, the temperature of coating deposition is relatively low (500°C and below). In addition PVD coatings do not embrittle the substrate material; this is very important for the majority of substrate materials, including steels and other engineering materials. Finally, the PVD process is extremely flexible. A wide variety of coating compositions and multilayer combinations, having unique structure characteristics and properties, can be synthesized. Included among these are nanostructured materials, which usually are difficult to obtain for other surface-engineering techniques.

9.3 PVD COATINGS FOR GENERAL-PURPOSE MACHINING

9.3.1 MONOLITHIC AND MULTILAYERED PVD TITANIUM NITRIDE COATINGS

9.3.1.1 Frictional Wear Behavior and Self-Organization of Ti–N PVD Coatings

The most widely used approach for the surface engineering of tools is based on titanium nitride coatings deposited by PVD methods. These coatings display a favorable combination of properties, such as good adhesion to the substrate, elevated hot hardness, and improved chemical and oxidation stability (up to 550 to 600°C), resulting in an increased resistance to solution wear. The ability to improve the contact conditions at the cutting edge (i.e., a reduction of the tool–chip contact length) has been reported, leading to lower friction and decreased temperatures at the surface of the tool [1]. However, monolithic Ti–N coatings have a critical weakness. Unfortunately, it is almost impossible to combine such divergent properties as good adhesion to steel or a hardmetal substrate coupled with minimal work-piece interactions in the monolithic Ti–N coating. In addition, high hardness and the possibility of energy dissipation, without coating failure, are often mutually exclusive properties [4]. One of the solutions to this problem is to adopt a classical metallurgical design approach. Coatings whose properties can be tailored specifically from the substrate to the top surface are required. Multilayered coatings or coatings with a metal-based sublayer could be applied [1]. However, technical problems have been encountered in the deposition of these coatings. An alternative approach is to vary the parameters of deposition, using a regular PVD unit, to optimize the coating properties and coating design.

A study designed to optimize the deposition parameters for Ti–N coatings Reference 5 determined that the nitrogen pressure is the most critical process parameter responsible for changes in the coating structure and properties. For the deposition conditions described in Reference 5, an increase in the nitrogen pressure up to 0.4 to 0.6 Pa leads to stoichiometric TiN (53 at. % N_2). Further increases in the nitrogen pressure lead to a decrease in the nitrogen concentration of the film (to 43 at. % N_2, Figure 9.1a). This is caused by a decrease in intensity of the plasma-chemical reaction as a result of a reduction in the flux and energy of the impinging ions. The phase composition changes from α – Ti + Ti_2N at very low nitrogen pressures to TiN at higher nitrogen pressures (0.4 to 0.6 Pa).

The structural parameters of the film also depend on the nitrogen concentration in the coating. The lattice parameter and subgrain size (or equivalently the dislocation density) are related to the nitrogen concentration in the coating (Figure 9.1b to Figure 9.1d) and have maximum values at a composition corresponding to stoichiometric TiN. An increase in the nitrogen pressure leads to a pronounced axial texture (Figure 9.1e), with (111) planes in the TiN layer being parallel to the substrate surface. The (111) axial texture increases to 95% as the gas pressure is raised to 0.6 Pa and remains practically unchanged with subsequent pressure increases. The residual compressive stress in the coating shows similar trends (Figure 9.1f), with the stress increasing from 200 to 1300 MPa as the gas pressure is raised to 0.6 Pa. The rate of increase in the compressive stress slows with a further increase in the nitrogen pressure up to 2.6 Pa, reaching a maximum value of 1600 MPa. It is important to realize that the trends shown by these structural-dependent parameters depend primarily on the deposition conditions. An increase in the nitrogen pressure over 1.3 Pa decreases the ion energy, and the effective temperature at the TiN crystallization front decreases. At a nitrogen pressure in excess of 1.3 Pa, the coatings form under conditions similar to those encountered in the balanced magnetron sputtering process (for the PVD method). These conditions can be indirectly characterized by the deposition rate, which should not exceed 5 to 6 μmh^{-1}. A decrease in the deposition rate enhances both the axial texture and the magnitude of the residual stresses. The most probable cause of the high compressive residual stress found in thin, condensed films deposited at nitrogen pressures greater than 1.3 Pa is a high density of point defects [6]. In addition, an increase in the nitrogen pressure also decreases the crystallite dimensions (Figure 9.1f).

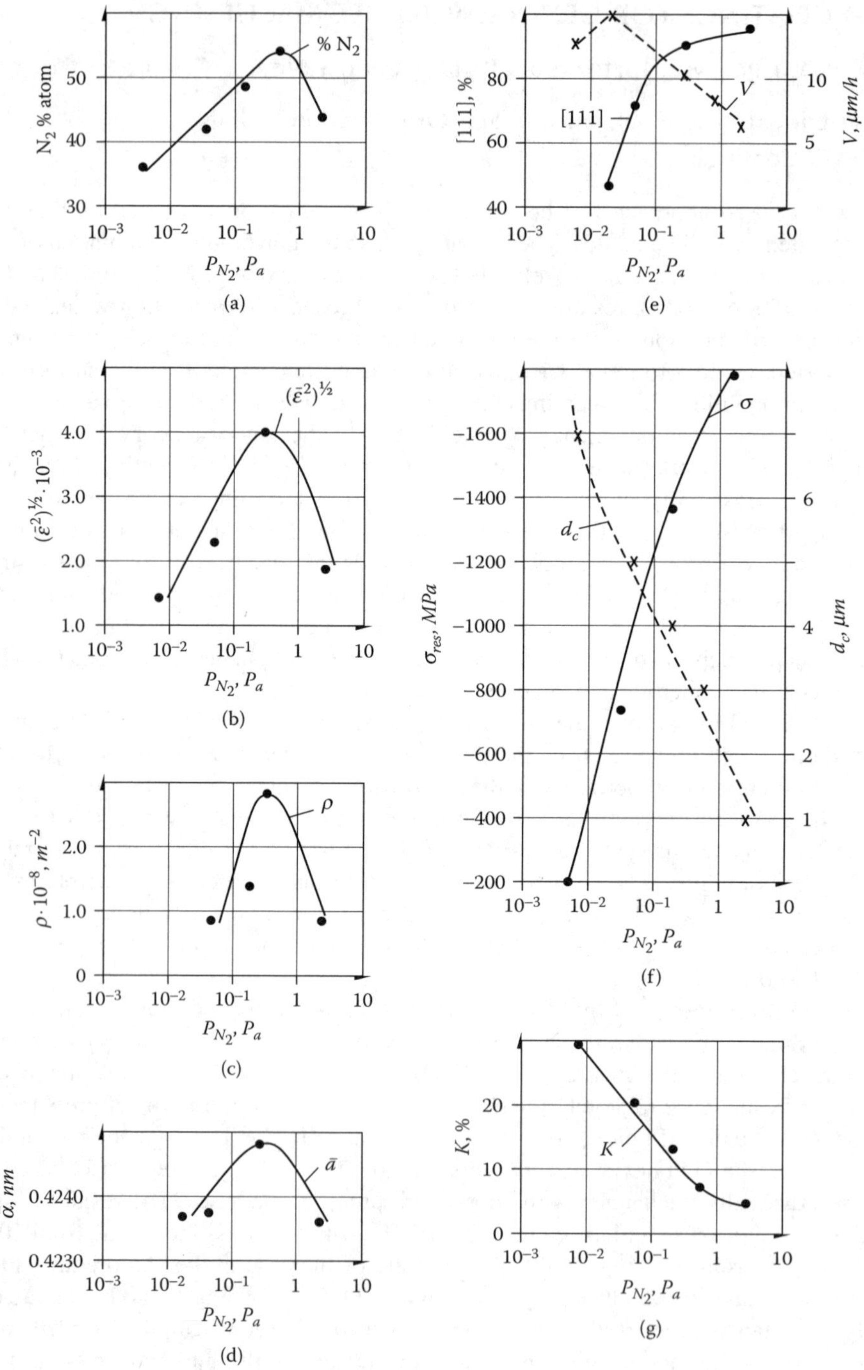

FIGURE 9.1 The dependence of the structural characteristics of Ti–N PVD coatings on the nitrogen pressure: (a) P_{N2} — nitrogen pressure in the chamber during Ti–N coating deposition, N_2 — nitrogen content in the coating layer in at. %; (b) ε — microdeformation of the TiN lattice; (c) $\rho \times 10^{-8}$ — density of dislocations; (d) *a* — lattice parameter; (e) [111] — axial texture, *v* — deposition rate; (f) σ_{res} — residual stress, dc — grain size; (g) *K* — quantity of "droplet" phase. (From Gershman, J.S., Bushe, N.A. Thin films and self-organization during friction under the current collection conditions. *Surf. Coat. Technol.* 2004, 186, *3*, 405–411. With permission.)

The microhardness of the coatings depends on their phase composition. The maximum microhardness ($H_{0.5}$ = 45 GPa) is achieved when the two-phase α-Ti + TiN composition changes into the three-phase composition α-Ti, Ti_2N, and TiN. The Palmquist toughness of the coatings (Figure 9.2a) is a structure-sensitive characteristic. The maximum toughness and microhardness dissipation parameter (see Chapter 6) correspond to a single-phase coating having a stoichiometric TiN composition. Any deviation from the stoichiometric composition causes a decrease in the Palmquist toughness, particularly when a second phase is formed in the coating (e.g., Ti_2N). This result seems unexpected at first, but we should remember that the hard Ti–N coating should be regarded as a quasi-brittle material, whose toughness is determined primarily by its crack-propagation resistance. The optimum nitrogen concentration corresponds to the largest lattice parameter [7], although the microhardness of the coating with this structure is not relatively high. In addition to the intrinsic mechanical properties of the coating, the level of the residual compressive stress is important for crack initiation and propagation. The value of the residual stress is about 800 MPa in stoichiometric Ti–N coatings (Figure 9.1f). The fracture resistance also appears to depend on the columnar grain size, which again can be controlled by deposition conditions [8]. The adhesion of the coating to the substrate and the shear load resistance (i.e., the cohesion of the coating) both decrease as the nitrogen pressure is increased. The nitrogen atoms (ions) in the plasma scatter the Ti ions, so that the net effect of the Ti-ion bombardment of the coating is reduced as the nitrogen gas pressure rises. As noted earlier, both the axial texture and the residual stress gradient at the coating–substrate interface increase as the pressure rises, and consequently the adhesion of the coating falls (Figure 9.2b). The coating wear resistance during dry sliding friction under high loads (Figure 9.2c) reaches its maximum value in the three-phase field area α-Ti + TiN + Ti_2N.

Turning inserts made of M2 high-speed steel with Ti–N coating were tested under adhesive wear conditions [9]. The tool life of TiN-coated tools depends on the friction coefficient values. The measurement of the coefficient of friction has been performed under conditions that mimic a seizure at the tool–workpiece interface, which is typical for cutting operations (Figure 9.2e).

The overall conclusion of this study is that the optimum combination of properties of the coating for adhesive wear is obtained at deposition rates (for the PVD method) of 5 μmh^{-1} and a stoichiometric composition of TiN. This can be achieved by optimizing the deposition parameters. In this case, the hardness and toughness increase, whereas the shear resistance decreases. A coating with the optimum structure will crack by shear failure at or near to the surface of the coating rather than by forming deep cracks leading to a catastrophic failure of the whole tool. However, at the same time the shear stress resistance of the coating should be strong enough to resist the flow of the chip. Because a monolithic Ti–N coating usually has a low adhesion to the substrate, adhesive sublayers are necessary to achieve high efficiency from this type of coating.

A number of principles guiding the selection of the processing of Ti–N multilayer coatings for adhesive wear conditions can be elicited from this study. The coating should have at least three sublayers:

1. An adhesion sublayer, deposited with substoichiometric nitrogen. These deposition conditions provide the maximum kinetic energy of the ions and a low nitrogen concentration in the layer (up to 35%). At the same time, the (111) axial texture should not exceed 50%, whereas the residual stress at the coating–substrate interface should be low (not more than 200 MPa). When this combination is achieved, the adhesion of the sublayer is high.
2. A transition layer deposited with a gradual increase in the nitrogen pressure to provide:
   a. Development of an axial (111) texture (from 48 to 100%) from the substrate to the top layer
   b. A residual compressive stress increase from 200 MPa in the adhesion layer to 1700 MPa
   c. A gradual transition from a three-phase structure α-Ti + TiN + Ti_2N to single-phase TiN

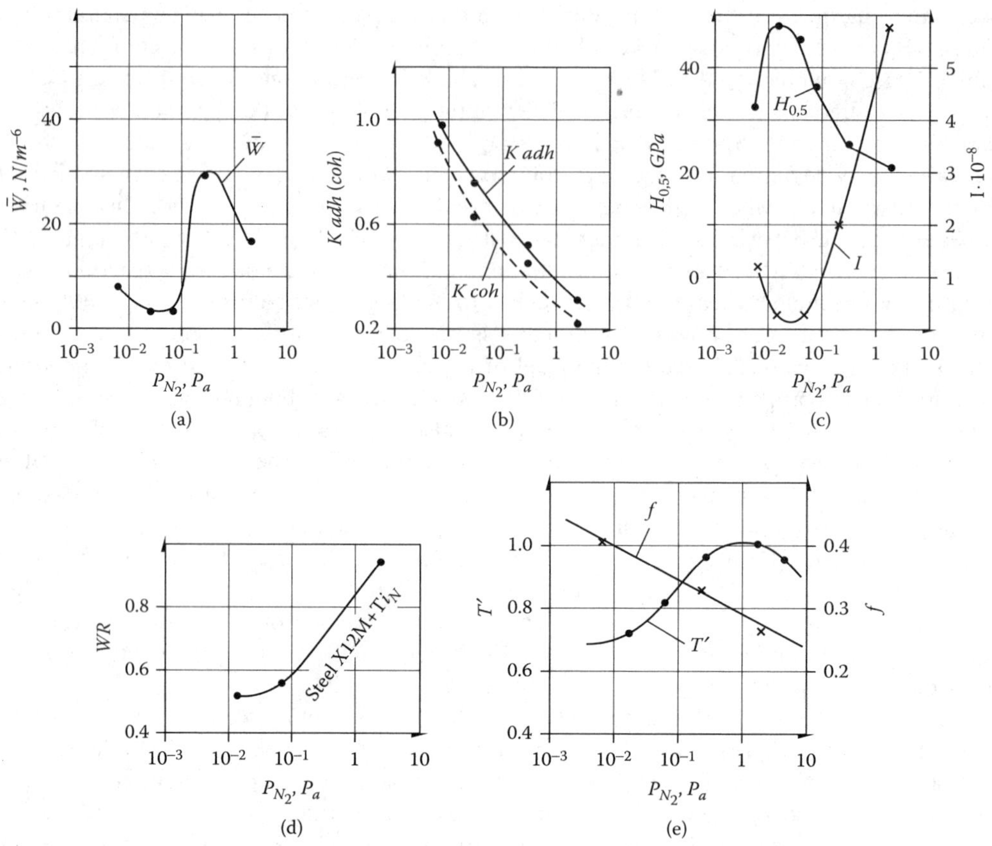

FIGURE 9.2 The dependence of the properties of Ti–N PVD coatings on the nitrogen pressure: (a) W — Palmquist toughness; (b) K_{adh}, K_{coh} — adhesion and cohesion of the Ti–N coating; (c) $H_{0.5}$ — microhardness, I — sliding wear intenstity; (d) W — relative wear resistance during stamping; (e) f — coefficient of friction, T — tool life during turning of 1040 steel. (From Gershman, J.S., Bushe, N.A. Thin films and self-organization during friction under the current collection conditions. *Surf. Coat. Technol.* 2004, 186, *3*, 405–411. With permission.)

3. A working (contact) layer deposited under high nitrogen pressure and balanced conditions (the deposition rate for the arc PVD method is 5 to 6.5 μm h^{-1}). In addition, the deposition conditions at this stage should be chosen to yield stoichiometric TiN, having a nearly perfect axial texture, a high residual compressive stress (more than 2000 MPa), and a fine, columnar, grain size containing minimal "droplet" phases.

A multilayer coating is required to meet these diverse requirements. These coatings offer many advantages (in comparison to monolithic coatings) in satisfying the broad range of mechanical properties needed. These coatings have a high adhesion to the substrate but low adhesion to the workpiece (i.e., a minimal friction coefficient), a high microhardness ($H_{0.5}$ = 35 GPa), and a high toughness (more than 50 J m^{-2}; see Table 9.2). This favorable blend of structural and mechanical properties has many advantages for wear resistance during cutting operations (Table 9.2).

The same multilayer coating could be used for filtered PVD coatings with additional advantages offered by this technology [10]. These systems not only eliminate the "droplet" phase from the coating, but can also be used to control the deposition conditions so that an excellent microstructure with the desired properties is obtained in the film. An extremely fine-grained structure (with the grain size is within a nanoscale range as compared to a grain size of microscale range in regular

TABLE 9.2
Comparative Characteristics of Ti–N Coatings Deposited by the PVD Method on the Coating Deposition Process

		Parameters			
PVD Method	Coating Design	Microhardness (GPa)	Palmquist Toughness (N m^{-2})	Coefficient of Adhesion to the Substrate	Relative Wear Resistance on Cutting
Regular arc deposition	Monolithic coating	25.0	26.0	0.5	1.0
	Multilayer coating	30–35	50–60	0.8	1.5–2.0
Filtered arc deposition	Multilayer coating	35–37	150–200	0.8	2.0–2.5

Source: Fox-Rabinovich, G.S. Structure of complex coatings. *Wear* 1993, 160, 67–76. With permission.

Ti–N coatings) can be achieved [11], with excellent mechanical properties. The hardness and Palmquist toughness of a Ti–N coating deposited by this method can also be increased up to 35 to 37 GPa (instead of 25 GPa) and 150 to 200 Jm^{-2} (instead of 26 Jm^{-2}), respectively. The adhesion of this coating is also very high (k_{adh} = 0.8, see Table 9.2). In addition, the wear resistance of filtered coatings is usually much better than regular coatings (Table 9.2). The principles outlined in the preceding text for a multilayered Ti–N coating can also be successfully applied to filtered coatings.

9.3.1.2 Wear Behavior and Self-Organization of Ti–N PVD Coatings

Hard Ti–N coatings inhibit an intensive adhesion of the workpiece to the tool surface [1]. The friction parameter of this coating is also low at the operating temperature (Figure 9.3). The wear behavior changes when Ti–N coatings are applied to cutting tools (Figure 9.4a) [12]. The initial rate of wear (during the running-in stage) is significantly lower, and the range of stable wear is expanded. This corresponds to the principles of wear control outlined in Chapter 1. To explain the enhanced wear characteristics imparted by Ti–N PVD coatings, a study of the self-organization phenomenon of the tool was performed [13]. Protective tribo-films of the Ti-O type form at the surface of hard PVD Ti–N coatings during cutting. The transition from the running-in stage to the normal wear stage is marked by the development of tribofilms which are a supersaturated solid solution of oxygen in titanium (Figure 9.5a to Figure 9.5c). The wear rate of the tool is reduced, and the process enters the steady state stage. Titanium oxide has a high resistance to friction [14] and cutting (see Chapter 7, Table 7.2) and readily fulfills a protective role for the underlying Ti–N coating.

The nature of the thin surface layer formed at the surface of the coatings was studied using EELFS analysis. An analysis of the fine structure obtained from the surface of the wear crater at different stages of operation in coated HSS cutting tools was used to follow the changes in the structure of the surface layers (see Figure 9.6a to Figure 9.6c). When the cutting period was 30 sec, i.e., at the running-in stage, the surface features in the EELFS spectrum agreed with crystalline titanium nitride. The characteristic signature of TiN is a peak at R_2 = 2.0 Å, as well as a peak at more remote interatomic distances (at about 4 to 5 Å, see Figure 9.6a). When the cutting lasts for 180 sec, titanium oxide develops in the coating, the degree of remote order in the crystal lattice is reduced, and the coating structure appears to amorphize. This is shown by the appearance of a peak at R_2 = 2.20 Å ($R_{Ti} + R_O$ = 1.45 + 0.73 = 2.18 Å) and by the attenuation of peaks at more remote interatomic distances (see Figure 9.6b). When catastrophic wear occurs (Figure 9.6c), at a cutting time of 2100 sec, the coating is destroyed while the steel surface is exposed. This is shown by the change in the form of the Fourier transform. The first peak is now located at a distance R_1 = 1.7 Å, whereas the second peak is at a distance R_2 = 2.65 Å. These peaks correspond approximately

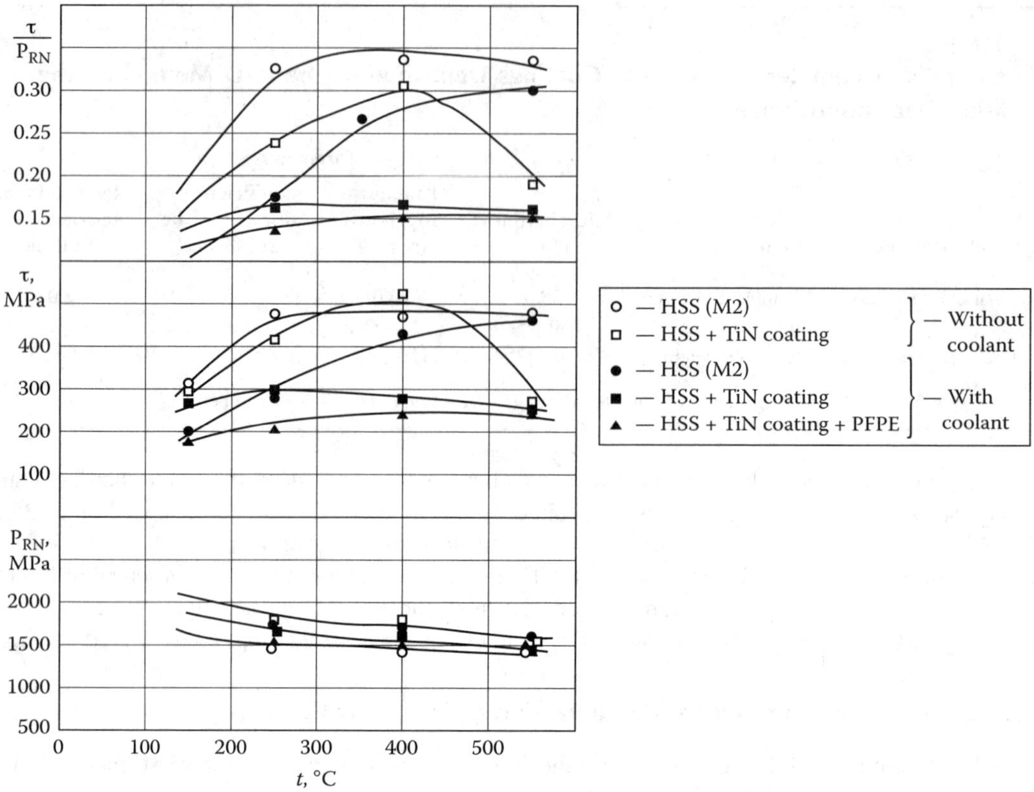

FIGURE 9.3 The dependence of the coefficient of friction of Ti–N PVD coatings on temperature.

to the length of C–Fe and Fe–Fe bonds ($R_C + R_{Fe} = 0.51 + 1.26 = 1.77$ Å); $R_{Fe} + R_{Fe} = 1.26 + 1.26 = 2.52$ Å). The spectrum shown in Figure 9.6c is typical of the BCC lattice of T15 high-speed steel.

The study of the wear resistance of coated tools demonstrates that the protective role of the coating is most efficient when the effects of the work of cutting can be localized in the near-surface region of the coating [12]. Current coating technologies achieve this goal by promoting the tribological compatibility of the tool. This is done in two ways:

1. By duplex and self-lubricated coatings for low- and moderate-speed machining
2. By the use of self-adaptive hard coatings, for high-performance machining applications

9.3.2 Duplex Coatings

The principal application of these coatings [15] is for cutting at low speeds, when HSS tools are used. It is desirable to deposit the hard coating, not directly onto the steel substrate but rather onto a sublayer, so that a gradual change in properties at the coating–substrate interface, i.e., a functionally graded material, is realized. This sublayer can be obtained by different technologies, e.g., ion nitriding. Usually, such coatings will then include both a nitrided sublayer and a hard PVD coating.

The nitrided sublayer has two roles. It prevents intensive plastic deformation of the substrate (HSS or DCPM) and cracking of the PVD coating that might be caused by deformation of the underlying substrate, while at the same time providing an additional thermal barrier [43]. The advantages of HSS cutting tools with duplex coatings are shown schematically in Figure 9.7.

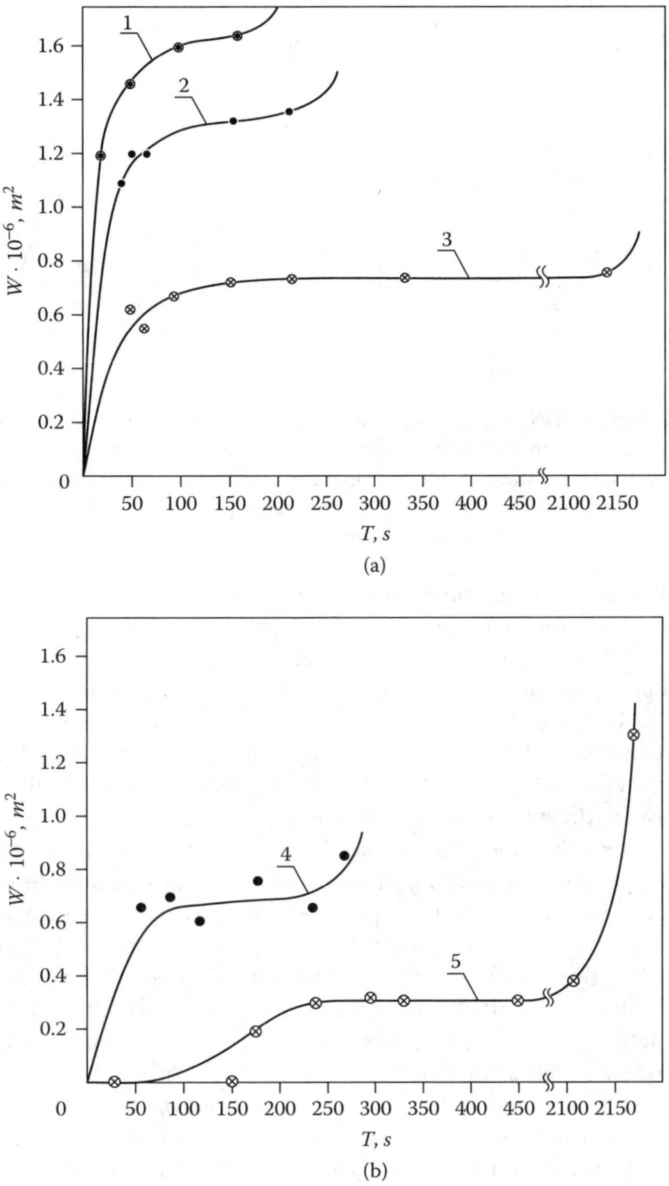

FIGURE 9.4 Tool flank wear vs. time. Turning test data (a — cutting speed = 70 m/min; b — cutting speed = 90 m/min; machined material — 1040 steel; depth (mm) — 1.0 mm; feed — 0.28 mm/rev). 1 — M2 HSS; 2 — M2 + ion nitriding; 3 — M2 + PVD Ti–N coatings; 4 — M2 + ion nitriding + PVD Ti–N coatings; 5 — T15 HSS + ion nitriding + PVD Ti–N coatings. (From Fox-Rabinovich, G.S., Kovalev, A.I., Afanasyev, S.N. Characteristic features of wear in tools made of high-speed steels with surface engineered coatings. Part I: Wear characteristics of surface engineered high-speed steel cutting tools. *Wear* 1996, 201, 38–44. With permission.)

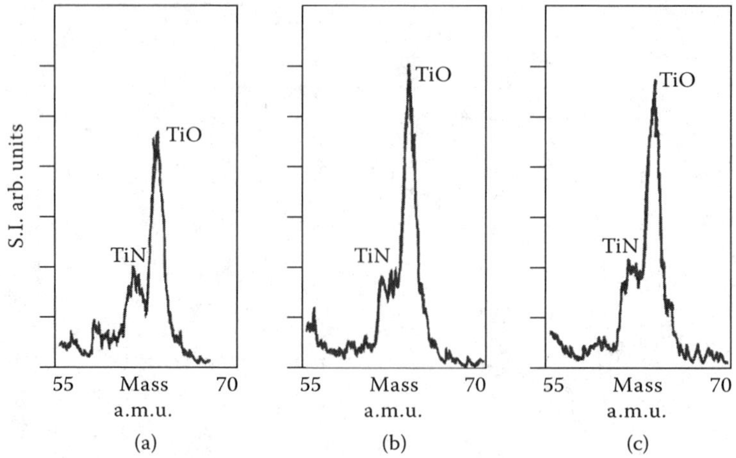

FIGURE 9.5 SIMS spectra of TiN and "quasi-oxide" Ti–O films as a function of the service time of M2 HSS tool with coatings: (a) cutting time = 15 sec; (b) cutting time = 90 sec; (c) cutting time = 120 sec. (From Fox-Rabinovich, G.S., Kovalev, A.I., Afanasyev, S.N. Characteristic features of wear in tools made of HSS with surface engineered coatings. Part II: Study of surface engineered HSS cutting tools by AES, SIMS and EELFAS methods. *Wear* 1996, 198, 280–286. With permission.)

However, the structure of the nitrided sublayer must be optimized in duplex coatings to achieve the best tool life. The duration and temperature of the process are the most important parameters in ion nitriding [5]. The ion current density should not be high, preferably about 3 Am^{-2}. This should be noted when an ion nitriding unit with combined heating is used.

The experimental data given in the following was obtained when the surface temperature during nitriding was between 500 and 530°C. At this temperature, rapid nitrogen diffusion occurs. The dependence of the structure and properties of M2 and D2 tool steels on the nitriding time is shown in Figure 9.8. Ion bombardment leads to the formation of a defective structure in the surface layers, which enhances nitrogen diffusion. During the first 10 to 20 min of nitriding, a saturated solid solution of N is formed. After 30 min of nitriding, a supersaturated solid solution of N is obtained at the surface (Figure 9.8a). The most pronounced changes in the lattice parameter and line broadening of the (211) reflection occurs after 0.5 to 2.0 h of nitriding (Figure 9.8b). A further increase in the nitriding time from 2 to 4 h has little effect on both the lattice parameter and the line broadening. Nitrides are observed after about 2 to 4 h. The formation of a nitride using x-ray diffraction can be detected when the concentration of the nitride is approximately 5%. The first nitride to be detected by x-ray diffraction in this study is the ε-phase $(W,Fe)_{2-3}N$, whereas after 4 h of nitriding, the ε and γ′ $(W,Fe)_4N$ phases are detected. After the 4 h of nitriding the nitrides can clearly be detected by optical metallography as a network of thin, needle- or lath-shaped particles.

It is known that the presence of tungsten, molybdenum, and chromium in a solid solution of steel can lead to the formation of fine nitrides with a high density and a marked increase in hardness. When the nitriding time is increased to 2 h or more, mixed (Cr, W, and Mo) nitrides will also nucleate. These nitrides are very finely dispersed and, hence, are difficult to detect by x-ray diffraction, but they contribute significantly to the increased hardness as well as the plasticity index.

The plasticity index of nitrided M2 steel changes in a way that is opposite to the way the hardness vs. time curve changes (Figure 9.8d and Figure 9.8e). This parameter (determined from a nanoindentation test) is highest (52%) when the hardness is low, and conversely decreases (to 48%) when the hardness is high. (The Palmquist toughness for nitrided steels cannot be used to give a meaningful measure of the fracture resistance as the depth of the nitrided layer changes as nitriding proceeds.) The plasticity of the nitrided layer is also sensitive to the microstructure. When there are no nitrides in the layer, the plasticity index is proportional to the nitrogen saturation.

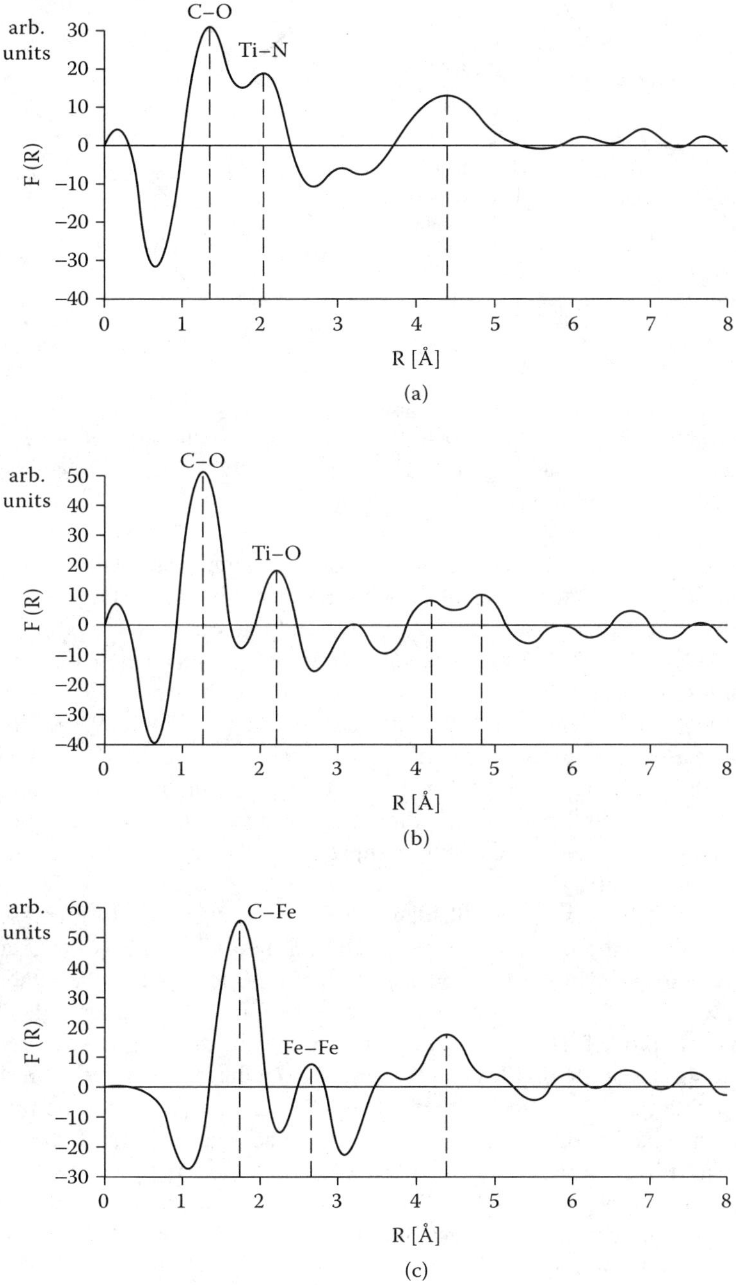

FIGURE 9.6 Fourier transform from EELFS analysis of a wear crater, Ti–Cr–N coating on nitrided T15 steel: (a) cutting time = 30 sec; (b) cutting time = 180 sec; (c) cutting time = 2100 sec. (From Fox-Rabinovich, G.S., Kovalev, A.I., Afanasyev, S.N. Characteristic features of wear in tools made of HSS with surface engineered coatings. Part II: Study of surface engineered HSS cutting tools by AES, SIMS and EELFAS methods. *Wear* 1996, 198, 280–286. With permission.)

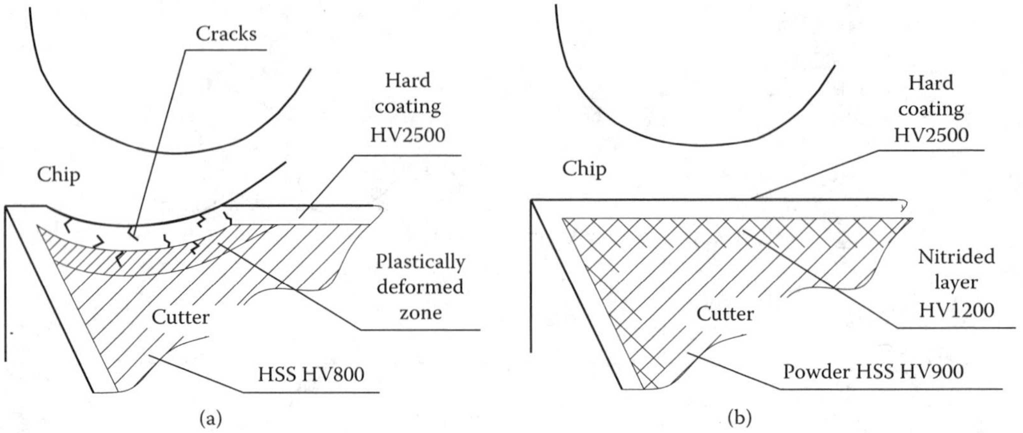

FIGURE 9.7 Schematic diagrams of composite cutting tools: (a) HSS + Ti–N PVD coating; (b) HSS + duplex coatings. (From Fox-Rabinovich, G.S., Kovalev, A.I., Afanasyev, S.N. Characteristic features of wear in tools made of high-speed steels with surface engineered coatings. Part I: Wear characteristics of surface engineered high-speed steel cutting tools. *Wear* 1996, 201, 38–44. With permission.)

The N content in this zone can be characterized by the lattice parameter of the α-phase (Figure 9.8a). As the nitrogen concentration (and lattice parameter) in the surface layer rises, there is a corresponding decrease in the plasticity, and *vice versa*. In an obvious way, it is related to the ion-nitrided layer of M2 and D2 steel. A low plasticity is correlated with an increased lattice deformation of the solid solution, associated with the dissolution of N into the iron lattice, as shown by the line broadening of the (211) reflection of the nitrided martensite (Figure 9.8b). In addition, some influence on the plastic properties is exerted by residual stresses that are formed on the surface layer during nitriding (Figure 9.8c). The residual stresses are high when the nitrogen content in the nitrided layer increases and extensive precipitation occurs on cooling. The volume of the surface layer increases on nitriding and, as a result, compressive residual stresses are formed. This effect is typical for M2 HSS steel.

High compressive stresses in the nitrided layer of M2 steel lead to increased hardness and plasticity, and inhibit flaking of the cutting edge during the tool life. It is important that the level and sign of stresses formed in the nitrided layer are similar to those in the adhesion sublayer of multilayer coatings. Then the stress gradient between the nitrided substrate and the coating is low, and the adhesion is improved. The service properties of the nitrided layer also have a high structural sensitivity. The longest tool life of nitrided HSS steels is obtained with an α-solid solution structure and is at least double that of unnitrided tools. The tool life increases with the nitrogen content in the layer, which, as noted earlier, can be monitored by the change in the lattice parameter of the nitrided martensite (Figure 9.8f). After nitrides have precipitated, the tool life decreases as a result of flaking at the cutting edge, caused by a decrease in the plasticity of the surface layer. The formation of a residual compressive stress also plays some role in flaking, as these stresses are highest with an N solid solution.

In addition to the structure of the duplex coating, the nature of the coating–substrate interface is also of great importance. The adhesion of the coating is one of the principal factors (together with the thermal stability) in determining tool life. The interface must be free from brittle compounds (such as oxides, nitrides, etc.) formed in the hardening process or during interaction with the environment. Several studies suggest that the surface of the tool be polished to remove surface nitrides formed after ion treatment [16]. Surface cleaning is also effective when ion etching is used, but the etching must be performed very carefully. The cutting edges of a sharp tool should not be rounded, the surface roughness should not increase, and the tool dimensions should be kept to a

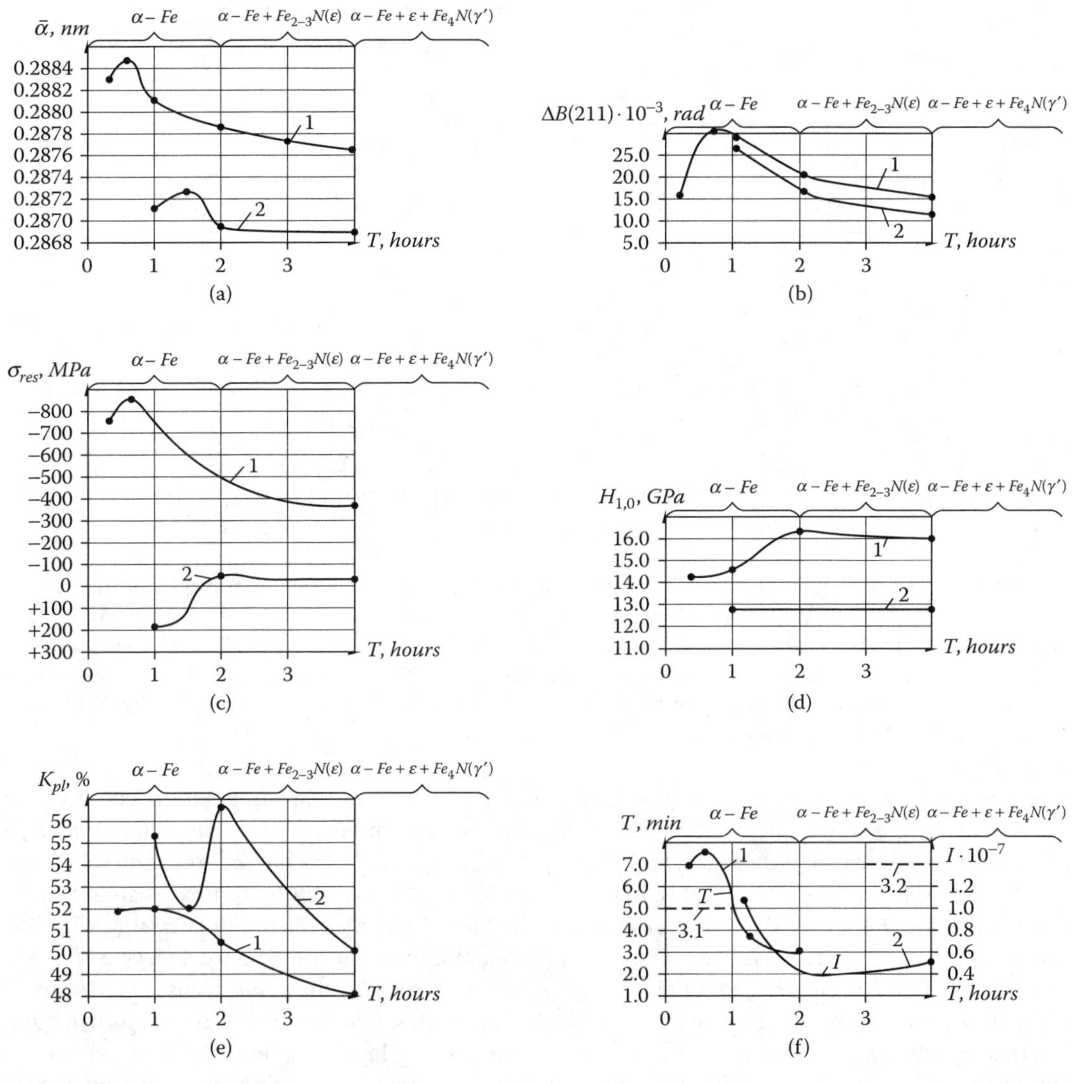

FIGURE 9.8 The time dependence of the structural characteristics and properties of the ion nitrided sublayer of a surface-engineered coating: 1 — M2 HSS; 2 — D2 tool steel, 3.1 nitrided layer of a cutting tool, 3.2 unnitrided layer of the die steel. (a) a — lattice parameter; (b) ΔB (211) — line broadening of the (211); (c) σ_{res} — residual stresses; (d) $H_{1.0}$ —microhardness; (e) K_{pl} — plasticity index; (f) T — turning inserts tool life during machining of 1040 steel, I — sliding wear intensity.

close tolerance. All of this is the subject of technological optimization, but with care, excellent results can be achieved [15].

9.3.2.1 Friction and Wear Behavior and the Features of Self-Organizing of Duplex Coatings

A duplex coating can act as a protective screen at the surface of a cutting tool (Figure 9.7). During steady state wear, a gradual but controlled wear of the coating takes place. All these advantages became even more obvious when duplex coatings are applied. Tests done at increased (90 m/min) cutting speeds (for HSS tools) enhance all the thermal processes associated with cutting. Under these conditions the heat-insulating effect of a hard Ti–N coating is diminished, the protective

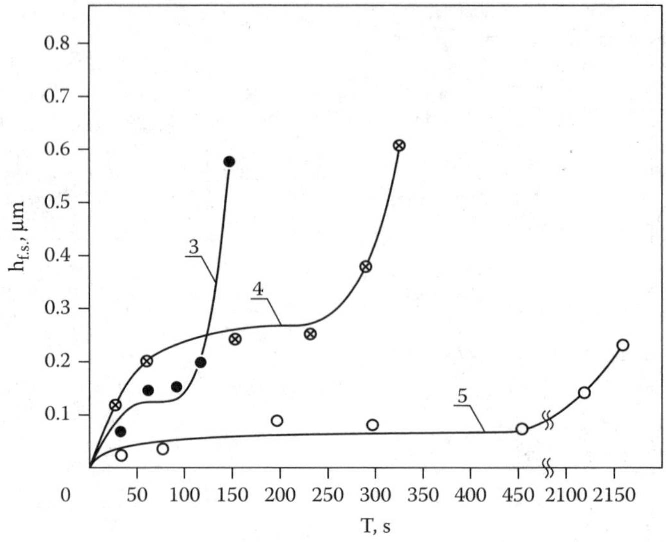

FIGURE 9.9 Tool flank wear vs. time. Turning test data (cutting speed = 90 m/min; machined material — 1040 steel; depth of cutting = 0.5 mm; feed rate = 0.28mm/rev): 3 — M2 + PVD Ti–N coatings; 4 — M2 + ion nitriding + PVD Ti–N coatings; 5 — T15 HSS + ion nitriding + PVD Ti–N coatings. (From Fox-Rabinovich, G.S., Kovalev, A.I., Afanasyev, S.N. Characteristic features of wear in tools made of high-speed steels with surface engineered coatings. Part I: Wear characteristics of surface engineered high-speed steel cutting tools. *Wear* 1996, 201, 38–44. With permission.)

function of the coating is reduced, plastic deformation of the steel substrate can occur, and the stability of cutting is disrupted. All of these trends can be seen in the data presented in Figure 9.4b and Figure 9.9. Hardening an M2 steel by a duplex coating can be employed to counteract these effects. The wear value is considerably lower, and the zone of stable cutting process is significantly broader (Figure 9.9, curve 4). The best results are achieved when a substrate material (T15 HSS) having a high heat resistance is used. In this case, compatibility of the tool and workpiece is realized to a great degree. The coating plays the role of a protective screen for the contact surfaces. However, it should be emphasized that the successful fulfillment of this function is possible only when the external thermomechanical effects are localized in the coating layer. Studies of coating wear have shown that this favorable condition occurs if a duplex coating was used because dissipation of energy is not channeled into processes other than friction (i.e., intensive surface damage).

The practical results of duplex coatings, i.e., a functionally graded tool material, are quite impressive. This material combines a high surface wear resistance (hard coating) and high core toughness (HSS). The tool life is increased by a factor of five to ten times [17], while at the same time the metalworking productivity can be increased by a factor of two to four. The cutting speeds of high-speed steel tools with duplex coatings (when cutting ordinary construction grades of steel) can be as high as 130 to 150 m min^{-1}. These cutting speeds are found with carbide tools only under certain limited cutting conditions.

9.3.3 Self-Lubricating Coatings

For transient or surface-damaging friction conditions (e.g., during the running-in or avalanche-like stages of wear, see Chapter 1, Figure 1.1), the efficiency of hard coatings becomes questionable owing to their brittleness. During the machining of several types of alloys (e.g., stainless steels- or nickel-based superalloys), unstable conditions of attrition wear starts to dominate [18] and surface damaging mechanisms become prevalent. In this case, the ability of a thin surface layer to dissipate most of the energy generated during cutting, thereby minimizing the chipping of the cutting-tool edge, becomes critically important.

For the most demanding cutting applications, a third type of coating — the self-lubricated hard coating — has been developed. Self-lubricating coating includes a lubricating compound in their structure that critically enhances tribological characteristics of wear-resistant coatings. A typical example of this type of development is the coating such as TiN–MoS_2 or TiAlN–MoS_2, with two energy-dissipating mechanisms built into the microstructure [19]. The first is associated with the formation of oxygen-containing tribo-films (most probably of Ti–O type, see earlier text) that readily form at the surface of the hard coating (Ti–N, Ti–Al–N) and play the role of a solid lubricant at elevated temperatures of cutting. The second is associated with the thin MoS_2 lubricating layer. This type of coating has the commercial name MoST. MoST coatings [20–21] are very efficient for cutting at low and moderate speeds as well as for stamping tools' application when intensive galling and seizure take place. The next generation of self-lubricating coating is the functionally graded self-lubricating coating. Grading the composition by drastic control of the deposition process has been achieved to combine a relatively hard TiAlN phase with a softer MoS_2 phase with an increase in the latter phase toward the top surface [22]. A similar design of coating is the multilayered coating with an outer layer of amorphous carbon on the TiAlN sublayer. This coating showed a high efficiency during drilling operations [23]. The limitation in the application of these coatings is thermal stability of the lubricating layer.

9.3.4 "Smart," Multilayered, Wear-Resistant Coatings

The next generation of the multilayer coatings is a "smart" coating. The "smart" wear-resistant coating concept implies the multilayered coatings that have a number of layers with programmable wear behavior that meet the requirements of the current stage of wear processes (Figure 9.10).

The service performance of multilayered smart coatings is characterized by the wear curves shown in Figure 9.11. Unfortunately, it is well known that not every mode of the running-in phase results in tribological compatibility of cutting tool–workpiece system [27,30], because damaging modes are also possible, especially during cutting. Thus, the goal of friction control is to prevent serious surface damage at the running-in stage and transform the tribosystem from its initial state into a self-organizing mode. If this can be achieved, the effective volume of interaction between the tool and the workpiece can drop by several orders of magnitude. For severe conditions of cutting, the effective thickness of the interaction volume at the self-organizing stage is in the range of 0.1 to 1 μm [24,27]. The high antifrictional nature of the surface layer is necessary to achieve

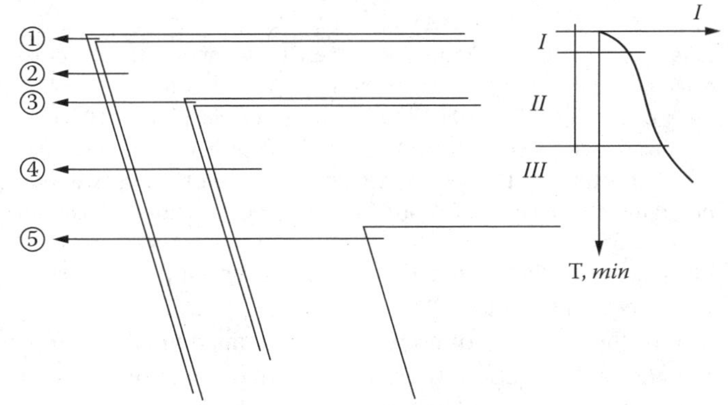

FIGURE 9.10 Schematic diagram of multilayered "smart" PVD coatings for cutting tools with a programmable change of properties: 1 — antifrictional layer (PFPE; see Chapter 12); 2 — hard Ti–N PVD coating; 3 — additional sublayer formed by (Ti + N) ion mixing; 4 — nitrided sublayer; 5 — HSS substrate.

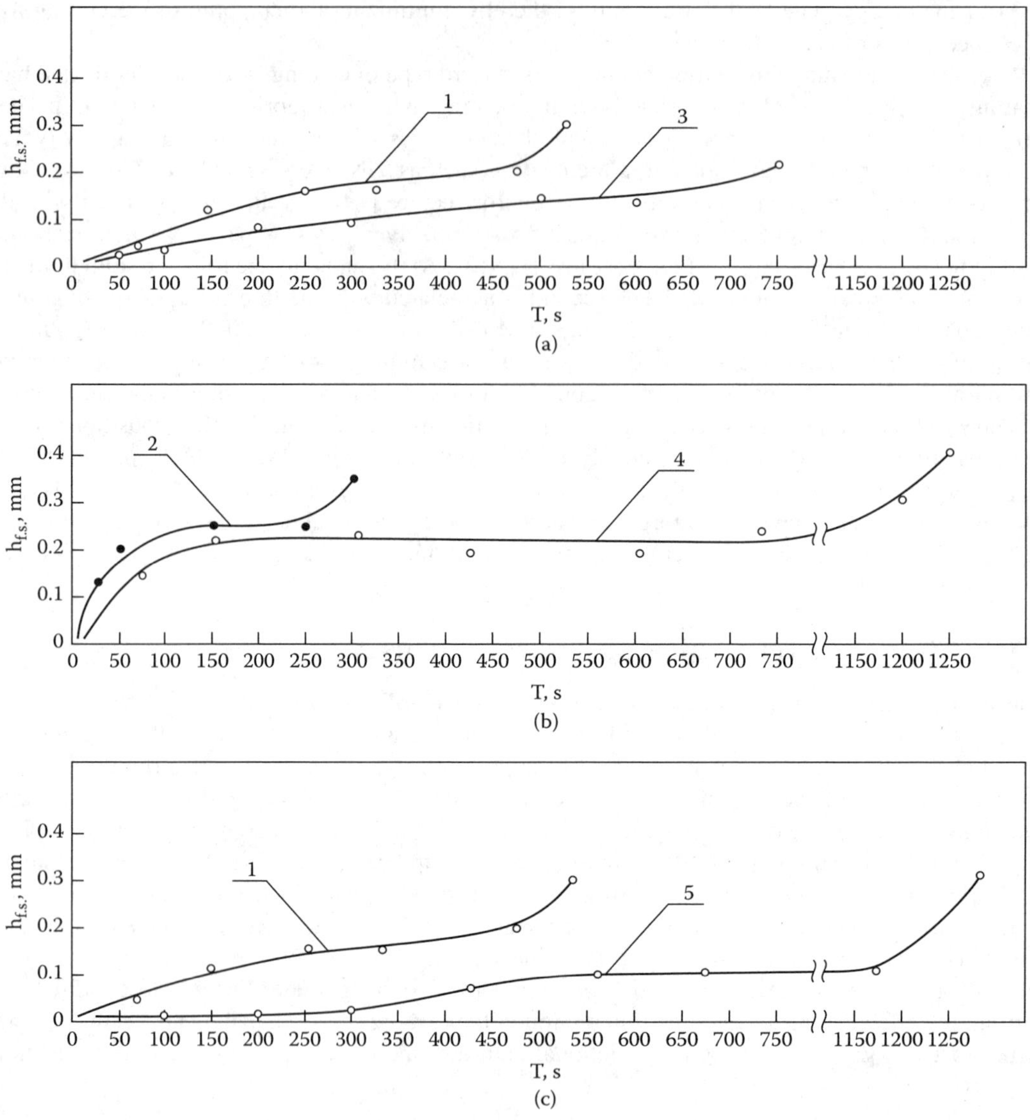

FIGURE 9.11 Tool flank wear vs. time. Turning test data (with and without coolant). Cutting speed = 90m/min; machined material — steel 1040; depth of cutting = 0.5 mm, feed rate = 0.28mm/rev: 1 — M2 + ion nitriding + PVD Ti–N coatings (with coolant); 2 — M2 + ion nitriding + PVD Ti–N coatings (without coolant); 3 — M2 + ion nitriding + PVD Ti–N coatings + PFPE antifrictional layer (with coolant); 4 — M2 + ion nitriding + PVD Ti–N coatings + Ti + N layer, modified by ion mixing (without coolant); 5 — M2 + "smart" coating with programmable change of properties, combining coatings 3 and 4 (with coolant).

these goals [31]. The details on the antifrictional layer (consisting of perfluorpolyether [PFPE] lubricating films) can be seen in Chapter 12.

The friction control of the cutting tool under surface-damaging conditions means the localization of the majority of external interactions at the maximum dissipation of energy, generated during friction within a thin surface layer. More channels of energy dissipation during the unstable running-in stage of wear ensure a higher tool life. This is a practical application of the universal principle of dissipative heterogeneity [30,32]. The multilayered smart coating includes a top (antifrictional) layer that leads to a decrease in flank wear as soon as the running-in stage is completed. As a result, the tool life is significantly increased (Figure 9.11). The application of the antifrictional layer is to prevent intensive surface damage of the hard coating owing to antifrictional properties of the layer, and to promote more protective and stable compound formation at the surface during the running-in stage of wear. Eventually, the stable stage of wear starts with a lower surface damage and corresponding to a tool life increase (Figure 9.11).

Similar problems of friction control at service conditions leading to surface damage arise when the wear process changes from stable to catastrophic or the avalanche-like stage. As noted earlier, cutting tools made of HSS usually operate under conditions of adhesive wear, where seizure occurs, accompanied by a rapid increase in the wear intensity [18,24]. Prolongation of the stable wear stage, however, is quite feasible even if seizure is a major problem. This can be achieved by applying an additional sublayer to the multilayered coating at the surface of the tool substrate. This layer should combine antifrictional properties with an ability to generate protective tribo-films at the coating–substrate interface.

One way to create these layers is by ion modification (ion alloying) of the surface of the tool. "Triplex" multilayered coatings have been studied [23,25]. The coating in these studies was deposited using three separate units. High-speed M2 steel was first nitrided using the glow discharge method. This was followed by ion implantation, prior to the application of a hard (Ti, Cr) N coating deposited by the PVD method [26]. Before applying the PVD coating, the samples were implanted at room temperature with 60-keV ions with a total flux of 4×10^{17} ions/cm^2. Sixteen different ions were chosen for the study. Prior to ion implantation, the surface of the samples were etched by argon ions, and surface contamination was controlled during implantation by the use of a cold trap that maintained a background pressure of about 2×10^{-6} Torr.

The sixteen elements selected for this work can be grouped as follows:

1. Elements forming stable protective surface films during friction [27], e.g., O, N, and Cl
2. Nonmetals (e.g., B, C, Si) forming compounds with good tribological properties when they interact with base material and elements from the environment
3. Metals including:
   a. Low-melting-point elements (in particular, In, Mg, Sn, and Ga) used as solid lubricants
   b. Cobalt-type metals with a hexagonal lattice and antifrictional properties owing to low shear strength [28–29]
   c. Metals (Al, Cr) that form stable oxide films during cutting, with a low coefficient of thermal conductivity
   d. Metals (Ag, Cu) known to have a low coefficient of friction, and low mutual solubility when in contact with steel, nickel, and titanium alloys (Figure 9.12) [29]

In addition, the surfaces were studied that have been subjected to treatments with:

- Four types of antifriction alloys used to improve conditions of sliding friction, viz., Zn + Al (9%) + Cu (2%), u + Pb (12%) + Sn (8%), Pb + Sn (1%) + Cu (3%), and Al + Sn (20%) + Cu (1%) + Si (0.5%) (see Chapter 3 [33])
- Zr + N, W + C, W + N, Ti + N, Al + O, to create layers with a high wear and oxidation resistance

The wear of these coatings were studied while turning 1045 carbon steels at a cutting speed of 70 m/min, a cutting depth of 0.5 mm, and a feed rate of 0.28 mm/revolution with and without a coolant. The flank wear of tetragonal, indexable HSS inserts with multilayered coatings was studied; when the flank wear exceeds 0.3 mm, the cutting tool loses its serviceability [18]. The effectiveness of ion modification was determined by comparing the cutting time to reach a specified flank wear of tools with multilayered coatings (i.e., those having both duplex coatings and ion modification)

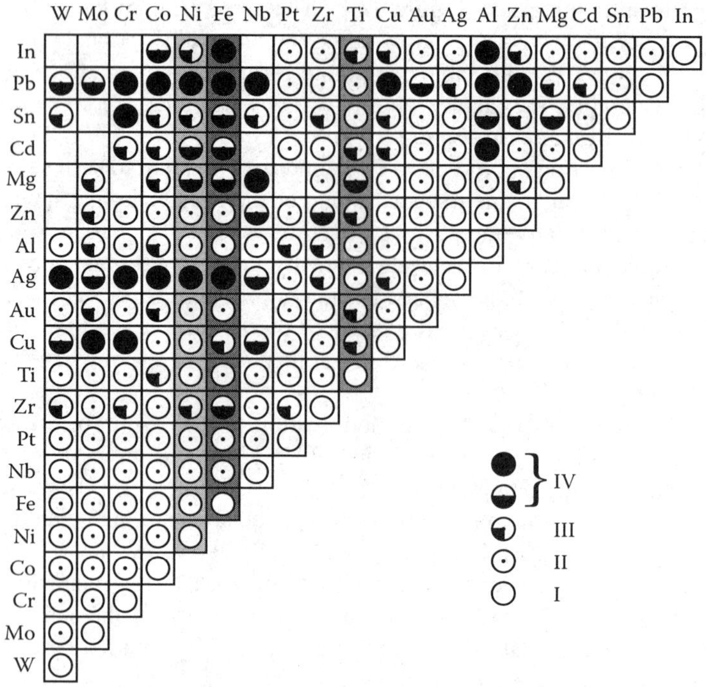

FIGURE 9.12 The mutual solubility of metals based on binary phase diagram data. The data are characterized in ranges running from IV (low mutual solubility) to I (high mutual solubility). (From Fox-Rabinovich, G.S., Bushe, N.A., Kovalev, A.I., Korshunov, S.N., Shuster, L.S., Dosbaeva, G.K. Impact of ion modification of HSS surfaces on the wear resistance of cutting tools with surface engineered coatings. *Wear* 2001, 249, 1051–1058. With permission.)

with identical duplex coatings prepared without the additional step of ion modification. Adhesion was determined using the scratch method. The coefficient of friction values were determined with the aid of a specially designed tribometer described in detail in Chapter 5 (see Figure 5.6).

The results of these tests, summarized in Table 9.3, demonstrate to a large extent that the influence of the implanted elements on the tool life is determined by the cutting conditions. The operational temperature during high-speed cutting is at least 600°C. If a coolant is used, the temperature is significantly reduced (by not less than 100°C) [18]. Thus, the effects of implantation vary, depending on whether the cutting is aided by coolants or not. Ion implantation significantly affects tool life [34], but the change in tool life is caused by a complex combination of interacting factors. The factors that are important in this context are:

- The formation of liquid phases or low-melting-point eutectics that act as lubricants
- The development of amorphous, oxygen-containing films with low coefficients of friction and thermal conductivity
- A reduction in the adhesion of the tool surface to the processed material and, at the same time, an increased adhesion of the hard PVD coating to the modified base material

The data from Table 9.3 shows that antifrictional alloys, widely used to improve the conditions of sliding friction [28,33], can double the tool life. However, this method of increasing the tool life, i.e., one that primarily depends on a reduction in the strength of the adhesion bonds between the tool and workpiece, is not the most efficient, as the adhesion of the coating to the modified surface was found to be rather low. This precludes their usage, as decohesion of a coating cannot be tolerated in practical applications.

TABLE 9.3
Tool Life of Cutters with a Modified Surface Layer (Ion Implantation and Ion Mixing)

N of Group (Subgroup)	Material	Element Composition	Coefficient of PVD-Coating Adhesion to Modified Surface Base	Relative Tool Life Compared to the Duplex Coating: Without Coolant	Relative Tool Life Compared to the Duplex Coating: With Coolant
		Surface Modified by Ion Implantation			
1	Elements with high affinity for oxidation	0	0.25	0.9	1.25
		N	0.41	2.0	1.83
		I	0.7–0.8	3.2	0.7
		Cl		1.8	
2	Nonmetals	B	0.6	1.2	0.65
		C	0.6	1.7	0.83
		Si		0.7	0.6
3	Metals				
	Low melting	In	0.6	2.4	2.1
		Mg	0.25	3.0	0.08
		Sn	0.6	0.8	0.7
		Ga		2.0	
b	With hexagonal lattice	Co	0.5	1.8	0.13
c	Forming stable oxides	Al	0.4	0.15	1.3
		Cr	0.6	0.2	1.2
d	With low coefficient of friction	Cu	0.55	1.0	2.5
		Ag	0.4	3.1	2.7
		Surface Modified by Antifriction Materials			
4	Zn-Al-Cu 9-1,5 GOST 21437-75 (Russia)	Zn + Al (9%) + Cu (2%)	0.44	1.98	—
	Bronze 8-12	Cu + Pb (11%) + Sn (9%)	0.4	0.95	—
	Babbitt BK2 GOST 1320-74 (Russia)	Pb + Sn (1.5%)	0.35	0.6	—
	Al-Sn-Cu 20-1 GOST 14113-69 (Russia)	Al + Sn (20%) + Cu (1%) + Si (0.5%)	0.3	0.4	—
		Surface Modified by Ion Mixing			
5		Al + O	0.4	3.0	—
		Ti + N	0.6	4.0	2.5
		Zr + N	—	0.53	—
		W + N	0.4	0.4	—
		W + C	0.4	1.33	—

Source: Gershman, J.S., Bushe, N.A. Thin films and self-organization during friction under the current collection conditions. *Surf. Coat. Technol.* 2004, 186, *3*, 405–411. With permission.

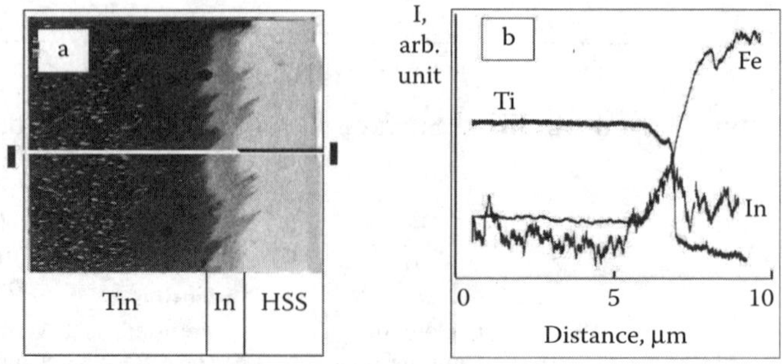

FIGURE 9.13 Microstructure of the multilayered HSS-based (Ti,Cr)N coating with an In-modified surface (ion implantation; 600× magnification): (a) microstructure of the angle lap section of the multilayered coating (SEM image); (b) distribution of elements along the II direction (x-ray microanalysis). (From Fox-Rabinovich, G.S., Bushe, N.A., Kovalev, A.I., Korshunov, S.N., Shuster, L.S., Dosbaeva, G.K. Impact of ion modification of HSS surfaces on the wear resistance of cutting tools with surface engineered coatings. *Wear* 2001, 249, 1051–1058. With permission.)

Implanting elements such as indium, silver, and nitrogen enhances tool life by a factor of 2 to 3 (see Table 9.3) for a range of cutting conditions (with and without cooling). These results are consistent with the observation that indium and silver show little interaction with iron, and find use as solid-state lubricants (Figure 9.12). Nitrogen implantation probably leads to the formation of an amorphous film with improved tribological characteristics [34]. Ion modification of the tool surface with the other elements studied led to unstable or negative effects, i.e., a reduction in tool life and/or poor adhesion between the hard coating and the substrate.

The most beneficial element in this study was indium. The maximum life of the tool was found, with or without the use of a coolant (see Table 9.3). At the same time, the adhesion between the coating and indium-modified surface of the tool was sufficient to ensure a reliable tool performance. Indium is a surface-active metal and usually displays a tribological compatibility with traditionally machined alloys based on steel, nickel, and titanium [29]. Because of this, the wear peculiarities of In-containing coatings have been comprehensively investigated [35].

Scanning electron microscopy and x-ray microanalysis were used to study surface-engineered cutting tools, composed of an ion-doped HSS surface, nitrided by a glow discharge technique, with a hard PVD coating over the In-modified layer (Figure 9.13a). Figure 9.13 shows the microstructure of a 5° angle lap specimen (including the surface-engineered coating), taken in the SEM with the back-scattered electron signal, which is sensitive to the mean atomic number. Separate layers of the multilayered coating (dark for TiN and gray for the In-rich sublayer) can be seen in the back-scattered electron image. The thickness of this zone is about 6 μm, so that the true depth of the modified (gray) layer is about 0.3 μm. It is probably an Fe-layer containing implanted Ar (as a result of etching by Ar^+ after nitriding) and In. The presence of W in the tool steel increases the intensity of the x-ray In K_α radiation and the background emission. This matrix effect influences the apparent emission volume of In K_α radiation and degrades the accuracy of measurement of the In-distribution. In addition, surface heating (up to 500°C) during (Ti,Cr)N deposition will modify the as-implanted In profile, which is expected to be about 0.3 μm in depth [36].

Following the x-ray microanalysis, the intensity ratios of the characteristic lines L_β/L_α for an In standard (99.99% purity) and the nitrided specimen were found to be 0.63 and 0.97, respectively. Changes in the intensity of the characteristic x-ray fluorescence are frequently observed when pure elements and their chemical compounds are compared [37]. In this study, clusters of In–N are thought to develop in the zone of In implantation. SIMS data and mass spectrometry (Figure 9.14) demonstrated that the ratio of the In concentration in a free state or present as clusters was approximately 10:1.

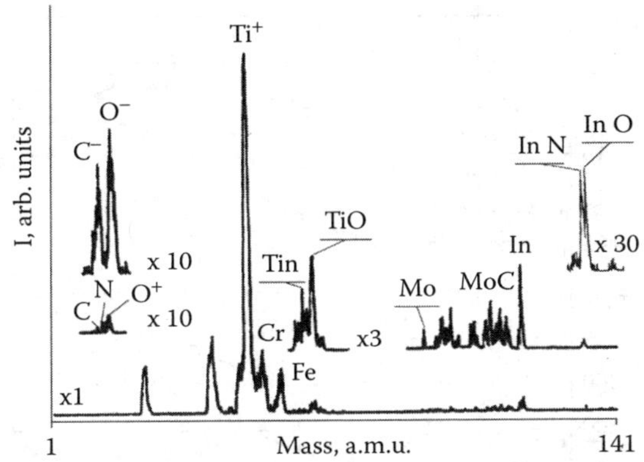

FIGURE 9.14 Secondary ion mass spectra from the wear zone of the cutting tool (cutting time is 30 min). (From Fox-Rabinovich, G.S., Bushe, N.A., Kovalev, A.I., Korshunov, S.N., Shuster, L.S., Dosbaeva, G.K. Impact of ion modification of HSS surfaces on the wear resistance of cutting tools with surface engineered coatings. *Wear* 2001, 249, 1051–1058. With permission.)

To explain how the implanted indium influences the tool life, the following factors were investigated:

1. The dependence of the friction coefficient on temperature
2. The distinctive features of indium oxidation in the wear zone (as investigated by SIMS)
3. The development of oxides on heating specimens with an In-modified surface

The temperature dependence of the friction coefficient demonstrated that In improves the frictional properties of HSS (Figure 9.15), by acting as a lubricant and reducing the shear strength (τ) of the adhesion bonds developed in the tribo-couples. This factor, however, is probably insufficient to account for the twofold increase in the tool life of cutters having an In-modified surface. Mass spectrometric analysis of the wear zone (Figure 9.14) suggests that the role of In is more complicated. Apart from metallic Indium, the wear zone reveals the presence of indium oxide, coming from both In and In–N dissociation and reaction during the wear process.

X-ray photoelectron spectroscopy (XPS) was used to study the changes in the shape of the In $3d_{5/2}$ lines in the electron spectra after oxidation. Figure 9.16a to Figure 9.16d presents the spectra obtained before and after heating the specimens to 823 K, with exposure times of 0, 0.5, 15, and 20 min, respectively. The position of the In $3d_{5/2}$ peak in the starting sample corresponds to a binding energy of 444.8 eV. Deconvolution of the spectra from the oxidized sample gave an additional peak, initially located at about 445.7 eV and, after a 25-min exposure, at 445.8 eV. These higher binding energies correspond to the formation of the oxide, In_2O_3. The relative intensity of this line compared to In $3d_{5/2}$ (the ratio of I_{In2O3}/I_{In}) was 23% in the initial state (Figure 9.16a), increasing to 41% after 25 min (Figure 9.16d).

Figure 9.17 presents the change in the relative concentration of In_2O_3 on the surface of HSS specimens during heating at 423, 623, and 823 K for times up to 25 min. At 823 K oxidation of the implanted indium rises quickly and saturates after about a 25-min exposure. The edge of a cutting tool runs at a temperature of about 773 K (500°C) during normal operations. These conditions suffice for oxidation of a fraction of the implanted indium. However, not all of the indium is oxidized, as part remains dissolved in solid solution in the iron matrix.

As the hard overlay (Ti,Cr)N coating is worn away, typically at the transition from the normal to catastrophic wear stage [12], the In-modified layer becomes exposed to the friction surface. This usually coincides with the point at which the protective PVD coating detaches from the contact face of the tool. Under conditions of high load and high temperature, partial oxidation of In will

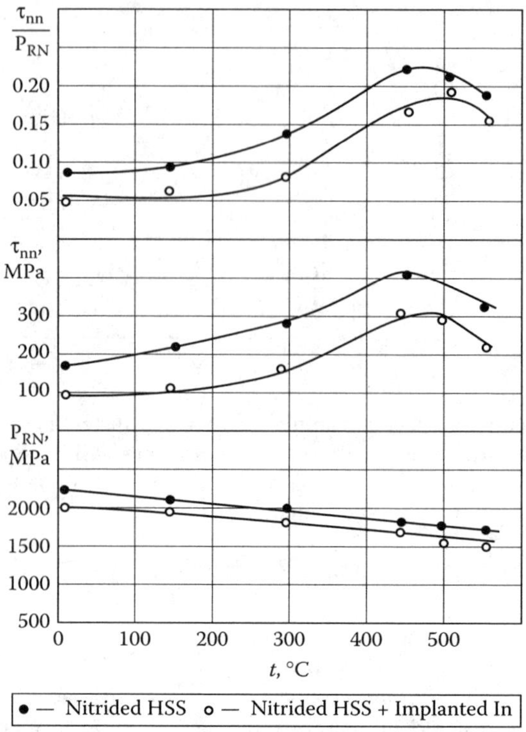

FIGURE 9.15 Impact of test temperature on the frictional properties of surface modified HSS cutting tools. (From Fox-Rabinovich, G.S., Bushe, N.A., Kovalev, A.I., Korshunov, S.N., Shuster, L.S., Dosbaeva, G.K. Impact of ion modification of HSS surfaces on the wear resistance of cutting tools with surface engineered coatings. *Wear* 2001, 249, 1051–1058. With permission.)

probably start before the complete destruction of the PVD coating. Because stable friction is characterized by minimal depth of damage of the contact surface [32], even a relatively thin ion-modified layer can enhance the tool life. Indium improves the frictional properties of the surface and reduces the sticking intensity over the friction surface. In addition, an oxygen-containing amorphous In–O film, formed by interaction with the environment, is likely to enhance favorable friction conditions in the contact zone of a cutting tool. Indium lies in the same group as Al in the periodic table, and probably forms oxygen-containing phases having low coefficients of thermal conductivity. These protect the tool surface, enhance the thermal conditions of cutting, and delay the onset of catastrophic wear. Thus, the influence of In is twofold: on the one hand, it acts as a metal lubricant, and on the other, it forms protective oxygen-containing phases. Indium enhances both the adaptability of the tribosystem and extends the stage of stable wear, in accordance with the principle of friction control [32].

The data presented in Table 9.3 show that the highest wear resistance after the triple surface treatment is achieved when transition metals together with nitrogen are used to modify the surface by ion mixing. Owing to the nonequilibrium character of the surface-modification process, stoichiometric compounds such as TiN, ZrN, WN, WC, and Al_2O_3 do not form. Only solid solutions of nonmetallic elements (O, C, and N) in corresponding metals are formed. The best wear resistance is achieved by a 1-μm-thick layer modified with Ti and N (Table 9.3; Figure 9.18a and Figure 9.18c). A metastable phase, which is most probably a solid solution of nitrogen in titanium, is formed on the surface of the ion-modified HSS substrate (Figure 9.18a). Minimal substrate heating during ion mixing is necessary to form the surface layer with a nanocrystalline structure and, in our case, an amorphous-like structure (see Figure 9.19) [38]. The ion mixing heats the growing

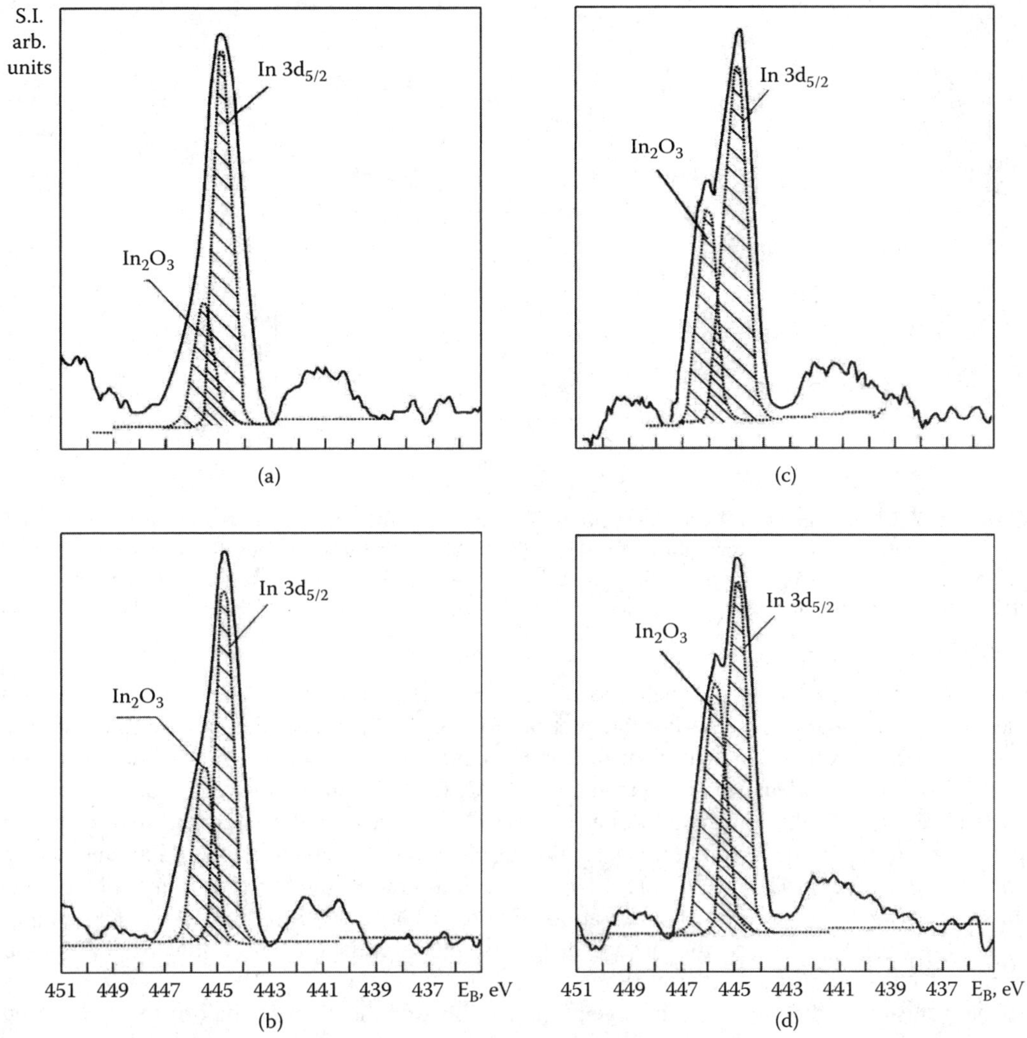

FIGURE 9.16 Change in the shape of the In 3d $_{5/2}$ line from the photoelectron spectrum taken from the HSS surface after ion implantation and oxidation at 823 K for: (a) 0 min, (b) 5 min, (c) 15 min, and (d) 20 min. Pressure of oxygen in the chamber = 2.5×10^{-6}. (From Fox-Rabinovich, G.S., Bushe, N.A., Kovalev, A.I., Korshunov, S.N., Shuster, L.S., Dosbaeva, G.K. Impact of ion modification of HSS surfaces on the wear resistance of cutting tools with surface engineered coatings. *Wear* 2001, 249, 1051–1058. With permission.)

films or surface layers at the atomic level. Therefore, it is called atomic scaled heating (ASH) [39–40]. ASH is caused by the condensation of sputtered atoms and subsequently by extremely fast cooling at the atomic level. We can assume that the formation of amorphous-like structures at the substrate surface is a result of ASH during strong nonequilibrium conditions of ion mixing. The amorphization of the surface layer (see Figure 9.19), which modifies the wear mechanism, is due to a prevention or delay in surface crack propagation [39].

Nanoindentation testing of a surface-modified layer has been performed for an ion nitrided HSS sample and a sample with additional ion mixing [41]. Microhardness and the work of elastic–plastic deformation were evaluated from load vs. displacement data measured using a computer-controlled nanoindentation tester MTI-3M. A Berkovich-type diamond indenter was used. The details of the experimental technique are shown in Chapter 5. Using the experimental data obtained, the elastic recovery parameter, K_e, was calculated as the ratio of the work of elastic

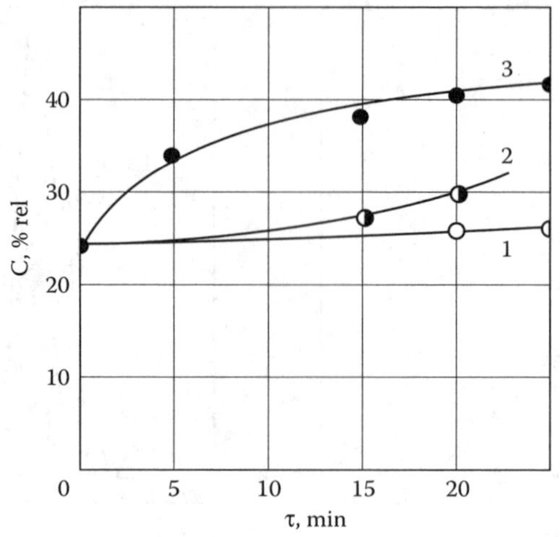

FIGURE 9.17 Change in the relative concentration of In_2O_3 on the surface of a HSS specimen after In-implantation and oxidation at temperatures: (1) 423, (2) 623, and (3) 823. (From Fox-Rabinovich, G.S., Bushe, N.A., Kovalev, A.I., Korshunov, S.N., Shuster, L.S., Dosbaeva, G.K. Impact of ion modification of HSS surfaces on the wear resistance of cutting tools with surface engineered coatings. *Wear* 2001, 249, 1051–1058. With permission.)

deformation to the total work of indentation. This parameter characterizes the ability of the surface to accumulate or store energy during indentation. Figure 9.20 presents the data on microhardness (Figure 9.20a), as well as the coefficient K_e (Figure 9.20b), as a ratio of work of elastic deformation to the total work of indentation vs. indentation depth for HSS specimens after ion nitriding and for ion nitriding with the ion mixed surface layer. We can see that the microhardness of the thin 1-μm layer, modified by ion mixing, has a value of HV 9.8 GPa, which is lower than the value for the ion nitrided layer at HV 11 GPa. However, there is a minor difference in the microhardness of both samples for bigger depths of indentation (Figure 9.20a). In Figure 9.20, b the K_e parameter is much higher, up to 0.64, at the surface of the ion-modified layer, with a lower microhardness compared to the ion nitrided layer at 0.49.

The diffusion of the implanted nitrogen into the chip and the reversed flux of oxygen from the environment into the tool surface leads to tribo-oxidation during cutting (Figure 9.18). The rapid formation of a protective tribo-film takes place (Figure 9.21) because the initial structure of the surface after mixing is similar to the structure of the films formed at the friction surface as a result of the self-organizing process. Ion mixing can produce thin surface layers with a nanocrystal, close to amorphous structure [39]. As noted earlier, the tribo-films or secondary structures (Chapter 1) have a similar microstructure containing an amorphous supersaturated solid solution of oxygen coming from the environment. These structures are formed by the reaction of the metal component with the tool material [27, 42]. Ion mixing enhances this process, which naturally evolves in the tribosystem during the self-organizing stage and results in the formation of stable tribo-films. In the final stage of wear, oxygen from the environment penetrates through the numerous pores and cracks in the PVD coating to the surface of the modified layer. Because this layer contains a very high density of point defects [39–40], the reaction with oxygen is rapid (Figure 9.18). As the PVD coating wears, this oxygen-rich layer can act to screen the ion-modified (Ti + N) surface and protect it against subsurface damage. Thus, when the hard coating is completely worn away, protective tribo-films would have already formed on the near-surface layers. These tribo-films delay the transformation to the avalanche-like stage of tool wear and, in addition, can again revert the tribosystem to a stable state (i.e., to a stable pattern of wear). From our point of view, this is the most beneficial effect of the modified Ti + N layer on the wear behavior. The transformation from

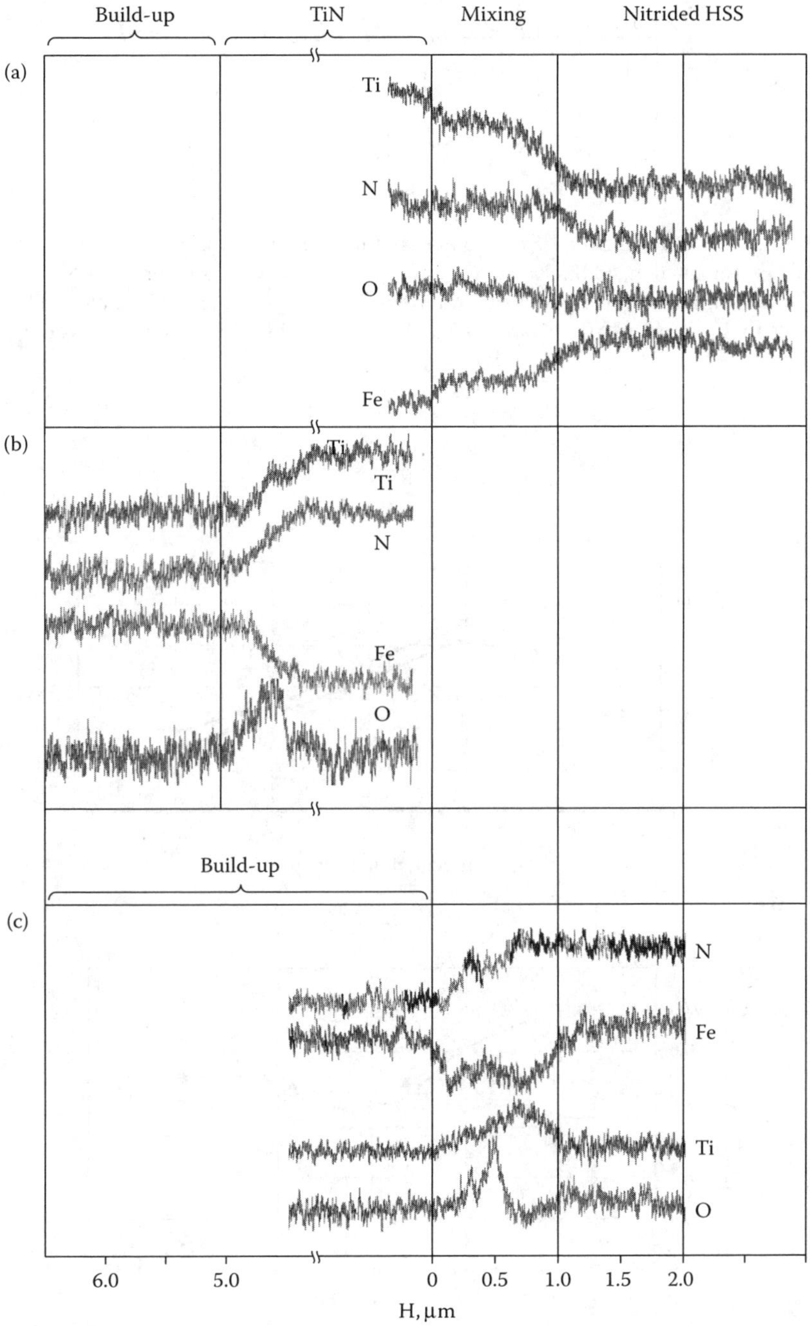

FIGURE 9.18 Distribution of chemical elements close to the "built-up–wear crater" interface: (a) initial stage of the Ti + N layer, as modified by ion mixing; (b) after a cutting time of 120 sec. Surface of (Ti,Cr)N PVD coating; (c) after a cutting time of 600 sec. Surface of the Ti + N layer, modified by ion mixing.

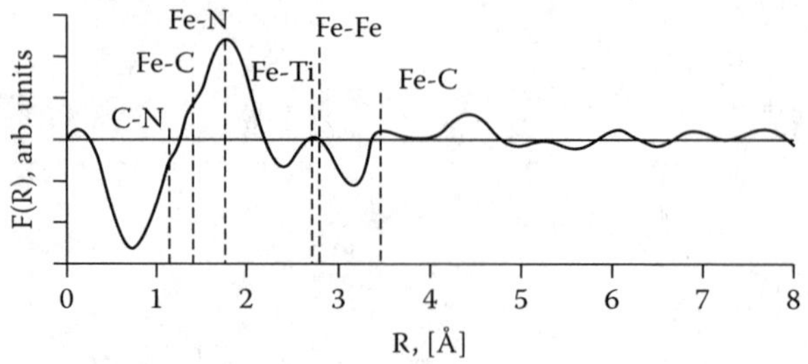

FIGURE 9.19 Fourier transform from EELFS analysis of the surface of the Ti + N layer, as modified by ion mixing. (From Fox-Rabinovich, G.S., Kovalev, A.I., Weatherly, G.C., Korshunov, S.N., Shuster, L.Sh.S., Veldhuis, C., Dosbaeva, G.K., Scvortsov, V.N., Wainstein D.L. Improvement of "duplex" PVD coatings for HSS cutting tools by Ion mixing. *Surf. Coat. Technol.* 2004, 187, 230–237. With permission.)

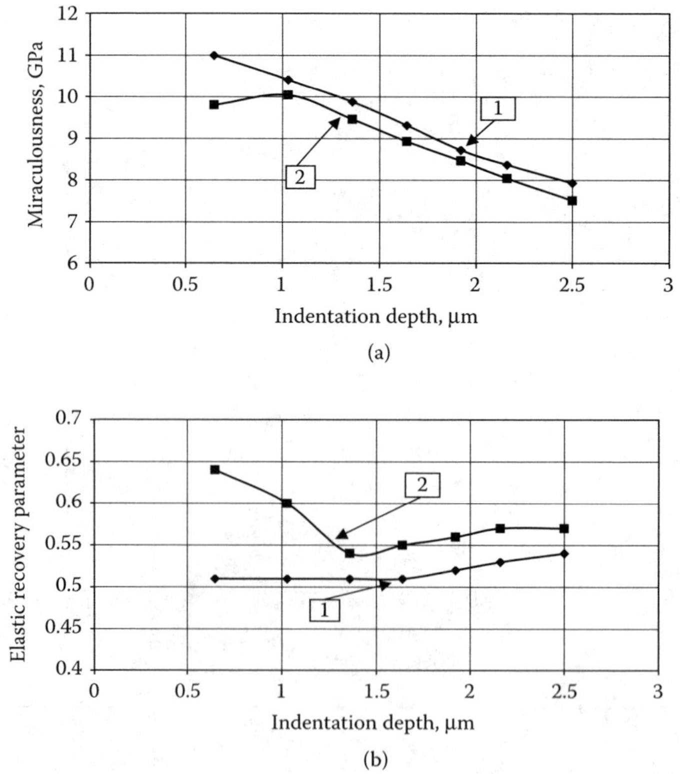

FIGURE 9.20 (a) Microhardness and (b) an elastic recovery parameter (ratio of elastic to total work of indentation K_e) vs. depth of indentation for the HSS specimens after (1) ion nitriding and (2) ion nitriding with the ion mixed surface layer. (From Fox-Rabinovich, G.S., Kovalev, A.I., Weatherly, G.C., Korshunov, S.N., Shuster, L.Sh.S., Veldhuis, C., Dosbaeva, G.K., Scvortsov, V.N., Wainstein D.L. Improvement of "duplex" PVD coatings for HSS cutting tools by Ion mixing. *Surf. Coat. Technol.* 2004, 187, 230–237. With permission.)

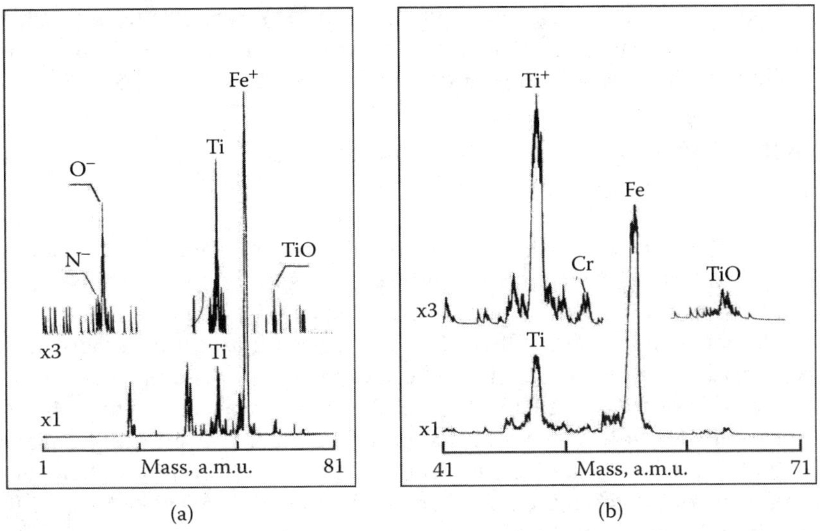

FIGURE 9.21 SIMS spectra of the surface of the Ti + N layer, modified by ion mixing: (a) initial stage of the Ti + N layer, as modified by ion mixing; (b) after a cutting time of 600 sec. (From Fox-Rabinovich, G.S., Kovalev, A.I., Weatherly, G.C., Korshunov, S.N., Shuster, L.Sh.S., Veldhuis, C., Dosbaeva, G.K., Scvortsov, V.N., Wainstein D.L. Improvement of "duplex" PVD coatings for HSS cutting tools by Ion mixing. *Surf. Coat. Technol.* 2004, 187, 230–237. With permission.)

stable to unstable wear modes (stable wear transforms to catastrophic one) is a typical deviation from a linear thermodynamic area to a nonlinear one. This is due to the accumulation of defects within the surface layer during friction, as was shown for stamping-tool wear behavior (see Chapter 7). To prevent this undesirable transformation, an additional Ti + N ion-modified layer with a given structure should be deposited on the surface to move the processes back to the linear thermodynamic area and decrease the wear rate. This is typical friction control with a positive feedback loop that results in a significant improvement to tool life.

The behavior of multilayer coatings illustrates an important principle that can be used to design effective materials for cutting tools. The more energy dissipation channels that can be built into the microstructure for the transitional (nonsteady) stage of tool wear, the longer the tool life will be. These channels can operate simultaneously in the same stage of the wear (e.g., during the running-in stage) and also subsequently using multilayer coatings when the wear process transforms from one stage to another. After completion of the first (running-in) stage of wear and exhaustion of the corresponding channel of energy dissipation in the top layers, the next layer of a multilayered coating can be used to control the "avalanche-like" wear with alternative channels of energy dissipation [4]. In this way a multilayered "smart" coating can be developed in which each layer fulfills a given function at a definite stage of wear (Figure 9.10), leading to high serviceability over a wide range of operating conditions. One of the novel areas of smart coating applications is used for dry machining conditions (see Chapter 12, Table 12.4).

In general, smart materials or systems respond and adapt to changes in conditions or the environment by integrating the functions of action and control [43–44]. This concept has been widely used for protective coatings for corrosion control [45], and as shown in the earlier text this concept can be extended to wear-resistant coatings. At the stable stage of wear, the coating must have adequate strength and toughness at actual service temperatures and adaptability, i.e., ability to form stable protective tribo-films during machining. In the unstable stages, the coating must have sufficient energy dissipation channels to prevent surface damage. A multilayer coating that includes a top antifrictional or lubricating layer, a working layer of a hard coating, and a sublayer that expands the range of stable wear will form the basis for future "smart" coatings. "Smart" coatings can adapt to rapidly changing conditions such as transformation from a stable wear stage to a catastrophic one. The coating with the design described in Reference 46 and shown in Figure 9.10 has the ability to adapt to the changing conditions outlined in the preceding text.

9.3.5 Wear-Resistant Coatings for High-Performance Machining Applications

Historically a major trend in materials' development for tooling applications is an improvement of their hardness and thermal stability. Starting from high-speed steel, the materials, which were developed, moved forward to hardmetals such as cemented carbide and further to ultrahard materials such as cubic boron nitride (CBN) and ceramics with exceptional thermal stability. The developments of hard coatings are mainly going the same way. As compared to the traditional Ti–N coating with microhardness 23 to 25 GPa, the most advanced coatings have a hardness almost three times higher [47] and, probably more importantly, a critically enhanced thermal stability (from 550°C to 1100°C and above). But cutting is a very complex phenomena in which one or two parameters of the tooling material cannot be totally responsible for tool life improvements. As was outlined in Chapter 5, a variety of characteristics are needed. In fact, the major improvements to the life time of cutting tools, compared to traditional Ti–N coating, have been achieved because of the development of titanium aluminum nitride, i.e., (Ti,Al)N, coatings, which have a similar or slightly higher hardness (up to 30 to 35 GPa). Ti–Al–N coatings currently replaced Ti–N on the market, and now it is the most widely used coating for machining application. Coatings such as Ti–Al–N with a Ti/Al ratio of 1.0 [48,49] display a unique combination of properties such as hot hardness together with relatively low thermal conductivity as well as chemical stability (i.e., stability to diffusion, dissolution into the chip, and oxidation stability). As a result, considerably more heat is dissipated via chip removal. An extremely important advantage of (Ti,Al)N coatings is their oxidation stability up to 850 to 925°C [50]. This is due to the formation of stable oxide films when compared to TiN [51–52] (see Table 9.4). Stable tribo-ceramic films can then be formed on the surface during cutting [53] and limit the diffusion of the coating material into the workpiece. These compounds are a mixture of alumina and rutile, as found in the oxidation of titanium aluminides [54–55].

TABLE 9.4
Oxidation Stability of the Compound (PVD, CVD Coating)

Coating	Loss of Oxidation Stability (Maximum working, T°C)
Ti–C	400
Ti(C,N)	450
Ti–N	550
Zr–N	500/600
Cr–N	650
Cr–C	700
Ti–Al–N(50:50)	850
TiC + Al_2O_3	1200

Source: Munz, W-D.M, Ti-Al nitride films: a new alternative to Ti–N coatings. *J. Vac. Sci. Technol.* 1986, A4, *6*, 2717–2725. With permission.

But with productivity improvements and more wide applications of high-performance machining, the actual temperature that occurs during cutting increase drastically (up to 1000°C and above). In modern machining, one of the major problems for the most widely used Ti–Al–N coating with a Ti/Al ratio of 1.0 is the hardness reduction vs. temperature [56] as well as the relatively high coefficient of friction values of this family of coatings at high temperatures [1].

There are two major ways to improve wear resistance of the Ti–Al–N-based coatings: (a) increasing hardness [57] and (b) improvements of tribological characteristics or reduction of coefficient of friction at the actual temperatures of cutting.

These two ways could be justified using the tribological compatibility approach. Tribological compatibility of the coated tool–workpiece system could be improved in two ways as shown in Chapter 3. The first way to improve the tribological compatibility is to reduce the coefficient of friction at high temperatures of cutting. In other words lubricity at the tool–workpiece interface should be enhanced. This results in lower wear intensity during the running-in stage of the process. The second way (which is most widely used practically) is to improve the mechanical properties, the most important being an increase in the hot hardness. This affects the tribological compatibility during post running-in stage.

Low-friction characteristics at high temperatures and improved lubricity at the cutting tool–workpiece interface leads to the lowering of seizure intensity. Lower seizure intensity results in less intensive cutting-tool surface damage during the running-in stage of wear.

High hardness of the coatings at elevated temperatures of cutting could lead to the following: the seizure formation is accompanied by intensive (around 50%) plastic deformation of surface layers as was outlined in Chapter 3. Owing to high hardness of a coating the zone of plastic deformation is concentrated within the very thin (most probably nanoscale range) surface layer with no subsurface layer involvement in the process of intensive deformation. This could result in a decrease of seizure intensity because this process is closely associated with the plastic deformation of surface layers (see Chapter 3 for details).

Both characteristics' (the hardness and the lubricity) improvements shift the processes that take place at the workpiece–tool interface from micro to nanohierarchical level (see Chapter 1). This shift is critical because during this change of the hierarchical level of tool–workpiece interaction the formation of relatively thin protective tribo-films probably starts to play an especially important role. If the hierarchical level of interaction changes, the substrate material (in our case a coating) is either working under less severe conditions (lower temperatures or stresses at the interface in the case of lubricious tribo-films formation), or is less intensively plastically deformed. In the latter case (if the high hot hardness is achieved) the layer of the coating supports the tribo-films formed with higher efficiency and prevents damage during friction. These two phenomena improve the tribo-film stability and their protection function [79]. As a result the energy dissipation and entropy production drop down. Thus, the wear intensity will also drop down.

9.3.5.1 Hard Nanoscale Crystalline/Composite/Multilayered Coatings

It was recently discovered that, in Ti–Al–N coating with high Al content (around 65 at. %) a secondary phase transformation (age-hardening) takes place during annealing at temperatures around 900 to 950°C. As the hardness grows, thermal stability improves because of the formation of a nanocomposite structure that consists of c-AlN domains and Ti-enriched c-(Ti,Al)N domains formed by spinodal decomposition (Figure 9.22). At the same time the residual stress relaxation takes place through annihilation of lattice defects during annealing [57].

The annealing of nanocrystalline Al–Ti–N coatings at 900°C for 2 h results in significant tool life improvements under high-speed turning conditions for 1040 steel (Figure 9.23, Table 9.5). This is most probably associated with the microstructure and to a lesser degree with the improvement of properties of the coating (Table 9.6) despite the similar tribological characteristics of the coatings. The chip characteristics that are formed during cutting are presented in Table 9.6. These

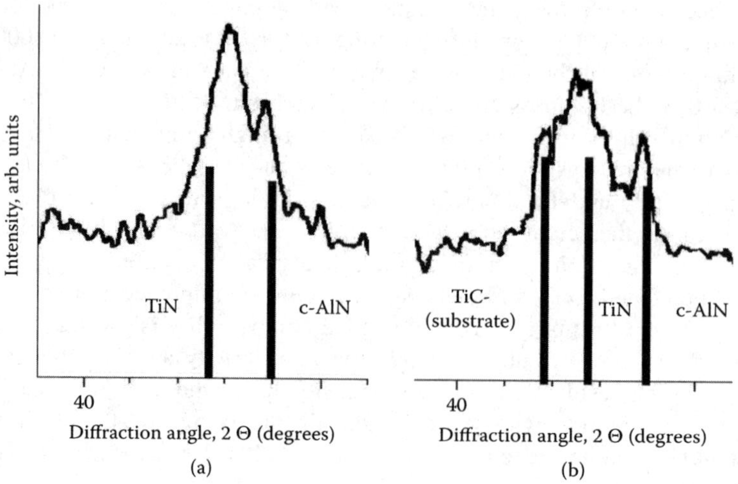

FIGURE 9.22 XRD data on Al–Ti–N coating (X.ceed) (a) before and (b) after annealing (900°C, 2 h, in vacuum, 1×10^{-6} mbar).

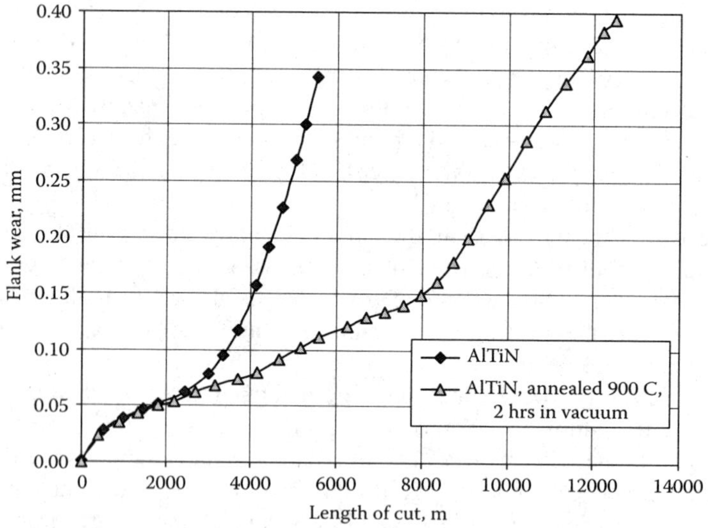

FIGURE 9.23 Tool life of Al–Ti–N coating (X.ceed) before and after annealing (900°C, 2 h, in vacuum, 1×10^{-6} mbar) under conditions of high-speed turning of 1040 steel (according to cutting conditions shown in Table 9.6).

characteristics are associated with *in situ* tribological characteristics of the coatings [18]. The values presented are far from optimal [18,111], and there is no significant difference in the chips' characteristics of the Al–Ti–N coatings before and after annealing.

Annealing of Al–Ti–N coatings significantly changes the microstructure (Figure 9.22 and Figure 9.24). Figure 9.24 presents the fine structure of the HREELS electron spectra in the proximity of the back-scattered electron line obtained from Al–Ti–N coatings (a) before and (b) after annealing. The P_1 and P_2 peaks on the spectra in Figure 9.24 are the plasmon peaks. The plasmon peak intensity in the spectrum of primary electrons energy loss is related to the excitation of collective oscillations among free electrons and determined by their concentration. On the spectra of the annealed coating, the lines are diffused and the intensity of the peaks is higher. It means that the density of the crystal structure imperfections in Al–Ti–N coating reduces after annealing. During friction the dislocations

TABLE 9.5
Cutting Data for Testing of Ti–Al–N Coatings

	Cutting Data					
Type of Cutting Operation	Speed (m/min)	Feed	Depth of Cut (mm)	Workpiece Material	Tooling	Tool Life Criteria
Turning	450	Feed 0.11 mm/rev	0.5	Steel 1040	Polished commercial indexable cutting inserts (SPG 422), CC H1P grade (Sandvik)	Flak wear = 0.3 mm

Source: Gershman, J.S., Bushe, N.A. Thin films and self-organization during friction under the current collection conditions. *Surf. Coat. Technol.* 2004, 186, *3*, 405–411. With permission.

outcrop on the surface as a result of plastic deformation. A surface with lower initial density of crystal structure imperfections shows more stable wear behavior during the post running-in stage of wear because the intensity of surface damage is reduced (see Figure 9.23). We can conclude that the equilibrium coating that formed as results of heat treatment promotes tool life improvement by expanding a range of stable wear. Moreover, the nanoscale tribo-films are embedded on less damaging surfaces and show better stability and protection ability during the stable stage of wear. The data presented in Figure 9.25 show that more stable tribo-films form on the surface of the annealed coating. The tribo-films that dynamically regenerate on the coated tool surface during cutting possess high thermal conductivity (see Chapter 1) and provide thermal barrier protection. Formation of more stable tribo-films leads to the beneficial heat redistribution at the cutting tool–chip interface. As a result more heat goes into the chips (Figure 9.25). A thicker zone of recrystallization with a coarser grain size forms within the contact zone of the chip. We believe that this is a major cause of tool life improvement when nanocomposite structure of the coatings is achieved.

It is known that in bulk materials as well as in coatings grain boundary hardening is one of the possibilities for hardness improvements. With a decrease in grain size, the multiplication and mobility of dislocation are hindered, and hardness of materials increases according to the widely known "Hall–Petch" relationship (Figure 9.26) [47,58]. This effect is especially prominent for grain sizes down to tens of nanometers. However, dislocation movement, which determines hardness of bulk materials, has little effect when the grain size is less then approximately 10 nm. A further reduction in grain size brings about a decrease in strength because of grain boundary sliding. Softening occurring along the grain boundary sliding is believed to be mainly caused by large amount of defects in grain boundaries. Therefore, further increase in hardness requires hindering of the grain boundary sliding. This could be achieved by proper metallurgical design, i.e., by increasing the complexity and strength of grain boundaries [59]. As different crystalline phases often exhibit different sliding systems and provide complex boundaries to accommodate a coherent strain, thus preventing voids or flaws, multiphase structures are expected to have interfaces with high cohesive strengths [60]. Apart from hardness, nanocomposite coatings also have very high cracking resistance [61]. In order to obtain superhardness plastic deformation is usually strongly prohibited and dislocation movement and grain boundary sliding are prevented. The high resistance obtained could be explained by the ability of the nanocomposite coating to absorb energy under loading. This is due to the increased resistance to plastic deformation (i.e., low-energy dissipation during plastic deformation, see Chapter 5) as well as the enhanced ability for amorphous structures to absorb energy during deformation (see Figure 9.20). This results in a very low intensity of crack initiation [61]. Crack propagation intensity could be achieved in these coats owing to optimization of the design to allow a certain degree of grain boundary sliding. This could be done by optimizing of the size, volume percentage, and distribution of the nanocrystals. [47].

TABLE 9.6
Characteristics of Al–Ti–N (X.ceed) Coatings with and without Annealing at 900°C and 2 h and Corresponding Chip Characteristics Formed during High-Speed Turning (speed = 450 m/min) of 1040 Steel

Coating	Grain size (nm) (AFM data)	Microhardness (GPa)	Reduced Elastic Modulus Er (GPa)	H/E ratio	Elastic Recovery Parameter ERP	Microhardness Dissipation Parameter	Chip Characteristics	
							Compression Ratio	Share Angle (degree)
Al–Ti–N	15–25	24.9 ± 4.1	279.2 ± 33.2	0.089	0.36 ± 0.04	0.450	1.47	34.23
Al–Ti–N + annealing 900°C	25–30	25.8 ± 4.4	274.2 ± 35.7	0.094	0.38 ± 0.05	0.418	1.40	36.54

Source: Gershman, J.S., Bushe, N.A. Thin films and self-organization during friction under the current collection conditions. *Surf. Coat. Technol.* 2004, 186, *3*, 405–411. With permission.

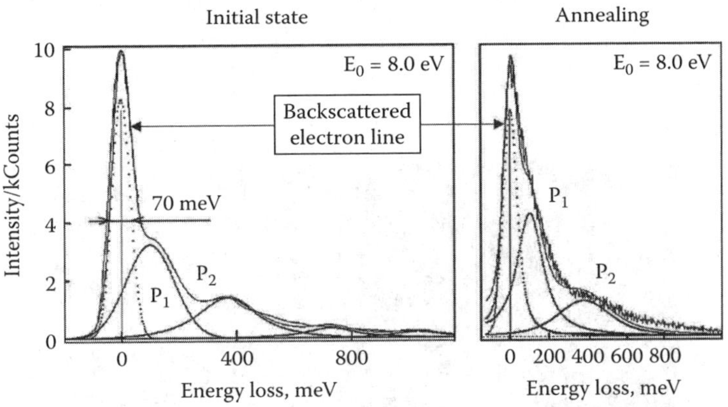

FIGURE 9.24 Spectrum of electron energy losses for the cutting tools with Al–Ti–N coating: (a) initial state; (b) annealing (900°C, 2 h, in vacuum).

A combination of two or more nanocomposite phases provides complex boundaries to accommodate coherent strain, which results in an increase in coating hardness. In this case, the phases involved must show a wide miscibility in the solid state, display thermodynamically driven spinodal phase segregation during deposition, and have a certain chemical affinity to each other to strengthen the grain boundaries. A typical example of this system is the (Ti–Al–Si)N coating. Thermal stability of the nanocomposites is probably the second most important advantage of these coatings. Significant reduction in the hardness of regular coatings takes place during annealing owing to relaxation of compressive stress generated during deposition. A way to improve thermal stability of the coatings is to modify the interface complexity, such as by using ternary systems that show immiscibility and undergo spinodal decomposition and segregation at high temperatures [60]. Strong segregation effects can lead to a thermodynamic stabilization of the grain boundaries, with high activation energy for grain coarsening [62]. Typical modern designs of nanocrystalline coatings include nanocrystalline phases embedded in an amorphous matrix [63]. Two different materials, namely, the crystalline and the amorphous phases, are deposited simultaneously, and a nanocomposite material forms by phase separation. A prerequisite for phase separation is complete immiscibility of the two phases. Nanocomposite materials could be obtained for certain material combinations. Diamond-like coating (DLC), amorphous carbon nitride, or other amorphous materials with high hardness and elastic modulus have been recognized as the primary candidates for the amorphous matrix, whereas nanosized refractory nitrides, such as TiN, TiAlN, AlN, etc., could be used as strengthening phases. The compounds with the metallic bonding are usually used as the strengthening phase, and compounds with covalent bonding are used as the amorphous matrix [4].

The nanocrystalline multicomponent coatings, usually named "superhard" coatings, have a hardness in excess of 40 GPa at room temperature, and the so-called "ultra hard coatings" have a hardness above 80 GPa [47]. Nanocomposite coatings have a hardness that significantly exceeds that given by the rule of mixture. A classic example of nanocomposite coatings is TiN or TiAlN crystallites of about 4 to 7 nm surrounded by an amorphous Si_3N_4 matrix. The hardness as well as crystalline size strongly depend on the Si content. The highest hardness corresponds to Si contents around 8 to 12% [63,64]. In nanocomposite materials, the dislocation motion and grain boundary sliding are suppressed. Thus, grain boundary sliding can replace dislocation climb and glide as the dominant plastic deformation mechanism.

Nanocrystalline materials can be prepared by methods that simultaneously ensure a high rate of nucleation and a low rate of growth. Magnetron sputtering and pulsed arc deposition can be used for the production of nanocrystalline films [65,66]. In both methods highly ionized plasma ensures rapid crystal nucleation on the one hand and very fast cooling rates on the other. The actual temperature of the coatings deposition must be relatively low to control the crystalline size.

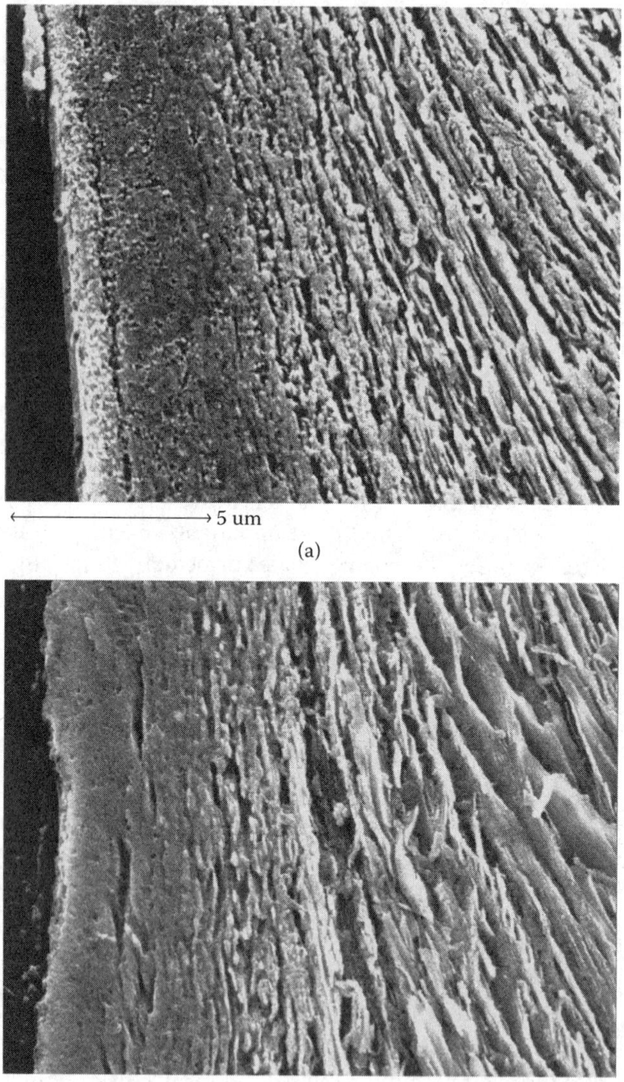

FIGURE 9.25 SEM images of the chips cross sections for the cutting tools with Al–Ti–N coatings before and after annealing (900°C, 2 h, in vacuum, 1×10^{-6} mbar (a); and without annealing (b). High-speed turning conditions (speed 450 m/min) of 1040 steel.

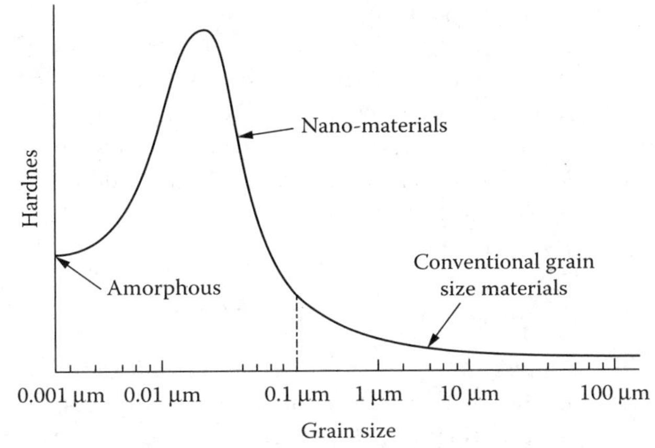

FIGURE 9.26 Hardness of the material as a function of the grain size. (From Zhang, S., Sun, D., Fu, Y., Du, H. Recent advances of superhard nano-composite coatings: a review. *Surf. Coat. Technol.* 2003, 167, 113–116. With permission.)

There are two types of nanocomposites that can be used for superhard coatings' fabrication. The first type of superhard coatings are nanocomposite -MeN/metals [66,68]. They rely on a combination of soft materials that do not form thermodynamically stable nitrides, such as Cu, Ni, Y, Ag, and Co, with hard transition metals nitrides [69]. These systems show low thermal stability due to relaxation of high compressive stresses that are formed during deposition under conditions of energetic ion bombardment. [61,70]. When the temperature of deposition is low, the kinetic energy of the bombarding ions is transferred into a very small volume of atomic dimensions while the cooling rate of the film is high [66]. These are highly nonequilibrium processes. The metastable nanocomposite layers begin to form under these conditions. This type of nanocomposite coating is usually not used for high-speed machining applications.

The second type of nanocomposite coatings such as nc-TiN/a-Si_3N_4, nc-TiAlSiN, and nc-TiN-ncBN [70–74] have improved tool life under high-speed machining conditions. They are deposited at high nitrogen pressures and relatively high temperatures (around 500 to 600°C) to enhance spinodal phase segregation [67].

A similar behavior exists in the superlattice or nanoscale multilayer coatings with a superlattice period ranging from 5 to 10 nm. The bilayers in these superlattice structures can be metal layers, nitrides, carbide, oxides of different materials, or a combination of these compounds such as TiAlN/NbN, TiAlN/CrN, or TiAlN/VN [75–77]. The mechanism of hardening in these coatings is associated with the restriction of dislocation motion across an interface or within the layer itself, owing to the suppression of the normal dislocation source and multiplication effects encountered in the bulk materials [75–77].

During the last few years, nanocrystalline or nanocomposite coatings have attracted the attention of different research groups all around the world owing to the unique combination of properties outlined in the preceding text, but the tribological and service behavior of these coating was somewhat questionable. Quite recently, nanocomposite Ti–Al–Si–N coatings have been successfully used for cutting-tool applications owing to the proper selection of Ti–Al–N coating as well as optimization of the Si content in the coating. Aluminum-rich Ti–A–N coatings with a high hardness at elevated temperatures (see earlier text) were used as a component of nanocomposite structures. Eventually a nanocomposite coating was formed with a very beneficial combination of properties such as high and even by increasing hardness as well as crack resistance at elevated temperatures [61,78]. At the same time, the formation of Al–O, Si–O tribo-films starts to control oxidation stability as well as tool life under high-speed machining conditions [80]. A significant tool life improvement when compared to the Al–Ti–N coating was reported [61]. AlCrN–Si_3N_4 nanocomposite coating also seem promising for specific cutting-tools applications [81].

Another way to improve the tool life of coatings is by the use of stable ceramic coatings (e.g., alumina or zirconia) or the alumina coating on the top of the TiAlN layer [82–83]. These ceramics are the most stable and wear-resistant materials for high-speed cutting applications. Unfortunately, these ceramics are brittle and when deposited as an outer layer, could be subjected to severe wear during initial running-in stage of wear. The deposition of the multilayered coating combining TiAlN and ceramic layers could be a possible solution.

9.3.5.2 Adaptive Coatings for Cutting-Tools Applications

An alternative way to improve the coating wear resistance is to enhance their adaptability [84–91]. Adaptive coatings are an emerging generation of coatings for specific applications. The metallurgical design of adaptive coatings strongly depends on the area of applications. Some researchers often try to present a "universal" solution, which does not exist in reality. But the trend of the development of commercial coatings during the last few years shows that the design of the coatings is quite different for specific applications. Instead of a "universal" Ti–N coating, which was used 20 years ago and almost the same "universal" Ti–Al–N a decade later, now we have a number of different coatings that are tailored for specific applications. From this viewpoint, the strategy of such world-leading coating companies as Balzers, Kobelco, Ion Bond, Teers, and others will offer the market a variety of different coatings for specific applications.

Adaptive coatings could be defined as coatings that are able to change their properties to meet the requirements of the operating conditions can change during their life time [98]. It is suggested that the whole family of adaptive coatings be divided into two groups: (1) adaptive coatings that change their structure and properties depending on the character of external impact and (2) self-adaptive coatings that are able to change their structure and properties as a result of the surface-engineered response to the constant environmental impact, owing to self-organization and dissipative structures' formation. Self-adaptive coatings possess the ability to generate tribo-films on a permanent basis, which protects the surface during operation and critically decreases the wear intensity. The major feature of the self-adaptive coatings is their ability of self-protection of the surface during friction. Adaptive coatings are a new generation of coatings that are similar in behavior to biolike systems.

9.3.5.2.1 Adaptive Coatings for General Tribological Applications

The first set of data on adaptive coatings was recently published in Reference 99 and Reference 100 and is related to a composite coating produced within the W–C–S system, consisting of 1- to 2-nm WC and 5- to 10-nm WS_2 grains embedded in an amorphous DLC matrix. The WC–DLD–WS_2 nanocomposite coatings are mainly used for aerospace tribosystems. They exhibited adaptive properties under operations of these systems. This adaptation was found in crystallization and reorientation of initially nanocrystalline and randomly oriented WS_2 grains. The graphitization of the initially amorphous DLC matrix and reversible regulation of the composition of the films at the WS_2–graphite interface follow environmental cycling from dry to humid air. Probably the DLC/WS_2 synergistic effect takes place, providing reduction of friction in oxidizing environments. Coefficient of friction values lower than 0.05 were recorded during a space simulation test of 2 million cycles, and recovery of low friction during the tests that simulates ambient or space environmental cycling was demonstrated. Another challenge for smart solid lubricants is the achievement of lubrication in the widest range of temperatures. No single material is known to be lubricious from ambient temperature up to 800°C. Thus, the way to produce a lubricating coating that can operate over a broad temperature range is to combine low- and high-temperature lubricants into a composite or a layered structure of the coating such as CaF_2 and WS_2, which interact during friction to form $CaSO_4$ [101]. A similar concept has been demonstrated by combining transition metal dichalcogenide (MoS_2 or WS_2) with oxides (ZnO or PbO) to form $PbMoO_2$ or $ZnWO_4$, which are lubricious at high temperature [102,103]. The future challenge of smart coatings is to achieve lubricious properties with reversibility under multiple cycles of temperature or environmental variation.

9.3.5.2.2 Multilayered Self-Adaptive Coatings

All known commercial coatings (e.g., TiN, TiCrN, and TiAlN) and even "state-of-the-art" nanocrystalline coatings mostly generate the tribo-ceramics such as rutile or mixtures of rutile and alumina that possess limited stability under high-speed cutting conditions [91–97]. The generation of a tribo-film that could improve lubrication properties of the coated tools at high temperatures is a plausible goal for the development of new coatings for high-speed cutting applications.

The most important phase of the structure adaptation and self-organization process is associated with the running-in stage of wear. During this stage, the wear process gradually stabilizes and finally transforms to a stable phase [104]. It is very important to prevent surface damage and promote formation of the protective tribo-films at the surface during the running-in stage of wear using the phenomenon of screening (see Chapter 1) [27, 104]. The less the surface damage at the beginning of the stable stage of wear, the longer the tool life (Figure 1.1).

Hard coatings are brittle and susceptible to extensive surface damage during the running-in stage of wear. Frequently, much of the hard coating is destroyed at this phase, prior to the beginning of the stable stage of wear, at which the wear rate can be lowered by an order of magnitude as a result of the self-organizing of the tribosystem. The initial surface damage often leads to a dramatic decline in the wear resistance of the coating. For this reason, a top layer with high antifrictional properties is a critical component and can be used to protect the surface of hard coatings. This is one of the most important goals of wear-resistant coatings, especially at low and moderate cutting speeds, and for handling hard-to-machine materials when adhesive and attrition wear modes dominate. This can be achieved by applying multilayer coatings that have self-adaptive features owing to the formation of a protective tribo-film during friction. Historically this was a first generation of self-adaptive coatings that appeared in the market around a decade ago. On the other hand, these coatings have adaptive features only during the initial stages of tool life. Because the outer adaptive layer is gone the coating does not show any feature of adaptability any more. One of the most effective commercial coatings of this type is the multilayered TiAlN–WC–C coating developed by Balzers [105]. The main advantage of this coating is a low coefficient of friction that results in a low initial wear rate, during the running-in stage of wear and leads to a significant increase in the tool life [23]. The VN outer layer on the TiAlN sublayer plays a similar role during cutting at moderate speeds [106]. The major feature of these coatings is the formation of W–O or V–O tribo-films as a result of tribo-oxidation during cutting. Both these oxides are known as high-temperature lubricants [98, 107].

A second example of a similar technology is the use of nanocomposite nc-TiN–BN coatings [108]. These coatings give good results at moderate cutting speeds. Following the discussion in Chapter 7 (see Figure 7.17), it seems likely that high-alloyed Ti–B–O tribo-films with amorphous-like structure and lubricious B_2O_3 films both form. The boron oxide plays the role of a liquid lubricant at the elevated temperatures of cutting [98].

Recently, several oxides [109] were found to exhibit good tribological properties at elevated temperatures. All of these oxides contain crystallographic shear planes with low shear strengths at high temperature [109]. They are promising materials as solid lubricants for elevated temperature applications.

The emerging generations of self-adaptive coating appeared recently. The major feature of these coating is the ability to form continuous layers of protective and lubricious tribo-oxides during the coating service life. This could be done by means of synergistic alloying as described in detail in Chapter 11.

9.3.5.2.3 Nanocrystalline Self-Adaptive Coatings for Severe Conditions of High-Speed Machining

All developments of self-adaptive coating for high-speed machining applications are associated with the enhancement of the coating layer to produce oxide films with high tribological properties at elevated temperatures.

The developments related to the self-adaptability of coatings under severe conditions of high-speed machining are one of the most challenging areas of recent research activity in the field of surface engineering. There are two major ways to improve adaptability of the coating for severe applications of high-speed machining:

- Microstructure improvement, i.e., development of nanocrystalline coatings with high resistance to chemical (crater) wear
- Optimization of the coating composition as a result of synergistic alloying (see Chapter 11)

Improvements in the life of cutting tools have been achieved by grain size refinement down to the nanoscale (grain size less than 100 nm) level [110]. This has been done by means of Ti–Al–N coating deposition, in strongly nonequilibrium conditions (see later text), through advanced deposition techniques (such as pulsed FAD) [111].

The current temperature at the crystallization front of the coatings is low because of the pulsed FAD deposition process application. The coating deposition conditions are close to nanoscale heating [66]. Thus, the cooling rate during a strongly nonequilibrium FAD process is very high. Pulsed ion bombardment of the growing films can restrict the grain growth and permit the formation of nanocrystalline films. This results in the formation of nanocrystalline coatings that could be very beneficial in some cases for wear-resistant coatings especially for high-speed machining applications.

The goal of the research presented in the following text is to study the friction and wear behavior of the cutting tools with Ti–Al–N PVD coatings and determine the impact of the FAD coatings' nanocrystalline structure on their adaptability under high-speed machining conditions.

The nanocrystalline coatings studied were deposited using a FAD PVD unit. The unit had up to three removable targets, also referred to as FAD modules. A sketch of the FAD module is shown in Figure 9.27.

The FAD system was shown to affect the physics, as well as the chemistry, of the plasma processes significantly when depositing refractory compounds such as TiAlN. This is because of an increase in the ionization rates of both metals and reactive gases [112]. The system included a hollow steel tube approximately 300 mm in diameter, which was a quarter segment of a torus (Figure 9.27). The tube was installed on the body of a standard unit of the arc evaporation PVD instead of on a conventional arc evaporator. An induction coil was then placed inside the tube to create the magnetic field needed to control the plasma path.

Regulation of the plasma flow in the FAD System is based on plasma optics principles. The influence of the magnetic field on the plasma flow allows the radial electric field to appear, and as a result regulation of the plasma path inside the internal volume of the FAD system can be achieved. Whereas only electrically charged particles (i.e., ions) are focused in the FAD system, the uncharged particles (so called "droplet" phase) are not affected by the magnetic and electric fields. The uncharged particles flying from the cathode surface parallel to the cathode axis do not reach the substrate surface and are thus deposited on the inside of the tube. However, the ions follow the bend and are focused on the substrate surface. The value of the continuous current running through the magnetic coil affects the distribution of the ion flow density. The body of the FAD system is biased with regard to the PVD unit's body. This is required if substrate biasing is not enough to achieve a suitable value for the deposition rate. The FAD system allows the substrate to be heated up to the deposition temperature by use of Argon ions, which are rarely achieved when applying

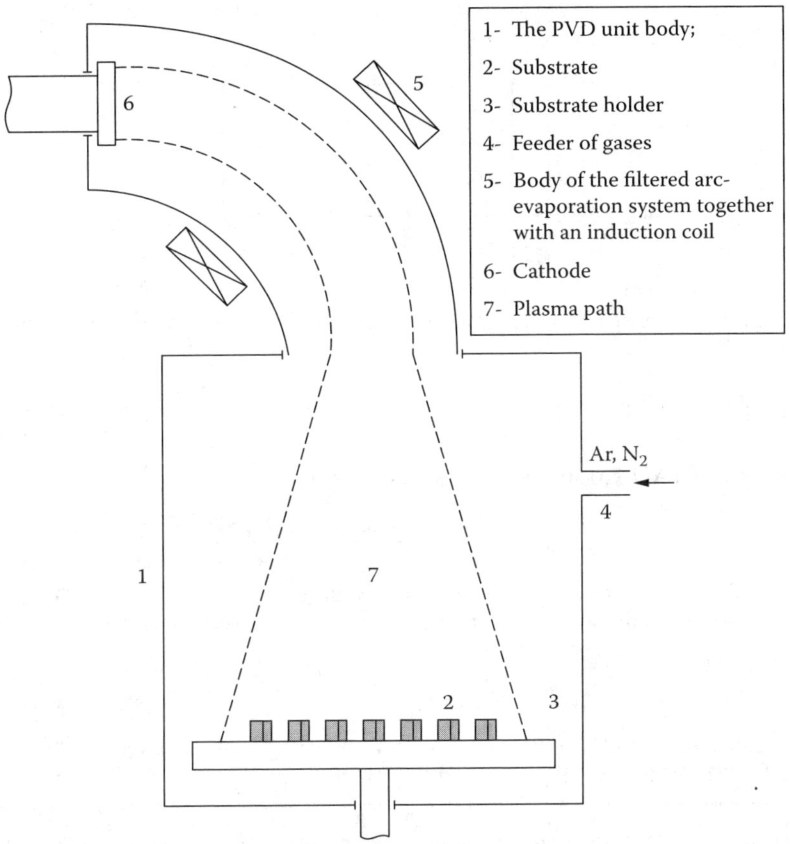

FIGURE 9.27 Scheme of the filtered arc evaporation system. (From Fox-Rabinovich, G.S., Weatherly, G.C., Dodonov, A.I., Kovalev, A.I., Veldhuis, S.C., Shuster, L.S., Dosbaeva, G.K., Wainstein, D.L., Migranov, M.S. Nano-crystalline FAD (filtered arc deposited) TiAlN PVD coatings for high-speed machining application. *Surf. Coat. Technol.* 2004, 177–178, 800–811. With permission.)

conventional arc evaporation PVD technology. This was shown to occur owing to the high ionization rate of the gases obtained when using the FAD system with elevated energy [111]. The filtering efficiency of the FAD technology for "droplet" phase deposition prevention is well known [111]. But probably a more important advantage of the FAD technique is the possibility to synthesize nanocrystalline coatings under nonequilibrium conditions.

The parameters of the FAD Ti–Al–N coatings' depositions strongly affect the microstructure and properties of the coatings. The coatings were deposited using accelerated ion bombardment (radiant annealing). Because the FAD technique application results in a very high plasma ionization rate, the ion bombardment could be performed using pulsed bias voltage during the coating deposition. In this case the ions accelerate up to the corresponding energy level and bombard the coating layer that was deposited between the impulses. The graph of the impulse bias voltage is shown in Figure 9.28.

Parameters of the PVD coating deposition process are shown in Table 9.7. Temperatures on the surface of the sample were measured using an optical pyrometer. Heating the substrate up to the deposition temperature was performed by argon ions during FAD at the substrate biasing voltage of 1 kV. After obtaining the required substrate temperature, the substrate biasing voltage value was reduced and nitrogen was fed into the chamber for deposition of the TiAlN films.

A TiAl target with Al/Ti ratio 1.0 was used. The target was made by GfE Company (Germany) using arc-melting technique in argon. The 2.5- to 3-μm-thick coatings were deposited on Sandvik

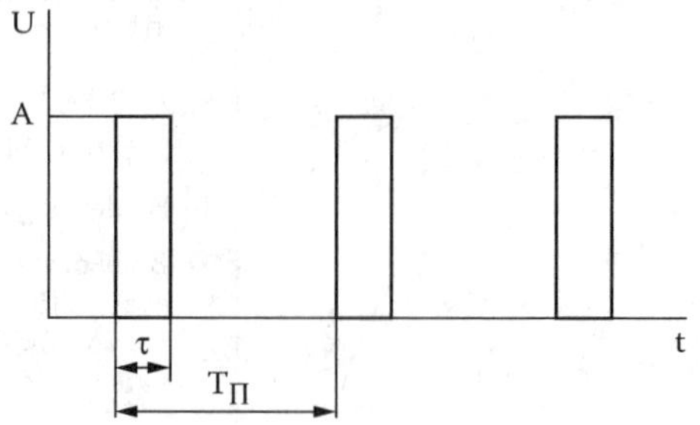

FIGURE 9.28 Scheme of pulsed FAD process. **A** — impulse amplitude, τ — impulse time, $\mathbf{T_n}$ — time between impulses.

TABLE 9.7
Parameters of FAD Coatings Deposition Microseconds

		Parameters of Deposition					
				High Voltage Impulses			
Number	Target	Arc Current (A)	Separator's Voltage (V)	Voltage (V)	τ (μsec)	T (μsec)	Substrate Temperature T°C
1	TiAl	200	16	1000	50	300	480–500
2	TiAl	200	16	1000	50	600	340–370

Source: Munz, W.-D.M. Ti-Al nitride films: a new alternative to TiN coatings. *J. Vac. Sci. Technol.* 1986, A4, *6*, 2717–2725. With permission.

H1P cemented carbide substrate. The chemical composition of the cemented carbides is the following: WC — 85.5; TiC — 7.5; TaC — 1.0; Co — 6.0 wt.%.

The microstructures of the coatings were investigated by transmission electron microscopy (TEM) using the JEOL JEM-2010F microscope with an acceleration voltage of 200 kV. TEM samples were prepared using a focused-ion-beam (FIB) technique on a JEOL JFIM-2100 system. The samples were thinned to approximately 0.1 μm by Ga ions with the acceleration voltage and current of the Ga ion source being 30 kV and 2.0 μA, respectively.

The chemical composition of secondary phases emerging on the tool surface during cutting was studied by means of a secondary ion mass spectroscopy (SIMS). This was carried out with the aid of an ESCALAB-MK2 (VG) electron spectrometer equipped with an SQ300 ion analyzer of quadruple type and AG-61 scanning ion gun, which allows the flow of argon primary ions with energies up to 5 keV to be focused on a spot up to 200 μm in diameter on the surface of a sample. The ion etching speed was on the order of 0.2 monolayer min^{-1}; the analysis was carried out in the static mode. The average chemical composition of the wear zone of the coating was then studied. To do this, the argon primary-ion beam was directed at the target spot on the surface of the sample. In some cases, an ion beam was moved along the direction chosen.

The atomic structure of the film on the tool surface during cutting was studied by means of EELFS with the aid of an ESCALAB-MK2 (VG) electron spectrometer. The friction surface was studied in the zones free of adhesion of workpiece material. High magnification (2000×) was applied, and the primary electron energy was E_p = 1000 eV. The fine structure of the spectrum energy loss was recorded close to the line of elastically scattered electrons in the range 250 eV. Analysis conditions were chosen in such a way as to ensure the best energy resolution with a good signal to noise ratio. The fundamentals of the method's details are presented in Chapter 4.

Wear of tetragonal indexable insert with $Ti–Al_x–N$ coatings were studied while turning 1040 steels under cutting conditions shown in Table 9.5. It is known that, as the flank wear exceeds 0.3

mm, the cutting tool loses serviceability [18]. The coefficient of friction values vs. temperature for the coating under analysis were determined using the method described in details in Chapter 5. The coefficient of friction at the rake face during cutting was also determined using method described in Reference 113. Cutting forces were measured *in situ* with a force dynamometer. Metallography analysis of the chips formed during cutting was made using an SEM. The chip compression ratio and chip shear plane angle were determined using standard methods [113].

To characterize the thermal conductivity of the coating *in situ*, the tool–chip contact length was measured in the very beginning of the cutting process using optical microscopy as well as an SEM [114].

The coating's characterization data are shown in Table 9.8. The results presented in Table 9.8 show that the coatings under analysis are close to a stoichiometric composition; they have a similar thickness (around 3 μm) and microhardness values (around 30 GPa).

Detailed TEM studies of Ti–Al–N coatings with Al/Ti ratio of 1.0 were performed. The major feature of the filtered coatings is the nanocrystalline grain size. The TEM images of regular arc as well as FAD Ti–Al–N coatings together with their diffraction patterns are presented in Figure 9.29. The light field images together with the diffraction patterns show that both of the coatings have column-like structures. The columns grow perpendicular to the surface of the substrate during coating deposition. The FAD coating has a significantly finer grain size as compared to the monolayered arc coating (Figure 9.29a, Figure 9.29c, Figure 9.29d, and Figure 9.29f correspondingly). Based on the dark field images presented (Figure 9.29b and Figure 9.29e) we can estimate the actual grain size. The grain size of a monolayered arc Ti–Al–N coating is around 50 nm, whereas the grain size of the FAD coating is around 15 nm.

Wear resistance of the Ti–Al–N coatings under analysis during high-speed turning mainly depend on two characteristics: (1) the grain sizes and (2) the thermodynamic state of the coating. The Ti–Al–N FAD coating with a grain size of 15 nm has a tool life 4 times higher than the Ti–Al–N coating with the coarsest grain size of 50 nm. It is also 1.4 times higher than that of the Ti–Al–N FAD with slightly coarser grain size of 25 nm (Figure 9.30b; Table 9.8). It is know from the literature [127] that the deposition of coatings result in generation of a high amount of lattice imperfections within a layer of the growing coating because the defects annihilation rate is not enough to compensate the defects creation rate under intensive ion bombardment. The lower the temperature of deposition, the higher the density of lattice imperfections within a coating layer. The energy that supplies the coating layer during deposition accumulates within these imperfections and shifts the system from equilibrium. That is why the FAD coatings deposited at a lower temperature (340°C) are in a lower equilibrium state when compared to the FAD coatings deposited at a higher temperature (500°C; see Table 9.7) Grain size refinement together with the thermodynamic state controls the coated tool life. We have studied in detail both the characteristics and the wear behavior of the FAD coating with the finest grain size.

The values of the coefficient of friction vs. temperature determined for the Ti–Al–N PVD coatings using the method described in detail in Chapter 5 correspond to their tool life (Figure 9.30 and Figure 9.31). The friction parameter values go up to 0.25 for the monolayered Ti–Al–N coatings at the elevated temperatures above 600°C. The higher values of the coefficient of friction correspond to the intensive (linear) wear of the cutting tools (Figure 9.30). In contrast, a critical decrease in the frictional parameter values down to 0.1 takes place for Ti–Al–N FAD coatings. This could be associated with protective tribo-films' formation (see later text).

Additional investigations of the cutting tool–workpiece interface have been done to further our understanding of this phenomenon.

Figure 9.32a and Figure 9.32b display the SEM images of the worn rake surface of the cutting inserts with Ti–Al–N coatings. Formation of the relatively large adhesion zones of the workpiece material on the tool surface takes place in the case of monolayered Ti–Al–N coating (Figure 9.32a). The wear behavior of the filtered coatings is different (Figure 9.32b). The low intensity sticking of workpiece materials takes place, and numerous small sticking zones with dimension of 10 to100

TABLE 9.8
Ti–Al–N PVD Coatings Characterisation Data

Ti–Al–N PVD Coatings	Chemical Composition (at. %)		Properties				
	Chemical Formula (Quantitative AES data)	Al/Ti Ratio	Microstructure Grain sizes (nm) (TEM, AFM data)	Thickness (μm)	Microhardness (GPa)	Tool–chip Contact Length (TCCL) (μm)	Coefficient of Friction on the Rake Surface
Uncoated carbide tool	—	—			15		
Ti–Al–N	$Ti_{0.25}Al_{0.22}N_{0.53}$	0.88	50	3.0	33	1450	0.986
Filtered Ti–Al–N	$Ti_{0.22}Al_{0.22}N_{0.56}$	1.0	15	2.8	35	450	0.857
Filtered Ti–Al–N		1.0	25	3.0	32	—	—

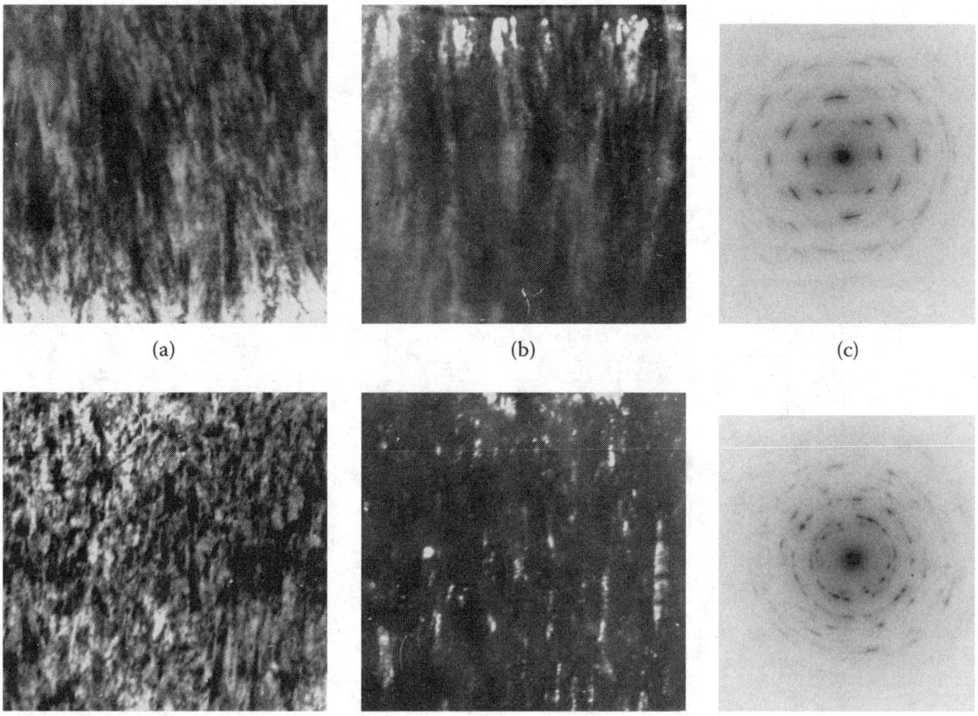

(a) (b) (c)

(d) (e) (f)

FIGURE 9.29 TEM images of Ti–Al–N coatings with their diffraction patterns (magnification 24000): (a)–(c) monolayered Ti–Al–N coating: (a) light field image, (b) dark field image, (c) diffraction pattern; (d)–(f) Ti–Al–N FAD coating: (d) light field image, (e) dark field image, (f) diffraction pattern.

μm are formed. The sticking zones are generated on the asperities that formed as a result of tool surface grinding.

Measurement of the tool–chip contact length (TCCL) was made for the cutting tools with the Ti–Al–N coatings in the study. The tool–chip contact length data was measured at the very beginning of the cutting for monolayered arc Ti–Al–N coating (1400 μm) and compared to the FAD coating (450 μm; Table 9.8). The tool–chip contact length measured in the very begging of cutting, prior to tribo-films formation, mainly relates to the thermal conductivity of the coatings [114] and, to a lesser degree, to its frictional characteristics. FAD coatings show lower thermal conductivity when compared to the arc monolayered TiAlN ones. SEM images of the chips cross section presented in Figure 9.32c and Figure 9.32d show that the lower value of the tool–chip contact length corresponds to more intensive bending of the chip. The angle α_1 of the chip flow lines within the extended deformation zone (see later text and Figure 9.32) is lower for the cutting tools with the monolayered arc Ti–Al–N coatings when compared to the angle α_2 for cutting tools with FAD Ti–Al–N coatings [111]. The lower thermal conductivity of FAD Ti–Al–N coatings leads to more beneficial heat distribution at the workpiece–cutting tool interface, and more heat leaves with the chip. Together with friction parameter enhancement (Figure 9.31) the lower thermal conductivity of the FAD coatings improves the wear performance as well as tool life.

An intensive tribo-oxidation of the cutting-tool surface takes place during high-speed machining. Figure 9.33 shows the Auger electron spectra for worn tools with (1) monolayered Ti–Al–N and (2) FAD coatings. The oxidation of the rake surface is obviously due to the high amount of oxygen present in both spectra. Intensive iron peaks correspond to the sticking zones of the workpiece material. The line of Ti is significantly diffused (Figure 9.33b) for this zone because of the tribo-oxidation process. An increased amount of alumina is observed on the spectrum of FAD

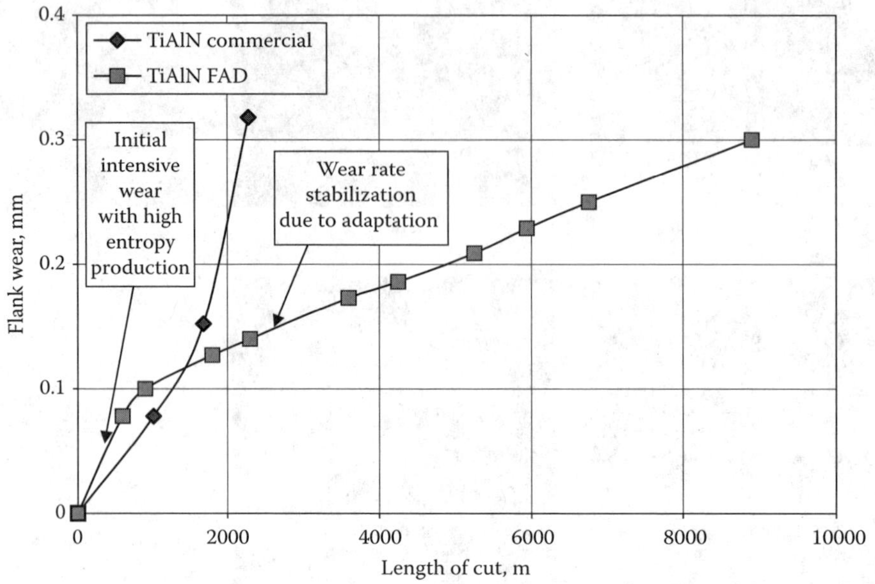

(a)

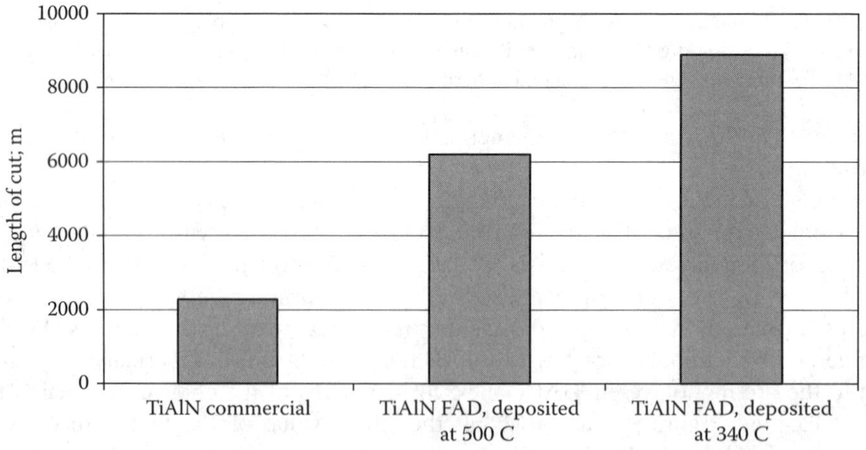

(b)

FIGURE 9.30 Tool life of Ti–Al–N coating under conditions of high-speed turning of 1040 steel (according to cutting conditions shown in Table 9.6): (a) flank wear vs. length of cut for Ti–Al–N commercial and FAD coatings; (b) tool life of the Ti–Al–N coating depending on the deposition parameters (Table 9.8). (From Fox-Rabinovich, G.S., Weatherly, G.C., Dodonov, A.I., Kovalev, A.I., Veldhuis, S.C., Shuster, L.S., Dosbaeva, G.K., Wainstein, D.L., Migranov, M.S. Nano-crystalline FAD (filtered arc deposited) TiAlN PVD coatings for high-speed machining application. *Surf. Coat. Technol.* 2004, 177–178, 800–811. With permission.)

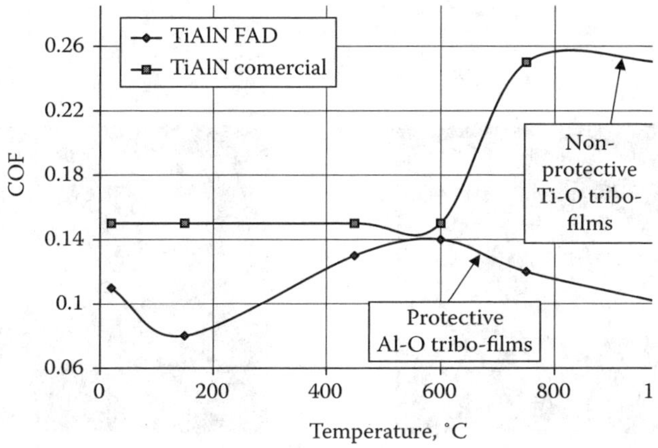

FIGURE 9.31 Coefficient of friction of Ti–Al–N coating vs. temperature in a contact with 1040 steel.

coatings that is displayed as a shift in the aluminum line to the zone of lower energy (60-eV). At the same time the intensity of metallic Al LMM line near the 68 eV level decreases. Presented in Figure 9.34a and Figure 9.34b, the series of positive secondary ion spectra for both arc monolayered and FAD Ti–Al–N coatings can be seen. Figure 9.34c and Figure 9.34d show the spectra of negative secondary ions for the same coatings. On both positive SIMS spectra the intensity of the TiO line is high. This is due to intensive tribo-oxidation that forms rutile-like films, whereas some amount of alumina forms only on the surface of FAD coatings (Figure 9.34c). This effect can be observed at the spectrum of negative secondary ions (Figure 9.34c and Figure 9.34d).

Formation of alumina films on the cutting-tool surface drastically changes the heat dissipation associated with chip removal. This can be illustrated through the SEM images of the cross sections of the chips as shown in Figure 9.35. In general, chips consist of three different zones (Figure 9.35) [18]. It is known that dynamic recrystallization of the chip contact area takes place during cutting (Zone 3) [18]. More heat flux goes into the chips, causing more intensive recrystallization to occur. This is exhibited in the chip grain coarsening within the contact zone (Figure 9.35). Figure 9.35a and Figure 9.35b shows recrystallization of the chip contact zone for cutting tools with monolayered Ti–Al–N coatings, whereas Figure 9.35c and Figure 9.35d present similar images for tools with FAD coatings. We can see that more intensive recrystallization of the contact zone takes place for FAD coatings.

When chips slide along the rake face of the cutting tool, curved flow lines are formed because of friction. An extended deformation zone (Zone 2) can be observed close to the contact area of the chip (Zone 3). Zone 1 is located at the chip outer surface [18]. The chips that are formed with Ti–Al–N arc monolayered coatings show a deformation zone, Zone 2. This zone was found to be thicker than the corresponding one with the FAD Ti–Al–N coatings. At the same time, the tool–workpiece interfaces are different. In the case of the arc monolayered coating, a severe seizure was observed in the cutting tool–chip interface. This corresponds to the morphology of the rake face with large sticking zones, as presented in Figure 9.32a. The interface of the chips formed using the FAD coating is very smooth and corresponds to Figure 9.32b with small, pointlike sticking zones. This data exhibits the reduction of friction for the tool with FAD coatings.

Atomic structure of the films that generate on the surface of the wear zone during tribo-oxidation of the Ti–Al coatings was compared to the oxide layer that is formed during oxidation of the binary TiAl alloy under equilibrium conditions (isothermal oxidation test; see Table 9.9 and Table 9.10). Figure 9.36 depicts the Fourier transforms of the fine structure of electron spectra close to the line of elastically scattered electrons. The position of the peaks on the Fourier transform corresponds to the interatomic distances for the nearest coordination spheres. The data interpretation was based on the analysis of known crystal characteristics of the oxides forming on the surface.

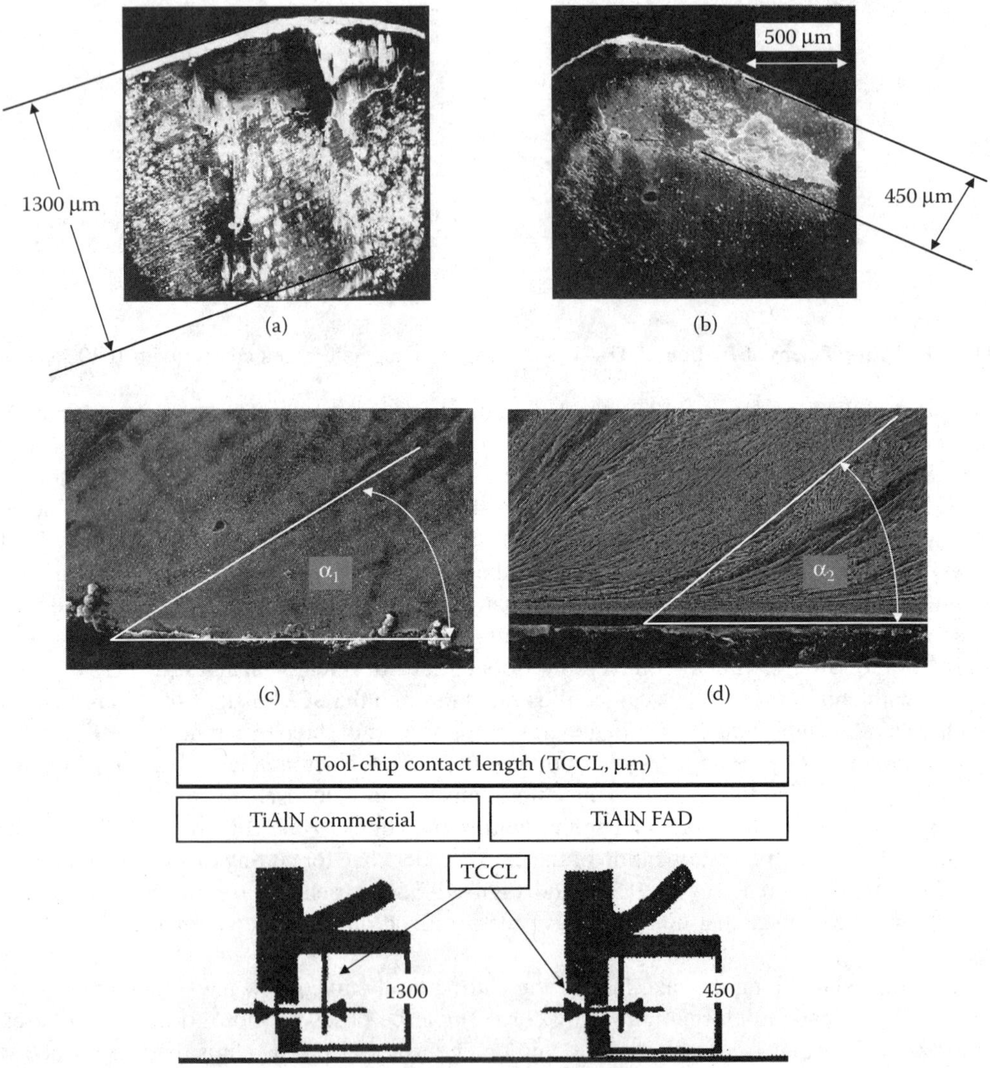

FIGURE 9.32 SEM images of the worn turning inserts with Ti–Al–N coatings and the corresponding chips cross sections: (a) morphology of the insert with monolayered coating; (b) morphology of insert with FAD coating. Length of cuts are 2280 and 8900 m, correspondingly; (c)–(d) corresponding chips cross sections. Lower value of the tool–chip contact length corresponds to more intensive bending of the chips: the angle α_1 of the chip flow lines within extended deformation zone is lower for cutting tools with the arc monolayered Ti–Al–N coatings compared to the same parameter α_2 for cutting tools with FAD Ti–Al–N coatings. (From Fox-Rabinovich, G.S., Weatherly, G.C., Dodonov, A.I., Kovalev, A.I., Veldhuis, S.C., Shuster, L.S., Dosbaeva, G.K., Wainstein, D.L., Migranov, M.S. Nano-crystalline FAD (filtered arc deposited) TiAlN PVD coatings for high-speed machining application. *Surf. Coat. Technol.* 2004, 177–178, 800–811. With permission.)

TABLE 9.9
Interpretation of Fourier Transforms for FAD (TiAl)N Coating at the Wear Zone of Cutting Tool

Phase	Bonds	Distances (Theoretical, Å)	Phase	Phase Composition	Distances (Experimental, Å)	Number of Lines
Al_2O_3	Al–O	1.89	Al_2O_3	Al_2O_3	1.61	1
Al_2O_3	Al–O	1.93	Al_2O_3		1.85	2
Al_2O_3	O–O	2.57	Al_2O_3		2.45	3
Al_2O_3	O–O	2.64	Al_2O_3		2.55	4
Al_2O_3	O–O	3.62	Al_2O_3		3.46	5
Al_2O_3	Al–Al	4.25	Al_2O_3		4.25	6
Al_2O_3	O–O	4.60	Al_2O_3		4.60	7
Al_2O_3	Al–Al	5.12	Al_2O_3		5.12	8
Al_2O_3	2 × O–O	2.64 × 2 = 5.28	Al_2O_3		5.30	9
Al_2O_3	2 × O–O	3.62 × 2 = 7.24	Al_2O_3		7.30	10
				TiO_2		
TiO_2	O–O	3.55	TiO_2		3.51	11
TiO_2	Ti–O	4.37	TiO_2		4.40	12
TiO_2	Ti–O	5.26	TiO_2		5.15	13
TiO_2	O–O	6.46	TiO_2		6.30	14

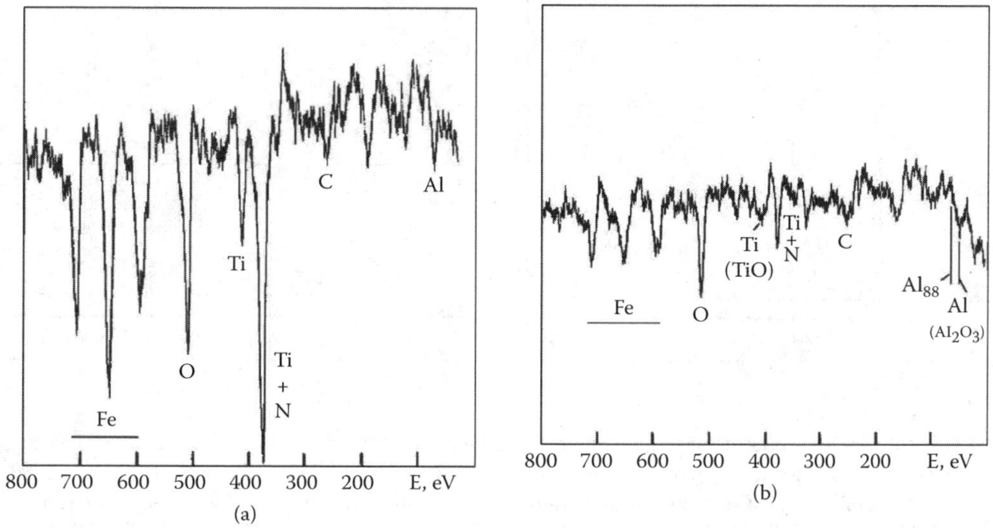

FIGURE 9.33 Auger electron spectra of turning inserts rake surface (cratering zone) for (a) monolayer Ti–Al–N coating and (b) FAD coating. Length of cuts are 2280 and 8900 m, correspondingly. (From Fox-Rabinovich, G.S., Weatherly, G.C., Dodonov, A.I., Kovalev, A.I., Veldhuis, S.C., Shuster, L.S., Dosbaeva, G.K., Wainstein, D.L., Migranov, M.S. Nano-crystalline FAD (filtered arc deposited) TiAlN PVD coatings for high-speed machining application. *Surf. Coat. Technol.* 2004, 177–178, 800–811. With permission.)

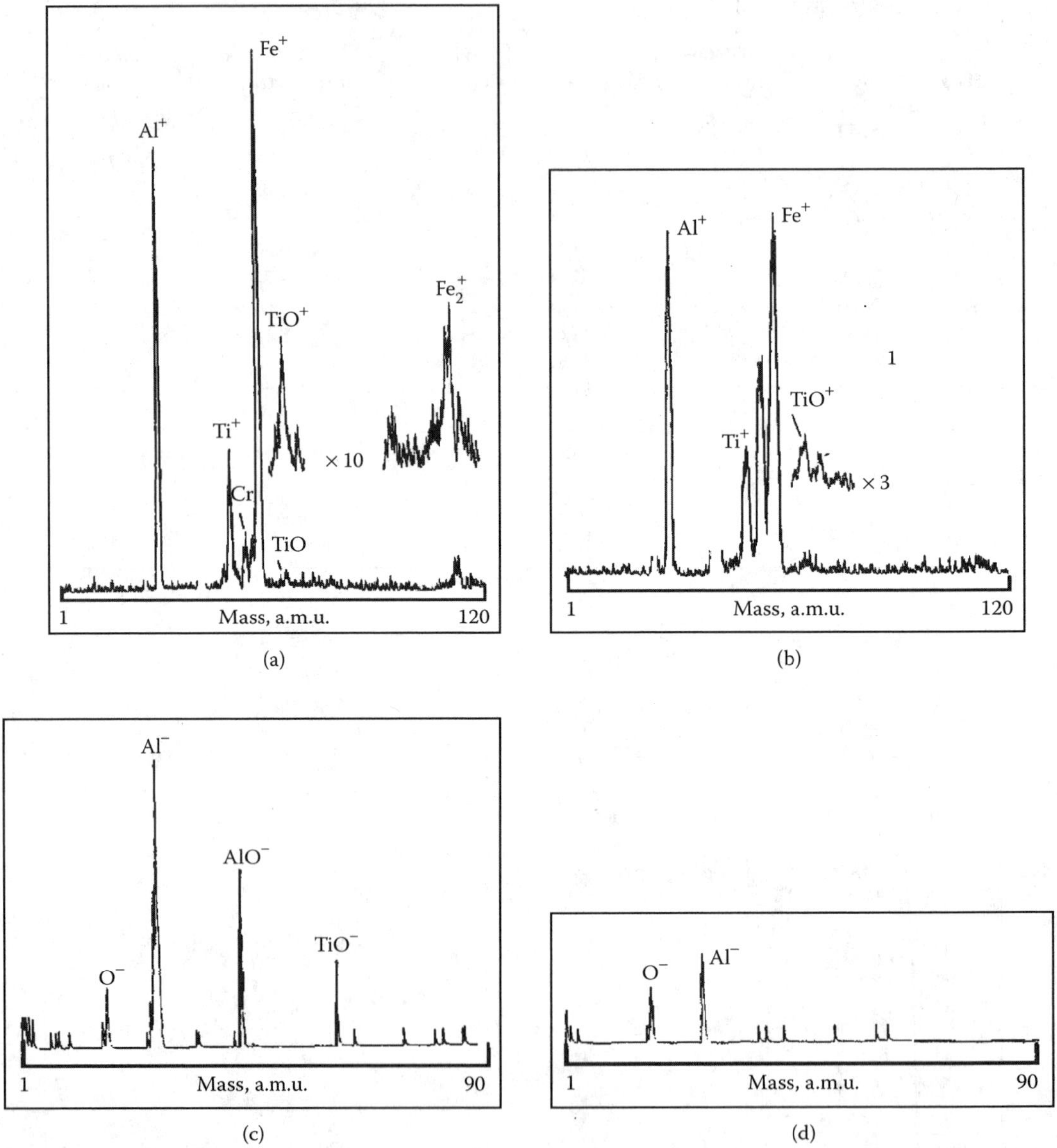

FIGURE 9.34 SIMS spectra of turning inserts rake surface (cratering zone) for $Ti–Al_x–N$ ($x = 0.5$) coating: (a)–(b) positive secondary ion spectra; (c)–(d) negative secondary ion spectra. Length of cuts are 2280 and 8900 m, correspondingly. (From Fox-Rabinovich, G.S., Weatherly, G.C., Dodonov, A.I., Kovalev, A.I., Veldhuis, S.C., Shuster, L.S., Dosbaeva, G.K., Wainstein, D.L., Migranov, M.S. Nano-crystalline FAD (filtered arc deposited) TiAlN PVD coatings for high-speed machining application. *Surf. Coat. Technol.* 2004, 177–178, 800–811. With permission.)

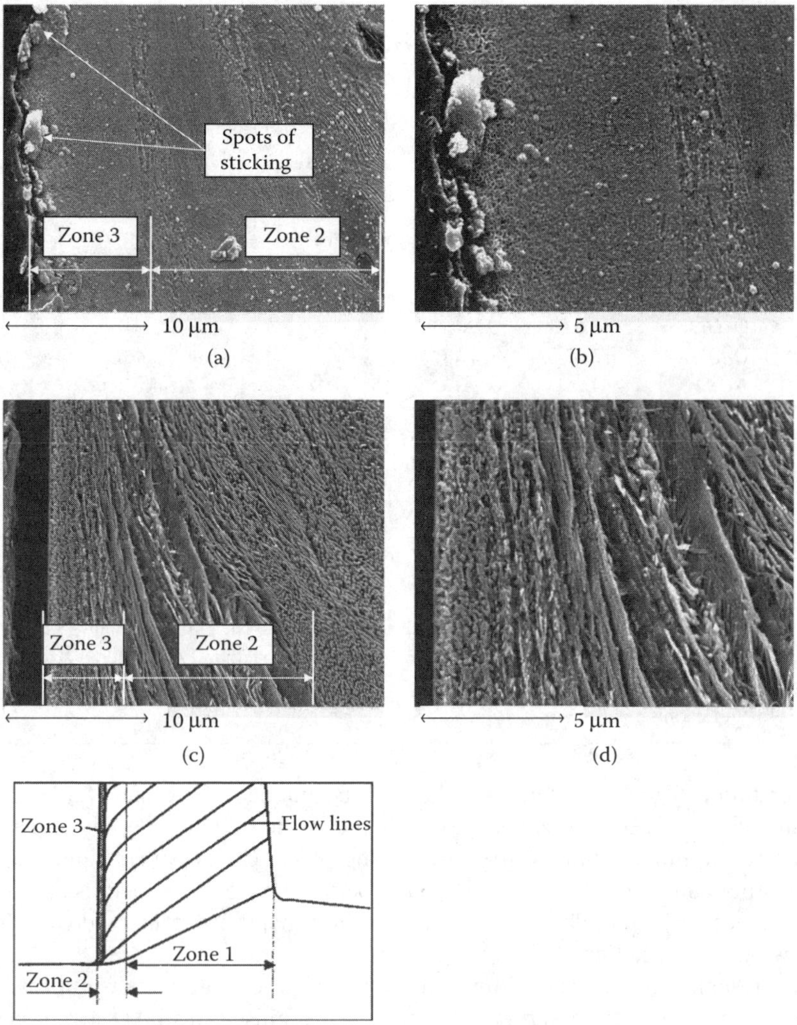

FIGURE 9.35 SEM images of the chips cross-section for Ti–Al$_x$–N (x = 0.5) coatings: (a)–(b) arc monolayered (c)–(d) FAD coating. Initial stage of cutting. Zone 3 — chip contact area; Zone 2 — extended deformation zone; Zone 3 — chip outer surface. (From Fox-Rabinovich, G.S., Weatherly, G.C., Dodonov, A.I., Kovalev, A.I., Veldhuis, S.C., Shuster, L.S., Dosbaeva, G.K., Wainstein, D.L., Migranov, M.S. Nano-crystalline FAD (filtered arc deposited) TiAlN PVD coatings for high-speed machining application. *Surf. Coat. Technol.* 2004, 177–178, 800–811. With permission.)

The interpretation of the data obtained is shown in Table 9.9 and Table 9.10. The positions of the main peaks on the Fourier transforms are in good agreement with the interatomic distances associated with Al_2O_3 and TiO_2 lattices (Figure 9.36). The peaks intensity of the Fourier transforms is connected to the amount of interatomic bonds of the type for the nearest atomic neighbors. The peak's intensity at the Fourier transform is higher for a larger coordination number for a sphere of definite radius and for a higher-ordered crystal structure. Major peak positions in Figure 9.36a and Figure 9.36b are similar because of the formation of titanium and aluminum oxide films in both cases. However, experimental distances in Figure 9.36b (Table 9.10) are closer to the theoretical data when compared to similar distances in Figure 9.36a. This means that equilibrium oxides form on the surface of binary TiAl alloy during the process of oxidation in air. The oxide films that form during tribo-oxidation have atomic structures with higher levels of defects. In this case

TABLE 9.10
Interpretation of Fourier Transformation of the Oxidized Surface for the Binary TiAl Alloy

Phase	Bonds	Distances (Theoretical, Å)	Phase	Phase Composition	Distances (Experimental, Å)	Number of Lines
Al_2O_3	Al–O	1.89	Al_2O_3	Al_2O_3	1.89	1
Al_2O_3	Al–O	1.93	Al_2O_3		1.93	2
Al_2O_3	O–O	2.57	Al_2O_3		2.57	3
Al_2O_3	O–O	2.64	Al_2O_3		2.64	4
Al_2O_3	O–O	3.62	Al_2O_3		3.46	5
Al_2O_3	Al–Al	4.25	Al_2O_3		4.25	6
Al_2O_3	O–O	4.60	Al_2O_3		4.60	7
TiO_2	Ti–O	1.91	TiO_2	TiO_2	1.97	8
TiO_2	Ti–O	1.97	TiO_2		2.01	9
TiO_2	O–O	2.46	TiO_2		2.44	10
TiO_2	O–O	3.55	TiO_2		3.54	11
TiO_2	O–O	4.37	TiO_2		4.49	12

Source: Fox-Rabinovich, G.S., Weatherly, G.C., Dodonov, A.I., Kovalev, A.I., Veldhuis, S.C., Shuster, L.S., Dosbaeva, G.K., Wainstein, D.L., Migranov, M.S. Nano-crystalline FAD (filtered arc deposited) TiAlN PVD coatings for high-speed machining application. *Surf. Coat. Technol.* 2004, 177–178, 800–811. With permission.

the actual interatomic distances differs from the equilibrium theoretical data (Table 9.9). The peaks' intensity at the distances above 4 Å decrease; this means that the degree of order is reduced and amorphization of the films atomic structure is taking place. On the other hand, the peaks of the Fourier transforms could be observed for small (below 4 to 5 Å) and large (above 5 to 8 Å) interatomic distances. Based on this data we can conclude that the surface oxide film has a complex amorphous–crystalline structure.

The data presented in the preceding text show that there are a few improvements in the characteristics of coating, which can be attributed to the pulsed arc FAD technique application. The first, and most widely known one is a filtering of the "droplet phase" [10,115]. This results in a better surface finish, which lowers the workpiece material adherence to the tool surface. It also improves the friction and wear behavior of the filtered coatings. This is very important for low-speed machining conditions within the domain of build-up formation [17, 18]. But for high-speed machining conditions in which cratering or (chemical) and oxidation wear dominates (see Chapter 6, Figure 6.1) [1,23] the ability of the coatings to form a protective surface during friction becomes more important. This ability is significantly improved for the Ti–Al–N coating with nanocrystalline structure (grain size around 15 nm) deposited at nonequilibrium conditions using pulsed FAD technique. The nanocrystals with grain size around 10 to 20 nm are free of defects [116]. So, the majority of defects are concentrated within the grain boundaries and start to play a decisive role in the beneficial mass transfer during tribo-oxidation [39]. A coating with a finer grain size has more grain boundaries; thus, more diffusion paths are available for outward diffusion of Al and inward diffusion of oxygen. This results in an increase of the oxidation resistance of the friction surface [111] because of the formation of a protective aluminum oxide film [117] and leads to significant tool life improvement during high-speed machining (Figure 9.31).

From this we can assume that tribo-oxidation is a very important and beneficial process for high-speed machining conditions. Tribo-oxidation of the cutting tools is very far from an equilibrium state. Tribo-oxidation of the Ti–Al–N coating results in structural adaptation of the surface

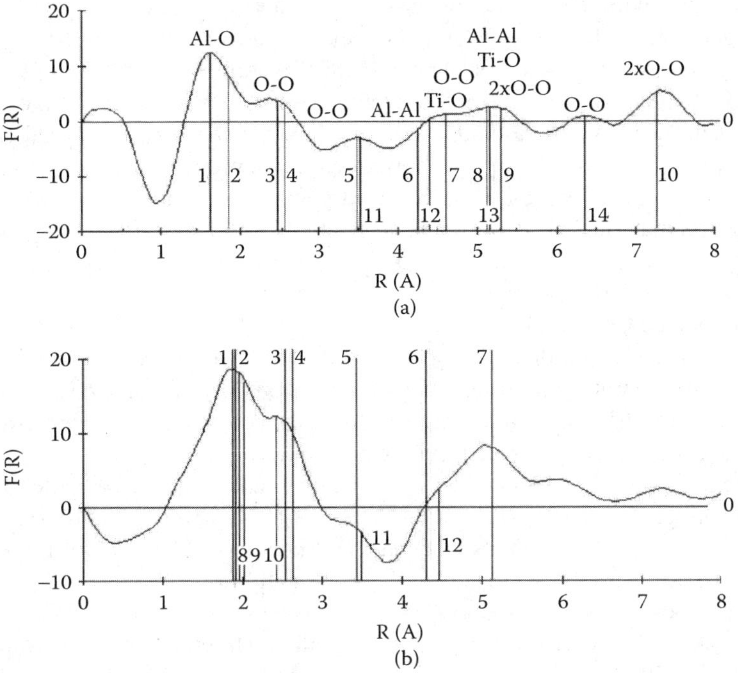

FIGURE 9.36 Fourier transforms of EELFS close to the line of elastic scattered electrons for the films forming on the surface of: (a) Ti–Al–N FAD coating after cutting; (b) binary TiAl alloy after isothermal oxidation in air. (From Fox-Rabinovich, G.S., Weatherly, G.C., Dodonov, A.I., Kovalev, A.I., Veldhuis, S.C., Shuster, L.S., Dosbaeva, G.K., Wainstein, D.L., Migranov, M.S. Nano-crystalline FAD (filtered arc deposited) TiAlN PVD coatings for high-speed machining application. *Surf. Coat. Technol.* 2004, 177–178, 800–811. With permission.)

layers to severe conditions of high-speed machining. Adaptation is a beneficial process based on the self-organizing phenomenon [95,118] that eventually results in increase of the life of the coated cutting tool. The coating that has this ability could be described as a self-adaptive one, and the FAD Ti–Al–N coating exhibits this adaptive characteristics under high-speed machining conditions. The metal-based oxygen-containing compounds that are formed during cutting can act as shields that protect the tool surface. Based on the data presented in Figure 9.33, Figure 9.34, and Figure 9.36, we can conclude that the oxide films that are formed on the tool surface with Ti–Al–N coatings are a mixture of alumina and rutile, but the only protective layer is the alumina one [119]. As soon as a significant portion of the alumina film starts to form, a beneficial phenomenon occurs at the cutting tool–workpiece material interface. During high-speed machining, alumina films formed at the surface limit the interaction of the underlying coating with the workpiece material. The data presented in Figure 9.36 shows that there are two types of protective aluminum-based oxygen-containing films formed at the surface during cutting: (1) amorphous-like and (2) crystalline. Previous detailed studies of protective film formation during cutting at low and moderate cutting speeds show that there is only one type of protective film formed on the surface as a result of the self-organizing phenomena. These films have amorphous-like structure with high plasticity and improved lubricity [42,95]. During high-speed machining, a more complex phenomenon take place. These are low-intensity peaks on the Fourier transform that are found at remote atomic distances, which are presented in Figure 9.36. From this we can assume that the majority of the films that form during tribo-oxidation under high-speed machining conditions are also amorphous-like. Only a small amount of crystalline films are forming as a result of the intensive thermal activation process [27]. The crystalline alumina films additionally improve the wear behavior because they have a low thermal conductivity [119] that prevents intensive heat generation during cutting from being conducted into the cutting-tool surface. Therefore, a significant part of the heat leaves with the chips (Figure 9.35). At the same time, alumina as a chemically stable material prevents intensive interaction at the workpiece–tool interface during cutting and depresses the adherence of the machined material (Figure 9.32) to the cutting-tool surface. It is worth noting that the adaptation process during high-speed machining exhibits the formation of alumina films with amorphous-like structure in a very clear manner. Spots of workpiece material sticking to the tool–chip interface could not be observed, and a perfectly smooth contact zone was formed (Figure 9.35b to Figure 9.35d). This is a result of the chip-flow improvement most probably due to the increased lubricity of the amorphous-like alumina tribo-films (see Chapter 1).

Ultimately, it leads to a significant improvement of the frictional characteristics (Figure 9.31). The coefficient of friction at high temperatures is as low as 0.1 (Figure 9.31). As a result of the favorable change in the friction conditions, the wear rate of the process was found to stabilize (Figure 9.30), thus significantly increasing the cutting-tool life.

As was mentioned earlier, the Ti–Al–N family of coatings has great potential owing to its favorable combination of mechanical and thermal properties as well as oxidation resistance [56], as outlined earlier, owing to the adaptability of these coatings. To improve the adaptability, and consequently the wear resistance of these coatings, formation of both types of protective alumina films (crystalline tribo-ceramics and amorphous-like) should be promoted. We can conclude that adaptive FAD Ti–Al–N PVD coatings with nanocrystalline structures exhibit promising combinations of properties ideally suited for high-speed machining applications.

We have to emphasize that a special thermodynamic state of the surface-engineered layer is needed to improve adaptability of the coated cutting tools. It is known that metal cutting is a strongly nonequilibrium process [23]. The tribo-films (dissipative structures) are formed as a result of nonequilibrium processes. Based on the irreversible thermodynamics analysis of entropy production during friction, a conclusion has been drawn that the formation of protective tribo-films results in the accumulation of energy generated during friction within the layer of these films (see Chapters 1 and 2). In this case, the entropy of the system is reduced compared to a condition without the protective films. This leads to a wear rate reduction [120]. The explanation of this phenomenon is that the nonequilibrium process needs plenty of energy. Therefore, energy of friction is spent for nonequilibrium processes instead of the wear.

The goal of surface engineering is to "help" the system to form the dissipative structures faster. The nanocrystalline coatings developed with the highest possible amount of energy accumulated during deposition and having a far-from-equilibrium thermodynamic state are the best accelerators of beneficial surface processes during severe high-temperature or high-stress friction conditions. This microstructure and thermodynamic state of the coatings gives them some decisive advantages to improve adaptability under severe conditions of high-speed machining.

Together with grain size refinement, down to the nanoscale level, the pulsed FAD techniques application results in an increase in the density of lattice imperfections within the coatings layer. Figure 9.37 presents the fine structure of the electron spectra in the proximity of the backscattered electron line obtained from TiAlN commercial and FAD coatings. The pronounced peaks on the spectrum in Figure 9.37 are the plasmon peaks. The plasmon losses are tied together with free surface electrons. On the spectra of FAD coatings, the intensity of the peaks is negligible, and the lines are diffused. The theoretical spectral intensity of each plasmon peak is related to its energy E and energy of primary electrons E_p. The plasmon peak intensity in the spectrum of primary electrons energy loss is related to the excitation of collective oscillations among free electrons and determined by their concentration. This permits us to determine the effective concentration of conduction electrons on the basis of the principles of the dielectric theory of energy losses by using the integral intensity of plasmon peaks (I_m) [121]:

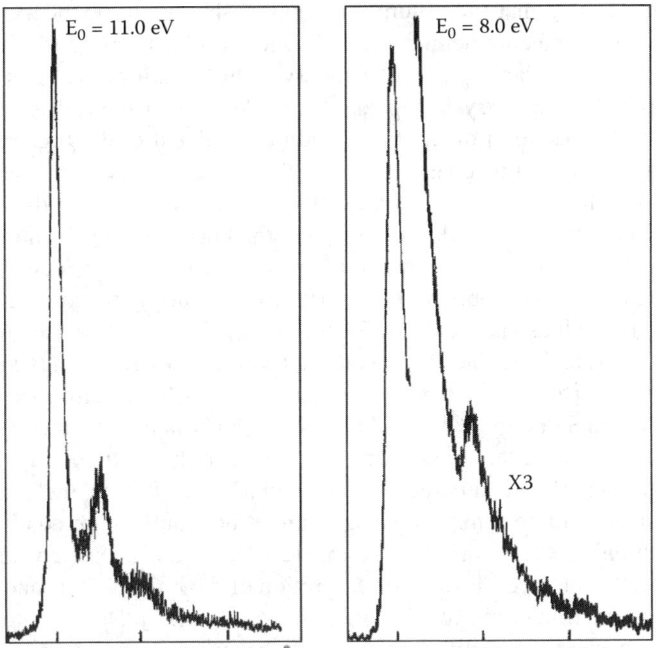

FIGURE 9.37 HREEL spectra for Ti–Al–N coatings: (a) monolayered; (b) FAD. Energy of the primary electrons is 9.0 eV.

$$n_{eff} = \frac{2\varepsilon_0 M}{\pi\hbar^2 e^2 n_a}\int I_m\left[-\frac{1}{\varepsilon}\right]E'dE' \tag{9.1}$$

After normalizing an integral intensity for plasmon peaks to the respective intensities of back-scattered electron lines, we can determine a change in the effective density of conduction electrons in Ti–Al–N commercial and FAD PVD coatings [121]:

$$\Delta n_{eff} = \frac{n_e(\text{TiAlN commercial}) - n_e(\text{TiAl NFAD})}{n_e(\text{TiAlN commercial})}\cdot 100\% \tag{9.2}$$

The density of the states of free surface electrons in FAD coating is 55% lower than the regular monolayered one. Free electrons form the atmospheres around the crystal structure defects. The concentration of the free electrons drops when the density of the lattice imperfections increases.

One of the advantages of the nanocrystalline Ti–Al–N coatings synthesized in nonequilibrium conditions using pulsed FAD technique is their ability to accelerate the beneficial mass transfer during friction within the initial stage of wear when the process of self-organization is taking place (Figure 9.30).

The mechanism of mass-transfer acceleration is as follows: the formation of the nanocrystalline structure of the Ti–Al–N coatings leads to anomalously enhanced grain boundary diffusion [122–126]. As we can see in Figure 9.28, the grain sizes of the coatings lie in the nanoscale range. For calculation purposes we can conditionally consider the nanocrystal as a double-phase structure, including "crystal" and "grain boundary" phases. According to the calculations presented in Reference 122, we can estimate that the volume fraction of the "grain boundary" in FAD coating is 40% as compared to 10% in a commercial coating. One must be aware of the fact that the diffusion coefficient of the "grain boundary" phase of nanocrystalline coatings is higher by 4 to 5 orders of magnitude as compared to the "crystal" phase [123]. We can deduce that the accelerated mass transfer during the tribo-oxidation of the FAD coating is related to the higher volume fraction of the "grain boundary" phase in this coating.

High-power plasma flow "pumps" a significant amount of energy into the surface layer during the FAD process. This energy accumulates within the thickness of the growing layer of the coating and is predominantly localized at the grain boundaries with a high density of imperfections [122,123]. The energy that was supplied to the surface layer during the coating deposition is stored in these imperfections, such as vacancies [127]. So the layer with an elevated density of imperfections could strongly activate the surface layers during friction to accelerate nonequilibrium processes of protective tribo-film formation. It is worth mentioning that the electron structure (Figure 9.37) also shows that the internal energy of the FAD's coating layer is increased compared to the regular coating [128]. It means that, if the FAD coating is considered as a thermodynamic system, it has more nonequilibrium states as compared to the regular one. Heating under stress during tribo-oxidation of the nanocrystalline films synthesized under nonequilibrium conditions also results in the intensive migration of the point defects [58,129], i.e., diffusion carriers [130] as well as dislocations, and greatly enhances diffusivity. Migration of the point defects and dislocations during heating leads to the formation of microvoids and micropores [131]. During intensive heating under high-speed machining conditions, disassociation of the nitride coating and a competitive metal diffusion of the metallic atoms takes place. An accelerated outward Al diffusion to the friction surface takes place owing to the lower activation energy needed for mass transfer of the more mobile and lighter atoms of aluminum than for the heavier atoms of titanium [133]. At the same time, it is easier for the smaller-radius ions (in our case, Al) to penetrate the micropores as well as the point and line defects of the coating during tribo-oxidation [132]. This probably could explain the well-known fact that nanocrystals promote selective oxidation of alloys and form desirable protective oxide films with low growth rate. The high-density grain boundaries also improve the oxides scale spallation resistance [134].

The level of the deviation of the coatings from a nonequilibrium state could be characterized by a parameter that we can tentatively call φ. A higher value of the φ parameter promotes the acceleration of beneficial mass transfer during friction. The parameter φ is related to the density of defects within the coating layer. The value of φ has to be as high as possible to enhance the desirable mass transfer to the outer surface of the coatings on the one hand, but is limited, on the other hand, to prevent the coating degradation at high-temperature and high-stress conditions generated during cutting. These limitations are obvious. First of all, they are associated with the hardness, especially hot hardness, of the coatings. It is known that the Ti–Al–N coatings have a maximal hardness values when the grain size is around 30 nm [135]. Hardness reduces when grain sizes fall below this range (see Figure 9.26). We have to keep in mind that input of residual stress in the hardness of the nanocrystalline coating is very important because these coatings have a very high density of lattice imperfections [17]. So, during heating under stress, the hardness of the coating should decrease significantly. Other issues such as poor adhesion or even buckling during coating deposition and under operation should also be taken into consideration. Outcropping of the dislocations on the surface during friction as a result of intensive plastic deformation should also be considered (see Figure 9.24 and Chapter 7) because it could lead to excessive surface damage. We can conclude that technological optimization or multilayered coating deposition is critical.

A thermodynamical approach [136] could be applied for the analysis of tribo-oxidation of Ti–Al–N coatings under extreme conditions of high-speed machining. It is known that the entropy production in the thermodynamics of irreversible processes is defined as the sum of multiplications of the connected thermodynamic forces (X_i) and flows (J_i):

$$dS_i / dt = \sum J_i X_i \tag{9.3}$$

We can divide the change of entropy production of the open system $\frac{dS_i}{dt} = \Phi$ into two parts [136] and connect one part with the change of the thermodynamic forces $X(d_x\Phi)$ and the other with the change of thermodynamic fluxes $I(d_i\Phi)$, i.e.,

$$d\Phi = d\sum_i I_i X_i = \sum_i I_i dX_i + \sum_i X_i dI_i = d_x\Phi + d_i\Phi \tag{9.4}$$

The value $d_x\Phi$ will be less than or equal to 0 when the system evolves to a steady state (owing to the formation of protective tribo-films). We cannot say anything about the sign of $d\Phi$, but we can say that, if the entropy production change is caused only by the rate of the thermodynamic forces, this change will be negative. During the tribo-film formation (especially during the initial stage of the process), on the surface of cutting tools X and I are changing. So, in our case, when the entropy production is caused by the rate of change in both the forces and fluxes, the entropy production will decrease with time as soon as the steady state with $d_i\Phi < 0$ is reached.

Intensive mass transfer due to the acceleration of the aluminum diffusion results in an increase in the flux of matter, I, in the initial stage of the tribo-oxidation process. This intensive diffusion leads to an entropy production increase at the very beginning of the process. This results in a change in X due to the protective oxide tribo-film formation (see Figure 9.33, Figure 9.34, and [111]). This oxide is a thermodynamically stable one (Al_2O_3), and thus implies that the rate of entropy production decreases until the moment a stable state is achieved. According to the Prigogine theorem [137], in a system with minimal entropy production, the flux of matter and, consequently, the wear rate of the tools have to drop significantly as a flux of a nonlimited thermodynamic force (Figure 9.30). This must occur not at the cost of the intensive wear during the running-in stage but rather due to the enhancement of the nonequilibrium processes of tribo-oxidation. As was mentioned earlier, the formation of the protective tribo-films requires a considerable level of energy. A layer of the nanocrystalline coatings with a high level of accumulated energy serves as a reservoir for the formation of protective tribo-films. It should be emphasized that the dissipative structure formation is a sudden-change type of process. Thus, on the curve of the tribosystem's properties vs. any corresponding parameter, a sudden change in the function takes place as the minor argument changes. Figure 9.30 shows that the wear rate (i.e., slope of the curve or its first derivative) for the filtered coating drops down by almost three times in a short period of time. In this case, the total time of the tribo-films formation is only a few minutes of cutting (Figure 9.30). This phenomenon could be described as a "trigger" effect. The similar effect was discovered previously for intermetallic systems [138] and has also been observed in biosystems [139].

Further considerations of the self-organizing process could be made based on the detailed thermodynamic analysis cutting tool–workpiece tribosystem [120].

As was shown in details in Chapter 2, the change of entropy of a frictional body (dS) will be as follows:

$$dS = dS_i + dS_e + dS_m + dS_f - dS_w \tag{9.5}$$

After differentiation on time, Equation 9.1 for a stationary condition will be:

$$dS/dt = 0 = dS_i/dt + dS_e/dt + dS_m/dt + dS_f/dt - dS_w/dt$$

or

$$dS_w/dt = dS_i/dt + dS_e/dt + dS_f/dt + dS_m/dt \tag{9.6}$$

where dS_i/dt is the entropy production, dS_e/dt is the flow of entropy, dS_m/dt is the change of entropy owing to entropy the matter from the frictional body in contact (in our case, because of cutting tool–chip interactions), dS_f/dt is the change of entropy due to tribo-films formation, and dS_w/dt is the change of entropy due to wear process (sign "–" is used in the equation because the products of the wear process leaves the frictional body with their own entropy). The change of entropy, dS_f/dt, could be negative if nonequilibrium processes are taking place on the surface, and it is positive if the equilibrium processes are occurring. However, the sum of these terms ($dS_i/dt + dS_f/dt > 0$) is supposed to be positive as the total entropy production. Following from Equation 9.6, the lower entropy production corresponds to the lower dS_w/dt value. The value of dS_w/dt characterizes entropy of the wear products. Bearing in mind that entropy is an additive value, we can consider that the lower value of dS_w/dt decreases the wear rate. Therefore, the decrease of entropy production leads to reduction of wear rate. Nonequilibrium processes on the surface ($dS_f/dt < 0$), other conditions being equal, can also lead to wear rate decrease.

Nonequilibrium processes develop steadily when the dissipative structures form during the process of self-organization. Self-organization can begin only after the system passes the point of instability. We will analyze the probability of stability loss in the tribosystem.

As it was shown experimentally, the major differences in the microstructure of regular and FAD coating are the grain size refinement (Figure 9.29) and a higher density of lattice imperfections in nanocrystalline coating (Figure 9.37). The FAD coating has higher φ parameter value (see earlier text). Owing to this, accelerated relaxation processes will take place during friction of the nonequilibrium FAD coating. The relaxation processes by themselves shift the system to the equilibrium state and lead to an increase in entropy. However, the interaction of the process with the other ones (in our case, with the process of friction) could result in self-organization. This is specifically interesting for the nonlinear area, which is typical for the conditions of high-speed machining. We will try to compare the thermodynamic stability of the regular and nanocrystalline FAD coatings.

If a single process is taking place within the system (friction in our case), the entropy of production has the following form:

$$\frac{dS_i}{dt} = X_h J_h = \frac{(kpv)^2}{\lambda B T^2} \tag{9.7}$$

where J_h is a heat flow, X_h is a thermodynamic force that is causing the thermodynamic flow and equal to $gradT/T^2$, $J_h = -\lambda BgradT$ (B is a contact area). At the same time, $J_h = kpv$ (k is a coefficient of friction, p is a load applied, v is a cutting speed), T is a temperature, and λ is a thermal conductivity.

It is shown in Reference 140 that the function $\delta^2 s$ could be used as Lapunov's function to analyze the condition of stability loss for the thermodynamic system. If the state of the thermodynamic system is far from equilibrium, a Lapunov's function $\delta^2 s$ is introduced [140], and it is shown that

$$\delta^2 s < 0 \tag{9.8}$$

where δ is a fluctuation and s is a local entropy.

This value is negative as the determined function of the increments in independent variables, which Gibbs' formula includes for local conditions of the dissipative system. Therefore, the system stability can be characterized based on the function $\delta^2 s$ as a Lapunov's one. Thus, the local stability condition will be [140]:

$$\frac{\partial}{\partial t}(\delta^2 s) \geq 0 \tag{9.9}$$

The inequalities of Equation 9.8 and Equation 9.9 are the sufficient conditions according to Lyapunov's theorem.

The local equilibrium principle implies the validity of Equation 9.9 at any point of the volume and at any moment. That is why owing to the continuity of $\delta^2 s$ function, the integration of Equation 9.9 by volume will give a sufficient condition for stability of a specified volume of the system:

$$\frac{\partial}{\partial t}(\delta^2 S) \geq 0 \text{ at} : \delta^2 S < 0 \tag{9.10}$$

where S is entropy of a specified volume of the system.

The time derivative for $\delta^2 S$ in Equation 9.10 is related, as is shown in Reference 140, to the entropy production generated by disturbance of the system, i.e.,

$$\frac{1}{2}\frac{\partial}{\partial t}(\delta^2 S) = \sum_n \delta X_n \delta J_n \geq 0 \tag{9.11}$$

The sum on the right-hand side of the equation is called *excessive entropy production*. The values of δX_n and δJ_n are the deviation of the corresponding thermodynamic flows and forces at the stationary state of the system. The state of the system is stable if the inequality Equation 9.10 conditions are satisfied from the beginning of disturbance of the system. However, during specific processes or during these processes' interactions, one can get negative input in excessive entropy production, which increases during development of the disturbance. Under these conditions this state could become unstable because the positive entropy production is a sufficient but not prerequisite condition of stability. Only after passing instability, self-organization can begin.

Let us consider friction as the only independent source of energy dissipation within the system. In this case the excessive entropy production according to Equation 9.10 and Equation 9.11 will be the following:

$$\frac{\partial}{2\partial t}\delta^2 S = \delta X_h \delta J_h \tag{9.12}$$

During the deviation of the system from the stable state owing to the "nonequilibrium" coating application instead of the "equilibrium" one, i.e., owing to parameter (φ) increase, the excessive entropy production, based on equation (Equation 9.7), will be:

$$\frac{\partial}{2\partial t}\delta^2 S = \frac{(pv)^2}{T^2 B\lambda}\left(\frac{\partial k}{\partial \varphi}\right)^2 (\delta\varphi)^2 \tag{9.13}$$

Equation 9.13 is based on the condition that friction is the only source of energy dissipation and the parameter φ depends only on friction. The right-hand side of the quadratic equation can only be positive (because λ is always greater than 0). Hence, if friction is the only source of energy dissipation within the tribosystem, this system does not lose stability and self-organization does not begin. Therefore, more than one independent process should take place within the system to start self-organization. In "real" systems some additional independent sources of energy dissipation exist, such as interaction of the frictional bodies and interaction with the environment. In our case the latter interaction (tribo-oxidation) plays a decisive role.

Based on the data obtained, we can say that physicochemical interactions of the frictional bodies and the interaction with the environment lead to changes in chemical composition and structure of the surface layers. In the equation of entropy production, this will be considered depending on the thermal conductivity and φ parameter values. In this case the equation for the excessive entropy production will be as follows:

$$\frac{\partial}{2\partial t}\delta^2 S = \frac{(pv)^2}{T^2 B}\left(\frac{1}{\lambda}\left(\frac{\partial k}{\partial \varphi}\right)^2 - \frac{k}{\lambda^2}\frac{\partial k}{\partial \varphi}\frac{\partial \lambda}{\partial \varphi}\right)(\delta\varphi)^2 \tag{9.14}$$

The right-hand side of the Equation 9.14 could be negative due to the sign of the second multiplier. For this to become negative, the following condition should be satisfied for Equation 9.14:

$$\frac{\partial k}{\partial \varphi}\frac{\partial \lambda}{\partial \varphi} > 0 \tag{9.15}$$

The condition in Equation 9.15 is realized if the coefficient of friction and the thermal conductivity increase or decrease at the same time with the φ parameter. The thermal conductivity as well as the coefficient of friction in FAD coatings is lower compared to the regular coating. Thus, it was proven experimentally that the condition Equation 9.11 is realized as the parameter φ grows. When the negative part of Equation 9.14 is higher than the positive one, the condition in Equation 9.9 could be violated and the system could loss stability. After stability loss, the process of the self-organization with dissipative structures formation could begin. Nonequilibrium processes and the structures that are formed as a result of these processes lead to the wear rate decrease.

As was outlined in the preceding text, the "equilibrium" coating will differ from the "nonequilibrium" one because of the acceleration of the relaxation processes. The relaxation processes in our case are mass transfer during tribo-oxidation. Mass transfer cannot follow the linear laws for the systems that are far from equilibrium. This is why a part of entropy production due to mass transfer $\left(\frac{dS}{dt}\right)_m$ can be presented in a generic form:

$$\left(\frac{dS}{dt}\right)_m = X_m \rho_m W(X_m) \tag{9.16}$$

where X_m is the thermodynamic force causing the mass transfer (gradients of chemical potentials during diffusion), ρ_m is the average material density involved in the mass transfer, and $W(X_m)$ is the average mass transfer rate, which is depending on Xm. We have to emphasize that $W(X_m)$ will grow as X_m increase.

The excessive entropy production due to systematic fluctuation of the φ parameter (that is related to the use of the "nonequilibrium" coating instead of the "equilibrium" one) will be the following:

$$\frac{\partial}{2\partial t}\delta^2 S_m = \frac{\partial X_m}{\partial \varphi}\left(W\frac{\partial \rho_m}{\partial \varphi} + \rho_m \frac{\partial W}{\partial X_m}\frac{\partial X_m}{\partial \varphi}\right)(\delta\varphi)^2 \tag{9.17}$$

The thermodynamic force of the mass transfer will increase with φ parameter growth, i.e., $\frac{\partial X_m}{\partial \varphi} > 0$. Because $\frac{\partial W}{\partial X_m} > 0$, the negative input in excessive entropy production could be only due to the term $\frac{\partial \rho}{\partial \varphi}$. Because of the density of the matter involved, the mass transfer should drop with increase of the φ parameter. Thus, the excessive entropy production Equation 9.17 could became negative and the system could loss stability under the conditions:

$$\frac{\partial \rho}{\partial \varphi} < 0 \tag{9.18}$$

A decrease of the material's density that takes place during mass transfer could be caused by the increase in the density of lattice imperfection in nanocrystalline FAD coatings [131]. Another (and most widely used in practice) way to reduce density is to deposit Al-rich Ti–Al–N coatings because Al is a lighter metal as compared to Ti. But in the "equilibrium" coatings that dominate in practice, the nonequilibrium processes do not develop or develops with a lower intensity. The probability that the system does pass instability in these coatings is much lower, and the process of self-organization cannot start in this case as it does for the regular Ti–Al–N coating (Figure 9.30).

Thus, the system with nonequilibrium coatings could lose stability and the wear rate will be reduced if conditions Equation 9.15 and Equation 9.18 are held. These conditions are interrelated because the decrease in the coefficient of friction as well as the thermal conductivity take place most intensively because of the change in surface layer's composition and the protective tribo-films formation. The surface layer composition changes if the condition in Equation 9.18 is realized.

Oxidation in equilibrium conditions results in competitive mass transfer of Ti and Al to the outer surface of the coating with subsequent formation of the corresponding oxides. Under conditions of high-performance machining, the surface temperatures are around 1000°C and above [18]. Intensively growing nonprotective rutile films are a regular feature of the equilibrium oxidation process in regular Ti–Al–N coatings under high-temperature conditions. Dominating formation of the protective films on the surface (in our case, alumina-like) is probably a feature of the nonequilibrium process for the high-temperature conditions. We could consider the protective alumina-like films formation as dissipative structures. Formation of dissipative structures is a spontaneous and sudden processes [140]. The mechanism of this process is very complex. It could be related to (1) selective oxidation of nanocrystalline coatings [134] and (2) selective wear of the tribo-films with preferential nonprotective rutile tribo-films removal. The tribosystem with the nanocrystalline nonequilibrium coating enhances formation of the protective alumina-like layer because both types of processes (i.e., tribo-oxidation and wear) are governed by reduction of entropy production as a result of the self-organizing process. This is associated with the "trigger" effect described earlier for nanocrystalline coatings.

A similar speculation and analytical modeling could be applied for the ion-modified (by means of the ion mixing) surface layers of the smart coatings described in Section 9.3.4. This layer, synthesized under nonequilibrium conditions, has similar features and wear behavior as compared to the FAD coatings. Acceleration of mass transfer in this case leads to the enhancement of inward oxygen diffusion with protective tribo-films formation. It means that the principles outlined here are generic.

If we summarize all the known data of nanoscale films such as nanocomposite superhard coatings, nanocrystalline Ti–Al–N coatings, and ion-mixed layers, it is apparent that a metastable material with some outstanding characteristics is created on the surface. There are two major types of the nanoscale coating designs currently presented. These types of coatings could affect the tool

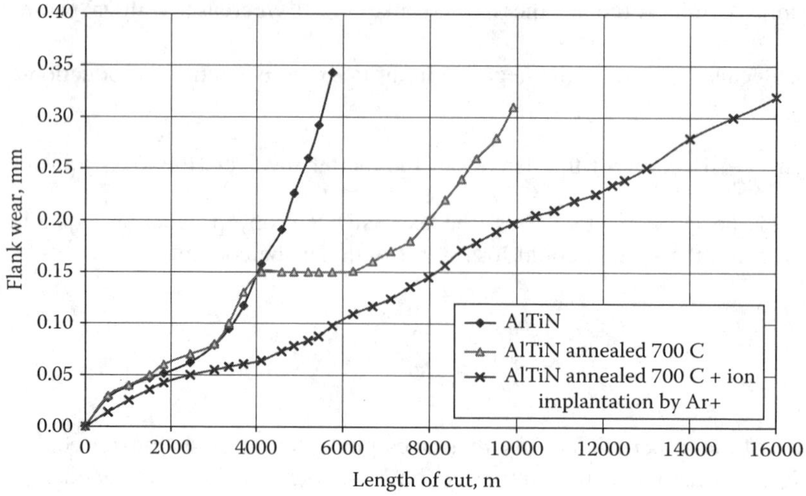

FIGURE 9.38 Tool life of Al–Ti–N coating (X.ceed) before and after annealing (700°C, 2 h, in vacuum, 1×10^{-6} mbar) and annealing with further ion implantation by Ar^+ ions under conditions of high-speed turning of 1040 steel (according to cutting conditions shown in Table 9.6).

life during different phases of wear associated with the principles of tribological compatibility improvement outlined in Chapter 1 and Chapter 3. The first one is the nanocomposites that are initially fabricated under nonequilibrium conditions and further heat-treated or deposited under controlled conditions to achieve a quasi-stable structure with exceptional thermal stability (see Figure 9.22 and Figure 9.24). The structure of the coating mainly improves the wear behavior of the coated tools during stable, post running-in stage of wear (Figure 9.23). The second one is the adaptive coatings and films with nanocrystalline microstructure that are fabricated under strongly nonequilibrium conditions (Figure 9.29). The major feature of nanocrystalline coatings is improvement of their tribological properties. We have to emphasize that both types of the state-of-art hard coatings could be successfully used for specific applications. Moreover, a combination of these two major types of coatings in one multilayered coating appears very promising. This idea has been realized by means of the Al–Ti–N-coating posttreatment using a combination of annealing with ion implantation by heavy Ar ions. The ion implantation performed results in the formation of a thin outer layer with a high density of lattice imperfections (see Figure 9.37) that shifts the surface into a nonequilibrium state. The multilayered adaptive or smart coating is able to ensure the tribological compatibility during overall periods of the tool life owing to accelerated formation of the protective or lubricious tribo-films during the running-in stage and their stable regeneration as well as prevention of excessive surface-damage during post running-in (stable) stage of wear. Eventually the tool life is increased significantly as a result of this "duplex" posttreatment (Figure 9.38). This novel generation of functionally graded or smart coatings combines nanocrystalline, nonequilibrium structures of the outer layer to enhance adaptability and nanocomposite structure to ensure stability of the coating during cutting under severe conditions of high-performance machining. Thus, it is possible to create an "artificial" material, possessing unique but previously unattainable properties, with a new level of stability under severe and even extreme environmental conditions and, as a result, a new level of tool performance. This is an exciting yet practical realization of the underlying principles of friction control.

REFERENCES

1. Holmberg, K., Matthews, A. *Coating Tribology: Principles, Techniques, and Application in Surface Engineering*. Elsevier Science: Amsterdam, 1994, 257–309.
2. Astakov, V.P. Tribology in metal cutting. In *Mechanical Tribology: Materials, Characterization, and Application*, Eds., Liang, H., Totten, G. Marcel Dekker: New York, 2004, pp. 307–389.
3. Stoiber, M., Perlot, S., Mitterer, C., Beschliesser, M., Lugmair, C., Kullmer, R. PACVD TiN/Ti–B–N multilayers: from micro- to nano-scale. *Surf. Coat. Technol.* 2004, *177–178*, 348–354.
4. Holleck, H.W. Advanced concepts of PVD hard coatings. *Vacuum* 1990, 41, *7–9*, 2220–2222.
5. Fox-Rabinovich, G.S. Structure of complex coatings. *Wear* 1993, 160, 67–76.
6. Palatnilk, L.S. *Pores in the Films*. Energizdat: Moscow, 1982, pp. 121–214.
7. Goldsmith, H.J. *Interstitial Alloys*. Butterworth: London, 1967, pp. 14–23.
8. Fox-Rabinovich, G.S., Veldhuis, S.C., Scvortsov, V.N., Shuster, L.Sh. Elastic and plastic work of indentation as a characteristic of wear behavior for cutting tools with nitride PVD coatings. *Thin Solid Films* 2004, *469–470,* 505–512.
9. Trent, E.M., Suh, N.P. *Tribophysics*. Prentice-Hall: Englewood Cliffs, NJ, 1986, pp. 125–489.
10. Gorokhovsky, V.I., Bhat, D.G., Shivpuri, R., Kulkarni, K., Bhattacharya, R., Rai, A.K. Characterization of large area filtered arc deposition technology. Part II: Coating properties and applications. *Surf. Coat. Technol.* 2001, *140*, 215–224.
11. Konyashin, I., Fox-Rabinovich, G.S. Nanograined titanium nitride thin films. *Adv. Mater.* 1998, 10, *12*, 952–955.
12. Fox-Rabinovich, G.S., Kovalev, A.I., Afanasyev, S.N. Characteristic features of wear in tools made of high-speed steels with surface engineered coatings. Part I: Wear characteristics of surface engineered high-speed steel cutting tools. *Wear* 1996, *201*, 38–44.
13. Fox-Rabinovich, G.S., Kovalev, A.I., Afanasyev, S.N. Characteristic features of wear in tools made of HSS with surface engineered coatings. Part II: Study of surface engineered HSS cutting tools by AES, SIMS and EELFAS methods. *Wear* 1996, *198*, 280–286.
14. Storz, O., Gasthuber, H., Woydt, M. Tribological properties of thermal-sprayed Magnéli-type coatings with different stoichiometries (Ti_nO2_{n-1}). *Surf. Coat. Technol.* 2001, 140, 76–81.
15. Fox-Rabinovich, G.S. The Method of Tool Hardening. Russian Patent 2026419, 1995.
16. Nosovskiy, I.G. On the mechanism of seizure of metals at friction. *J. Friction and Wear* 1993, 14, *1*, 19–24.
17. Fox-Rabinovich, G.S. Scientific Principles of Material Choice for Wear-Resistant Cutting Tools and Dies from the Point of View of Surface's Structure Optimization. D.Sc. dissertation, All-Russian Railway Transport Research Institute: Moscow, Russia, 1993.
18. Trent, E.M., Wright, P.K. *Metal Cutting,* 4th ed. Butterworth-Heinemann: Woburn, MA, 2000.
19. Lahres, M., Doerfel, O., Neumüller, R. Applicability of different hard coatings in dry machining an austenitic steel. *Surf. Coat. Technol.* 1999, 120–121, 687–691.
20. Fox, V.C., Renevier, N., Teer, D.G., Hampshire, J., Rigato, V. The structure of tribologically improved MoS_2-metal composite coatings and their industrial applications. *Surf. Coat. Technol.* 1999, 116–119, 492–497.
21. Fox, V.C., Jones, A., Renevier, N.M., Teer, D.G. Hard lubricating coatings for cutting and forming tools and mechanical components. *Surf. Coat. Technol.* 2000, 125, 347–353.
22. Savan, A., Pfluger, E., Goller, R., Gissler, W. Use of nanoscaled multilayer and compound films to realize a soft lubrication phase within a hard, wear-resistant matrix. *Surf. Coat. Technol.* 1999, 126, 159.
23. Fox-Rabinovich, G.S., Kovalev, A.I., Weatherly, G.C. Tribology and the design of surface engineered materials for cutting tool applications. In *Modeling and Simulation for Material Selection and Mechanical Design*, Eds., Totten, G., Xie, L., Funatani K. Marcel Dekker: New York, 2004, pp. 301–382.
24. Bushe, N.A. Solved and unsolved problems of tribosystems compatibility. *J. Friction and Wear* 1993, 14, *1*, 26–33.
25. Fox-Rabinovich, G.S., Bushe, N.A., Kovalev, A.I., Korshunov, S.N., Shuster, L.S., Dosbaeva, G.K. Impact of ion modification of HSS surfaces on the wear resistance of cutting tools with surface engineered coatings. *Wear* 2001, 249, 1051–1058.

26. Panckow, A.N., Steffenhagen, J., Wegener, B., Dübner, L., Lierath, F. Application of a novel vacuum-arc ion-plating technology for the design of advanced wear resistant coatings. *Surf. Coat. Technol.* 2001, 138, 71–76.
27. Kostetsky, B.I. An evolution of the materials' structure and phase composition and the mechanisms of the self-organizing phenomenon at external friction. *J. Friction and Wear* 1993, 14, *4*, 773–783.
28. Bushe, N.A. Tribo-engineering materials. In *International Engineering Encyclopedia: Practical Tribology — World Experience*, Vol.1. Science and Technique Center: Moscow, Russia, 1994, pp. 21–29.
29. Rabinowicz, E. *Wear Control Handbook*. ASME: New York, 1980, pp. 475–476.
30. Bershadskiy, L.I. On self-organizing and concept of tribosystem self-organizing. *J. Friction and Wear* 1992, 13, *6*, 1077–1094.
31. Karasik, I.I. *Methods of Tribological Tests in Standards of Various Countries*. Science and Technique: Moscow, 1993, p. 325.
32. Kostetsky, B.I. *Surface Strength of the Materials at Friction*. Technika: Kiev, 1976, 76–154.
33. Beliy, V.A., Ludema, K., Mishkin, N.K. *Tribology: Studies and Applications: USA and USSR Experience*. Allerton Press: New York, 1993, pp. 202–452.
34. Manory, R.R., Perry, A. J., Rafaja, D., Nowak, R. Some effects of ion beam treatments on titanium nitride coatings of commercial quality. *Surf. Coat. Technol.* 1999, 114, 137–142.
35. Vladimirov, B.G., Guseva, M.I. Carbide cutting tool life improvement by ion implanting methods. *J. Friction and Wear* 1993, 14, 544–551.
36. Liau, Z.L., Mayer, J.W. Influence of ion bombardment on material composition. In *Treatise on Materials Science and Technology,* Hirvonen, J.K., Ed. Academic Press: New York, 1980, 18, pp. 49–57.
37. Komkonder, L., Sahin, E., Buyukksap, E. The effect of the chemical environment on the K_β/K_α x-ray intensity ratio. *Nuovo cimento*1993, 5, *10*, 1215–1299.
38. Fox-Rabinovich, G.S., Kovalev, A.I., Shuster, L.Sh., Bokiy, Yu.F., Dosbaeva, G.K. Characteristic features of wear in HSS-based compound powder materials with consideration for tool self-organization at cutting. Part 2: Cutting tool friction control due to the alloying of the HSS-based deformed compound powder material. *Wear* 1998, 214, 279–286.
39. Musil, J. Hard and superhard nanocomposite coatings. *Surf. Coat. Technol.* 2000, 125, 322–330.
40. Musil, J., Vlcek, J. A perspective of magnetron sputtering in surface engineering. *Surf. Coat. Technol.* 1999, 112, 162–169.
41. Fox-Rabinovich, G.S., Kovalev, A.I., Weatherly, G.C., Korshunov, S.N., Shuster, L.Sh.S., Veldhuis, C., Dosbaeva, G.K., Scvortsov, V.N., Wainstein D.L. Improvement of "duplex" PVD coatings for HSS cutting tools by Ion mixing. *Surf. Coat. Technol.* 2004, 187, 230–237.
42. Kovalev, A.I., Wainstein, D.L., Mishina, V.P., Fox-Rabinovich, G.S. Investigation of atomic and electronic structure of films generated on a cutting tool surface. *J. Electron Spectrosc. Relat. Phenomena* 1999, 105, 63–75.
43. Hardwicke, C.U. Recent developments in applying smart structural materials. *J. Mater.* 2003, 12, 15–16.
44. Wax, S.C., Fisher, G.M., Sands, R.R. The past, present and future of DARPA's investment strategy in smart materials. *J. Mater.* 2003, 12, 17–23.
45. Nicholls, J.R., Simms, N.J. Smart overlay coating-concept and practice. *Proc. Int. Conf. Metallurgical Coatings and Thin Films*. San Diego, CA 2001, p. 96.
46. Fox-Rabinovich, G.S., Kovalev, A.I., Wainstein, D.L. Intelligente Oberflachenbeschinchtung fur Schneidwerkzeuge. Germany Patent 20116404.3, October 10, 2001.
47. Zhang, S., Sun, D., Fu, Y., Du, H. Recent advances of superhard nano-composite coatings: a review. *Surf. Coat. Technol.* 2003, 167, 113–116.
48. Tönshoff, H.K., Karpuschewski, B., Mohlfeld, A., Seegers, H. Influence of subsurface properties on the adhesion strength of sputtered hard coatings. *Surf. Coat. Technol.* 1999, 116–119, 524–529.
49. Wu, S.K., Lin, H.C., Liu, P.L. An investigation of unbalanced-magnetron sputtered TiAlN films on SKH51 high-speed steel. *Surf. Coat. Technol.* 2000, 124, 97–103.
50. Wang, D-Y., Li, Y-W., Ho, W-Y. Deposition of high quality (Ti, Al)N hard coatings by vacuum arc evaporation process. *Surf. Coat. Technol.* 1999, 114, 109–113.
51. Munz, W.-D.M. Ti-Al nitride films: a new alternative to TiN coatings. *J. Vac. Sci. Technol.* 1986, A4, *6*, 2717–2725.

52. Woo, J.H., Lee, J.K., Lee, S.R., Lee, D.B. High-temperature oxidation of $Ti_{0.3}\ Al_{0.2}\ N_{0.5}$ thin films deposited on steel substrate by ion plating. *Oxidation Met.* 2000; 53, *5/6*, 529–537.
53. Bouzakis, K.-D., Vidakis, N., Michailidis, N., Leyendecker, T., Erkens, G., Fuss, G. Quantification of properties modification and cutting performance of $(Ti_{1x}\ Al_x)N$ coatings at elevated temperatures. *Surf. Coat. Technol.* 1999, 120–131, 34–43.
54. Doychak, J. Oxidation behavior of high-temperature intermetallics. In *Intermetallic Compounds*, Vol. 2, Eds., Westbrook, J.H., Fleisher, R.L. John Wiley & Sons: New York, 1994, pp. 73–90.
55. Okafor, I.C.I., Reddy, R.C. The oxidation behavior of high-temperature aluminides. *J. Mater.* 1999, 6, 35–39.
56. PalDey, S., Deevi, S.C. Single layer and multilayer wear resistant coatings of (Ti,Al)N: a review. *Mater. Sci. Eng. A* 2003, 342, 1–2, *15*, 58–79.
57. Horling, A., Hultman, L., Oden, M., Sojolen, J., Karlsson, L. Mechanical properties and machining performance of TiAlN coated cutting tools. *Surf. Coat. Technol.* 2005, 191, *2–3*, 384–392.
58. Schiotz, J. *Proc. 22nd Riso Int. Symp. Materials Science*. Roskilde, Denmark, 2001, p. 127.
59. Henk, H., Westphal, M. Recent advances of superhard nanocomposite coatings: a review. *Euro PM99 Conference on Advances in Hard Materials Production*. Turin, Italy, 1999.
60. Gallister, W.C. *Material Science and Engineering: An Introduction*. John Wiley & Sons: New York, 1985.
61. Veprek, S. Superhard and functional nanocomposites formed by self-organisation. *Rev. Adv. Mater. Sci.* 2003, 5, 6–16.
62. Neiderhofer, A., Nesladek, P., Mannling, H-D., Moto, K., Veprek, S., Jilek, M. Structural properties, internal stress and thermal stability of nc-TiN/a-S_3N_4, nc-TiN/$TiSi_x$ and nc-$(Ti_{1y}\ Al_y\ Si_x)N$ superhard nanocomposite coatings reaching the hardness of diamond. *Surf. Coat. Technol.* 1999, 120–121, 173–178.
63. Hauert, R., Patscheider J. From alloying to nano-composites: improved performance in hard coatings. *Adv. Eng. Mater.* 2000, 2, *5*, 247–259.
64. Yau, B.-S., Huang, J.-L., Lii, D.-F., Sajgalik, P. Investigation of nanocrystal-$(Ti,Al)N_x$/amorphous-SiN_y composite films by co-deposition process. *Surf. Coat. Technol.* 2004, 177–178, 209–214.
65. Konyashin, I., Fox-Rabinovich, G.S. Nanograined titanium nitride thin films. *Adv. Mater.* 1998, 10, *12*, 952–955.
66. Musil, J. Hard and superhard nanocomposite coatings. *Surf. Coat. Technol.* 2000, 125, 322–330.
67. Levchuk, D., Maier, H. *The Influence of Plasma Variables on the Characteristics of Superhard Nano-Composite Coatings*. Max-Plank Institute fur Plasmaphysik: Munchen, 2004.
68. Musil, J., Zeman, P., Hrubý, H., Mayrhofer P.H. ZrN/Cu nanocomposite films: novel superhard material. *Surf. Coat. Technol.* 1999, 120–121, 179–183.
69. He, J.L., Setsuhara, Y., Shimizu, I., Miyake, S. Structure refinement and hardness enhancement of titanium nitride films by addition of copper. *Surf. Coat. Technol.* 2000, 137, 38–42.
70. Holubar, P., Jilek, I., Sima, M. Nanocomposite nc-TiAlSiN and nc-TiN-BN coatings: their applications on substrates made of cemented carbide and result of cutting tests. *Surf. Coat. Technol.* 1999, 120–121, 184–188.
71. Mitterer, C., Mayrhofer, P.H., Beschliesser, M., Losbichler, P., Warbichler, P., Hofer, F. Microstructure and properties of nanocomposite Ti-B-N and Ti-B-C coatings. *Surf. Coat. Technol.* 1999, 120–121, 405–411.
72. Veprek, S., Nesládek, P., Niederhofer, A., Glatz, F., Jílek, M., Síma, M. Recent progress in the superhard nanocrystalline composites: towards their industrialization and understanding of the origin of the superhardness. *Surf. Coat. Technol.* 1998, 108–109, 138–147.
73. Wong, M.S., Lee, Y.C. Deposition and characterization of Ti-B-N monolithic and multilayer coatings. *Surf. Coat. Technol.* 1999, 120–121, 194–199.
74. Munz, W-D.M, Ti-Al nitride films: a new alternative to TiN coatings. *J. Vac. Sci. Technol.* 1986, A4, *6*, 2717–2725.
75. Zeng, X.T. TiN/NbN superlattice hard coatings deposited by unbalanced magnetron sputtering. *Surf. Coat. Technol.* 1999, 113, 74–79.
76. Munz, W.-D., Donohue, L.A., Hovsepian, P.E. Properties of various large-scale fabricated TiAlN- and CrN-based superlattice coatings grown by combined cathodic arc-unbalanced magnetron sputter deposition. *Surf. Coat. Technol.* 2000, 125, 269–277.

77. Lembke, M.I., Lewis, D.B., Münz, W.-D. Localized oxidation defects in TiAlN/CrN superlattice structured hard coatings grown by cathodic arc/ unbalanced magnetron deposition on various substrate materials. *Surf. Coat. Technol.* 2000, 125, 263–268.
78. Parlinska-Wojtan, M., Karimi, A., Coddet, O., Cselle, T., Morstein, M. Characterization of thermally treated TiAlSiN coatings by TEM and nanoindentation. *Surf. Coat. Technol.* 2004, 188–189, 344–350.
79. Shevela,V.V. Internal friction as a factor of wear resistance of the tribosystems. *J. Friction and Wear* 1990, 11, *6*, 979–986.
80. Carvalho, S., Rebouta, L., Ribeiro, E., Vaz, M., Denannot, F., Pacaud, J., Rivière, J.P., Paumier, F., Gaboriaud, R.J., Alves, E. Microstructure of (Ti,Si,Al) N nano-composite coatings. *Surf. Coat. Technol.* 2004, 177–178, 369–375.
81. Cselle, T. *Proc. 3rd Int. Conf. Coatings and Layers*. Rosnvov, CZ, 2004, p. 7.
82. Larsson, A., Halvarsson, M., Ruppi, S. Microstructural changes in CVD-Al_2O_3 coated cutting tools during turning operations. *Surf. Coat. Technol.* 1999, 111, 191–198.
83. Bolt, H., Koch, F., Rodet, J.L., Karpov, D., Menzel, S. Al_2O_3 coatings deposited by filtered vacuum arc characterization of high temperature properties. *Surf. Coat. Technol.* 1999, 116–119, 956–962.
84. Hollek, H. *Surface Engineering: Science and Technology*. The Minerals, Metals and Materials Society, 1999, pp. 207–231.
85. Holleck, H. Properties of titanium based hard coatings. *J. Vac. Sci. Technol.* 1986, A4, 2661–2676.
86. Löffler, F.H.W. Systematic approach to improve the performance of PVD coatings for tool applications. *Surf. Coat. Technol.* 1994, 68/69, 729–740.
87. Yamamoto, K., Sato, T., Takahara, K., Hanaguri, K. Properties of (Ti, Cr, Al) N coatings with high Al content deposited by new plasma enhanced arc-cathode. *Surf. Coat. Technol.* 2003, 174/175, 620–626.
88. Anderson, W.P., Edwards, W.D., Zerner, M.C. Electronic spectra of hydrated ions of the first row transition-metals from semi-empirical calculations. *Inorg. Chem.* 1986, 25, 2728.
89. Zerner, M.C. Semi-empirical molecular orbital methods: reviews. In *Computational Chemistry II*, Eds., Lipkowitz, K.B., Boyd, D.B. VCH: New York, 1991, 313–366, chap. 8.
90. Nemoshkalenko, V.V., Kucherenko, Y. *The Computing Methods of Physics in Solid State Theory*; Naukova Dumka: Kiev, 1986, p. 128.
91. Seung-Hoon, Jhi., Jisoon, Ihm., Steven Louie, G., Cohen, M.L. Electronic mechanism of hardness enhancement in transition-metal carbonitrides. *Nature* 1999, 399, 132.
92. Kovalev, A.I., Barskaya, R.A., Wainstein, D.L. Effect of alloying on electronic structure of nickel aluminid. *Surf. Sci.* 2003, 35, 532–535.
93. Trent, E.M., Wright, P.K. *Metal Cutting,* 4th ed. Butterworth-Heinemann: Woburn, MA, 2000.
94. Trent, E.M., Suh, N.P. *Tribophysics*. Prentice-Hall: Englewood Cliffs, NJ, 1986.
95. Fox-Rabinovich, G.S., Kovalev, A.I., Weatherly, G.C. Tribology and the design of surface engineered materials for cutting tool applications. In *Modeling and Simulation for Material Selection and Mechanical Design,* Eds., Totten, G., Xie, L., Funatani, K., Marcel Dekker: New York, 2004, 301–382.
96. Fox-Rabinovich, G.S., Veldhuis, S.C., Scvortsov, V.N., Shuster, L.Sh., Dosbaeva, G.K., Migranov, M.S. Elastic and plastic work of indentation as a characteristic of cutting tool life with nitride PVD coatings. *Thin Solid Films* 2004, 469–470, 505–512.
97. Rass, I., Leyendecker, T., Feldehege, M., Erkens, G. TiAlN-Al_2O_3 PVD multi-layer for metal cutting operation. *Proc. 5th Eur. Conf. Adv. Mater. Proc. Appl.* Zeedijk 1997, pp. 23–27.
98. Donnet C., Erdemir, A. Historical developments and new trends in tribological and solid lubricant coatings. *Surf. Coat. Technol.* 2004, 180–181, 76–84.
99. Voevodin, A.A., O'Neull, J.P., Zabinsky, J.S. Nanocomposite tribological coatings for aerospace applications. *Surf. Coat. Technol.* 1999, 116–119, 36–45.
100. Voevodin, A.A., Zabinsky, J.S. Supertough wear-resistant coatings with "chameleon" surface adaptation. *Thin Solid Films* 2000, 370, 223–231.
101. John, P.J., Prasad, S.V., Voevodin, A.A., Zabinsky, J.S. Calcium sulfate as a high temperature solid lubricant. *Wear* 1998, 219, 2, 155–161.
102. Walck, S.D., Donley, M.S., Zabinsky, J.S., Dyhouse,V.J. Synthesis and characterization of TiC and TiCN coatings. *J. Mater. Sci.* 1994, 9, 236–239.
103. Walck, S.D., Zabinsky, McDevitt, N.T., Bultman, J.E. Characterization of air-annealed, pulsed laser deposited ZnO-WS_2 solid film lubricants by transmission electron microscopy. *Thin Solid Films* 1997, 305, 130–143.

104. Ivanova, V.S., Bushe, N.A., Gershman, I.S. Structure adaptation at friction as a process of self-organization. *J. Friction and Wear* 1997, 18, *1*, 74–79.
105. Derflinger, V., Brändle, H., Zimmermann, H. New hard/lubricant coating for dry machining. *Surf. Coat. Technol.* 1999, 113, 286–292.
106. Mayrhofer, P.H., Hovsepian, P.Eh., Mitterer, C., Münz, W.-D. Calorimetric evidence for frictional self-adaptation of TiAlN/VN superlattice coatings. *Surf. Coat. Technol.* 2004, 177–178, 341–347.
107. Lugscheider, E., Bärwulf, S., Barimani, C. Properties of tungsten and vanadium oxides deposited by MSIP-PVD process for self-lubricating applications. *Surf. Coat. Technol.* 1999, 120–121, 458–464.
108. Holubár, P., Jílek, M., Síma, M. Nanocomposite nc-TiAlSiN and nc-TiN-BN coatings: their applications on substrates made of cemented carbide and results of cutting tests. *Surf. Coat. Technol.* 1999, 120–121, 184–188.
109. Erdemir, A. A crystal-chemical approach to lubrications by solid oxides. *Tribol. Lett.* 2000, 8, 97–102.
110. Mayrhofer, P.H., Mitterer, C., Musil, J. Structure-property relationship in single- and dual-phase nanocrystalline hard coatings. *Surf. Coat. Technol.* 2003, 174–175, 725–731.
111. Fox-Rabinovich, G.S., Weatherly, G.C., Dodonov, A.I., Kovalev, A.I., Veldhuis, S.C., Shuster, L.S., Dosbaeva, G.K., Wainstein, D.L., Migranov, M.S. Nano-crystalline FAD (filtered arc deposited) TiAlN PVD coatings for high-speed machining application. *Surf. Coat. Technol.* 2004, 177–178, 800–811.
112. Martin, P., Netterfield, R., Bernard, A., Kinder, T. *Annual Technical Conference Proceedings of the Society of Vacuum Coaters*. Society of Vacuum Coaters, Boston, MA, 1993, p. 375
113. Shaw, M.S. *Metal Cutting Principles*. Oxford University Press: New York, 1996, p. 594.
114. Erkens, R., Cremer, T., Hamoudi, K.-D., Bouzakis, I., Mirisidis, S., Hadjiyiannis, G., Skordaris, A., Asimakopoulos, S., Kombogiannis, J., Anastopoulos, A., Efstathiou, K. Properties and performance of high aluminum containing (Ti,Al)N based supernitride coatings in innovative cutting applications. *Surf. Coat. Technol.* 2004, 177–178, 727–734.
115. Harris, S.C., Doyle, E.D., Vlasveld, A.C., Dolder, P.G. Dry cutting performance of partially filtered are deposited TiAlN coating with various metal nitride base coatings. *Surf. Coat. Technol.* 2001, 149–147, 305.
116. Karnakova, P., Veprek-Heijman, M.G., Zindulka, J., Bergmaier, O., Veprek, S. Superhard nc-TiN/a-BN and nc-TiN/a-TiB$_x$/a-BN coatings prepared by plasma CVD and PVD: a comparative study of their properties. *Surf. Coat. Technol.* 2003, 163–164, 149–153.
117. Pint, B.A., Leibowitz, J., DeVan, J.K. Characterization of the breakaway Al content in Alumina-forming alloys. *Oxid. Met.* 1999, 51, 45.
118. Prigogine, I.R. *From Being to Becoming*. W.H. Freeman and Company: San Francisco, CA, 1980, pp. 84–87.
119. Schulz, H., Dorr, J., Rass, I.J., Schulze, M., Leyenecker, T., Erkens, G. Performance of oxide PVD coatings in dry cutting operations. *Surf. Coat. Technol.* 2001, 146–147, 480–485.
120. Gershman, J.S., Bushe, N.A. Thin films and self-organization during friction under the current collection conditions. *Surf. Coat. Technol.* 2004, 186, *3*, 405–411.
121. Criscenzi, M.D., Lozzi, L., Piccozzi, P., Santicci, S. Electron energy-loss fine structure of a polycrystalline surface. *Phys. Rev.* 1989, B 39, *12*, 8409.
122. Palumbo, G., Erb, U., Aust, K.T. Interfacial geometry in simulated rolling and recrystallization structure. *Scr. Met.* 1990, 24, 2347.
123. Dickenscheid, W., Birringer, R., Gleiter, H. Diffusion in nanocrystalline metals and alloys — a status report. *Solid State Commun.* 1991, 79, 683.
124. Ovidko, I.A., Reizis, A.B. Grain-boundary dislocation climb and diffusion in nanocrystalline solids. *Phys. Solid State* 2001, 43, *1*, 35–38.
125. Horvath, J., Birringer, R., Gleiter, H. Diffusion in nanocrystalline material. *Solid State Commun.* 1987, 62, 319–322.
126. Ovidko, I.A., Reizis, A.B., Masumura R.A. Effects of transformations of grain boundary defects on diffusion in nanocrystalline. *Mater. Phys. Mech.* 2000, 1, 103–110.
127. Palatnik, L.S. *Pores in the Films*. Energoizdat: Moscow, 1982, p. 214.
128. Kostetskaya, N.B. Mechanism of deformation, fracture and wear particles formation during friction. *J. Friction and Wear* 1990, 11, *1*, 108–115.
129. Schiotz, J., Vegge, T., Di Tolla, F.D. Atomic scale simulations of the mechanical simulation of nanocrystal metals. *Phys. Rev.* 1999, B 60, 17, 11971–11983.

130. Perevezentsev, V.N. Grain-boundary diffusion and properties of nanostructural materials. *Tech. Phys.* 2001, 46, *11*, 1481.
131. Yoshida, N., Yasukawa, M., Muroga, T. Evolution of microstructure in nickel by low-energy deuterium ion irradiation. *J. Nucl. Mater.*, 1993, 205, 385–393.
132. Fischer, W.R. Properties of and heavy metal complexation by aqueous humic extracts. In *Bonner Bodenkundliche Abhandlungen*, Eds., Brümmer, G.W., Skowronek, A. Institut für Bodenkunde: Bonn, 1999, pp. 205–219.
133. Levchuk, D., Maier, H. *The Influence of Plasma Variables on the Characteristics of Superhard Nano-Composite Coatings*. Max-Planck institute fur Plasmaphysik: Munchen, 2004, p. 31.
134. Gao, W., Liu, Z., Li, Z., Gong, H. Nano-crystal alloy and alloy-oxide coating and their high temperature corrosion properties. *Int. J. Mod. Phys.: Appl. Phys.* 2002, 16, 1–2, 128–136.
135. Musil, J., Hruby, H. Superhard nanocomposite $Ti_{1x}Al_xN$ films prepared by magnetron sputtering. *Thin Solid Films* 2000, 365, *1*, 104–109.
136. Gershman, I.S., Bushe, N.A., Mironov, A.E., Nikiforov, V.A. Self-organization of secondary structures at friction. *J. Friction and Wear* 2003, 24, *3*, 323–334.
137. Prigogine, I., Kondepudi, D. *Modern Thermodynamics*. John Wiley & Sons: New York, 2002, pp. 84–86.
138. Fox-Rabinovich, G.S., Weatherly, G.C., Wilkinson, D.S., Kovalev, A.I. The role of chromium in protective alumina scale formation during the oxidation of ternary TiAlCr alloys in air. *Intermetallics* 2004, 12, 2, 165–180.
139. Ebeling, A., Engel, R., Feistel, A. *Physik der evolutionsprozesse*. Akademie-Verlag: Berlin, 1990, pp. 73–74.
140. Glansdorff, P., Prigogine, I. *Thermodynamic Theory of Structure, Stability and Fluctuations*. John Wiley & Sons: New York, 1971, p. 268.

10 Synergistic Alloying of Self-Adaptive Wear-Resistant Coatings

German S. Fox-Rabinovich, Kenji Yamamoto, and Anatoliy I. Kovalev

CONTENTS

10.1 Introduction........297
10.2 Principles of Proper Coatings' Selection for Specific Application of High-Performance Machining........299
10.2.1 Band Structure........302
10.2.2 Electrostatic Potential........303
10.2.3 Coated-Tool Properties........304
10.3 Synergistically Alloyed Self-Adaptive Wear-Resistant Coatings........306
10.3.1 Monolayered Self-Adaptive Coatings: Ternary Nitride Coatings........307
10.3.2 Nanoscale Multilayered Self-Adaptive Coatings: Quaternary Nitride Coatings........315
References........333

10.1 INTRODUCTION

Owing to the huge variety of PVD coatings available in the scientific and industrial market, a natural question arises: "What is the major trend in wear-resistant-coating development for future applications?" Based on the literature and experimental data presented in Chapter 9, we can assume that nanoscale structured coatings will dominate in the future. This includes nanocomposite, nanocrystalline, and nanolaminate coatings. These coatings have excellent hot hardness and oxidation stability. But as we have emphasized many times, friction is a complex phenomenon, and a comprehensive approach to the coating characterization and development is critical to solve a variety of problems associated with machining operations. A generic tribological approach is needed, and therefore, tribological characteristics of the coating, primarily their adaptability, have to be considered in more detail. To confirm this theory, a simple experiment was performed. The tool life of different coatings was compared under extreme conditions of dry, high-performance end milling of hardened H13 tool steel (HRC 50). Major state-of-the-art coatings were compared, including nanocomposite Al–Ti–N/Si_3N_4, nanocrystalline Al–Ti–N, as well as self-adaptive Al–Ti–Cr–N and Ti–Al–Cr–N/WN ones. All the coatings had the same cubic crystal structure and were based on Al-rich (around 65 at. %) Ti–Al–N compositions. The results are presented in Figure 10.1. The data presented shows that the coatings studied have quite different tool lives and wear behavior. All the coatings wear intensively during the running-in stage of wear, but once this stage is completed, the wear rate drops and transforms to the stable or post running-in stage. But the

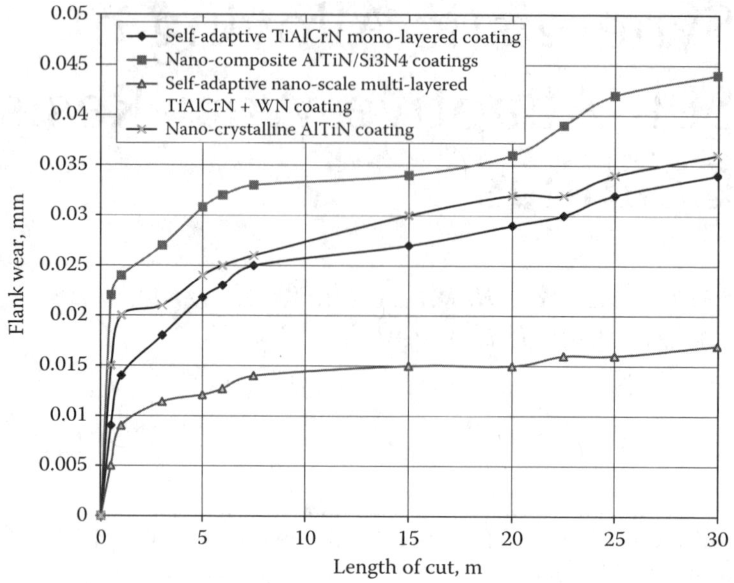

FIGURE 10.1 Comparative tool life data for the state-of-the-art coatings under high-performance dry machining conditions: Ball-nose E/M — 5-mm radius, 2 flutes (C-2MB, Mitsubishi); workpiece material — AISI H13 (HRC50); cutting conditions: speed — 200 m/min; feed — 0.06 mm/flute = 840 mm/min; axial depth — 5.0 mm; radial depth — 0.6 mm; cutting length, total — 30 m.

level of surface damage after the running-in stage of wear is very different for the coatings studied. The nanocomposite Al–Ti–N/Si_3N_4 coating had the most intensive surface damage during the running-in stage. This is due to extra high hardness and the relatively high coefficient of friction [1]. Nanocrystalline Al–Ti–N coating showed better wear behavior, but the coefficient of friction for this coating is also relatively high. The self-adaptive monolayered Al–Cr–Ti–N coating had less intensive surface damage during the running-in stage owing to improved lubricity at elevated temperatures (see the following text). The best tool life and wear behavior has the self-adaptive nanoscale, multilayered Ti–Al–Cr–N+WN coating. This coating has a less intensive wear rate during the running-in stage of wear, and the wear process stabilizes very quickly at the lowest level of surface damage. This wear behavior corresponds to the generic principles of friction control outlined in Chapter 1. Based on the data presented in Figure 10.1, it can be included that self-adaptive coatings have a great future potential.

This chapter focuses on the emerging self-adaptive coatings. The major feature of this coating is its ability to form protective and lubricious layers of tribo-oxides during periods of service. To perform this task, a proper metallurgical design of the coatings is critically needed. This can be achieved in a few ways: first, by the coating design; second, by optimizing the microstructure; and third, by means of proper composition selection. The first two methods have been considered in detail in Chapter 9. In this chapter, we will mainly focus on composition optimization. Synergistic alloying of the coatings will be considered in detail for specific applications such as high-performance turning and end-milling operations.

Synergistic alloying of the materials generically means that the elements in the material work synergistically if used in combination. They support each other and multiply the effectiveness to enhance formation of the phases that critically improve a specific property, or a set of properties, to achieve desired workability of the metallurgical system (alloy, coating, etc.).

The first step in this process is the proper selection of the compounds to be used as a coating material for specific applications.

10.2 PRINCIPLES OF PROPER COATINGS' SELECTION FOR SPECIFIC APPLICATION OF HIGH-PERFORMANCE MACHINING

The proper selection of hard-coating compositions are limited by the number of compound combinations available to form specific crystal structures (Figure 10.2, Table 10.1). The composition of coatings is related to interatomic bonding that is formed in the compound selected for the coating deposition. The majority of coatings used for cutting tools' applications are based on titanium nitride coatings with cubic crystal lattices. Improvement of properties and tool life of Ti–N-based coatings can be achieved by means of alloying. It is well known that the chemical bonding between the atoms involved is changed by the element added (Table 10.1 [3], Figure 10.2 [2,4]). A number of authors have already outlined the opportunities that exist for alloying of hard coatings to achieve the range of characteristics that are needed for demanding applications [3,4].

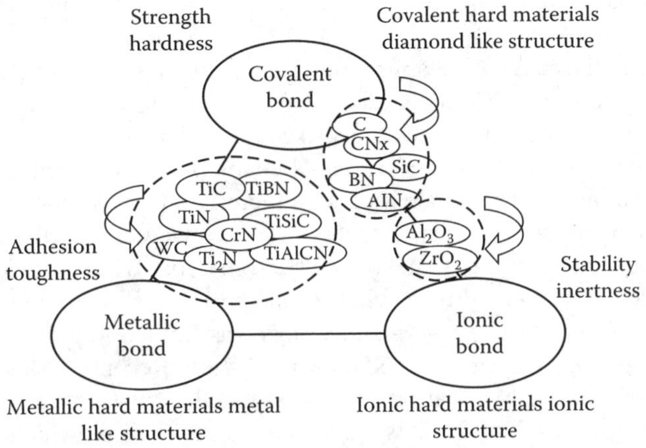

FIGURE 10.2 Hard materials for nanocomposite coatings in the bond triangle and changes in properties with the change in chemical bonding. (From Patscheider, J., Zehnder, T., Diserense, M. Structure-performance relations in nanocomposite coating. *Surf. Coat. Technol.* 2001, 146–147, *5*, 201–208; Holleck, H. Material selection for hard coatings. *J. Vac. Sci. Technol.* 1986, 4, 2661.)

TABLE 10.1
General Classification of (Binary) Hard Coatings

Type	Materials		Examples
Metallic	Borides	of transition metals	TiB_2, Mo_2B_5
	Carbides		TiC, VC, WC
	Nitrides		TiN, CrN, ZrN
Covalent	Borides	of Al, Si, and B	SiB_6, AlB_{12}
	Carbides		SiC, B_4C
	Nitrides		c-BN, AlN, Si_3N_4
	Carbon		Diamond
Ionic (ceramics)	Oxides	of Al, Zr, Ti, Be	Al_2O_3, ZrO_2, BeO, TiO_2

Source: Holleck, H. Material selection for hard coatings. *J. Vac. Sci. Technol.* 1986, 4, 2661.

Recently, new methods have become available to characterize the interatomic bonding within the hard-coating layer by means of detailed electronic structure calculations. Such calculations can be used to design new complex compound coatings for specific applications. Change in the composition of a coating because of alloying within the same crystal structure (as in the case of Ti–N cubic crystal lattice) gives us a chance to present the impact of interatomic bonding on the properties and wear behavior of a coating.

The latest trend in the last decade in hard-coating development for cutting tools has been in utilizing the variability of the design of the coating for specific domains of application. Thus, at present, the composition of a coating and, ultimately, its tool life strongly depends on the specific domain of application. The goal of this study is, therefore, to identify a relationship between the composition of the coating, as well as its interatomic bonding and the cutting-tool life under different domains of application.

To achieve this goal, Ti–Al–N and two compositions of Ti–Al–Cr–N coatings were deposited. The initial concept behind the metallurgical design of the Ti–Al–Cr–N coatings was to increase their hardness and oxidation resistance as compared to the Ti–Al–N coating. This concept is briefly described in the following text.

Ti–Al–N coating replaced Ti–N coating after its discovery by Münz [5]. Ti–Al–N coatings were characterized by improved high hardness and oxidation-resistance properties and were extremely beneficial for dry cutting applications. Ikeda et al. investigated the effect of Al content on the hardness and oxidation behavior of Ti–Al–N coatings with different Al contents [6]. They reported that the oxidation resistance increases with increase in the Al content. The hardness also increased as the Al content was increased to 60 at. %. However, a further Al-content increase yielded an undesirable change in the crystal structure from cubic B1 to hexagonal B4 that resulted in drastic hardness decrease. Owing to inevitable change in crystal structure and corresponding hardness decrease, the Al content in Ti–Al–N coatings was limited to around 60 at. % in spite of obvious merits such as improvement in oxidation resistance. Recently, Makino et al. calculated maximum solubility of cubic AlN in various transition metal nitrides and reported that CrN could dissolve in cubic AlN up to 77.2 at. %. This is much higher than Ti-N, which was calculated to be 65.3 at. % [7]. This result suggests that by introduction of Cr in the Ti–Al–N system, the cubic B1 structure can be stabilized at an Al content above 60 at. %. Several compositions of Ti–Al–Cr–N coatings were tested, and the result validated the ideas presented in preceding text. By substituting a part of Ti with Cr, the cubic B1 structure was stable up to an Al content of 73 at. %. As a result, these Ti–Al–Cr–N coatings had much higher oxidation resistance and improved hardness compared to conventional Ti–Al–N [8].

Hot pressed Ti–Cr–Al targets (100 mm diameter) with different compositions ($Ti_{0.5}Al_{0.5}$, $Ti_{0.25}Al_{0.65}Cr_{0.10}$, $Ti_{0.1}Al_{0.70}Cr_{0.20}$) were prepared. The deposition of these coatings was conducted in a laboratory-type arc coater (AIP SS002, Kobe Steel Ltd.) equipped with a plasma-enhanced cathode [8]. Samples were ultrasonically cleaned in an ethanol solvent and introduced into the chamber. Following the evacuation and preheating process, the samples were cleaned with Ar-ion bombardment. During the deposition process, only nitrogen gas was fed into the chamber at a pressure of 2.66 Pa and the arc current was typically set at 150 A. The substrate temperature was within the range of 400 to 500°C. The substrate bias was 30 V for Ti–Al–N and 150 V for the Ti–Al–Cr–N coatings. The thickness of the coating was approximately 3 μm for the film characterization and cutting test work. Ti–Al–N and the two Ti–Al–Cr–N coatings have a cubic B1-phase crystal structure. The lattice constants of Ti–Al–Cr–N coatings are slightly smaller than that of Ti–Al–N coatings, reflecting the smaller ionic radii of the Al and Cr atoms. The harnesses of $Ti_{0.5}Al_{0.5}N$, $Ti_{0.1}Al_{0.7}Cr_{0.2}N$, and $Ti_{0.25}Al_{0.65}Cr_{0.1}N$ were 28, 35, and 32 GPa, respectively, as measured by nanoindentation technique (see Figure 10.6 for details).

A theoretical study has been performed on the interatomic bonding of the Ti–N-based coatings. The energy band structure of the system considered has been calculated by means of the self-consistent density functional methods (ZINDO1). The calculations associated with this study were

TABLE 10.2
Cutting Data for Testing of the Self-Adaptive Coating under Conditions of High-Performance Machining

Type of Cutting Operation	Cutting Data					
	Speed (m/min)	Feed	Depth of Cut (mm)	Workpiece Material	Tooling	Tool Life Criteria
Turning	450	0.11 mm/rev	0.5	Steel 1040	Polished commercial indexable cutting inserts (SPG 422), CC H1P grade (Sandvik)	Flak wear = 0.3 mm
End milling	220 = 7000 rpm	0.06 mm/flute = 840 mm/min	Axial depth: 5.0 mm; radial depth: 0.6 mm	Tool steel H13 (HRC50)	The carbide ball-nose end mills (2 flutes, 10-mm diameter)	Comparative flank wear after 30 and 250 m length of cut

performed using the "ÄrgusLab" and "HyperChem 7.5" software (www.planaria-software.com; www.hyper.com). The ZINDO method is the most suitable semiempirical method to determine the structure and energy of the molecules or crystals for the first- and second-transition row metals. These metals have a wide range of valences, oxidation states, and spin multiplicities, and possess unusual bonding situations. In addition, the nondirectional nature of the metallic bonding is less amenable to a ball-and-spring interpretation [9–10]. These factors determine that the molecular orbital calculations on metals yield less reliable results compared to the results obtained from organic compounds. Nevertheless, these quantum mechanical calculations are very useful to predict the wide spectrum of interatomic bonding and interpretation of the physical properties of the metals and their compounds. These properties include the dipole moment, polarizability, total electron density, total spin density, electrostatic potential, heat of formation, orbital energy levels, vibrational normal modes and frequencies, strength, and plasticity.

The cutting-tool life was tested under conditions of turning and end-mill cutting (Table 10.2). The coated cemented carbide cutting inserts (SPG 422; Sandvik grade H1P) with mirror-polished top surfaces were used for the turning tests. End-milling tests were performed during dry machining of tool steel H13 (HRC50, Table 10.2). The flank wear of the milling cutters was compared after a 30-m linear length was cut.

Nanoindentation testing of a surface-modified layer has been performed. Microhardness and the work of elastic–plastic deformation were evaluated from load vs. displacement data measured using a computer-controlled nanoindentation tester MTI-3M [11]. A Berkovich-type diamond indenter was used. The indentation depth was measured electronically and the indentation curves were evaluated. Thirty measurements were performed on each sample. Standard statistical methods were used to calculate the average indentation curves. The scatter of microhardness measurements was around 2%. The experimental indentation curves closest to the calculated average ones were used for the analysis. The total work of elastic and plastic deformation during nanoindentation testing of the modified surface layers was calculated using the loading and unloading indentation curves. To measure the plastic and total work for each specimen, the loading and unloading indentation curves were approximated by a third-order polynomial. The parameter of determination was $R^2 = 1$. The total work of elastic and plastic deformation was estimated by the integration of the loading curve [11–12]. The integration step is equal to the step of the current loading increase, which was 200 mN in our case. The range of loads was from 0 to 2500 mN. The work of plastic deformation was determined as the difference between the total work of elastic and plastic deformation and the work of elastic deformation. The latter parameter was determined by integrating the unloading curve. Using the data obtained, the microhardness dissipation parameter [13–16] was calculated as the ratio of the plastic work to the total work of deformation during indentation.

10.2.1 Band Structure

The TiN crystal has a simple cubic lattice (space symmetry group O_h^5). The cubic lattice constant is a = 0.422 nm, and the interatomic distance is d = 0.211 nm. The nearest surrounding Ti atoms form a tetrahedron of 4N atoms. The cubic unit cells contain 8 or 27 atoms. Several Ti atoms in the TiN lattice have been replaced by Al and Cr atoms.

The electronic band structures of TiAlN, TiCrN, and TiAlCrN nitrides were investigated. Figure 10.3 demonstrates the energy diagram of the calculated molecular orbitals (MO) of (1) TiN, (2) TiAlN, (3) TiCrN, and (4) TiAlCrN. The resulting set of levels includes a set of valence and conduction bands. The electronic band structure of the TiN crystal is well known and described in detail in many sources. Therefore, we do not present a detailed discussion of our calculated results (Figure 10.3). However, they are in agreement with known results that were obtained by means of other calculation methods [17]. Our calculations (Figure 10.3a) show that the total density of the states of TiN comprises approximately three fine structures. The lowest energy structure is centered on –16.6 to –7.1 eV and is composed mainly of electrons from the 2s bands of nitrogen atoms and the 3p and 3d bands of titanium. The middle energy structure is centered on –2.7 eV and primarily comprises of electrons from the 2p (valence) bands of the nitrogen atoms and electrons from the 3d (valence) and 3p bands of the titanium atoms. The narrow highest energy structure is centered on 1.2 to 4.2 eV and is primarily composed of electrons from the 3d bands of the titanium atoms.

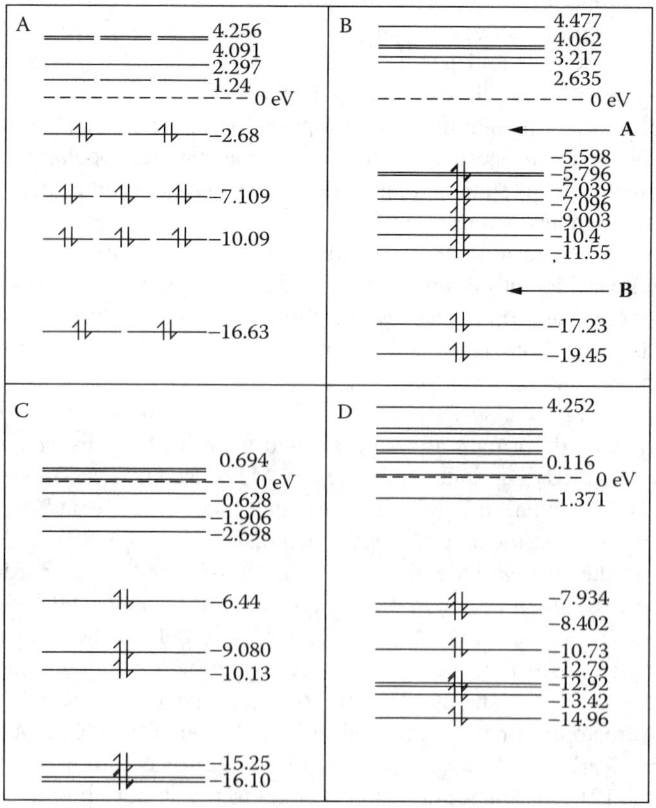

FIGURE 10.3 The energy diagram of calculated molecular orbitals (MO) of (a) TiN, (b) Ti–Al–N, (c) TiCrN, and (d) Ti–Al–Cr–N.

If the valence band is full, the conduction band has to accommodate 1N electrons. Three of the four valence electrons from Ti complete the filling of the valence band. The one remaining electron of the Ti atom populates the conduction band, thus making TiN a metallic conductor.

The alloying of TiN by Al transforms the electronic band structure of the complex nitride TiAlN (Figure 10.3b). The aluminum addition strongly increases the energy gap between the conduction and valence bands (A in Figure 10.3b). The metallic character of the interatomic bonds is thus weakened in the TiAlN lattice. As a result, this nitride obtains properties similar to that of a semiconductor. The valence band is divided by the energy-empty zone (B in Figure 10.3b). It is known that the thickness of this zone growth increases with the chemical bonds' polarity. The thickness of this zone in TiAlN is higher compared to that of Ti–N. Growth in the thickness of this zone in TiAlN and the electron state localizing within the energies –5.5 to –11.5 eV means that the ion covalent bonds are increased in the short-range interatomic interactions.

On the contrary, the chromium addition to the complex TiCrN nitride increases the density of states near to the Fermi level and decreases the empty zone B in the valence band (see Figure 10.3c). Chromium significantly enhances the metallic character of the interatomic attraction in the nitride lattice. This is due to chromium atoms having five electrons on the 3d level as compared to the two electrons on the 3d Ti level. These common features are kept in the structure of TiAlCrN nitride (Figure 10.3d). The metal–metal interaction is considered predominant when compared to the nitrogen atoms as electron donors. The empty zone of the valence band disappears and this decreases the covalent component of the interatomic attraction. The shift in the lowest energy structure of the MO diagram to the Fermi level and the constraint in the levels reflect the high hybridization of the d π-electronic states. These states provide the necessary intermetallic atom interaction and enhance the metallic properties of the nitride.

The type of interatomic bonds in the crystal lattice of the transition-metal carbides and nitrides strongly affects their basic mechanical properties such as hardness and plasticity [3,18]. Strong ion covalent types of bonds between the metallic and nitrogen atoms of the nitride result in a high hardness. The d π-electrons bring together the metallic bonds within the crystal lattice of the compound and ensure its metallic properties. The dislocation mobility significantly affects the material plasticity and relates to Peierls and stacking fault energies. A correlation between these factors and the electron concentration within the contact zone and the π-level electron density is then determined. With a decrease in the covalent component of the interatomic bonds, the Peierls energy U_p, and the corresponding Peierls barrier of plastic deformation, σp, are decreased:

$$\sigma_p = \frac{2\pi Up}{ba}$$

where a is the lattice parameter and b is the Burgers vector.

The Peierls stresses in the transition-metal nitrides are very high. The strengths of these compounds are tied more by the difficulty to nucleate and move the dislocation through the crystal lattice than by the difficulty associated with moving the dislocation through microstructural obstacles. Also, dropping the σp leads to an increase in the plasticity of the compound under analysis [19].

10.2.2 Electrostatic Potential

The electrostatic potential is the potential energy felt by a positive "test" charge at a particular point in space. The distribution of the electrostatic potential determines the packing of the units in the crystal lattice. Figure 10.4 presents electrostatic potential maps on the ground-state surfaces for the 27 atom clusters of the TiN-based nitrides. High polar interatomic bonds are typical of TiN and TiCrN nitrides, which is distinct in comparison to the TiAlN and TiAlCrN where the aluminum atoms are poor in electrons. The chromium-containing nitride TiCrN (Figure 10.4c) has a high concentration of valence electrons as compared to TiN. The maps also show that for each TiN or

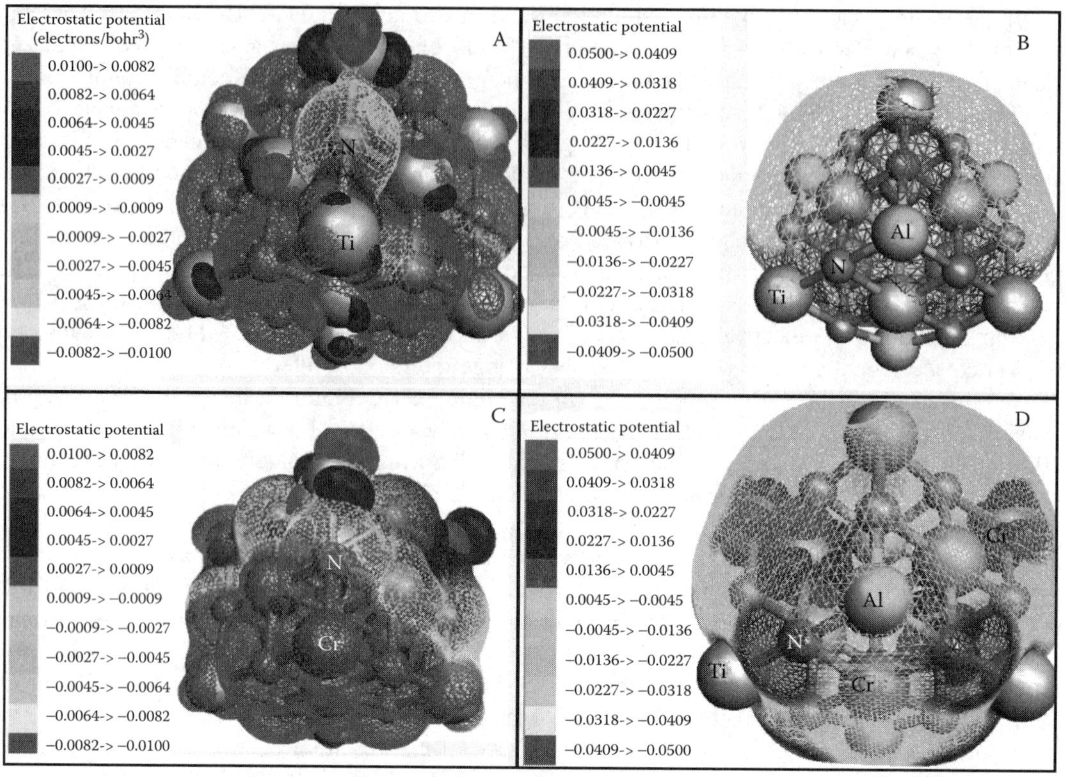

FIGURE 10.4 Electrostatic potential maps on the ground-state surfaces for atom clusters of the TiN-based nitrides.

TiCrN cluster, a complementary pattern of positive and negative potentials is generated so that stabilizing interfaces will be created in the crystal lattices of these high-strength nitrides. The general topology of the positive and negative isosurfaces in Figure 10.4a and Figure 10.4c makes it clear why these units pack face to face as they do in the crystal, with regions of negative potential in one polyhedron overlapping with regions of positive potential in a neighboring polyhedron. These interfaces in the TiAlN and TiAlCrN nitrides are characterized by small long-range attractions. The plasticity of the Cr-containing nitrides is somewhat higher than that of the TiN-based nitrides without Cr.

10.2.3 Coated-Tool Properties

The wear resistance of the coated cutting tools was investigated under different high-performance machining applications. Cutting data is shown in Table 10.2. Tool life data collected during turning is presented in Figure 10.5. Under conditions of continuous cutting (turning), crater (chemical) wear controls the tool life [20]. Three types of coating compositions, Ti–Al–N and two types of Ti–Al–Cr–N (see the experimental section for details) were compared. The Ti–Al–N-coated inserts exhibit better wear resistance under high-speed turning conditions but the inserts with the Ti–Al–Cr–N showed a lower tool life.

However, for interrupted cutting conditions (end milling, Table 10.2), the coatings richest in Al (Ti10–Al70–Cr20) N (Figure 10.6) display a tool life almost two times better than the Ti–Al–N coating. This can be explained as follows. Under end-milling conditions, the cutting tool undergoes intensive adhesive-fatigue wear accompanied by oxidation attack [20–22]. A hard coating working under conditions of adhesive-fatigue wear should have some reserve in plasticity to dissipate part

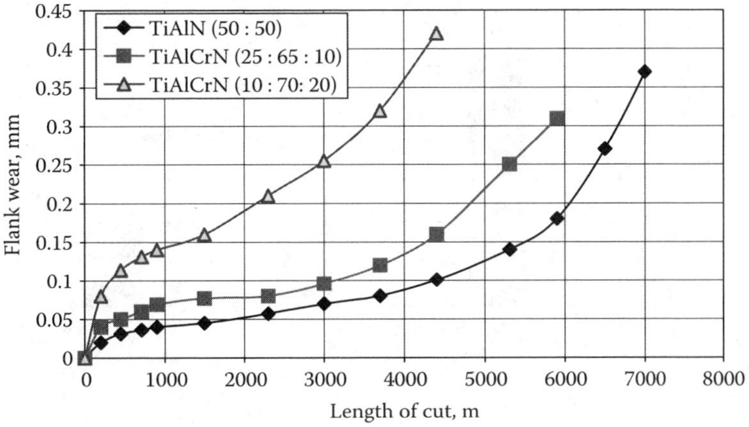

FIGURE 10.5 Tool life data of Ti–Al–N and Ti–Al–Cr–N coatings during turning of 1040 steel.

of the energy generated during friction [11]. As published earlier [10], the microhardness dissipation parameter (MDP) of the coating measured during nanoindentation can be used to characterize this property of a coating.

Figure 10.6 presents the flank wear of the end mills with Ti–Al and Ti–Al–Cr–N coatings plotted vs. MDP values measured by a nanoindentation method. The microhardness of the coating is also presented. Figure 10.6 shows that the wear rate for an end mill depends on the plasticity of the coating. In this case, a coating with better plasticity shows an improved tool life. The Ti–Al–Cr–N coatings have better oxidation stability at high temperatures as compared to the Ti–Al–N coating (see Figure 10.7). This also extends the tool life under end-milling conditions.

The ESP map aids in understanding the differences in tribo-chemical activity of the investigated coatings. If the ESP is negative, this is a region of stability for the positive test charge. Conversely, if the ESP is positive, this is a region of relative instability for the positive test charge. Thus, an ESP-mapped density surface can be used to show regions of a molecule that might be more favorable to nucleophilic or electrophilic attack, thus making these types of surfaces useful for qualitative interpretations of chemical reactivity. As shown in Figure 10.4a to Figure 10.4c, Al addition to titanium nitrides dramatically decreases the heterogeneity of the molecular electrostatic potential. It means that the addition of Al to the Ti nitride significantly reduces the inhomogeneity of the

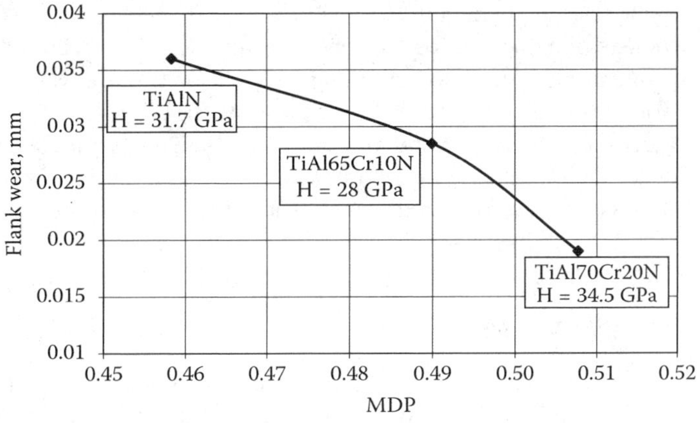

FIGURE 10.6 Flank wear of the coated end-milling cutters vs. microhardness and plasticity index of Ti–Al–N and Ti–Al–Cr–N coatings.

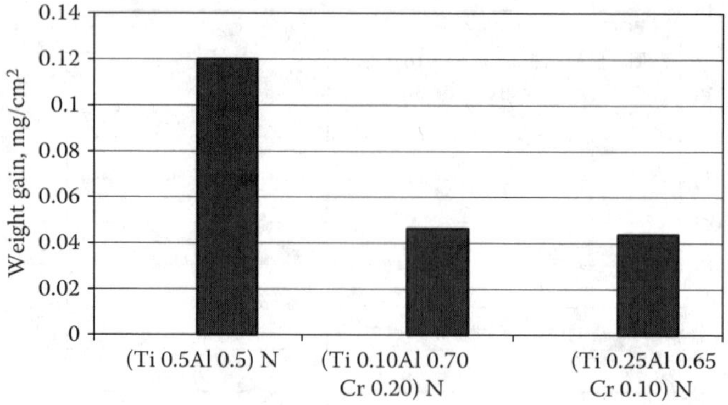

(a)

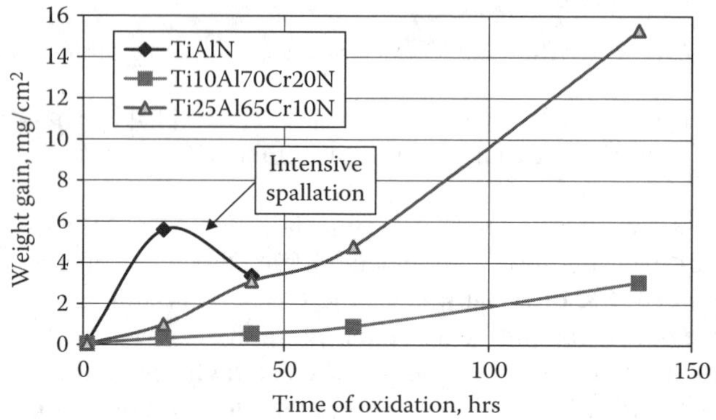

(b)

FIGURE 10.7 Weight gain data of the Ti–Al–N and Ti–Al–Cr–N PVD coatings deposited on the Ti–Al coupons after: (a) short-term oxidation in air at 900°C during 1 h; (b) long-term oxidation during 140 h. (From Fox-Rabinovich, G.S., Yamamoto, K., Veldhuis, S.C., Kovalev, A.I., Dosbaeva, G.K. Tribological adaptability of TiAlCrN PVD coatings under high performance dry machining. *Surf. Coat. Technol.* 2005, 200, 1804–1813. With permission.)

electron density distribution within the volume of the molecule and, as a result, drops the chemical reactivity of the Ti–Al–N coating. That is why this coating has a high tool life during turning when the chemical wear mechanism is dominant [20–22].

As outlined previously, the addition of Cr content in the Ti–Al–N coating adds an enhanced metallic character of interatomic attraction resulting in a plasticity improvement that leads to an increase in tool life for end-milling applications. Conversely, the Al-content Ti–Al–Cr–N coatings have sufficiently strong polar interatomic bond to resist, with a high degree of efficiency, the oxidation attack experienced under the elevated temperatures associated with cutting. We can conclude that this type of composition is tailored for specific application of high-performance end milling.

10.3 SYNERGISTICALLY ALLOYED SELF-ADAPTIVE WEAR-RESISTANT COATINGS

High-performance dry machining is one of the major trends in modern manufacturing. This is very topical for machining, particularly end milling, of hardened tool steels for die and mold applications. This advanced technology application results in a significant productivity increase, labor cost savings, and also solves environmental issues during the manufacturing process. High-performance dry machining generates severe cutting conditions associated with high temperature and stress within the cutting zone. In this application, the use of advanced coated tools is critical to realizing the benefits of high-performance machining. As emphasized in the literature, high-performance coatings are needed to withstand the harsh environment associated with this specific application [23]. The tool life data presented in Figure 10.1 show that the adaptability of the coatings, i.e., ability to generate protective and lubricious oxide tribo-films, controls the tool life under these severe wear conditions. These films play a vital role as high-temperature protective and lubricating layers. If stable tribo-films are formed on the surface during the initial stage of friction, the wear process will achieve a higher degree of stability without experiencing intensive surface damage over a longer period of time.

As it was shown in Chapter 9, two major types of oxygen-containing tribo-films are formed during high-speed machining. The first type is tribo-ceramics with high thermodynamic stability, and the second type is surface film with enhanced lubricity. The generation of these two types of tribo-films during high-performance machining is the goal of synergistic alloying of wear-resistant coatings.

The elements in the synergistically alloyed coating working under severe conditions of high-speed machining should enhance the formation of both types of oxide tribo-films during cutting. To perform this task, the synergistically alloyed coating should enhance the formation of protective oxide tribo-films with high thermodynamic stability (see Chapter 11 for details) but, on the other hand, drop friction forces by means of lubricity improvement at the cutting tool–workpiece interface.

The aluminum-rich Ti–Al–Cr–N-based coatings synergistically alloyed by several elements have shown promising results for high-performance machining applications because of the combination of high hardness and oxidation stability [8,24]. Ti–Al–Cr–Y–N and Ti–Al–N/Cr–N superlattice coatings show excellent oxidation resistance [22,25–26], and the former demonstrates promising wear resistance at elevated temperatures [26]. The addition of Y drastically reduces the grain size [26] and leads to superhard coatings.

10.3.1 Monolayered Self-Adaptive Coatings: Ternary Nitride Coatings

Ti–Al–Cr–N coatings having different atomic compositions were deposited by a commercial-type arc ion plating apparatus (Kobe Steel Ltd., AIP-SS002) equipped with a magnetically steered arc source. Ball-nose end mills and γ-TiAl intermetallic coupons were used as the substrates. Interrupted oxidation tests were performed at 900°C in air for 1 h and 140 h.

The coefficient of friction vs. temperature was determined with the aid of a specially designed apparatus described in Chapter 5 [22]. A rotating sample of the coated substrate was placed between two polished specimens made of tool steel H13 (hardness HRC 50). To simulate tool friction conditions, the specimens were heated by resistive heating in the temperature range of 150 to 1000°C. A standard force of 2400 N generated plastic strain in the contact zone. The friction parameter value was determined as the ratio of the shear strength induced by the adhesion bonds between the tool and the workpiece to the normal contact stress developed on the contact surface at the test temperatures. Three tests were performed for each kind of coating and the scatter for the friction parameter measurements was found to be around 5%.

Cutting tests have been performed using a ball-nose end mill under severe conditions of dry high-performance machining of the hardened H13 tool steel (Table 10.2). Surface morphology of the oxidized and worn tool samples, as well as the chips formed during cutting, were studied using an SEM. The composition of the oxide films formed during the oxidation tests was also investigated using EDS.

To understand the physical mechanism associated with phenomena that occur at the cutting tool–workpiece interface, the chemical composition of the tribo-films were studied by means of Auger electron spectroscopy (AES). The chemical bonds of Cr within the layer of tribo-films were investigated by means of photoelectron spectroscopy (XPS) using a VG ESCALAB-MK2 spectrometer. Scanning Auger spectroscopy was used to analyze the composition on the worn surface of the end mills. Several 15 × 15 μm sectors were chosen for analysis. A TV sweep-speed for the primary electrons with a 2000-Å beam diameter was used. The Auger signal was recorded in the mode $CRR = 2V$ at a speed of 2.1 eV/sec and primary electrons energy $E = 2500$ eV. The XPS studies were carried out using monochromated Al Kα radiation with an energy of 1486.6 eV. The hemispherical electron analyzer was operated with a B3 slit at constant pass energy of 20 eV. Under these conditions, the full width at half-maximum (FWHM) of the Ag $3d_{5/2}$ line recorded from a clear Ag standard was 1.0 eV. The binding energy of the Ag $3d_{5/2}$ line was 368.4 eV. The spectrum acquisition parameters were: $\Delta E = 0.05$ eV, 100 msec/channel, 100 scans. Curve fitting of the photoelectron spectra was performed utilizing a nonlinear least-squares algorithm with Shirley background subtraction and a Gaussian–Lorentzian convolution function [25].

The oxidation stability of the hard Ti–Al–Cr–N coatings in Reference 8 has been studied. The oxidation of the Ti–Al coupons with the Ti–Al–Cr–N coatings has been performed at 900°C in air. Weight gain data is presented in Figure 10.7 after the short-term isothermal oxidation (1 h) and after long-term isothermal interrupted oxidation (140 h). Figure 10.7 shows that the weight gain of the two Ti–Al–Cr–N coatings under analysis in this study is similar at the initial stages of oxidation (after 1 h), but the coating with the higher Al content (70%) shows much better oxidation stability during long-term oxidation (Figure 10.7b). Surface morphology of the oxidized samples after short-term oxidation is presented in Figure 10.8. The oxide scales that form on the surface of the $(Ti_{0.10}Cr_{0.20}Al_{0.70})$ N coating are more protective compared to the $(Ti_{0.25}Cr0_{.1}Al_{0.65})$ N ones. The EDS spectra shows that only a minor amount of rutile scales are formed on the surface of the $(Ti_{0.10}Cr_{0.20}Al_{0.70})$ N coatings (Figure 10.8c), but less protective films that are richer in rutile oxide films are formed on the surface of $(Ti_{0.25}Cr_{0.1}Al_{0.65})$ N coatings (Figure 10.8b). This results in the better oxidation stability of the $(Ti_{0.10}Cr_{0.20}Al_{0.70})$ N coating after long-term oxidation (Figure 10.7b).

The wear resistance of the coated cutting tools was investigated under high-speed dry milling conditions. The best tool life was found in the coating richest in aluminum $(Ti_{0.10}Cr_{0.20}Al_{0.70})$ N (Figure 10.6). This coating has tool life twice as long as the Ti–Al–N coating. This coating also has the best oxidation stability at high temperature (Figure 10.7).

This result corresponds to the coefficient of friction values vs. temperature in contact with the H13 tool steel (HRC 50) at the actual temperature range of cutting (around 900°C, Figure 10.9).

SEM images of the worn end mills with different coatings (Figure 10.10) show that a build-up edge forms under these cutting conditions. The adherence of the workpiece material is more intensive on the rake as well as the flank surfaces of the end-mill cutters with the Ti–Al–N coatings (Figure 10.10a to Figure 10.10c) as compared to the $(Ti_{0.10}Cr_{0.20}Al_{0.70})$ N coatings (Figure 10.10b to Figure 10.10d).

The shapes of the chips formed during end milling are shown in Figure 10.11. There is a minor difference in the shape of the chips for the end-milling cutters with Ti–Al–N and $(Ti_{0.10}Cr_{0.20}Al_{0.70})$ N coatings at the initial stage of cutting (Figure 10.11a and Figure 10.11b), but when the length of cut is long, approaching 30 m, the nature of the chips changes. The chips from the end mill with Ti–Al–N coating are dull and have a large curl diameter (Figure 10.11c) and the undersurface of the chips are relatively rough (Figure 10.11e). In contrast, the chips from the cutter with the $(Ti_{0.10}Cr_{0.20}Al_{0.70})$ N coatings are more intensively curled into a spiral (Figure 10.11d) and the undersurface of these chips is smoother (Figure 10.11f).

Cross sections of the chips are shown in Figure 10.12. The microstructure of the chips strongly depends on the type of coating being applied. The chips that are formed on the surface of the end-milling cutters with the Ti–Al–N coating have a wide contact zone of secondary hardening (Zone 1) [20,27]. The intensive secondary hardening of the contact area leads to initial crack formation (Figure 10.12a to Figure 10.12c). In the case of the Ti–Al–N coating, a severe seizure was observed at the end mill–chip interface. This corresponds to the morphology of the rake surface experiencing intensive sticking of the workpiece material (Figure 10.11a).

The application of the $(Ti_{0.10}Cr_{0.20}Al_{0.70})$ N coatings was found to critically change the microstructure of the chips (Figure 10.12b to Figure 10.12d). The width of the secondary hardened zone is much narrower than the one that corresponds to the image shown in Figure 10.12c. The smaller size of the sticking zone is caused by the higher temperature stability of the coating. As published

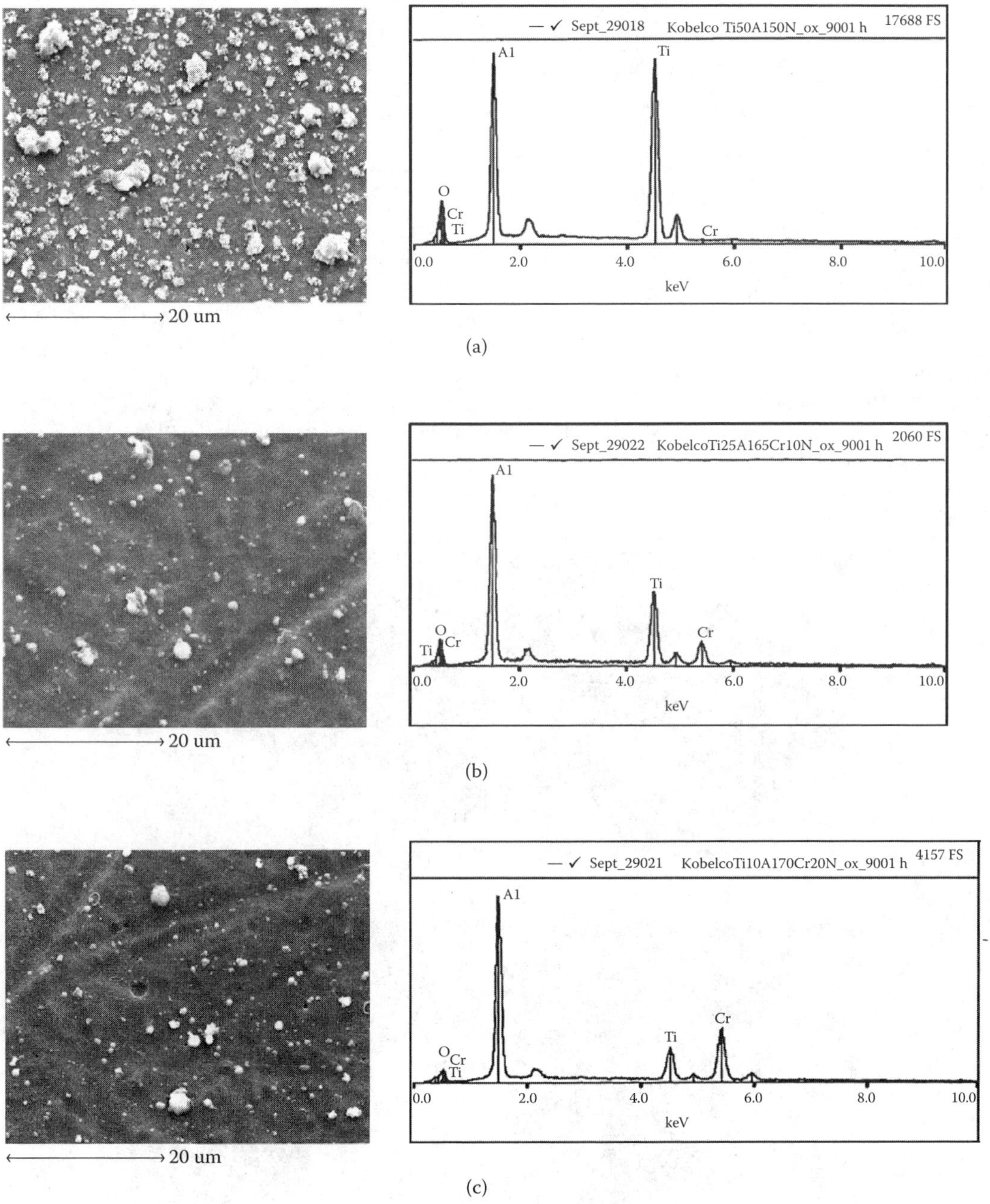

FIGURE 10.8 SEM micrographs of the surface morphology together with EDS spectra showing the relative abundance of Al, Ti, and Cr in the oxide films: (a) $Ti_{0.50}Al_{0.50}N$; (b) $Ti_{0.25}Cr_{0.10}Al_{0.65}$ N; (c) $Ti_{0.10}Cr_{0.20}Al_{0.70}$ N after short-term oxidation in air during 1 h. (From Fox-Rabinovich, G.S., Yamamoto, K., Veldhuis, S.C., Kovalev, A.I., Dosbaeva, G.K. Tribological adaptability of TiAlCrN PVD coatings under high performance dry machining. *Surf. Coat. Technol.* 2005, 200, 1804–1813. With permission.)

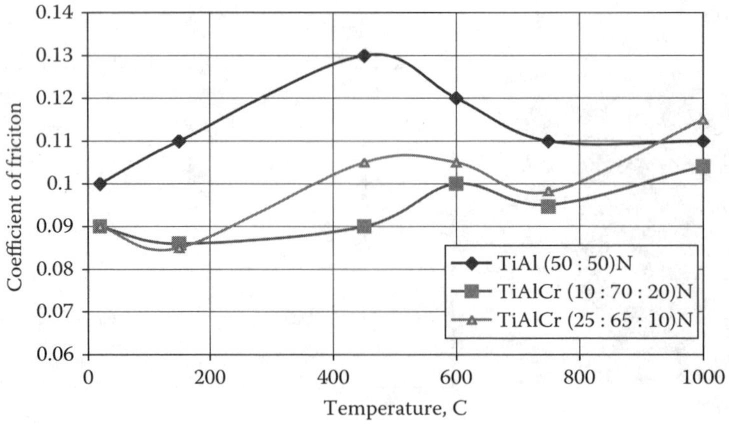

FIGURE 10.9 Coefficient of friction vs. temperature for Ti–Al–N and Ti–Al–Cr–N coatings in contact with H13 steel (HRC 50). (From Fox-Rabinovich, G.S., Yamamoto, K., Veldhuis, S.C., Kovalev, A.I., Dosbaeva, G.K. Tribological adaptability of TiAlCrN PVD coatings under high performance dry machining. *Surf. Coat. Technol.* 2005, 200, 1804–1813. With permission.)

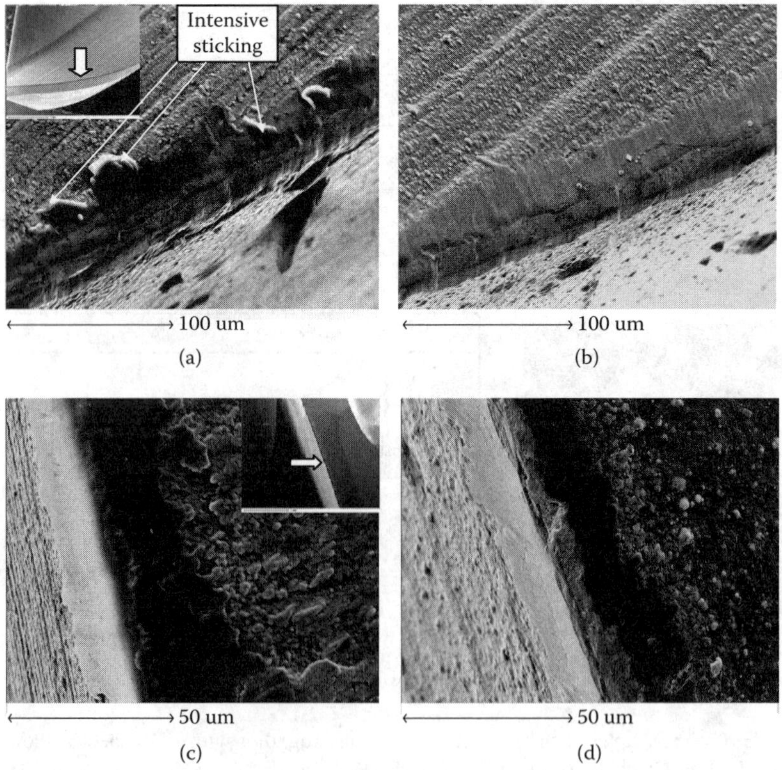

FIGURE 10.10 SEM images of the worn ball-nose carbide end mills; length of cut: 30 m; (a)–(b) rake surface of the mills; (c)–(d) flank surface of the mills; (a)–(c) milling cutters with $Ti_{0.50}Al_{0.50}N$ coating; (b)–(d) milling cutters with $Ti_{0.10}Cr_{0.20}Al_{0.70}N$ coating. (From Fox-Rabinovich, G.S., Yamamoto, K., Veldhuis, S.C., Kovalev, A.I., Dosbaeva, G.K. Tribological adaptability of TiAlCrN PVD coatings under high performance dry machining. *Surf. Coat. Technol.* 2005, 200, 1809–1813. With permission.)

FIGURE 10.11 Types of chips for end-mill cutters with (a, c, e) $Ti_{0.50}Al_{0.50}N$ coatings; (b, d, f) $Ti_{0.10}Cr_{0.20}Al_{0.70}$ N coatings. (a)–(b) length of cut: 0.9 m; (c)–(f) length of cut: 30 m; (a)–(d)-chip shape; (e)–(f) undersurface morphology. (From Fox-Rabinovich, G.S., Yamamoto, K., Veldhuis, S.C., Kovalev, A.I., Dosbaeva, G.K. Tribological adaptability of TiAlCrN PVD coatings under high performance dry machining. *Surf. Coat. Technol.* 2005, 200, 1804–1813. With permission.)

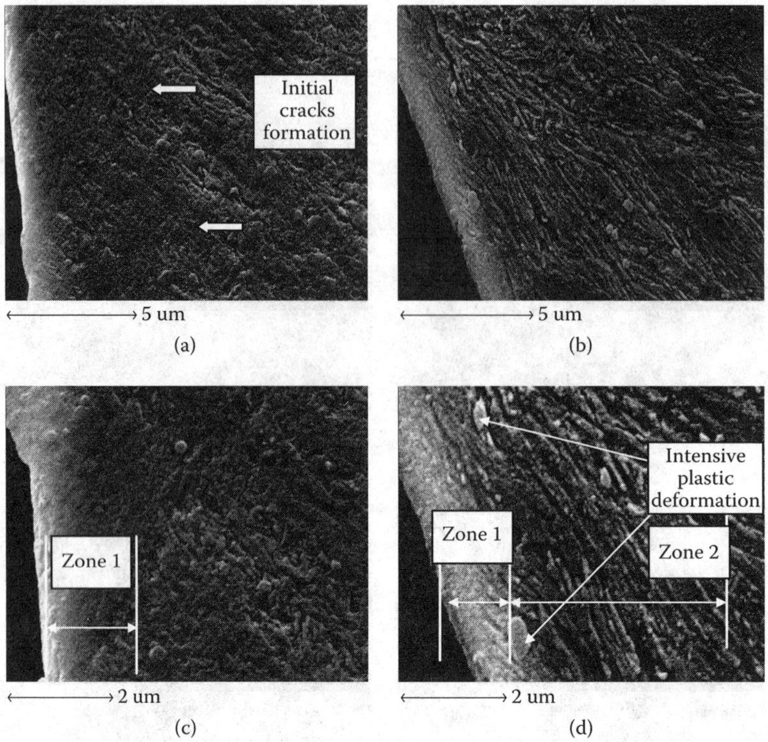

FIGURE 10.12 SEM images of the chips cross-sections for (a)–(c) $Ti_{0.50}Al_{0.50}N$ and (b)–(d) $Ti_{0.10}Cr_{0.20}Al_{0.70}$ N coating after end-mill cutting (length of cut: 30 m). Zone 1 — the contact zone of secondary hardening; Zone 2 — extended deformation zone. (From Fox-Rabinovich, G.S., Yamamoto, K., Veldhuis, S.C., Kovalev, A.I., Dosbaeva, G.K. Tribological adaptability of TiAlCrN PVD coatings under high performance dry machining. *Surf. Coat. Technol.* 2005, 200, 1804–1813. With permission.)

previously, the coating has a high hardness value at elevated temperatures [8]. When a chip slides along the rake surface of the end mill, the curved flow lines are formed because of friction. An extended deformation zone (Zone 2) [20,27] could be observed close to the contact area of the chip. The plastic deformation within this zone is so intensive that even the carbides that are present in the structure of the H13 steel are involved in this process. This means that the actual tribological characteristics of the interface have been significantly improved.

Figure 10.13 shows the Auger spectrum of the worn surface of the end mill with the Ti–Al–N coating, and Figure 10.14 exhibits the Auger lines AES spectra of the end mill with the ($Ti_{0.10}Cr_{0.20}Al_{0.70}$) N coating for the following zones of the surface: (1) far from the wear zone, (2) near the wear zone, and (3) in the wear zone. Figure 10.14d to Figure 10.14f present detailed AES spectra of aluminum (Al LMM Auger lines) associated with the (a–c) zones, respectively.

Figure 10.14a to Figure 10.14c show that the intensity of the oxygen and carbon peaks increase when compared to titanium and nitrogen, as we move toward the wear zone. At the same time, the peak of aluminum shifts down to a lower level of energy (Figure 10.14d to Figure 10.14e). It is known that the Auger line of pure aluminum (Al LMM) corresponds to 68.0 eV and it shifts to 64.5 eV for the Ti–Al–Cr–N coating (Figure 10.14d). This is known as a "chemical shift." During the tribo-oxidation process, the level of ionization grows within the layer of the compound formed on the surface. This leads to a "chemical shift" increase (Figure 10.14e to Figure 10.14f). Combining this data with the data presented in Figure 10.14f as well as with the higher intensity of the oxygen line, we can conclude that an aluminum–oxygen compound is being formed during friction, most probably Al_2O_3. Comparative Auger analysis of the worn surfaces shows (see Figures 10.13 and

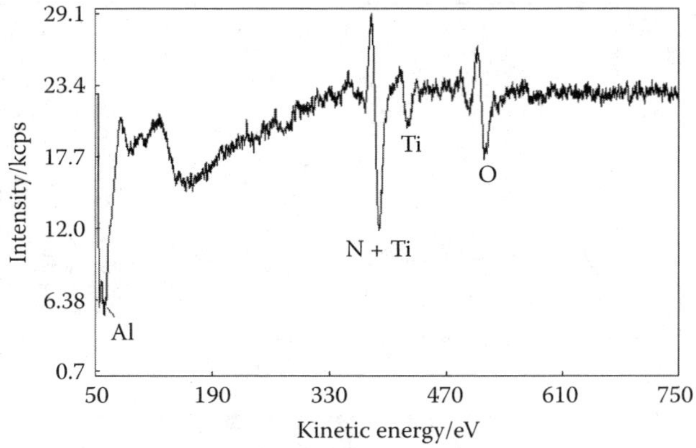

FIGURE 10.13 AES spectra of the end-mill worn surface with Ti–Al–N coatings in the wear zone. Length of cut: 30 m. (From Fox-Rabinovich, G.S., Yamamoto, K., Veldhuis, S.C., Kovalev, A.I., Dosbaeva, G.K. Tribological adaptability of TiAlCrN PVD coatings under high performance dry machining. *Surf. Coat. Technol.* 2005, 200, 1804–1813. With permission.)

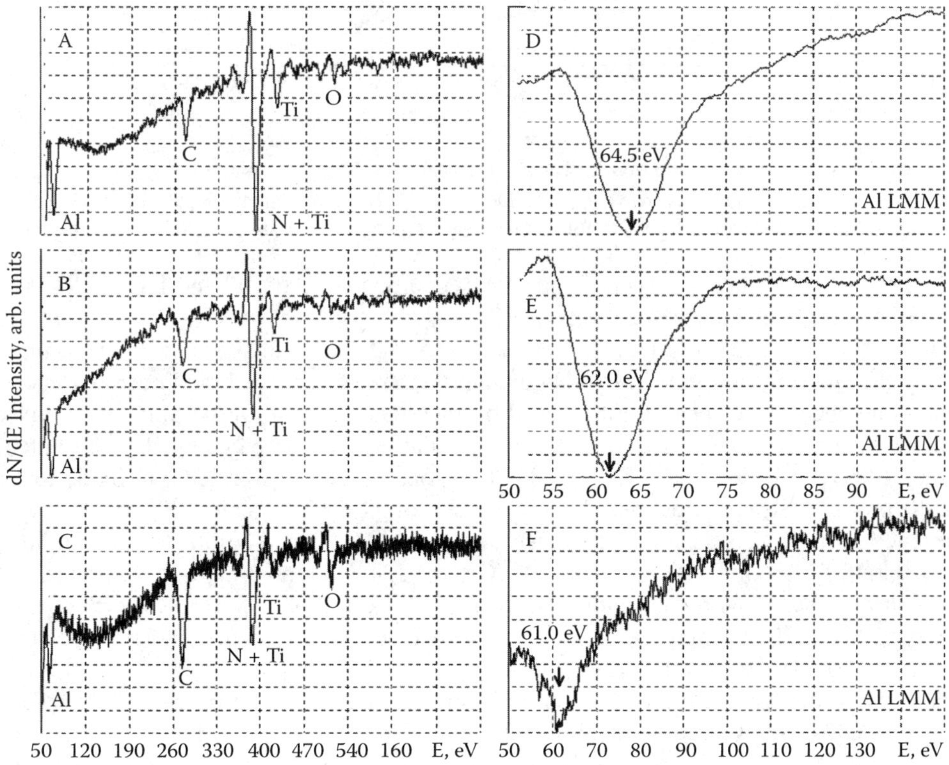

FIGURE 10.14 AES spectra of the end-mill worn surface with Ti–Al–Cr–N coatings for: (a) far from wear zone; (b) near wear zone; (c) in the wear zone; (d)–(f) detailed AES spectra (LMM Auger lines of aluminum) associated with (a)–(c) zones correspondingly. Length of cut: 30 m. Figure 10.14 a should be considered together with Figure 10.14d; Figure 10.14b, corresponds to the detailed spectrum in Figure 10.14e; and Figure 10.14c corresponds to Figure 10.14f. (From Fox-Rabinovich, G.S., Yamamoto, K., Veldhuis, S.C., Kovalev, A.I., Dosbaeva, G.K. Tribological adaptability of TiAlCrN PVD coatings under high performance dry machining. *Surf. Coat. Technol.* 2005, 200, 1804–1813. With permission.)

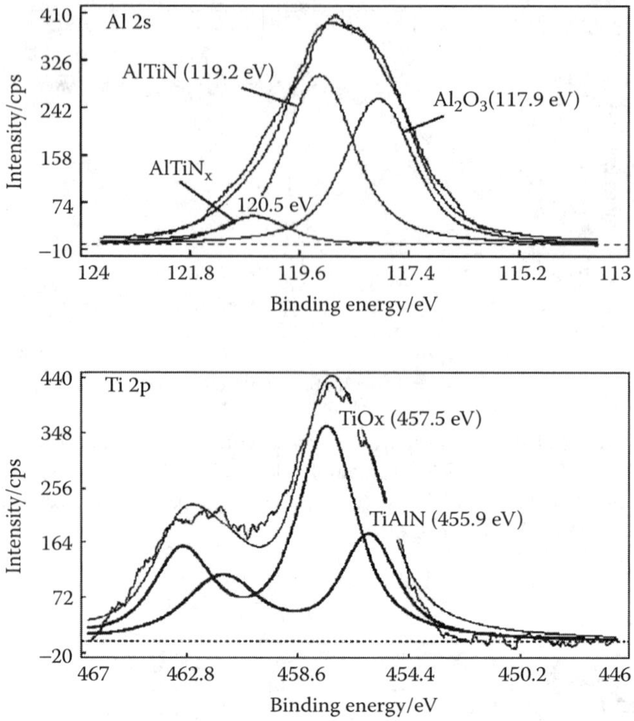

FIGURE 10.15 XPS spectra of the end-mill worn surface with Ti–Al–N coatings. (From Fox-Rabinovich, G.S., Yamamoto, K., Veldhuis, S.C., Kovalev, A.I., Dosbaeva, G.K. Tribological adaptability of TiAlCrN PVD coatings under high performance dry machining. *Surf. Coat. Technol.* 2005, 200, 1804–1813. With permission.)

Figure 10.14) that the relative concentration of Al/Ti corresponds to 2.47 and 1.11 for Ti–Al–Cr–N and Ti–Al–N-based coatings, respectively.

Figure 10.15a and Figures 10.15b represent XPS Ti2p and Al2s spectra of the worn crater surface of the end mills with the Ti–Al–N coating. We have determined the integrated intensity of separate components after peak fitting, based on the standard data on the binding energy of the titanium and aluminum oxides and nitrides. The calculations performed on the data presented show that around 65% of Ti and 56% of Al in Ti–Al–N coating are transformed into oxide-like layers during the process of tribo-oxidation. The XPS spectra of the Cr $2p_{3/2}$ region (Figure 10.16) indicate the presence of three Cr states on the surface of the worn coating. The fitted spectrum has peaks at 575.9 (CrN), 577.3(Cr^{+4}–O), and 578.7 eV (Cr^{+3}–O). This interpretation of chromium bonds is obtained in accordance with data published in other sources [30, 31]. The data, which has been calculated from the XPS components area, show that 86% of Cr in the Al–Ti–Cr–N coating is transformed to oxide-like layers during cutting.

The flank wear rate determines the tool life during an end-milling operation [19]. This wear intensity, under aggressive cutting conditions, strongly depends on the oxidation resistance of the coating. The coating with the best oxidation stability at elevated temperatures (Figure 10.7) has the best tool life. The higher oxidation stability of the $(Ti_{0.10}Cr_{0.20}Al_{0.70})$ N coating leads to the better tool life of the end-milling cutter. This higher oxidation stability is a result of the formation of a protective oxide film (a mixture of alumina and chromia [24]) on the surface during oxidation (Figure 10.8). Similar films are formed during the nonequilibrium tribo-oxidation process (Figure 10.13 to Figure 10.16). Based on electron spectroscopy data, the Ti–Al–Cr–N coatings are more likely to produce aluminum tribo-oxides during cutting compared to the Ti–Al–N coatings.

The formation of the alumina-like tribo-films with high chemical stability reduces the adherence of the workpiece material to the cutting-tool surface (Figure 10.10). Reduced seizure intensity results

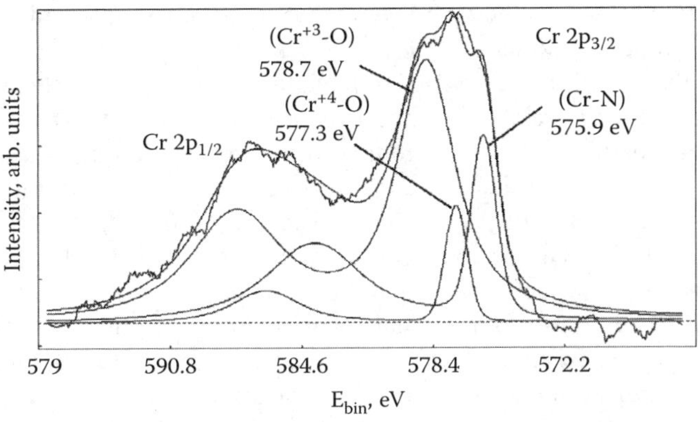

FIGURE 10.16 XPS spectra of the end-mill worn surface with Ti–Al–Cr–N coatings. (From Fox-Rabinovich, G.S., Yamamoto, K., Veldhuis, S.C., Kovalev, A.I., Dosbaeva, G.K. Tribological adaptability of TiAlCrN PVD coatings under high performance dry machining. *Surf. Coat. Technol.* 2005, 200, 1804–1813. With permission.)

in lower heat generation at the tool–workpiece interface. The addition of Cr to the Ti–Al–N-based coating leads to the formation of Cr oxide-like layers (Figure 10.16). Owing to the improved lubricity of this oxide under elevated temperatures of cutting [32], the value of the coefficient of friction drops (Figure 10.9). The type of chips that are formed changes critically from the dull to the curled (Figure 10.11), and the metal flow along the tool or chip interface is improved (Figure 10.12). All of these phenomena serve to enhance the adaptability of the coatings, thus resulting in better tool life.

The role of Cr in improving the tribological adaptability of the aluminum-rich Ti–Al–N-based coating comes from the formation of an additional lubricating oxide layer. This layer works in conjunction with the alumina-like surface films [24] and thus enhances the surface protection under friction. We can consider the Ti–Al–Cr–N coating as a synergistically alloyed one because all the metallic elements of the coating act together to achieve enhanced adaptation under friction.

Furthermore, based on the data presented in Chapter 9 [22,27] the tribo-films that form under conditions of high-performance machining most likely have amorphous-like structures, which again provide improved lubricity.

Based on the data presented we can conclude that the application of the Ti–Al–Cr–N coating improves the tribological adaptability of end-milling cutters. Thus, we can consider the response of the $(Ti_{0.1}Cr_{0.20}Al_{0.70})$ N coating to this harsh environment as adaptive.

We have to note that there is some similarity in synergistic alloying of oxidation-resistant metals and coating that leads to collaborative action of elements to improve protective alumina or alumina–chromia film formation. But there is a significant difference in the synergistic alloying of the wear-resistant coatings for severe dry high-performance machining applications. The tribo-films that form during cutting act in synergy. The wear behavior is improved because both types of tribo-films, protective and lubricious, are formed as outlined in Chapter 1 and Chapter 9. They protect the surface as thermodynamically stable oxide films on the one hand and improve the surface lubricity on the other hand. This is a very important difference because more complex mechanisms of adaptability occur during cutting and more complex synergistic action of the alloying elements should be taken into consideration. This idea can be illustrated by the wear behavior of the synergistically alloyed quaternary nitride nanolaminated coatings.

10.3.2 Nanoscale Multilayered Self-Adaptive Coatings: Quaternary Nitride Coatings

Based on the data obtained for the ternary nitride Ti–Al–Cr–N coatings, two different groups of the quaternary coatings have been studied. The general idea was to further improve the coatings lubricity by the addition of the elements (or layers) that could be transformed into high-temperature lubricious oxides during tribo-oxidation.

We have to note that a traditional practice of metallurgical design is associated with the specific characteristics that are directly related to one or more major volumetric properties. For example, the composition of the coating (with regard to their metallurgical design) is usually directly related to their major properties such as hardness or oxidation resistance (see preceding text). The principles of synergistic alloying for tribological materials considered in this chapter are associated with the tribo-films that form on the surface during friction (see Chapter 1). The development of new generations of self-adaptive surface-engineered materials promises to be a more complicated task for material scientists. They must improve major volumetric properties, but also keep in mind the composition and properties of the tribo-films formed during friction because this is critically important to enhance the service characteristics. Under these circumstances, the novel adaptive materials development becomes more complicated on the one hand because both stages of the design should be planned in advance. On the other hand, this could be paid back owing to critical enhancement of the service characteristics, as will be described in following text.

Two groups of coating were selected for the studies. Both of them are able to generate high-temperature lubricating oxides during the tribo-oxidation. Previously, we considered only adaptive monolayered Ti–Al–Cr–N coating. In this section, we will focus on two types of nanoscale multilayered coatings [33]. The first is the Ti–Al–Cr–N/WN nanomultilayered coating and the second is Ti–Al–Cr–N/BCN. A typical structure of the Ti–Al–Cr–N/WN coating is shown in Figure 10.17 [33]. The $Ti_{0.25}Cr_{0.10}Al_{0.65}$ powder metallurgical target has been used to fabricate the layers of the nitride coatings with cubic structure to ensure high mechanical properties of the multilayered coating. Metallic tungsten and B_4C targets have been used to fabricate nitride layers that further transform to lubricious oxides during friction.

Both types of the coatings are hard with microhardness within the range of 25 to 35 GPa. The design of the coating is also somewhat different. The total amount of tungsten in Ti–Al–Cr–N + WN multilayered coating, measured by EDS, increases from 5 to 20 at. % depending on the UBM power applied (within the range of 0.5 to 2.0 kW). The thickness of the Ti–Al–Cr–N layers in these coatings is around 10 to 12 nm. The thickness of the WN layers is around 3 to 5 nm and grows linearly with the UBM power applied (Figure 10.17). This results in a tungsten-content increase within the overall layer of the coating.

The Ti–Al–Cr–N + BN coating has a different design. All the coatings have the same amount of B (around 5 at. %) but the multilayer period varies depending on the different substrate rotation speeds (within 3 to 10 r/min).

The oxidation behavior of the quaternary coatings studied is quite different. It is known that short-term oxidation resistance of the coatings has a major impact on the wear resistance of the cutting tools under high-performance end-milling conditions [20]. Short-term oxidation resistance of the Ti–Al–Cr–N/WN coatings is slightly lower as compared to Ti–Al–Cr–N because of the formation of tungsten oxide with intermediate thermodynamic stability (Figure 10.18 and Figure 10.19). Short-term oxidation resistance of Ti–Al–Cr–N/BCN coatings is significantly lower as compared to the Ti–Al–Cr–N because of the formation of nonprotective boron oxide (Figure 10.20). We have to note that the reduction of oxidation stability of the quaternary nitride coatings could be balanced by lubricity improvements, and thus the tool life is enhanced. As outlined earlier, the modern approach to advanced coating design should involve more than one stage of property optimization.

To prove this hypothesis, the measurement of the coefficient of friction (Figure 10.21) and the wear resistance of the coatings were studied under conditions of dry high-performance end milling of the 1040 steel and hardened H13 tool steel (Figure 10.23).

The coefficient of friction parameters values for Ti–Al–Cr–N/WN coatings are as low as 0.05 to 0.06 and remain almost constant within a wide range of the temperatures (Figure 10.21).

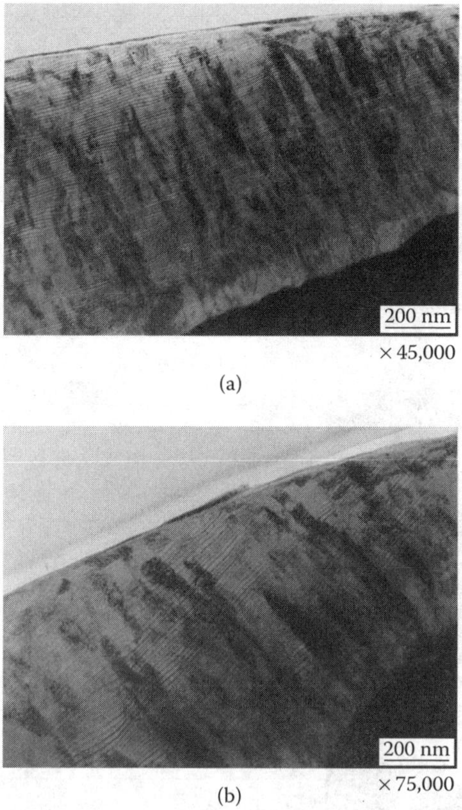

FIGURE 10.17 TEM images of the ball-nose end-mill surface with Ti–Al–Cr–N/WN nanomultilayered coating (FIB cross-sections): (a) before service; (b) after service; length of cut: 30 m.

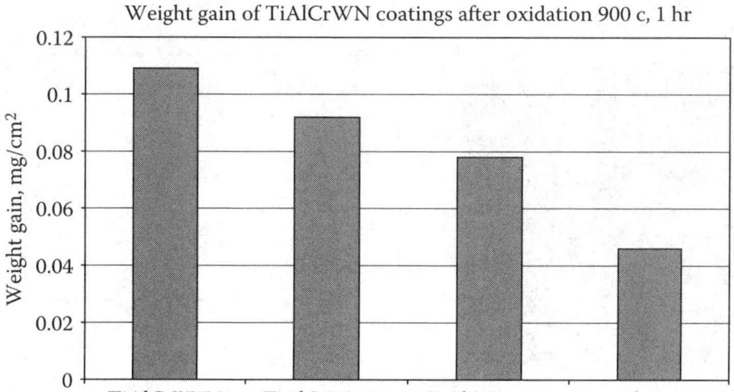

FIGURE 10.18 Weight-gain data of the Ti–Al–Cr–N/WN PVD coatings deposited on the TiAl coupons after short-term oxidation in air at 900°C during 1 h.

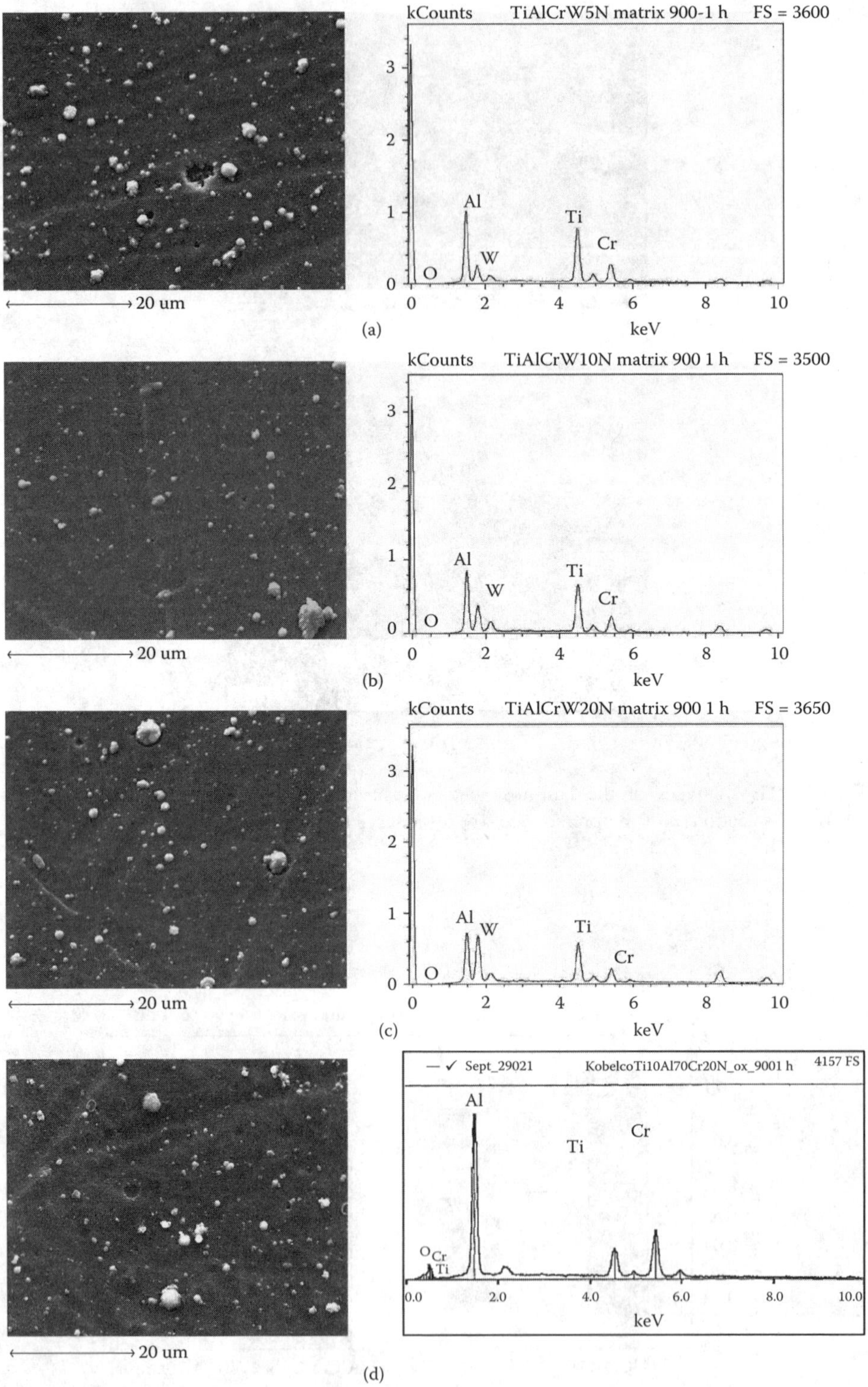

FIGURE 10.19 SEM micrographs of the surface morphology together with EDS spectra of the quaternary nitride coatings after oxidation at 900°C for 1 h: (a) with addition of 5 % W; (b) with addition of 10 % W; (c) with addition of 20% W; (d) $TiAl_{70}Cr_{20}N$ coating.

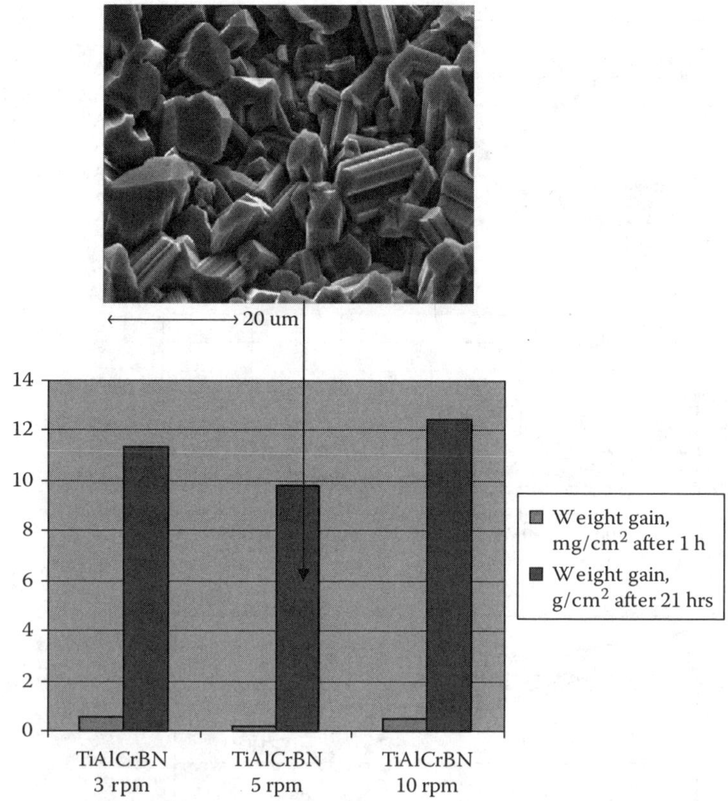

FIGURE 10.20 Weight-gain data of the Ti–Al–Cr–N/BCN PVD coatings deposited on the Ti–Al coupons after short-term oxidation in air at 900°C during 1 h.

Low coefficient of friction values at elevated temperatures for the Ti–Al–Cr–N/WN coatings impact on the major service characteristics of the coated cutting tools.

It is known that one of the major methods to improve wear behavior during cutting is to reduce: (1) cutting forces (2) heat generated during cutting [38]. The cutting forces have been measured *in situ* during ball-nose end milling of H13 steel. The data presented in Figure 10.22 exhibit relatively low cutting forces for the ball-nose end mills with Ti–Al–Cr–N/WN coating as compared to the other types of coatings studied.

Application of the Ti–Cr–Al–N/WN coatings also reduces heat generation at the workpiece–tool interface during cutting. It is also known that the color the chips corresponds directly to the temperature generated at the tool–workpiece interface (Table 10.3 [38]). Comparing the color of the chips formed and the temperature estimation presented in Table 10.3 [38], we can assume that the temperature at the workpiece–tool interface drops by around 100°C from 980 to 880°C for end-mill cutters with Ti–Al–Cr–N coating and for Ti–Al–Cr+WN coatings, correspondingly (Table 10.4).

These improvements in the surface characteristics of the coatings result in increase in coated-tool life. It was shown that the wear resistance of the Ti–Al–CN/WN nanoscale multilayered coating is almost twice as high when compared to Ti–Al–Cr–N coatings (Figure 10.23a and Figure 10.23b). This corresponds directly to the improvements in the coefficient of friction.

The seizure process is typical of many cutting operations [20]. In contrast, there is a very low seizure intensity on the flank and rake surfaces of worn end mill with nanoscale multilayered $Ti_{0.25}Cr_{0.10}Al_{0.65}$ N/WN coatings (Figure 10.24a and Figure 10.24b). SEM elemental mapping shows

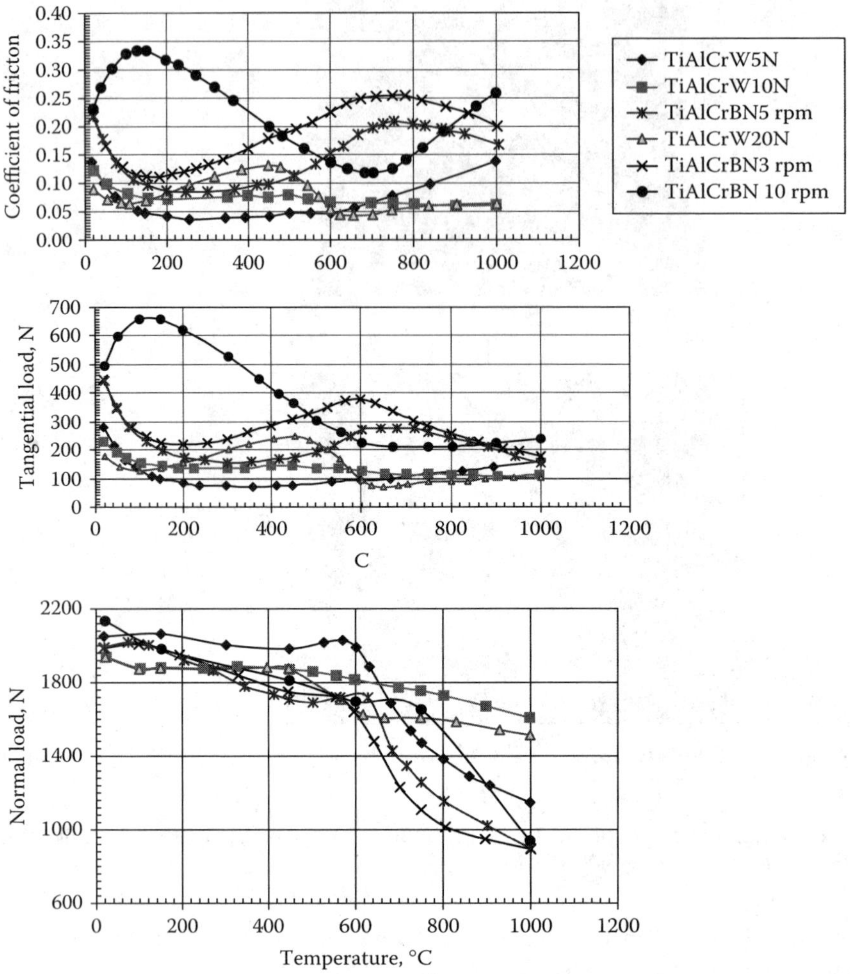

FIGURE 10.21 Coefficient of friction for Ti–Al–Cr–N/WN and Ti–Al–Cr–N/BCN coatings vs. temperature in contact with 1040 steel.

that it could be explained by the formation of the protective alumina on the flank (Figure 10.24a) and lubricious W–O tribo-films on the rake surface (Figure 10.24b).

Figure 10.25 shows the intensive photoelectrons lines on the electron spectrum for Ti, Al, and W. After deconvolution and fitting the main components, it is possible to identify the nature of the tribo-films formed on the worn ball-nose end-mill surface. The calculations performed on the data presented show that around 60% Ti, 50% Al, and 60% W was transformed into the tribo-oxide layers during cutting within the surface layer of Ti–Al–Cr–N/WN coating. The ability of the tungsten nitride nanolayer to form the tribo-oxide seems to be greater than that of the other components of the coating (Figure 10.25). Chromium tribo-oxides were not found in these experiments.

As inferred from Figure 10.24b, the formation of W oxide tribo-films is caused by a local sticking of the workpiece material to the coated tool surface. This leads to heat generation and rapid transformation of the WN nanolayer closest to the friction surface into the oxide tribo-film. It is known that W oxide, such as WO_3, is a high-temperature lubricant owing to its low shear strength. At the same time, this oxide has a relatively high hardness of around 15 to 22 GPa at elevated temperatures (See Chapter 12 for details) [34–35]. Due to tungsten tribo-oxide formation, intensity of seizure and coefficient of friction values drop (Figure 10.21), cutting forces decrease (Figure

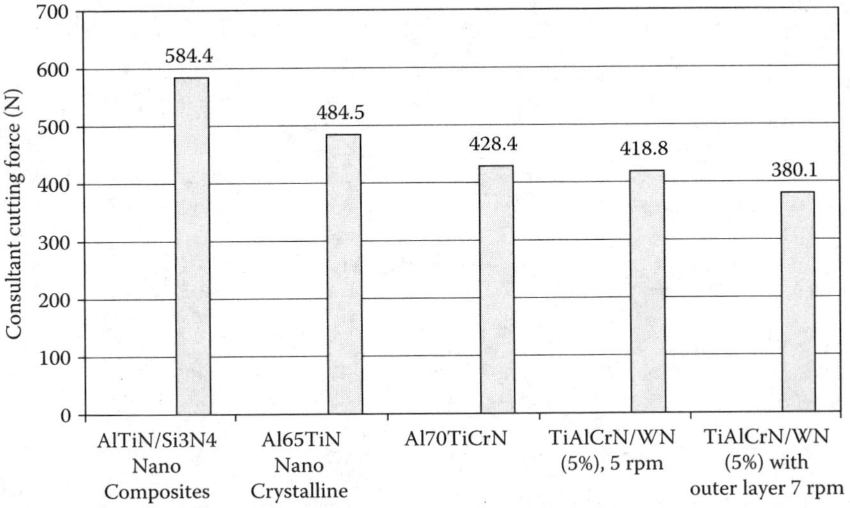

FIGURE 10.22 Cutting forces for the coated end-mill cutters measured *in situ* during machining of H13. Cutting conditions: speed — 200 m/min; feed — 0.06 mm/flute = 840 mm/min; axial depth — 5.0 mm; radial depth — 0.6 mm; cutting length, total – 30 m.

TABLE 10.3
Chip Temperature vs. Chip Color for End Milling of H13 Steel with Hardness HRC 50

Chip Temperature (°C)	Chip Color
981	Dark blue
900	Dark blue + brown
881	Brown
837	Light brown

Source: Ning, Y., Rahman, Y., Wong, S. Investigation of chip formation in high- speed end milling. *J. Mater. Proc. Technol.* 2001, 113, 360–367.

TABLE 10.4
Chip Color vs. Length of Cut for Ball-Nose End Mills with Ti–Al–Cr–N and Ti–Al–Cr–N+WN Coatings[a]

Length of Cut (m)	Coating	Chip Color	Temperature Estimated According to Reference 38	Coating	Chip Color	Temperature, Estimated According to Reference 38
7.5	Ti–Al–Cr–N	Blue	981	Ti–Al–Cr–N+WN	881	Brown
15		Blue	981		881	Brown
30		Blue	981		881–900	Brown or blue + brown

[a] Workpiece material: H13 steel with hardness HRC 50.

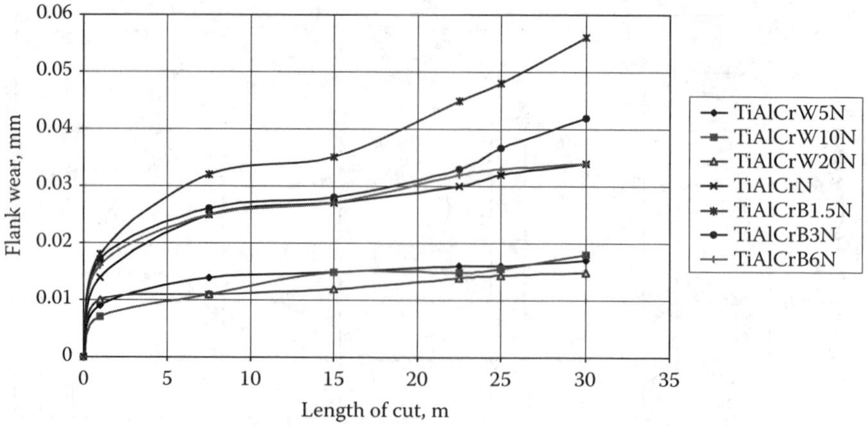

(a)

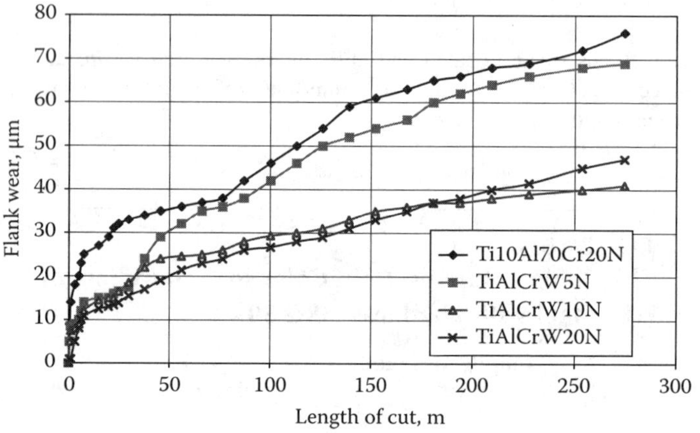

(b)

FIGURE 10.23 Tool life of the Ti–Al–Cr–N as well as Ti–Al–Cr–N/WN and Ti–Al–Cr–N/BCN PVD coatings under high performance dry machining conditions: (a) Ball-nose E/M 5mm radius, 2 flutes (C-2MB, Mitsubishi); workpiece material — AISI H13 (HRC50); cutting conditions: speed — 200 m/min; feed — 0.06 mm/flute = 840 mm/min; axial depth — 5.0 mm; radial depth — 0.6 mm; cutting length — 30 m; (b) 250 m; (c) End mills, D = 12 mm; 4 flutes; workpiece material — AISI 1040; cutting conditions: 1400 rpm; Ad — 3 mm, Wd — 10 mm, feed — 63 mm/min; Ti–Al–Cr–N/WN coating — 540 r/min.

Continued.

10.22), and heat generation at the tool–workpiece interface also reduces (Table 10.4). Eventually, it results in wear intensity reduction (Figure 10.23). This is an example of friction control with a positive feedback loop [22]. In synergy with the other tribo-oxides (in this case, alumina) that form during friction of Ti–Al–Cr–N nanolayers [33], this lubricious tungsten oxide formation leads to significant tool life improvement. Combining this data with the tool life data (Figure 10.23), we can conclude that the lubricious and protective tribo-oxide films that form on the tool surface during cutting mostly isolate the workpiece from the cutting-tool surface. The surface-damaging wear mechanism (seizure) comes close to the classical case of external wear with critically lower wear intensity [37]. That is why dry machining of the hardened H13 steel becomes quite efficient if the nanomultilayered Ti–Al–Cr–N/WN coating has been applied on the cutting-tool surface.

Nanomultilayered structures of coatings (Figure 10.17) play a significant role in wear behavior improvement. Most probably, the nanoscale laminar structure critically enhances two beneficial

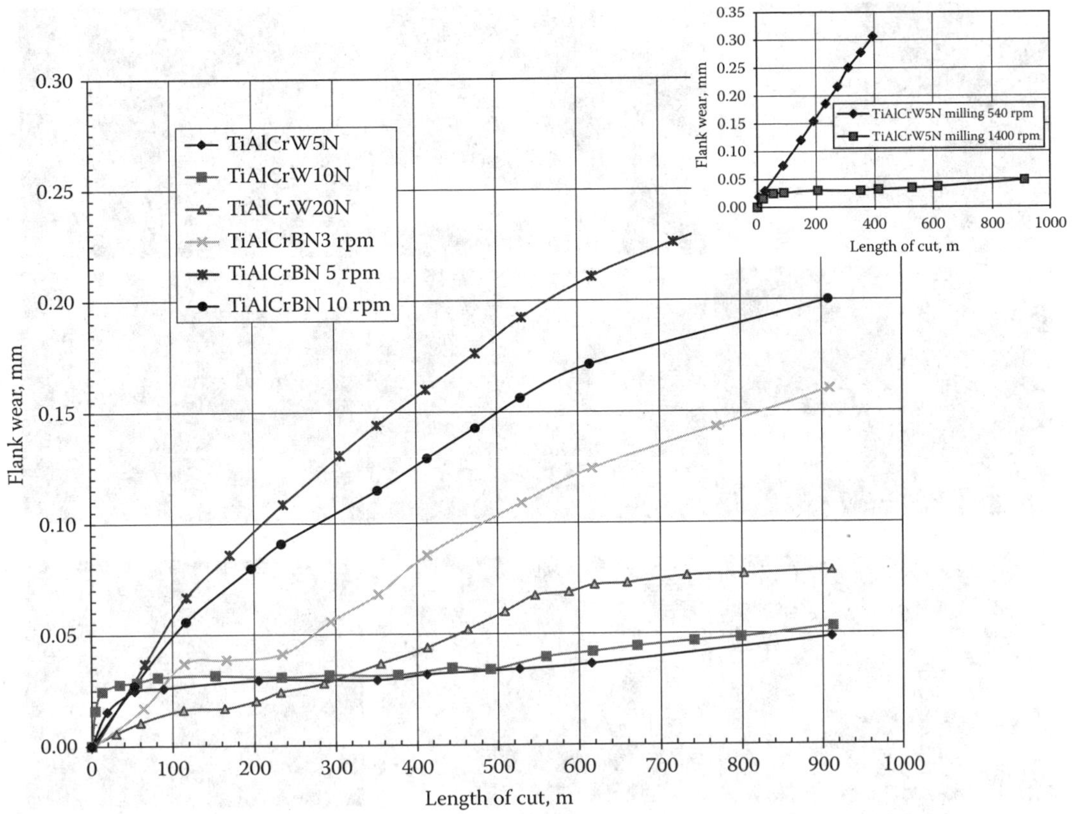

FIGURE 10.23 *Continued.*

phenomena: (1) mass transfer within the very beginning of the running-in stage owing to formation of so-called bridges that result in quick formation of the tribo-oxides with the required composition and microstructure [36]; (2) selective oxidation within the islandlike seizure zones with further formation of the tribo-oxides that critically reduces intensity of this surface-damaging process and results in tool life improvement.

However, a significant transformation of initial nanolaminar structure of Ti–Al–Cr–N/WN coating occurs during cutting (Figure 10.17 and Figure 10.26). The microstructure of the worn cutting-tool surface with Ti–Al–Cr–N/WN coating within an island seizure of the workpiece material has been studied using TEM. The electron diffraction patterns (Figure 10.26) show that an intensive recrystallization process occurs within the seizure zone and close to the surface areas (Zone A and Zone B). Electron diffraction spots in the electron diffraction patterns of three zones have been indexed. Zone A presents close to perfect (110) Fe (bcc) structure of the recrystallized workpiece material (tool steel H13); Zone B shows cubic WN structure; Zone C indicates a nanocrystalline structure far from the surface of the Ti–Al–Cr–N layer.

A few observations can be made with regard to the wear behavior of the self-adaptive synergistically alloyed coatings.

If we compare the curves of wear for cutting tools with Ti–Al–Cr–N/WN coatings that have been tested under different machining conditions (Figure 10.23c), a very remarkable conclusion can be made. Wear behavior of the same coating critically depends on the cutting parameters used. Less aggressive cutting (cutting speed 540 r/min) results in significantly higher intensity of wear. In contrast, the wear intensity under higher cutting speeds (1400 r/min) leads to wear rate stabilization at very low levels (flank wear is around 0.02 to 0.025 mm). This can be explained by

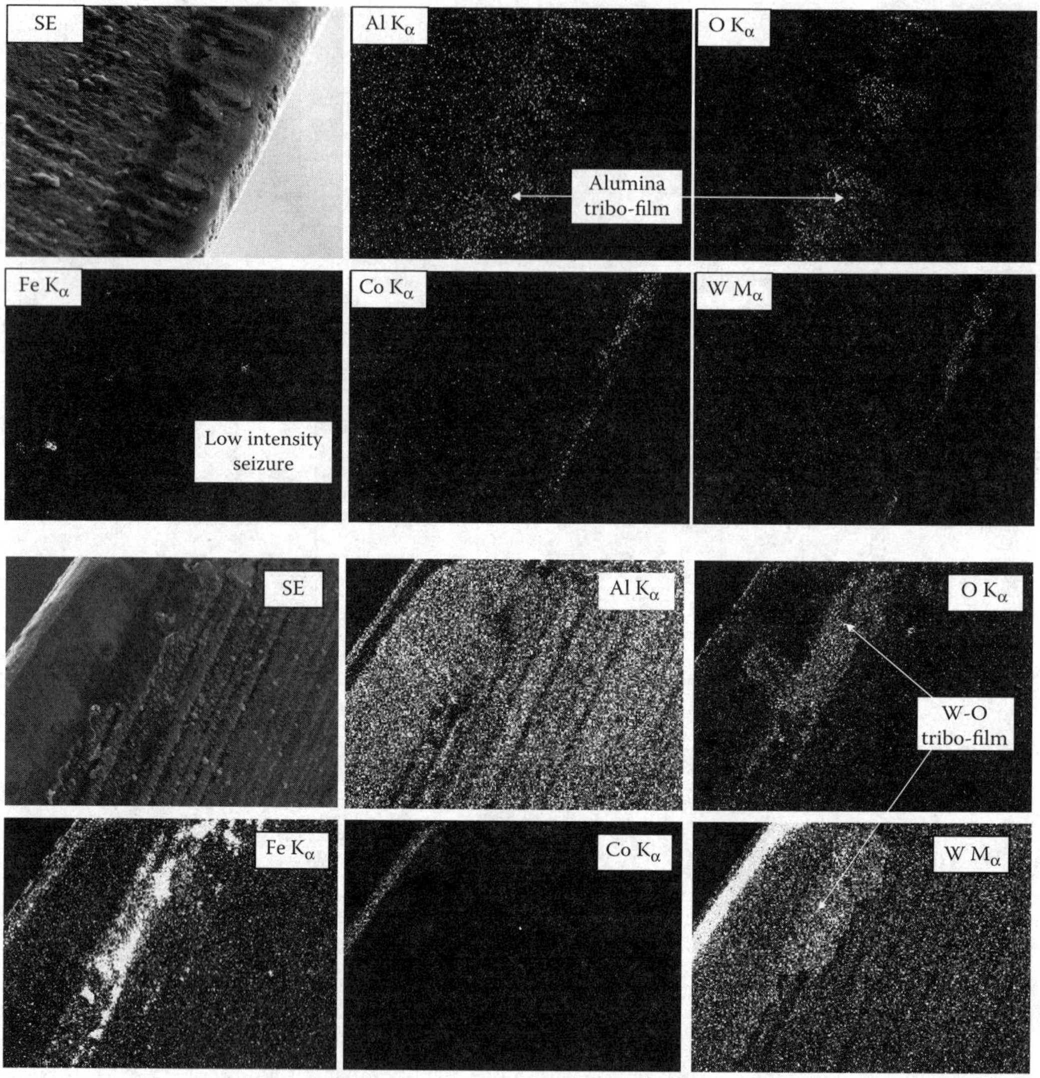

FIGURE 10.24 SEM images and EDX elemental maps of the worn surface of ball-nose end mill with Ti–Al–Cr–N/WN PVD coatings. Machining of H13 steel, length of cut: 30 m: (a) flank surface; (b) rake surface.

formation of the synergistically acting tribo-films at elevated temperatures of cutting that almost arrest the wear-intensity growth. The cause of this phenomenon is discussed in Chapter 11. We have to emphasize that, according to Chapter 2, to achieve fully protective or lubricious conditions, the tribo-films have to be formed under nonequilibrium conditions of high-speed machining with significant gradients of temperature and other parameters that control the friction.

An observation of these wear curves is also very important from a practical viewpoint. These wear curves mean that we can significantly increase productivity of machining and simultaneously critically reduce cutting-tool wear intensity. This is an extremely beneficial situation from the point of view of the productivity improvements and manufacturing process cost reduction.

The data presented shows that tool life improvement during end-milling operations is controlled mainly by tribological properties rather than hot hardness values. This corresponds to the data in Reference 41. But the improvement of tribological characteristics of the coating affects not only the tool life but also critically changes the manufacturing processes. The temperatures at the tool–chip interface during cutting are very high (around 1000°C; see Table 10.4) and close to the

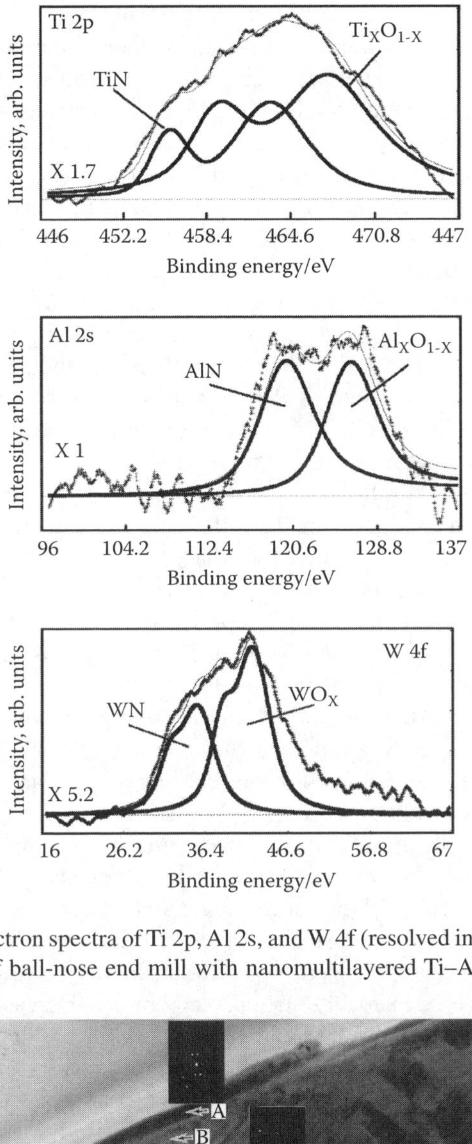

FIGURE 10.25 The photoelectron spectra of Ti 2p, Al 2s, and W 4f (resolved into oxide and nitride component peaks) for the worn surface of ball-nose end mill with nanomultilayered Ti–Al–Cr–N/WN coating.

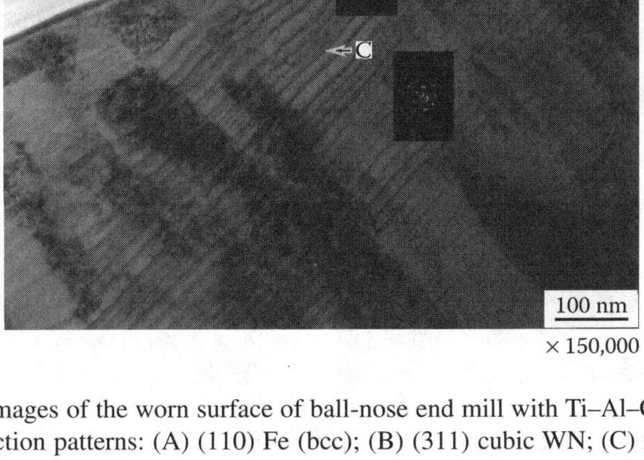

FIGURE 10.26 TEM images of the worn surface of ball-nose end mill with Ti–Al–Cr–N/WN nanomultilayered coating with diffraction patterns: (A) (110) Fe (bcc); (B) (311) cubic WN; (C) nanocrystalline structure of Ti–Al–Cr–N (cubic).

temperature of the $\alpha \rightarrow \gamma$ transformation for the machined tool steel H13. The shapes of chips formed have been studied for the end-mill cutters with different state-of-the-art coatings (see Figure 10.1). The SEM images (Figure 10.27) show that the chips from the end mill with nanocrystalline as well as nanocomposite coatings with relatively high coefficient-of-friction values [1] are very dull and have almost no curling (Figure 10.27a and Figure 10.27b). In contrast, the chips from the cutter with the self-adaptive coating are intensively curled into a spiral (Figure 10.27c). The undersurface of the chips from the cutters with the self-adaptive coating is much smoother (compare Figure 10.27b and Figure 10.27c). Cross sections of the chips are shown in Figure 10.28. The microstructure of the chips strongly depends on the type of coating used. The chips that are formed on the surface of the end-milling cutters with the $Al_{0.65}Ti_{0.35}N$ nanocrystalline coating have a very wide contact zone of secondary hardening (Figure 10.28a) [20,27]. This corresponds to the temperature data presented in Table 10.4. The chips that are formed on the surface of the end mill with nanocomposite $Al_{0.65}Ti_{0.35}N/\ Si_3N_4$ coatings have a narrower zone of secondary hardening but no intensive flow zone could be observed (Figure 10.28b). In contrast, the chips that are formed on the surface of the end-mill cutters with self-adaptive Ti–Al–Cr–N/WN coatings have a very narrow zone of secondary hardening and a flow zone with intensive plastic deformation (Figure 10.28c). The microhardness measurements performed on the chip cross section vs. the distance from the tool–chips interface (Figure 10.29) show that secondary hardening of the workpiece material occurs if nanocomposite and nanocrystalline coatings are applied. In contrast, a softening of the chip surface area occurs if a self-adaptive coating is applied (Figure 10.29). The phase composition of the chips' undersurface corresponds to the SEM images as well as microhardness measurements of the chips' cross sections. Roentgen analysis shows that a mixture of retained austenite and martensite forms on the undersurface of chips from the milling cutter with a nanocrystalline coating (Table 10.5). There is only a minor amount of the retained austenite on the surface of the chips from the milling cutters with the nanocomposite coating and no austenite on the surface of the chips from the self-adaptive coating. Based on the data obtained, we can conclude that during machining of hardened H13 steel, a secondary hardening of the thin surface layer of workpiece material could occur if coatings with relatively poor tribological characteristics are applied. This could result in secondary hardening of the surface layer and leads to more severe friction conditions. This could be one of the causes of intensive wear of these coatings during the running-in stage of wear (Figure 10.1). The self-destructive tribosystem is probably created in this case. In contrast, if the self-adaptive coating is applied, the actual temperature decreases; no $\alpha \rightarrow \gamma$ transformation occurs and a beneficial softening takes place at the tool–workpiece interface (Figure 10.29). Besides, it is known that secondary hardening results in formation of tensile residual stresses on the surface of the workpiece (see Chapter 6). We can assume that owing to the beneficial phase transformation observed, a quality of the machine part can be improved if a self-adaptive coating is used.

Ultimately, the self-adaptive coating application could result both in the improvement of productivity of machining process as well as better quality of the machined part.

Application of the Ti–Al–N/WN self-adaptive coating is very challenging for cutting of nickel-containing aerospace alloys. Poor machinability of these aerospace alloys is related to the formation of high forces and high temperatures at the tool–workpiece interface during cutting [20]. A major method to improve tool life and enhance productivity of machining aerospace alloys is similar to that described earlier for the hardened tool steels, i.e., to reduce cutting force and heat generation during cutting.

Ti–Al–Cr–N/WN coating also improves tool life as compared to Ti–Al–N coatings during turning of austenitic stainless steels (γ-SS) 316 (Figure 10.30a). During machining of these workpiece materials, a few phenomena occur that are associated with specific properties of the stainless steel [21]. On the one hand, an intensive seizure of workpiece material takes place because of high ductility of the austenitic stainless steel that significantly increases the tool–chip contact length. Moreover, the γ-SS has a tendency to work-harden that intensifies adhesive interaction. On the other hand, the heat that is generated during friction mainly goes into the body of the cutter because

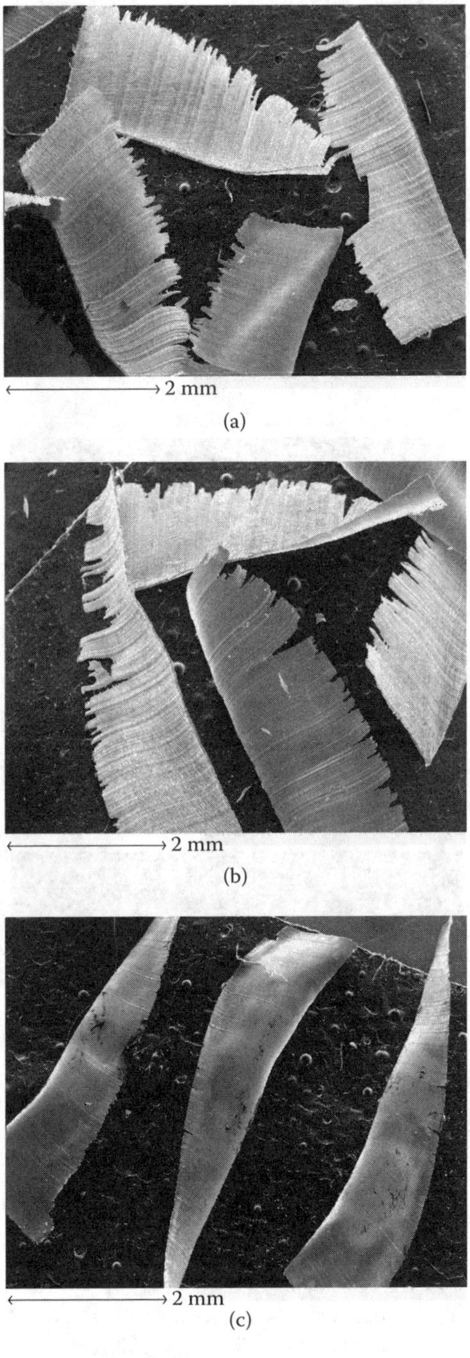

FIGURE 10.27 Types of chips for end-mill cutters with (a) $Al_{0.65}Ti_{0.35}N$ nanocrystalline coatings; (b) $Al_{0.65}Ti_{0.35}N/\ Si_3N_4$ nanocomposite coatings; (c) Ti–Al–Cr–N/WN self-adaptive coating. Length of cut — 30 m.

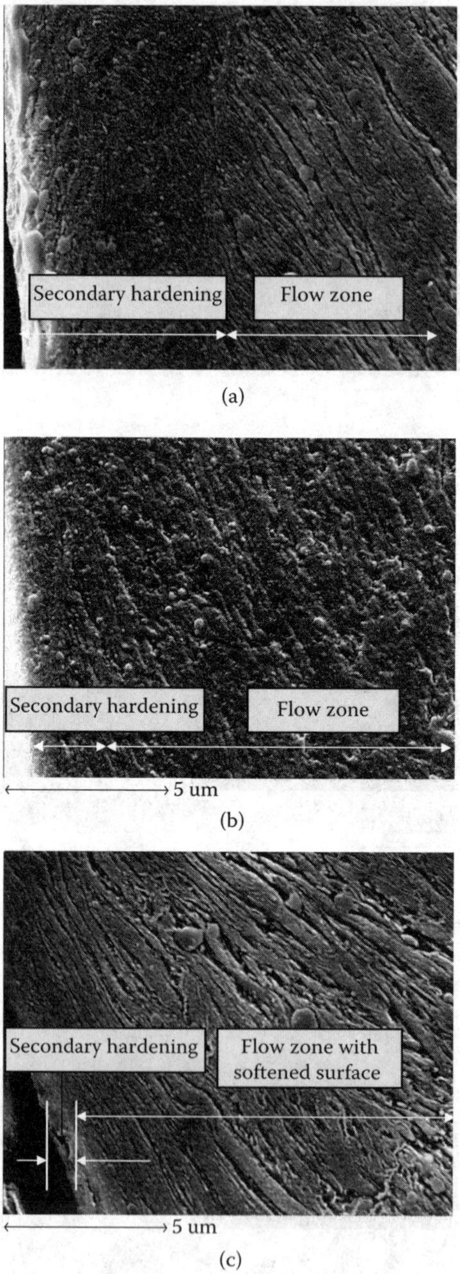

FIGURE 10.28 SEM images of the chips' cross-sections for (a) $Al_{0.65}Ti_{0.35}N$ nanocrystalline coatings; (b) $Al_{0.65}Ti_{0.35}N/Si_3N_4$ nanocomposite coatings; and (c) $Ti_{0.25}Cr_{0.10}Al_{0.65}N + WN$ coating after end-mill cutting (length of cut — 30 m). Zone 1 — the contact zone of secondary hardening; Zone 2 — extended deformation zone.

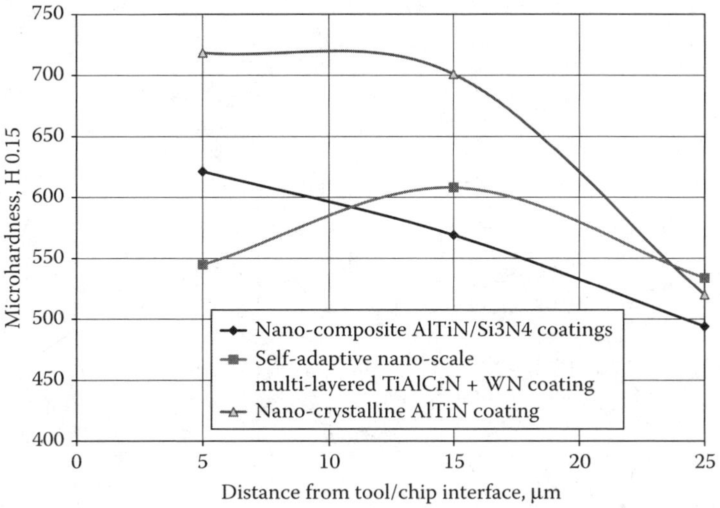

FIGURE 10.29 Microhardness of the chips vs. the distance from the tool or chips interface.

TABLE 10.5
Phase Composition on Chip Undersurface (XRD Data)[a]

Coating	Phase Composition of the Chips Undersurface
Nanocrystalline Al–Ti–N	$\alpha + \gamma$ (some amount)
Nanocomposite Al–Ti–N/Si_3–N_4	$\alpha + \gamma$ (traces)
Self-adaptive Ti–Al–Cr–N/WN	α only

[a] Ball-Nose End Milling of H13 Steel (HRC 50).

of the low thermal conductivity of stainless steel. In this case, application of the self-adaptive Ti–Al–Cr–N/WN coating with a low coefficient of friction at elevated temperatures results in less intensive sticking of the workpiece material and lowers actual temperatures at the tool–workpiece interface. This is why Ti–Al–Cr–N/WN coatings show tool life improvements.

Creep-resistant superalloys have improved mechanical characteristics at elevated temperatures that have made these materials difficult to machine. These materials also intensively work-hardening during cutting. However, application of the self-adaptive Ti–Al–Cr–N/WN coating results in a significant increase in tool life when compared to other types of state-of-the-art coatings during machining of these alloys (Figure 10.31 and Figure 10.32). More aggressive cutting conditions result in better tool life (Figure 10.31; Figure 10.32a and Figure 10.32b). A similar phenomenon was observed for end milling of 1040 steel (Figure 10.23b). This phenomenon is associated with the enhanced formation of protective and lubricious tribo-films (see the preceding text) and further explained in Chapter 11 based on a thermodynamics approach.

As outlined in the beginning of this chapter, the metallurgical design of coatings should be strongly tailored to specific applications. In contrast, the Ti–Al–N coating shows a better tool life when compared to Ti–Al–Cr–N/WN coatings during high-speed turning of structural 1040 steel (Figure 10.30b). This means that strong interatomic bonds in the Ti–Al–N coating (see Section 10.1 of this chapter) also control tool life in this case.

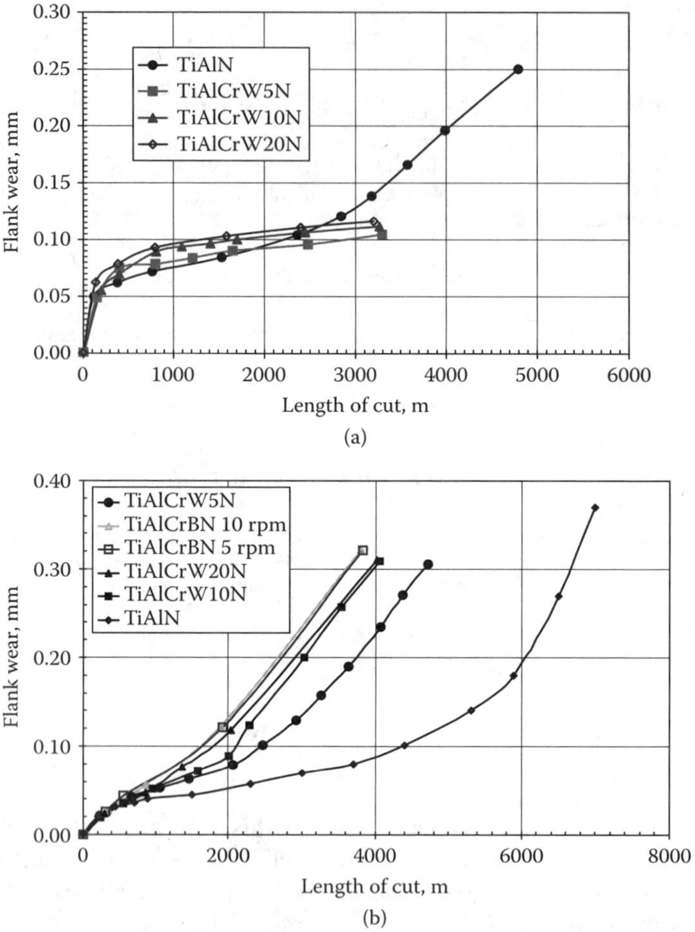

FIGURE 10.30 Tool life of the Ti–Al–Cr–N/WN and monolayered Ti–Al–N PVD coatings under turning conditions: (a) machining of austenitic stainless steel 316; speed — 246 m/min, feed — 0.11 mm/rev, depth of cut — 0.5 mm; (b) machining of 1040 steel, annealed; speed — 450 m/min, feed — 0.11 mm/rev, depth of cut — 0.5 mm.

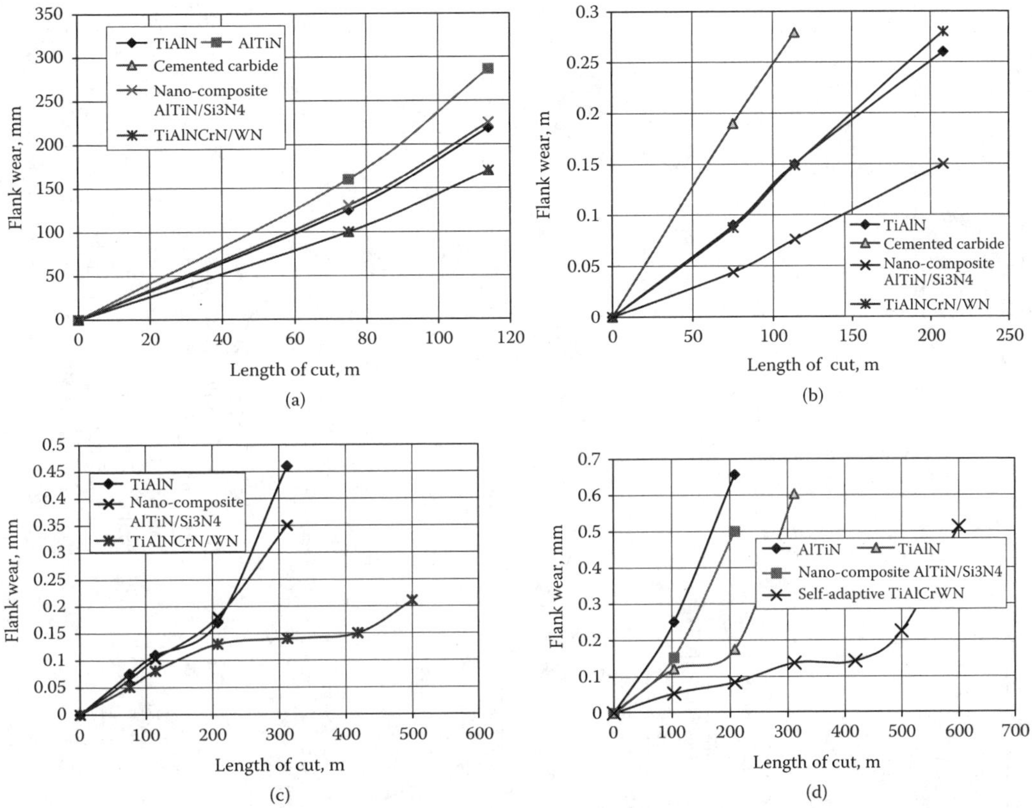

FIGURE 10.31 Tool life of the cemented carbide inserts with state-of-the-art PVD coatings during dry turning of the Waspalloy discs (aged, HRC 41). Cutting data: feed — 0.2 mm/rev; depth of cut — 0.250 mm: (a) speed 40 m/min; (b) speed 100 m/min; (c) speed 125 m/min; (d) speed 150 m/min.

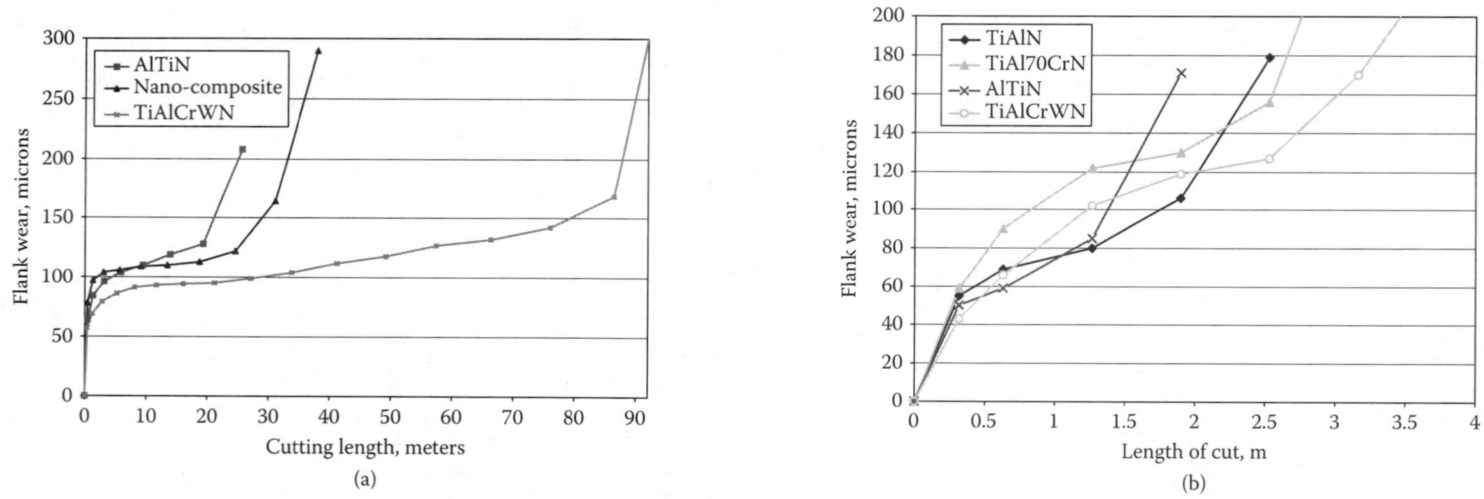

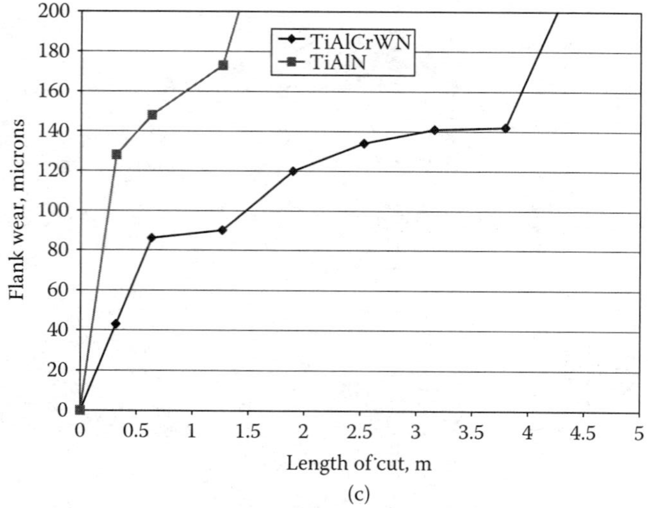

FIGURE 10.32 Tool life of the cemented carbide end-mill cutters with state-of-the-art PVD coatings during dry machining of the Waspalloy blocks (aged, HRC 41): (a) End mills, D — 12 mm; 4 flutes; cutting conditions: 1500 rpm; Depth of cut — 1 mm, Width of cut — 5 mm, feed — 150 mm/min. (b) Ball-nose E/M 5mm radius, 2 flutes (C-2MB, Mitsubishi); cutting conditions: speed — 47 m/min; feed — 0.025 mm/flute — 150 mm/min; axial depth — 5.0 mm; radial depth — 1 mm. (c) Ball-nose E/M 5mm radius, 2 flutes (C-2MB, Mitsubishi); cutting conditions: speed — 47 m/min; feed — 0.083 mm/flute — 250 mm/min; axial depth — 5.0 mm; radial depth — 1 mm.

REFERENCES

1. Ma, S., Procházka, J., Karvánková, P., Ma, Q., Niu, X., Wang, X., Ma, D., Xu, K., Veprek, S. Comparative study of the tribological behavior of superhard nanocomposite coatings nc-TiN/a-Si_3N_4 with TiN. *Surf. Coat. Technol.* 2005, 194, *20*, 143–148.
2. Patscheider, J., Zehnder, T., Diserense, M. Structure-performance relations in nanocomposite coating. *Surf. Coat. Technol.* 2001, 146–147, *5*, 201–208.
3. Holleck, H. Material selection for hard coatings. *J. Vac. Sci. Technol.* 1986, 4, 2661.
4. Knotek, O., Löffler, F., Wolkers, L. Phase stability and formation of Ti-Al-C-N PVD coatings. *Surf. Coat. Technol.* 1994, 68/69, 176–180.
5. Munz, W.-D. Titanium aluminum nitride films: a new alternative to TiN coatings. *J. Vac. Sci. Technol.* 1986, 4, *6*, 2717–2725.
6. Ikeda, T., Satoh, H. Phase formation and characterization of hard coatings in the Ti-Al-N system prepared by the cathodic arc ion plating method. *Thin Solid Films* 1991, 195, 99.
7. Sugiyama, A., Kajioka, H., Makino, Y. Phase transition of pseudobinary Cr-Al-N films deposited by magnetron sputtering method. *Surf. Coat. Technol.* 1997, 97, 590.
8. Yamamoto, K., Sato, T., Takahara, K., Hanaguri, K. Properties of (Ti, Cr, Al) N coatings with high Al content deposited by new plasma enhanced arc-cathode. *Surf. Coat. Technol.* 2003, 174/175, 620–626.
9. Anderson, W.P., Edwards, W.D., Zerner, M.C. Calculated spectra of hydrated ions of the first transition series. *Inorg. Chem.* 1986, 25, 2728.
10. Zerner, M.C. Semi-empirical molecular orbital methods, reviews. In *Computational Chemistry*, Vol. II, Eds., Lipkowitz, K.B., Boyd, D.B. VCH: New York, 1991, pp. 313–324.
11. Fox-Rabinovich, G.S., Veldhuis, S.C., Scvortsov, V.N., Shuster, L.Sh., Dosbaeva, G.K., Migranov, M.S. Elastic and plastic work of indentation as a characteristic of wear behavior for cutting tools with nitride PVD coatings. *Thin Solid Films* 2004, 469–470, 505–512.
12. Sjostrom, H., Hultman, L., Sundgren, J.-E., Hainsworth, S.V., Page, T.F., Theunssen, G.S.A.M. Structural and mechanical properties of carbon nitride CN_x (0.2 × 0.35) films. *J. Vac. Sci. Technol.* 1996, 14, *1*, 56–62.
13. Brisccoe, B.J., Fiori, L., Pelillo, E. Nano-indentation of polymeric surfaces. *Appl. Phys.* 1998, 31, 2395.
14. Andrievskii, R.A., Kalinnikov, G.V., Hellgren, N., Sandstorm, P., Stanskii, D.V. Nanoindentaiton and strain characteristics of nanostructured boride/nitride films. *Phys. Solid State* 2000, 42, 1624–1671.
15. Catledge, S.A., Vohrå, Y., Woodard, S., Venugopalan, R. Structural and mechanical properties of nanostructured metalloceramic coatings on cobalt chrome alloys. *Appl. Phys. Lett.* 2003, 82, 1625.
16. Catledge, S.A., Vohra, Y., Woodard, S., Venugopalan, R. Structure and mechanical properties of functionally-graded nanostructured metalloceramic coatings. In *Mechanical Properties Derived from Nanostructuring Materials*, Vol. 778, Eds., Kung, H., Bahr, D.F., Moody, N.R., Wahl, K.J. MRS Proceedings: Pittsburgh, PA, 2003, p. 781.
17. Nemoshkalenko, V.V., Kucherenko, Y. *The Computing Methods of Physics in Solid State Theory.* Naukova Dumka Publisher: Kiev, 1986.
18. Seung-Hoon, Jhi., Jisoon, Ihm., Louie, G. Steven, Cohen, M.L. Electronic mechanism of hardness enhancement in transition-metal carbonitrides. *Nature* 1999, 399, 132.
19. Kovalev, A.I., Barskaya, R.A., Wainstein, D.L. Effect of alloying on electron structure, strength and ductility characteristics of nickel aluminide. *Surf. Sci.* 2003, 532–535, 35.
20. Trent, E.M., Wright, P.K. *Metal Cutting.* 4th ed. Butterworth-Heinemann: Woburn, MA, 2000.
21. Trent, E.M, Suh, N.P. *Tribophysics.* Prentice-Hall: Englewood Cliffs, NJ, 1986.
22. Fox-Rabinovich, G.S., Kovalev, A.I., Weatherly, G.C. Tribology and the design of surface engineered materials for cutting tool applications. In *Modeling and Simulation for Material Selection and Mechanical Design*, Eds., Totten, G., Xie, L.K., Funatani, K. Marcel Dekker: New York, 2004, p. 301.
23. Erkens, G., Cremer, R., Hamoudi, T., Bouzakis, K.-D., Mirisidis, I., Hadjiyiannis, S. Properties and performance of high aluminum containing (Ti,Al)N based supernitride coatings in innovative cutting application. *Surf. Coat. Technol.* 2004, 177–178, 724–734.
24. Fox-Rabinovich, G.S., Weatherly, G.C., Wilkinson, D.S., Kovalev, A.I. Oxidation resistant Ti-Al-Cr alloy for protective coating applications. *Intermetallics* 2004, 12, *2*, 165.

25. Hörling, A., Hultman, L., Odén, M., Sjölén J., Karlsson, L. Mechanical properties and machining performance of $Ti_{1x}Al_xN$-coated cutting tools. *Surf. Coat. Technol.* 2005, 191, *2–3*, 384–392.
26. Munz, W.-D. Large-Scale manufacturing of nano-scale multilayered hard coating deposited by Cathodic Arc/Unbalanced Magnetron Sputtering. *MRS Bull.* 2003, 28, *3*, 173–180.
27. Fox-Rabinovich, G.S., Weatherly, G.C., Dodonov, A.I., Kovalev, A.I., Veldhuis, S.C., Shuster, L.S., Dosbaeva, G.K. Nano-crystalline filtered arc deposited (FAD) TiAlN PVD coatings for high-speed machining applications. *Surf. Coat. Technol.* 2004, 177–178, 800–811.
28. Veprek, S., Veprek-Heijman, M.G.J., Karnakova, P., Prochzka, J. Different approaches to superhard coatings and nano-composites. *Thin Solid Films* 2005, 476, 1–29.
29. Briggs, D., Seach, M.P. *Practical Surface Analysis by Auger and X-Ray Photoelectron Spectroscopy.* John Wiley & Sons: New York, 1983.
30. Tsitsumi T., Ikemoto I., Namikawa T. Interatomic bonds in nitrides. *Bull. Chem. Soc. Japan* 1981, 54, 3, 913.
31. Romand H., Robin M. Electronic and structural properties of transition metals compounds. *Analyses* 1974, 4, *7*, 308.
32. Ho, W.Y., Huang, D.-H., Hsu, C.-H., Wang, D.-Y. Study of characteristics of Cr_2O_3/CrN duplex coatings for aluminum die casting applications. *Surf. Coat. Technol.* 2004, 177–178, 172.
33. Yamamoto, K., Kujime, S., Takahara, K. Properties of nano-multilayered hard coatings deposited by a new hybrid coating process: combined cathodic arc and unbalanced magnetron sputtering. *Surf. Coat. Technol.* 2005, 200, 435–439.
34. Lugscheider, E., Bärwulf S., Barimani, C. Properties of tungsten and vanadium oxides deposited by MSIP–PVD process for self-lubricating applications. *Surf. Coat. Technol.* 1999, 120–121, 458–464.
35. Erdemir, A. A crystal-chemical approach to lubrication by solid oxides. *Tribol. Lett.* 2000, 8, 97–102.
36. Gachon, J.-C., Rogachev, A.S., Grigorian, H.E., Illarionova, E.V., Kuntz, J.-J., Kovalev, A.N., Nosyrev, D.Y., Sachkova, N.V., Tsygankov, P.A. On mechanism of heterogeneous reaction and phase formation in Ti/Al multilayered nano-films. *Acta Mater.* 2005, 53, *4*, 1225–1231.
37. Kragelski, I.V., Dobychin, N.M., Kombalov, V.S. *Foundations of Calculations of Friction and Wear.* Mashinostroenie: Moscow, 1977.
38. Ning, Y., Rahman, Y., Wong, S. Investigation of chip formation in high-speed end milling. *J. Mater. Proc. Technol.* 2001, 113, 360–367.
39. Geller, Yu. *Tool Steels.* Metallurgy: Moscow, 1978.
40. Gershman, S., Bushe, N.A. Thin films and self-organization during friction under the current collection conditions. *Surf. Coat. Technol.* 2004, 186, *3*, 405–411.
41. Daitzemberg, J.H., Jaspers, S.P.F.C., Taminai, D.A. The workpiece material in machining. *Int. J. Adv. Manuf. Technol.* 1999, 15, 383–386.

11 Development of the Ternary and Higher-Ordered Protective or Wear-Resistant Materials and Coatings for High-Temperature Applications and Thermodynamics-Based Principles of their Synergistic Alloying

German S. Fox-Rabinovich, Iosif S. Gershman, Anatoliy I. Kovalev, and Kenji Yamamoto

CONTENTS

11.1 Introduction....336
11.2 Synergistically Alloyed Ti–Al–Cr Intermetallics....336
11.2.1 Ternary Intermetallic Systems....336
11.2.1.1 General Oxidation Behavior....342
11.2.1.2 The First Stages of Oxidation of the Laves-Phase Alloy at 900°C....343
11.2.1.3 The Long-Term Oxidation of the Laves-Phase Alloy at 900°C....346
11.2.1.4 Oxidation at Temperatures Greater than 1000°C....347
11.3 Quaternary and Higher-Ordered Intermetallic Systems....351
11.4 Ternary and Higher-Ordered Hard Coatings....358
11.5 Thermodynamic-Based Principles of Synergistic Alloying of Protective and Wear-Resistant Materials and Coatings....366
11.5.1 The Trigger Effect for Synergistically Alloyed Oxidation-Resistant Materials and Wear-Resistant Coatings....367
11.5.2 Thermodynamic Principles of Synergistic Alloying of Quaternary Wear-Resistant Coatings for Severe Applications....368
11.6 Conclusions....377
References....378

11.1 INTRODUCTION

This chapter focuses on the principles of synergistic alloying of intermetallics and oxidation-resistant coatings, as well as wear-resistant coatings. The principles developed are based on the concept of irreversible thermodynamics. Although the composition of the coating differs for specific applications, the principles are similar.

The set of metals and compounds available for surface protection against severe environmental attack is small. These metals include transitional metals, such as Ti, Zr, Hf, V, Nb, Ta, Cr, Mo, and W, as well as Al and Si. The compounds used in protective-coating applications are the following: carbides, nitrides, oxides, and, less frequently, borides (see Chapter 10, Figure 10.2). The majority of novel materials are based on combinations of these elements and compounds. To improve the advanced materials' development, it is extremely important to find elements and compounds that can act synergistically under specific service conditions.

The main focus of this chapter is on Ti–Al–Cr-based intermetallics and coatings with high-temperature protective properties. The major ability of these materials is self-protection against severe environmental attack, as will be described in detail.

11.2 SYNERGISTICALLY ALLOYED TI–AL–CR INTERMETALLICS

11.2.1 Ternary Intermetallic Systems

One of the most desirable features of Ti–Al-based binary and ternary alloys is their ability to form protective surface oxides during oxidation. It is known that about 60 to 70 at. % Al is needed for binary Ti–Al alloys to form a protective alumina layer during oxidation in air [1–2], but these binary alloys are extremely brittle, limiting their practical use. In recent years, a large amount of data has been accumulated concerning the effect of alloying additions on the oxidation of TiAl ductile compounds. Several authors have shown that certain additional elements are beneficial in this context [3–5]. Ternary chromium-containing alloys (when the Cr addition is above 8 to 10 at. %) [6–7] exhibit excellent oxidation resistance when compared to γ-TiAl alloys owing to the formation of a protective alumina layer during oxidation in air (Figure 11.1). These alloys display promising properties for oxidation-protective-coating applications. Excellent oxidation stability over a wide range of temperatures (750 to 1100°C) has shown $Ti_{0.28}Al_{0.44}Cr_{0.28}$ (Laves-phase-based alloy) to be one of the more promising alloys. The mechanism of oxidation of this alloy is very complicated and not well understood. The oxidation behavior has been linked to synergistic alloying effects on the surface and at the oxide–matrix interface during oxidation.

The goal of this study is to further understand the mechanism of the so-called "chromium effect" [6–7] in ternary Ti–Al–Cr alloys. To research this effect, the oxidation behavior and oxide scale formation for two ternary Ti–Al–Cr alloys has been studied. They are the $Ti_{0.25}Al_{0.67}Cr_{0.08}$ τ-phase alloy with a L12 crystal structure and the $Ti_{0.28}Al_{0.44}Cr_{0.28}$–Ti (Al, Cr)$_2$ Laves-phase alloy with a C14 crystal structure. Their behavior has been compared with the alumina-forming binary Ti–Al_3 phase alloy having a tetragonal crystal structure DO_{22} (Figure 11.1 to Figure 11.3).

The chemical composition of the alloys under study is given in Table 11.1. All Ti–Al melts were prepared in-house. The starting materials were either in wire, pellet, or ribbon form (Ti, Al, and Cr, respectively) with a minimum of 99.5% purity. The pure metals were weighed to the desired composition and arc-melted in an argon atmosphere in a water-cooled hearth. The alloy button (5.0 to 5.5 cm^3) was remelted several times to promote homogeneity and was then heat-treated in argon at 1000°C for 100 h to complete the homogenization of the sample. Alloys were verified by x-ray diffraction to have the desired crystal structure. In addition, energy dispersive spectroscopy (EDS) was performed on polished sections of the alloys to check the uniformity of the composition and provide a semiquantitative analysis. To improve the accuracy of the EDS measurements, calibrating Ti–Al and Ti–Al–Cr samples were made, and their chemical composition was determined using

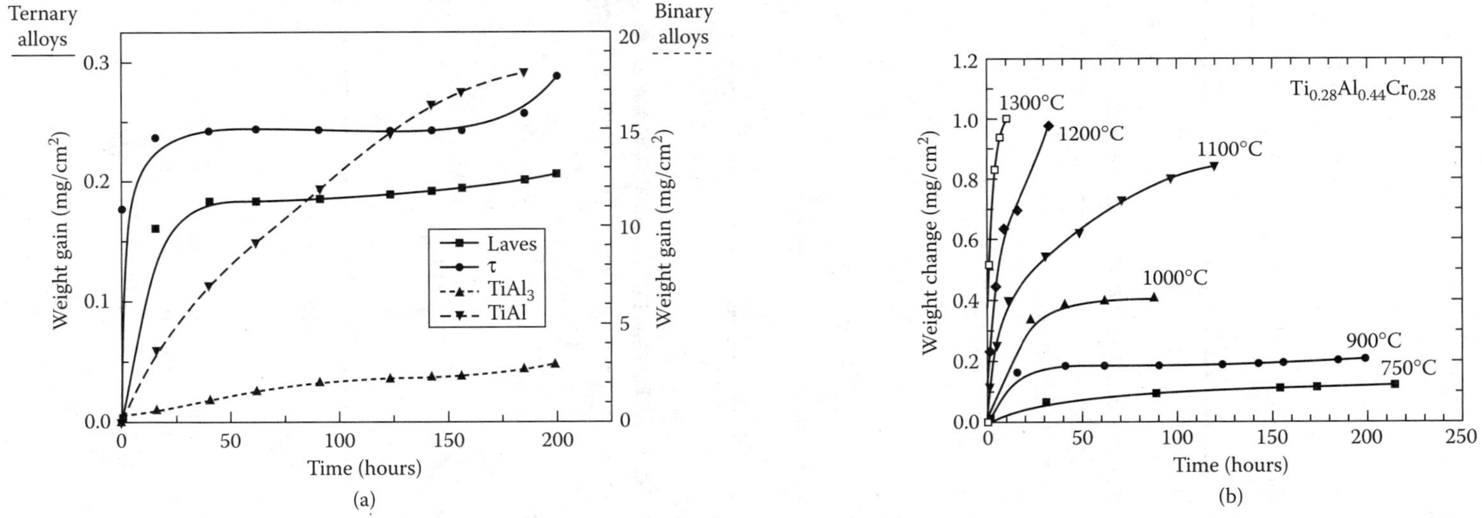

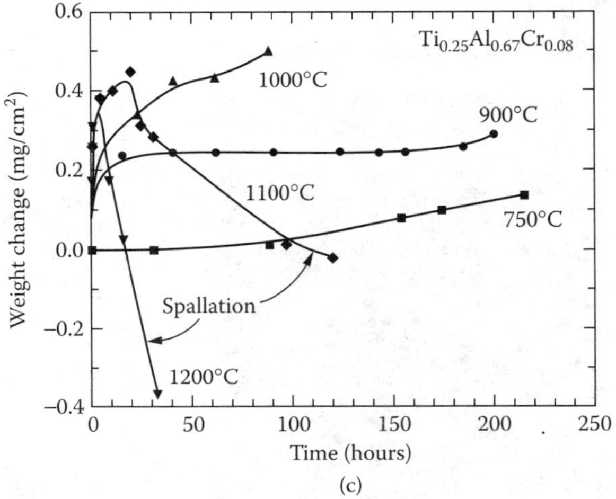

FIGURE 11.1 (a) Comparison of oxidation behavior of binary γ-TiAl and Al_3Ti alloys with ternary Ti–Al–Cr Laves and τ alloys at 900°C in air. Interrupted isothermal oxidation tests data, (b) oxidation behavior of $Ti_{0.28}Al_{0.44}Cr_{0.28}$ (Laves) alloy, and (c) $Ti_{0.25}Al_{0.67}Cr_{0.08}$ (τ) alloy in temperatures ranging from 750 to 1300°C. (From Fox-Rabinovich, G.S., Weatherly, G.C., Wilkinson, D.S., Kovalev A.I. The role of chromium in protective alumina scale formation during the oxidation of ternary TiAlCr alloys in air. *Intermetallics* 2004, 12, 2, 165–180. With permission.)

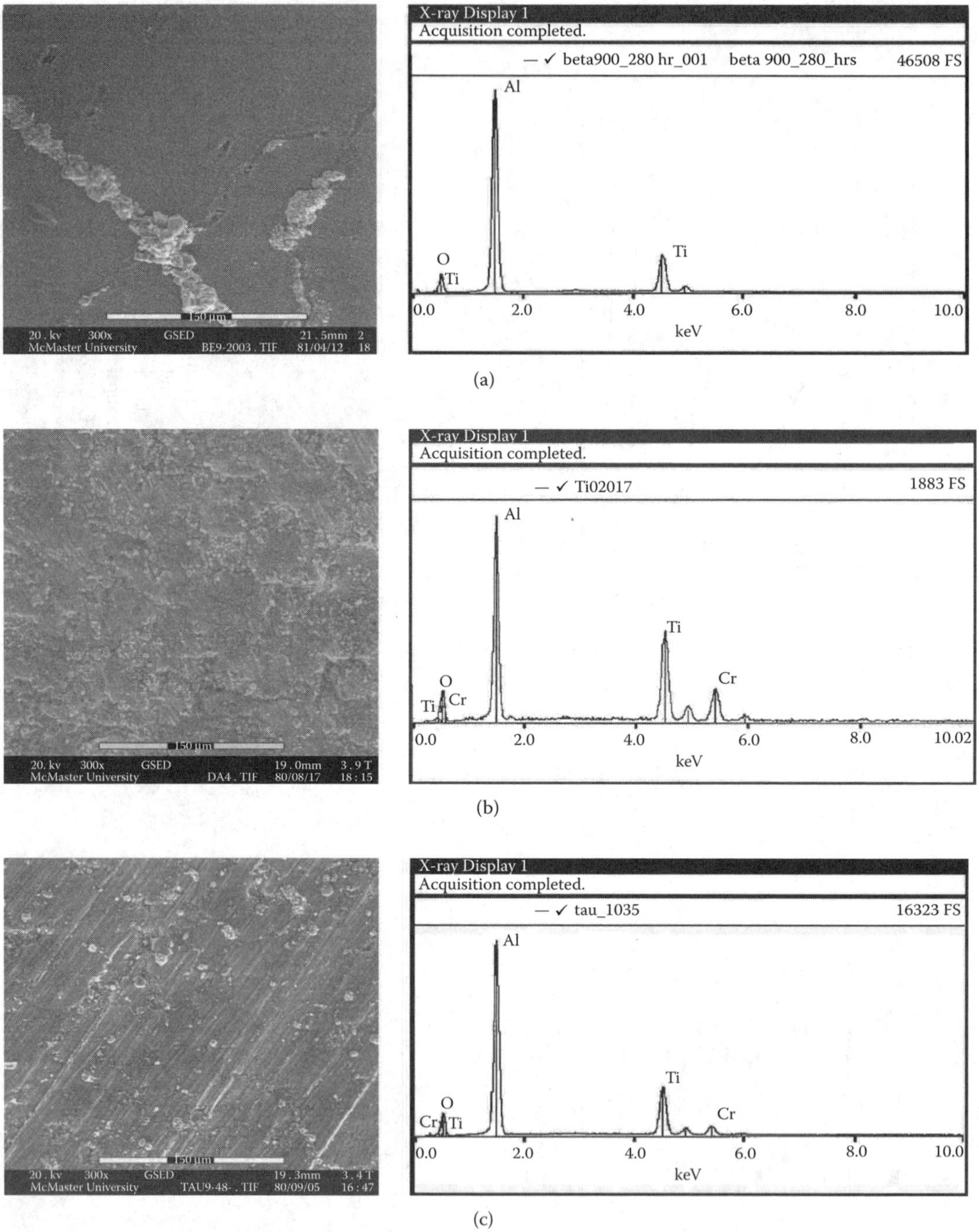

FIGURE 11.2 SEM micrographs of the oxide morphology, together with EDX spectra showing the relative abundance of Al, Ti, and Cr in the oxide films (a)–(c) oxide scales morphology of $TiAl_3$, Laves, and τ alloys at 900°C; (d)–(f) the same alloys at 1100°C. (From Fox-Rabinovich, G.S., Weatherly, G.C., Wilkinson, D.S., Kovalev A.I. The role of chromium in protective alumina scale formation during the oxidation of ternary TiAlCr alloys in air. *Intermetallics* 2004, 12, 2, 165–180. With permission.)

Continued.

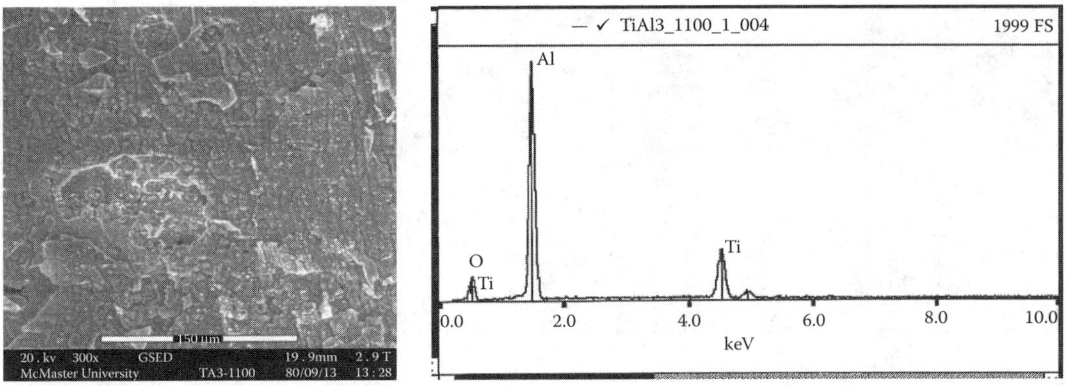

(d)

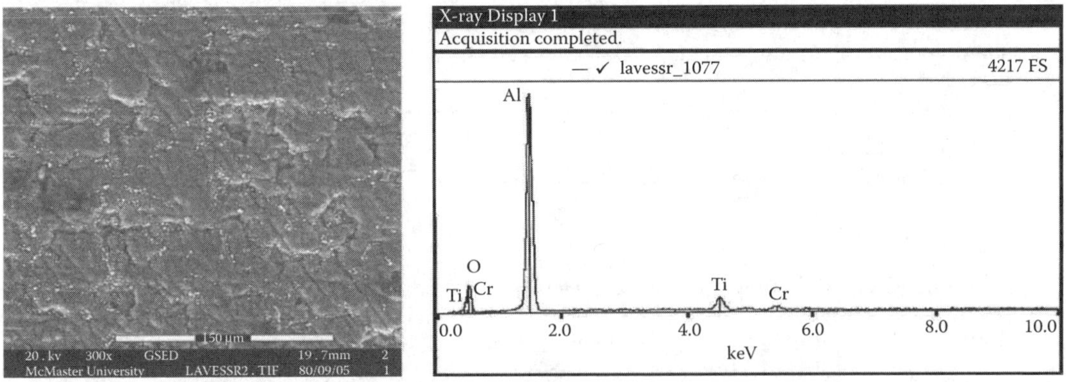

(e)

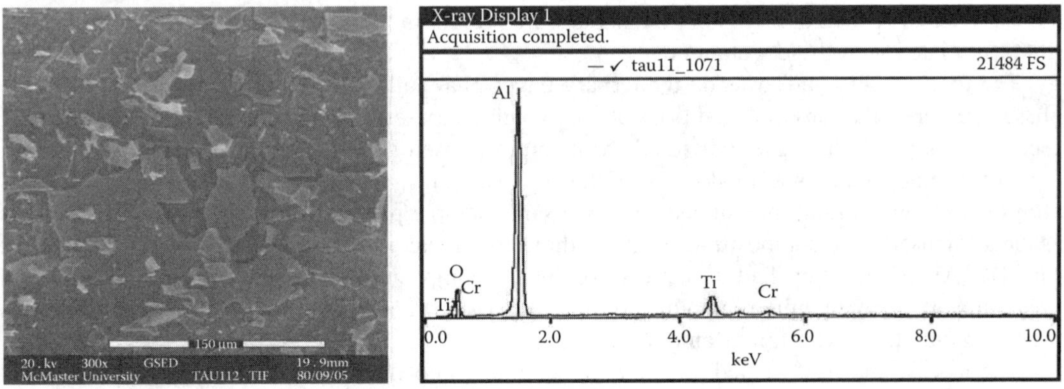

(f)

FIGURE 11.2 *Continued.*

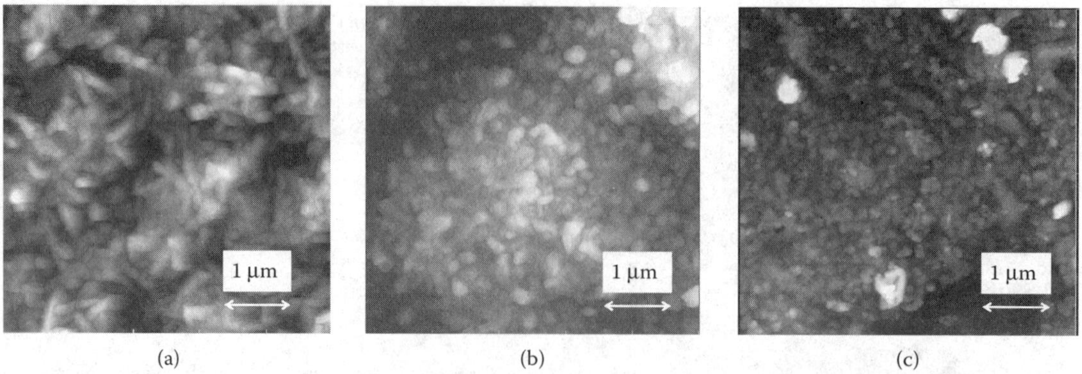

FIGURE 11.3 Atomic force microscope (AFM) images showing the oxide (alumina) morphology and grain size in (a) β, (b) τ, and (c) Laves-phase alloys after 200 h at 900°C. (From Fox-Rabinovich, G.S., Weatherly, G.C., Wilkinson, D.S., Kovalev A.I. The role of chromium in protective alumina scale formation during the oxidation of ternary TiAlCr alloys in air. *Intermetallics* 2004, 12, 2, 165–180. With permission.)

TABLE 11.1
Chemical Compositions of Binary Ti–Al and Ternary Ti–Al–Cr Alloys under Study

Number	Alloy	Measured Composition (at. %)	Phases Presented (XRD Data)
1	$TiAl_3$	Ti–74.5Al	β
2	$Ti_{0.25}Al_{0.67}Cr_{0.08}$	Ti–67.3Al–8.2Cr	τ (major), γ (traces)
3	$Ti_{0.28}Al_{0.44}Cr_{0.28}$	Ti–43.4Al–27.2Cr	Laves (major), τ, (β-Cr) traces

the wet-chemistry method. No other elements were present in the alloys, at or above the detectable limit associated with these techniques.

The focus of this study was on the ternary Laves phase alloy. It was found to consist of three-phase structures, the Laves, τ, and β-Cr phase, which are described in more detail in the following section. In order to study the individual oxidation response of each of the microstructural constituents of the alloy, buttons were also cast of the $Ti_{0.33}Al_{0.36}Cr_{0.31}$, $Ti_{0.21}Al_{0.31}Cr_{0.48}$, and $Ti_{0.30}Al_{0.47}Cr_{0.23}$ alloys (see Figure 11.4a) and subjected to the same sample preparation and oxidation treatments as the alloy itself. The compositions of these three additional alloy buttons are listed in Table 11.2. The $Ti_{0.33}Al_{0.36}Cr_{0.31}$ alloy had a single Laves phase composition. The $Ti_{0.21}Al_{0.31}Cr_{0.48}$ alloy was predominantly τ with minor amounts of Laves phase, whereas the $Ti_{0.30}Al_{0.47}Cr_{0.23}$ alloy was predominantly β-Cr with minor amounts of Laves.

The as-cast and heat-treated alloys were sectioned into flat specimens having an area of 0.3 cm^2 and a thickness of 1.0 mm using the electro-discharge wire cutter. The surfaces were ground down using a number 600 SiC paper and cleaned with ethanol. Each specimen was placed in an open alumina crucible and, after testing at a given temperature, was removed with the crucible from the furnace. A lid was placed on the crucible immediately after the oxidation treatment to recover any spalled scales.

Isothermal interrupted oxidation tests were performed in air over a wide range of temperatures from 750 to 1200°C. The exposures were conducted in an open-"room" atmosphere. The surface morphologies of the oxidized specimens were studied using the scanning electron microscope (SEM) and EDS, as well as the atomic force microscopy (AFM). The cross sections of scales (fractured sections) were studied using the same techniques. The thickness of the oxide films was

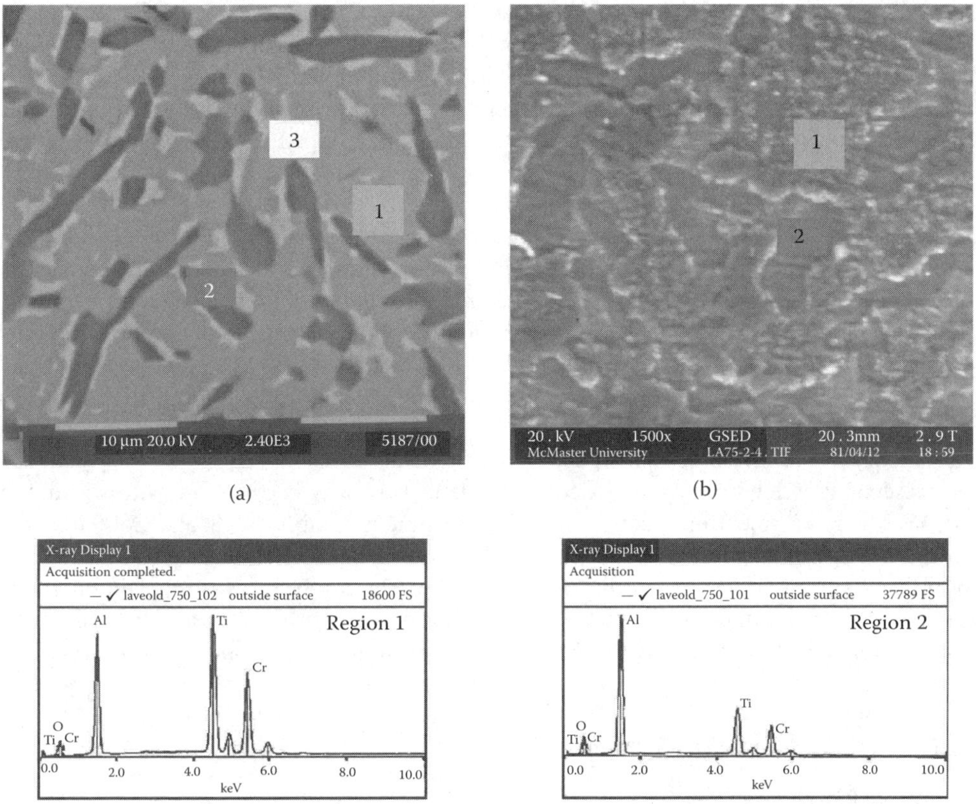

(c)

FIGURE 11.4 (a) (BSE)SEM image of $Ti_{0.28}Al_{0.44}Cr_{0.28}$ alloy showing the three-phase microstructure. The matrix phase is Laves (Mark 1). The two minor phases are predominantly τ (Mark 2) and predominantly β-Cr (Mark 3). (b) SEM image of alloy surface after 200 h at 750°C. (c) The Laves matrix (area 1) is covered by a mixture of alumina (containing Cr) and rutile, whereas the area 2 (τ)) is predominantly covered by alumina with minor amounts of rutile. (From Fox-Rabinovich, G.S., Weatherly, G.C., Wilkinson, D.S., Kovalev A.I. The role of chromium in protective alumina scale formation during the oxidation of ternary TiAlCr alloys in air. *Intermetallics* 2004, 12, 2, 165–180. With permission.)

TABLE 11.2
Chemical and Phase Composition of Multiphase $Ti_{0.28}Al_{0.44}Cr_{0.28}$ Alloy

Phases (Figure 11.3b)	Chemical Composition of the Multiphase Alloy and the Constituent Phases (at. %)			Phase Volume Content (%)	Phase Composition (XRD Data)
	Al	Ti	Cr		
$Ti_{0.28}Al_{0.44}Cr_{0.28}$ multiphase alloy (general)	41.6	29.3	29.1	—	Laves+τ+(γ+$AlCr_2$)-traces
Matrix phase $Ti_{0.33}Al_{0.36}Cr_{0.31}$ (see Mark 1, Figure 11.4a)	36.1	33.4	30.5	66.5	Laves
Phase $Ti_{0.30}Al_{0.47}Cr_{0.23}$ (see Mark 2, Figure 11.4a)	47.0	30.5	22.5	23.7	τ+Laves
Phase $Ti_{0.21}Al_{0.31}Cr_{0.48}$ (see Mark 3, Figure 11.4a)	31.4	21.0	47.6	9.8	β-Cr+Laves

measured. The transmission electron microscopes (a Philips CM12 TEM and JEOL 2010F STEM) were used to study the microstructure of the oxidized surface. An EDS analytical system was then used to identify the phase composition in the STEM micrographs.

The chemical composition of the surface layer was studied by scanning Auger electron spectroscopy (AES) using an ESCALAB-MK2 (VG) spectrometer equipped with a LEG 200 electron gun and an AG2 ion gun for preliminary ion cleaning of the specimens. An angle lap cross section at 5 degrees was fabricated at the surface of the sample. This geometry permits the investigation of any heterogeneity of the composition of the surface. Prior to AES, the degreased samples were placed inside the spectrometer preparation chamber and subjected to 5 min of weak etching by argon ions. The etching was carried out at a gas pressure of 10^{-4} Pa, an accelerating voltage of 4 kV, and an etching speed of the order of 2.0 monolayers/min. Such specimen preparation methods rule out the possibility of cross contamination on the analysis results. In previous investigations, under these conditions of ion etching, selective sputtering or amorphization of the surface layers were not observed [8–12].

The chemical and phase compositions of the oxidized Ti–Cr–Al alloys were investigated by means of secondary ion mass spectroscopy (SIMS). This was carried out with the aid of an ESCALAB MK2 (VG) electron spectrometer. The machine is equipped with an SQ300 ion analyzer of quadrupole type and an AG-6 argon ion gun that allows energy flows of up to 5 keV on a spot size of up to 0.5 μm in diameter on the sample surface. The ion beam was scanned on an area of 20×20 μm during accumulation of the spectra. The ion etching speed was of the order of 0.2 monolayers min^{-1}. The time of spectra recording was 10 min, and the analysis was carried out in the static mode.

11.2.1.1 General Oxidation Behavior

The interrupted isothermal oxidation test data for Ti–Al binary, as well as the Ti–Al–Cr ternary alloys at 900°C, are shown in Figure 11.1a to 11.1c. The data for the ternary Ti–Al–Cr Laves- and τ-phase alloys are shown in Figure 11.1b and 11.1c for a range of oxidation temperatures from 750 to 1200°C. The Laves-phase alloy shows stable oxidation behavior (with a parabolic or near-parabolic weight gain vs. time response) over the whole temperature range (Figure 11.1b). On the other hand, the τ-phase alloy, which has an excellent oxidation resistance at lower temperatures, shows extensive spallation at higher temperatures (Figure 11.1c). Although all three alloys predominantly form films of alumina, the oxidation rate of the $TiAl_3$ alloy is faster by about an order of magnitude at 900°C (Figure 11.1a). The EDS results show that the $TiAl_3$-, τ-, and the $Ti(Al,Cr)_2$ Laves-phase-based alloys form a layer of stable α-alumina (Al_2O_3) on the surface over the range of temperatures 900 (Figure 11.2a to Figure 11.2c) to 1100°C (Figure 11.2d to Figure 11.2f). This layer protects the surface of the Laves alloy against oxidation with high efficiency up to 1200°C [4]. The main difference in the oxidation behavior of these alloys lies in their spallation behavior. The binary $TiAl_3$ alloy forms a discontinuous alumina layer associated with intensive spallation. This is seen in Figure 11.2a, in which rutile crystals have formed in the spallation cracks, interrupting the formation of the continuous alumina layer, thereby increasing the oxidation rate. The morphology of the alumina on the surface of the binary $TiAl_3$ alloy is whisker shaped (Figure 11.3a) with a low adherence to the substrate and a low protective ability [13]. This morphology is typical of many other binary aluminides (e.g., nickel aluminides [4,13]).

Cr additions change the morphology of the oxidized surfaces. The alumina that forms on the surface of the ternary Cr-containing alloys has a smoother morphology (Figure 11.2b and Figure 11.2c) with fine-grained oxide scales that are compact and uniformly adherent (Figure 11.3b and Figure 11.3c).

Ternary Cr-containing titanium aluminides (i.e., τ and Laves) show excellent oxidation resistances up to 1000°C. These alloys form a continuous thin protective alumina layer during prolonged oxidation at temperatures up to 1000°C (Figures 11.1b and Figure 11.1c). However, the τ phase is unstable and decomposes at temperatures greater than 1000°C [14,15]. This is thought to be reason for the weakness of τ-phase-based alloys in spallation at temperatures above 1000°C (Figure 11.1c and Figure 11.2f), which accounts for the weight gain and oxide thickness changes observed.

The best oxidation stability over a wide temperature range was found with Laves-phase-based Ti $(Al,Cr)_2$ alloys, especially at high temperatures (1000 to 1200°C). This has been associated with the high stability of the Laves phase at these temperatures and a very low affinity of the Laves phase for oxygen [1–2,7]. However, the protective role of the Laves phase has definite limitations. Under severe conditions of high temperature oxidation (above 1000°C), Laves-phase alloys continue to exhibit a parabolic shape of oxidation curve (Figure 11.1b), but the alumina scales show evidence of some spalling (Figure 11.2e).

11.2.1.2 The First Stages of Oxidation of the Laves-Phase Alloy at 900°C

The microstructure of the Laves-phase alloy is seen in Figure 11.4a. Three phases are found, corresponding to the atomic number contrast image (taken with back-scattered electrons in the SEM) shown in Figure 11.4a. The continuous matrix phase (Mark 1), accounting for a volume fraction of 0.67, is a pure Laves phase ($\sim Ti_{0.33}Al_{0.36}Cr_{0.31}$). The second phase (Mark 2), accounting for a volume fraction of 0.24 consists predominantly of $\tau(\sim Ti_{0.30}Al_{0.47}Cr_{0.23})$, whereas the third phase (Mark 3) consists of predominantly β-Cr ($\sim Ti_{0.21}Al_{0.31}Cr_{0.48}$). In the initial stage of oxidation of the multicomponent alloy, each of these three phases exhibited independent oxidation behavior (see Figure 11.4b), i.e., different oxides initially formed (at different rates) at the surface of each of the three phases. However, on continued oxidation, the surface of the alloy is protected everywhere by a continuous alumina scale, a phenomenon to be described in greater detail in the following text.

The different oxidation response of the three phases in the Laves-phase alloy was also clearly demonstrated by the behavior of the alloys listed in Table 11.2. The results (see Figure 11.5) show that the rate of oxidation is highest for the "$Ti_{0.21}Al_{0.31}Cr_{0.48}$/predominantly β-Cr phase" and least for the "$Ti_{0.30}Al_{0.47}Cr_{0.23}$/predominantly τ phase." Surprisingly, the $Ti_{0.33}Al_{0.36}Cr_{0.31}$ alloy (pure Laves) has an oxidation rate about ten times greater than the $Ti_{0.30}Al_{0.47}Cr_{0.23}$ alloy (predominantly τ), which in turn shows slightly better oxidation resistance at 900°C than the multicomponent Laves alloy. The overall rates of oxidation of the four alloys, which can be seen in Figure 11.5, correlate well with the rutile content of the oxide, this being highest for the β-Cr alloy (where no alumina forms occur) and least for the τ alloy (Figure 11.6).

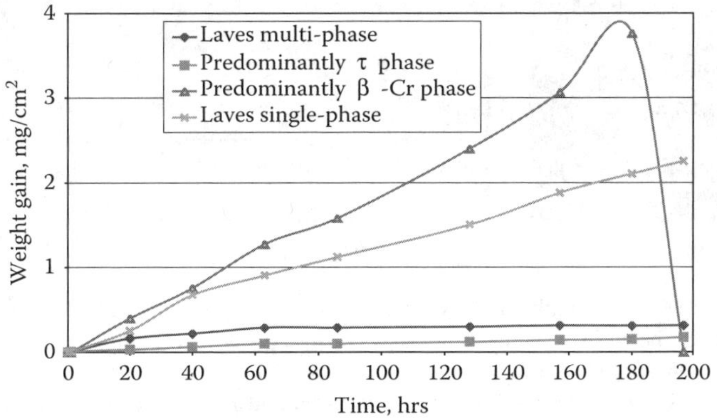

FIGURE 11.5 Oxidation behavior at 900°C of the Laves, predominantly τ, and predominantly β-Cr phases, compared to the multiphase Laves alloy. (From Fox-Rabinovich, G.S., Weatherly, G.C., Wilkinson, D.S., Kovalev A.I. The role of chromium in protective alumina scale formation during the oxidation of ternary TiAlCr alloys in air. *Intermetallics* 2004, 12, 2, 165–180. With permission.)

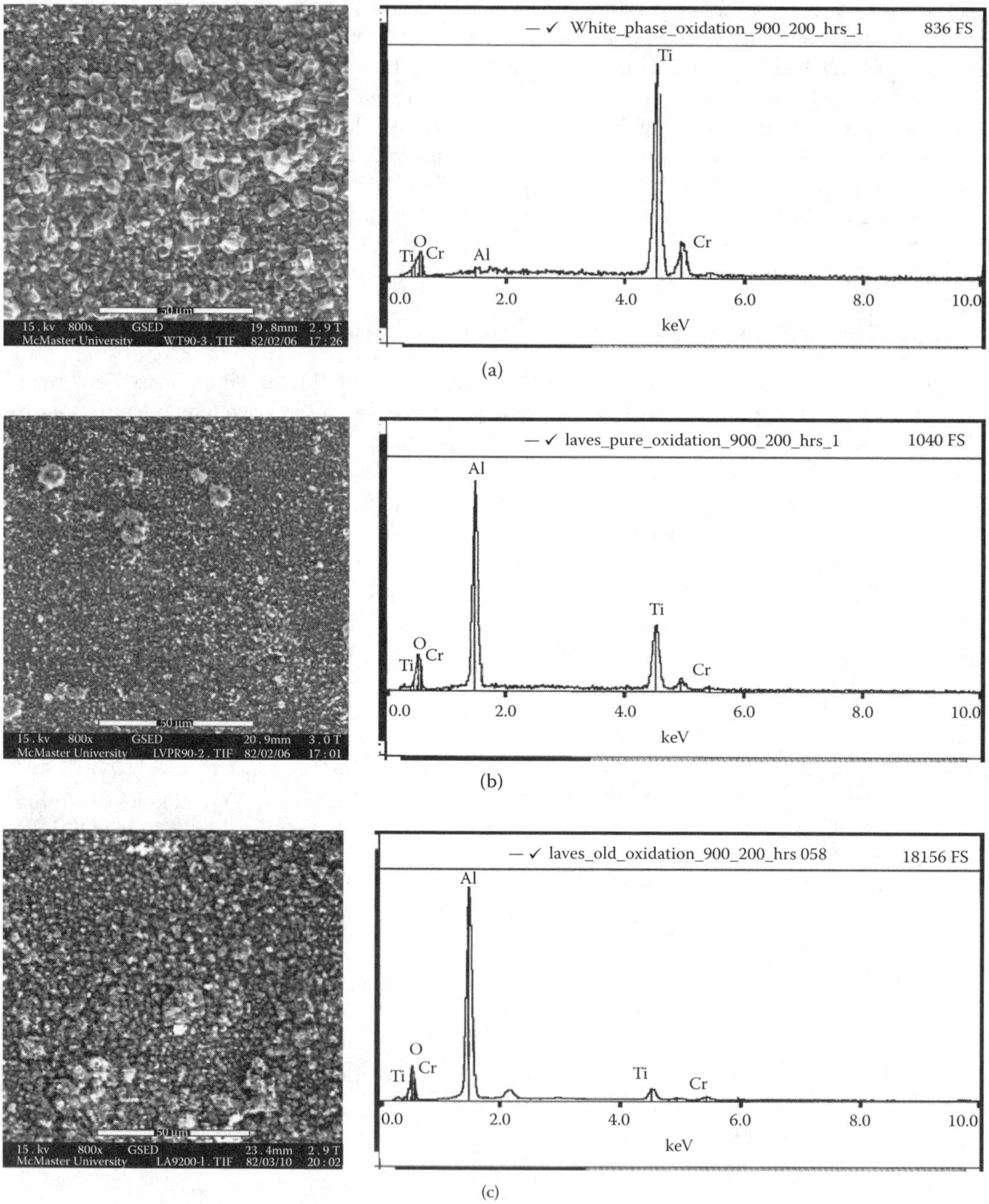

FIGURE 11.6 The SEM micrographs and associated EDX spectra show that the highest (lowest) rates of oxidation correlate with the rutile content of the oxide films: (a) predominantly β-Cr phase; (b) Laves phase; (c) multiphase Laves alloy; (d) predominantly τ phase alloy. (From Fox-Rabinovich, G.S., Weatherly, G.C., Wilkinson, D.S., Kovalev A.I. The role of chromium in protective alumina scale formation during the oxidation of ternary TiAlCr alloys in air. *Intermetallics* 2004, 12, 2, 165–180. With permission.)

Continued.

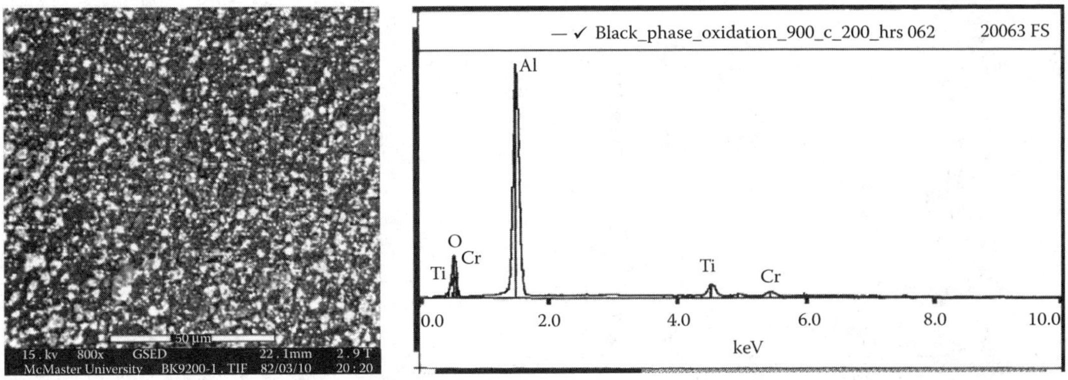

(d)

FIGURE 11.6 *Continued.*

Figure 11.7 presents the results of the chemical inhomogeneity study of the oxidized samples. The chemical inhomogeneity at the oxide–matrix interface has been studied using the procedures described previously for Auger analysis. The Auger data (Figure 11.7a) confirm that the surface oxide is richer in aluminum and poorer in titanium and chromium compared to the matrix alloy. The vertical lines in Figure 11.7 mark the approximate positions at which the analysis procedure samples either the metal or oxide phases. The peaks marked as P1 and P2 are rich in chromium and oxygen and depleted in aluminum (as shown by the Al_{60} signal that comes from the oxide), i.e., they are Cr-rich oxides.

Figure 11.8a presents the results of a SIMS study of the Laves-phase-based alloy after 1 h at 900°C. The ion beam scanned over an area 20 × 20 µm on the sample surface during the accumulation of the spectra. The mass spectrum of the positive and the negative secondary ions for the samples oxidized at 900°C are presented. It should be noted that the coefficients of the secondary ion egress for the refractory metal compound of the different metals vary. For the experimental conditions used in this study, the figure for chromium is 1.5, for aluminum it is 1.8,

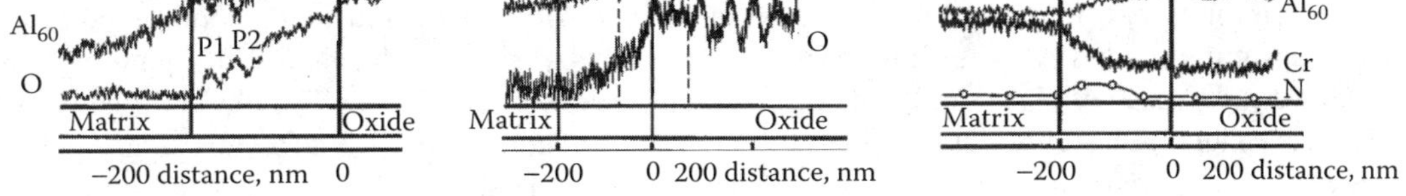

(a) (b) (c)

FIGURE 11.7 Scanning Auger electron spectra recorded across the metal-oxide interface region of multiphase Laves alloy after (a) 1 h at 900°C, (b) 200 h at 900°C, and (c) 50 h at 1100°C. (From Fox-Rabinovich, G.S., Weatherly, G.C., Wilkinson, D.S., Kovalev A.I. The role of chromium in protective alumina scale formation during the oxidation of ternary TiAlCr alloys in air. *Intermetallics* 2004, 12, 2, 165–180. With permission.)

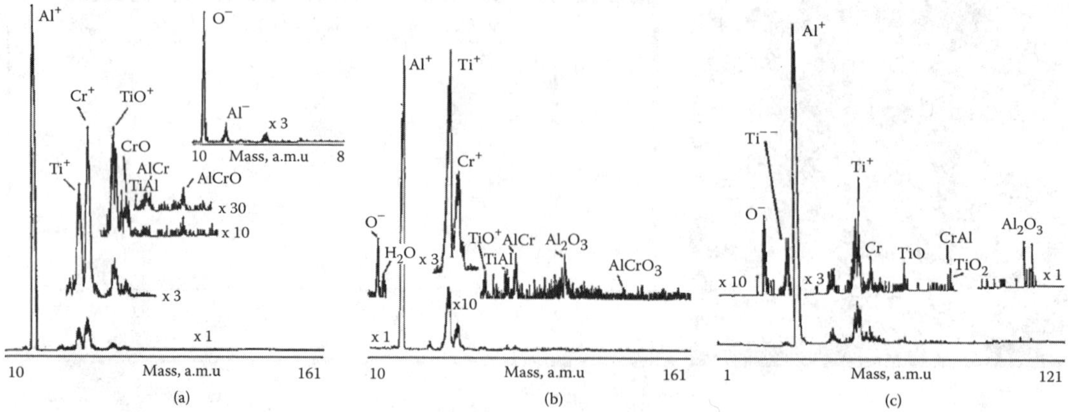

FIGURE 11.8 Mass spectra of secondary ions of the oxide film on the surface of the Laves alloy after (a) 1 h at 900°C, (b) 200 h at 900°C, and (c) 50 h at 1100°C. (From Fox-Rabinovich, G.S., Weatherly, G.C., Wilkinson, D.S., Kovalev A.I. The role of chromium in protective alumina scale formation during the oxidation of ternary TiAlCr alloys in air. *Intermetallics* 2004, 12, 2, 165–180. With permission.)

and for titanium it is 2.0 [16]. This means that even if the Ti content in the oxide is low, the intensity of its lines on the SIMS spectrum is significant. During short-time oxidation, the peaks of chromium and oxygen are pronounced. The presence of the CrO peak in the spectrum in Figure 11.8a implies that chromium actively takes part in the oxide formation during the initial stage of oxidation. The AlCrO (95 amu) peak is observed at all times of oxidation. This suggests that even during the initial stages of oxidation, the formation of complex aluminum and chromium oxides takes place.

11.2.1.3 The Long-Term Oxidation of the Laves-Phase Alloy at 900°C

A STEM elemental map of the surface after longer-term oxidation at 900°C shows a 1000-nm-thick alumina film (Figure 11.9a). The alumina layer contains only a small amount of chromium (Figure 11.9e). Several rutile islands can be seen, principally on the outer oxide surface (Figure 11.9d). The surface inhomogeneity study using Auger analysis confirms the STEM results (Figure 11.7b). There is an aluminum-depleted layer under the oxide (Figure 11.9c) at a relatively fixed concentration of titanium and chromium (Figure 11.7b). The outer layers of the oxide are poorer in chromium and titanium at a roughly constant concentration of aluminum (Figure 11.7b).

The SIMS study of the oxide film after 200 h at 900°C (Figure 11.8b) shows the $AlCrO_3$ cluster, indicative of the formation of a complex oxide of chromium and aluminum. In addition, we no longer observe the CrO peak that is found after short-term oxidation (Figure 11.8a). We can assume that chromium oxide-rich nuclei form only at the initial stages of oxidation and later in the oxidation process chromium oxide dissolves into the alumina. The AlCrO (95 amu) peak is observed at all times during the oxidation. After long-term oxidation, the Al_2O_3 peak forms, which again implies a significant decrease of chromium content in the later stages of oxidation. The oxide formed after long-term oxidation has more titanium (i.e., a higher rutile) content than that found after the short-term oxidation, in agreement with the STEM data (Figure 11.9d).

At this stage of oxidation, the phase next to the oxide is a coarse-grained Laves phase, which can be identified by the electron diffraction pattern seen in Figure 11.9g. A single Laves phase is found at the surface after about 20 h at 900°C, reaching a maximum width of about 30 μm after 70 h, but thereafter the width slowly decreases (see Figure 11.10). The reason for this decline is probably associated with the formation of thermal cracking at the surface of the Laves phase (see the inserts in Figure 11.10), which lead to enhanced local spalling. The single Laves phase is much

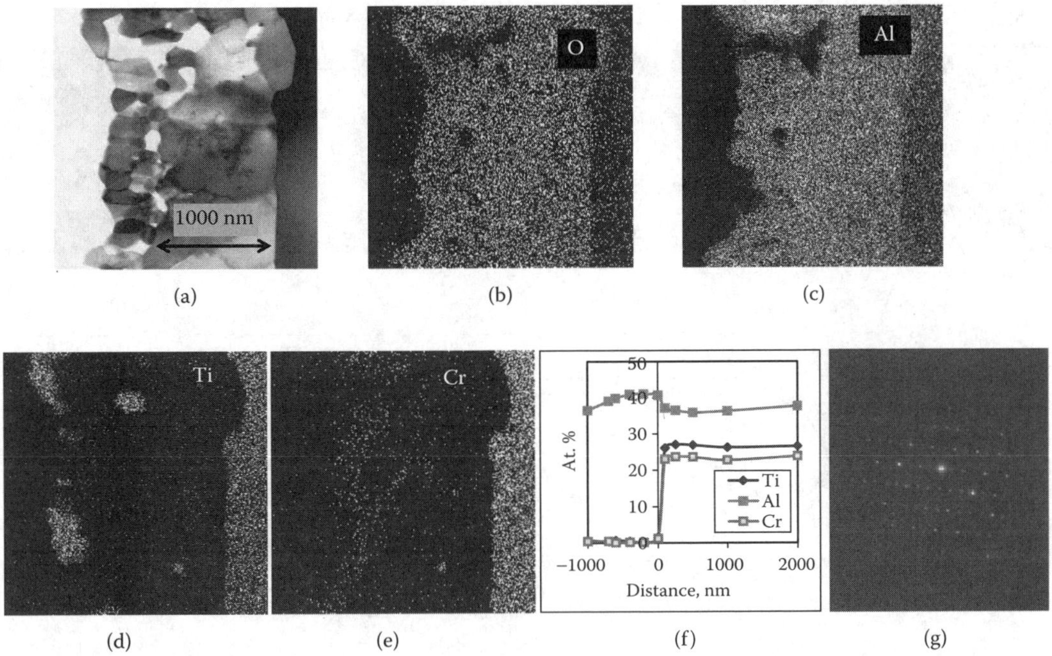

FIGURE 11.9 (a) STEM micrograph, (b)–(f) EDX elemental maps and line spectra of oxide film on Laves alloy after 200 h at 900°C, (g) diffraction pattern from the substrate next to the oxide film shows that it is predominantly a single Laves phase in a [10-10] orientation with respect to the electron beam. (From Fox-Rabinovich, G.S., Weatherly, G.C., Wilkinson, D.S., Kovalev A.I. The role of chromium in protective alumina scale formation during the oxidation of ternary TiAlCr alloys in air. *Intermetallics* 2004, 12, *2*, 165–180. With permission.)

less resistant to cracking than the multiphase alloy, as demonstrated by the crack lengths detected in microhardness impressions in the single and multiphase alloys (Figure 11.11).

11.2.1.4 Oxidation at Temperatures Greater than 1000°C

At high temperatures (1100°C), the process of oxidation of the Laves-phase-based alloy becomes more complicated. Figure 11.7c presents the Auger lines intensity change in the direction from the core metal to the oxidized surface using an angle lap cross section for a specimen oxidized at 1100°C for 50 h. A significant decrease in chromium content accompanied by oxygen enrichment is observed. The intensity of the titanium signal rises slightly in the intermediate region, and this correlates with the pickup of nitrogen at the interface. It is known that at high temperatures TiN can form in the air during oxidation experiments of the Laves phase. TiN is unstable on exposure to oxygen and forms rutile [1,2]. The oxidation of the aluminum in the Laves alloy and alumina formation proceeds at the same time as nitride and rutile formation.

Based on the SIMS data presented in Figure 11.8c, we can see that the oxide forming on the surface of the Laves-phase-based alloy during high-temperature oxidation (1100°C) consists of alumina, rutile, and a mixed Al- and Cr-containing oxide. The $CrAl^+$ (79 amu) cluster is also present in the spectrum. Comparing the change of the Ti and Cr peak intensities for the samples oxidized at 900 and 1100°C, we can see that the ratio I_{Ti}/I_{Cr} changes from 2.25 to 3.44. As soon as the temperature of the oxidation is increased, the formation of rutile also increases. This dependency varies noticeably with temperature, with the rutile content increasing most rapidly in the temperature range of 1000 to 1100°C.

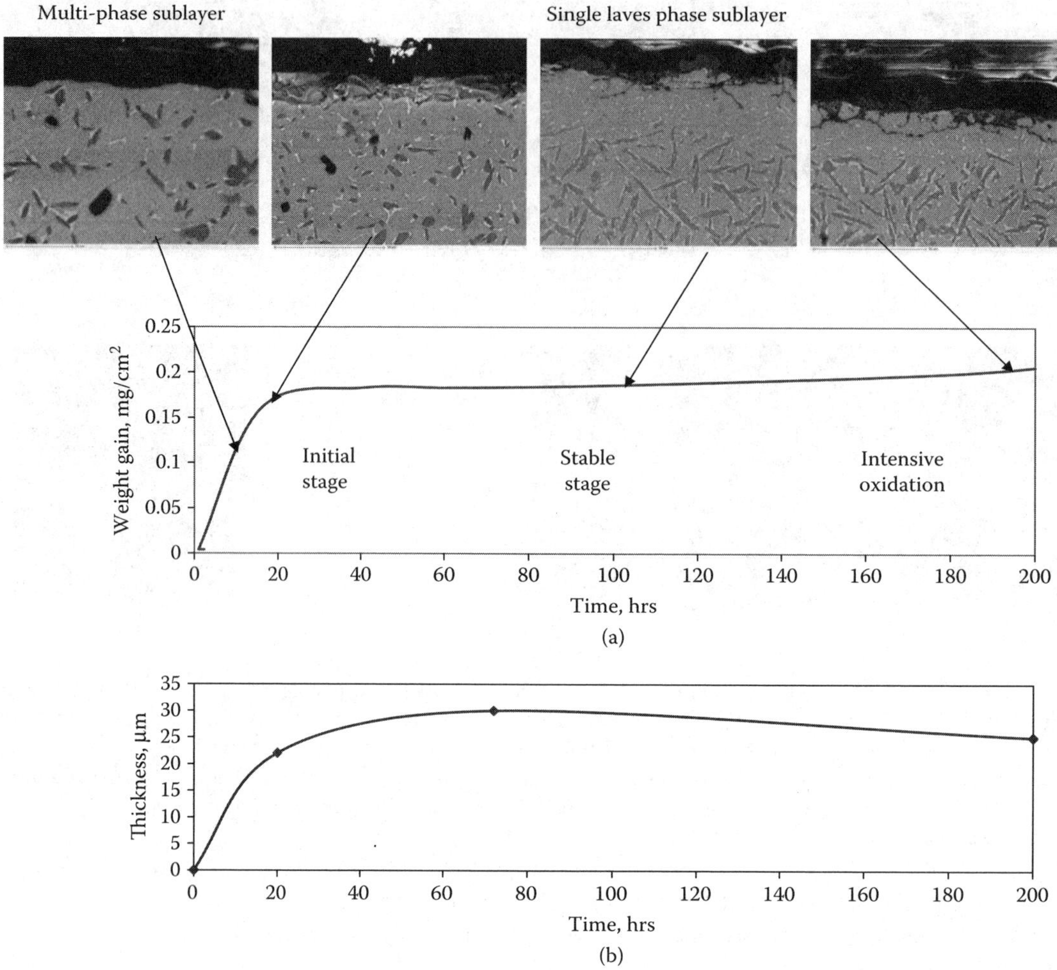

FIGURE 11.10 The formation of the single Laves layer at the surface of the multiphase Laves alloy at 900°C. Note the progressive disappearance of the τ phase at the surface and the formation of spall cracks in the Laves layer after 200 h. (From Fox-Rabinovich, G.S., Weatherly, G.C., Wilkinson, D.S., Kovalev A.I. The role of chromium in protective alumina scale formation during the oxidation of ternary TiAlCr alloys in air. *Intermetallics* 2004, 12, 2, 165–180. With permission.)

The Auger spectroscopy data also shows that the chemical composition of the surface oxide layer depends on the temperature and time of oxidation. The change in the atomic concentration ratio (as measured by AES) of Cr and Al as a function of the time and temperature of oxidation can be seen in Table 11.3.

Our results on the oxidation behavior of Ti–Cr–Al alloys are in broad agreement with previous studies of this subject [1–2,6–7,14,17]. The critical observations are:

1. At temperatures below 1000°C, excellent oxidation resistance is displayed by the ternary τ alloy containing 67 at. % Al. Although it contains less Al than the binary β alloy (75 at. % Al), the finer alumina grain size of the ternary alloy appears to prevent spalling and the formation of rutile, which limits the oxidation resistance of the binary alloy.
2. At temperatures above 1000°C, the best cyclic oxidation resistance is displayed by the multiphase Laves alloy. Although this alloy contains less Al than the τ alloy (44

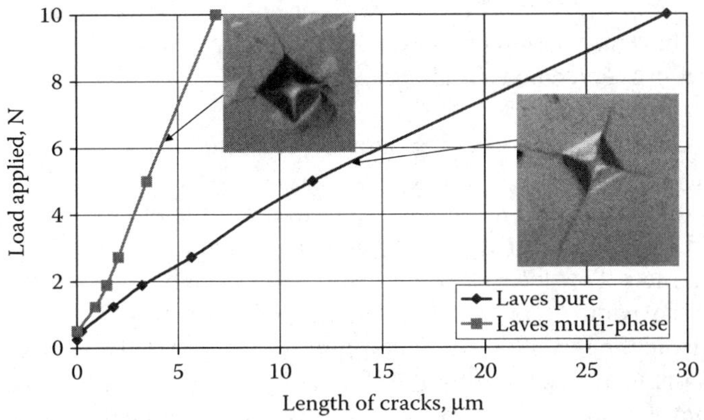

FIGURE 11.11 Microhardness indentation study of multiphase and single-phase Laves alloy showing the maximum crack length observed as a function of the applied load. The inserts compare the appearance of the cracked indentations at a load of 500 g. (From Fox-Rabinovich, G.S., Weatherly, G.C., Wilkinson, D.S., Kovalev A.I. The role of chromium in protective alumina scale formation during the oxidation of ternary TiAlCr alloys in air. *Intermetallics* 2004, 12, 2, 165–180. With permission.)

TABLE 11.3
Composition of $(Al,Cr)_2O_3$ Oxide Film vs. Time and Temperature of Oxidation

Oxidizing Parameters		Composition of	AES
Temperature	Time (h)	$(Al,Cr)_2O_3$	I_{Cr}/I_{Al}
900	1	43% Cr_2O_3; 57% Al_2O_3	0.78
900	200	39% Cr_2O_3; 61% Al_2O_3	0.65
1100	53	16% Cr_2O_3; 84% Al_2O_3	0.20

vs. 67 at. %), it is stable up to the melting point of the Laves phase, whereas is unstable [14].

3. The multiphase Laves alloy has superior oxidation resistance compared to an alloy that contains a single Laves phase. This is related to the reservoir effect [7], in which the high-Al-content τ phase in the multicomponent alloy dissolves and "feeds" the growth of a stable alumina film at the surface of the alloy. At the same time, a coarse-grained single Laves phase develops immediately below the surface of the alloy. The formation of this phase cannot account for the good oxidation resistance of the multicomponent alloy as the single Laves-phase alloy has relatively poor oxidation resistance. The effectiveness of the alumina film in protecting the multiphase alloy during long oxidation periods appears to be limited by the low fracture toughness of the Laves phase under thermal cyclic conditions.
4. Cr has an active role in the formation of alumina. Both the Auger spectroscopy and SIMS results point to the formation of Cr-rich oxides in the first stages of oxidation. In the later stages of oxidation, the average Cr content in the oxide declines and a mixed $(Al,Cr)_2O_3$ form. This observation might explain the "Cr effect" [1–2] and is discussed in more detail in following text.
5. In alloys in which rutile forms extensively, the oxidation resistance is low. The ratio of alumina to rutile in the oxide (following the completion of any transient oxidation effects as found in the multiphase Laves alloy) provides a good measure of the oxidation

> resistance of the alloy. A critical Al content is needed in the base alloy to suppress rutile formation in the first stages of oxidation of the alloy, in addition to elements that promote the formation of a fine equiaxed alumina structure.

The origin of the "chromium" effect remains to be explained. Of all the elements that have been added to binary Ti–Al alloys to enhance their oxidation resistance, chromium is unique in that relatively small amounts (e.g., 8 at. % in τ phase) are sufficient to form a protective alumina film and suppress the formation of rutile. A number of factors are probably important. All the components of the ternary Ti–Al–Cr alloy are involved in the oxidation process one after another. We can assume that the oxidation rates of each element of the ternary Ti–Al–Cr alloy are different. We can speculate that in the series Cr, Al, Ti chromium has the highest oxidation rate and titanium has the lowest. This is mainly caused by the diffusivity of the element in the ternary alloy. We know that chromium cannot form an oxide more thermodynamically stable than alumina and titania or change the thermodynamic activity of aluminum and titanium [18] in the ternary alloy during oxidation. Our results show that in the early stages of oxidation, a chromium-rich mixed oxide forms. The diffusion rate of oxygen in Cr_2O_3 is about 100 times higher than in Al_2O_3 [19], so the initial growth rate of the oxide will be enhanced. Expansion of chromium oxide nuclei takes place because of intensive oxygen inward diffusion. At the same time, aluminum is displaced to the neighbor sites and the alumina nuclei are formed but at a lower rate. Chromium oxide is completely soluble in alumina and is followed by the formation of complex $(Al,Cr)_2O_3$ oxide. This very fine-grained oxide will also promote alumina-based oxide film growth by grain boundary diffusion. During the early stage of oxidation, kinetic parameters control the processes that take place on the surface. The oxidizing surface is an open thermodynamic system reacting with the environment via heat, matter, and entropy [20–21]. The formation of an oxide film occurs under conditions that are close to equilibrium but are conditions in which the concepts of irreversible thermodynamics apply. Under these conditions, the rate of entropy production in the system will be minimized [21–22].

Following the arguments presented in the following text (see Part 3), the rate of entropy production can be shown to be the sum of two terms, one associated with the thermodynamic forces and the other with the fluxes [23]. If the oxide film that forms is protective, the change in entropy production decreases, leading to a decrease in the flux of matter and hence a rapid drop in the rate of oxidation, as observed in Figure 11.1. The same arguments could be used in Laves single sublayer formation under the oxide film. This sublayer has a low affinity to oxygen and prevents its intensive inward diffusion during oxidation [7]. This layer formation most probably promotes stability of the oxidation process.

The "chromium effect" is most evident in the early stages of oxidation in the lower temperature range we have studied. On prolonged oxidation at lower temperatures, or at higher temperatures, the Cr content of the oxide film declines. Under these conditions, the most thermodynamically stable Al-rich $(Al,Cr)_2O_3$ forms at the metal–oxide interface. This observation reinforces the idea that the "Cr effect" is associated with the initial nucleation and growth of the Cr-doped alumina.

We have to emphasize that to achieve an oxidation resistance greater than the binary alloy, the components in a ternary Ti–Al–Cr alloy must collaborate synergistically during oxidation. The initial stage of the oxidation of the ternary Ti–Al–Cr alloy is most important. During this initial stage, the kinetics of the competitive multi-component diffusion of the alloy is the major factor that controls the oxidation process. Cr has the highest diffusivity when compared to other components of the ternary alloy and forms its own oxide at the onset of oxidation. This impacts both the interaction of the alloy's components with oxygen and alumina formation during the initial stage of oxidation. Owing to the formation of a fine-grained complex $(AlCr)_2O_3$ oxide film, the ability of the ternary Ti–Al–Cr alloy to produce the protective oxide layer is enhanced. During longer-term oxidation, the chromium oxide dissolves in alumina, which reduces the interdiffusion flow because of the formation of a thermodynamically stable oxide.

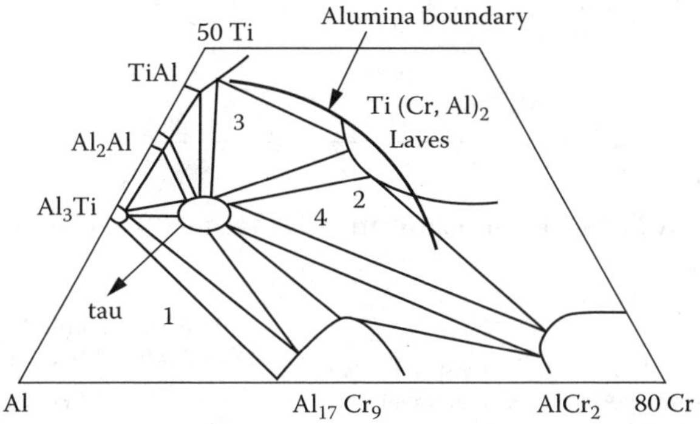

FIGURE 11.12 Schematic Ti–Al–Cr ternary phase diagram. Marks: 1 — τ-phase ($Ti_{0.25}Al_{0.67}Cr_{0.08}$) alloy, 2 — Laves + τ + β-Cr ($Ti_{0.28}Al_{0.44}Cr_{0.28}$) alloy, 3 — γ + Laves ($Ti_{0.37}Al_{0.51}Cr_{0.12}$) alloy, 4 — τ + Laves ($Ti_{0.30}Al_{0.50}Cr_{0.20}$) alloy. (From Brady, M.P., Gleeson, B., Wright, I.G. Alloy design strategies for promoting protective oxide-scale formation. *J. Met.* 2000, 52, *1*, 16–21.) (From Fox-Rabinovich, G.S., Wilkinson, D.S., Veldhuis, S.C., Dosbaeva, G. K., Weatherly, G.C. Oxidation resistant Ti-Al-Cr alloy for protective coating applications. *Intermetallics*, 2006, 14, 2, 189–192. With permission.)

11.3 QUATERNARY AND HIGHER-ORDERED INTERMETALLIC SYSTEMS

There are a number of Ti–Al–Cr alloys with high oxidation stability that assure alumina formation during oxidation (Figure 11.12) [24]. The first group comprises aluminum-rich, τ-phase-based alloys. This alloy exhibits excellent oxidation stability in the temperature range of 750 to 900°C [25,26]. On the other hand, the τ-phase alloy shows extensive spallation at temperatures above 1000°C [27,28]. The second group is the chromium-rich $Ti(Al,Cr)_2$ Laves-phase alloy. These alloys have high oxidation stability during oxidation at temperatures above 900°C. To improve the oxidation stability within a wide range of temperatures, a combination of Laves and τ phases in a multiphase of Ti–Al–Cr alloy seems a promising solution.

The goals of this work are, therefore, as follows:

1. To compare oxidation stability of two types of the ternary Ti–Al–Cr multiphase alloys, i.e., Laves- and τ-phase-based alloys
2. To optimize the phase and chemical composition of the multiphase Laves- and τ-phase-based alloys to improve their oxidation stability by means of synergistic alloying using active elements and Si

The chemical compositions of the ternary Ti–Al–Cr alloys under the study are given in Table 11.4 and Table 11.5. All the Ti–Al–Cr melts were prepared in-house.

The main feature of the ternary Ti–Al–Cr alloys under this study is their ability to form alumina on the surface during oxidation in air (Figure 11.12). To understand the oxidation mechanism in such alumina-forming alloys, isothermal interrupted oxidation tests were performed in air from 750 to 1100°C. The oxidation kinetics data at temperatures 900 and 1100°C are presented in Figure 11.13 and Figure 11.15. Oxide scale morphology at the corresponding temperatures is presented in Figure 11.14 and Figure 11.16. The SEM micrographs of the microstructure of the studied materials are shown in Figure 11.18.

The oxidation behavior of these alloys strongly depends on the temperature of oxidation. The oxidation stability of τ phase is excellent at a temperature of 900°C (Figure 11.13 and Figure 11.14). However, at 1000°C and above, the oxidation behavior deteriorates owing to an island-like (discontinuous) alumina oxide film formation and intensive spallation of this layer (Figure 11.15

TABLE 11.4
Chemical and Phase Composition of the Ternary Ti–Al–Cr Alloys under Study

Number	Nominal Alloy Composition	Measured Alloy Composition (at. %)	Phases Presented (XRD and Quantitative Metallography Data)			
			τ	Ti (Al,Cr)$_2$ Laves	γ	β-Cr
1	$Ti_{0.25}Al_{0.67}Cr_{0.08}$	Ti–67.3Al–8.2Cr	100		traces	
2	$Ti_{0.25}Al_{0.60}Cr_{0.15}$	Ti–60.3Al–15.2Cr	94.23	5.77		
3	$Ti_{0.28}Al_{0.44}Cr_{0.28}$	Ti–43.4Al–27.2Cr	29.15	62.47		8.38
4	$Ti_{0.37}Al_{0.51}Cr_{0.12}$	Ti–51.2Al–11.8Cr		26.1	73.9	
5	$Ti_{0.30}Al_{0.50}Cr_{0.20}$	Ti–50.2Al–20.8Cr	54.27	42.33		3.44
6	$Ti_{0.30}Al_{0.55}Cr_{0.15}$	Ti–54.2Al–15.8Cr	63.11	36.89		
7	$Ti_{0.25}Al_{0.55}Cr_{0.20}$	Ti–55.1Al–19.9Cr	75.12	19.92		4.96

Source: Fox-Rabinovich, G.S., Wilkinson, D.S., Veldhuis, S.C., Dosbaeva, G. K., Weatherly, G.C. Oxidation resistant Ti-Al-Cr alloy for protective coating applications. *Intermetallics*, 2006, 14, 2, 189–192. With permission.

TABLE 11.5
Chemical Compositions and Weight Gain after Isothermal Oxidation Tests of the Ternary Ti–Al–Cr Alloys Doped with Hf, Y, Si, and Zr

Nominal Alloy Composition	Measured Alloy Composition (at. %)	Weight Gain (mg/cm^2)	
		900°C, 400 h	1100°C, 100 h
$Ti_{0.25}Al_{0.60}Cr_{0.15}$	Ti–60.3Al–15.2Cr	0.175	1.25
$Ti_{0.24}Al_{0.60}Cr_{0.15}Si_{0.01}$	Ti–60.3Al–15.2Cr–0.9Si	0.25	1.73
$Ti_{0.245}Al_{0.60}Cr_{0.15}Si_{0.005}$	Ti–60.3Al–15.2Cr–0.55Si	0.168	0.77
$Ti_{0.25}Al_{0.55}Cr_{0.20}$	Ti–55.1Al–19.9Cr	0.165	0.78
$Ti_{0.248}Al_{0.55}Cr_{0.20}Zr_{0.002}$	Ti–55.1Al–19.9Cr–0.2 Zr	0.16	0.58
$Ti_{0.248}Al_{0.55}Cr_{0.20}Hf_{0.002}$	Ti–55.1Al–19.9Cr–0.2 Hf	0.13	0.41
$Ti_{0.248}Al_{0.55}Cr_{0.20}Y_{0.002}$	Ti–55.1Al–19.9Cr–0.2 Y	0.149	0.66
$Ti_{0.247}Al_{0.55}Cr_{0.20}Hf_{0.002}Y_{0.001}$	Ti–55.1Al–19.9Cr–0.2 Hf–0.1Y	0.09	0.45
$Ti_{0.243}Al_{0.55}Cr_{0.20}Hf_{0.002}Si_{0.005}$	Ti–55.1Al–19.9Cr–0.2 Hf–0.5Si	0.1	0.42
$Ti_{0.242}Al_{0.55}Cr_{0.20}Hf_{0.002}Si_{0.005}Y_{0.001}$	Ti–55.1Al–19.9Cr–0.2Hf–0.5Si–0.1Y	0.046	0.4

Source: Fox-Rabinovich, G.S., Wilkinson, D.S., Veldhuis, S.C., Dosbaeva, G. K., Weatherly, G.C. Oxidation resistant Ti-Al-Cr alloy for protective coating applications. *Intermetallics*, 2006, 14, 2, 189–192. With permission.

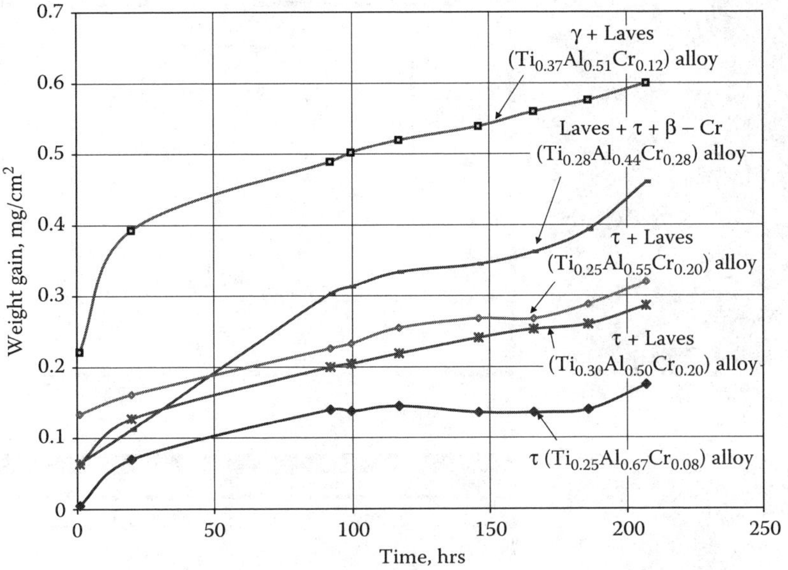

FIGURE 11.13 Oxidation behavior at 900°C of the Ti–Al–Cr based alloys in air. (From Fox-Rabinovich, G.S., Wilkinson, D.S., Veldhuis, S.C., Dosbaeva, G. K., Weatherly, G.C. Oxidation resistant Ti-Al-Cr alloy for protective coating applications. *Intermetallics*, 2006, 14, 2, 189–192. With permission.)

and Figure 11.16). It is known from the literature that the τ phase is unstable at temperatures above 1000°C [27–28]. The stability of the τ-phase alloy at 1100°C may be improved slightly by increasing the Cr content from 8 to 15% (Figure 11.17) [29]. The major phase of these alloys is τ ($Ti_{0.25}Al_{0.67}Cr_{0.08}$ and $Ti_{0.24}Al_{0.66}Cr_{0.10}$ alloys; see Table 11.4 and Figure 11.18a), but alloys richer in Cr (10 to 15 at. %) contain a small amount of the Laves-phase $Ti_{0.25}Al_{0.60}Cr_{0.15}$ alloys; see Table 11.1 and Figure 11.18b).

The Laves-phase-based alloys exhibit promising oxidation resistance (Figure 11.13 to Figure 11.15), especially at high temperatures because of the low permeability of this phase to oxygen [7] and the high adherence of the oxide layer to the substrate matrix. High oxidation resistance at high temperatures shows the Laves-phase-based $Ti_{0.28}Al_{0.44}Cr_{0.28}$ alloy. This alloy contains the $Ti(Al,Cr)_2$ Laves phase, but also τ and β-Cr (Table 11.4; Figure 11.18c). This data suggests that alloys containing a mixture of Laves and other phases might be promising in terms of oxidation stability. A similar conclusion can be made for an alloy containing a mixture of Laves and γ phases (Figure 11.18d) [26], although this does not offer the most beneficial oxidation rates (Figure 11.13 and Figure 11.15), i.e., Laves + τ phase alloys offer better oxidation stability than γ + Laves alloys [30]. Indeed, this family of alloys shows the lowest oxidation rate from 750 to 1100°C (Figure 11.13 and Figure 11.15) because of continuous alumina layer formation (Figure 11.14 and Figure 11.16). We can then conclude that the alloys containing several phases (Laves, τ, and a small amount of a β-Cr) show the best oxidation stability (Figure 11.13 to Figure 11.16). Thus, further optimization of the phase and chemical compositions (in the range of compositions $Ti_{0.25}Al_{0.60}Cr_{0.15}${τ (mainly) + Laves}/$Ti_{0.28}Al_{0.44}Cr_{0.28}${Laves (mainly) + τ}) of the ternary Cr-containing alloys for oxidation-protective-coating applications has been done (Table 11.4 and Figure 11.19).

The data presented in Table 11.4 together with micrographs shown in Figure 11.19 demonstrate that the alloys containing τ phase as a major phase and the $Ti(Al,Cr)_2$ Laves phase in the range of 25 to 45% have the highest oxidation stability (Figure 11.18a to Figure 11.18c). We believe that the presence of Laves phase in the τ-phase-based alloy stabilizes oxidation resistance of the multiphase alloy at high temperatures. The best oxidation stability over a wide range of temperatures

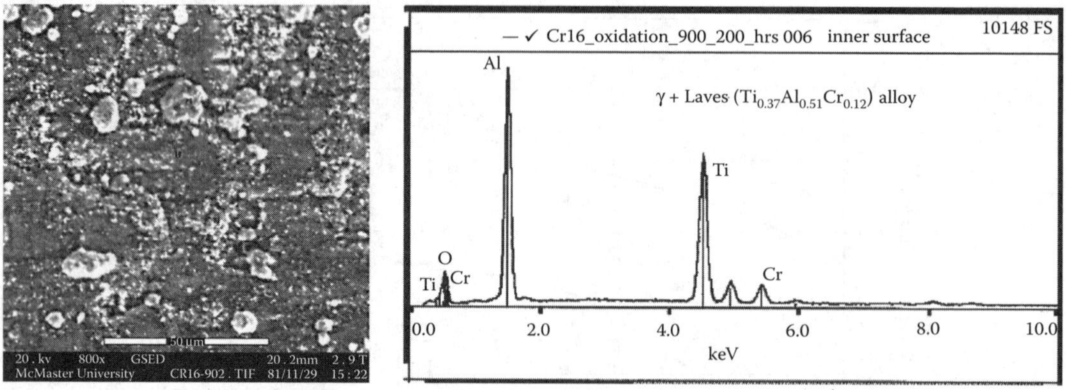

(a)

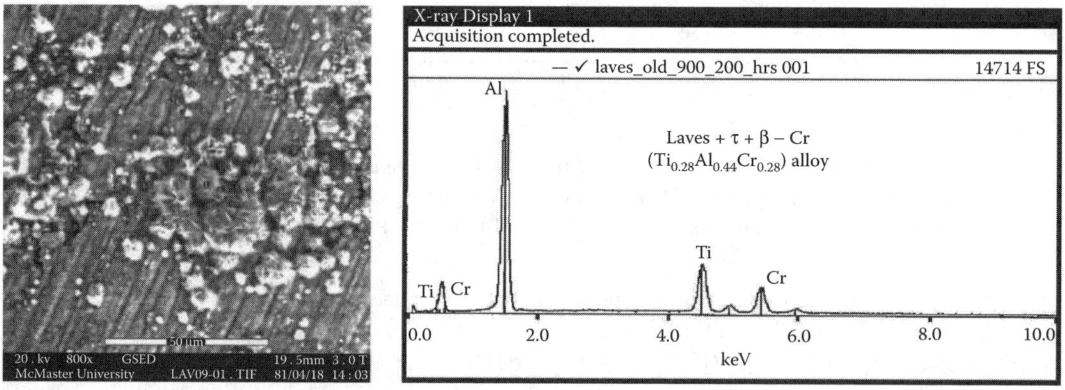

(b)

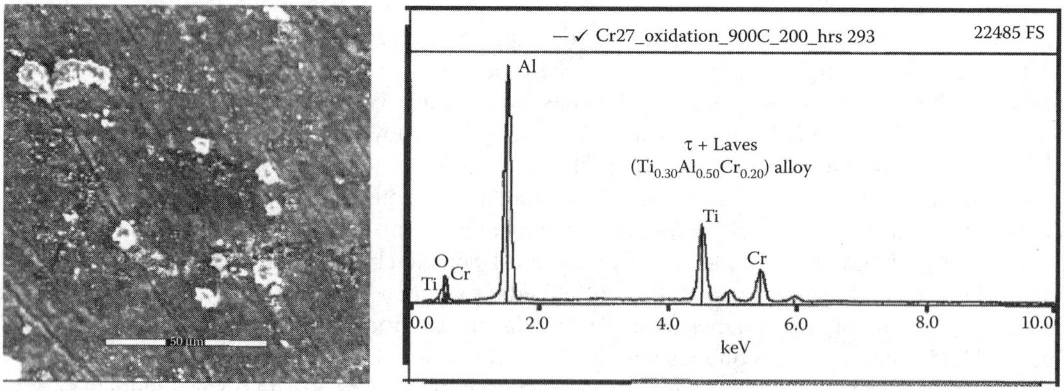

(c)

FIGURE 11.14 SEM micrographs of the oxide morphology, together with EDX spectra showing oxide scales morphology of Ti–Al–Cr based alloys at 900°C, in air: (a) γ + Laves (($Ti_{0.37}Al_{0.51}Cr_{0.12}$) alloy, (b) Laves + τ + β-Cr ($Ti_{0.28}Al_{0.44}Cr_{0.28}$) alloy, (c) τ + Laves ($Ti_{0.30}Al_{0.50}Cr_{0.20}$) alloy, (d) τ + Laves ($Ti_{0.25}Al_{0.55}Cr_{0.2}$) alloy, (e) τ-($Ti_{0.25}Al_{0.67}Cr_{0.08}$) phase alloys. (From Fox-Rabinovich, G.S., Wilkinson, D.S., Veldhuis, S.C., Dosbaeva, G. K., Weatherly, G.C. Oxidation resistant Ti-Al-Cr alloy for protective coating applications. *Intermetallics*, 2006, 14, 2, 189–192. With permission.)

Continued.

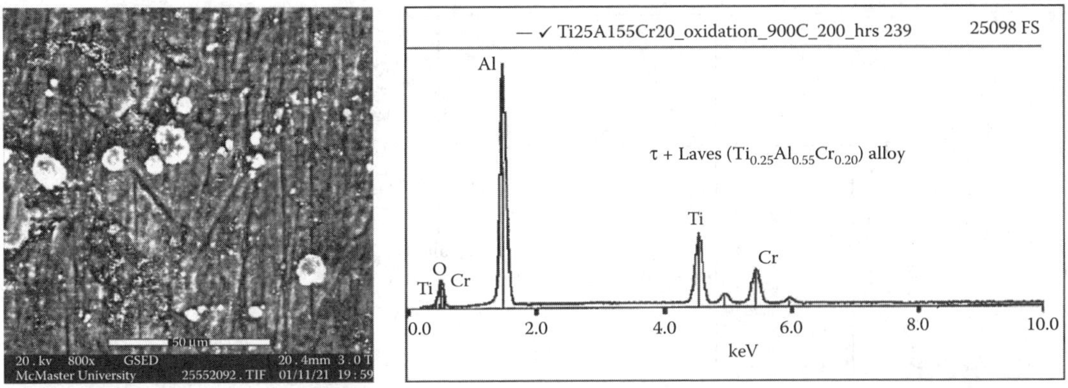

(d)

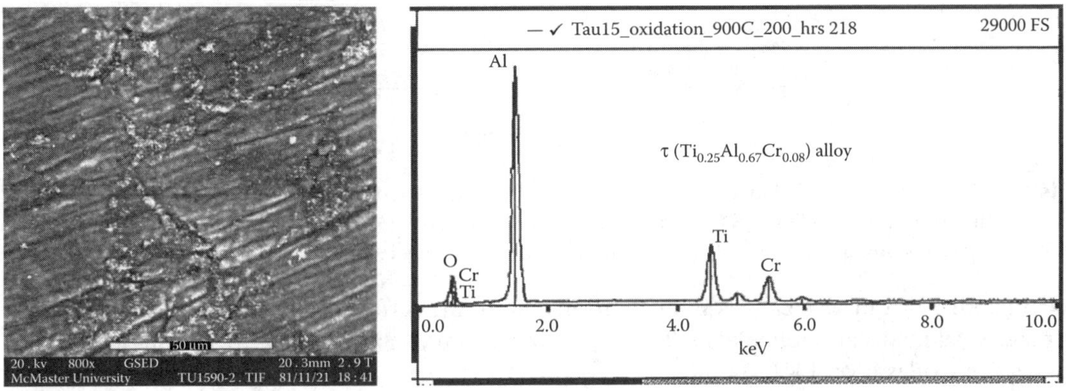

(e)

FIGURE 11.14 *Continued.*

was for the τ-phase-based $Ti_{0.25}Al_{0.55}Cr_{0.20}$ alloy that contains around 20% of the Laves phase (see Table 11.4) in the structure (Figure 11.13, Figure 11.15, and Figure 11.19c). This alloy has no spallation at high temperatures (1100°C) and exhibits very smooth morphology (Figure 11.16) with minimal amount of rutile nodules (Figure 11.20a).

One of the problems of Ti–Al–Cr intermetallics that contain the Laves phase is their brittleness compared to the binary γ-TiAl alloys (see Table 11.5) [7]. This property is very important for the thermal cyclic oxidation conditions. To improve the brittleness of the Laves-phase-based alloys, the multiphase compositions with a certain amount of ductile phases such as γ and τ were used. The compositions to be examined are the γ + Laves Ti–Al–Cr alloy and the τ + Laves Ti–Al–Cr alloy (Table 11.6). The ternary Ti–Al–Cr multiphase alloy containing τ + Laves phases is more resistant to cracking than the multiphase γ + Laves alloy. This is demonstrated by the crack lengths detected in microhardness impressions in the single and multiphase alloys (Table 11.6 and Figure 11.21). Based on the data presented, we can conclude that τ + Laves Ti–Al–Cr alloys have a more beneficial combination of properties (oxidation resistance as well as resistance to cracking) than known γ + Laves Ti–Al–Cr alloys [7].

Further improvement of the oxidation resistance of the τ + Laves multiphase alloys can be achieved by addition of small amounts (in the range of 0.1 to 0.5%) of such active elements as Hf and Y [31], as well as Zr and Si (Table 11.5). The mechanisms by which these elements impact the oxidation stability of the Ti–Al–Cr alloys are different. Hf and Y elements form segregations at the oxide–matrix interface, which prevent oxide grain coarsening during oxidation [32]. It is

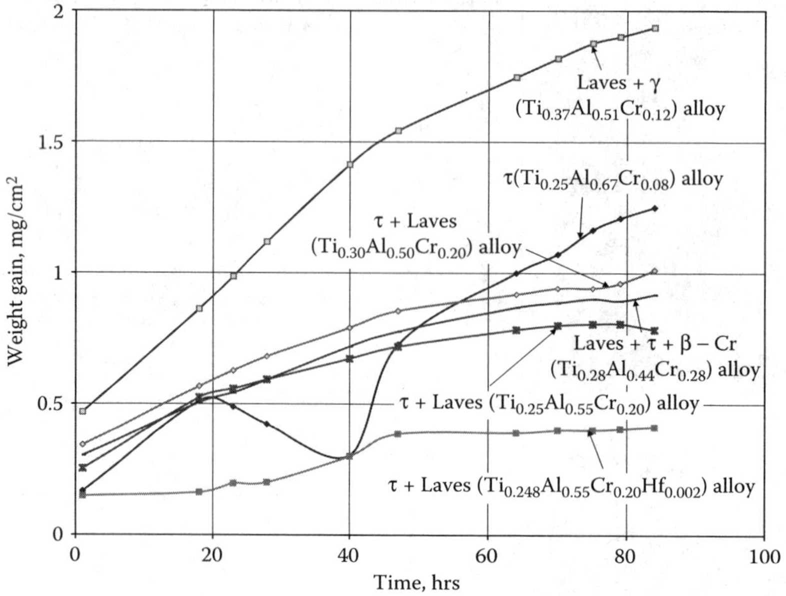

FIGURE 11.15 Oxidation behavior at 1100°C of the Ti-Al–Cr based alloys in air. (From Fox-Rabinovich, G.S., Wilkinson, D.S., Veldhuis, S.C., Dosbaeva, G. K., Weatherly, G.C. Oxidation resistant Ti-Al-Cr alloy for protective coating applications. *Intermetallics*, 2006, 14, 2, 189–192. With permission.)

worth noting that these elements promote formation of the surface oxide columnar structure [33]. Zr and Si additions also refine oxide film grain sizes and probably promote formation of the stable protective oxides layer [34].

The efficiency of every elements (Hf, Y, Zr, and Si) addition has been evaluated. Only a small amount of these elements (in the range of 0.1 to 0.5%) has a beneficial impact on the oxidation stability. Based on the literature and our own data, the addition of larger amounts of the elements is not beneficial, especially at high temperatures (Table 11.5).

The main element that critically changes oxidation stability of the ternary Ti–Al–Cr alloys (from several times to almost an order of magnitude) is Hf [35]. Hf additions to the τ-phase-based alloys is very beneficial at elevated (900°C) and at high (1100°C) temperatures of oxidation (Table 11.5). Hf improves the adherence of alumina to the substrate metal and prevents whiskerlike morphology formation, which is very important at high temperatures of oxidation (Figure 11.20). The addition of Hf changes the equiaxed structure of the surface oxide film to a columnar one (Figure 11.22). The additions of Zr, Y, and Si are less beneficial to oxidation resistance (Table 11.5). The addition of two elements such as Hf–Y and Hf–Si results in the alloy having a higher oxidation resistance (Table 11.5). Ternary Ti–Al–Cr τ-phase-based alloys doped with Hf–Si and Hf–Y exhibit improved oxidation resistance at 900°C, but at high temperatures (1100°C) the beneficial effect is not obvious when compared to the Hf-doped alloy. The addition of three elements (Hf, Y, and Si) results in significant grain refinement (Figure 11.20), formation of a columnlike structure in the oxide layer (Figure 11.22c), and the best oxidation stability during the alloy oxidation at 900°C. The oxidation resistance of this multicomponent alloy at high temperatures (1100°C) is close to that of the Ti–Al–Cr alloy doped by Hf. All the alloys of this type exhibit very smooth surface morphology. The multicomponent alloying results in excellent adherence of the surface oxide layer at high temperatures. Absolutely no spallation has been observed (Figure 11.20). Based on the data obtained, we can conclude that elements such as Hf, Y, and Si added to the Ti–Al–Cr intermetallic, in the amount defined in Table 11.5, act synergistically during oxidation.

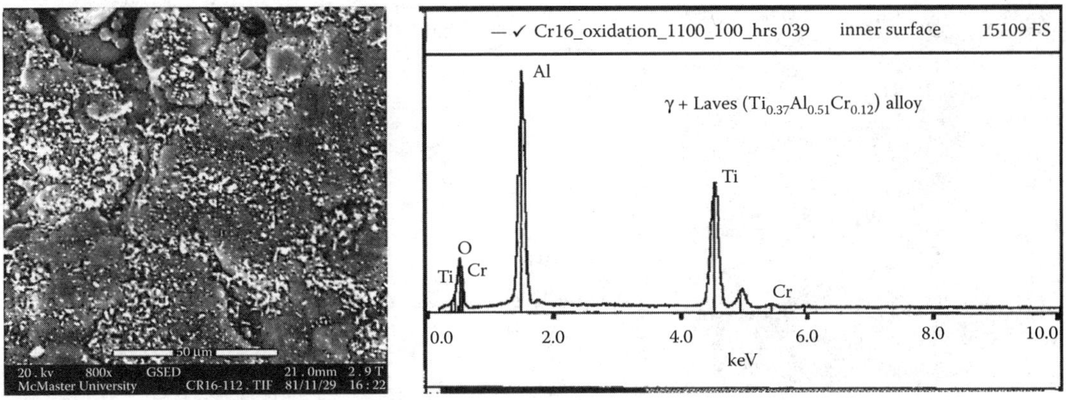

(a)

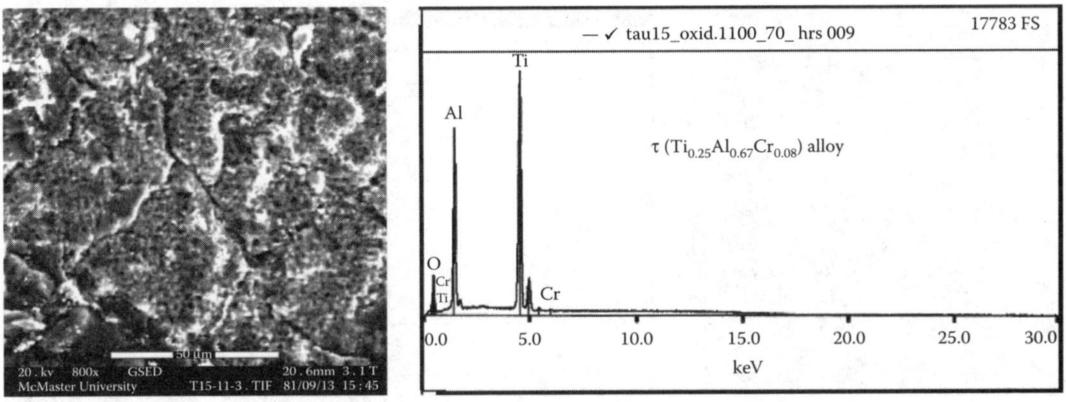

(b)

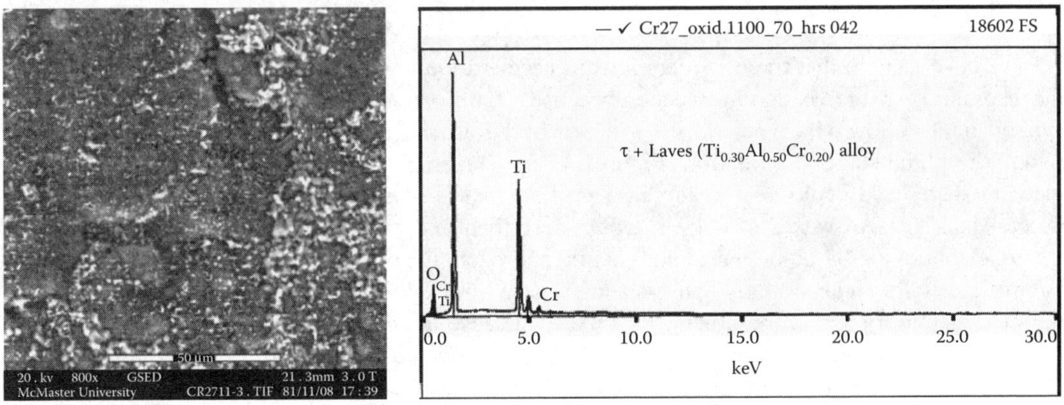

(c)

FIGURE 11.16 SEM micrographs of the oxide morphology, together with EDX spectra showing oxide scales morphology of Ti–Al–Cr based alloys at 1100°C, in air: (a) τ + Laves(($Ti_{0.37}Al_{0.51}Cr_{0.12}$) alloy, (b) τ-($Ti_{0.25}Al_{0.67}Cr_{0.08}$) phase alloys, (c) τ + Laves ($Ti_{0.30}Al_{0.50}Cr_{0.20}$) alloy, (d) Laves + τ + β-Cr ($Ti_{0.28}Al_{0.44}Cr_{0.28}$) alloy, e) τ + Laves ($Ti_{0.25}Al_{0.55}Cr_{0.2}$) alloy.

Continued.

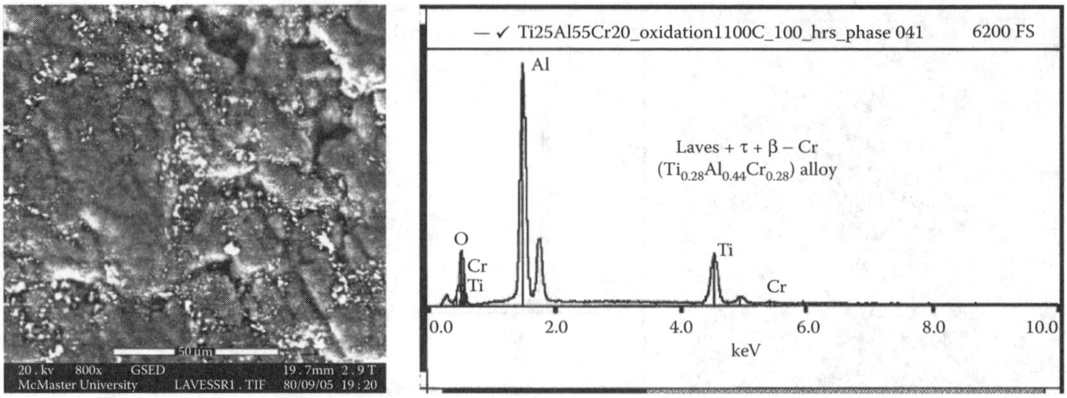

(d)

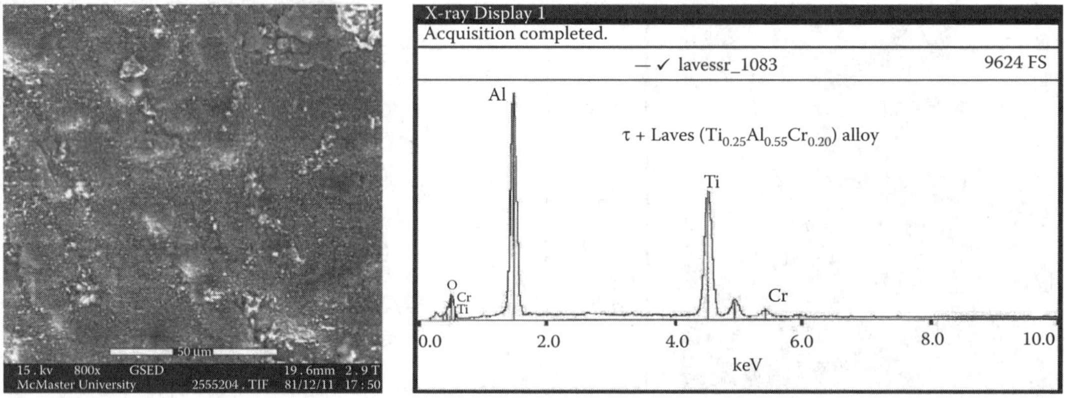

(e)

FIGURE 11.16 *Continued.*

We have to note that there is a significant acceleration of the aluminum outward diffusion in the beginning of the oxidation process because of the very-fine-grained microstructure of the synergistically alloyed intermetallics. Formation of the surface alumina film takes place very quickly within the initial stage of oxidation (Figure 11.15). Thus, this process passes through many intermediate stages and could be considered a process that is under nonequilibrium conditions. This process is associated with the "trigger" effect described in detail later. If we consider the effect of synergistic alloying based on a thermodynamic approach, the change of entropy production for the complex metallurgical systems can be shown as a sum of two terms, one associated with the thermodynamic force and the other with fluxes. Because the oxide film that forms on the surface is a protective alumina layer, the rate of entropy production decreases and the weight-gain oxidation curve shows a parabolic shape (Figure 11.15). This means that the flux of matter also decreases.

We have to emphasize that the alloying specifics strongly depend on the applications of the material, but the generic approach presented in preceding text allows us to give an example of how to achieve the properties needed, in particular oxidation resistance, by means of synergistic alloying.

11.4 TERNARY AND HIGHER-ORDERED HARD COATINGS

Based on the data obtained, a metallic powder metallurgical target was made from synergistically alloyed $Ti_{0.242}Al_{0.55}Cr_{0.20}Hf_{0.002}Si_{0.005}Y_{0.001}$ intermetallics (Table 11.5) and a protective PVD coating deposited in the nitrogen atmosphere. An oxidation test was performed, and the data obtained can

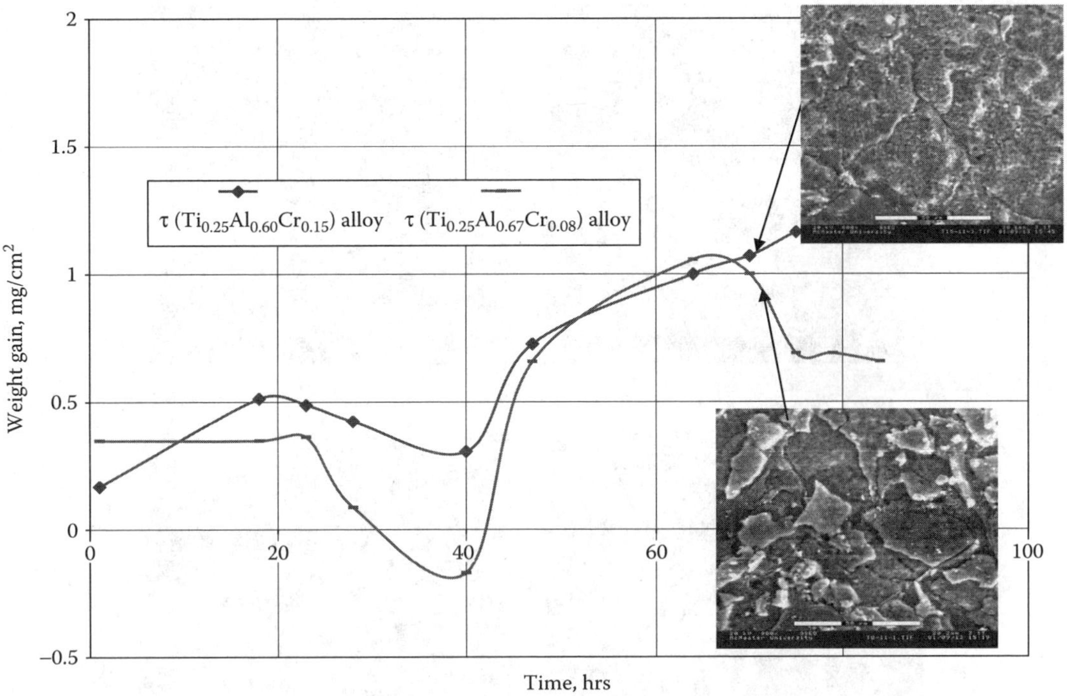

FIGURE 11.17 Oxidation behavior at 1100°C of the τ-phase Ti–Al–Cr based alloys in air with different Cr content (8 to 15 at. %).

be seen in Figure 11.23. We have to note that the oxidation resistance of the coating can vary significantly because its properties critically depend on the metallurgical design. The metallurgical design of the coating includes composition, microstructure and architecture (mono-, multilayered, and so on).

A coating can be considered a metallurgical system. It is known that a coating is condensed matter and, therefore, has a very high density of the lattice imperfections [36]. These lattice imperfections are largely concentrated at the grain boundaries. A coating with higher amounts of lattice imperfections possesses a more nonequilibrium state because of the energy stored in these imperfections within the coating layer during deposition [37]. The smaller the coating's grain size, the higher the density of the lattice imperfections. These two characteristics are dependent and related to the φ parameter value, which was suggested in Chapter 9 to characterize the nonequilibrium state of a coating. A critical value of the φ parameter exists within the coating that corresponds to a specific breakpoint. When the metallurgical system passes this breakpoint, an interaction with the external environment starts to follow the laws of irreversible thermodynamics (see Figure 2.1, Chapter 2). For the coatings described in this chapter, this critical φ parameter value probably corresponds to nanocrystalline coatings that have grain sizes around 10 nm and below. Starting at this point, the density of lattice imperfections is so high and the deviation from equilibrium is so substantial that the break point on the curve presented in Figure 2.1 is passed, and the process of self-organization occurs during the interaction of the coating with the environment. It does not matter whether the external conditions of this interaction are: (1) nonequilibrium during friction or (2) equilibrium during oxidation. In both cases, the self-organization process can begin for the nanostructured coatings. It is also unimportant how this state of the nanostructured coating is achieved, i.e., by means of alloying, deposition conditions, or any other way. The process of self-organization can start, shifting the metallurgical system to a behavior that is driven by reduction of entropy production. This can be illustrated by the examples shown in the following text.

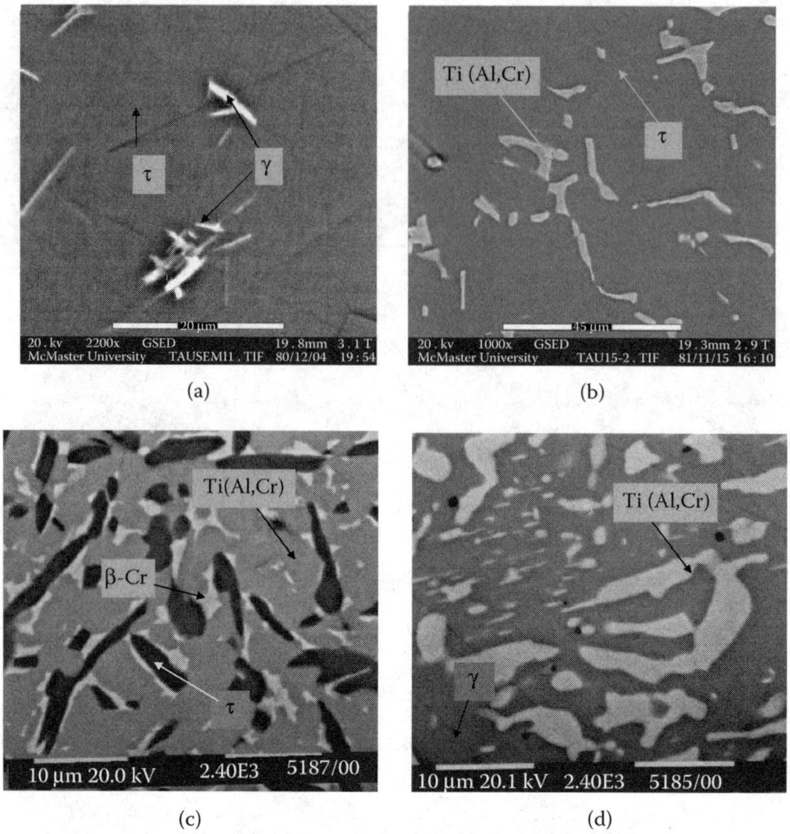

FIGURE 11.18 SEM micrographs of selected samples with different phase compositions: (a) $Ti_{0.25}Al_{0.67}Cr_{0.08}$, $Ti_{0.24}Al_{0.66}Cr_{0.10}$ alloys containing predominantly τ phase (Mark 1 in the Ti–Al–Cr ternary diagram; Figure 11.12), (b) $Ti_{0.25}Al_{0.60}Cr_{0.15}$ alloy containing τ phase + Laves phase, (c) $Ti_{0.28}Al_{0.44}Cr_{0.28}$ alloy containing a mixture of Laves + τ + β-Cr (traces) phases (Mark 2 in the Ti–Al–Cr ternary diagram; Figure 11.12), (d) $Ti_{0.37}Al_{0.51}Cr_{0.12}$ alloy containing a mixture of γ + Laves phases (Mark 3 in the Ti–Al–Cr ternary diagram; Figure 11.12). (From Fox-Rabinovich, G.S., Wilkinson, D.S., Veldhuis, S.C., Dosbaeva, G. K., Weatherly, G.C. Oxidation resistant Ti-Al-Cr alloy for protective coating applications. *Intermetallics*, 2006, 14, 2, 189–192. With permission.)

We will consider three different types of coatings:

- Nanocrystalline coatings deposited under relatively low temperatures in nonequilibrium conditions
- Nanocrystalline coatings with a structure formed because of the addition of amorphization-enhancing elements
- Nanolaminated coatings

The importance of deposition conditions of the coatings regarding oxidation resistance improvement can be illustrated by the data presented in Figure 11.24. Coatings with the same $(Ti_{0.242}Al_{0.55}Cr_{0.20}Hf_{0.002}Si_{0.005}Y_{0.001})N$ composition, but deposited under different conditions (see Chapter 9, Table 9.7) using a pulsed FAD technique, show different oxidation behaviors. The nanocrystalline coatings, deposited at low temperatures (340 to 370°C; see Chapter 9, Table 9.7) in a nonequilibrium state (when lattice imperfections formed during deposition do not annihilate because of recrystallization), form a protective alumina film during oxidation. In contrast, the coating with the same composition, but with a coarser grain size and a more stable equilibrium

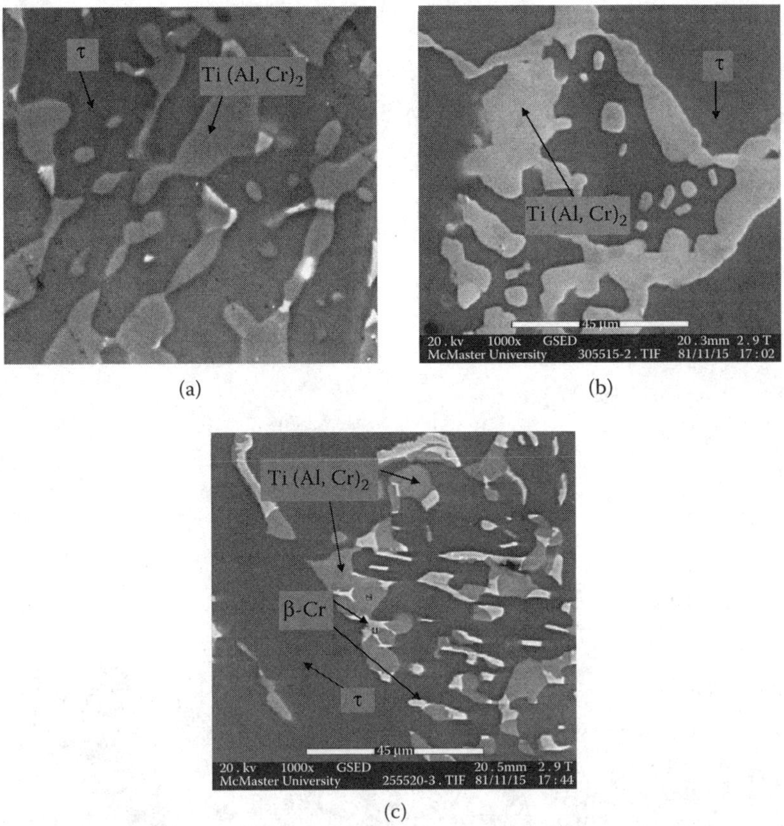

FIGURE 11.19 SEM micrographs of selected samples with τ + Laves + β-Cr (traces) phase composition (Mark 4 on the Figure 11.12): (a) $Ti_{0.30}Al_{0.50}Cr_{0.20}$ alloy, (b) $Ti_{0.30}Al_{0.55}Cr_{0.15}$ alloy, (c) $Ti_{0.25}Al_{0.55}Cr_{0.20}$ alloy. (From Fox-Rabinovich, G.S., Wilkinson, D.S., Veldhuis, S.C., Dosbaeva, G. K., Weatherly, G.C. Oxidation resistant Ti-Al-Cr alloy for protective coating applications. *Intermetallics*, 2006, 14, 2, 189–192. With permission.)

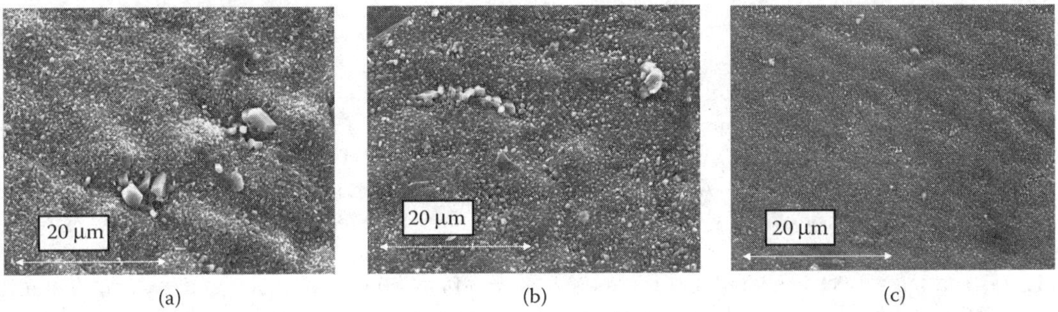

FIGURE 11.20 SEM micrographs of the Ti–Al–Cr alloys after oxidation at 1100°C during 200 h: (a) $Ti_{0.25}Al_{0.55}Cr_{0.20}$, (b) $Ti_{0.25}Al_{0.55}Cr_{0.20}$ doped by Hf (0.2 at. %), (c) $Ti_{0.25}Al_{0.55}Cr_{0.20}$ doped by Si, Hf, and Y (0.5, 0.2, and 0.1 at. %, correspondingly). (From Fox-Rabinovich, G.S., Wilkinson, D.S., Veldhuis, S.C., Dosbaeva, G. K., Weatherly, G.C. Oxidation resistant Ti-Al-Cr alloy for protective coating applications. *Intermetallics*, 2006, 14, 2, 189–192. With permission.)

TABLE 11.6
Microhardness Indentation Study of a Few Binary Ti–Al and Ternary Ti–Al–Cr Alloys Showing the Crack Length Observed vs. Load Applied

Type of Ti–Al-Based Alloy	Hardness ($HV_{0.1}$)	Length of Cracks (μm) under the Loads Applied	
		20 N	3 N
Ti–Al–Cr Laves single phase	824		20
Ti–Al–Cr Laves multiphase	764		8.7
γ-Ti–Al	330	0	
Ti–Al–Cr γ+Laves	400	2.91	
Ti–Al–Cr τ single phase	200	0	
Ti–Al–Cr τ+Laves	190	0	

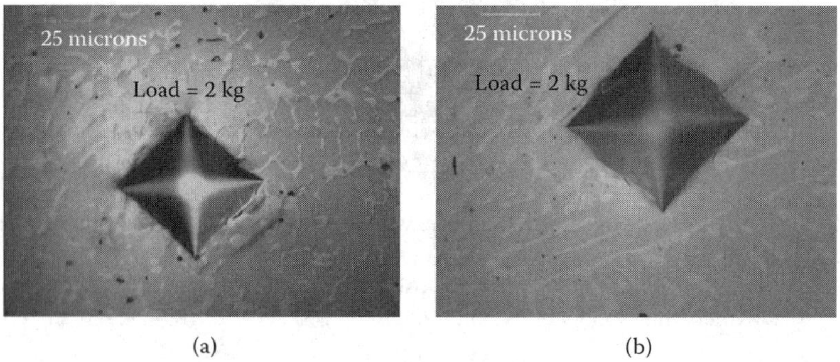

FIGURE 11.21 Microhardness indentation study of multiphase Laves-based alloys showing the appearance of the cracked indentations at a load of 20 N: (a) γ + Laves phase alloy; (b) τ + Laves phase alloy.

state due to a higher temperature of deposition (500°C; see Chapter 9, Table 9.7), does not show any protective abilities at all.

From a thermodynamic point of view, the nonequilibrium state of the nanocrystalline coatings can critically change the oxidation phenomena. During oxidation of the "equilibrium" coatings (that are deposited, for instance, at a higher temperature of 500°C; Figure 11.24b), only equilibrium processes occur on the surface. According to the conclusions made in Chapter 2, these processes cannot cause instability of the system and consequently cannot lead to self-organization. During the oxidation of "nonequilibrium" coatings (deposited at a lower temperature of 340°C; Figure 11.24a), at least two irreversible processes occur: (1) a chemical reaction between metallic components of the coating and oxygen from the environment, and (2) the relaxation process of a nonequilibrium coating associated with diffusion. These two irreversible processes start to interact with each other within a nonlinear area, i.e., far from equilibrium. As a result, the rates of these processes begin to depend on each other. In particular, the oxidation rate decreases with the value of the φ parameter. Starting with a definite value of the φ parameter, this dependence indicates a decrease in entropy (illustrated in Figure 11.23) based on the parabolic shape of the weight-gain curve with time. Thus, we can make a very important conclusion that the interaction of equilibrium

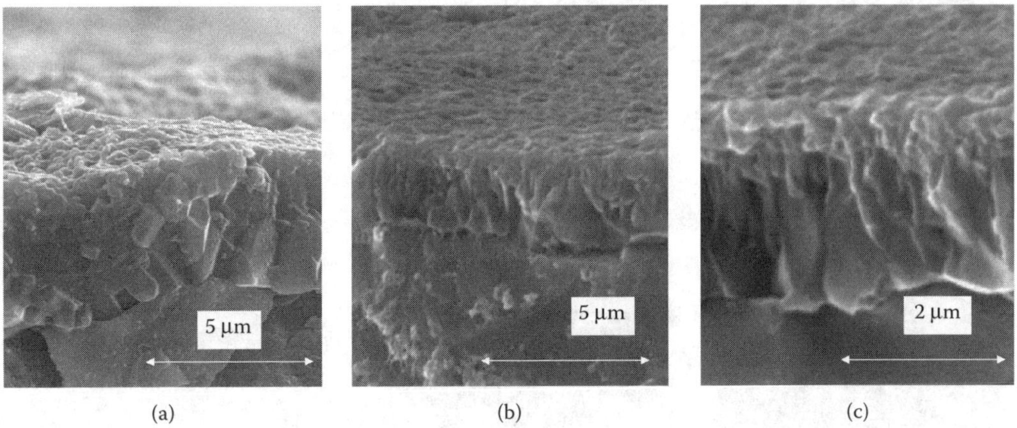

FIGURE 11.22 Fracture surfaces of (a) $Ti_{0.25}Al_{0.55}Cr_{0.20}$, (b) $Ti_{0.25}Cr_{0.55}Al_{0.20}$ + Hf, (c) $Ti_{0.25}Al_{0.55}Cr_{0.20}$ + Hf + Y + Si showing the oxide morphology in the cross-sectional view. (From Fox-Rabinovich, G.S., Wilkinson, D.S., Veldhuis, S.C., Dosbaeva, G. K., Weatherly, G.C. Oxidation resistant Ti-Al-Cr alloy for protective coating applications. *Intermetallics*, 2006, 14, 2, 189–192. With permission.)

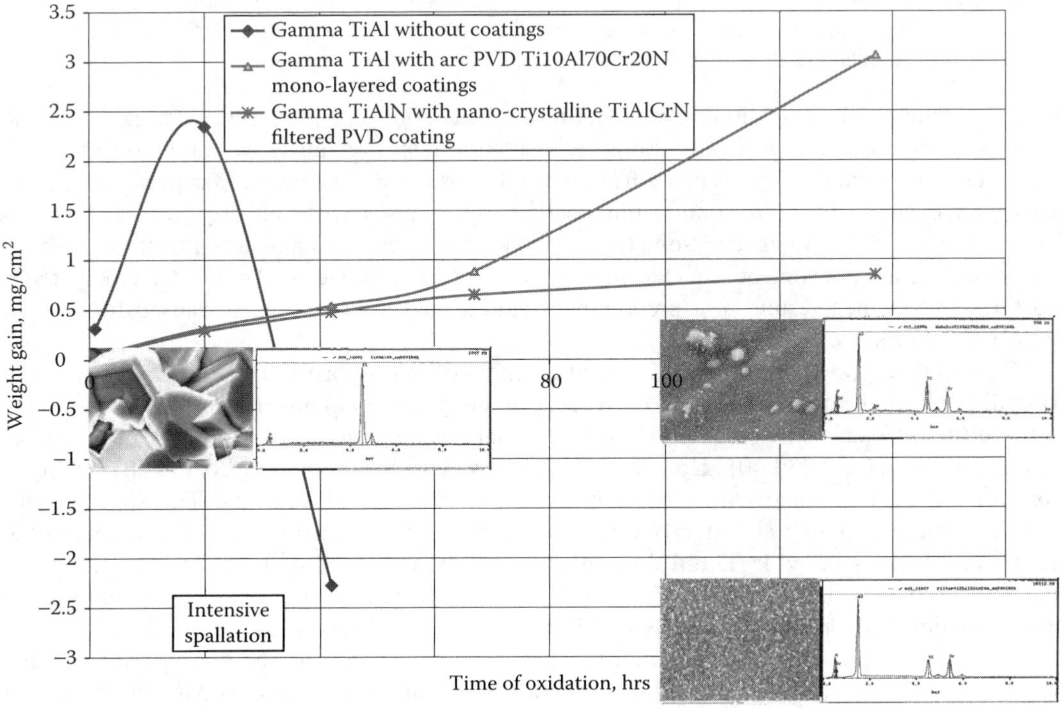

FIGURE 11.23 Weight gain vs. time of oxidation for Ti–Al coupons with protective Ti–Al–Cr–N-based coatings with different composition and microstructure (regular and nanoscale grain size).

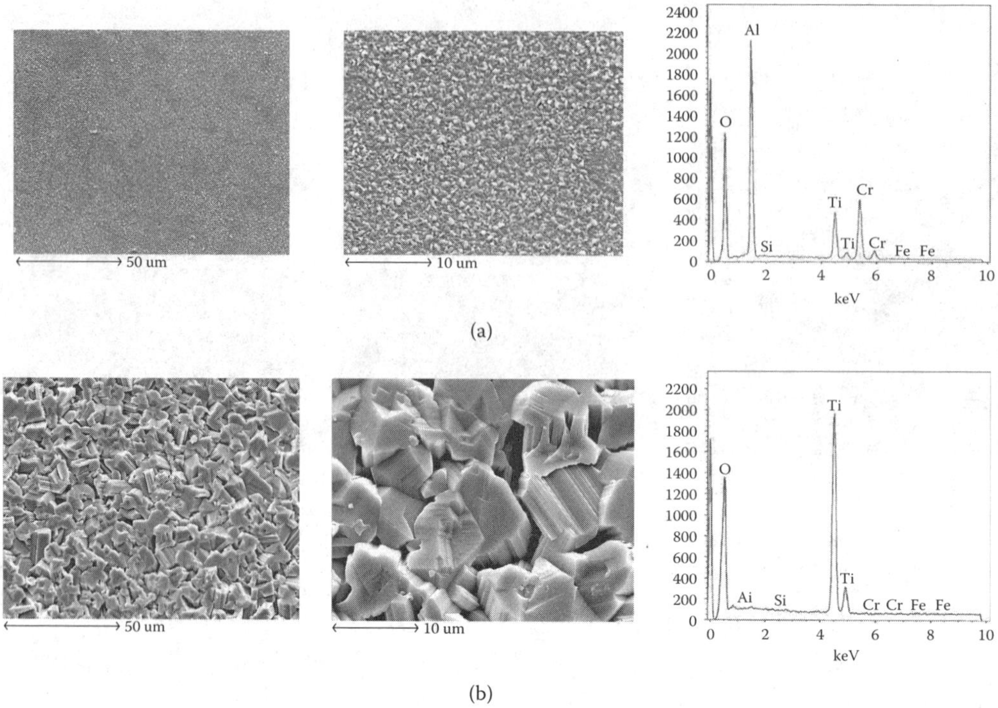

FIGURE 11.24 SEM micrographs and EDS spectra of the oxide films morphology formed on the surface of the synergistically alloyed Ti–Al–Cr–Si–Hf–Y–N pulsed FAD PVD-coating, deposited at (a) 340 to 360°C and (b) deposited at 500°C after oxidation in air at 900°C and 100 h.

processes, such as the oxidation of "nonequilibrium" material (in our case the coating), can result in self-organization. This critically drops the oxidation rate. Previously, we only considered the interaction of different materials under friction conditions that are themselves far from equilibrium. Using an example of a nonequilibrium coating, the approach of self-organization could be expanded to a much wider, and sometimes unexpected, area of practical applications. We can conclude that the phenomenon named in the literature as selective oxidation of nanocrystalline coatings [38], which results in a predominant alumina layer formation, is caused by the self-organization process.

The second type of nanocrystalline coating is fabricated by a PVD method when a nanocrystalline structure forms as a result of synergistic alloying. It is known from the literature that the addition of elements to the Ti–Al–N-based coatings, such as Si or Y, results in significant grain size refinement [39,40]. However, the ternary Ti–Al–Cr–N-based coating has a higher oxidation stability compared to similar intermetallic alloy-based coatings. The specifics of a nitride coating's alloying are slightly different compared with intermetallics. The amount of Al in the Ti–Al–Cr–N-based PVD nitride coatings could be increased up to 65 to 70% along with the amount of Si (within a range of 5 at. %) to achieve grain sizes of around 8 to 10 nm [39,41]. The oxidation data on nanocrystalline (Ti15–Al60–Cr20–Si5)N coatings presented in Reference 39 indicates the predominant formation of alumina films on the surface during oxidation (so-called "selective oxidation" [38]) within a wide range of temperatures from 800 to 1200°C (Figure 11.26) [39]. An "equilibrium" (Ti25–Al65–Cr10)N coating deposited at 500°C [41] does not show protective abilities during oxidation at temperatures above 900°C (Figure 11.25e). In contrast, the nanocrystalline (Ti15–Al60–Cr20–Si5)N coatings show excellent protective ability up to 1200°C, which also corresponds to the data presented in Figure 11.25f,g for the nanocrystalline coatings.

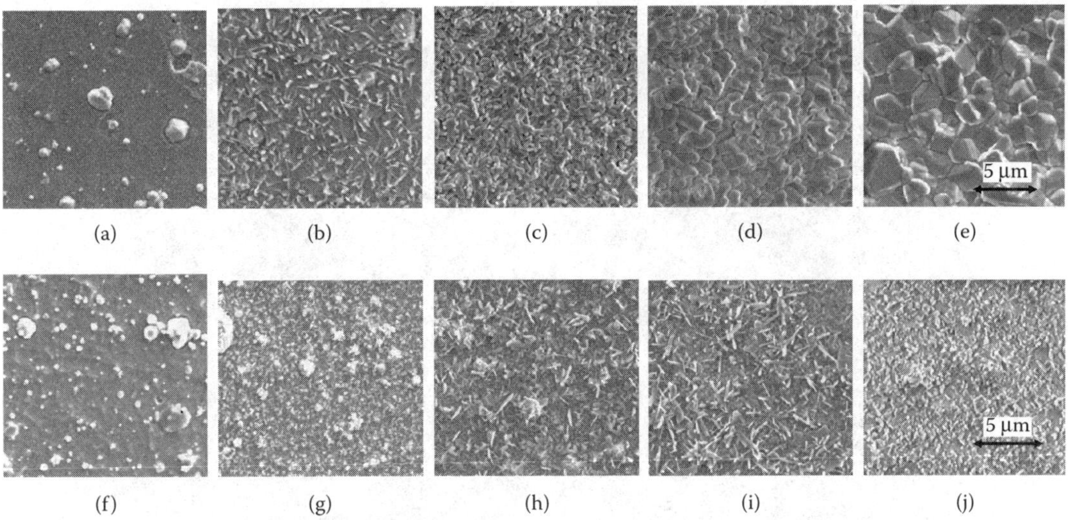

FIGURE 11.25 (a–e) SEM micrographs of the oxide films morphology formed on the surface of the monolayered (Ti25–Al65–Cr10)N and (f–j) nanocrystalline synergistically alloyed (Ti15–Al60–Cr20–Si5)N PVD coatings after oxidation in air at 800 to 1200°C and 0.5 h, correspondingly.

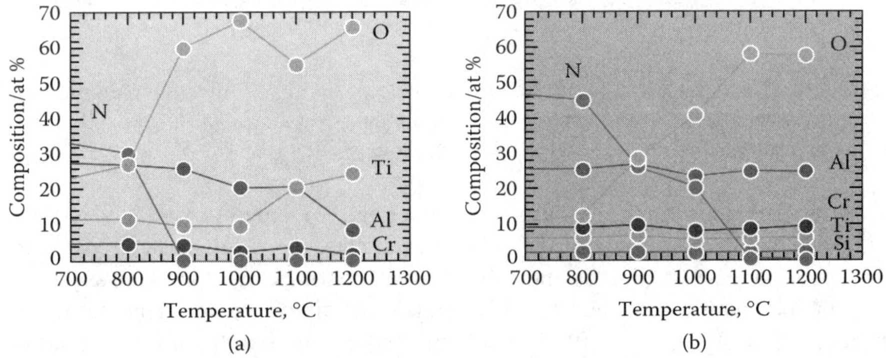

FIGURE 11.26 EDS spectra of the oxide films formed on the surface of the (a) monolayered (Ti25–Al65–Cr10)N and (b) nanocrystalline synergistically alloyed (Ti15–Al60–Cr20–Si5)N PVD coating after oxidation in air at 800 to 1200°C and 0.5 h, correspondingly.

Stability of the nanocrystalline (Ti15–Al60–Cr20–Si5)N coatings under increasing oxidation temperatures (up to 1100°C) is related to self-organizing of the nonequilibrium coating that occurs during oxidation. It is known that after self-organization has begun, the intensity of the corresponding processes that occur within the system (in our case, the intensity of oxidation) will decrease [42]. Thus, the metallurgical system enters a state in which the entropy production is constant within a range of given external conditions [42]. If the external conditions become more severe (the temperature of oxidation increases), some of the system's characteristics (most likely the intensity of mass transfer) have to adjust to this change in order to prevent an increase in entropy production. However, if self-organization does not start, a similar interaction of processes (in our case, chemical reaction on the surface and coating relaxation) cannot happen, because it only appears after the self-organization process initiates.

Similar observations can be made for nanolaminated Ti–Al–Cr–N/WN coatings studied in detail in Chapter 10. One of the major features of nanolaminated structures is the increased density of lattice imperfections located at the Ti–Al–Cr–N/WN interface and grain boundaries [43]. A smaller

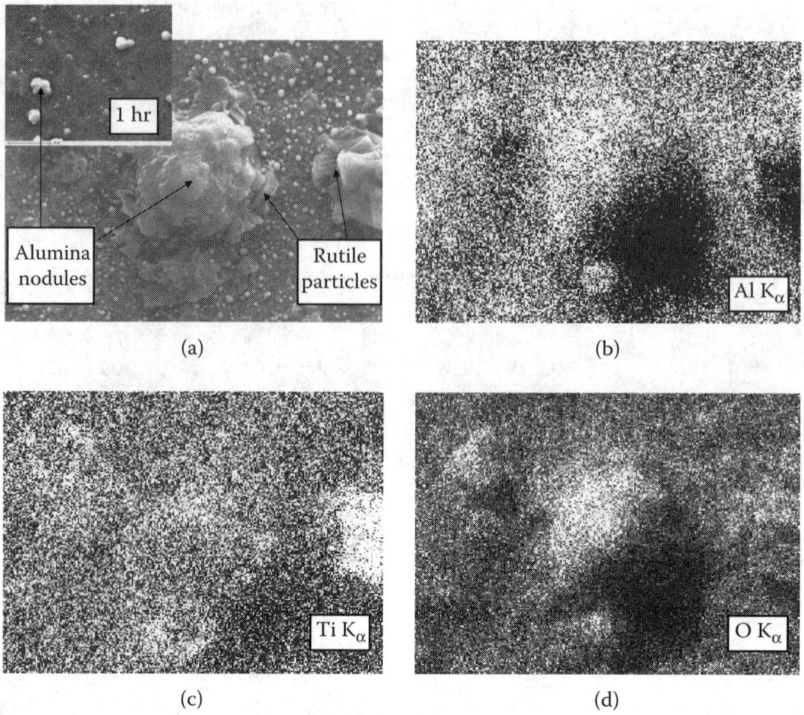

FIGURE 11.27 SEM micrographs and EDS elemental map of the oxide films morphology formed on the surface of the nano-laminated Ti–Al–Cr–N/WN coating oxidation in air at 900°C and 100 h: (a) SE image. (An image in the left upper corner presents surface morphology after 1 h), (b) Al K_α, (c) Ti K_α; (d) O K_α.

grain size of the nanolaminated Ti–Al–Cr–N/WN coatings compared to monolayered Ti–Al–Cr–N should also be taken into consideration [43–45]. We can view nanoscale-structured coatings (either nanolaminated or nanocrystalline) as "nonequilibrium" coatings. We can also assume that during oxidation of the nanolaminated coating, self-organization can take place. The main feature of the oxidation behavior of Ti–Al–Cr–N/WN nanolaminated coatings is the ability of alumina–oxide films to prevent (or, at least, limit) growth of rutile particles. Many oxide nodules are formed during oxidation of the Ti–Al–Cr–N/WN coating. At the initial stage of the process, they consist of pure alumina (Figure 11.27a), likely due to the enhanced outward diffusion [43] of aluminum to the surface, as was shown for the nonequilibrium nanocrystalline TiAlN coatings in Chapter 9. The alumina nodules transform with time into a mixture of alumina and rutile. However, these particles are relatively small due to the coating's grain size refinement (Figure 11.27a). The alumina films form shell-like nodules on the top of the rutile particles (Figure 11.27b–d) that limit the growth of these nonprotective particles, which results in significant improvement of oxidation protection. Based on the preceding speculation regarding nanocrystalline coatings, we can assume that the process of oxidation of the nanolaminated coating is also controlled by the self-organization phenomena.

11.5 THERMODYNAMIC-BASED PRINCIPLES OF SYNERGISTIC ALLOYING OF PROTECTIVE AND WEAR-RESISTANT MATERIALS AND COATINGS

The principles of synergistic alloying for protective and wear-resistant coatings are developed to enhance the formation of specific films during service. There are two major types of films that are formed on the surface of hard coatings as a response to external impact during oxidation and wear at elevated temperatures.

The first type of film is protective, mainly performing passivating and barrier functions to prevent against interaction with the environment during oxidation, or when in contact with a frictional body during wear at elevated temperatures. This type of surface film formation is governed by the trigger effect (Chapter 9). However, under severe friction conditions such as high speed or dry cutting of hard-to-machine alloys, where seizure starts to play a dominating role, lubricity of the tribo-films at high temperatures controls the entropy production during friction and the wear intensity. The principles of synergistic alloying of hard coatings with more sophisticated designs tailored for severe tribological applications have also been developed.

11.5.1 The Trigger Effect for Synergistically Alloyed Oxidation-Resistant Materials and Wear-Resistant Coatings

A typical feature of synergistically alloyed materials and coatings is stabilization of the wear or oxidation rate within a short period of time, at a very low level of intensity for the corresponding process (Figure 11.15) [46]. We can consider this process based on the thermodynamic approach, as done previously for nanocrystalline Ti–Al–N PVD coatings (see Chapter 9).

Oxidation and tribo-oxidation processes of the ternary and higher-ordered Ti–Al–Cr-based alloys and coatings pass through a number of intermediate stages. Nonuniform composition of the oxide or tribo-oxide films in the very beginning of the oxidation or wear process results in fluxes of matter and energy, as well as changes in chemical potentials. These features of oxidation- and tribo-oxidation-related wear processes can be explained using the approach of irreversible thermodynamics.

According to the laws of irreversible thermodynamics [21], if a stationary state is achieved, the entropy of production due to an irreversible process inside the system (such as diffusion and chemical reactions) will be $\frac{dS_i}{dt} \rightarrow \min$.

Generally, the entropy of production can be expressed as [22]:

$$\frac{dS_i}{dt} = \sum_i I_i X_i$$

where X is the thermodynamic force associated with chemical potential and the I are the thermodynamic fluxes related to heat and mass. We can divide the change of the entropy production of the open system $\frac{dS_i}{dt} = \Phi$ into two parts [23], connecting one part with the change of thermodynamic forces $X(d_x\Phi)$ and the other with the change of thermodynamic fluxes $I(d_i\Phi)$, i.e.:

$$d\Phi = d\sum_i I_i X_i = \sum_i I_i dX_i + \sum_i X_i dI_i = d_x\Phi + d_i\Phi$$

The value $d_x\Phi$ will be less than or equal to 0 when the system evolves to stationary state because of the formation of (1) protective oxide films (in the case of oxidation) and (2) protective or lubricious oxide films (in the case of wear). It is impossible to conclude anything about the sign of $d\Phi$, however, when the entropy production variation is caused only by the change in thermodynamic forces. This deviation will be negative according to the evolution criterion [22,47]. During the initial stage of oxidation or tribo-oxidation of the ternary Ti–Al–Cr-based compound or nitride coatings, X and I change. So, in this case, when the variation of entropy production is caused by the changes in both the forces and fluxes with $d_X\Phi$ and $d_i\Phi < 0$ at the same time, entropy production will vary as a stationary state with dS_i/dt = min is reached. After that, entropy production ceases to change [42]. This occurs after the stabilization of the composition, structure, and properties of surface films. For example, in the Ti–Al–Cr-based alloys, it occurs after the chromium oxide dissolves in alumina films during oxidation.

Synergistic alloying by the addition of Cr to titanium aluminides or to Ti–Al–N-based coatings or by a number of elements such as Hf, Si, and Y increases the flux of matter I in the initial stage of the process by the acceleration of metal-surface oxidation or tribo-oxidation. This occurs because of the formation of oxides such as chromium oxide or complex oxides, which have lower thermodynamic stability than alumina [46]. At the same time, oxidation of aluminum is also accelerated in intermetallics and in nanoscaled crystalline or multilayered coatings by different mechanisms [43,46]. This can be seen in the Ti–Al–Cr-based intermetallics synergistically alloyed by Hf, Si, and Y with fine-grained columnar structure (Figure 11.22). This intensive diffusion leads to an increase in entropy production at the very beginning of the process, resulting in a change of X owing to the formation of protective oxide films. These oxides (Al_2O_3) are thermodynamically stable and capable of forming a continuous protective layer and limiting surface oxidation or wear rate. This implies that the entropy production decreases when a stable state is achieved. According to the Prigogine theorem [22], in a system with minimal entropy production, the flux of matter and, consequently, the oxidation or wear rate have to decrease significantly, which has been confirmed by the parabolic shape of the weight gain or wear curve (Figure 11.15 and Chapter 10, Figure 10.1).

If we compare the aforementioned phenomena, wear rate stabilizing within the running-in stage (Chapter 3), oxidation of Ti–Al–Cr intermetallics, and the tribo-oxidation of wear-resistant coatings (Chapter 9 to Chapter 11), we can see that sudden changes in the function values happen during a minor alteration of argument on the curve of the material's properties with the corresponding parameter (see Chapter 2, Figure 2.1). Such a "catastrophe" of the function (see Chapter 2) shows that, at a specific point, the system could lose its stability, thus resulting in self-organization, which corresponds to the drastic decrease of the wear or oxidation rate. In fact, the wear or oxidation rate (i.e., slope of the curve or its first derivative) for the synergistically alloyed metals or coatings critically decreases in a short period of time.

It is possible to conclude that a trigger effect could be described as a complex phenomena that is associated with intensive mass transfer at the beginning of the wear or oxidation process, which leads to a beneficial change of chemical potential of a metallurgical system and eventually results in a decrease of entropy production. The trigger effect is a feature of metallurgical systems that are designed so as to stabilize the wear or oxidation processes during the shortest period of time under a very low level of intensity of the corresponding process. Based on the data obtained for a variety of the materials (intermetallics; binary, ternary, and higher-ordered coatings), as well as external conditions (oxidation and wear), we can conclude that the "trigger" effect is commonly observed. The trigger effect can be observed in metals and surface-engineered materials and relates to the surface phenomena that occur under severe environmental impact associated with high-temperature and high-stress conditions. An optimized composition, as well as special structure of the surface layer such as nanoscale crystalline structures with highly nonequilibrium states (Chapter 9), strongly enhances the trigger effect (Figure 11.23). This can result in a critical improvement of service properties of the advanced materials.

11.5.2 Thermodynamic Principles of Synergistic Alloying of Quaternary Wear-Resistant Coatings for Severe Applications

In this section, we will chiefly focus on wear-resistant coatings for severe cutting applications (see Chapter 10). Self-organization starts at the initial stage of the wear process, causing the decrease in wear intensity. This could prevent a wear intensity increase if friction conditions become more severe. Therefore, a state of the tribosystem occurs in which the entropy production can be constant within the range of given cutting conditions (see preceding text) [42]. If certain cutting parameters become more aggressive, for instance, the cutting speed increases, some of the tribosystem's characteristics such as the coefficient of friction have to adjust to this change to prevent growth in entropy production. As stated before, a similar interaction of processes cannot happen until the self-organization process begins.

As outlined in Chapter 10 (Figure 10.23b), wear behavior of self-adaptive quaternary Ti–Al–Cr–N/WN coatings critically depends on the cutting parameters employed. In contrast to regular coatings, more aggressive conditions do not lead to higher wear rates (see Figure 10.23b and Figure 10.31). Under higher cutting speeds, the wear rate stabilizes at a very low level because the self-organization process has started.

Intensification of cutting parameters leads to beneficial change of wear behavior of the cutting tools with Ti–Al–Cr–N/WN coatings, as was mentioned earlier in Chapter 10.

Under cutting conditions overviewed earlier in Chapter 10 (ball-nose end milling of the hardened H13 steel with cutting speed 200 m/min under dry friction), the major difference in wear behavior of cutting tools with monolayered Ti–Al–Cr–N coating and nanolaminated Ti–Al–Cr–N/WN coating is the following. It is known that cutting conditions on rake and flank surfaces of the cutting tools significantly differ, and more intensive heat generation occurs on the rake surface than on the flank surface [48]. SEM images of the surface morphology of worn end-mill cutters are presented in Figure 11.28b. The micrographs show that, after long-term cutting tests (length of cut being 250 m; Figure 11.30a), sliding of chips along the rake face of cutting tools with a Ti–Al–Cr–N coating leads to intensive adhesive interaction and subsequent heat generation at the tool–workpiece interface [49]. This means that alumina–chromia-like tribo-films that form on the rake surface under operation (see Chapter 10) cannot efficiently protect the surface during long-term cutting when intensive seizure controls the friction process. This results in intensive diffusion and chemical wear, which leads to further crater formation on the rake surface (Figure 11.28b). Owing to lower temperature on the flank surface of an end-mill cutter [48], the alumina-like tribo-films formed on the flank surface are efficient enough, and lower intensity of workpiece material pickup was observed (Figure 11.28a).

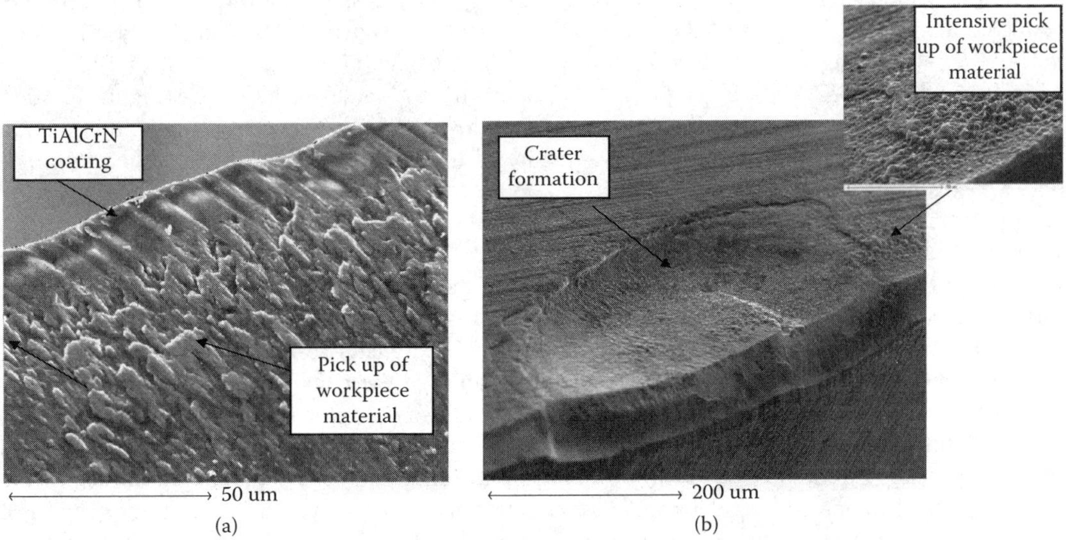

FIGURE 11.28 SEM images of the worn surface of ball-nose end mill with Ti–Al–Cr–N. Machining of H13 steel; cutting speed 200 m/min; length of cut = 250 m; (a) flank surface, (b) rake surface.

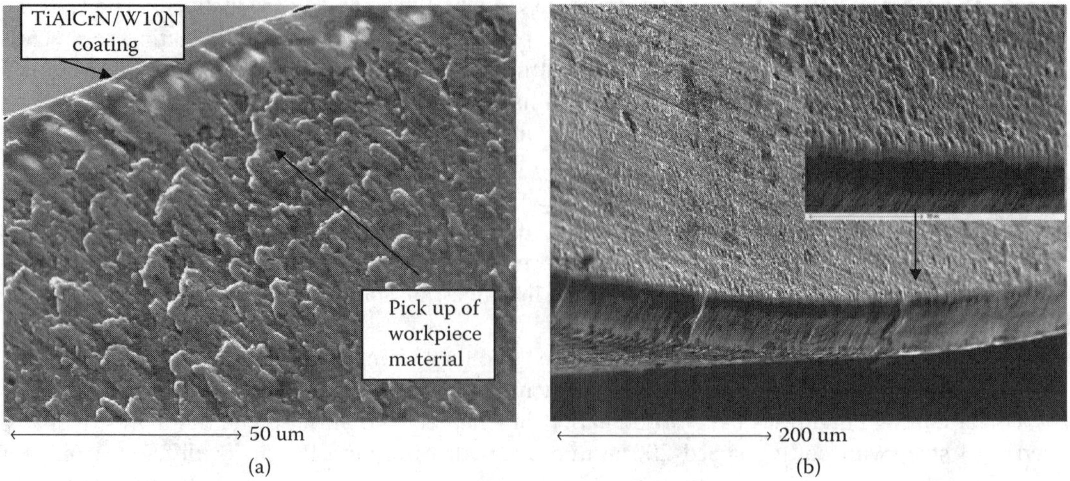

FIGURE 11.29 SEM images of the worn surface of ball-nose end mill with Ti–Al–Cr–N/WN. Machining of H13 steel; cutting speed 200 m/min; length of cut = 250 m; (a) flank surface, (b) rake surface.

Simultaneous formation of protective alumina-like tribo-films and lubricious W–O tribo-films on the friction surface of the nanolaminated coating at elevated temperatures (see Chapter 10) result in milder friction conditions at the tool–workpiece interface (Figure 11.29). A lower seizure intensity takes place on the rake surface of the end-mill cutter, where major adhesive interaction between sliding chips and the cutting tool occurs (Figure 11.29b). No cratering is observed on the rake face after long-term tests (length of cut being 250 m) that lead to an improvement of tool life compared to the Ti–Al–Cr–N coating (Figure 11.30a). However, intensity of the workpiece material pickup is similar on the flank surface for both studied coatings (Figure 11.29a). It is necessary to note that end-mill cutters with a Ti–Al–N coating present a near-linear dependence of the wear rate on the length of cut [50], because the self-organization process did not begin for these coatings under the severe operating conditions.

Intensification of the cutting parameters, primarily cutting speeds, from 200 to 300 m/min, results in a wear rate increase of the coated end-mill cutters. Whereas the wear of end mills with a Ti–Al–Cr–N coating rapidly transforms to catastrophic stage after length of cut of 170 m, the end mills with Ti–Al–Cr–N/WN coatings still remain within a stable stage of wear (Figure 11.30b).

SEM images of worn end-mill cutters show that the Ti–Al–Cr–N coating is partially gone on the flank surface (Figure 11.31a). Intensive seizure of the workpiece material, together with oxidation, takes place on the coating-free areas.

In contrast, a low-intensity seizure could be observed on the flank surface of the end-mill cutters with nanolaminated Ti–Al–Cr–N/WN coatings (Figure 11.32a). The coating protects the surface with high efficiency on both the rake and flank faces of the cutting tools (Figure 11.32b). To explain this phenomenon, we have to remember that tungsten is a heavy metal. Higher temperatures are needed to activate this metal and to initiate intensive mass transfer to the friction surface. During operation at cutting speeds of 200 m/min, the temperature at the flank surface is not high enough to initiate this process, and only alumina-like tribo-films form (Chapter 10, Figure 10.24). However, at higher cutting speeds of 300 m/min, the temperature raises and tungsten begins to diffuse toward the flank surface. Lubricious W–O tribo-films begin to form, and the friction conditions shift to a milder mode. In contrast, a catastrophic wear rate was observed for a Ti–Al–N coating. The thermodynamic aspects of this phenomenon are considered in the following text. Thus, once the self-organization has begun, more severe friction conditions do not lead to significant wear rate increase, as was outlined in Chapter 2. However, we have to note that this effect has obvious

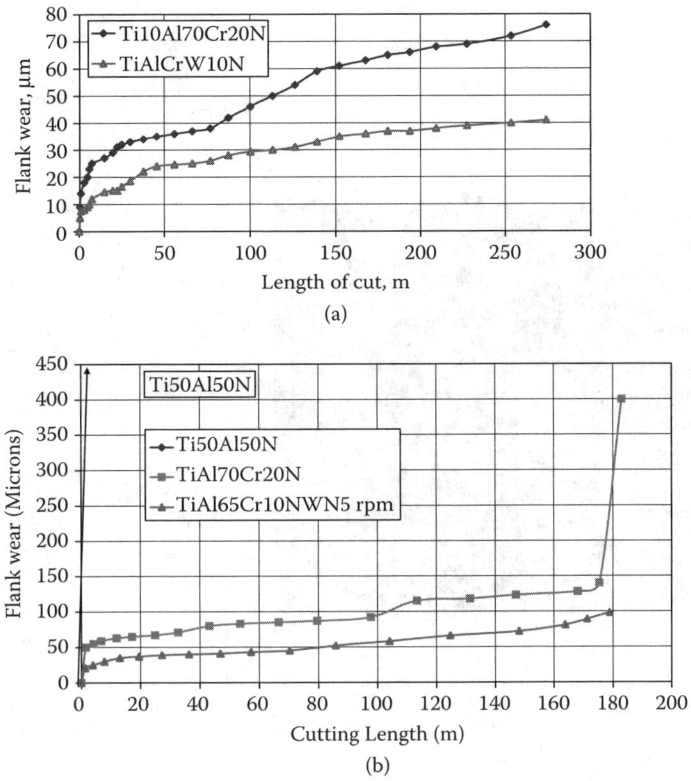

FIGURE 11.30 Tool life of the Ti–Al–Cr–N and Ti–Al–Cr–N/WN PVD coatings under high-performance dry-machining conditions: (a) ball-nose E/M 5-mm radius, 2 flutes (C-2MB, Mitsubishi); workpiece material — AISI H13 (HRC50); cutting conditions — feed 0.06 mm/flute = 840 mm/min; axial depth — 5.0 mm; radial depth —0.6 mm; speed —200 m/min; cutting length — 250 m; (b) speed — 300 m/min; length of cut — 150 m.

limitations. Under even more severe cutting conditions (speed of 400 m/min), tool life is lower than could be expected.

A similar observation was witnessed during the turning of hard-to-machine nickel-based Waspalloy. It is known that the machining of Waspalloy results in intensive seizure during cutting. These are extremely severe cutting conditions. Under low cutting speeds (40 to 100 m/min), only linear wear could be observed for different types of cutting tools and self-organization does not occur (Chapter 10, Figure 10.31). At 40 m/min, uncoated cutting tools exhibit the best tool life (Chapter 10, Figure 10.31a). At higher cutting speeds, i.e., 125 m/min and above, self-organization of the turning inserts with self-adaptive Ti–Al–Cr–N/WN coatings initiates. This can be seen by the shape of the wear curve with low initial wear rate (during the running-in stage) and with expanded range of stable wear (Chapter 10, Figure 10.31c and Figure 10.31d). Overall wear rate of the turning inserts with Ti–Al–Cr–N/WN coatings dramatically decreases. From a thermodynamics point of view, this is caused by the reduction of entropy production. This condition is stable within a range of cutting speeds 125 to 150 m/min. The only explanation of this phenomenon is the decrease in the coefficient of friction during high-temperature cutting, as can be seen from data presented in Chapter 10, Figure 10.21. As shown, the coefficient of friction decrease leads to the reduction of the cutting forces, as well as heat generation at the interface. All these changes result in tool life improvement.

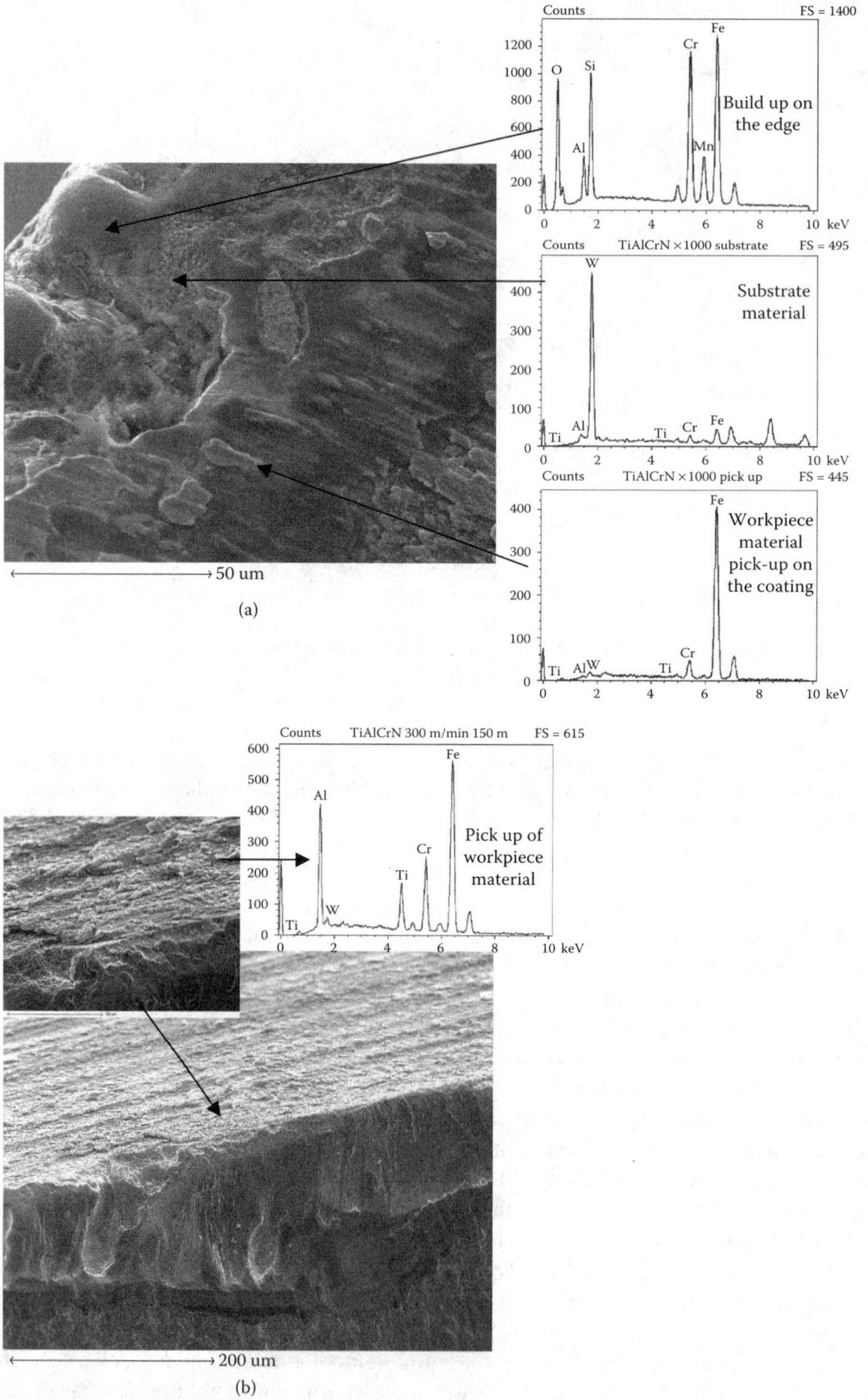

FIGURE 11.31 SEM images of the worn surface of ball-nose end mill with Ti–Al–Cr–N. Machining of H13 steel; cutting speed — 300 m/min; length of cut = 150 m; (a) flank surface; (b) rake surface.

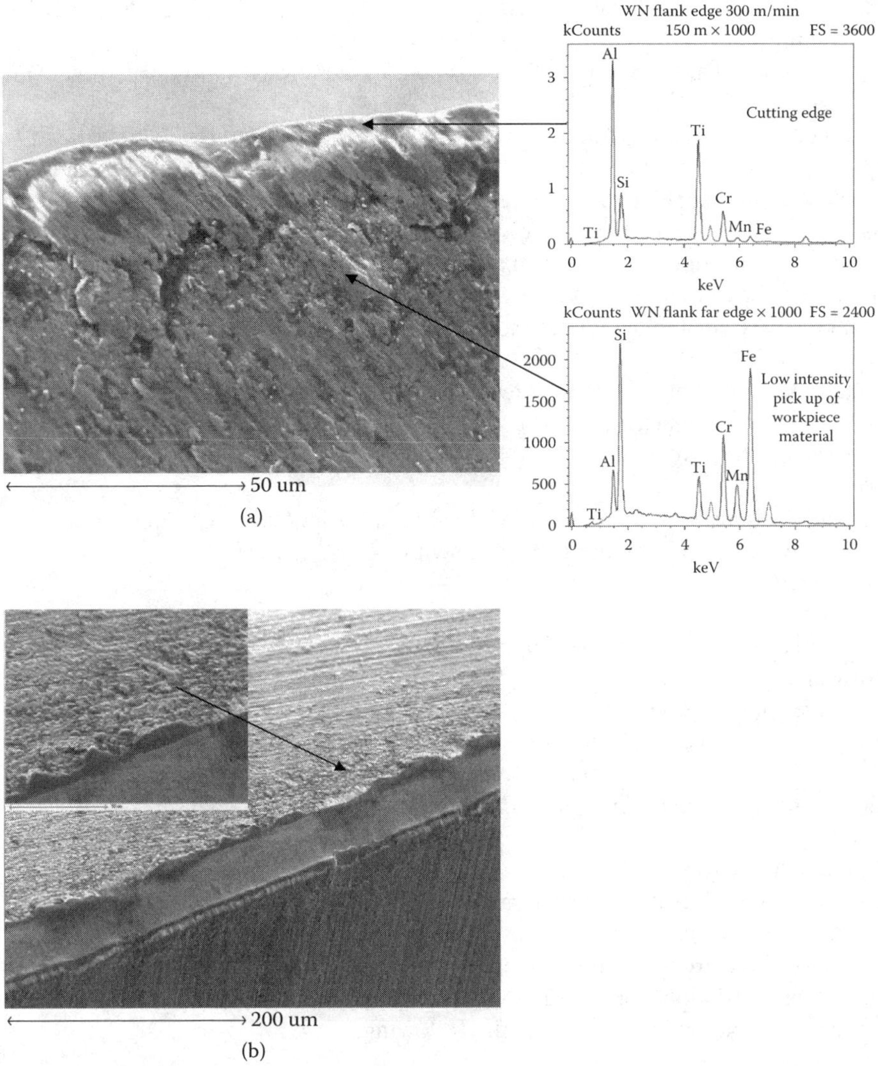

FIGURE 11.32 SEM images of the worn surface of ball-nose end mill with Ti–Al–Cr–N/WN. Machining of H13 steel; cutting speed — 300 m/min; length of cut = 150 m; (c) flank surface; (b) rake surface.

The cause of these beneficial transformations can be explained based on the conclusions of analytical modeling presented here. Energy transformation, as well as the physical–chemical processes that occur during high-performance machining conditions, proceeds under strong nonequilibrium conditions [47–51]. Owing to these circumstances, these processes can be analyzed using methods of nonequilibrium thermodynamics. Only preliminary results relating to this approach have been obtained to date. It should be noted that much more research needs to be performed in this field. However, we believe it is reasonable to present our preliminary results in order to indicate further challenging incentives for future research. The results of thermodynamic modeling evaluated from the experimental data obtained can create a general understanding of the considered complex phenomena.

We can assume that a combination of two processes is taking place during cutting. The first is intensive external friction and the second is seizure of the workpiece material. These processes are simultaneous, but they occur at different sites of the cutting tool's friction surface. This corresponds to the SEM observations of the worn ball-nose end mills with Ti–Al–Cr–N/WN coatings (See Chapter 10, Figure 10.24).

The simultaneous character of the processes considered allows us to create a general equation of entropy production.

We assume that there is no sliding within the zones of seizure. That is why the velocity parameter can be represented by the speed of deformation of welded asperity junctions between the chip and the tool face. We can consider the seizure as just a given spontaneous process with no energy (activation) required in order to start. The zones of seizure can be characterized by the area of seizure (in future, by the amount of welded asperity junctions). Intensity of the relaxation processes such as mass transfer can characterize the nonequilibrium state of the material.

The symbols used for the analytical modeling performed are the following:

Coefficient of friction at the seizure-free areas: K
General area of the contact: G (constant)
Relative seizure-free area: n.
Relative seizure area: $1 - n$
Stress generated during tension of the welded asperity junctions: σ (constant).
Displacement during seizure up to the moment of fracturing: l
Sliding velocity: v
Load applied: F
Density of matter involved in mass transfer: ρ
Mass-transfer intensity: w
Mass-transfer intensity within a zone of external friction: w_f
Mass-transfer intensity within a zone of seizure: w_s
Density of matter involved in mass transfer within the zone of external friction: ρ_f
Density of matter involved in mass transfer within the zone of seizure: ρ_s
Temperature: T
Thermal conductivity: λ
Nonequilibrium state of the material: φ (see Chapter 9 for details)

The equation of entropy production includes three terms that characterize three different phenomena: external friction, seizure, and mass transfer.

The equation of entropy production is the following:

$$\frac{dS_i}{dt} = \frac{(kFv)^2}{\lambda TnG} + \frac{\sigma v(1-n)G}{T} + \frac{\rho wG}{T} \tag{11.1}$$

It is possible that mass transfer within zones of external friction and seizure will progress in different ways. That is why we have to separate the term in Equation 11.1, thus describing mass transfer in two terms:

$$\frac{\rho wG}{T} = \frac{\rho_S w_S(1-n)G}{T} + \frac{\rho_f w_f Gn}{T} \tag{11.2}$$

It is worth noting that the temperatures within the zone of external friction and the zone of seizure will be different. However, we are not going to consider this difference at the moment.

The self-organization process begins if a system loses its thermodynamic stability. The φ parameter could be used to characterize a systematic deviation of the system (in our case, a wear-resistant coating) from a stable mode.

Sufficient conditions of the system's stability are controlled by the negative or positive sign of excessive entropy production. In our case, excessive entropy production is equal to:

$$\sum_i \delta J_i \delta X_i = \frac{(Fv)^2}{\lambda TG}\frac{\partial k}{\partial \varphi}\left(\frac{\partial k}{n\partial \varphi} - \frac{k}{n^2}\frac{\partial n}{\partial \varphi}\right)(\delta\varphi)^2 - \frac{vG}{T}\frac{\partial \sigma}{\partial \varphi}\frac{\partial n}{\partial \varphi}(\delta\varphi)^2 + \frac{G}{T}\frac{\partial \rho_f}{\partial \varphi}\times$$
$$\left(\frac{\partial w_f}{\partial \varphi}n + w_f\frac{\partial n}{\partial \varphi}\right)(\delta\varphi)^2 + \frac{G}{T}\frac{\partial \rho_S}{\partial \varphi}\left(\frac{\partial w_S}{\partial \varphi}(1-n) - w_S\frac{\partial n}{\partial \varphi}\right)(\delta\varphi)^2 \tag{11.3}$$

We can consider the characteristics K (coefficient of friction), ρ_f and ρ_s (density of matter involved in mass transfer), w_f and w_s (mass-transfer intensity) depending on the φ parameter.

The system can lose its stability if the sign of excessive entropy production changes from positive to negative. There are four terms in Equation 11.3. Excessive entropy production can be negative if every term of the equation is negative. An additional requirement has to be established that the area of external friction grows with the area of seizure:

$$\frac{\partial n}{\partial \varphi} > 0 \tag{11.4}$$

The condition in the Equation 11.4 can be followed, and the first term of the equation will be negative if:

$$\frac{\partial k}{\partial \varphi} > 0 \tag{11.5}$$

The second term could be negative if:

$$\frac{\partial \sigma}{\partial \varphi} > 0 \tag{11.6}$$

The third term could be negative if:

$$\frac{\partial w_f}{\partial \varphi} > 0, \quad \frac{\partial \rho_f}{\partial \varphi} < 0 \tag{11.7}$$

or:

$$\frac{\partial w_f}{\partial \varphi} < 0, \quad \frac{\partial \rho_f}{\partial \varphi} > 0 \tag{11.7a}$$

The fourth term could be negative if:

$$\frac{\partial w_S}{\partial \varphi} < 0, \quad \frac{\partial \rho_S}{\partial \varphi} > 0 \tag{11.8}$$

or:

$$\frac{\partial w_S}{\partial \varphi} > 0, \quad \frac{\partial \rho_S}{\partial \varphi} < 0 \tag{11.8a}$$

If the condition in Equation 11.4 is mandatory, there are four options for the tribosystem to lose stability. The combinations of either Equation 11.5, Equation 11.6, Equation 11.7, and Equation 11.8; Equation 11.5, Equation 11.6, Equation 11.7, and Equation 11.8; Equation 11.5, Equation 11.6, Equation 11.7, and Equation 11.8; or Equation 11.5, Equation 11.6, Equation 11.7, and Equation 11.8 has to take place simultaneously.

If the combination Equation 11.5, Equation 11.6, Equation 11.7, and Equation 11.8 takes place simultaneously, the density of matter that is involved in the mass transfer to the outer friction surface has to drop within the zone of external friction and grow within the zone of seizure with the φ parameter. Ti–Al–Cr–N/WN coating applications can fit this combination of conditions. As a matter of fact, intensive mass transfer of aluminum from the internal layers of the Ti–Al–Cr–N/WN coating to the cutting-tool surface results in a decrease in the density of material at the friction surface (see Chapter 10, Figure 10.24a). Within a zone of seizure, because of local heat generation a tungsten tribo-oxide forms as a result of transformation of the current WN nanolayer into an oxide tribo-film (see Chapter 10, Figure 10.25 and Figure 10.26b). Because tungsten is a heavy metal, the density of the matter involved in mass transfer within the zone of seizure grows.

When intensive seizure occurs, the tribosystem faces a choice. The first option is to develop a process of friction with an intensive wear rate. The second option is self-organization and formation of dissipative structures. It is not accidental that outward diffusion of a heavy metal such as tungsten (or similar heavy metals such as Nb and Ta) [48] leads to the reduction of seizure intensity. Seizure is associated with severe friction conditions. The tribosystem under severe friction conditions begins to involve a nonequilibrium process that requires the consumption of plenty of energy, because a high level of energy is required to activate mass transfer of the heavy elements. In contrast, energy that has to be otherwise spent on the process of wear begins to be used in these nonequilibrium processes. As a result, the wear rate reduces. We can assume that this is directly related to the process of self-organization and dissipative structure formation. Initiation of the mass transfer of heavy tungsten atoms to the outer surface during friction requires plenty of energy. Therefore, the tribosystem spends a major portion of available energy on this process instead of developing wear. It can be concluded that outward diffusion of a heavy element to the friction surface is a nonequilibrium process that is chiefly responsible for wear rate decrease.

Thus, the results of analytical modeling that was performed allow us to understand why self-organization of the tools with Ti–Al–Cr–N/WN coatings begins only at elevated cutting-speed conditions. We have to note that for conditions of high-performance machining, the heavy-metal-based oxide tribo-films (such as W, Nb, Ta, and others; see Chapter 7) look the most promising because they have lubricous properties [48] and possibly high hardness at elevated temperatures due to strong interatomic bonds in a compound with a heavy atomic nucleus [52]. These oxides also have a complex electronic structure, and a variety of tribo-oxide films with different stoichiometry and structures could form during friction that is typical for these heavy-metal-based compounds [52]. This could be considered as a second nonequilibrium process. Therefore, the tribosystem has a higher possibility of selecting an oxide with a better combination of properties suitable for specific conditions. This could lead to additional improvements in tool life and wear behavior. This is confirmed by XPS data presented in Figure 11.33.

Figure 11.33 shows the W 4d XPS line after peak deconvolution. The interpretation of phase composition was close to the values reported for WN and Tungsten oxides in [70]. The XPS W4d5/2 line at 243.1 eV corresponds to WN. The Binding Energy values of the oxide XPS lines are higher than metallic or WN due to the ionic character of the W-O bond. The positions of W4d 5/2 peak

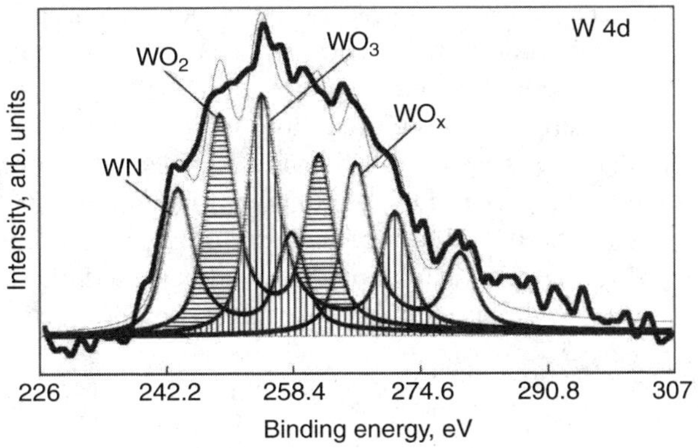

FIGURE 11.33 The photoelectron spectra of W 4 d line after peak deconvolution for the worn surface of ball-nose end mill with nano-multilayered TiAlCrN/WN coating.

TABLE 11.7
Binding Energy of W4d$_{5/2}$ and Volume Concentration of WO Tribo-Films; XPS Data

Phases	Binding Energy of W4d5/2, eV	Volume Concentration, at.%
WN	243.1	19
WO_2	248.1	27
WO_3	254.7	32
WO_X	265.4	22

at binding energies of 248.1 and 254.7 eV belong to WO_2 and WO_3 correspondingly. The W4d5/2, 4d3/2 doublet at 166.5–282.7 eV has a very high binding energy (Table 11.7). This increase in binding energy is probably a result of charging of the amorphous WOx films, which are forming under severe friction conditions associated with cutting, as it was shown earlier in Chapter 9 (Figure 9.36 [53–55]). Concentration of the phases on the friction surface is shown in Table 11.7. This data is obtained basing on analysis of the peaks areas. It is worth noting that the most intensive peak belongs to WO_3 tribo-oxide with high lubricious properties [56]. This tribo-oxide together with amorphous films is forming preferentially within the area of wear crater. It means that as a result of self-organization and adaptation process the tribo-system selects a specific type of the tribo-films tailored to given frictional conditions.

Based on the concept of irreversible thermodynamics presented earlier, we can predict that synergistically alloyed nanocrystalline or laminated coatings can be considered the next generation of wear-resistant coatings because they can enhance the trigger effect and promote the formation of both continuous protective tribo-ceramic films and high-temperature lubricious tribo-films.

11.6 CONCLUSIONS

We can conclude that the thermodynamics-based principles of synergistic alloying of oxidation- and wear-resistant coatings have to be tailored to their specific operating conditions. For an oxidation- and wear-resistant coating, in which a surface protective (passivate) oxide layer is needed to prevent intensive interaction with the environment or a frictional body, the trigger effect outlines the principles of synergistic alloying for hard coatings. This effect also determines the role of every element acting in synergy, i.e., the element that performs protective functions (such as aluminum) or the element that accelerates beneficial mass transfer (such as chromium, hafnium, or other surface-active elements in titanium aluminides and Ti–Al-based nitride coatings).

For wear-resistant coatings working under severe conditions, such as high-performance dry machining, the principles of synergistic alloying are more sophisticated, as outlined in the analytical model. Double surface protection is needed, which includes the formation of passivating tribo-ceramic layers as well as high-temperature lubricious films. The friction process itself governs the efficiency of the coating elements. In the case of high-performance machining, thermal activation plays a critical role. Only if the thermal activation threshold is passed, the elements begin to work in synergy, leading to a given effect, i.e., the decrease of entropy production.

REFERENCES

1. Brady, M.P., Smialek, J.L., Smith, J.W., Humphrey, D.L. The role of Cr in promoting protective alumina scale formation by γ-based Ti-Al-Cr alloys. Part I: Compatibility with alumina and oxidation behavior in oxygen. *Acta Mater.* 1997, 45, *6*, 2357–2369.
2. Brady, M.P., Smialek, J.L., Humphrey, D.L., Smith, J.W. The role of Cr in promoting protective alumina scale formation by γ-based Ti-Al-Cr alloys. Part II: Oxidation behavior in air. *Acta Mater.* 1997, 45, *6*, 2371–2382.
3. Shida, Y., Anada, H. The effect of ternary additives on the oxidation behavior of Ti-Al in high-temperature air. *Oxidation Met.* 1996, 45, *1*, 197.
4. Westbrook, J.H., Fleischer, R.L. Principles. In *Intermetallic Compounds: Principles and Practice*, Vol. 1, Eds., Westbrook, J.H., Fleischer, R.L. John Wiley and Sons: London, 1995, p. 73.
5. Young-Won, K. Ordered intermetallic alloys. Part III: Gamma titanium aluminides. *J. Met.* 1994, 46, *7*, 30.
6. Brady, M.P., Brindley, W.J., Smialek, J.L., Locci, I.E. The oxidation and protection of gamma titanium aluminides. *J. Met.* 1996, 8, *11*, 46–50.
7. Brady, M.P., Gleeson, B., Wright, I.G. Alloy design strategies for promoting protective oxide-scale formation. *J. Met.* 2000, 52, *1*, 16–21.
8. Fox-Rabinovich, G.S., Kovalev, A.I., Wainstein, D.L. Investigation of atomic and electronic structure of films generated on cutting tools surface. *J. Electron. Spectrosc. Relat. Phenomena* 1997, 85, 65.
9. Kovalev, A.I., Scherbedinsky, G.V. *Modern Methods of Metals and Alloys Surface Study.* Metallurgia: Moscow, 1989, p. 191.
10. Proctor, A., Sherwood, P.M.A. Data analysis techniques in x-ray photoelectron spectroscopy. *Anal. Chem.* 1982, 54, 13.
11. Derrier, J., Chainef, E., De Crescenzi, M., Noguero, C. Evidence of fine structures in the Auger spectra: a new approach for surface structural studies. *Solid State Commun.* 1986, 57, *7*, 473–552.
12. Kovalev, A.I., Mishina, V.P., Scherbedinsky, G.V., Wainstein, D.L. EELFS method for investigation of equilibrium segregations on surfaces in steel and alloys. *Vacuum* 1990, 41, *7–9*, 1794–1795.
13. Brumm, M.W., Grabke, H.J. The oxidation behaviour of NiAl-I. Phase transformations in the alumina scale during oxidation of NiAl and NiAl-Cr alloys. *Corros. Sci.* 1992, 33, *11*, 1677–1692.
14. Jewett, T.J., Ahrens, B., Dahms, M. Phase equilibria involving the γ-$L1_2$ and $TiAl_2$ phases in the Ti-Al-Cr system. *Intermetalics* 1996, 4, 543.
15. Wagner, Y.-W.R., Yamaguchi, M. Oxidation behavior of structure-controlled TiAl. In *Gamma Titanium Aluminides*, Ed., Yoshihara K. TMS: Warrendale, PA, 1995, p. 93.
16. Beninghoven, A., Muller, A. AES and SIMS investigations of the oxide films. *Phys. Lett.* 1979, A40, 169.
17. Tang, Z., Shemet, V., Niewolak, L., Singheiser, L., Quadakkers, W.J. Effect of Cr addition on oxidation behavior of Ti-48Al-2Ag alloys. *Intermetalics* 2003, 11, *1*, 1–8.
18. Benard, J. Theory of metal oxidation. In *Oxidation des Metaux*, Ed., Gauthier-Villars. Paris, 1964, p. 68.
19. Doychak, J. Oxidation behavior of high-temperature intermetallics. In *Intermetallic Compounds*, Vol. 1, Eds., Westbrook, J.H., Fleischer, R.L. John Wiley & Sons: Chichester, 1994, pp. 977–1016.

20. Sneggil, J.G. The role of chromium in protective alumina scale formation. *Mater. Sci. Eng.* 1987, 87, 261.
21. Yao, Y.L. *Irreversible Thermodynamics*. Science Press: Beijing, 1981, p. 273.
22. Prigogine, I. *From Being to Becoming*. W.H. Freeman and Company: San Francisco, CA, 1980, p. 84.
23. Gershman, I.S., Bushe, N.A., Mironov, A.E., Nikiforov, V.A. Self-organization of secondary structures at friction. *J. Friction Wear* 2003, 24, *3*, 329–334.
24. Brady, M.P., Smialek, J.L., Terepka, F. Thermodynamics of Selected Ti-Al and Ti-Al-Cr Alloys. *Scr. Metall. Mater.* 1995, 32, 10, 1659–1665.
25. Klansky, J.L., Nic, J.P., Mikkola, D.E. Structure/property observation for AlTiCr intermetallic alloys. *J. Mater. Res.* 1994, 2, 555–563.
26. Mabuchi, H., Tsuda, H., Matsui, T., Morii, K. Fabrication and structural control of nano-structured thin films by solid-state reaction of compositionally modulated multilayers. *J. Crystl. Growth* 2002, 1, 237–239.
27. Jewett, T.J., Ahrens, B., Dahms, M.J. Phase equilibria involving the γ-$L1_2$ and $TiAl_2$ phases in the Ti-Al-Cr system. *Intermetallics* 1996, 4, 543–551.
28. Wagner, Y-W.R., Yamaguchi, M. In *Gamma Titanium Aluminides*, Ed., Yoshihara, K. TMS: Warrendale, PA, 1995, pp. 93–123.
29. Lee, D.B., Kim, S.H., Niinobe, K., Yang, C.W., Nakamura, M. Effect of Cr on the high temperature oxidation of $L1_2$-type Al_3Ti intermetallics. *Mater. Sci. Eng. A* 2000, 290, 1–7.
30. Lee, J.K., Lee, H.N., Oh, M.H., Wee, D.M. Effects of Al–21Ti–23Cr coatings on oxidation and mechanical properties of Ti-Al alloy. *Surf. Coat. Technol.* 2000, 155, 59–66.
31. Mrowec, S., Jedlinsky, J., Gil, A. The influence of certain reactive elements on the oxidation behavior of chromia- and alumina-forming alloys. *Mater. Sci. Eng. A* 1989, 120–121, 169–174.
32. Stringer, J. The reactive element effect in high-temperature corrosion. *Mater. Sci. Eng. A* 1989, 120–121, *1*, 129–136.
33. Mennicke, G., Shumann, E., Ruhle, M., Hussey, R.J., Sproule, I., Graham, M.J. The influence of reactive-element coatings on the high-temperature oxidation of pure-Cr and high-Cr-content alloys. *Oxidation Met.* 1998, 49, 455–467.
34. Tanaguchi, S., Juso, H., Shibata, T. The influence of the alloy microstructure on the oxidation behavior of Ti–46Al-1Cr-0.2Si alloy. *Mater. Trans., JIM*, 1996, 37, *3*, 245–251.
35. Tanaguchi, S., Juso, H., Shibata, T. Effect of Hf additions on the isothermal-oxidation behavior of TiAl at high temperatures. *Oxidation Met.* 1998, 49, *3–4*, 325–337.
36. Palatnik, L.S. *Pores in the Films*. Energoizdat: Moscow, 1982, p. 214.
37. Escudeiro Santana, A., Karimi, A., Derflinger, V.H., Schütze, A. The role of hcp-AlN on hardness behavior of $Ti_{1x}Al_xN$ nanocomposite during annealing. *Thin Solid Films*, 2004, 22, 469–470, 339–344.
38. Gao, W., Liu, Z., Li, Z., Gong, H. Nano-crystal alloy and alloy-oxide coating and their high temperature corrosion properties. *Int. J. Mod. Phys: Appl. Phys.* 2002, 16, *1–2*, 128–136.
39. Yamamoto, K., Kujime, S., Taharara, K. Structural and mechanical property of Si incorporated (TiAlCr) N coating deposited by arc ion plating process, *Surf. Coat. Technol.* 2006, 5–6, 1383–1390.
40. Leyens, C., Peters, M., Hovsepian, P.Eh., Lewis, D.B., Luo, Q., Münz, W.-D. Novel coating systems produced by the combined cathodic arc/unbalanced magnetron sputtering for environmental protection of titanium alloys, *Surf. Coat. Technol.* 2002, 155, *2–3*, 103–111.
41. Yamamoto, K., Sato, T., Takahara, K., Hanaguri, K. Properties of (Ti, Cr, Al) N coatings with high Al content deposited by new plasma enhanced arc-cathode. *Surf. Coat. Technol.* 2003, 174–175, 620–626.
42. De Grote, S., Mazur, P. *Irreversible Thermodynamics*. Mir: Moscow, 1964.
43. Gachon, J.-C., Rogachev, A.S., Grigorian, H.E., Illarionova, E.V., Kuntz, J.-J., Kovalev, D.Y., Nosyrev, A.N., Sachkova, N.V., Tsygnkov, P.A. On mechanism of heterogeneous reaction and phase formation in Ti/Al multi-layer nano-films. *Acta Mater.* 2005, 53, 4, 1225–1231.
44. Lewis A.C., Josell D., Weihs T.P., Stability in thin film multi-layers and micro-laminates: the role of free energy, structure, and orientation at interfaces and grain boundaries. *Scripta Mater.* 2003, 48, 1079–1085.
45. Boeth, M., Ulrich, A., Pompe, W. Destratification mechanisms in coherent multi-layers. *J. Metastable Nanocrystalline Mater.*, 2004, 19, 153–178

46. Fox-Rabinovich, G.S., Weatherly, G.C., Wilkinson, D.S., Kovalev A.I. The role of chromium in protective alumina scale formation during the oxidation of ternary TiAlCr alloys in air. *Intermetallics* 2004, 12, 2, 165–180.
47. Prigogine, I.R. Dilip, K. *Modern Thermodynamics*. John Wiley & Sons: Chichester, 2002, p. 459.
48. Brookes, K.J.A. *Hardmetals and Other Hardmaterials*, 3rd ed. International Carbide Data: Hertfordshire, U.K., 1998, p. 220.
49. Ning, Y., Rahman, M., Wong, Y.S. Investigation of chip formation in high speed milling. *J. Mater. Proc. Technol.* 2001, 113, 360–367.
50. Yamamoto, K., Sato, T., Takahara, K., Hanaguri, K. Properties of (Ti,Cr,Al)N coatings with high Al content deposited by new plasma enhanced arc-cathode. *Surf. Coat. Technol.* 2003, 174–175, 620–626.
51. Gershman, J.S., Bushe, N.A. Thin films and self-organization during friction under the current collection conditions. *Surf. Coat. Technol.* 2004, 186, *3*, 405–411.
52. Crist, B.V. *XPS Handbook of the Elements and Native Oxides*. John Wiley & Sons, New York, 2000, p. 458.
53. Fox-Rabinovich, G.S., Weatherly, G.C., Dodonov, A.I., Kovalev, A.I., Veldhuis, S.C., Shuster, L.S., Dosbaeva, G.K., Wainstein, D.L., Migranov, M.S. Nano-crystalline FAD (filtered arc deposited) TiAlN PVD coatings for high-speed machining application. *Surf. Coat. Technol.* 2004, 177–178, 800–811.
54. Fox-Rabinovich, G.S., Kovalev, A.I., Weatherly, G.C. Tribology and the design of surface engineered materials for cutting tool applications. In *Modeling and Simulation for Material Selection and Mechanical Design*, Eds., Totten, G., Xie, L., Funatani, K. Marcel Dekker: New York, 2004, pp. 301–382.
55. Fox-Rabinovich, G.S., Weatherly, G.C., Dodonov, A.I., Kovalev, A.I., Veldhuis, S.C., Shuster, L.S., Dosbaeva, G.K., Wainstein, D.L., Migranov, M.S. Nano-crystalline FAD (filtered arc deposited) TiAlN PVD coatings for high-speed machining application. *Surf. Coat. Technol.* 2004, 177–178, 800–811.
56. Erdemir, A. A crystal-chemical approach to lubrication by solid oxides, *Tribology Letters* 2000, 8, 97–102.

12 Coolants and Lubricants to Enhance Tribological Compatibility of the "Tool–Workpiece" Tribosystem

Stephen C. Veldhuis, German S. Fox-Rabinovich, and Lev S. Shuster

CONTENTS

12.1 Introduction 382
12.2 Classification of the Coolants and Lubricants 382
12.2.1 Gaseous Lubricants 383
12.2.2 Cutting Fluids 383
12.2.2.1 Cutting Oils 383
12.2.2.2 Water-Based Cutting Fluids 383
12.2.2.3 Metallic Liquids 384
12.2.3 Plastic Lubricants 384
12.2.4 Solid Lubricants 385
12.3 Fundamentals of Fluids and Lubricants' Application for Cutting Operations 385
12.3.1 Functionality and Service Properties of Cutting Fluids 388
12.3.1.1 Lubrication 388
12.3.1.2 Cutting Fluids as Coolants 389
12.3.1.3 Washing Ability of Cutting Fluids 390
12.3.1.4 Impact of Cutting Fluids on Tool Life, Cutting Forces, Surface Finish, and the Other Parameters of Cutting 392
12.4 Future Trends 392
12.4.1 New Lubrication Techniques 392
12.4.2 New Lubrication Materials: Perfluorpolyether Lubricating Nanofilms 393
12.4.3 Dry Machining 404
12.4.4 High-Temperature Lubricious Oxides 405
References 407

12.1 INTRODUCTION

Cutting is a process of friction associated with intensive heat generation. That is why coolants, as well as lubricant application, are critically important to enhance the cutting process, especially under high performance conditions. Lubricants also play a critical role in the adaptation-process enhancement during cutting.

Modern machinery has some specific features. First of all, owing to environmental concerns, cutting fluids are beginning to be considered as hazardous materials. To solve this problem new techniques such as minimal quantity lubrication (MQL) and new lubricious polymer-based films are being used more and more widely in advanced industrial practice. Therefore, there is a trend toward the development of dry machining techniques. For extremely severe conditions of high performance, new types of lubricating compounds, such as oxides, could be used for dry machining where temperatures within the cutting zone can reach 1000°C. All these issues are considered in the present chapter.

The proper selection and application of cutting fluids is one of the major economical as well as engineering issues for high performance machining [1,2]. Rational application of coolant and lubricants results in cutting-tool life improvements by 1.2 to 4 times, cutting parameters intensifying by 20 to 60% and productivity increase by 10 to 50%. The majority of the coolants and lubricants that are used in practice presently are fluids. It is well known from the practice of machining that a flood of liquid is directed over the tool to act as a coolant or/and a lubricant. The reasons for using these fluids are to:

- Prevent overheating of the machine, tool and workpiece during cutting (cooling effect)
- Reduce friction and heat generation that result in tool life and surface finish improvements (lubricating effect)
- Reduce seizure phenomena (antifrictional properties)
- Enhance fluid penetration to the cutting zone (good wetting properties)
- Clear chips from the cutting zone (flushing effect)
- Ensure corrosion protection
- Allow chip and metal debris settle out (low viscosity)
- Prevent the formation of a sticky or gummy residue on parts as well as machine tools
- Provide stability for a safe work environment
- Effect economical efficiency

Application of efficient coolants and lubricants results in better surface finish of the machined component especially during machining of the hard-to-machine alloys and materials. A large variety of cutting fluids are available on the market. Fluids should improve cutting performance and meet the numerous requirements outlined in the preceding text. Selection and development of the advanced fluids are possible with a deep understanding of the mechanism of their interaction with the tools and workpiece materials during cutting. But at present, selection of the fluids mainly depends on testing and practical experience. Fundamentals of fluid impact on friction in general are based on the well-known tribological studies [3–10]. The results of these studies have lead to the wide use of surface-active materials in the cutting fluids for machining applications.

12.2 CLASSIFICATION OF THE COOLANTS AND LUBRICANTS

All coolants and lubricants can be divided into groups based on the state of aggregation of matter: gaseous, fluids, plastic, and solids.

Coolants and lubricants could also differ depending on the application either universal or specific. Universal coolants and lubricants could be used for a wide range of machining operations.

Specific coolants and lubricants, such as gaseous, plastic, solid, and liquid lubricants, are developed and used in a relatively narrow domain of application.

12.2.1 Gaseous Lubricants

The lubricants of this type are different (nitrogen, argon, helium) or active (oxygen containing: air, oxygen, carbon dioxide) ones. Active gases both cool and protect the surface due to the formation of oxide films. Machining (turning, drilling) of hard-to-machine materials such as Inconel in oxygen environments gives some beneficial results. Use of CO_2 is also effective and clean but expensive. Gaseous coolants are not widely used. The limitation of the gaseous coolants can be illustrated by application of liquid nitrogen [11–12]. The temperature at the cutting zone reduces slightly and still remains very high. Even extreme cooling action could not prevent overheating of the rake surface and thus intensive crater wear.

12.2.2 Cutting Fluids

There are two major types of cutting fluids:

- Oils
- Water-based fluids

Some liquid-state metals could be also considered as cutting fluids.

12.2.2.1 Cutting Oils

These cutting fluids consist of oils of petroleum of both animal and vegetable origin. These oils usually contain so-called extreme pressure or antiweld additives such as sulfur, chlorine and others. These additives react under pressure and heat to give the oil better lubricating properties. The obvious advantages of oils are good lubricity, antiseizure properties and high corrosion protection. The usage of oil is mandatory in many cutting operations such as tapping of hard-to-machine alloys, free machining, deep drilling, and gear hobbing. They are practically irreplaceable under these conditions. Sometimes synthetic oils or their mixture with mineral oils are used as fluids. However, the fluids that have this advantage could not be priced competitively in comparison to the mineral oils. Oil-based fluids increase cutting-tool life, improve workpiece surface finish, and protect the machine tools against corrosion.

Oils without additives are used for machining of magnesium, copper, and copper-based alloys (such as brass, bronze), and sometimes carbon steels under nonaggressive cutting conditions. These oils have low efficiency during machining of hard-to-machine materials and under high-performance machining conditions due to relatively poor cooling.

The disadvantages of these fluids are the following:

- Relatively low cooling ability
- Low thermal stability
- Increased evaporation
- High price

12.2.2.2 Water-Based Cutting Fluids

These fluids could contain emulsifiers, mineral oils, water, spirits, inhibitors to prevent corrosion and growth of bacteria, antigalling as well as antifoaming additives, electrolytes, and other organic and inorganic additives. The fluids could contain some fat or oils to improve lubricity. Recently some surface-active materials or chlorine additives are used. The lubricating properties of the fluids, i.e., water-based emulsions as well as oils, could be improved by the addition of chlorine and sulfur. These fluids are widely used as emulsions or water solutions during cutting of a wide range of steels and nonferrous metals. The main advantage of the water-based fluids is their higher efficiency as coolants. These fluids are also cheaper and less toxic. The disadvantage of the water-based fluids is the relatively low lubricity, low efficiency under specific cutting conditions, enhanced corrosion, and relatively low stability of the properties vs. time. Water-based fluids could be divided into the following groups: emulsions, semisynthetic, synthetic, and electrolyte solutions.

An emulsion is usually a mineral oil that is dissolved in water. These fluids are a suspension of oil droplets in water. These are the most popular cutting fluids because they combine the lubricity of oils and cooling properties of water and can be used in a wide range of cutting operations under various machining conditions. Emulsions are mixtures of mineral oil, emulsifiers, corrosion and bacteria growth inhibitors and other additives. The content of mineral oils in the cutting fluid could be up to 85%. Emulsions are used in 1 to 5% water solutions.

Semisynthetic cutting fluids are close to emulsions in composition but the concentration of the components is different. The basis of the semisynthetic cutting fluids is water (50%) and emulsifiers (up to 40%). The presence of mineral oil with low viscosity (3 to 10 vv/sec at 50°C) is mandatory. The semisynthetic cutting fluids could contain some additives. This type of fluid is used as a 1 to 10% water solution.

Synthetic cutting fluids are mixtures of water-soluble polymers, surface-active materials, additives, and water. To improve lubricity of these types of fluids some antiwear and antigalling additives are used. Synthetic coolants are used as a 1 to 10% water solution. Synthetic fluids that consist of inorganic or organic materials dissolved in water are used to improve the lubricity of these cutting fluids. The primary functions of these fluids are to cool and inhibit rust and corrosion. The synthetic fluids contain no oils and usually have low lubricating values. Their lubricating properties could be improved by the addition of active agents, which lower the surface tension of water (wetting ability) and form colloidal aggregates among the surface-active molecules. The disposal of these fluids is a relatively easy task compared to other lubricants. These fluids have a definite future potential. Semisynthetic fluids have recently become less popular because lubricity significantly improves for the state-of-the-art synthetic fluids.

12.2.2.3 Metallic Liquids

Metallic liquids such as tin, cadmium, bismuth, and zinc are used during machining of hard-to-machine metals such as Inconel, titanium, and titanium-based alloys. The application of these liquids results in significant tool life improvement but it is very difficult to use the liquid metals under conditions of mass production.

12.2.3 Plastic Lubricants

Plastic lubricants are used under conditions of small-lot production during taping, drilling, and reaming operations. The application range of these lubricants is narrow owing to the difficulties related to penetration of the cutting zone, cleaning, and recycling. Plastic lubricants can be divided as follows:

- Lubricants that are based on hydrocarbon thickening agents (paraffin, wax, and a few polymers)
- Lubricants that are based on soap thickening agents (sodium, lithium, calcium, barium, aluminum, lead-based, and others)
- Lubricants that are based on inorganic thickening agents (silica gel, clay, molybdenum disulfide, mica, and asbestos)

TABLE 12.1
General Classification and Examples of Solid Lubricants

Type	Principle of Lubrication	Examples
Soft materials	Pure metals Soft materials (at high temperatures); other inorganic solids	Pb, In, Sn, Ag, Au, CaF_2, or BaF_2 PbO, PbS, or CdO
Lamellar solids	Layered-lattice structure with strong, weak covalent bonding within the layers but bonding between the layers. Adsorbed vapors such as water or organic compounds are necessary to maintain easy shear (low friction). Layered structure with fluorination or intercalation used to insert atoms or molecules into the gap between planes, thus widening the spacing	Dichalcogenides: MoS_2, WS_2, TaS_2, etc., and similar di-tellurides and di-selenides; talc Graphite and hexagonal boron nitride (BN_{hex}) Graphite fluoride or Ag_xNbSe_2
Organic polymers	Polymers with smooth molecular structures (i.e., no side groups), that can slide easily over one another	PTFE, FEP, PF A, PTFCE, nylon, acetals, polyimides, metal soaps, waxes, solid fatty acids, esters
Chemical conversion layers	Surface oxides that reduce local cold-welding and can be more easily sheared than the underlying metal Porous surfaces that provide a reservoir for solid (or liquid) lubricants Surfaces with increased hardness to improve friction and wear characteristics	Oxide films Anodized surfaces Phosphated surfaces

Source: Erdemir, A. A crystal-chemical approach to lubrication by solid oxides, *Tribol. Lett.* 2000, 8, 97–102. With permission.

12.2.4 Solid Lubricants

The general classification of solid lubricants is shown in Table 12.1 [13].

Solid lubricants for cutting applications can be divided into groups (see Table 12.1) according to their chemical composition: inorganic products (talcum, graphite, mica, molybdenum disulfide, and other lamellar solids), organic compounds and polymers (wax, soap, solid fats), soft metals and materials (tin, lead, copper), and chemical conversion layers such as oxides. These lubricants are used under extreme cutting conditions where heavy loads and temperatures are applied or if other types of lubricants do not work well. Solid lubricants deposit coatings on the surface of the cutting tools as well as on the machined part. The deposition method of these lubricants depends on the type of solid lubricant. Recently, some of these coatings have been deposited using a plasma vapor deposition technique.

12.3 FUNDAMENTALS OF FLUIDS AND LUBRICANTS' APPLICATION FOR CUTTING OPERATIONS

Cutting is one of the most "ancient" methods of metalworking [14,15]. This includes several types: turning, milling, drilling, tapping, broaching, reaming, gear hobbing, grinding, and so on. The processes of surface plastic deformation, fracture, and chip removal take place during cutting. During machining a chip leaves the cutting zone with some particles of the worn tool. The actual tool's wear volume is small enough, but the cutting edge profile intensively changes vs. time and eventually the tool loses its workability. Both wear of rake and flank surfaces of the tool takes place during cutting. Wear is a very complicated phenomenon.

The major wear modes during cutting are the following:

- Adhesive: formation of welded asperity junctions between the chip and the tool face. The subsequent fracture of the junctions by shear leads to microscopic fragments of the tool material being torn out and subsequently adhering to the chip or the workpiece
- Diffusive or chemical wear: a complicated process of material transfer from one metal to another through the interface and their combination
- Abrasive: similar to grinding

There are some other mechanisms of the cutting-edge failure during service such as fatigue fracture (cracks formation and chipping) and plastic deformation of the cutting edge (see Chapter 7).

When a chip leaves the cutting zone, a juvenile surface of the workpiece is formed. This surface has some very special features. Ideally, a clean, juvenile surface has a very high chemical reactivity that intensively emits electron flow and has a very high coefficient of friction. These features of the juvenile surface enhance adhesive interaction and diffusion at the cutting tool–workpiece interface. Besides, the juvenile surfaces that generate during friction act as a catalyst to the chemical reaction at the cutting tool–workpiece interface, especially in the presence of lubricants. Interaction of the juvenile surface with active molecules of the cutting fluids such as oxygen, iodine, and chlorine results in the formation of chemical films, which decrease both the friction forces and the intensity of adhesive and diffusive processes during cutting-tool wear. Even a small amount of fluid reaching the cutting zone results in a significant change in the efficiency of the cutting process. Therefore, much attention is paid to the issues associated with the cutting fluids penetration to the cutting zone.

One of the major features of the cutting phenomena is intensive seizure at the cutting tool–workpiece interface. When intensive seizure is taking place, which is typical for machining operations such as turning of ductile materials, there is little or no access of the cutting fluids to the workpiece–cutting tool interface. Any fluid access is supposedly precluded especially along the plastic part of the tool–chip contact length [15]. In spite of this, cutting-fluid applications obviously improve critical parameters of cutting such as tool wear rate and the surface finish of the machined component. Direct observations were made that prove that cutting fluids sometimes reduce the tool–chip contact length [2]. This is why the theory considering cutting fluids as boundary lubricants has gained prominence [16]. Besides, boundary lubrication conditions can take place when seizure does not result in the formation of a continuous metal-to-metal contact layer at the workpiece–tool interface. The cutting fluid application could be very effective if the seizure at the chip–tool interface is island-like which is typical for some milling or gear-hobbing operations. A great number of publications relate to cutting fluids but only a few of them aim at understanding the role of cutting fluids on the mechanism of cutting [15].

The major paths of cutting fluid penetration to the contact zone are through the use of pores and capillary networks that form at the cutting tool–workpiece interface (Figure 12.1a) and the voids that form as a result of the build-up edge tearing off (Figure 12.1b). The vibration during machining also enhances the lubricant penetration to the contact zone (Figure 12.1c). Experimental studies show that the contact of the chip and cutting tool is discontinuous during cutting under low and moderate cutting speeds. A formation of around a micron-size capillary network during cutting ensures the permanent supply of fluids and their mist to the cutting zone under these conditions. Here fluid velocity reaches 3.5 to 4.0 m/sec and thus permanent supply of the cutting tool–workpiece interface by the lubricant is ensured. The depth of the cutting fluids' penetration depends on the size of the capillary at the cutting tool–workpiece material interface, surface tension at the phase boundary, and the lubricant density.

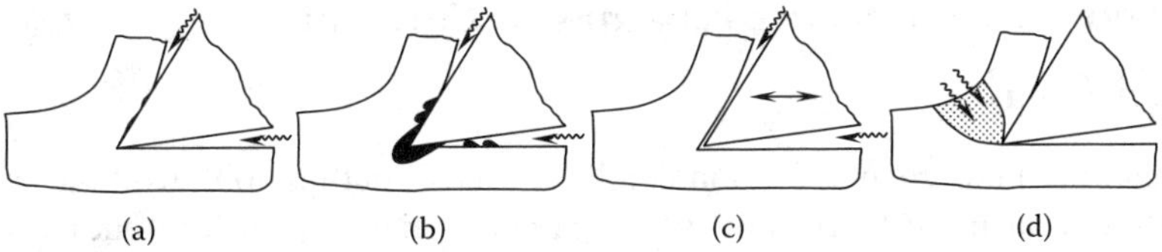

FIGURE 12.1 Major paths of cutting fluids penetration to the "workpiece–cutting tool" contact zone: (a) pores and capillary network that form at the cutting tool–workpiece interface; (b) voids that are forming as a result of the build-up edge tearing off; (c) vibration during machining; (d) through distorted lattice structure within the deformed layer of the chips.

Under high-performance-machining conditions, a close-to-continuous seizure zone forms on the rake surface of the cutting tools. The size of the capillary decreases dramatically, and there is only one possible way for fluid penetration of the cutting tool–workpiece interface, i.e., penetration of the fluids in the gaseous state. In this case the actual lubrication at the cutting tool–workpiece material interface is performed by the products of the fluids' thermal dissociation. These products are interacting with the friction surfaces or with oxygen from the environment and after that, with the machined metal surface. Another possible path of penetration for the particles of fluids, such as ions, atoms or molecules, is through the distorted lattice structure within the deformed layer of the chips (Figure 12.1d). It was determined that the cutting fluids penetrate through the bypath sides of the chips.

The driving force of the lubricant penetration to the cutting tool–workpiece interface is the force of chemical interactions, adsorption, vibrations, external electrical, and magnetic fields. Among the most important parameters that could determine the efficiency of the cutting fluids penetration are the following:

- Means of the fluid supply to the cutting zone (free fluids flow, forced lubrication, spraying)
- Viscosity
- Surface properties
- Chemical nature and size of the molecules, atoms, and ions

Cutting fluids act as coolants, lubricants, dispersants, and cleaning agents. The different role of the cutting fluids is dependent upon the specific segments of the cutting tool–workpiece contact zone (Figure 12.2). The segments I and II correspond to the cooling zones and zones of embrittlement, whereas the segments III and IV are lubricating zones due to the formation of the protective film [10].

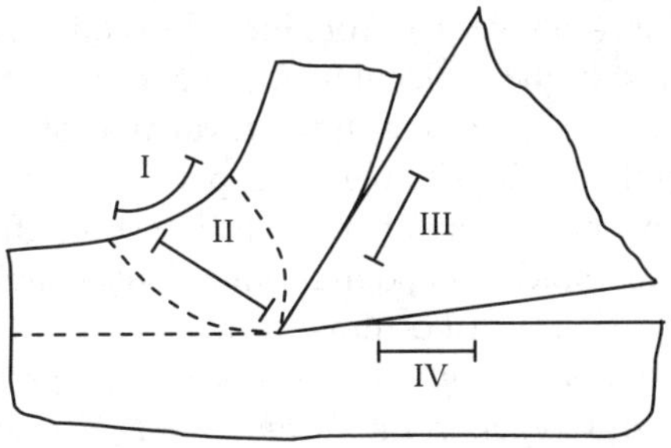

FIGURE 12.2 Specific segments of the "cutting tool–workpiece" contact zone. The segments I and II are the cooling zones and zones of embrittlement correspondingly, the segments III and IV are lubricating zones due to protective films formation.

12.3.1 Functionality and Service Properties of Cutting Fluids

12.3.1.1 Lubrication

Lubricants in cutting fluids change friction conditions at the cutting tool–workpiece interface. The major goal of the cutting fluids lubrication is to: (1) reduce seizure intensity at the cutting tool–workpiece interface, (2) generate heat due to friction, (3) change the tool–chip contact length, (4) decrease cutting forces, and (5) affect build-up formation.

Under high-performance-machining conditions, the efficiency of the cutting fluid is lower because any fluid access is supposed to be prohibited especially along the plastic part of the tool–chip contact length at the cutting tool–workpiece interface.

Enhanced lubricity of the cutting fluids leads to:

- Tool life improvements due to a decrease in the adhesion and diffusion at the interface
- Productivity increase
- Surface-finish improvements of the workpiece
- Lowering of the residual stresses in the machined component owing to a decrease of the adhesion intensity and localization of the shearing processes
- Cutting forces and heat-generating reduction

At the same time improvement of cutting fluids lubricity can lead to negative results:

- Decrease of the tool–chip contact length on the rake surface of cutting tools. This leads to a stress concentration at the cutting edge, as a result cutting-edge chipping could occur.
- Lowering the protective function of build-up.
- Growth of the cutting forces during machining. For example, cutting forces during machining of aluminum, lead, and copper using high-speed steel cutting tools are higher in air than in a vacuum. The cause of this phenomenon is formation of surface tribo-films whose strength is higher than the bulk material.

The formation of strongly reactive juvenile surfaces results from high temperatures at the cutting zone, heavy applied loads, and plastic deformation. The molecules of surface-active materials of the cutting fluids are adsorbed on the surface. The presence of moisture and oxygen from the environment accelerates the processes associated with chemisorption. Temperature plays a significant role in these processes. The adsorbing and chemisorbing films have low temperature stability. They have low efficiency during machining of hard-to-machine materials and the materials that generate significant amount of heat during cutting. The lubricating films added to the cutting fluids as antiwear and antigalling additives are more efficient for these conditions. The molecules of these additives interact with each other, with oxygen from the environment, and with the juvenile surfaces of the frictional bodies. As a result, the molecules of the additives dissociate and form atoms and radicals that interact with the metals, resulting in the formation of a surface-lubricating layer.

There are different data on the composition and properties of lubricants with high-temperature stability and efficiency at various machining operations. The most efficient are compounds such as oxides, sulfides, chlorides, and complex compounds whose shear strength is lower when compared to the contacting frictional bodies. The tribo-films' formation based on these compounds prevent direct interaction of the frictional bodies, which decreases adhesive and fatigue wear rates for the cutting tools. The positive role of oxygen in the formation of protective tribo-films with improved lubricity is known from the literature [17]. However, an excess of oxygen in the cutting zone leads to the formation of relatively thick and brittle oxide films on the surface of the cutting tools that could result in an increase in wear intensity. Therefore, it is important to control the process, aiming to accelerate or if necessary depress the reactions of frictional surfaces with oxygen, instead of reducing the natural air environment. The major goal of this control is to inhibit or more realistically to reduce the adhesion and diffusion processes at the cutting tool–workpiece interface. Besides, the oxygen, carbon, nitrogen, boron, and other elements that are supplied by fluids to the cutting zone could play a similar role. The amount of these elements, as well as the amount of oxygen and carbon within the cutting zone, should be optimized to get the best results.

In general the mechanism of the lubrication of cutting fluids is a result of the following processes:

- Depression of reactivity of the juvenile surfaces that form during cutting
- Formation of the boundary tribo-films, protecting materials of the cutting tool and workpiece against mechanical and physicochemical destruction
- Lowering of the surface energy of the workpiece material as well as decreasing of the shearing strength of the surface layer (Rebinder's effect [18])
- Lowering the tool–chip contact length and decreasing the frictional forces [19]
- Formation of the wedge pressure in microcracks that appear in workpiece material during cutting and inhibition of the formed cracks sides welding

Lubrication of the cutting fluids could be evaluated during cutting and tribological tests based on the parameters: (1) cutting forces and work, (2) friction parameter on the rake and flank surfaces, (3) chip compression ratio, (4) wear rate data, and (5) build-up formation. The results of the cutting test also depend on the cleaning, cooling and dispersant properties of the cutting fluids. The tribological tests' data though, are not sensitive to these properties. Sometimes lubricating properties of cutting fluids are evaluated using a special tribological apparatus that models the conditions of cutting. However, the data obtained using these apparatus could only indirectly characterize the lubricity as well as antiwear, antigalling, and antifrictional properties of the cutting fluids. This is because the parameters of the cutting process, such as cutting data, chip thickness, and chip compression ratio, strongly affect the experimental values obtained. That is why the lubricity of cutting fluids are evaluated most often using universal tribo-meters, where the efficiency of the lubricant is estimated based on the friction parameter and wear rate data.

Temperature-dependent properties of the cutting fluids are the most important parameters used to characterize lubricity of these materials. Based on the knowledge of the actual temperature in the cutting zone and the temperature dependent properties of the cutting fluids, the rational selection of the cutting fluids can be made for specific applications.

The role of antifrictional fat-based additives in cutting fluids consists of adsorbed lubricating films on the surface of cutting tools that lead to a wear rate decrease. The concentration of these fat-based additives in cutting fluids is in the range of 0.5 to 20%.

Antigalling additives inhibit the adhesion between the cutting tool and the workpiece material to promote a wear rate decrease under high temperatures and heavy loads. The mechanism of the action of these additives includes the chemical interaction of the products of these materials' dissociation with the machined surface during cutting. The chemical compounds form due to this process, resulting in lower shear strength and melting temperature compared to the bulk material. Chemical lubricating films decrease cutting forces and prevent intensive adhesion and diffusion. The concentration of the antigalling additives in cutting fluids is in the range of 0.5 to 50% depending on the cutting conditions.

12.3.1.2 Cutting Fluids as Coolants

There are two major sources of heat generation during cutting operations. The first one is the primary share plane, and the second one is the tool–workpiece interface. Usually the major portion of heat is generated because of shearing of the workpiece material. The heat that is generated during friction makes a minor contribution to the overall heating during cutting. The coolant has no direct access to the areas of the heat sources. Water-based coolants efficiently reduce the temperature in the cutting zone. The removal of the heat from the primary share zone slightly changes the cutting tools' performance. In contrast, the heat generated during friction at the tool–workpiece interface controls the tool life and wear behavior. During cutting, the tool is the only stationary component of the system, which is why in most cases cooling is most effective through the tool. It is a well-known fact that some types of cutting tools include nozzles for fluids. It should be mentioned here that the flow zone at the rake surface of the cutting tool has the highest actual temperature. It was proved by experiments that the coolant application could not prevent heat generation at the cutting tool–workpiece interface because the interface is unacceptable to the direct action of the fluids. However, coolant application significantly reduces the volume of the tool surface affected by overheating. In practice, overall cooling is most widely used. The major part of the mechanical energy generated during cutting is transformed to heat, and only a small part of the energy is used in the structural transformation of the workpiece surface layer. The heat generation during cutting leads to the temperature growth within the cutting zone that includes the tool, workpiece, chips, or environment. Heat generation is most influenced by the cutting speed but other cutting parameters such as feed and depth of cut are important. Heat generation during machining of ductile materials such as steels is higher compared to machining of brittle materials such as cast iron.

The cooling action of the cutting fluids impacts the:

- Cutting-tool life
- Accuracy of machining
- Surface finish
- Residual stress forming on the surface layer of the machined component
- Helical chip formation (tool–chip contact length changes)
- Machining productivity

Cutting forces, as was outlined in earlier text, are largely determined by diffusion and adhesion processes that take place during cutting. These processes are temperature dependent. Cutting-fluid application leads to a significant decrease of the workpiece, cutting tool, and chip temperature. Thus, cutting fluids can impact the temperature distribution during cutting. The enhancement of the cutting fluids' cooling properties leads to tool life improvements during cutting operations with intensive heat generation. Machining of steels is usually associated with usage of fluids. The use of a coolant is very important when HSS cutting tools are used. Cutting fluids decrease the temperature in the cutting zone when machining is carried out with speeds up to 150 m/min. At higher cutting speeds, cutting fluids stabilize the temperature of the workpiece.

Using a coolant is efficient for turning operations, but for some operations such as the drilling of shallow holes, the use of a coolant is critical because it significantly decreases the temperature in the cutting zone to perform cost-effective machining. For heavy loaded operations of cutting such as tapping and gear hobbing, the lubrication properties of the fluid controls the cutting performance.

On the other hand, the cooling of the cutting zone can have some negative results such as intensive cutting-edge chipping for interrupted cutting operations (milling, planing) especially at elevated cutting speeds. This is very typical when ceramic and cemented carbide tooling are used because the coolants' application result in intensive thermal cycling that leads to cutting-edge chipping. Besides, sometimes the intensive cooling of the workpiece leads to the formation of tensile internal stresses that result in poor surface properties.

12.3.1.3 Washing Ability of Cutting Fluids

During cutting, chips and other waste materials such as wear debris, oxide particles, dust, dirt, and products of cutting fluids' destruction are formed.

The washing ability of the cutting fluid is a combination of the chemical phenomena that leads to cleaning of the surface by cutting and machine tools, as well as the workpiece. This characteristic of the cutting fluids ensures the cleaning of chips and waste material from the workpiece, cutting tools, and machine tools. We have to emphasize that the term "washing ability" of cutting fluids covers two meanings: (1) washing ability in the proper sense of the word and (2) ability to wash out and evacuate chips or metal debris, as well as inhibit the formation of carbon build-up on the surface of cutting tools and workpiece. This carbon build-up is a result of two processes during cutting: (1) adsorption of carbohydrate molecules present in the coolant from heating the machined surfaces and (2) chemical reaction with oxygen that is solved in the coolant.

The initial stage of the washing action of the cutting fluids is to moisten the dirt surface. This includes metals as well as nonmetal particles that are formed during machining. The next stage of the washing action is the formation of a stable suspension of waste material that is formed in the fluids. The molecules of surface-active materials adsorption leads to the dispersion of the debris due to the surface strength. In addition, the wedging action decreases. Besides, owing to the surface-active action of the molecules, the so-called "colloidal solution" of the dirt takes place in the cutting fluids.

Washing and evacuation of the chips is one of the most important functional properties of the cutting fluids. The washing ability greatly depends on the amount of cutting fluid supplied to the cutting zone, the flow velocity, the method of fluid supply as well as the concentration of the washing components in the fluid. In industrial practice, the washing ability is evaluated visually. Quantitative estimation of this property is very time consuming and expensive.

Based on the properties of cutting fluids outlined earlier, we can similarly explain the mechanisms of action for cutting fluid efficiency enhancement, using the scheme suggested below (Figure 12.3).

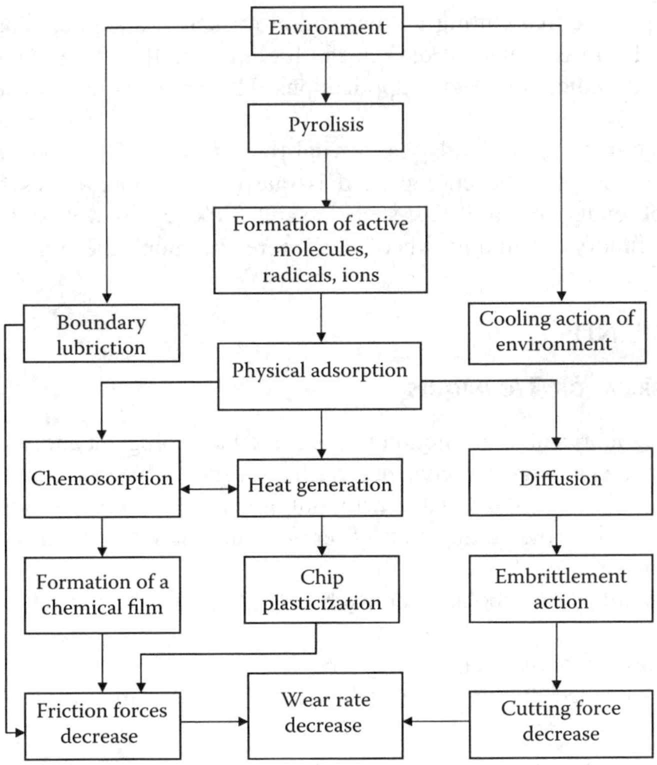

FIGURE 12.3 The scheme of cutting fluids action.

12.3.1.4 Impact of Cutting Fluids on Tool Life, Cutting Forces, Surface Finish, and the Other Parameters of Cutting

The main criteria for the major service properties evaluated for cutting fluids are the following:

- Wear of cutting tools
- Tool life
- Surface finish (dimension accuracy and roughness of the machined surfaces)

Additional criteria that can characterize the service properties of cutting fluids are the following:

- Torque
- Cutting forces
- Temperature distribution at the tool surface
- Microprofiles of the cutting tool–workpiece material interface
- Vibration-resistance during cutting
- Cutting-tool chipping

The most comprehensive evaluation of all the service properties of cutting fluids can be done through testing under real cutting conditions. However, these tests require a large number of machines to perform, significant consumption of cutting tools, workpiece materials and also cutting fluids. Therefore, industrial testing is very expensive and time consuming. To develop new cutting fluids for specific applications new methods of accelerated testing have been developed using different sets of apparatus (such as tribo-meters) as well as machine tools.

We have to emphasize that cutting fluid selection for specific applications is a very difficult technological task. There is plenty of data in the literature [20] that can be used as a guide to the selection of cutting fluids for specific applications. The choice is influenced by many machining parameters.

We can conclude that cutting fluids play a vital role in machining productivity. The environmental cost consideration for the cutting fluid is one of the major issues for the future. The development of biofriendly cutting fluids, such as synthetic ones, as well as intensive research in the field of dry or semidry machining is becoming more and more relevant.

12.4 FUTURE TRENDS

12.4.1 New Lubrication Techniques

Recently, minimizing lubricant consumption has become the ecological and economic objective of machining operations. As a long-term goal, dry machining should be considered, but dry machining has some obvious disadvantages associated with tool overheating, excessive seizure, and problems with chips' evacuation from the cutting zone. Therefore, alternative methods of cooling and lubrication should be considered.

Methods that use little or no coolants for machining [7,9,21] should also be mentioned, such as:

- Cooling with compressive air
- The use of so-called "minimum mist lubrication" or "minimum quantity lubrication" (MQL)

However, the use of MQL will only be acceptable if the main task of the cutting fluids [17], i.e., cooling, lubrication, and chip removal in the cutting process is resolved [22]. Mist cooling offers the option of combining gases having good penetration to the cutting zone with those of liquids having good lubricity. The substance of the MQL method is very small quantities of high-lubricity oil applied on to the point, creating a semidry lubricant-supplying system called near dry machining (NDM) or minimum quantity lubrication (MQL), which has its origin in the American aerospace industry with a view to cutting difficult materials [2]. The benefits of this process include dry metal chips that are ready to remelt and improved chip-removal capabilities using vacuum systems. Minimum quantities of oil provides a highly reduced risk of fire, significantly reduce the administration costs for lubricants, reduce the need for waste treatment, eliminate the problem of lubricating oil life, and greatly reduce waste oil disposal [23].

Another promising method is the "high-jet" technique. If a stream of compressed air or a cutting fluid is applied at high velocity to the cutting zone, the tool life improvement is significantly better compared to regular free-flow cutting fluids [23].

12.4.2 New Lubrication Materials: Perfluorpolyether Lubricating Nanofilms

The perfluorpolyether (PFPE) lubricating films are multicomponent systems that contain fluorine-organic surface-active substances (fluorine-SAS) and controlling agents in various solvents. Usually PFPE (a 0.5% solution of perfluorine polyester acid [R_f–CH_2OH] in freon 113) can be deposited by dipping the tool part into a boiling solution. PFPE has the following chemical structure: $HO–CH_2–CF_2O–(C_2F_4O)_6–(CF_2O)_{20}–CF_2–CH_2–OH$. The PFPE film was deposited by dipping into a solution maintained at the boiling point. The physicochemical properties of PFPE are shown in Table 12.2. The thin film consists of a close-packed molecular monolayer that provides an even coating to a rough tool surface. The process of PFPE films' deposition consists in applying the fluorine-SAS molecules on the surface of a frictional body. As a result of this process, a thin (40 to 80 Å thick) film of oriented fluorine-SAS molecules are formed, fixed on the surface by the chemosorption forces. This coating has high adsorption ability, and owing to its low thickness it also has high adhesion to the substrate and penetrates into the pores. The surface energy of oils contained in the typical coolant regularly used for machining is higher than the surface energy of the PFPE film. As a result of the molecular interaction of the oil and PFPE film, the latter film is not sheared from the surface of the cutting tool during the first stages of cutting.

TABLE 12.2
Physicochemical Properties of Z–DOL

Property	Value
Molecular mass	2194
Average number of units in the molecule	12
Molecular dimensions of SAM	5 nm
Density	1560 kg/m^3
Thickness of epilamon layer	5–2500 nm
Critical unit load	3 GPa
Maximum service temperature	723 K

Source: From Fox-Rabinovich, G.S., Kovalev, A.I., Weatherly G.C. Tribology and the design of surface engineered materials for cutting tool applications. In *Modeling and Simulation for Material Selection and Mechanical Design*. Eds., Totten G.E., Xie, L., Funatani, K. Marcel Dekker: New York, 2004, chap. 5.

The fluorine-SAS film has the following features [24]:

- Reduces the surface energy of the material by approximately 1000 times, that prevents spreading of the lubricant on the surface.
- Enhances tribological compatibility, owing to the reduction of the coefficient of friction and the lowering/preventing of adhesive interaction during friction.
- High temperatures are stable (up to 400 to 450°C; the short-term temperature flashes up to 700°C).
- High-load-bearing capacity (the approved unit pressure is 3000 MN/m^2).
- Changes the wear mode: due to reduction of adhesion at the frictional bodies interface, the fatigue wear mode starts dominating.
- Reduces the probability of a dry friction condition occurrence resulting from the prevention of the overcritical displacement of the lubricant in HLTS.
- Prevents electrochemical corrosion (prevents hydrogen embrittlement) due to high penetrability of the fluorine-SAS molecules, which fill up and de-gas all the pores and cracks; protects the surfaces from exposure to moisture and aggressive substances.

The properties of the fluorine-SAS films outlined in earlier text result in beneficial surface modification and give the modified surfaces antifriction, antisticking, anticorrosion, and some other specific properties.

The organic-based SAS are widely used in practice for specific industrial applications, such as designations for heavy loading applications. PFPE films could also be used for manufacturing applications such as cutting and stamping operations, as well as injection dies and molds. These SAS have improved thermal stability, resistance to oxidation, and very low fugacity that could expand the field of application for PFPE films. The PFPE films application results in tool life improvements, better surface finish, and dimensional accuracy.

A great advantage of SAS films is that the deposition processes of these films are relatively simple. There include cold, hot, and technological processes.

The hot process method is when the fluorine-SAS molecules are deposited at a temperature higher than the boiling point of the solvent. The cold process occurs when the fluorine-SAS molecules are deposited from the fluorine-SAS-containing solution at a temperature lower than the boiling point of the solvent. The technological medium process is when the fluorine-SAS molecules are deposited from the technological medium (oil, emulsion, grease), containing the fluorine-SAS.

The process of PFPE films deposition includes the following production operations:

- Surface preparation
- Applying the fluorine-SAS film
- Thermal fixing or drying
- Quality control

The surface preparation involves cleaning all kinds of dirt from the surface. Various solvents (acetone, benzene, alcohol, and chlorine-containing solvents) could be used for this purpose. The surface should be dried after cleaning. The drying time is around 0.5 to 3 h at a temperature of 90 to 100°C.

In the cold process, the part is immersed in the fluorine-SAS-containing solution, or applied to the surface by spraying. The film is formed after evaporation of the solvent. In the hot process, the part is placed in the boiling solution and runs in a sealed container.

In the technological medium process, some quantity of the oil-compatible (or water-solvable) fluorine-SAS is introduced into the technological medium (oil, emulsion, grease). The technological

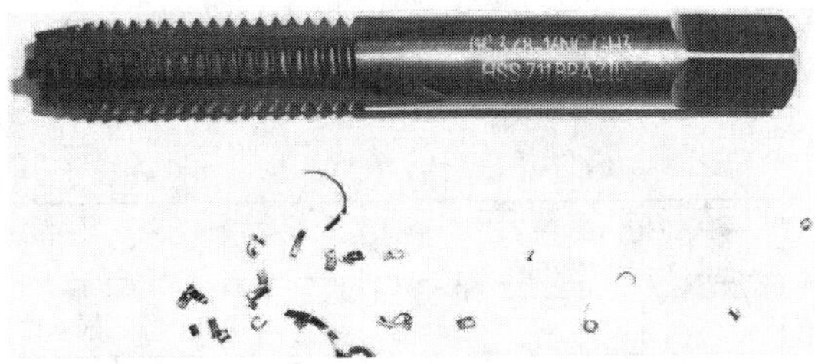

(a)

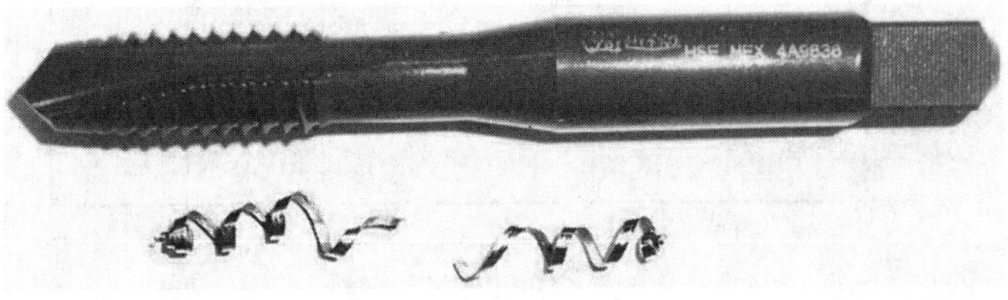

(b)

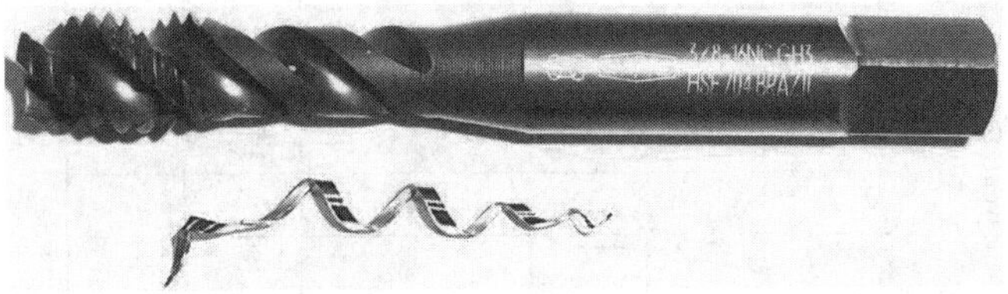

(c)

FIGURE 12.4 Images of different types of taps tested: (a) regular; (b) spiral point; (c) spiral.

medium transports the fluorine-SAS molecules to the friction surfaces. The most active process of the film-forming runs on the surfaces in contact.

The hot process usually provides better quality of the film when compared to the cold one but the deposition technique should be tailored to specific applications. As a rule this process is carried out in driers with the temperature around 90 to 150°C. The drying process strengthens the chemisorption bonds between the fluorine-SAS film and the treated surfaces.

Some standard methods of quality control such as the wetting angle of an oil drop are used. The angle should be equal to or more than 72° [25]. The oil drop should have a diameter of 2 to 4 mm and remain on the surface. More advanced methods based on of the surface energy of the material estimation can also be used.

Experimental data from other authors for different cutting applications such as tapping and end milling have been obtained.

HSS taps were tested during machining of P20 steel (Figure 12.4). Figure 12.5 presents the tool life data of the taps with and without PFPE films. The data show that the design of the taps critically affects the tool life of taps both with and without PFPE films. The highest tool life has spiral taps (Figure 12.5c) and the least is spiral pointed (Figure 12.5b).

Coefficient of friction data vs. temperature in a contact with the P20 steel with hardness HRC 30–35 presented in Figure 12.6. An application of PFPE films result in an improved coefficient of friction, within a temperature range of up to 500°C, that is typical for tapping operations.

SEM data on surface morphology of the taps with and without PFPE films is shown in Figure 12.7 and Figure 12.8. Figure 12.7 presents the surface morphology of the worn taps with "regular" design and Figure 12.8 presents similar data for the worn spiral taps with the best tool life as

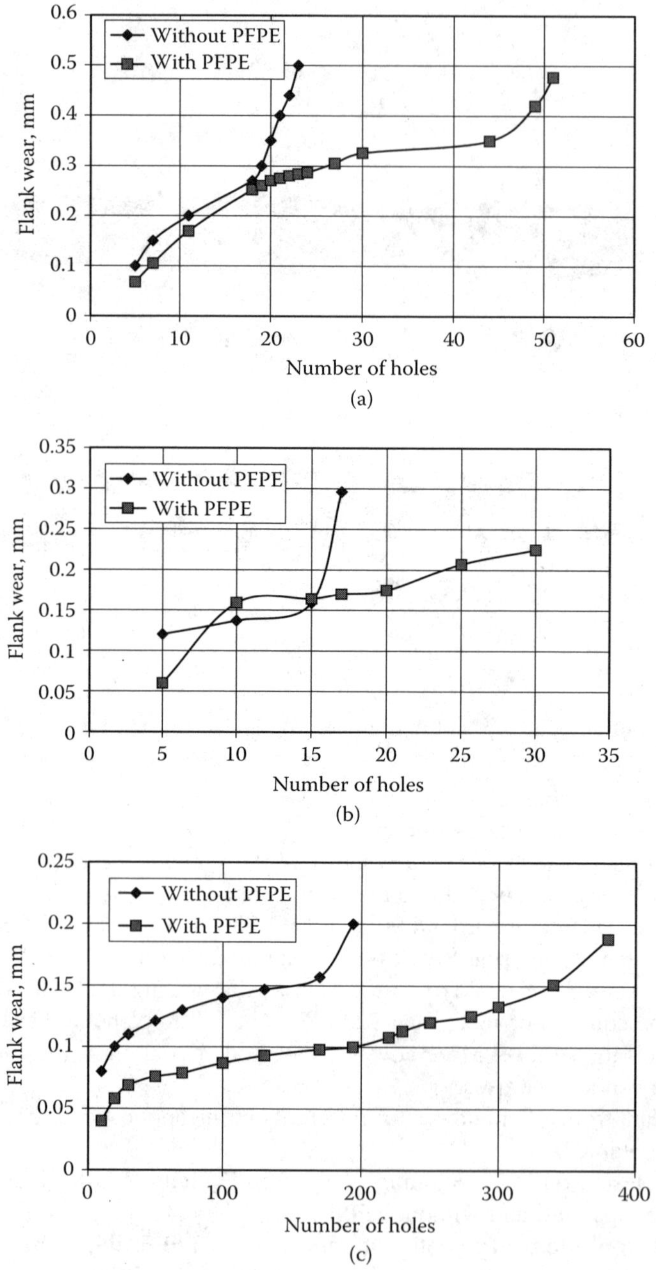

FIGURE 12.5 Flank wear of the taps with and without PFPE films vs. number of holes: (a) regular; (b) spiral point; (c) spiral.

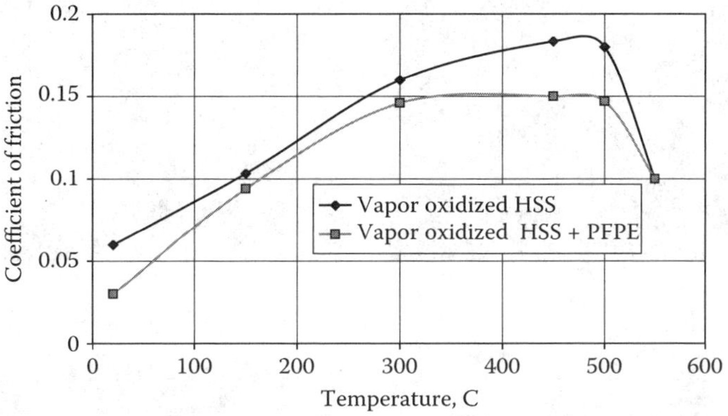

FIGURE 12.6 Coefficient of friction vs. temperatures for the vapor oxidized HSS steels samples with and without PFPE films.

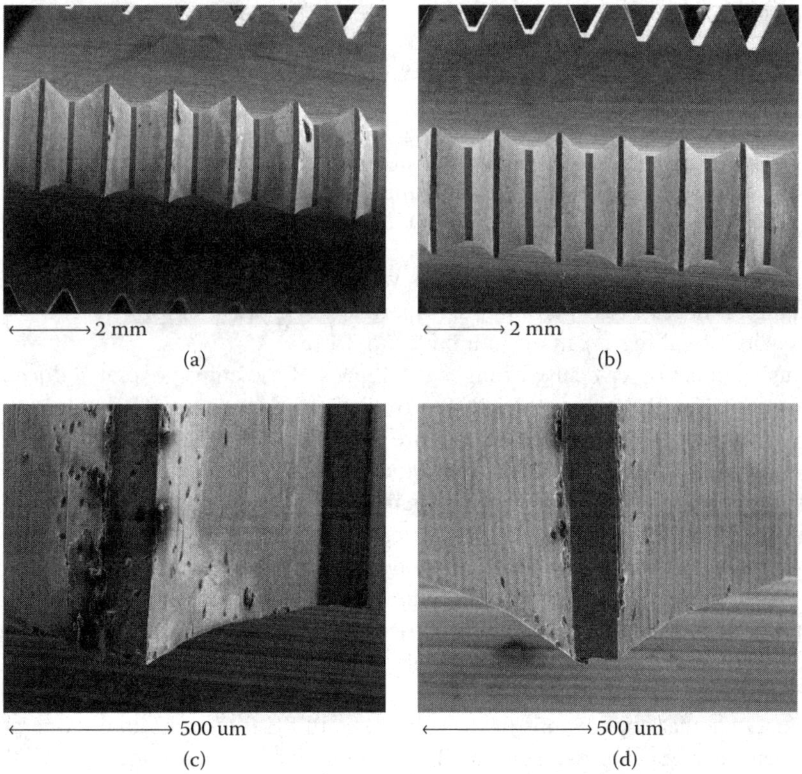

FIGURE 12.7 SEM images of the "regular" taps' worn surface with and without PFPE films: (a) without PFPE films after 23 holes, overview; (b) with PFPE films after 51 holes, overview; (c) detailed view of the image shown in (a); (d) detailed view of the image shown in (b).

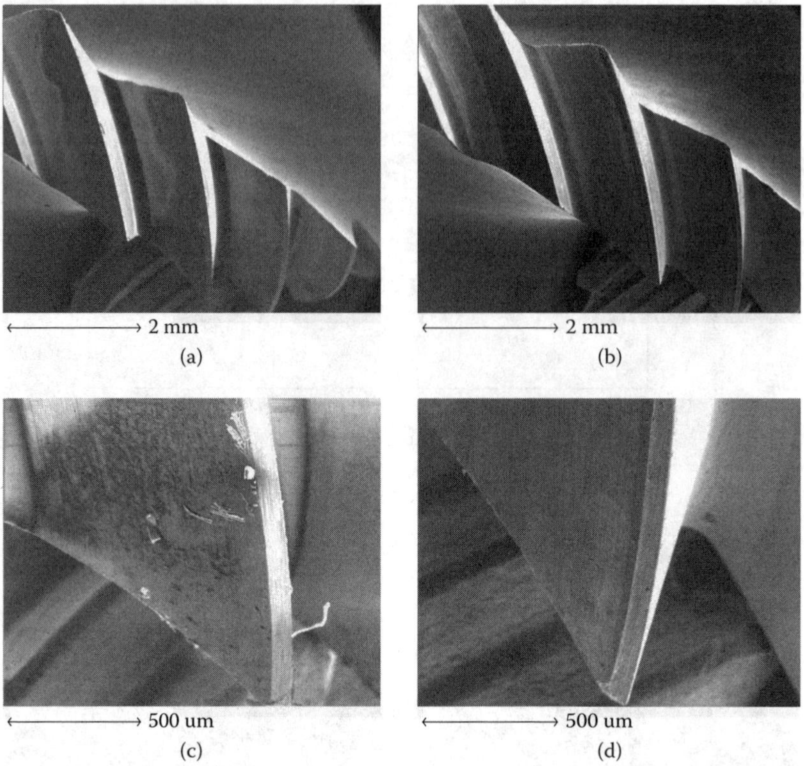

FIGURE 12.8 SEM images of the spiral taps' worn surface with and without PFPE films: (a) without PFPE films after 194 holes, overview; (b) with PFPE films after 380 holes, overview; (c) detailed view of the image shown in (a); (d) detailed view of the image shown in (b).

compared to the other types of taps tested. Taps without PFPE films have intensive galling of the workpiece material that is typical for tapping operations. Application of PFPE films significantly drops the intensity of galling resulting in a better tool life.

PFPE films' application critically changes the shapes of the chips generated during the tapping operation (Figure 12.9). Relatively thick chips from the taps without PFPE films (Figure 12.9b) showed a lower curling (Figure 12.9a and Figure 12.9c) and had a rougher undersurface (Figure 12.9e). The thinner chips (Figure 12.9d) from the taps with PFPE films were curled in a spiral (Figure 12.9b and Figure 12.9d) and had a smooth undersurface (Figure 12.9f).

The cross sections of the chips show two areas: a contact flow zone and a zone of extended deformation (Figure 12.10). There is a very wide flow zone for chips from taps without PFPE (Figure 12.10a). In contrast, the flow zone together with the zone of extended deformation is much narrower for the taps with PFPE (Figure 12.10b). This means that more intensive flow of the workpiece material is taking place during cutting owing to improved surface layer lubricity of the taps with PFPE films.

We can conclude that applications of the lubricious PFPE films for tapping operations leads to friction and wear behavior improvement due to the reduction of seizure intensity of the workpiece material to the cutting-tool surface. This results in tool life improvement by a factor of 2.

Another example of PFPE films application is the deposition of this layer on the sublayer of hard Ti–N coatings [26]. As is known, the main disadvantage of hard coatings is their high coefficient of friction. The principal function of the top antifrictional PFPE layer is to increase the lubricity of the cutting tools with hard nitride coatings (see Chapter 9, Figure 9.3; Chapter 10).

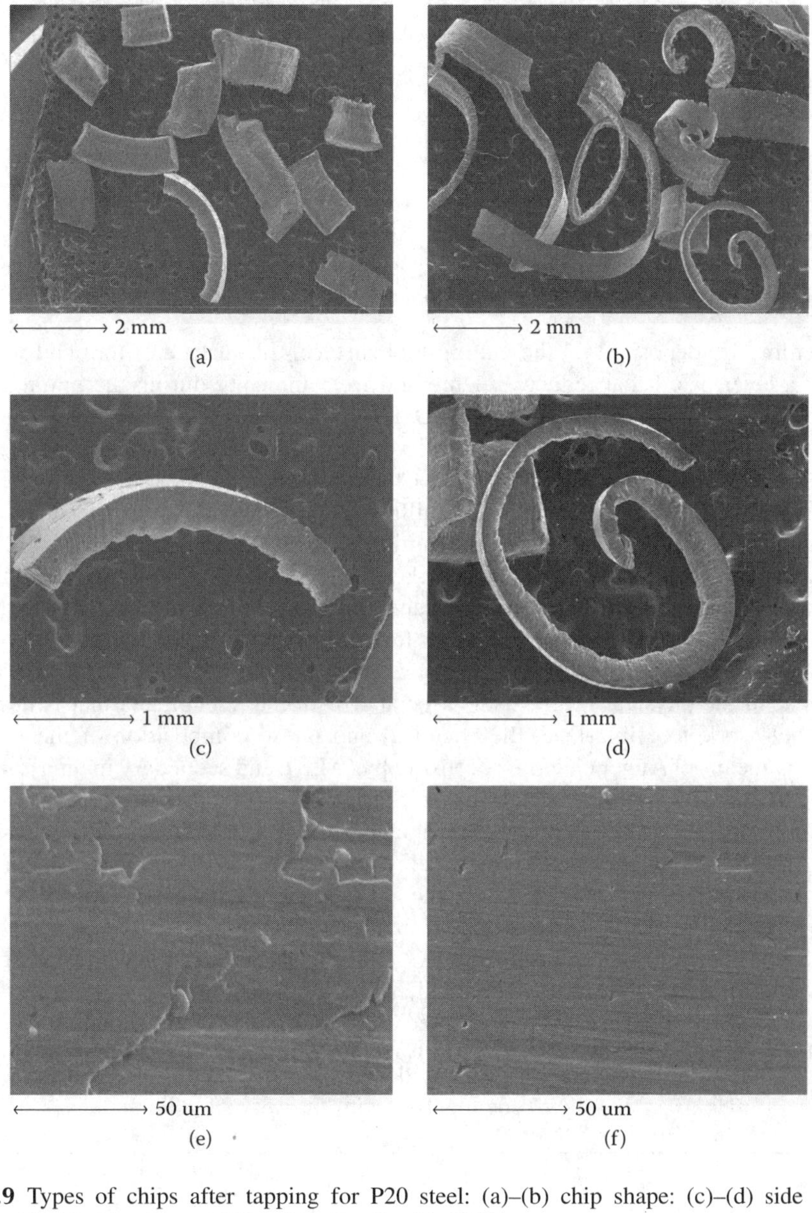

FIGURE 12.9 Types of chips after tapping for P20 steel: (a)–(b) chip shape: (c)–(d) side view: (e)–(f) undersurface. (a), (c), (e) without PFPE; (b), (d), (f) with PFPE.

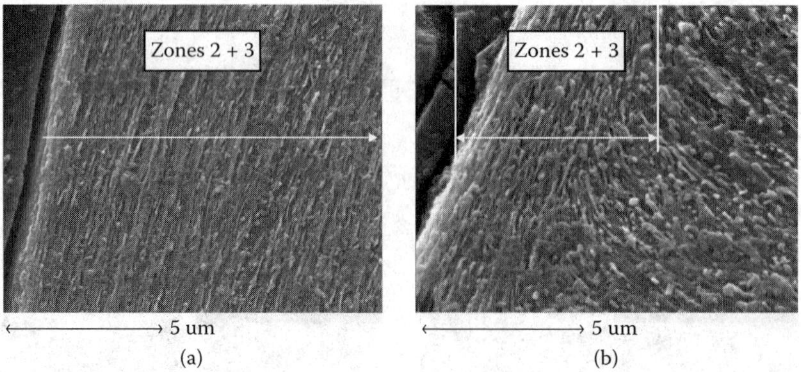

FIGURE 12.10 SEM images of the chips cross sections: (a) without PFPE; (b) with PFPE films. Zone 2 and Zone 3 — contact flow zone and a zone of extended deformation.

If PFPE films are deposited on the cutting-tool surface, the surface of tool and workpiece are separated by a layer of oil that reduces seizure and wear intensity during the initial stages of the tool service. Studies of duplex coatings (a PVD Ti–Cr–N hard coating and a top layer of PFPE) deposited on a HSS substrate in contact with a 1040 steel show that the friction characteristics are improved at the service temperature (500°C, see Chapter 9, Figure 9.3). The wear process of coated cutting tools was studied during turning and milling of 1040 steel. The cutting tools were tested when cutting with and without a coolant. The cutting tests data are shown in Table 12.3. The wear of tetragonal indexable HSS inserts with multilayered coatings was also studied. These service conditions can lead to a very intensive surface-damaging wear of the cutting tools. As it was shown in Chapter 7, adhesive wear dominates the work faces of the cutting tool during cutting at low and moderate speeds [15].

To understand the physical mechanism associated with the phenomena that is taking place at the cutting tool–workpiece interface, the chemical and phase composition of the surface layers were studied by means of Auger electron spectroscopy (AES), and secondary ion mass spectroscopy (SIMS) using VG ESCALAB-MK2 spectrometer. Scanning Auger spectroscopy was used to analyze the surface composition of the hollow on the face cutting-tool edge at different stages of its development, as well as for the analysis of the chip contact surface. The phase composition of the surface of the wear crater was analyzed with the aid of scanning mass spectroscopy of secondary positive ions. The analysis was made in approximately static mode.

HREELS spectra were acquired with a VG EMU–50. The spectra was recorded with 300 msec/mV dwell time and CAE = 1 eV. The incident electron beam energy was 7.0 eV. The FWHM of the peak of back-scattered electrons was 7 meV as from the graphite standard.

The tool life data (Figure 12.11a) reflects a very low pattern of surface damage at the running-in stage of wear, leading to a marked improvement in the overall tool performance.

TABLE 12.3
Cutting Tests Data

		Cutting Parameters					
Machined Material	Operation	Speed m/min	Depth, mm	Feed	Width, mm	Coolant	HSS Cutting Tool
Steel 1040	Turning	50–70	1.0	0.28 mm/rev		+	Indexable insert
	Milling	21	3	0.028 mm/flute	5		End mill

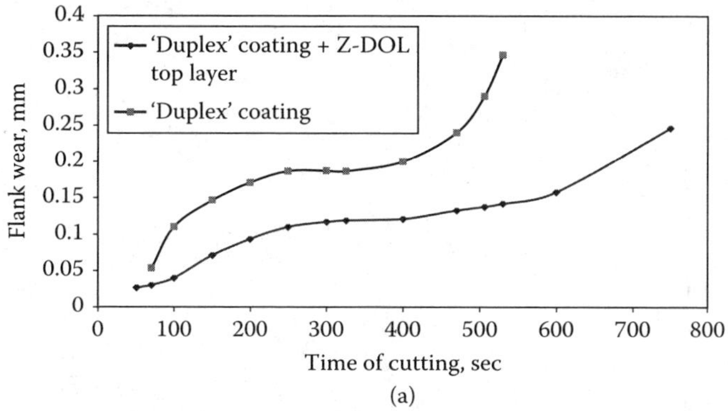

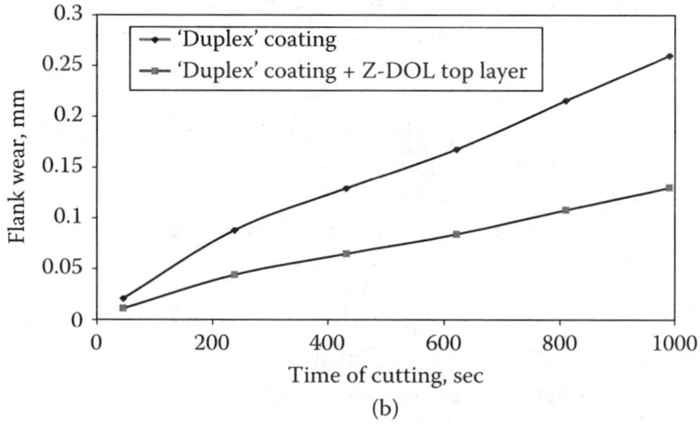

FIGURE 12.11 Tool flank wear vs. time. (From Fox-Rabinovich, G.S., Kovalev, A.I., Wainstein, D.L., Shuster, L.Sh., Dosbaeva, G.K. Improvement of "duplex" PVD coatings for HSS cutting tools by PFPE (perfluorpolyether "Z-DOL"). *Surf. Coat. Technol.* 2002, 160, *1*, 99–107. With permission.) (a) Turning test data (machined material — 1040 steel; cutting speed = 70 m/min; depth — 1.0 mm; feed — 0.28 mm/rev). Cutting tool: indexable inserts, tool material: M2 HSS with "Duplex" coatings; cutting with coolant. (b) Milling test data (machined material — 1040 steel; cutting speed — 21 m/min; depth — 3.0 mm; width — 5 mm; feed: 0.028 mm/flute). Cutting tool: end mills with "Duplex" coatings; cutting with coolant. (From Fox-Rabinovich, G.S., Kovalev, A.I., Wainstein, D.L., Shuster, L.Sh., Dosbaeva, G.K. Improvement of "duplex" PVD coatings for HSS cutting tools by PFPE (perfluorpolyether "Z-DOL"). *Surf. Coat. Technol.* 2002, 160, *1*, 99–107. With permission.)

The results of the friction parameter vs. temperature study for the multilayered coatings, consisted of PVD (Ti–N) hard coating plus PFPE's top layer, in contact with 1040 steel show (Chapter 9, Figure 9.3) that the friction parameter is slightly improved. The improvement of the friction parameter alone cannot explain a tool life increase of 1.5 to 2.0 times for the cutting tool with multilayered coatings (Figure 12.11).

Additional investigations of the cutting tool–workpiece interface have been done to understand this phenomenon. Figure 12.12a to Figure 12.12c presents the spectra of the positive and negative ions obtained after analyzing the chemical and phase composition of Ti–N + PFPE coating. The Ti–N coating gives the following peaks: Ti^+ (48); TiN^+ (62); TiO^+ (64 amu). The PFPE gives the following peaks: O^- (16); F^- (19); CF_2^{-2} (25); CF_2O^{-2} (33); F_2 (38 amu). The composition of the coating changes during cutting. The small intensity peaks at 64 and 86 amu appear after cutting for 200 sec. Obviously, we can attribute these peaks to the tribo-decomposition of PFPE and TiN, the formation of titanium–oxygen compounds and TiF_2.

Elemental composition analysis of the tool wear crater surface was carried out by means of AES. The results are presented as a series of Auger spectra obtained from the surface of wear crater

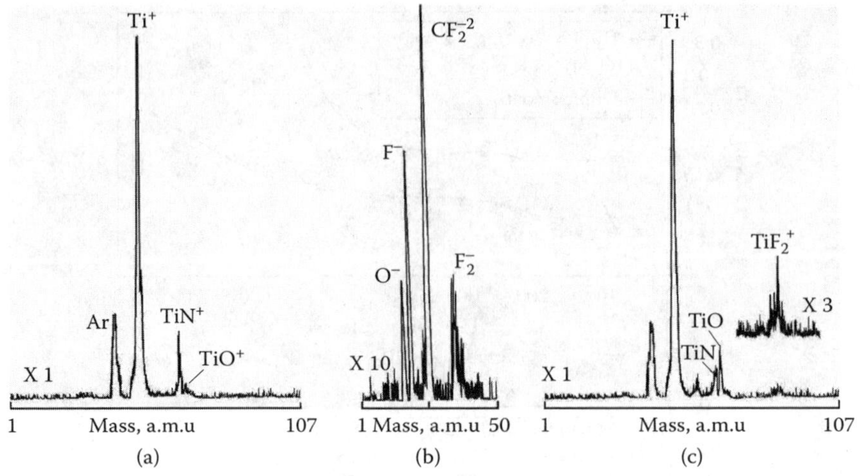

FIGURE 12.12 SIMS spectra of Ti–N coating with Z–DOL antifrictional top layer. (a) spectrum of positive ions (before service); (b) spectrum of negative ions (before service); (c) spectrum of positive ions (after 200 sec cutting). (From Fox-Rabinovich, G.S., Kovalev, A.I., Wainstein, D.L., Shuster, L.Sh., Dosbaeva, G.K. Improvement of "duplex" PVD coatings for HSS cutting tools by PFPE (perfluorpolyether "Z-DOL"). *Surf. Coat. Technol.* 2002, 160, *1*, 99–107. With permission.)

at different times of cutting (Figure 12.13). The results show that in the initial stages of wear (in the running-in phase), a gradual tribo-oxidation of the Ti–N PVD coating takes place under high local stress and temperatures generated during cutting. This process was observed in Ti–N coatings with and without a PFPE antifrictional top layer. A gradual reduction of nitrogen content was observed on the surface at the running-in stage. Simultaneously, oxygen replaces the nitrogen. A significant increase in intensity of Auger lines for iron after prolonged cutting is brought about by the adhesion of the workpiece material (1040 steel) to the surface of the cutting tool and by the destruction of the Ti–N coating at the final stage of the wear. There is a good correlation between the change in wear resistance of the cutting tool and the composition of the coating (Figure 12.11 and Figure 12.13). The depletion of the coating surface by nitrogen and its enrichment by oxygen during the transition from the running-in stage to the steady stage of wear is connected to the tribo-oxidation of TiN. Our previous studies show that the oxygen-containing compound developing on the surface is a nonequilibrium Ti–O solid solution [25].

Tribo-oxidation of titanium nitride in our case is thought to be a beneficial process. The formation of a titanium-based oxygen-containing compound acts as a shield that protects the surface. That is why the appearance of the oxygen-containing compound at the transition from the running-in stage to the normal wear testifies to the self-organization of the "tool–workpiece" tribosystem [25].

Figure 12.14 shows the vibration spectrum of Ti–N PVD coating (a) without and (b) with the PFPE lubricant deposited on the tool surface. The surface was cleaned prior to examination by heating in a high vacuum of 1×10^{-10} Torr at 473 K for 15 min. Vibration energy loss peaks for the PVD Ti–N coating deposited on HSS are observed at 110 and 370 meV. The first asymmetrical peak (110 meV) is observed at characteristic energies of the transverse optical phonons with out-of-plane displacements ($TO_{\perp}$) in the spectra. In addition, we detected a broad loss peak centered at 185 meV due to optical phonons with in-plane displacement. This feature in the high-energy loss portion of the spectrum at 370 meV is attributed to an overtone of the loss at 185 meV.

Vibration energy loss peaks for PFPE films deposited on TiN are observed at 100 and 159 meV (See Figure 12.14b). In addition, intensity loss spectra were detected at energies in the range of 350 to 1400 meV. The intense loss at 159 meV is typical for perfluoropolyethers and can be associated with the coupled stretching motion of the (CF_2–O–CF_2) linkage [18]. The peaks of losses

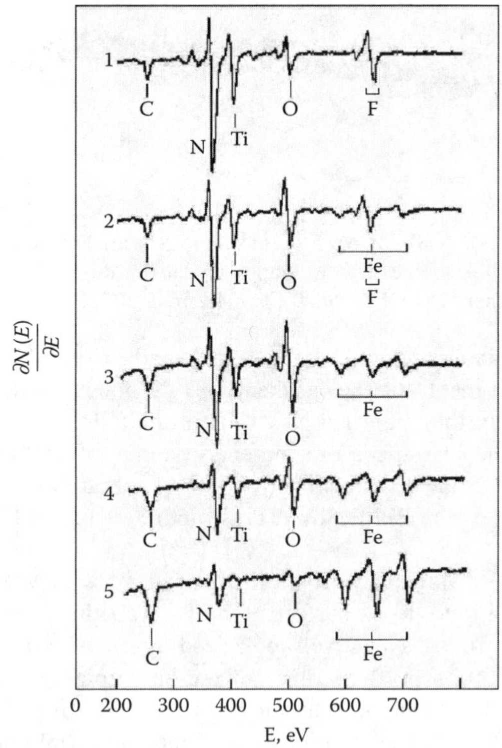

FIGURE 12.13 Auger spectra from cutting-tool surface: (1) Ti–N coating with Z–DOL antifrictional top layer before service; (2) after 100 sec; (3) 300 sec; (4) 500 sec; (5) 750 sec. (From Fox-Rabinovich, G.S., Kovalev, A.I., Wainstein, D.L., Shuster, L.Sh., Dosbaeva, G.K. Improvement of "duplex" PVD coatings for HSS cutting tools by PFPE (perfluorpolyether "Z-DOL"). *Surf. Coat. Technol.* 2002, 160, *1*, 99–107. With permission.)

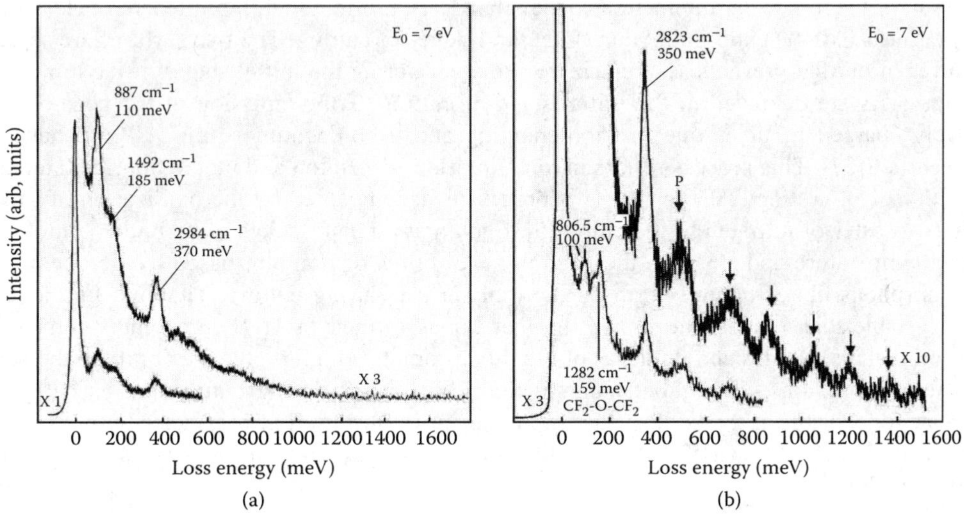

FIGURE 12.14 (a) HREELS spectrum of Ti–N coating on HSS; (b) HREELS spectrum of PFPE Z–DOL on Ti–N coating. In both cases the primary energy was 7.0 eV. (From Fox-Rabinovich, G.S., Kovalev, A.I., Wainstein, D.L., Shuster, L.Sh., Dosbaeva, G.K. Improvement of "duplex" PVD coatings for HSS cutting tools by PFPE (perfluorpolyether "Z-DOL"). *Surf. Coat. Technol.* 2002, 160, *1*, 99–107. With permission.)

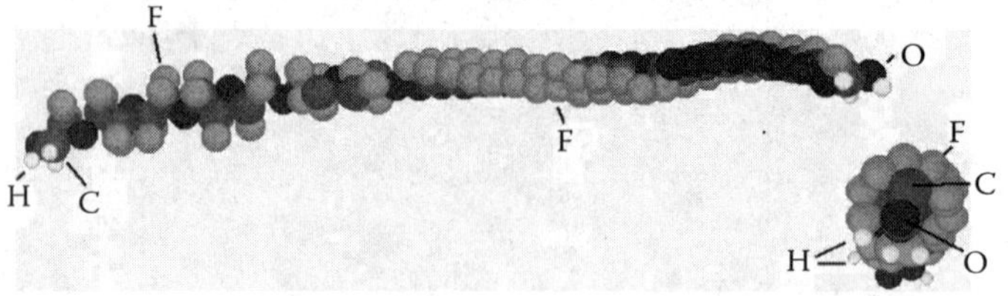

FIGURE 12.15 Side and end-on view of a PFPE molecule. (From Fox-Rabinovich, G.S., Kovalev, A.I., Wainstein, D.L., Shuster, L.Sh., Dosbaeva, G.K. Improvement of "duplex" PVD coatings for HSS cutting tools by PFPE (perfluorpolyether "Z-DOL"). *Surf. Coat. Technol.* 2002, 160, *1*, 99–107. With permission.)

at 100 meV and 350 meV are analogous to the 110 meV and 370 meV losses on the TiN spectrum. These softened phonons come from the weakened Ti–N bonds of the substrate, owing to the additional chemical bonds formed between N with H of the (CH_2–OH) groups of PFPE film. The periodic fine structure of the loss spectra in the energy range of 350 to 1400 meV is due to the overtone phonon spectra associated with collective surface vibrations and can be explained by the highly disordered structure of the PFPE film. The structure of the PFPE molecule is represented in Figure 12.15.

This study suggests the following wear mechanism of a hard PVD coating with a top PFPE layer. The polymer molecule binds to the surface of the Ti–N coating by positive polarized hydrogen atoms in the OH and CH groups. Negatively polarized atoms of nitrogen in the TiN compound can be the centers of adsorption on the surface of the hard coating. Negative polarized fluorine atoms in CF groups push back out of electron clouds surrounding the cutting tool–workpiece interface. A similar process for the generation of a surface potential under PFPE lubrication was proposed elsewhere [27]. Owing to this, the electrostatical effect of the PFPE lubricant decreases the friction parameter at the cutting tool–workpiece interface. In this way the total operating life of the tool increases.

This is not the only explanation of the PFPE film effect. Another explanation of PFPE films effect is outlined in earlier text and associated with the low surface energy of PFPE films [25,28]. As a result of the molecular interaction of the oil and PFPE film, the film of the latter is not sheared from the surface of the cutting tool during the first stages of cutting. The two surfaces are separated by a layer of oil that prevents severe seizure and wear during the initial stages of the tool service.

Lubricants are degraded at the initial stage of cutting. Tribo-emission of the negatively and positively charged particles due to tribo-charging and tribo-decomposition of PFPE have been shown earlier [27]. This process results in ionic fluorine adsorption and metal–fluoride interaction. After 200 sec of cutting, TiF_2 forms. This process is accompanied by the oxidization of TiN and the mass transfer of nitrogen to the chip [25]. Such mass transfer takes place under conditions of extreme temperature and stress in the friction zone. Films of titanium-based oxygen compounds with amorphous-like structures form on the tool surface during cutting. Titanium fluorides may be more stable than oxides due to the stronger bonds formed in Ti–F compounds compared to Ti–O compounds. Oxides and fluorides of titanium promote a high protective ability to the coating under wear conditions. The stable titanium fluorides formed during cutting are probably more important for tool life increase than the improvement of any friction parameter by the PFPE top layer. Earlier it was shown [25,29] that the nonequilibrium Ti–O films acts as a thermal and mechanical shield.

12.4.3 Dry Machining

Dry machining is a very challenging area. To achieve this goal the cutting tool has to meet a number of requirements. The major way to improve dry machining conditions is through a proper cutting-tool design [30–32], advanced tooling materials selection [33,34] as well as application of surface engineering [20,35–41]. Cutting-tool manufacturers such as Titex Plus Company http://www.titex.com/ have used these approaches. The cutting-tool designs used were:

- Small friction forces due to large rake angle and dispensing with heavily-rounded cutting edges
- Low friction via narrow cylindrical lands and strong back taper
- Good chip evacuation was obtained due to special flute profiles

To achieve the design features specified previously, special micro grain carbide with high cobalt content was used to improve toughness and edge strength. The advanced coatings used are mandatory for high speed machining operations. The coatings have a few critical functions for dry machining, i.e., thermal isolation and separation of the tool and workpiece and reduction of friction. Traditional hard coatings such as TiAlN or multilayered WC/C as an outer layer of Ti–Al–N coatings could be used with high efficiency for these conditions.

One of the most challenging areas of modern dry machining is the high-speed end milling of hardened tool steels. The end milling of hardened dies and molds results in significant labor costs saving up to 30%. The only lubricating agents that could be stable during this process are high temperature lubricating oxides. The overview of these solids is presented in the following text.

As was outlined in Chapter 9, one of the best ways to form oxides layers at the workpiece–cutting tools interface is a deposition of the advanced self-adaptive synergistically alloyed coatings such as Ti–Al–Cr–N-based ones (Chapter 9 and Chapter 10). We can assume that further improvements of the adaptive coating for high-performance machining applications will be based on the compositions that enhance formation of lubricious oxides with high-temperature stability during friction.

One of the most challenging areas of dry machining is the cutting of lightweight materials such as aluminum-based alloys. Cutting tools with DLC (diamondlike carbon) coatings look very promising for this type of machining applications. This hard coating has properties such as low coefficient of friction; high hardness, high chemical stability, and high wear resistance when in contact with aluminum. Additionally, DLC coatings can achieve high antiadherent properties without lubricant that make them excellent coatings for dry machining of aluminum. There are a variety of the different DLC coatings commercially available on the market. Applications of nonhydrogenated amorphous carbon film with low coefficients of friction (around 0.1) of the cemented carbide substrates results in improvement of surface finish of the machined part and prevent intensive build-up-edge formation that is critical for the machining of aluminum. This leads to maintaining the sharpness of the edge, the reduction of approximately 50% cutting resistance, and the improvement of the machinability of aluminum-based alloys with Mg and Si additions. The surface finish of the component after dry machining using cemented carbides with DLC coatings is equal to the uncoated inserts under wet conditions. At the same time, the tool life was also improved significantly. Even in dry machining, DLC coated tools have excellent advantages over uncoated tools during cutting of aluminum cutting [42].

A "smart" coating concept that was considered in detail in Chapter 9 could be successfully used for dry machining applications. A multilayered coating that includes an outer DLC layer and a sublayer of implanted In (see Chapter 9, Table 9.4) has been used for dry drilling of aluminum alloy. The drilling conditions are shown in Table 12.4. Tests have been performed until the drill breaks. The drills that were tested to failure typically resulted from fracture of the drill after its flutes became clogged with aluminum. The results of the test are shown in Table 12.4. The data obtained show that the smart coating application results in tool life improvement by four times.

12.4.4 High-Temperature Lubricious Oxides

The self-organization phenomenon that is considered in detail in this book is mainly related to the tribo-oxidation phenomenon and the oxygen-containing films' formation. The tribo-films formed

TABLE 12.4
Tool Life Data on the "Smart" DLC-Based Coatings

Cutting Tool	Tool Life Improvement
HSS uncoated	1.0
HSS+DLC coating	1.68
HSS+ smart DLC coating with In sublayer	4.06 as compared to uncoated drills and 2.4 as compared to DLC coatings

Note: Data presented by J. Dasch and C. Ang, R&D Center, General Motors. The drilling conditions are as follows: cutting tool: 1/4 in. HSS drills; workpiece material: B 319 (Al — base, Si — 5.5–6.5, Cu — 3–4, Fe, Mg, and Ni — < 1 wt.%) cast aluminum BHN500; hardness 80 HB; depth of cut, 19 mm; blind holes; speed 61 m/min; feed 0.13 mm/rev; no coolant (dry machining conditions).

during friction have high-surface-protection ability as well as improved lubricity at elevated temperatures. On the other hand, a permanently increasing demand for more aggressive machining conditions, especially severe dry machining conditions, requires the development of advanced adaptive coatings (see Chapter 10) that could generate protective/lubricious oxides on the surface during friction under high-temperature or high-stress conditions. It is critically important to overview the major properties of the lubricious oxides to understand the future trends of surface engineering for high performance machining applications.

Achieving and maintaining low friction at high temperatures is typical for high performance machining (700 to 1000°C). It is one of the toughest problems encountered in the field of tribology [43]. It is known from literature that at elevated temperatures above 500°C the solid lubricant coating such as silver, molybdenum disulfide, and hexagonal boron nitride quickly degrade from the surface during friction due to intensive oxidation and/or corrosion wear. Because of their low friction at elevated temperatures, oxides of certain metals and metalloids (i.e., Re, Ti, Mo, Cr, Zn, V, W, B, etc.) have been used as lubricants. Oxide-based self-lubricating materials can be prepared as alloys or by designing appropriate coatings [43]. At high temperatures, the oxide layer is depleted from the surface by wear, and the most useful alloying ingredients diffuse toward the surface where the oxygen potential is higher. They then oxidize again and replenish the consumed lubricious layer, which has low shear strength and/or surface energy to decrease friction.

Many oxides are abrasive or hard to shear, thus they lead to severe abrasion. Only oxides with low shear strength can be used as lubricants. One of the well-known lubricious oxides is rutile. Research performed by various authors (see Chapter 1, Chapter 5 to Chapter 10) show that Ti–O films that are formed on the surface of Ti-based nitride coatings. This oxide has unique shear properties. The crystal chemistry of oxides is strongly related to shear rheology. An important crystal chemical property was developed, i.e., ionic potential, which is responsible for many important characteristics of numerous oxides. The ionic potential is a ratio of the formal cationic charge to the cation radius. The ionic potential is related to the melting point of the oxides and the higher the ionic potential, the lower the coefficient of friction at elevated temperatures (see Table 12.5) [43].

Based on the data presented in Table 12.5, we can conclude that compounds that can generate oxides with relatively low melting point such as B_2O_3 and V_2O_3 can be used with high efficiency for low or moderate cutting speed conditions when the temperature on the surface does not exceed 700°C. For more aggressive cutting conditions, compounds generating such oxides as WO_3 can be used. This oxide does not melt during cutting and has low shear strength. Even aluminum oxide has quite good lubricious properties at high temperatures around 1000°C. Experimental data presented in Chapter 7 to Chapter 10 confirm these conclusions. The chemistry and crystallography

TABLE 12.5
Relationship between Coefficient of Friction, Melting Point, and Ionic Potential of Certain Oxides

Oxide	Ionic Potential	Melting Point (°C)	Coefficient of Friction (Temperature Range, °C)
B_2O_3	12	500	0.3–05 (550–730)
V_2O_3	8.4	670	0.32–0.3 (600–1000)
MoO_3	8.9	795	0.27–0.2 (600–800)
WO_3	8.8	1470	0.3–0.25 (600–800)
TiO_2	5.8	1850	0.55–0.35 (800–1000)
Al_2O_3	6	2040	0.5–03 (800–1000)
ZrO_2	5	2800	0.5 (800)
CoO	2.7		0.6–0.4 (300–700)
FeO	2.7		0.6 (300–800)

Source: Savan, A., Pflüger, E., Goller, R., Gissler, W. Use of nanoscaled multilayer and compound films to realize a soft lubrication phase with a hard, wear-resistant matrix. *Surf. Coat. Technol.* 2000, 126, *2–3*, 159. With permission.

of the oxide formed is critically important for new generation of the self-adaptive coatings that work under conditions of high-speed dry machining (see Chapter 10).

Future developments most probably will be related to the deposition of compounds that can generate complex lubricious oxides during high temperature friction [43,44]. As is shown in Chapter 7, complex oxides are more stable, and such oxides as Mo–Ti–O probably ensure better lubricity. It is also proven that oxide films that are generated during friction can work synergistically and significantly enhance the protective/lubricious ability of the surface under severe temperature/heavy stress conditions (see Chapter 10).

REFERENCES

1. Byrne, G., Scholta, E. Environmentally clean machining processes — a strategic approach. *Ann. CIRP* 1993, 42, *1*, 471.
2. Astakov, V.P. Tribology in metal cutting. In *Mechanical Tribology. Materials, Characterization, and Applications*, Eds., Liang, H., Totten, G. Marcel Dekker: New York, 2002, p. 307, chap. 9.
3. Braun, E.D., Bushe, N.A., Buyanovsky, I.A. *Fundamentals of Tribology (Friction, Wear and Lubricants).* Science and Technique: Moscow, 1995, 778.
4. Garkunov, E.D. *Tribo-Engineering (Design, Manufacturing and Machine Service)*, 5th ed. MSCA Publishing: Moscow, 2002, p. 632.
5. Goracheva, I.G. *Mechanics of Frictional Interaction.* Nauka: Moscow, 2001, p. 478.
6. Mang, T.A. *New Generation of Non Hazardous and Environmentally Safe Metal Working Oils.* ITC: Melbourne, 1987, p. 336.
7. Chichinadze, A.V. *Fundamentals of Tribology (Friction, Wear and Lubricants).* Mashinostroenie: Moscow, 2001, p. 664.
8. Shkolnikov, V.M. *Fuels, Lubricants, Engineering Fluids. Selection and Applications.* 2nd ed. Chemistry: Moscow, 1999, p. 596.
9. Sholom, V.J., Crioni, N.K., Shuster, L.S., Migranov, M.S. Influence of active additives on anti-galling properties of lubricants. In *Proc. Int. Conference Reliability, Quality in Industry, Power Generation and Transportation.* Samara State University Publishing: Samara, 1999, p. 64.
10. Shuster, L.S. *Adhesive Interaction in Solids.* Gilem: Ufa, 1999, p. 199.
11. Wang, Z.Y., Rajurkar, K.P. Cryogenic machining of hard-to-cut materials. *Wear* 2000, 238, 169.

12. Hong, S.Y., Ding, Y., Ekkens, R.G. Improving low carbon steel chip breakability by cryogenic cooling. *Int. J. Mach. Tool. Manuf.* 1999, 39, 1065.
13. Savan, A., Pflüger, E., Goller, R., Gissler, W. Use of nanoscaled multilayer and compound films to realize a soft lubrication phase with a hard, wear-resistant matrix. *Surf. Coat. Technol.* 2000, 126, *2–3*, 159.
14. Sutor, P. Solid lubricants: overview and recent developments. *MRS Bull.* 1999, 24, 45–52.
15. Trent, E. M., Wright, P.K. *Metal cutting*. Butterworth-Heinemann: Boston, MA, 2000, p. 446.
16. Bailey, J.A. Friction in metal machining-mechanical aspect. *Wear* 1975, 31, 243.
17. De Chiffre, L. Function of cutting fluids in machining. *Lubr. Eng.* 1988, 44, 514.
18. Rebinder, P.A., Likhtman, V.I. Effect of surface-acting media on strains and ruptures in solids. *Proc. 2nd Int. Conf. Surf. Activity.* Butterworth: London, 1947, p. 563.
19. De Chiffre, L. What can we do about chips formation mechanics? *Ann. CIRP* 1985, 34, 129.
20. Silliman, J.D. *Cutting and Grinding Fluids: Selection and Applications*. 2nd ed. SME: Dearborn, MI, 2002, p. 216.
21. Lahres, M., Muller-Hummel, P., Doerfel, O. Applicability of different hard coatings in dry milling aluminum alloys. *Surf. Coat. Technol.* 1997, 91, 116–121.
22. Chen, N.N., Pun, W.K. Stresses at cutting tool wear land. *Int. Mach. Tool. Manuf.* 1988, 28, 79.
23. Sreejith, P.S., Ngoi, B.K.A. Dry machining: machining of the future. *J. Mater. Process.* 2000, 101, 287.
24. Napreev, I.S. Control bearings tribological characteristics by PFPE films formation, Ph.D. thesis, Belarussian State University, Minsk, 1999, 210–218.
25. Fox-Rabinovich, G.S., Kovalev, A.I., Weatherly G.C. Tribology and the design of surface engineered materials for cutting tool applications. In *Modeling and Simulation for Material Selection and Mechanical Design*. Eds., Totten G.E., Xie, L., Funatani, K. Marcel Dekker: New York, 2004, chap. 5.
26. Fox-Rabinovich, G.S., Kovalev, A.I., Wainstein, D.L., Shuster, L.Sh., Dosbaeva, G.K. Improvement of "duplex" PVD coatings for HSS cutting tools by PFPE (perfluorpolyether "Z-DOL"). *Surf. Coat. Technol.* 2002, 160, *1*, 99–107.
27. Nakayama, K., Nguyen, S. Triboelectromagnetic phenomena in a diamond/hydrogenated-carbon-film tribosystem under perfluoropolyether fluid lubrication *Appl. Surf. Sci.* 2000, 158, 229.
28. Gulanskiy, L.G. Kinematic model of boundary lubrication layer formation in the cutting zone. *J. Friction Wear* 1992, 4, 695.
29. Fox-Rabinovich, G.S., Kovalev, A.I., Shuster, L.S., Bokiy, Y.F., Dosbaeva, G.K., Wainstein, D.L. Characteristic features of alloying HSS-based deformed compound powder materials with consideration for tool self-organization at cutting: 1. Characteristic features of wear in HSS-based deformed compound powder materials at cutting. *Wear* 1997, 206, 214.
30. Balzer, S.A., Haan, D.M., Rao, P.D., Olson, W.W., Sutherland, J.W. Minimizing the quantity of cutting fluid required for specific machining processes through the manipulation of process input parameters. *J. Mater. Proc. Technol.* 1998, 79, 72.
31. Diniz, A.E., Micaroni, R. Cutting conditions for finish turning process aiming: the use of dry cutting. *Int. J. Mach. Tool. Manuf.* 2002, 45, 899.
32. Knolcke, F., Eisenblatter, G. Dry cutting. *Ann. CIRP* 1997, 46, 2, 519–526.
33. Braga, D.U., Diniz, A.E., Miranda, G.W.A., Coppini, N.L. Using a minimum quantity of lubricant (MQL) and a diamond coated tool in the drilling of aluminum–silicone alloys. *J. Mater. Proc. Technol.* 2002, 122, 127–138.
34. Andrews, C.J.E., Feng, H.-Y., Lau, W.M. Machining of an aluminum/SiC composite using diamond inserts. *J. Mater. Proc. Technol.* 2000, 102, *1–3*, 25–29.
35. Tonshoff, H.K., Micaroni, R. PVD-Coatings for wear protection in dry cutting operations. *Surf. Coat. Technol.* 1997, 96, 88.
36. Lahers, M., Jorgensen, G. Properties and dry cutting performance of diamond-coated tools. *Surf. Coat. Technol.* 1997, 96, 198–204.
37. Kustus, F.M., Fehrehnbacher, L.L., Komandiri, R. Nanocoatings on cutting tools for dry machining. *Ann. CIRP* 1997, 46, *1*, 39.
38. Renevier, N.M., Liobindo, N., Fox, V.C., Teer, D.G., Hampshire, J. Coating characteristics and tribological properties of sputter-deposited MoS_2/metal composite coatings deposited by closed field unbalanced magnetron sputter ion plating. *Surf. Coat. Technol.* 2000, 123, 84.

39. Harris, S.G., Vlasveld, A.C., Doyle, E.D., Dolder, P.J. Dry machining — commercial viability through filtered arc vapor deposited coatings. *Surf. Coat. Technol.* 2000, 133–134, 383–388.
40. Gresik, W., Zalisz, Z., Nieslony, P. Friction and wear testing of multilayer coatings on carbide substrates for dry machining applications. *Surf. Coat. Technol.* 2002, 155, 37.
41. Klocke, F., Kreig, T. Coated tools for metal cutting — features and applications. *Ann. CIRP* 1999, 48, 2, 515–525.
42. Haruyo Fukui, Junya Okida, Naoya Omori, Hideki Moriguchi, Keiichi Tsuda. Cutting performance of DLC coated tools in dry machining aluminum alloys. *Surf. Coat. Technol.* 2004, 187, *1*, 70–76.
43. Erdemir, A. A crystal-chemical approach to lubrication by solid oxides, *Tribol. Lett.* 2000, 8, 97–102.
44. Fox-Rabinovich, G.S., Weatherly, G.C., Dodonov, A.I., Kovalev, A.I., Veldhuis, S.C., Shuster, L.S., Dosbaeva, G.K., Wainstein, D.L., Migranov, M.S. Nano-crystalline filtered arc deposited (FAD) TiAlN PVD coatings for high-speed machining applications. *Surf. Coat. Technol.* 2004, 177–178, 800–811.

13 Geometrical Adaptation of Cutting Tools

Stephen C. Veldhuis, Michael M. Bruhis, Lev S. Shuster, and German S. Fox-Rabinovich

CONTENTS

13.1 Introduction 411
13.2 Assessment of the Metal Flow and Temperature Rise in the Chip and the Chip–Cutting Tool Interface during Orthogonal Machining 412
13.2.1 Upper-Bound Analysis 413
13.2.2 Upper-Bound Analysis for Orthogonal Cutting with a Chamfered Tool 420
13.2.3 Assessment of Temperature Rise in Orthogonal Cutting 427
13.2.3.1 Temperature Rise within the Chip 427
13.2.3.2 Assessment of Temperature Rise on the Tool Face 435
13.2.3.3 Assessment of the Stress Distribution in the Edge of the Cutting Tool 436
13.3 Design Optimization for the Adaptive Cutting Tools with Hard PVD Coatings 442
13.3.1 HSS Cutters 446
13.3.2 CC Cutters 448
13.4 Conclusion 450
References 451

13.1 INTRODUCTION

This chapter focuses on the adaptation of cutting-tool geometry during machining. Major attention is paid to the surface-engineered cutting tools with hard PVD Ti–Al–N-based coatings.

It is widely known that the hard-coating deposition results in improvement of cutting-tool life. However, surface damage can be enhanced by the brittleness of the hard coatings. It is known from the literature that the design of cutting-edge improvements could significantly affect the coated-cutting-tool life and wear behavior [1–2]. As was outlined in Chapter 1, three levels of workpiece–cutting tool interactions exist, i.e., macro-, micro-, and nanolevel ones. The topics considered in this book are mainly covering the micro- and nanoscale interactions, i.e., the so-called structural adaptation process related to the nanoscale tribo-film formation on the surface during friction (Chapter 1). But micro- and even macroscopic adaptation is also extremely important during cutting. Almost no publication could be found in the literature on this topic, although the role of cutting-edge geometry on machining processes has been studied in detail [2]. In this chapter we present some data aimed to fill the gap in the understanding of this problem.

The major contribution to the understanding of the geometrical adaptability of tribosystems has been made by Shultz [3]. Thermodynamics-based tribology considers the generic adaptation process taking place during the initial running-in stage of wear. The adaptation is a very complex process related to the geometrical, chemical, phase, and structure transformations on the surface and within subsurface layers during friction. As shown in Chapter 3, the self-organization process of tribosystems (especially heavily loaded tribosystems) takes place at the costs of intensive wear rate, which is associated with surface damage [4]. Scientifically speaking, according to the principle of dissipative heterogeneity [5–7] the geometrical adaptation (changing of the shape of the frictional body coincides with the structural adaptation (protective tribo-films formation) [7]. Both of these processes compete with each other in a very complex manner.

Adaptation during cutting includes processes related to changes in the cutting-edge geometry within the friction zone. This type of adaptation is conditionally named *geometrical adaptation*. This is a very important process from a practical point of view. It eventually results in the formation of quasi-stable cutting-edge shapes, as well as wear rate stabilization. According to the ideas considered in Reference 3, the geometry of frictional bodies (in our case, cutting tools) can be modified in a way to mimic the changes that take place during cutting. Once the running-in stage of wear is completed, the tribosystem transforms to the stable stage of wear with a lower wear rate. Thus, the shape of the cutting edge that corresponds to the point of transformation from nonequilibrium (running-in) to the quasi-equilibrium (stable) stage of wear is a so-called shape of natural wear [3]. If we mimic this shape, we can significantly stabilize the wear rate. This hypothesis was used to develop an adaptive geometry of the cutting tool with hard PVD coatings.

The development of the cutting-tools adaptive geometry has been performed in two stages: (1) analytical modeling to justify the design of the cutting tool and (2) experimental optimization of cutting-edge design.

13.2 ASSESSMENT OF THE METAL FLOW AND TEMPERATURE RISE IN THE CHIP AND THE CHIP–CUTTING TOOL INTERFACE DURING ORTHOGONAL MACHINING

The design of cutting-edge optimization has been performed based on the analytical modeling described in the following text. A detailed upper-bound analysis is performed for orthogonal cutting with a chamfered-edge preparation tools under plane strain conditions. The solutions under consideration for the sharp tool were first proposed by Lee and Shaffer [4], which had proposed a shear angle solution based on the slip-line field theory. A new upper-bound model for the machining with a chamfered-edge tool is developed in this chapter. The temperature rise at the chip–tool interface has also been determined. This chapter presents variations of strain and flow stresses through the steady state cutting process. The results obtained were compared with finite element analysis (FEA) [5] using experimental data from Reference 6.

Understanding and characterizing the basic chip formation process of the orthogonal cutting is imperative in the analysis of modern tools with effective rake, clearance, and setting angles. The simplest deformation modes for machining are based on a single shear plane and slip-line field techniques. Ernst and Merchant [7] first presented that the chip is formed in a shear narrow zone, along an inclined line, which extends from the tool edge to the workpiece free surface. The slip-line field (SLF) solution to analyze the plastic deformation of the orthogonal cutting process was developed by Lee and Shaffer [4]. However, plastic deformation is assumed to occur on the inclined line. Kudo [8] proposed a corrected new slip-line solutions for orthogonal cutting. Oxley [9] developed a mathematical model, which is most significant for practical machining. The slip-line solution obtained includes strain hardening for orthogonal cutting process with a sharp cutting edge. Maity and Das [10] presented a new class of SLF for machining with slipping and sticking contact at the tool–workpiece interface. However, at the higher rake angle the agreement between theory and experimental data is not good. William and Endres [11] proposed an upper-bound solution for orthogonal cutting with an edge rounded tool. The author made an assumption that flow of the workpiece material is assumed to occur without a built-up edge and that there is no plastic deformation at the tool–chip interface. Zhang et al. [12] modeled the orthogonal cutting with a chamfered tool. Investigations related to the chamfered tool have shown that the shear angle is lowered by about 3° in comparison with a single-rake angle and the dead metal zone is not dependent on the chamfered angle. Ren and Altintas [13] developed SLF for the mechanics of machining with a chamfered tool. Their model derived that the plastic deformation in the secondary shear zone is assumed to occur within a thin layer of the work material. Toporov and Sung-Lim Ko [14] developed a solution for orthogonal cutting with an unrestricted rake of the tool face. However, the correctness of the SLF gives rise to doubts because: (1) at a free surface the α and β lines do not meet at a 45° angle and (2) at the tool–workpiece interface the incline angle of the slip line cannot have a discontinuous jump [15–20]. For the majority of the orthogonal cutting investigations, the tool–workpiece interface is flat with a single rake angle. Fang [21,22] presented a new slip-line model with a double-rake angled grooved tool. The double-rake angle tool had 5° primary rake angle and 10° to 25° secondary rake angle. The author made the following remarks: "the cutting forces increase with a double-rake angle tool in comparison to a single-rake angle, and there is no significant increase in chip thickness."

However, there is not enough data in the publications outlined to predict the chip back-flow with grooved two-rake face tools or with chamfered-edge tool or worn tool. Therefore, the appropriate modeling of the material flow and fracture during machining has significant practical interest, and can be useful for process planning and for the cutting-tool design optimization. We will also present other correct kinematically admissible upper-bound solutions for orthogonal cutting with a worn tool. A limit load expression was derived on the basis of a simplified deformation field constructed by the upper-bound technique and showed good agreement with the experimental results. Microstructural studies showed that crack formation occurred in the area of deformation bands with the shear strain of maximum value. It has been shown that the simplicity of the upper-bound method provides a more practical alternative to the finite element method under orthogonal cutting conditions, and may be useful for specific applications.

13.2.1 Upper-Bound Analysis

Simulations presented in this chapter are based on the upper-bound technique (UBT). The limit-load analysis is based on assumed deformation mechanisms and a rigid plastic workpiece material model. UBT have been used to estimate the limit load for orthogonal machining. The technique of a kinematically admissible velocity field is an efficient method to solve plane problems for orthogonal machining. The theoretical model investigates the workpiece material flow in the ortogonal machining (Figure 13.1). Figure 13.2 shows the upper-bound velocity field for the orthogonal cutting

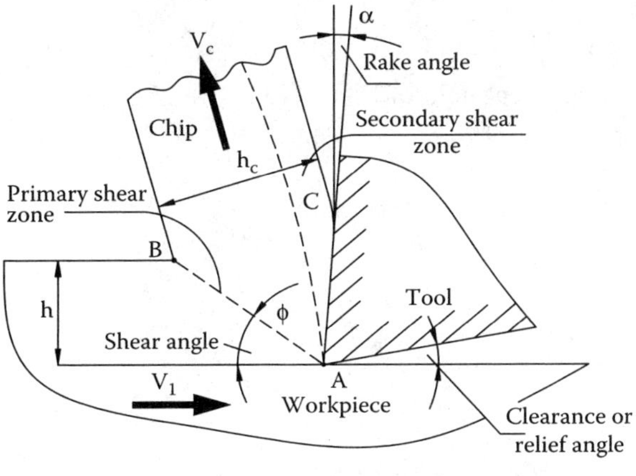

FIGURE 13.1 Schematic view of the orthogonal cutting process.

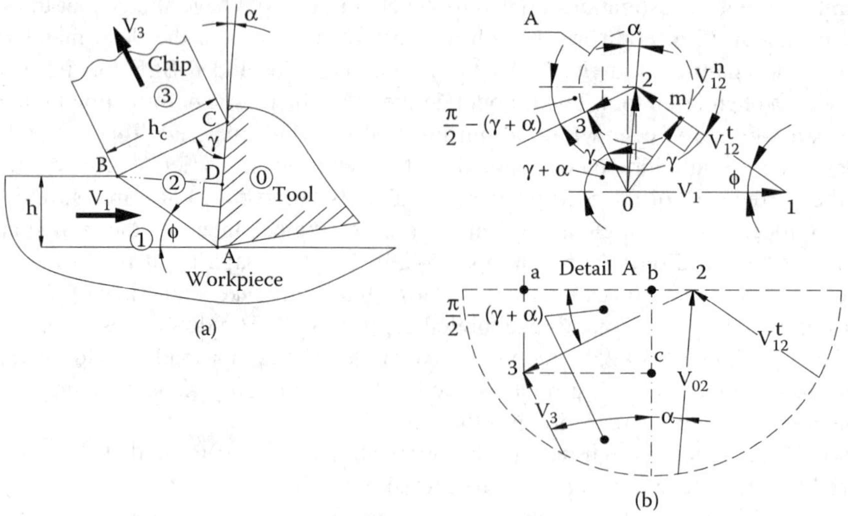

FIGURE 13.2 (a) Triangle upper-bound field for orthogonal cutting with a sharp tool and (b) corresponding hodograph and enlarged detail of region round point A.

process. The strain regions consist of the rigid zones separated by triangular elements. This condition is necessary for the given method. The relative velocity between adjacent zones is defined from the hodograph of velocities (Figure 13.2b). The principles of hodograph construction have been described in details in [16–20, 23]. The purpose of the hodograph is to verify if the chosen velocity pattern is kinematically admissible, as well as to define graphically the relative velocities between the adjacent triangles. The goal of the present work is to obtain analytically the force condition and parameters for plastic deformation during the machining process.

We have assumed that the plastic deformation occurs in the triangle zone ABC, and we have accepted the following boundary conditions for the considered task:

1. Zone 1 (workpiece material) is moved with a unity speed V_1 to the tool.
2. The tool is sharp, and there is no contact along relief angle (Figure 13.1).
3. The cut depth is equal to h and is constant.
4. The free workpiece surfaces are straight lines.
5. The triangle ABC is assumed to be an isosceles triangle.
6. The chip thickness is equal to h_c.

From geometry of the physical field of Figure 13.2a we obtained

$$AB = BC = \frac{h}{\sin\phi} \tag{13.1}$$

$$\gamma = \frac{\pi}{2} - (\phi - \alpha) \tag{13.2}$$

$$AC = l_{02} = 2AB\cos\gamma \overset{(1),(2)}{=} \frac{2h}{\sin\phi}\sin(\phi - \alpha) \tag{13.3}$$

$$\sin\gamma = \frac{BD}{AB} \rightarrow BD = AB\sin\gamma \overset{(1),(2)}{=} \frac{h}{\sin\phi}\cos(\phi-\alpha) \tag{13.4}$$

The hodograph shown in Figure 13.2b was designed in consideration of constancy of volume (i.e., the material flow obeys the overall mass conservation) or the velocity component normal to the line of discontinuity must be continuous to the material flow rate through the boundary between adjacent triangles. Thus, from geometry of the hodograph (Figure 13.2b), we obtain

$$V_1 h = V_{02} BD \overset{(4)}{=} V_{02}\frac{h}{\sin\phi}\cos(\phi-\alpha) \tag{13.5}$$

or

$$V_1 h = V_3 BC \overset{(1)}{=} V_3 \frac{h}{\sin\phi} \tag{13.6}$$

where V_1 is the cutting speed and it is equal to unity, V_{02} is the displacement speed of the workpiece material along tool surface, V_3 is the chip velocity vector, h is the cut depth, h_c is the chip thickness, ϕ is the shear angle, and α is the effective rake angle.

Equation 13.5 leads to

$$(5) \rightarrow V_{02} = \frac{V_1 h \sin\phi}{h\cos(\phi-\alpha)} = \frac{\sin\phi}{\cos(\phi-\alpha)} \tag{13.7}$$

Equation 13.6 leads to

$$(6) \rightarrow V_3 = V_{23}^n = \frac{V_1 h\sin\phi}{h} = \sin\phi \tag{13.8}$$

Equating Equation 13.5 and Equation 13.6 leads to

$$(5),(6) \rightarrow \frac{V_{02} h\cos(\phi-\alpha)}{\sin\phi} = \frac{V_3 h}{\sin\phi} \rightarrow V_3 = V_{02}\cos(\phi-\alpha) \tag{13.9}$$

From the geometry of the hodograph (Figure 13.2b), the following expression is obtained

$$\frac{V_{12}^n}{V_{02}} = \sin\gamma \rightarrow V_{12}^n \overset{(9)}{=} \frac{\sin\phi}{\cos(\phi-\alpha)}\sin\gamma \overset{(2)}{=} \frac{\sin\phi\sin\left[\frac{\pi}{2}-(\phi-\alpha)\right]}{\cos(\phi-\alpha)} = \sin\phi \tag{13.10}$$

$$\frac{2\bar{m}}{V_{02}} = \cos\gamma \rightarrow 2\bar{m} = V_{02}\cos\gamma \tag{13.11}$$

$$\frac{\overline{1m}}{V_1} = \cos\phi \rightarrow \overline{1m} = V_1 \cos\phi \tag{13.12}$$

$$V_{12}^t = \overline{2m} + \overline{1m} \overset{(11),(12)}{=} V_{02} \cos\gamma + V_1 \cos\phi \tag{13.13}$$

Inserting Equation 13.2 and Equation 13.7 into Equation 13.13. the tangential velocity V_{12}^t can be found:

$$V_{12}^t = \frac{\sin\phi}{\cos(\phi-\alpha)}\sin(\phi-\alpha) + \cos\phi = \sin\phi\tan(\phi-\alpha) + \cos\phi \tag{13.14}$$

It can be seen from the hodograph Figure 13.2b that

$$\frac{\overline{2b}}{V_{02}} = \sin\alpha \rightarrow \overline{2b} = V_{02}\sin\alpha \tag{13.15}$$

$$\frac{\overline{3c}}{V_3} = \cos\left[\frac{\pi}{2} - (\gamma+\alpha)\right] \rightarrow \overline{3c} = V_3 \sin(\gamma+\alpha) \tag{13.16}$$

$$\cos\left[\frac{\pi}{2} - (\gamma+\alpha)\right] = \frac{\overline{2b} + \overline{3c}}{V_{23}^t} \rightarrow V_{23}^t \overset{(17),(18)}{=} \frac{V_{02}\sin\alpha + V_3 \sin(\gamma+\alpha)}{\sin(\gamma+\alpha)} \tag{13.17}$$

Inserting Equation 13.7 and Equation 13.8 into Equation 13.17, the tangential velocity V_{23}^t can be found:

$$V_{23}^t \overset{(9),(10)}{=} \sin\phi\left[\frac{\sin\alpha}{\sin(\gamma+\alpha)} + \sin\gamma\right] \tag{13.18}$$

where V_{12}^t, and V_{23}^t are the velocity discontinuities along the shear lines ($l_{12} = AB$, and $l_{23} = BC$).

From the definition of the UBT, the energy balance [17, 20] can be written as

$$W_i = W_e \tag{13.19}$$

where W_i is the rate of energy dissipation (internal work) and W_e is the rate of the external work.

The work/rate of energy dissipation can be expressed (internal work W_i) as

$$W_i = \sum kV_{ij} l_{ij} = k\left(V_{12}^t l_{12} + +V_{23}^t l_{23}\right) \tag{13.20}$$

where l_{ij} is the length of the shear lines (Figure 13.2a), that is, $l_{12} \overset{(1)}{=} AB$ and $l_{23} \overset{(1)}{=} BC$ and k is the value of the applied shear stress, and according to the Von Mises yield condition k is equal to

$$k = \tau_y = \frac{\sigma_y}{\sqrt{3}} \tag{13.21}$$

where τ_y is the yield stress in pure shear and σ_y is the yield stress in tension.
The rate of external work [20, 23] is equal to

$$W_e = pV_1 l_{02} \cos\alpha - W_f = pV_1 l_{02} \cos\alpha - \tau_c V_c l_c = pV_1 l_{02} \cos\alpha - 2\mu k V_{02} l_{02} \tag{13.22}$$

where p is the normal cutting pressure on the tool face, $l_c = l_{02} \overset{(3)}{=} AC$ is the tool–chip contact length, $W_f = \tau_c V_c l_c = 2\mu k V_{02} l_{02}$ is the frictional work/rate of energy dissipation to overcome friction along the tool–chip interface, V_c is the sliding velocity of the workpiece material along the contact surface, and τ_c is the tangential stress on the tool–workpiece interface, and according to Reference 23 and Reference 24 this stress can be written as $\tau_c = 2\mu k$, and μ is the Coulomb's frictional coefficient $0 \le \mu \le 0.5$. Quoting Equation 13.22 and Equation 13.24, we obtain the critical average pressure as

$$pV_1 l_{02} \cos\alpha = k(V_{12}^t l_{12} + V_{23}^t l_{23}) + 2\mu k V_{02} l_{02} \tag{13.23}$$

Substituting this in Equation 13.23, the correlation (ratio) between the velocities and the value of the shear lines we obtain the critical load that is required for the orthogonal cutting:

$$p\overset{(3)}{\frac{2h}{\sin\phi}}\sin(\phi-\alpha)\cos\alpha = k\left\{\begin{array}{l}\left[\sin\phi\overset{(15)}{\tan(\phi-\alpha)}+\cos\phi\right]\overset{(1)}{\frac{h}{\sin\phi}}+\left[\overset{(19)}{\frac{\sin\alpha}{\sin(\gamma+\alpha)}}+\sin\gamma\right]\sin\phi\overset{(1)}{\frac{h}{\sin\phi}}+\\ 2\mu\overset{(9)}{\frac{\sin\phi}{\cos(\phi-\alpha)}}\overset{(3)}{\frac{2h}{\sin\phi}}\sin(\phi-\alpha)\end{array}\right\}$$

or

$$p\frac{2h}{\sin\phi}\sin(\phi-\alpha)\cos\alpha = kh\left[\tan(\phi-\alpha)+\cot\phi+\frac{\sin\alpha}{\sin(\gamma+\alpha)}+\sin\gamma+4\mu\tan(\phi-\alpha)\right]$$

The general solution for the limiting cutting pressure is

$$\frac{p}{2k} = \frac{\sin\phi}{4\cos\alpha\sin(\phi-\alpha)}\left[\tan(\phi-\alpha)+\cot\phi+\frac{\sin\alpha}{\sin(\gamma+\alpha)}+\sin\gamma+4\mu\tan(\phi-\alpha)\right] \tag{13.24}$$

It is easy to see that when the rake angle $\alpha = 0$ the limiting cutting pressure (Equation 13.24) reduces to

$$\frac{p}{2k} = \frac{1}{4}[\tan\phi(1+4\mu)+\cot\phi+\sin\gamma]$$

or

$$\frac{p}{2k} \overset{(2)}{=} \frac{1}{4}[\tan\phi(1+4\mu)+\cot\phi+\cos\phi] \tag{13.25}$$

According to the formal statement of the upper-bound theorem, the total power dissipation requires minimization with respect to all the unknown variables. It is obvious that the magnitude of $\frac{p}{2k}$ for this field will depend upon values of ϕ, μ, and α. In order to find the lowest upper-bound solution based on this velocity field, we have to minimize $\frac{p}{2k}$ with respect to ϕ. It is seen in Figure 13.3 that the lowest value of $\frac{p}{2k}$ occurs when $\phi = 50°$ ($\mu = 0$), $\phi = 44°$ ($\mu = 0.1$), and $\phi = 32°$ ($\mu = 0.5$). Substituting optimal values of ϕ and μ, Equation 13.25 can be rewritten in the form

$$\mu = 0 \rightarrow \frac{p}{2k} = 0.668 \rightarrow p \overset{(23)}{=} 0.771 * \sigma_y \tag{13.26}$$

$$\mu = 0.1 \rightarrow \frac{p}{2k} = 0.777 \rightarrow p \overset{(23)}{=} 0.897 * \sigma_y \tag{13.27}$$

The chip thickness (13.1) can be calculated knowing the angle ϕ (Figure 13.2a):

$$h_c = \frac{h}{\sin\phi} \overset{(\mu=0)}{=} \frac{0.5}{\sin 50} = 0.652mm \tag{13.28}$$

$$h_c = \frac{h}{\sin\phi} \overset{(\mu=0.1)}{=} \frac{0.5}{\sin 44} = 0.72mm \tag{13.29}$$

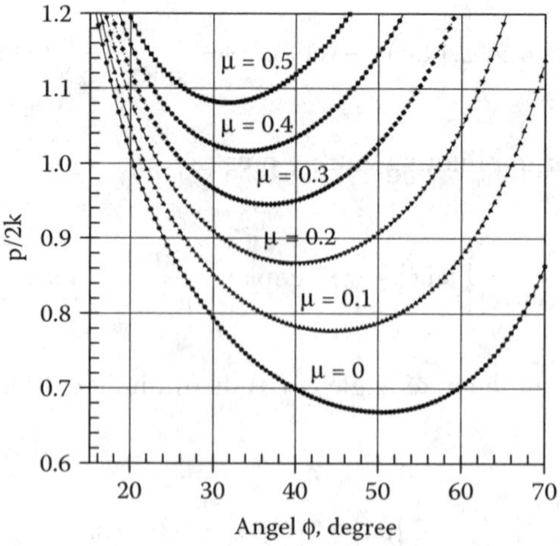

FIGURE 13.3 The limiting cutting pressure vs. angle ϕ.

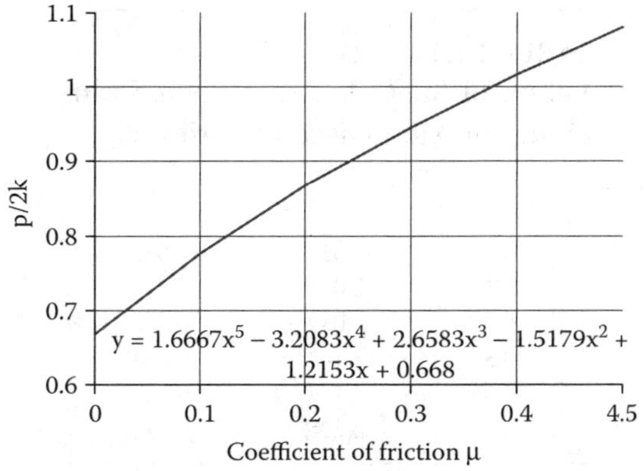

FIGURE 13.4 Limiting cutting pressure vs. coefficient of friction.

The tool–chip contact length ratio can be found as

$$\frac{h_c}{l_{AC}} \overset{(3),(30),(31)}{=} 0.652 \div 0.72 \tag{13.30}$$

The obtained expressions (Equation 13.26 and Equation 13.27) have a very simple and convenient form for evaluation of limit load in practice. The load per unit length of the contact surface is

$$P = pl_{AC} \cos\alpha \tag{13.31}$$

where $l_{AC} = AC \overset{(3)}{=} \frac{2h}{\sin\phi}\sin(\phi-\alpha)$ is the contact interface length.

It can be seen in Figure 13.4 that the limiting cutting pressure increases with increasing friction on the tool–workpiece interface. Using the velocity characteristics [17, 20], the shear strain of an element when it crosses a line of the velocity discontinuity can be found. This shear strain can be written as

$$\gamma_{ij} = \frac{V_{ij}^t}{V_{ij}^n} \tag{13.32}$$

where γ_{ij} is the shear strain on the boundary, V_{ij}^t is the velocity component tangential to the boundary, and V_{ij}^n is the velocity component normal to the boundary. From the hodograph of Figure 13.2b we can see that the maximum value of the shear strain occurs on the line $l_{12}(AB)$. From Equation 13.10, Equation 13.13, and Equation 13.32,

$$\gamma_{12} = \tan(\phi-\alpha) + \cot\phi$$

and from Equation 13.8, Equation 13.18, and Equation 13.32,

$$\gamma_{23} = \frac{\sin\alpha}{\sin(\gamma-\alpha)} + \sin\gamma \tag{13.33}$$

TABLE 13.1
Values of the Calculated Shear Strain along the Line Discontinuities

μ	γ_{12}	γ_{23}
0	2.03	0.643
0.1	2.0	0.719
0.2	2.032	0.766
0.3	2.102	0.809
0.4	2.157	0.829
0.5	2.225	0.848

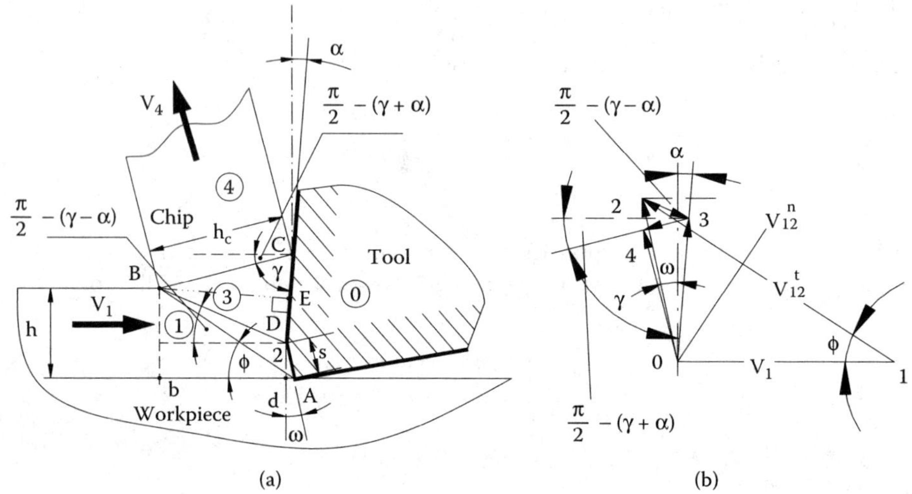

FIGURE 13.5 (a) Triangle upper-bound field for machining with a chamfered-edge preparation tool and (b) corresponding hodograph.

For the rake angle $\alpha = 0$ Equation 13.33 gives the simple relation:

$$\mu = 0\ (\phi = 50°) \rightarrow \gamma_{12} = 2.03 \text{ and } \gamma_{23} = 0.643 \tag{13.34}$$

$$\mu = 0.1\ (\phi = 44°) \rightarrow \gamma_{12} = 2.0 \text{ and } \gamma_{23} = 0.719 \tag{13.35}$$

A summary of the calculated effective shear strain along the line discontinuities is given in Table 13.1.

13.2.2 Upper-Bound Analysis for Orthogonal Cutting with a Chamfered Tool

The analytical solution for the orthogonal machining process with a chamfered tool is shown in Figure 13.5. We have assumed for this type of deformation that the plastic deformation occurs in the two triangle zones *ABD* and *DBC*. The boundary conditions of the considered task are as follows:

1. Zone 1 (workpiece material) is moved with a unity speed V_1 to the tool.
2. The tool is grooved with two rake faces (Figure 13.5).
3. The cut depth is equal to h and is constant.
4. The free workpiece surface is straight line.
5. The triangle DBC is assumed to be an isosceles triangle.
6. The chip thickness is equal to h_c.

From the analysis of the upper-bound field Figure 13.5a, it can be shown that

$$AB = l_{12} = \frac{h}{\sin\phi} \tag{13.36}$$

$$\frac{h}{Ab} = \tan\phi \rightarrow Ab = \frac{h}{\tan\phi} \tag{13.37}$$

$$\frac{\overline{dA}}{DA} = \sin\omega \rightarrow \overline{dA} = DA\sin\omega = s*\sin\omega \tag{13.38}$$

$$\frac{\overline{dD}}{DA} = \cos\omega \rightarrow \overline{dD} = DA\cos\omega = s*\cos\omega \tag{13.39}$$

where s is the length of chamfer edge and ω is the chamfer angle or second rake angle.

$$\frac{h-\overline{dD}}{BD} = \sin\left[\frac{\pi}{2}-(\gamma-\alpha)\right] \rightarrow BD \overset{(36)}{=} \frac{h-s*\cos\omega}{\cos(\gamma-\alpha)} = A \tag{13.40}$$

where $A = \dfrac{h-s*\cos\omega}{\cos(\gamma-\alpha)}$, h is the cut depth, γ is the angle of the isosceles triangle DBC, and α is the effective rake angle.

The length of a straight boundary $BD = BC = l_{23} = l_{34}$

$$\tan(\gamma-\alpha) = \frac{\dfrac{h}{\tan\phi} - s*\sin\omega}{h - s\cos\omega} \rightarrow (\gamma-\alpha) = \arctan\left[\frac{\dfrac{h}{\tan\phi} - s*\sin\omega}{h - s\cos\omega}\right] \tag{13.41}$$

$$\cos\gamma = \frac{DE}{BD} \rightarrow DE = BD\cos\gamma \overset{(40)}{=} A\cos\gamma \tag{13.42}$$

$$DC = l_{03} = 2DE \overset{(41)}{=} 2A\cos\gamma \tag{13.43}$$

$$\sin\gamma = \frac{BE}{BD} \rightarrow BE \overset{(40)}{=} A\sin\gamma \tag{13.44}$$

The magnitude of the chip velocity V_4 and the magnitude of the displacement velocity V_{03} of the workpiece material along tool surface can be expressed in terms of the cutting speed V_1:

$$V_1 h = V_{03} BE \overset{(44)}{=} V_{03} A \sin\gamma \tag{13.45}$$

or

$$V_1 h = V_4 BC \overset{(40)}{=} V_4 A \tag{13.46}$$

Equating Equation 13.45 and Equation 13.46, V_4 and V_{03} can be expressed as

$$V_{03} A \sin\gamma = V_4 A \tag{13.47}$$

$$V_{03} \sin\gamma = V_4 \tag{13.48}$$

From Equation 13.47,

$$V_{03} = \frac{V_1 h}{A\sin\gamma} \tag{13.49}$$

From Equation 13.46,

$$V_4 = \frac{V_1 h}{A} \tag{13.50}$$

From the geometry of the hodograph (Figure 13.5b), the following expressions are obtained:

$$V_{02} \cos\omega = V_{12}^t \sin\phi \tag{13.51}$$

$$V_{02} = V_{12}^t \frac{\sin\phi}{\cos\omega} \tag{13.52}$$

$$V_1 = V_{12}^t \cos\phi - V_{02} \sin\omega \tag{13.53}$$

and after rearrangement of Equation 13.51 and Equation 13.53, the following expression is obtained:

$$V_{12}^t = \frac{V_1}{\cos\phi - \sin\phi\tan\omega} = \frac{1}{\cos\phi - \sin\phi\tan\omega} \tag{13.54}$$

The tangential velocities V_{23}^t and V_{34}^t of the shear discontinuities are

$$\sin\left[\frac{\pi}{2}-(\gamma-\alpha)\right]=\frac{V_{12}^t\sin\phi-V_{03}\cos\omega}{V_{23}^t} \tag{13.55}$$

or

$$V_{23}^t=\frac{V_{12}^t\sin\phi-V_{03}\cos\omega}{\cos(\gamma-\alpha)} \tag{13.56}$$

$$\sin\left[\frac{\pi}{2}-(\gamma+\alpha)\right]=\frac{V_{03}\cos\alpha-V_4\cos(\gamma+\alpha)}{V_{34}^t} \tag{13.57}$$

$$V_{34}^t=\frac{V_{03}\cos\alpha-V_4\cos(\gamma+\alpha)}{\cos(\gamma+\alpha)} \tag{13.58}$$

The energy dissipation rate of the plastic work W_i is

$$W_i=\sum kV_{ij}l_{ij}=k(V_{12}^tl_{12}++V_{23}^tl_{23}+V_{34}^tl_{34}) \tag{13.59}$$

where l_{ij} is the length of the shear lines of the physical plane in Figure 13.5a ($l_{12} \overset{(36)}{=} AB$, and $l_{23}=l_{34} \overset{(40)}{=} BD=BC$).

Substituting the component of the tangential velocities and the length of the shear lines into Equation 13.59, one obtains

$$W_i=k\left[V_{12}^t\frac{\overset{(36)}{h}}{\sin\phi}+\overset{(56)}{\frac{V_{12}^t\sin\phi-V_{03}\cos\omega}{\cos(\gamma-\alpha)}}\overset{(40)}{A}+\overset{(58)}{\frac{V_{03}\cos\alpha-V_4\cos(\gamma+\alpha)}{\cos(\gamma+\alpha)}}\overset{(40)}{A}\right]$$

After mathematical transformations, this equation can be reduced to a short form

$$W_i=k\left\{\begin{array}{l}\dfrac{1}{\cos\phi-\sin\phi\tan\omega}\left[\dfrac{h}{\sin\phi}+\dfrac{A\sin\phi}{h\cos(\gamma-\alpha)}\right]+\\[2ex]\dfrac{h}{\sin\gamma}\left[\dfrac{\cos\alpha}{\cos(\gamma+\alpha)}-\dfrac{\cos\omega}{\cos(\gamma-\alpha)}\right]-h\end{array}\right\} \tag{13.60}$$

The rate W_e of the external work can be found as

$$W_e=pV_1l_{02}\cos\omega-W_{f02}+pV_1l_{03}\cos\alpha-W_{f03} \tag{13.61}$$

where $W = pV_1 l_{02}\cos\omega + pV_1 l_{03}\cos\alpha$ is the cutting energy of the workpiece, p is the normal cutting pressure on the tool face, $W_{f\,02}$ is the frictional energy rate on the chamfer edge of the tool surface, and $W_{f\,03}$ is the frictional work/rate of energy dissipation to overcome friction along the tool contact lengths, which are equal to l_{03}, $l_{02} = AD = s$ and $l_{03} \overset{(43)}{=} DC$.

The frictional work rates, $W_{f\,02}$ and W_{f03} are equal to

$$W_{f\,02} = \tau_c V_{02} l_{02} = 2\mu_{02} k V_{02} l_{02} \tag{13.62}$$

$$W_{f\,03} = \tau_{c_1} V_{03} l_{03} = 2\mu_{03} k V_{03} l_{03} \tag{13.63}$$

Substituting Equation 13.62 and Equation 13.63 into Equation 13.61, simplification of the external work is as follows:

$$W_e = pV_1(l_{02}\cos\omega + l_{03}\cos\alpha) - 2\mu_{02}V_{02}l_{02}k - 2\mu_{03}V_{03}l_{03}k \tag{13.64}$$

where $l_{02} = AD = s$ and $l_{03} \overset{(45)}{=} DC$ are the tool contact lengths, V_{02} and V_{03} are the sliding velocities of the workpiece material along the contact surfaces, τ_c is the tangential stress on the interface AD, and τ_{c_1} is the tangential stress on the tool–workpiece interface DC. These stresses can be written as $\tau_c = 2\mu_{02}k$ and $\tau_{c_1} = 2\mu_{03}k$; $0 \le \mu_{02} \le 0.5$ and $0 \le \mu_{03} \le 0.5$ are the Coulomb's frictional coefficients between the workpiece material and tool contact surfaces.

A power law (Hollomon) type of constitutive equation was used to describe plastic behavior of the workpiece material:

$$\sigma_y = K(\varepsilon)^n \tag{13.65}$$

where K is the strength coefficient, ε is the true plastic strain, and n is the strain-strengthening exponent.

Thus, the rate of the external work in Equation 13.64 can be written as

$$W_e = p(s*\cos\omega + \overset{(42)}{2A\cos\gamma}\cos\alpha) - 2\mu_{02}V'_{12}\overset{(51)}{\frac{\sin\phi}{\cos\omega}}ks - 2\mu_{03}k\overset{(48)}{\frac{h}{A\sin\gamma}}\overset{(42)}{2A\cos\gamma} \tag{13.66}$$

$$W_e = p(s*\cos\omega + 2A\cos\gamma\cos\alpha) - 2\mu_{02}k\overset{(53)}{\frac{1}{\cos\phi - \sin\phi\tan\omega}}\frac{\sin\phi}{\cos\omega}s - \mu_{03}kh\cot\gamma \tag{13.67}$$

Equating 13.60 and Equation 13.67, we obtain the critical average pressure for the chamfered tool

$$p(s*\cos\omega + 2A\cos\gamma\cos\alpha) = k\left\{\begin{array}{l}\dfrac{1}{\cos\phi - \sin\phi\tan\omega}\left[\dfrac{h}{\sin\phi} + \dfrac{A\sin\phi}{\cos(\gamma-\alpha)} + \dfrac{2\mu_{02}s*\sin\phi}{\cos\omega}\right] + \\ \dfrac{h}{\sin\gamma}\left[\dfrac{\cos\alpha}{\cos(\gamma+\alpha)} - \dfrac{\cos\omega}{\cos(\gamma-\alpha)}\right] + h(\mu_{03}\cot\gamma - 1)\end{array}\right\} \tag{13.68}$$

The limiting cutting pressure for the chamfered tool is

$$\frac{p}{2k}=\frac{1}{2\left(s^*\cos\omega+2A\cos\gamma\cos\alpha\right)}$$
$$\left\{\begin{array}{l}\dfrac{1}{\cos\phi-\sin\phi\tan\omega}\left[\dfrac{h}{\sin\phi}+\dfrac{A\sin\phi}{\cos(\gamma-\alpha)}+\dfrac{2\mu_{02}s^*\sin\phi}{\cos\omega}\right]+\\ \dfrac{h}{\sin\gamma}\left[\dfrac{\cos\alpha}{\cos(\gamma+\alpha)}-\dfrac{\cos\omega}{\cos(\gamma-\alpha)}\right]+h\left(\mu_{03}\cot\gamma-1\right)\end{array}\right\} \tag{13.69}$$

It is easy to see when the rake angle $\alpha = 0$ then the limiting cutting pressure for the chamfered tool becomes

$$\frac{p}{2k}=\frac{1}{2(s^*\cos\omega+2A\cos\gamma)}$$
$$\left\{\begin{array}{l}\dfrac{1}{\cos\phi-\sin\phi\tan\omega}\left[\dfrac{h}{\sin\phi}+\dfrac{A\sin\phi}{\cos\gamma}+\dfrac{2\mu_{02}s^*\sin\phi}{\cos\omega}\right]+\\ \dfrac{h}{\sin\gamma}\left[\dfrac{1-\cos\omega}{\cos\gamma}\right]+h(\mu_{03}\cot\gamma-1)\end{array}\right\} \tag{13.70}$$

Substituting component A (Equation 13.40) into Equation 13.70, one obtains

$$\frac{p}{2k}=\frac{1}{2(2h-s^*\cos\omega)}$$
$$\left\{\begin{array}{l}\dfrac{1}{\cos\phi-\sin\phi\tan\omega}\left[\dfrac{h}{\sin\phi}+\dfrac{(h-s^*\cos\omega)\sin\phi}{\cos^2\gamma}+\dfrac{2\mu_{02}s^*\sin\phi}{\cos\omega}\right]+\\ \dfrac{h}{\sin\gamma}\left[\dfrac{1-\cos\omega}{\cos\gamma}\right]+h(\mu_{03}\cot\gamma-1)\end{array}\right\} \tag{13.71}$$

According to the formal statement of the upper-bound theorem, the optimal cutting pressure has to be minimized with respect to all the unknown variables ϕ, γ, ω, μ_{02}, and μ_{03}. When the rake angle $\alpha = 0$, the corresponding value of γ (Figure 13.5a) can be easily found from Equation 13.41:

$$\gamma=\arctan\left[\frac{\dfrac{h}{\tan\phi}-s^*\sin\omega}{h-s^*\cos\omega}\right] \tag{13.72}$$

Figure 13.6 shows the relationship of the calculated values of $\frac{p}{2k}$ vs. the angle ϕ (the cutting conditions were $h = 0.5$ mm, $s = 0.1 * h$, and $\omega = 5°$). Thus, the lowest value of the limiting cutting pressure for the preceding conditions occurs when $\phi = 43°$ and $\gamma = 49.75°$.

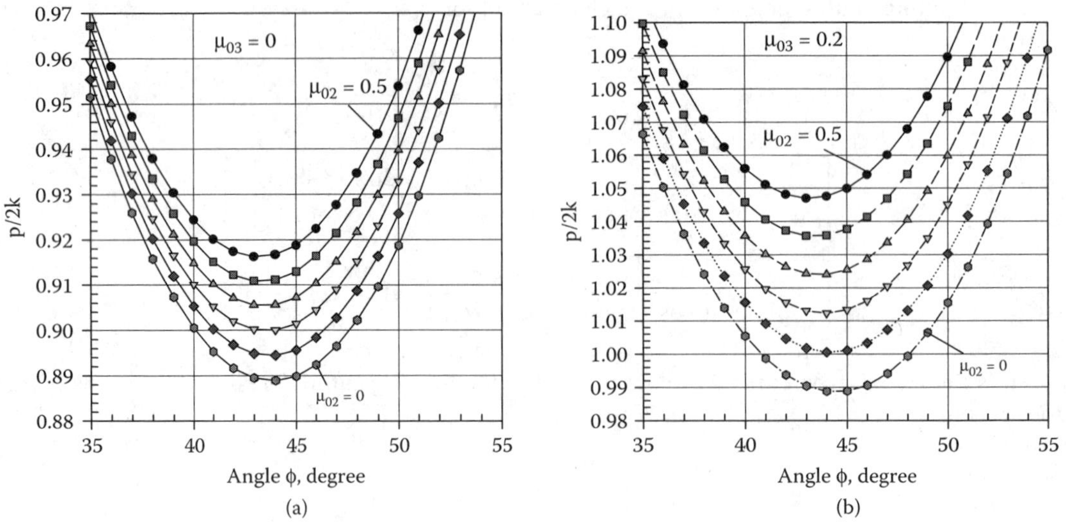

FIGURE 13.6 The limiting cutting pressure vs angle ϕ for (a) $\mu_{03} = 0$ and (b) $\mu_{03} = 0.2$.

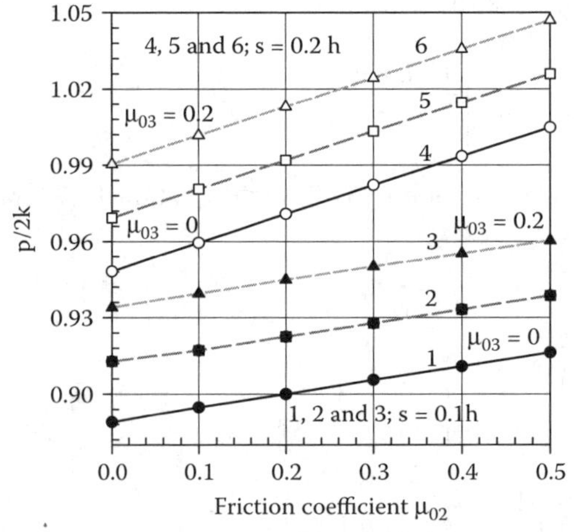

FIGURE 13.7 The limiting cutting pressure vs. friction coefficient μ.

The effect of friction on the limiting machining pressure is shown in Figure 13.7. We can see that the cutting pressure rises as the coefficient of friction increases. Let us assume that $\mu_{03} = 0.1$, $\mu_{02} = 0.4$ (in the Zone 2, Figure 13.5a, there is intensive seizure), and $s = 0.1h$; then the limiting cutting pressure is

$$\frac{p}{2k}^{(73)} = 1.0146 \tag{13.73}$$

When the chamfer angle is $\omega = 10°$, the limiting cutting pressure is only 12% larger than that for $\omega = 5°$. Figure 13.8 shows the relationship of the calculated values of $\frac{p}{2k}$ vs. the angle ϕ (the cutting conditions were $h = 0.5$ mm, $s = 0.1 * h$, and $\omega = 10°$). We can see that the lowest value of the limiting cutting pressure for the preceding conditions occurs when $\phi = 41°$ and $\gamma = 51.49°$.

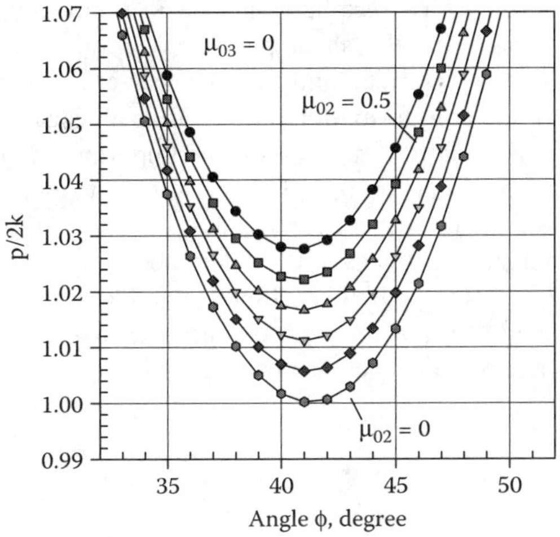

FIGURE 13.8 The limiting cutting pressure vs angle ϕ ($\omega = 10°$).

For the cutting conditions $h = 0.5$ mm, $s = 0.1 * h$, $\phi = 43°$, $\gamma = 49.75°$, and $\omega = 5°$, the chip thickness can be calculated using Equation 13.40:

$$h_c \overset{(40)}{=} \frac{h - s * \cos\omega}{\cos(\gamma - \alpha)} = 0.697mm \tag{13.74}$$

and for the cutting conditions $h = 0.5mm$, $s = 0.1 * h$, $\phi = 41°$, $\gamma = 51.49°$, and $\omega = 10°$,

$$h_c \overset{(40)}{=} \frac{h - s * \cos\omega}{\cos(\gamma - \alpha)} = 0.722mm \tag{13.75}$$

The tool–chip contact-length ratio for the cutting conditions $\gamma = 51.49°$ and $\omega = 10°$ can be found as

$$\frac{h_c}{l_{03}} \overset{(43)}{=} 0.802 \tag{13.76}$$

It can be seen in the results obtained from Equation 13.30 and Equation 13.76 that the tool-chip contact ratio for the chamfered-edge tool is 12% higher. The results obtained for the cutting pressure in Equation 13.27 and Equation 13.73 show that cutting with a chamfered tool needs a higher pressure. At the same time, a double-rake angle tool in comparison with a single-rake angle tool shows best wears resistance.

13.2.3 Assessment of Temperature Rise in Orthogonal Cutting

13.2.3.1 Temperature Rise within the Chip

Assessment and prediction of temperature rise has been a major focus of research on orthogonal cutting during the last few years. Plastic flow in machining may be characterized with a large value of shear strain and increase in temperature. Pekelharing [25] analyzed the failure mechanism of the carbide tools using FE analysis. He shows that high compressive stresses at the tool's tip and within the plastic zone of the chip can explain the rise in temperature in the zone of the chip formation. Shaw [26] has shown that the shear plane temperature has a major influence on the temperature rise on the tool–chip interface and on the wear rate. According to the conservation law, all-mechanical energy of the chip formation during adiabatic steady cutting processes converts into heat [20,27]. Komaduri and Hou [28] reviewed some important theoretical solutions to analyze the temperature rise in orthogonal machining. They showed that researchers mainly used a single shear plane model of the deformation zone for the assessment of the temperature increase first presented by Ernst and Merchant [7]. However, the calculated temperature rise under these assumptions gives a significant error (~ 50% or higher). Johnson and Kudo [20,27] extended the kinematically admissible velocity fields to find the temperature rise in the plastic flow zone. Using the velocity characteristics the temperature increase can be found as

$$\Delta T_{ij} = \frac{k\gamma_{ij}}{Jc\rho} \overset{(23)}{=} \frac{\sigma_y \gamma_{ij}}{\sqrt{3}Jc\rho} \tag{13.77}$$

where ΔT_{ij} is the rise in temperature, $k \overset{(21,65)}{=} \frac{K(\varepsilon)^n}{\sqrt{3}}$ is the value of the applied shear stress, γ_{ij} is the shear strain on the boundary, J is the Jouleís mechanical equivalent of heat, c is the specific heat of the workpiece material, and ρ is the density of the workpiece material.

In this equation, it is desired that the temperature rise depend on the flow stress. From the geometry of the hodograph (Figure 13.5b), we obtain the normal components of the velocities at the lines of the shear discontinuities.

$$\sin\phi = \frac{V_{12}^n}{V_1} \rightarrow V_{12}^n = V_1 \sin\phi \tag{13.78}$$

$$\sin\gamma = \frac{V_{23}^n}{V_{03}} \rightarrow V_{23}^n = V_{03}\sin\gamma \overset{(48),(40)}{=} \frac{h\cos\gamma}{h - s\cos\omega} \tag{13.79}$$

$$\sin\gamma = \frac{V_{34}^n}{V_{03}} \rightarrow V_{34}^n = V_{03}\sin\gamma \overset{(48),(40)}{=} \frac{h\cos\gamma}{h - s\cos\omega} \tag{13.80}$$

The shear strain at the lines AB, BD, and BC can be found using Equation 13.32. From Equation 13.32, Equation 13.54, and Equation 13.78,

$$\gamma_{12} = \frac{V_{12}^t}{V_{12}^n} = \frac{1}{\sin\phi(\cos\phi - \sin\phi\tan\omega)} \overset{(\omega=5^\circ)}{=} 2.181 \tag{13.81}$$

From Equation 13.79,

$$\gamma_{23} = \frac{V_{23}^t}{V_{23}^n} = \frac{\dfrac{\sin\phi}{\cos\gamma(\cos\phi - \sin\phi\tan\omega)} - \dfrac{h\cos\omega}{\sin\gamma(h - s\cos\omega)}}{\dfrac{h\cos\gamma}{h - s\cos\omega}} \overset{(\omega=5^\circ)}{=} 0.186 \tag{13.82}$$

where

$$V_{23}^{t} \overset{(49),(54),(56),(40)}{=} \frac{\sin\phi}{\cos\gamma(\cos\phi - \sin\phi\tan\omega)} - \frac{h\cos\omega\cos\gamma}{\sin\gamma\cos\gamma(h - s\cos\omega)}$$

From Equation 13.58, Equation 13.49, Equation 13.50, Equation 13.40, and Equation 13.80,

$$\gamma_{34} = \frac{V_{34}^{t}}{V_{34}^{n}} = \frac{\dfrac{h}{A\sin\gamma} - \dfrac{h}{A}\cos\gamma}{\dfrac{h\cos\gamma}{h - s\cos\omega}} \overset{(\omega=5°)}{=} 0.664 \tag{13.83}$$

From the von Mises criterion, the effective strain can be found as

$$\varepsilon_{ij} = \frac{\sqrt{2}}{3}[(\varepsilon_1 - \varepsilon_2)^2 + (\varepsilon_2 - \varepsilon_3)^2 + (\varepsilon_3 - \varepsilon_1)^2]^{1/2} \tag{13.84}$$

where ε_1, ε_2, and ε_3 are the principal strain components, and ε_{ij} is the effective strain.

Following Reference 27, the effective shear strain is defined as follows:

$$\gamma_{ij} = \sqrt{\frac{2}{3}}[(\varepsilon_1 - \varepsilon_2)^2 + (\varepsilon_2 - \varepsilon_3)^2 + (\varepsilon_3 - \varepsilon_1)^2]^{1/2} \tag{13.85}$$

where γ_{ij} is the effective shear strain.

Solving Equation 13.84 and Equation 13.85, one obtains the relationship between effective strain and effective shear strain as

$$\varepsilon_{ij} = \frac{\gamma_{ij}}{\sqrt{3}} \rightarrow \varepsilon_{12} \overset{(81)}{=} \frac{\gamma_{12}}{\sqrt{3}} = 1.259 \tag{13.86}$$

where ε_{12} is the effective strain on *AB* (Figure 13.5a).

A summary of the calculated effective strain and effective shear strain along the line discontinuities for sharp and chamfered tools is given in Table 13.2.

Jaspers and Dautzenberg [6], using the Rastergaev compression tests, found experimentally the values of the strength coefficient, *K*, and the strain-strengthening exponent, *n*, for AISI 1045 steels and AA 6082-T6 aluminum alloys, which are shown in Table 13.3 (see also Figure 13.9).

Examples of the calculated values of the shear stress and temperature rise, which are acting along the lines l_{12} for AISI 1045 steel, are

$$k_{12} \overset{(21,65,86)}{=} \frac{1120 * 1.259^{0.12}}{\sqrt{3} * 9.807 * 0.01} = 6778.55\frac{kg}{cm^2} \tag{13.87}$$

$$\Delta T_{12} = \frac{k\gamma_{12}}{Jc\rho} = \frac{6778.55 * 2.181}{42700 * 0.11 * 0.0078} = 403.53°C \tag{13.88}$$

TABLE 13.2
Values of the Calculated Effective Strains along the Line Discontinuities

Coefficient of Friction	Sharp Tool (Figure 13.2)		Chamfered Tool (Friction Conditions: $\mu_{0.3}$ = 0.4 and $\mu_{0.3}$ = 0.1, Figure 13.5)	
μ	ε_{12}	ε_{23}	ε_{ij}	γ_{ij}
0	1.172	0.371	$l_{12} \rightarrow \varepsilon_{12} = 1.259$	$\gamma_{12} = 2.181$
0.1	1.155	0.415	$l_{23} \rightarrow \varepsilon_{23} = 0.107$	$\gamma_{23} = 0.186$
0.2	1.173	0.442	$l_{34} \rightarrow \varepsilon_{34} = 0.383$	$\gamma_{34} = 0.664$
0.3	1.214	0.467		
0.4	1.245	0.4768		
0.5	1.2846	0.489		

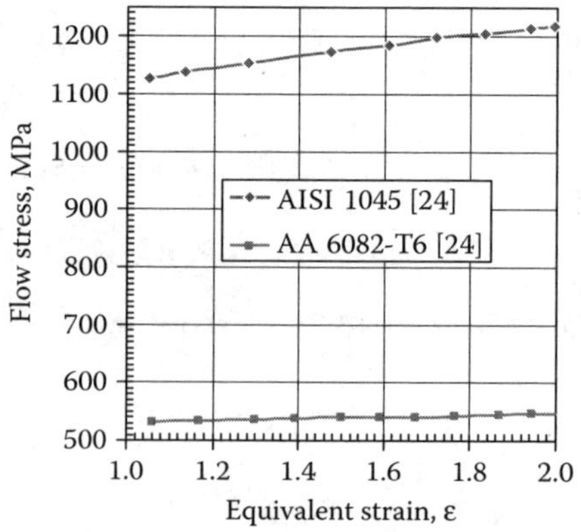

FIGURE 13.9 True stress–strain behavior of the carbon steel AISI 1045 and aluminum alloy AA6082-T6. (From Zhang, H.T., Liu, P., Hu, R.S. A three zone model and solution of shear angle in orthogonal machining. *Wear* 1991, 143, 29–43. With permission.)

TABLE 13.3
Experimental Values of Strength Coefficient, *K*, and the Strain-Strengthening Exponent, *n*

	AISI 1045	AA 6082-T6
Strength coefficient, K	1120 MPa	530 Pa
Strain-strengthening exponent, n	0.12	0.042

Substituting Equation 13.81, Equation 13.82, Equation 13.83, and Equation 13.87 into Equation 13.77, one obtains the temperature rise in the triangle plastic zones 23 and 34 in Figure 13.5a

$$\Delta T_{23} = \frac{k\gamma_{23}}{Jc\rho} = \frac{5042.66 * 0.186}{42700 * 0.11 * 0.0078} = 25.6°C \tag{13.89}$$

$$\Delta T_{34} = \frac{k\gamma_{34}}{Jc\rho} = \frac{5876.49 * 0.664}{42700 * 0.11 * 0.0078} = 106.5°C \tag{13.90}$$

where $J = 427\frac{kg * m}{kcal}$, $c = 0.11\frac{kcal}{kg * °C}$, and $\rho = 7.8\frac{g}{cm^3}$ for mild steel The investigations performed [26] show that, if the tool temperature increases above 735°C, the strength of the carbide tool can be reduced by 35 to 45%. As the yield stress of the cobalt binder decreases vs. temperature, the carbide becomes more ductile and plastic flow of the tool material can occur.

In Table 13.4 and Table 13.5, the values of the calculated shear stresses and temperature rise that are acting along the line discontinuities for sharp and chamfered tools can be found. The total flow rate and the specific work along any boundary in the zone of chip formation [29] of length l_{ij} can be found as

TABLE 13.4
Values of the Shear Stresses and Rise in Temperature Calculated along the Line Discontinuities (Sharp Tool)

Coefficient of Friction	Sharp Tool (Figure 13.2)			
μ	k_{12}	k_{23}	ΔT_{12}, °C	ΔT_{23}, °C
0	6720.45	6051.23	372.37	106.2
0.1	6708.67	5933.25	366.23	116.44
0.2	6721.13	5978.3	372.78	125
0.3	6748.9	6017.9	387.21	132.89
0.4	6769.35	6032.9	398.55	136.51
0.5	6794.8	6051.2	412.66	140.06

TABLE 13.5
Values of the Shear Stresses and Rise in Temperature Calculated along the Line Discontinuities (Chamfered Tool)

Chamfered Tool (Friction Conditions: $\mu_{0.3}$ = 0.4 and $\mu_{0.3}$ = 0.1, Figure 13.5)					
k_{12}	k_{23}	k_{34}	ΔT_{12}, °C	ΔT_{23}, °C	ΔT_{34}, °C
6778.55	5042.66	5876.49	403.53	25.6	106.5

$$\bar{V}_{ij} = V^n_{ij} l_{ij} \tag{13.91}$$

$$w_{ij} = \frac{\bar{W}_{ij}}{\bar{V}_{ij}} \overset{(90,)}{=} \frac{2\mu k V_{ch} l_{ij}}{V^n_{ij} l_{ij}} \tag{13.92}$$

where $\bar{V}_{ij}$ is the flow rate of the workpiece material across the boundary, V^n_{ij} is the normal component of velocity to the boundary, l_{ij} is the boundary length, w_{ij} is the specific work of workpiece material, and $\bar{W}_{ij}$ is the dissipated power and is defined as

$$\bar{W}_{ij} = F l_{ij} V^t_{ij} \tag{13.93}$$

The tangential stress along the tool–chip interfaces is $\tau = 2\mu k$ and V^t_{ij} is the tangential component of velocity along any boundary. Thus, the temperature rise caused by the tool–chip friction (tool-face temperature) can be found as

$$\Delta T_f J c \rho \approx w \overset{(92),(93)}{\rightarrow} \Delta T_f \overset{(89)}{=} \frac{\bar{W}_{ij}}{\bar{V}_{ij} J c \rho} \tag{13.94}$$

where ΔT_f is the tool-face temperature rise. For two surfaces in friction, the rate of the frictional works, $\bar{W}_{ij}$, will be

$$\bar{W}_{ij} = W_{f02} + W_{f03} \tag{13.95}$$

Equating the two expressions (Equation 13.62 and Equation 13. 63) to Equation 13.95, one obtains

$$\bar{W}_{ij} = 2\mu_{02} k V_{02} l_{02} + 2\mu_{03} k V_{03} l_{03} \tag{13.96}$$

$$\begin{aligned} \bar{W}_{ij} &= 2\mu_{02} k V^t_{12} \overset{(52)}{\frac{\sin\phi}{\cos\omega}} s + 2\mu_{03} k \overset{(49)}{\frac{V_1 h}{A \sin\gamma}} \overset{(43)}{2A\cos\gamma} = \\ &2k\left(\mu_{02} V^t_{12} \frac{\sin\phi}{\cos\omega} s + 2\mu_{03} V_1 h \cot\gamma\right) \end{aligned} \tag{13.97}$$

or

$$\bar{W}_{ij} \overset{(54)}{=} 2kV_1 h\left(\mu_{02} \frac{\sin\phi}{\cos\omega(\cos\phi - \sin\phi\tan\omega)} 0.1 + 2\mu_{03}\cot\gamma\right) \tag{13.98}$$

The flow rate, $\bar{V}_{ij}$, of the workpiece material across the frictional tool–chip interface may be estimated as follows:

$$\bar{V}_{ij} = V_1 s \cos\omega + V_1 l_{03} \cos\alpha \overset{(40,43)}{=} V_1 h\left(0.1\cos\omega + 2\frac{1-0.1\cos\omega}{\cos(\gamma-\alpha)}\cos\gamma\cos\alpha\right) \tag{13.99}$$

Equating the two expressions (Equation 13.98 and Equation 13. 99) to the Equation 13.94, we can estimate the temperature rise caused by the tool–chip friction (tool-face temperature):

$$\Delta T_f = \frac{2k\left(\mu_{02}\dfrac{0.1\sin\phi}{\cos\omega(\cos\phi - \sin\phi\tan\omega)} + 2\mu_{03}\cot\gamma\right)}{Jc\rho\left(0.1\cos\omega + 2\dfrac{1-0.1\cos\omega}{\cos(\gamma-\alpha)}\cos\gamma\cos\alpha\right)} \tag{13.100}$$

Thus, for the cutting conditions $\alpha = 0°$, $h = 0.5$ mm, $s = 0.1 * \mathrm{h}$, $\omega = 0.5°$, $\mu_{03} = 0.1$, $\mu_{02} = 0.4$, $\phi = 43°$, and $\gamma = 49.75°$, the temperature rise caused by friction is

$$\Delta T_f \overset{(\omega=5°)}{=}$$

$$\frac{2*6778.55\left(0.4\dfrac{0.0682}{0.996(0.646-0.763*0.0875)} + 2*0.1*0.847\right)}{42700*0.11*0.0078\left(0.0996 + 2\dfrac{0.9}{0.646}0.646\right)} = 42.2°\mathrm{C} \tag{13.101}$$

Using the results from Table 13.5, an average rise in temperature in the stress field for the machining with a chamfered tool will be the sum of the shear temperatures in the plastic zones 1, 2, and 3 and by frictional forces at the tool–face interface [26] (Figure 13.5a).

$$\Delta T_c = \Delta T_{12} + \Delta T_{23} + \Delta T_{34} + \Delta T_f \overset{(88,89,90,101)}{=} 403.53 + 25.6 + 106.5 + 42.2 \approx 578°\mathrm{C} \tag{13.102}$$

If we assume that the coefficient of friction has the value of 0.1, then for the sharp tool (Table 13.4) the temperature rise can be found as

$$\Delta T_c = \Delta T_{12} + \Delta T_{23} + \Delta T_f = 366.5 + 116.44 + 35.4 \approx 518.34°\mathrm{C} \tag{13.103}$$

where

$$\Delta T_f = \frac{\bar{W}_{ij}}{\bar{V}_{ij} Jc\rho} = \frac{2\mu k_{12} V_{02} l_{02}}{V_{02}^n l_{02} Jc\rho} \overset{(3,9)}{=} \frac{2*0.1*0.966k*2h}{2hJc\rho} = \frac{0.1932*6708.67}{42700*0.11*0.0078} = 35.4°\mathrm{C}$$

$k_{12} \overset{\mu=0.1}{=} 6708.67$ (Table 13.6), and $V_{02}^n = V_1\cos\alpha \overset{\alpha=0}{=} V_1$ (Figure 13.2).

Table 13.6 shows the values of the calculated temperature rise in the friction area, which is acting along the tool–chip interface for sharp and chamfered tools. Figure 13.10 shows the calculated rise in temperature due to the plastic work done for the orthogonal cutting. The upper-bound theory of plasticity is applied to orthogonal machining to describe the distribution of the temperature rise in the plastic deformation zone. For the sharp tool it was found that the temperature in the plastic zone for annealed carbon steel AISI 1045 is 518.34°C, and it increases up to 668.7°C when the

TABLE 13.6
The Values of Calculated Temperature Rise in the Friction Area, Which Are Acting along the Tool–Chip Interface for Sharp and Chamfered Tools

Sharp Tool (Figure 13.2)		Chamfered Tool
Coefficient of Friction	**Calculate Temperature Rise**	**Friction Conditions: $\mu_{0.3} = 0.4$ and $\mu_{0.3} = 0.1$, Figure 13.5**
μ	ΔT_f °C	ΔT_f °C
0	0	42.2
0.1	35.4	
0.2	61.58	
0.3	80.24	
0.4	99.62	
0.5	118.99	

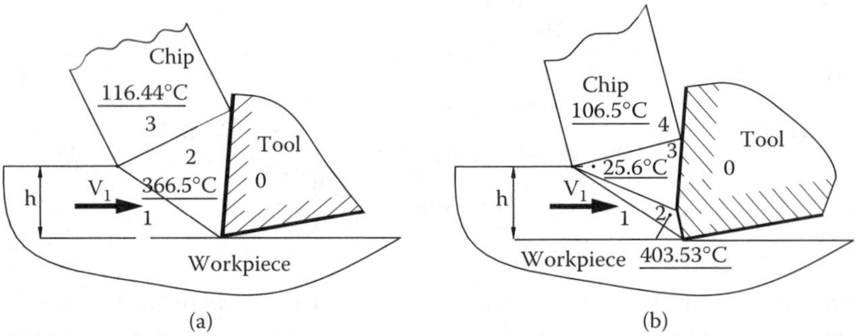

FIGURE 13.10 Theoretical predictions of the temperature distribution for (a) sharp tool and (b) chamfered tool.

maximum coefficient of friction is reached (value of 0.5). Al Hauda et al. [30] used the two-color pyrometer to investigate the temperature rise at the tool-chip interface for machining of annealed carbon steels (AISI 1045). They found that the temperature of the tool–chip interface is 840°C at a cutting speed of 100 m/min, and it increases up to 970°C when cutting speed grew to 200 m/min. FE simulations obtained for three different cutting speeds showed that the temperature in the plastic deformation zone is constant at a maximum of 500°C (Figure 13.11).

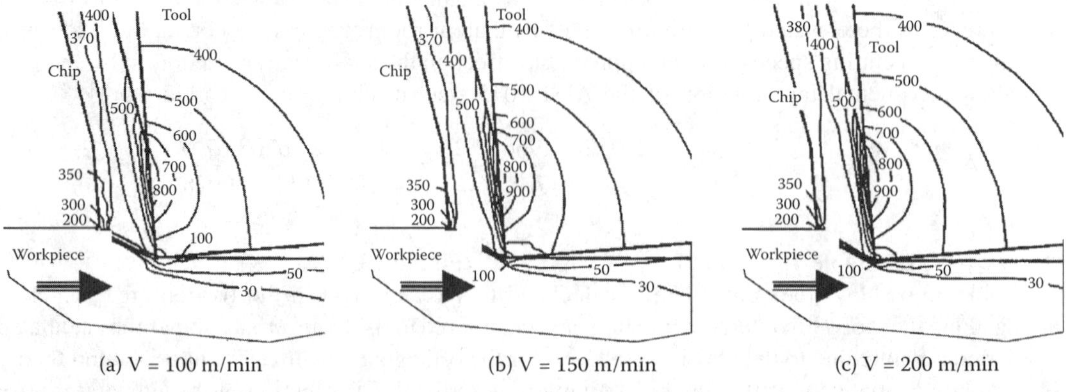

FIGURE 13.11 FEM calculated tool–chip interface temperature rise during machining of AISI 1045 steel. (From Al Huda, M., Yamada, K., Hosokawa, A., Ueda, T. Investigation of temperature at tool-chip interface in turning using two-control pyrometer. *J. Manuf. Sci. Eng.* 2002, 124, 200–207. With permission.)

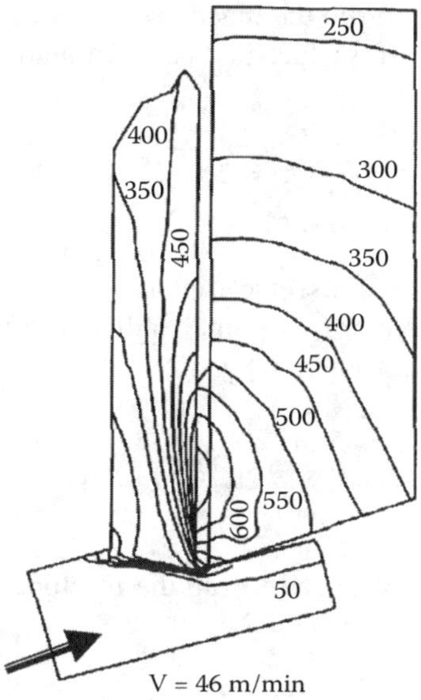

FIGURE 13.12 Tool–chip interface temperature rise during machining of AISI 1045 steel. (From Childs, T.H.C., Maekawa, K., Maulik, P. Effects of coolant on temperature distribution in metal cutting. *J. Mater. Sci. Technol.* 1998, 4, 1006–1019. With permission.)

The upper-bound method was used for calculations of the rise in temperature due to the plastic work done by orthogonal cutting. The calculated average increases in temperature in the chip for the experimental data [6] and FE simulations [30] are in a good agreement.

Comparing the limiting load of the sharp and chamfered tools, we are able to conclude that the critical average pressure on the tool edge is almost the same for both types of tool. Thus, the bearing capacity for the chamfered tool is increased.

13.2.3.2 Assessment of Temperature Rise on the Tool Face

During orthogonal cutting, temperature plays a key role in the cutting-tool life. The most important factors that control the tool life are temperature, friction, and dynamic compression of the tool material. As frictional forces on the tool–workpiece material interface increase, the temperature on the tool face increases also [26]. Bowden [32] investigated friction between two sliding molybdenum plates. Under dry conditions, when one plate slides against another, the friction coefficient is 1.8 times higher at the elevated temperature (1960°F) than at room temperature. In additional, it was found that high cutting speeds have an appreciable effect on the coefficient of friction and wear rate.

The experimental investigation of the AISI 1045 steel machining with carbide tools (Figure 13.12) shows that the maximum temperature rise on the tool face is approximately 600°C [31]. During the cutting process, the hot chip surface slides over the tool face, leading to intensive tool wear. The main sources of wear are the groove formations, cratering, and chipping on the tool interface [37–40]. When the tool is worn, the clearance tool face intensively rubs along the surface of the workpiece material. Thus, mechanical energy develops owing to the compressive stress and rubbing at the tool–workpiece interface over a long cutting pass. This mechanical energy then converts into heat that raises the temperature of the workpiece. The temperature raise at the worn surface may be higher than on the chip–tool surface. The power [38] expended to rub the machined material along the clearance tool face can be found as

$$\bar{W}_{cl.f} = \tau_t l_{cl.f} V_1 \tag{13.104}$$

where $\bar{W}_{cl.f}$ is the power expended to rub the machined material along the clearance tool face, and τ_t is the tangential stress on the worn surface and can be found as

$$\tau_t = 2\mu \frac{\sigma_{TR}}{\sqrt{3}} \tag{13.105}$$

where σ_{TR} is the tensile strength or transverse rupture stress (TRS) of the carbide tool material (which is usually measured in a three or four point bending tests), $l_{cl.f}$ is the contact length of worn surface, and V_1 is the cutting speed. Hence,

$$\bar{W}_{cl.f} = 2\mu \frac{\sigma_{TR}}{\sqrt{3}} l_{cl.f} * V_1 \tag{13.106}$$

On the other hand, the power expended to rub the machined material leads to the dissipation of heat and can be found as

$$\bar{W}_{cl.f} = \Delta T_{cl.f} Jc\rho * l_{cl.f} * V_1 \tag{13.107}$$

where $\Delta T_{cl.f}$ is the tool-face temperature rise on the worn surface.

If we assume that the adiabatic conditions prevail on a worn tool material surface along a path, then the energy balance of the frictional heat source can be written as

$$(106), (107)\ \Delta T_{cl.f} Jc\rho * l_{cl.f} * V_1 \approx 2\mu \frac{\sigma_{TR}}{\sqrt{3}} l_{cl,f} * V_1 \tag{13.108}$$

where $c = 0.0321 \frac{\text{kcal}}{\text{kg*°C}}$ (for carbide tool insert).

When the carbide tool temperature increases above 735°C, the strength of carbide tool is reduced by 25 to 35% as was outlined earlier [40]. Astakhov [41] has investigated the thermal softening of the cutting edge of the carbide tools. It was shown that, if the temperature rises up to 1000 to 1200°C, then the cutting edges deform as a result of high-temperature creep.

The temperature rises when the cutting speed increases, when there is no heat dissipated from the tool surface over time, and when there is heat transfer from the deformation shear zone. In this case, the tool edges soften, and wear volume increases with the sliding distance. Therefore, with a worn tool, there are three main mechanisms of the energy dissipation, which are converted into heat: plastic deformation of the workpiece material, friction at the chip–tool interface, and the friction on the clearance surface.

13.2.3.3 Assessment of the Stress Distribution in the Edge of the Cutting Tool

In this section, we are considering the solution of the problem of the cutting-edge failure. Fracture of tool materials under tension has been developed for various types of material behavior. However, relatively little is known about chipping under compression in carbide materials. Trent [38], using a photoelastic method, showed that the maximum stress acts on the tip of the cutting-tool edge.

The applied normal compressive force on one side of the wedge-shaped tool causes failure of the tool edge. At high temperatures, the wedge of the carbide tool insert deforms plastically. Astakhov [41] defined this cutting-tool failure mechanism as high-temperature creep.

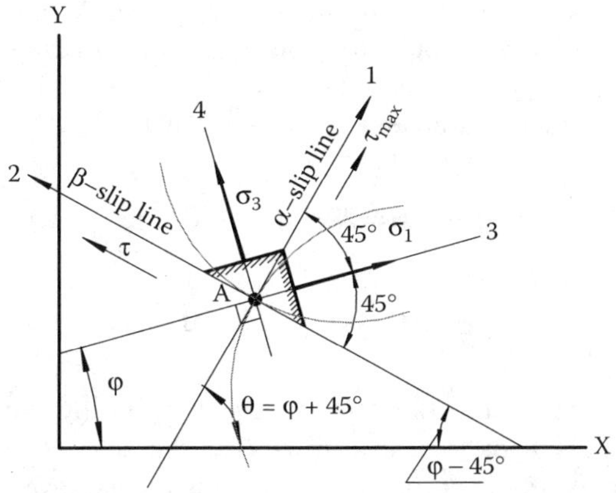

FIGURE 13.13 Slip lines (1, 2) and trajectory of principal normal stresses (3, 4).

The preceding analysis shows that the highly concentrated loading is applied on the rake face of the cutting tool. These pressures on the wedge increase with larger tool motion, with larger positive rake angle and an increased cutting depth. The excessive rise in pressure on one side of the wedge can result in plastic deformation of the tool edge and tool chipping. Therefore, it is important to define the limiting load for the tool wedge. The approach of the bearing capacity of the wedge-shaped tool was considered to become an instrument for the cutting-tool designers. Simulations presented in this section are based on the slip-line field theory (SLF). Green [43–44] first presented the SLF solution to analyze plastic deformation of the bending of cantilevers and fixed ended beams. Kachanov [15] proposed SLF solutions for the problem of upsetting a wedge under a uniform pressure. The effects of friction and nonhomogeneous deformation, however, were ignored.

In this study, investigations were carried out using SLF to simulate short cantilever wedge-shaped beams under concentrated loading. The slip-line fields were designed by a graphical method. Basically, the SLF method consisted of the following information. The plastic flow for the plane problem is described by the equilibrium conditions:

$$\frac{d\sigma_x}{dx} + \frac{d\tau_{xy}}{dy} = 0 \qquad \frac{d\sigma_y}{dy} + \frac{d\tau_{xy}}{dx} = 0 \tag{13.109}$$

The plasticity conditions

$$(\sigma_x - \sigma_y)^2 + 4\tau_{xy}^2 = 4k^2 \tag{13.110}$$

The trajectory of principal stresses σ_1 and σ_3 passing through the arbitrary point A are shown in Figure 13.13. From Figure 13.2, it follows that for the slip lines of family α and β it is possible to write

$$\frac{dy}{dx} = \tan\theta; \qquad \frac{\sigma}{2k} - \theta = \text{constant} = \xi \tag{13.111}$$

$$\frac{dy}{dx} = -\cot\theta; \qquad \frac{\sigma}{2k} + \theta = \text{constant} = \eta \tag{13.112}$$

Equation 13.111 and Equation 13.112 are the differential equations of slip lines. Hencky first derived these relations for the plane problem in the theory of plasticity. In passing from one slip line to another of the α family the parameter ξ changes in general. Likewise, in passing from one line to another of the β family, the parameter η changes.

$$\sigma - 2k\theta = 2k\xi \quad \text{along } \alpha \text{ line} \tag{13.113}$$

$$\sigma + 2k\theta = 2k\eta \quad \text{along } \beta \text{ line} \tag{13.114}$$

where $2k\eta$ and $2k\xi$ are arbitrary functions that are constant during movement along one of slip lines, and change only as result of transition from one characteristic to another. The plasticity condition is satisfied if the expressions for the stress components have the form:

$$\sigma_x = \sigma_n - k \sin 2\theta \tag{13.115}$$

$$\sigma_y = \sigma_n + k \sin 2\theta, \tag{13.116}$$

$$\tau_{xy} = k \cos 2\theta \tag{13.117}$$

where σ_n is the mean normal stress, and θ is the slope angle of the maximum tangential stress to the X axis. The major properties of slip lines are the following:

1. The pressure along a slip line varies directly as the angle of the slip line with the X axis. This property is obvious because $\sigma = 2k\theta$ + constant along the α line, and $\sigma = -2k\theta$ + constant along the β line.
2. If we pass from one slip line to another of the β family along any slip line of the α family, the angle α and the pressure σ change by a constant amount (Hencky's first theorem). From the relations in Equation 13.111 and Equation 13.112, it follows that

$$\sigma = k(\xi + \eta) \tag{13.118}$$

$$\theta = \frac{1}{2}(\eta - \xi) \tag{13.119}$$

If the slip lines (SL) and the stress at a point on the characteristic are know, the stress at any point on that characteristic can be determined by using Equation 13.113 and Equation 13.114. In distinction to the Green's models [42,43], our cantilever beams have a wedge shape and are loaded under a uniform pressure as shown in Figure 13.14. On the free surface *AE* (Figure 13.14b) the stress $\sigma_3 = 0$ and the compression stress $\sigma_1 = -2k$ (where σ_3 is the largest and σ_1 is the algebraically smallest principal stress). Thus, from the yield criterion the magnitude of the mean normal stress at point *E* can be expressed as

$$\sigma_3 - \sigma_1 = 2k \Rightarrow \sigma_{yE} - \sigma_{xE} = 2k \rightarrow \sigma_{xE} = -2k \tag{13.120}$$

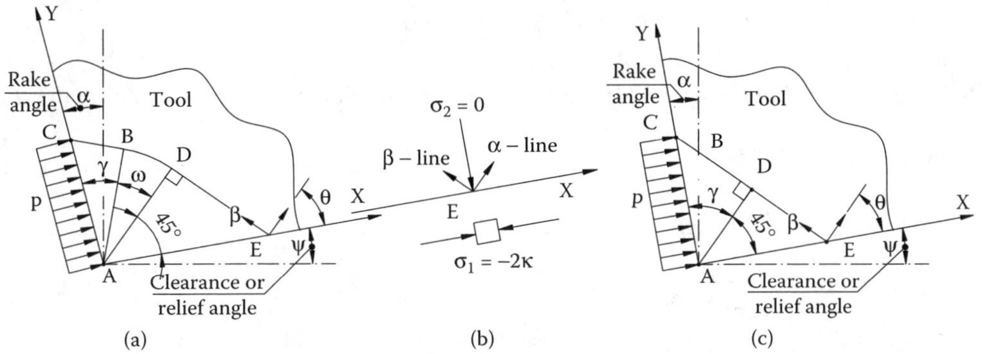

FIGURE 13.14 Slip-lines field for the wedge short cantilevers beams: (a) $\frac{\pi}{2}+\alpha-\psi>\frac{\pi}{2}$, (b) directions of principal stresses at point E, (c) $\frac{\pi}{2}+\alpha-\psi=\frac{\pi}{2}$.

$$\sigma_{nE}=\frac{\sigma_{yE}+\sigma_{xE}}{2}=-k \tag{13.121}$$

The angle θ_E of the α line at point E is equal to $\frac{\pi}{4}$. Using Equation 13.115,

$$\frac{\sigma_{nE}}{2k}+\frac{\pi}{4}=\eta_E=-\frac{k}{2k}+\frac{\pi}{4}=-\frac{1}{2}+\frac{\pi}{4} \tag{13.122}$$

At point C

$$\frac{\sigma_{nC}}{2k}+\theta_c=\eta_C=\eta_E=-\frac{1}{2}+\frac{\pi}{4} \tag{13.123}$$

or

$$\frac{\sigma_{nC}}{2k}=-\frac{\pi}{4}-\omega-\frac{1}{2}+\frac{\pi}{4}=-\frac{1}{2}-\omega\rightarrow\sigma_{nC}=-k(1+2\omega) \tag{13.124}$$

where $\theta_C=\frac{\pi}{4}+\omega$.

It is possible to write down (from plane strain condition for point C)

$$\sigma_{nC}=\frac{\sigma_{yC}+\sigma_{xC}}{2} \tag{13.125}$$

and

$$\sigma_1 - \sigma_3 = Y, \quad \sigma_{yC} - \sigma_{xC} = 2k \tag{13.126}$$

Solving together Equation 13.124, Equation 13.125, and Equation 13.126, we have the following:

From Equation 13.125,

$$2\sigma_{nC} = \sigma_{yC} + \sigma_{xC} \text{ or } \sigma_{xC} = 2\sigma_{nC} - \sigma_{yC} \tag{13.127}$$

From Equation 13.126,

$$\sigma_{xC} = \sigma_{yC} - 2k \tag{13.128}$$

From Equation 13.123, Equation 13.127, and Equation 13.128,

$$\sigma_{xC} = \sigma_{nC} + k = -2k(1+\omega) = -2k\left(1 + \frac{\pi}{4} + \alpha - \psi - \gamma\right) \tag{13.129}$$

From Equation 13.126,

$$\sigma_{yC} = -2k\omega \tag{13.130}$$

where $\omega = \frac{\pi}{4} + \alpha - \psi - \gamma$.

Thus, the pressure, p, on the contact surface, AC, is uniformly distributed and is equal to $p = -\sigma_{xC}$.

$$\frac{p}{2k} \overset{(129)}{=} 1 + \frac{\pi}{4} + \alpha - \psi - \gamma \tag{13.131}$$

From Equation 13.134, it follows that, if $\psi + \gamma = \frac{\pi}{4} + \alpha$, then the limiting pressure is $\frac{p}{2k} = 1$.

Maximum of the friction at the interface is when $\tau = k = \mu p$. It follows from the Equation 13.117 that

$$\tau_{xy} = k\cos 2\theta = \mu p = k \tag{13.132}$$

Therefore, the value of the friction angle can be represented in the form of the expression:

$$\theta = \frac{1}{2}\arccos 2\mu \tag{13.133}$$

where $0 \le \mu \le 0.5$.

In our case, the β line of velocity discontinuity $CBDE$ intersects the tool face at the angle γ; thus

$$\gamma = \frac{1}{2}\arccos 2\mu \quad (13.134)$$

When the shear stress is equal to $\tau_{xy} = 0$, the friction angle γ is equal to $\frac{\pi}{4}$, and when $\tau_{xy} = k$, the value of the friction angle is equal to 0.

Using Equation 13.134, Equation 13.131 can be rewritten as

$$\frac{p}{2k} = 1.785 + \alpha - \psi - \frac{1}{2}\arccos 2\mu \quad (13.135)$$

If we assume that the flank angel is equal to $\psi = 10°$, then the angle $\omega = \frac{7}{36}\pi + \alpha - \gamma$ and Equation 13.135 can be written as

$$\frac{p}{2k} = 1.611 + \alpha - \frac{1}{2}\arccos 2\mu \quad (13.136)$$

Using Equation 13. 135, the load per unit width acting at the tool surface is

$$P = AC * p \overset{(3)}{=} \frac{2kh}{\sin\phi}\sin(\phi + \alpha)\left[1.611 + \alpha - \frac{1}{2}\arccos 2\mu\right] \quad (13.137)$$

The effect of the tool rake angle and the coefficient of friction on the limiting load is shown in Figure 13.15. It can be seen in Figure 13.15 that the limiting upsetting pressure on the wedge-shaped tool rise as the tool rake angle increases. On the other hand, with an increase of the negative rake angle, the bearing capacity of the tool also increases.

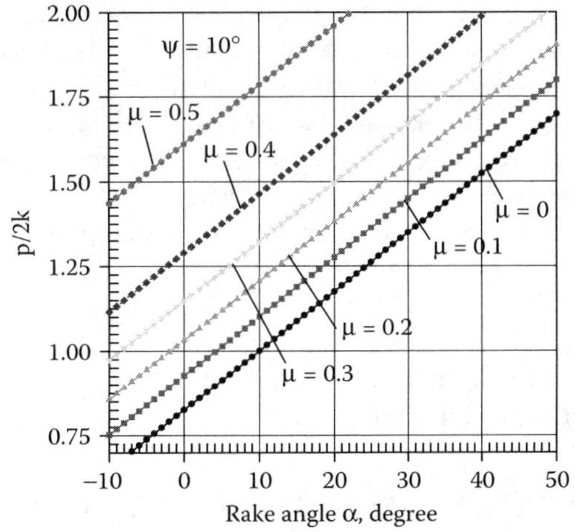

FIGURE 13.15 The limiting compression pressure $\frac{p}{2k}$ vs. tool rake angle α.

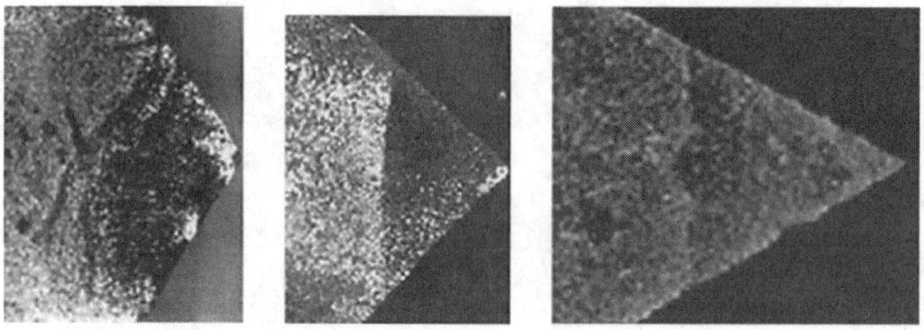

(a) (b) (c)

FIGURE 13.16 The experimental evidence of the theoretical prediction of the SLF for different angles of the wedge short cantilever beams: (a) $\frac{\pi}{2}+\alpha-\psi>\frac{\pi}{2}$, (b) $\frac{\pi}{2}+\alpha-\psi=\frac{\pi}{2}$, (c) $\frac{\pi}{2}+\alpha-\psi<\frac{\pi}{2}$.

Using the etching method in Fry's reagent [16], we have validated the assumed SLF for the wedge short cantilever beams. Figure 13.21 shows the SLF for different angles of the wedges, which are similar to the theoretical predictions.

The following example illustrates the principle of law choice for the tool rake angle. We assume that the friction coefficient is equal to $\mu = 0.4$ for high-speed machining (with a sharp cutting tool) and that the maximum of the cutting pressure is equal to $\frac{p}{2k} = 1.02$ (Figure 13.3). In this case, the wedge-shaped tools are loaded under a uniform pressure $\frac{p}{2k} = 1.02$. Using the data shown in Figure 13.16 (for friction coefficient $\mu = 0.4$), we obtain that the negative rake angle of the tool has to be equal to 10°. We can choose a larger negative rake angle, but the preceding analysis shows that, with an increase in the angle, the cutting pressure increases. With the chosen tool rake angle of $\alpha = 10°$, the limit pressure on the wedge-shaped tool is less, and correspondingly the bearing capacity of the tool is increased. Therefore, it is necessary to specify the technological parameters of the process to prevent the tool failure.

The following conclusion could be drawn based on analytical modeling performed:

1. Plastic limit analysis for the machining with sharp and chamfered tools has been presented. The rises in temperature in the deformation zone have been examined.
2. The velocity, strain, and temperature fields are analytically described.
3. The upper-bound technique was used for the calculation of the temperature rise due to the plastic work done for the orthogonal cutting. High-speed machining causes more plastic work and an increase in friction. This results in energy dissipation into heat and a greater temperature rise in the cutting-tool material.
4. The calculated average rises in temperature at the chip–tool interface is in a good agreement with experimental data [6] and FE simulations [5].
5. An experimental verification has been proposed. The maximum of the shear strain occurs at the boundary line AB, which is the primary shear plane [9].
6. To prevent the tool wear by plastic deformation, the tool rake angle has to be chosen using the nomogram shown in Figure 13.15.

13.3 DESIGN OPTIMIZATION FOR THE ADAPTIVE CUTTING TOOLS WITH HARD PVD COATINGS

The tetragonal indexable cutting inserts were made of T15 HSS, as well as Sandvik H1P WC/TiC/Co cemented carbide grade that are tailored for high-speed machining applications. Two types of

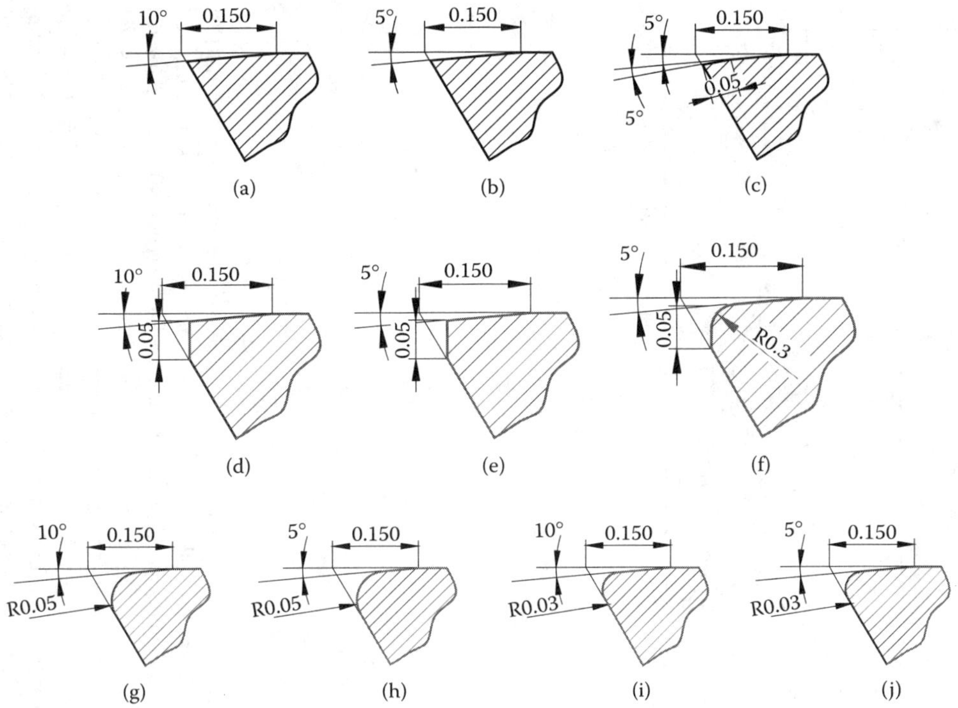

FIGURE 13.17 Design of the facets investigated. The length of lend is 50 to 300 μm.

Balzers Ti–Al–N PVD coatings were deposited on the cutting-tool substrates. Ti–Al–N monolayered coatings tailored for high-speed machining applications were deposited on cemented carbide substrates. Multilayered adaptive WC/C + Ti–Al–N coating were deposited on HSS substrates to reduce the intensity of the build-up edge formation that is typical for HSS cutting tools [38].

The results of the analytical modeling were used to justify the design of the adaptive cutting tools. Based on the calculations performed, chamfered cutting tools were used with a negative rake angle of 5 to 10° (Figure 13.17a and Figure 13.17b) and a varying size of facet (from 50 up to 350 μm). The basic, i.e., chamfered rake face design was compared with a variety of different cutting-edges designs (Figure 13.17).

The cutting tests were performed under the turning conditions using coated high-speed steel, as well as cemented carbide SPG 422 insert. The indexable inserts were tested under corresponding cutting conditions, i.e., under moderate (for HSS substrate) and high-speed (for CC substrate) turning conditions of the structural steels (Table 13.7). The data obtained was plotted as the flank wear value vs. the length of cut (Figure 13.18a to Figure 13.18c).

Based on the literature data [37–38], we can assume that the type of cutting-tool failure strongly depends on the cutting conditions. HSS cutting tools under conditions of turning at low and moderate cutting speeds, when attrition wear mode dominates [38], exhibit stochastic deep surface damage of the cutting edge (Figure 13.19c). At high-speed machining conditions, though, the cutting edge of the CC tool undergoes plastic deformation, or the so-called plastic lowering (Figure 13.20b and Figure 13.20c) [41]. The processes of plastic deformation and fracture switch the tool–workpiece tribosystem to conditions that are very far from the equilibrium state and their mimicking is practically impossible. The process of wear should be transformed to a quasi-stable mode due to modifications of the cutting-edge design using the results of the analytical modeling performed. Because the process of wear is transformed to a quasi-stable mode, we can mimic the edge shape that corresponds to the point of transformation from the running-in to the post running-in stage of wear with significantly lower wear intensity.

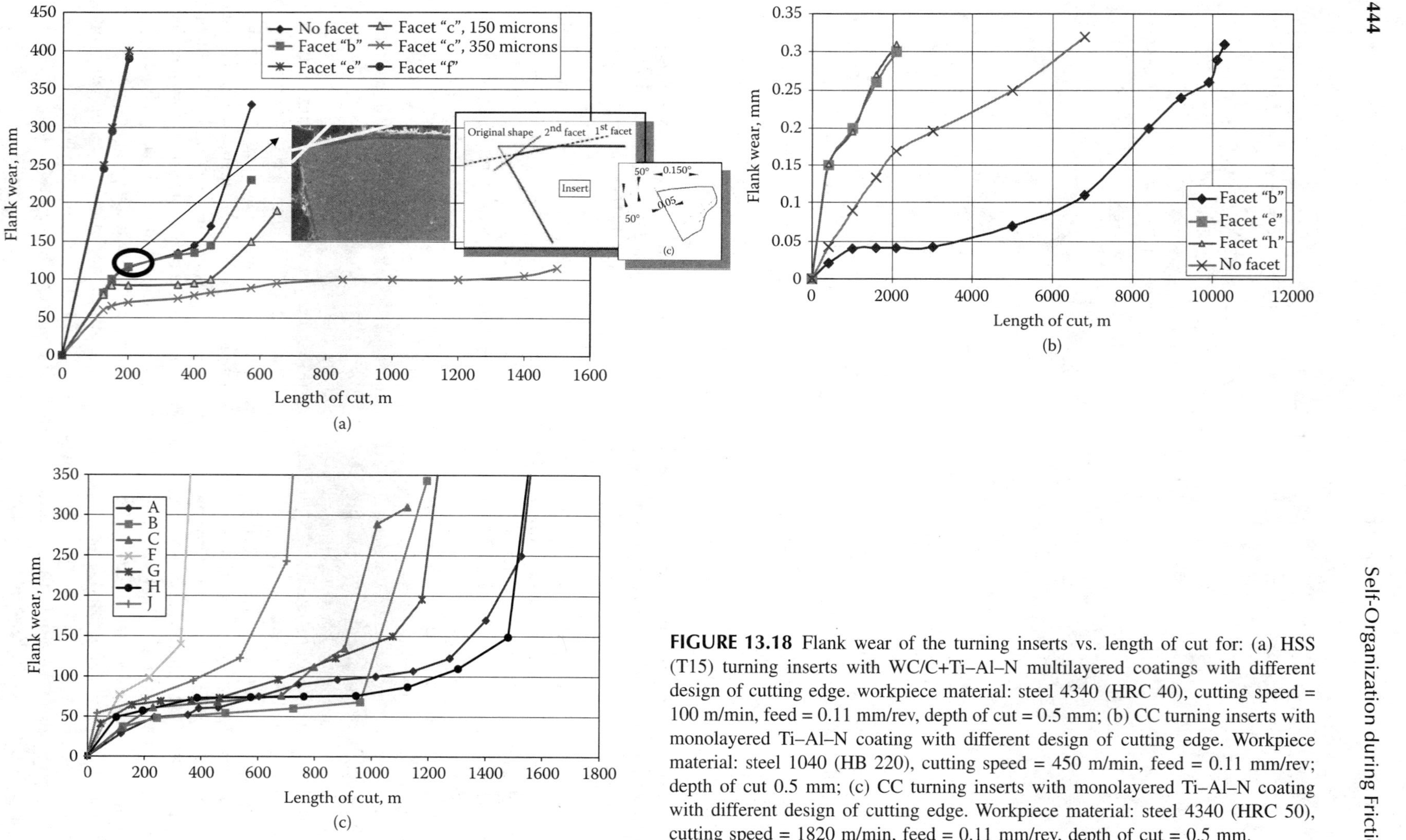

FIGURE 13.18 Flank wear of the turning inserts vs. length of cut for: (a) HSS (T15) turning inserts with WC/C+Ti–Al–N multilayered coatings with different design of cutting edge. workpiece material: steel 4340 (HRC 40), cutting speed = 100 m/min, feed = 0.11 mm/rev, depth of cut = 0.5 mm; (b) CC turning inserts with monolayered Ti–Al–N coating with different design of cutting edge. Workpiece material: steel 1040 (HB 220), cutting speed = 450 m/min, feed = 0.11 mm/rev; depth of cut 0.5 mm; (c) CC turning inserts with monolayered Ti–Al–N coating with different design of cutting edge. Workpiece material: steel 4340 (HRC 50), cutting speed = 1820 m/min, feed = 0.11 mm/rev, depth of cut = 0.5 mm.

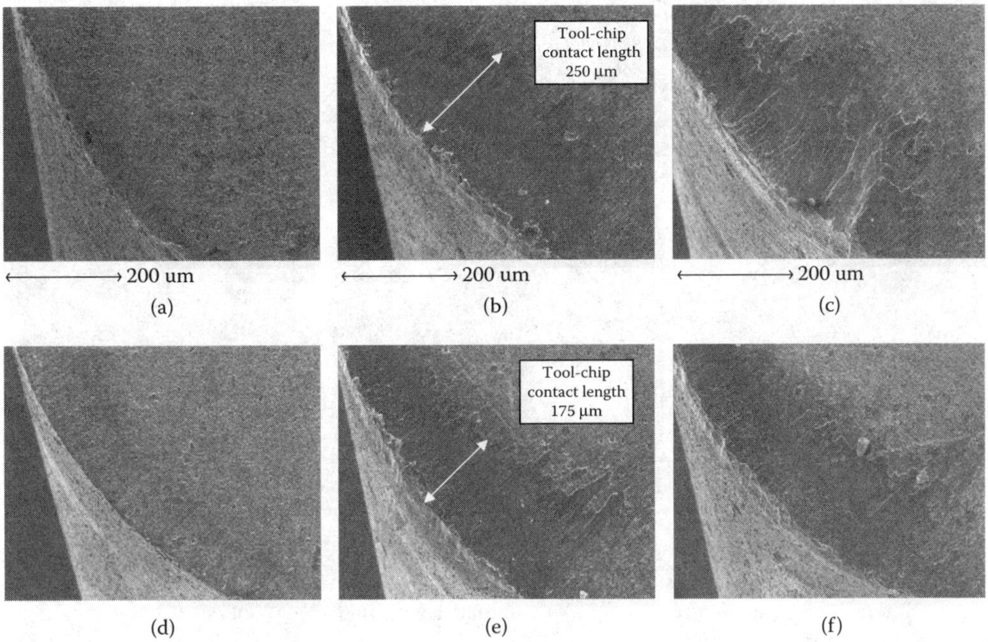

FIGURE 13.19 SEM images of the HSS turning inserts with WC/C+Ti–Al–N adaptive multilayered coatings. The rake face morphology changes vs. length of cut: (a)–(c) no facet; (d)–(f) with facet, type "a" (Figure 13.1) on a rake face; (a), (d) initial stage; (b), (e) length of cut = 200 m; (e), (f) length of cut = 500 m.

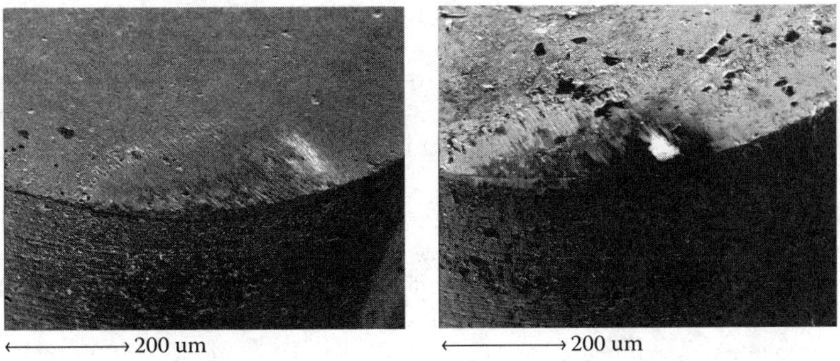

FIGURE 13.20 SEM images of the turning inserts made of cemented carbides (CC) with monolayered PVD Ti–Al–N coating. The rake face morphology changes vs. length of cut.

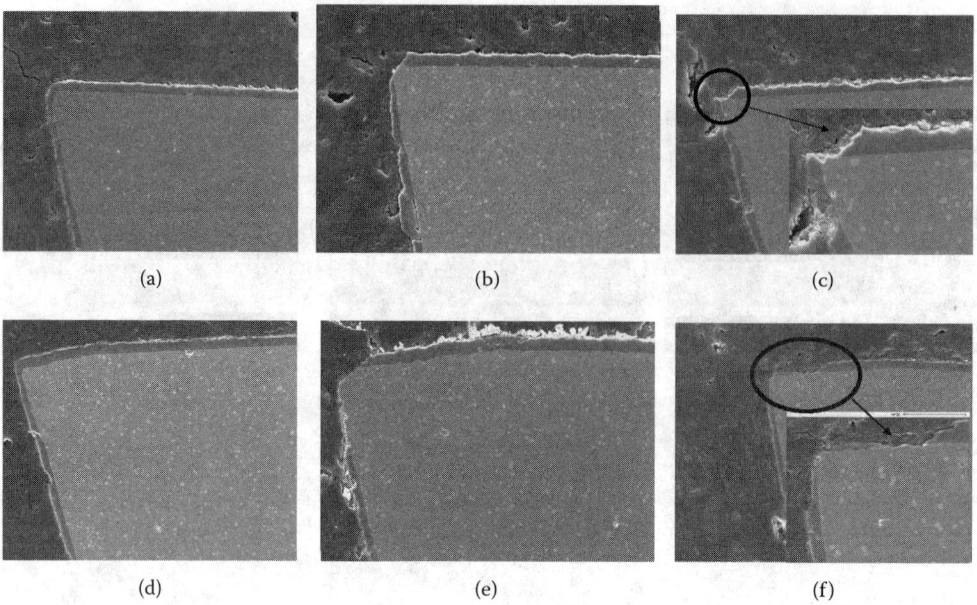

FIGURE 13.21 SEM images of the cross-sections of the turning inserts made of HSS with Balzers Hardlube multilayered coatings (outer layer has WC/C coating; sublayer has Ti–Al–N coating): (a)–(c) no facet; (d)–(f) with chamfered rake face (see Figure 13.1a); (a), (d) initial stage; (b), (e) length of cut = 200 m; (c), (f) length of cut = 500 m.

Owing to the difference in the failure mechanisms for the HSS cutters under moderate cutting speed conditions and the CC cutters under high-speed conditions, the impact of the cutting edge design on their adaptability is considered separately for specific applications.

13.3.1 HSS Cutters

HSS cutting-tool morphology has been observes (Figure 13.19a to Figure 13.19c), and the worn area cross sections were studied (Figure 13.21c) vs. the length of the cut for the turning inserts with and without a chamfered rake surface.

The cutting-edge chipping intensity within the initial stage of wear (Figure 13.19a to Figure 13.19c) could be eliminated or at least significantly reduced by strengthening it (Figure 13.19d to Figure 13.19d f). This could be done by means of the facet fabrications [2]. A few types of facet were made (Figure 13.17) [2].

It was found that the only type of the facet, i.e., "a" type, on a rake face has beneficial impact on the cutting-tool life (Figure 13.18a). This facet strengthens the cutting edge and prevents a deep surface damage (Figure 13.19d to Figure 13.19f). This leads to wear rate stabilization (Figure 13.19d to Figure 13.19f) and improves the overall wear behavior and chip formation during cutting.

The rake surface chamfering for the HSS turning inserts leads to significant changes in the chips shape (Figure 13.22). The chips collected from the unchamfered cutting inserts were curled (Figure 13.22a). The chips that formed when the chamfered cutting inserts were used had almost no curling (Figure 13.22c). In contrast, the chip's undersurface morphology investigation shows that a smoother surface is formed in this case (Figure 13.22b and Figure 13.22d). This corresponds to the analytical modeling data showing that the chamfering of the turning cutters rake surface affects the surface damage intensity, the metal flow, and the heat redistribution at the tool–workpiece interface. The SEM metallography studies of the chips cross section also prove this idea (Figure 13.23). It is known that chips consist of a few typical zones [38–39]. A zone of dynamic recrys-

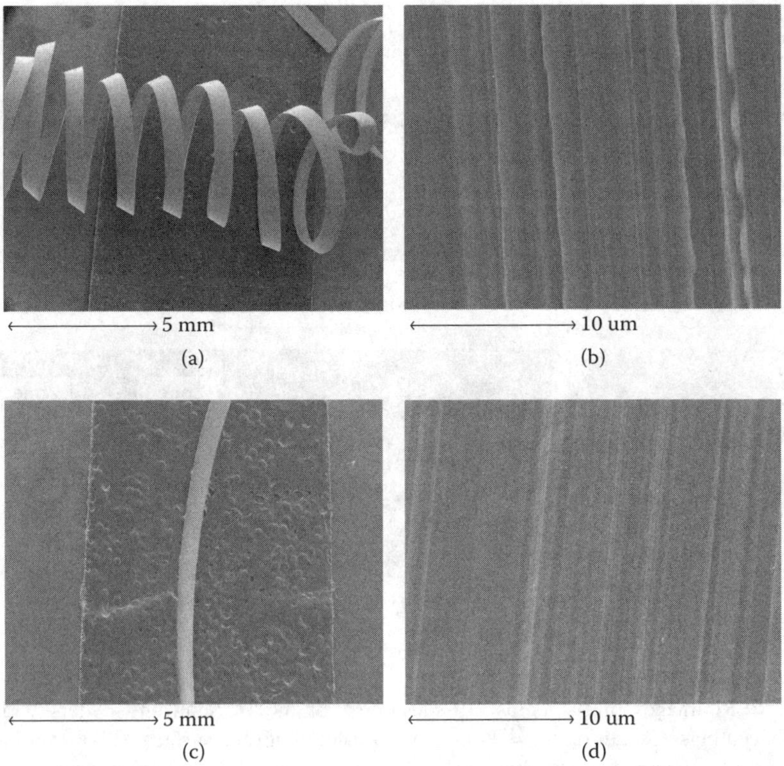

FIGURE 13.22 Types of chips for HSS turning cutters with multi-layered WC/C+Ti–Al–N coatings. Length of cut = 96 m; (a)–(c) chip shape; (b)–(d) undersurface morphology; (a)–(b) without facet on the rake surface; (c)–(d) with facet on a rake surface.

tallization is located at the chip–tool interface (Zone 1), and the extended deformation zone (Zone 2) is located far away from the interface. More heat flux goes into the chips, causing more intensive recrystallization to occur. This is exhibited in the chips' grain coarsening within the contact zone (Figure 13.23c). Figure 13.23b and Figure 13.23c show a zone of dynamic recrystallization for the chips collected from the inserts with the chamfered rake surface. Figure 13.23e to Figure 13.23f present similar images for the chips collected from the tools without the facet. We can observe more intensive recrystallization of the contact zone takes place from the chips collected from the inserts with a chamfered rake surface because the thickness of Zone 1 is significantly greater and the grains are coarser.

When the chips slide along the rake face of the turning inserts, curved flow lines are formed because of friction. More intensive metal flow results in more deformation within the extended deformation zone (Zone 2). Owing to the higher intensity of deformation, a thinner Zone 2 is formed (Figure 13.23a and Figure 13.23d) for the chips collected from the tools with chamfered rake surface.

Moreover, the tool–chip contact length measured in Figure 13.19b to Figure 13.19f is higher for the tools without the facet. This means that *in situ* frictional characteristics on the rake surface is presumed to be better and the curling of the chips should be more intensive for the cutting tools with facets. However, owing to intensive heat flux into the chip (thicker zone of dynamic recrystallization and coarser grain sizes; Figure 13.23), dull chips are formed. The possible cause of this beneficial heat redistribution is the formation of protective tribo-films on the surface of the cutter. If inserts without facet are used, the protective films cannot form in time because of the deep surface damage (Figure 13.19c). When inserts with the chamfered rake surface were used, deep

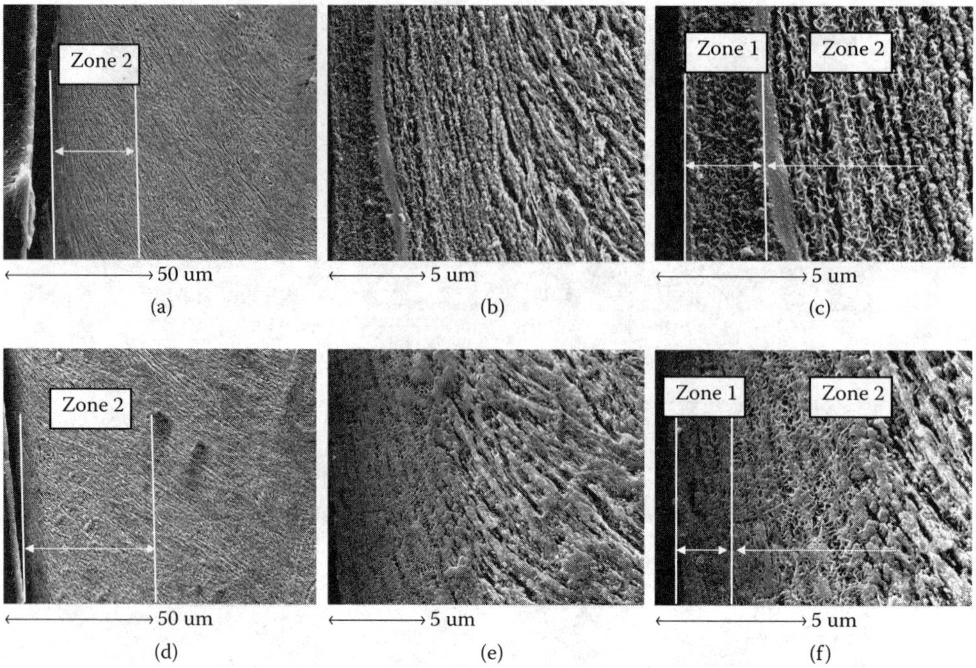

FIGURE 13.23 SEM images of the chips cross-sections for the HSS turning cutters with multilayered WC/C+Ti–Al–N coatings. Length of cut = 96 m; (a)–(c) facet on a rake surface; (d)–(e) no facet. Zone 1 — the contact zone of the dynamic recrystallization; Zone 2 — extended deformation zone.

surface damage is prevented, which results in a wear rate reduction due to the formation of the stable protective tribo-films (Figure 13.19g to Figure 13.19h).

Using a chamfered cutting tool, we can create quasi-stable wear conditions and eventually observe an evolution of the cutting-edge shape vs. length of cut (Figure 13.21c and Figure 13.21d). Intensive wear takes place on both faces of the inserts (Figure 13.21e and Figure 13.21f), and eventually the hard coating intensively wears on the rake surface. The area of wear is wide owing to lower stress concentration as compared to the tool without facets (Figure 13.21b and Figure 13.21c). We can assume that the "natural" wear shape corresponds to the image presented in Figure 13.21e and Figure 13.21f. Based on the cross section shown in Figure 13.21e and Figure 13.21f, we can suggest a few variants of the facets design that mimic the shape of natural wear (Figure 13.17). However, the cutting tests performed show that only the double chamfered rakes surface (Figure 13.17c) results in tool life improvement (Figure 13.18a). This shape corresponds to the shape of natural wear of the cutting edge at the specific point of transformation from the running-in to the post running-in stage, where the wear rate drops significantly (Figure 13.18a). The size of the chamfered zone is another important parameter. An increase in the total length of the double chamfered zone from 150 to 350 μm results in critical tool life improvement because of the reduced surface damage during the running-in stage, as well as wear rate stabilization (Figure 13.18a).

We can assume that the wear rate stabilization of the double chamfered tools is also associated with the microstructure of the tribo-films formed. Low wear intensity could indicate the formation of a more stable tribo-film. Thus, the geometrical adaptability could affect the structural adaptability. Further research is needed to confirm this hypothesis.

13.3.2 CC Cutters

The tool life data on the turning inserts made of cemented carbides is presented in Figure 13.18b and Figure 13.18c and evolution of surface morphology is presented in Figure 13.20. The cutting-

TABLE 13.7
Cutting Data

Cutting Tools	Workpiece Material/Hardness	Speed, m/min	Feed, mm/rev	Depth of Cut, mm
Cemented carbides with Ti–Al–N coating	1040 steel/HB 220	450	0.11	0.5
	4340 steel/HRC 50	182		
HSS with multilayer WC/C + Ti–Al–N coatings	4340 steel/HRC 40	100		

tool life has been studied under different cutting conditions (Table 13.7): (1) high-speed machining of the annealed (HB 220) 1040 steel (Figure 13.18b) and (2) turning of 4340 steel (HRC 40) (Figure 13.18c). Thus, the tool life was studied under two significantly different cutting conditions. First type of cutting condition was high-speed machining when thermal processes dominate, but stresses at the workpiece–tool interface are moderate because the annealed workpiece material has been machined. In this case, creep resistances of the cutting wedge mainly controls the tool life and wear behavior [41]. The second type of cutting condition was the machining of hardened steel (hardness HRC 50) under moderate cutting speeds. In this case, the stresses generated at the workpiece–tool interface play a decisive role. The load-bearing capacity of the cutting wedge, as well as stress concentration at the cutting edge most probably controls the tool life.

Analytical modeling performed predicts that the rake face chamfering of CC turning inserts results in an improved load-bearing capacity of the cutting edge and stress and temperature distribution at the tool surface. This prevents intensive creep of the cutting wedge during friction and results in wear rate stabilization during high-speed machining conditions (Figure 13.18b). The data presented show that all other types of cutting-edge design exhibits linear wear curves. It means that the wear rate is high. In contrast, the rake face chamfering significantly improves the tool life because of wear rate stabilizing at a low level after a short period of time (Figure 13.18b). The rake face chamfering leads to the cutting-wedge strengthening, which results in lowering the platic prevention and wear rate stabilization. The length of the facets is also important. Impact of the facets length on the turning inserts tool life is shown in Table 13.8. A length increase from 50 to 150 μm leads to tool life growth by 10%. Lower impact of the length value on the tool life of the CC chamfered tool could be explained by the more stable wear mode of the CC cutters, as compared to the stochastic surface damage of HSS cutters under attrition wear conditions. During machining of the hardened 4340 steel, when stress-related processes dominate, the most efficient design of the cutting edge is similar to the design in Figure 13.18c. Chamfering of the rake surface stabilizes the wear rate at a lower level and significantly widens the stable stage of wear. Slightly better wear behavior of the chamfered rake surface with the cutting-edge rounding could be explained by the intensive stress concentration of the cutting edge during machining of hardened steel.

The optimal design of the facet that stabilizes the wear rate strongly depends on the type of cutting operation and the design of the cutting tools. For turning operations, the best tool life was shown by the inserts with the chamfered rake surface (Figure 13.17 and Figure 13.18). The other designs of the cutting edge are not beneficial (Figure 13.18). In contrast, the facet fabricated on the flank surface of the end-mill cutter (Figure 13.24) improves tool life (Figure 13.25a). As shown for HSS turning inserts, facet fabrication is especially beneficial when the attrition wear mechanism dominates. An intensive chipping of the cutting edge occurs (see preceding text as well as Chapter 5), especially during end milling of SS 304 stainless steels, because of intensive build-up formation. There are two ways to prevent, or at least to reduce, the intensity of this undesired phenomenon. One way is by the development and application of multilayered Ti–Al–N/Ti–N coating with the optimized ability to accumulate the energy of deformation (see Chapter 5). But there is another and possibly a cheaper way to achieve this goal. The fabrication of a facet on the flank surface of

TABLE 13.8
Relative Tool Life of the Cemented Carbides with Ti–Al–N Coating vs. the Length of the Facets on the Rake Face

Length (μm)	Relative Tool Life
30	0.9
50	1.0
100	1.04
150	1.1

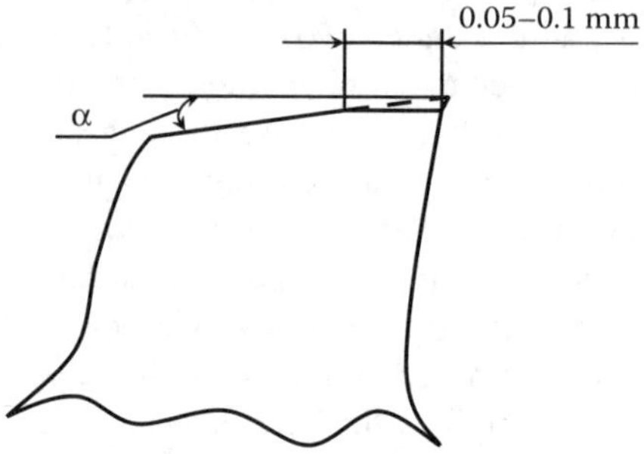

FIGURE 13.24 Design of the facet on the end-mill cutter.

the end-mill cutters with a regular monolayered Ti–Al–N coating leads to tool life improvement that is comparable to the result achieved if an advanced nanolayered coating is applied (Figure 13.25b). Most probably, a combination of adaptive cutting-edge designs with advanced PVD multilayered coating could be a suitable solution for this specific application.

13.4 CONCLUSION

We can conclude that owing to the variety of cutting conditions and complexity of the cutting phenomena in general, the adaptability of the cutting edge behaves differently depending on the cutting conditions. For cutting processes driven by intensive plastic deformation of the cutting wedge, a design of the cutting edge with improved load bearing capacity could result in desirable tool life and wear behavior improvements. But, for other applications, the evolution of the cutting wedge during friction could be monitored and successfully mimicked, which would result in significant tool life improvement due to the adaptability enhancement of the tool–workpiece tribosystem.

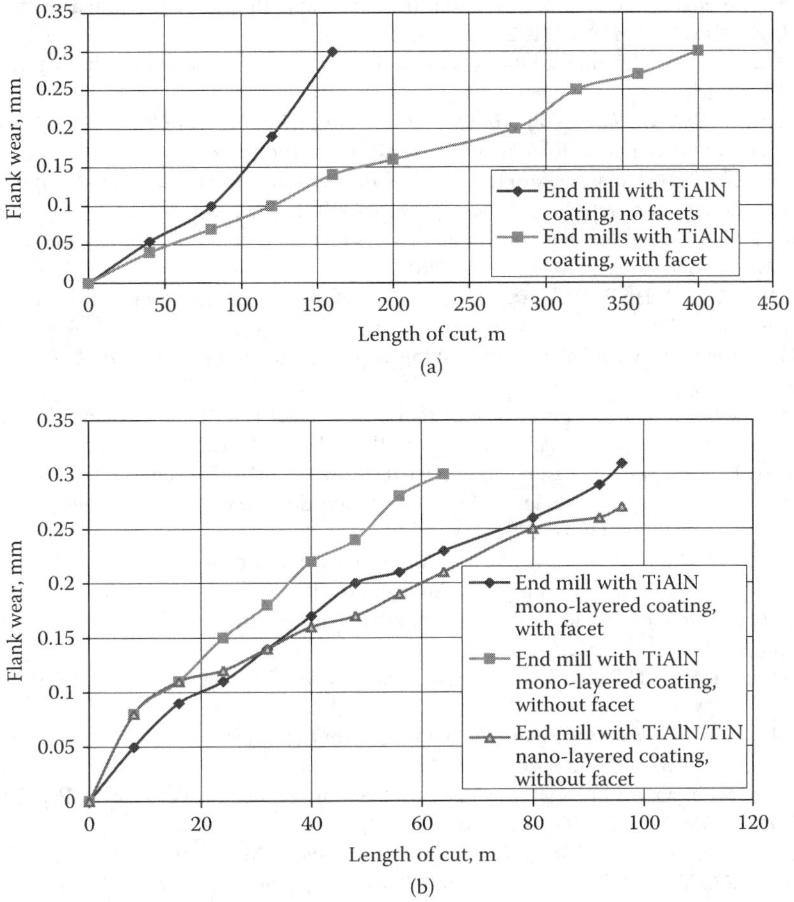

FIGURE 13.25 Flank wear of the end-mill cutter vs. length of cut for cemented carbide end mills (1-in.diameter, 4 flutes) with monolayered Ti–Al–N coating. Cutting data: n = 560 rev/min, depth of cut = 3 mm, width of cut = 10 mm; feed = 63 mm/min. (a) Machining of 1040 steel; (b) machining of SS 304 stainless steel.

REFERENCES

1. Rechberger, J. Geometry, substrate, coatings and applications: complex mechanisms. *CemeCon Tools* 2002, 17, 6–8.
2. Bouzakis, K.-D., Michailidis, N., Skordaris, G., Kombogiannis, S., Hadjiyiannis, S., Efstathiou, K., Pavlidou, E., Erkens, G., Rambadt, S., Wirth, I. Optimization of the cutting edge roundness and its manufacturing procedures of cemented carbide inserts, to improve their milling performance after a PVD coating deposition. *Surf. Coat. Technol.* 2003, 163–164, 625–630.
3. Schultz, V.V. *The Shape of Natural Wear of Machine Parts and Tools*. Mashinostroenie: Leningrad, 1990, p. 208.
4. Lee, E.H., Shaffer, B.W. The theory of plastic applied to a problem of machining, *Trans. ASME, J. Appl. Mech.* 1951, 18, 405–413.
5. Al Huda, M., Yamada, K., Hosokawa, A., Ueda, T. Investigation of temperature at tool-chip interface in turning using two-control pyrometer. *J. Manuf. Sci. Eng.* 2002, 124, 200–207.
6. Jaspers, S.P.F.C., Dautzenberg, J.H. Material behavior in conditions similar to metal cutting: flow stress in the primary shear zone. *Int. J. Mater. Process. Technol.* 2004, 146, 72.

7. Ernst, H. Merchant, M. E. Chip formation, friction, and high quality machined surfaces. *Surf. Treat. Met., Trans. Am. Soc. Met.* 1941, 29, 299–328.
8. Kudo, H. Some new slip-line solutions for two-dimensional steady-state machining. *Int. J. Mech. Sci.* 1965, 17, 43–55.
9. Oxley, P.L.B. *The Mechanics of Machining: An Analytical Approach to Assessing Machinability.* Halsted Press (a division of John Wiley & Sons): New York, 1989.
10. Maity, K., Das, N.S. A class of slip line field solutions for metal machining with slipping and sticking contact at the chip-tool interface. *Int. J. Mech. Sci.* 2001, 43, 2435–2452.
11. William, J.M., Endres, J. A new model and analysis of orthogonal machining with an edge-radiused tool. *J. Manuf. Sci. Eng.* 2000, 122, 384-390.
12. Zhang, H.T., Liu, P., Hu, R.S. A three zone model and solution of shear angle in orthogonal machining. *Wear* 1991, 143, 29–43.
13. Ren, H., Altintas, Y. Mechanics of machining with chamfered tools. *J. Manuf. Sci. Eng.* 2000, 122, 650–659.
14. Toporov, A., Sung-Lim Ko. Prediction of tool-chip contact length using a new slip-line solution for orthogonal cutting, *Int. J. Mach. Tool. Manuf.* 2003, 43, 1209–1215.
15. Kachanov, L.M. *Fundamentals of the Theory of Plasticity.* Mir: Moscow, 1974.
16. Johnson, W., Sowerby, R., Venter, R.D. *Plane-Strain Slip-Line Fields for Metal-Deformation Processes.* Pergamon Press: Oxford, 1982.
17. Backofen, W.A. *Deformation Processing*. Addison-Wesley: Reading, MA, 1972.
18. Prager, W. *An Introduction to Plasticity.* Addison-Wesley: Reading, MA, 1959.
19. Bishop, J.F.W. On the complete solution to problems of deformation of plastic rigid material. *J. Mech. Phys. Solids* 1953, 2, 43–53.
20. Johnson, W., Kudo, H. *The Mechanics of Metal Extrusion.* Manchester University Press: Manchester, 1962.
21. Fang, N. Machining with tool-chip contact on the tool secondary rake face — Part I: New slip-line model. *Int. J. Mech. Sci.* 2002, 44, 2355–2368.
22. Fang, N. Machining with tool-chip contact on the tool secondary rake face — Part II: New slip-line model. *Int. J. Mech. Sci.* 2002, 44, 2337–2354.
23. Tomlenov, A.D. *Theory of Plastic Deformations in Metals.* Metallurgy: Moscow, 1972.
24. Thomsen E.G., Yang, C.T., Kobayashi, Sh. *Plastic Deformation in Metal Processing*, Collier-MacMillen: London, 1965.
25. Pekelharing, J. The exit failure in interrupted cutting. *Ann. CIRP — Proc. Int. Prod. Eng. Res.* 1978, 27, 5–10.
26. Shaw, M.C. *Metal Cutting Principles.* Oxford University Press: Oxford, 2005.
27. Johnson, W., Mellor P.B. *Engineering Plasticity.* Ellis Horwood: Chichester, 1985.
28. Komaduri, R., Hou, Z.B. Thermal modeling of the metal cutting processes. Part I: Temperature rise distribution due to shear plane heat source. *Int. J. Mech. Sci.* 2000, 42, 1715–1752.
29. Backofen, W.A. *Deformation Processing*. Addison-Wesley: Reading, MA, 1972.
30. Al Huda, M., Yamada, K., Hosokawa, A., Ueda, T. Investigation of temperature at tool-chip interface in turning using two-control pyrometer. *J. Manuf. Sci. Eng.* 2002, 124, 200–207.
31. Childs, T.H.C., Maekawa, K., Maulik, P. Effects of coolant on temperature distribution in metal cutting. *J. Mater. Sci. Technol.* 1998, 4, 1006–1019.
32. Bowden, F.P. Recent studies of solid friction. *Friction and Wear Symposium*, Detroit, R. Davies, Elsevier: New York, 1959.
33. Gershman, J.S., Bushe, N.A. Thin films and self-organization during friction under the current collection conditions. *Surf. Coat. Technol.* 2004, 186, *3*, 405–411.
34. Kostetsky, B.I. An evolution of the materials' structure and phase composition and the mechanisms of the self-organizing phenomenon at external friction. *J. Friction Wear* 1993, 14, *1*, 773–783.
35. Bershadskiy, L.I. On self-organizing and concept of tribosystem self-organizing. *J. Friction Wear* 1992, 13, *6*, 1077–1094.
36. Kostetsky, B.I. Structural-energetic adaptation of materials at friction. *J. Friction Wear* 1985, 6, *2*, 201–212.
37. Astakov, V.P. Tribology of metal cutting. In *Mechanical Tribology. Materials, Characterization, and Applications*, Eds., Liang, H., Totten, G. Marcel Dekker: New York, p. 307, chap. 9.

38. Trent, E.M., Wright, P.K. *Metal Cutting,* 4th ed. Butterworth-Heinemann: Woburn, MA, 2000.
39. Fox-Rabinovich, G.S., Weatherly, G.C., Dodonov, A.I., Kovalev, A.I., Veldhuis, S.C., Shuster, L.S., Dosbaeva, G.K., Wainstein, D.L., Migranov, M.S. Nano-crystalline FAD (filtered arc deposited) TiAlN PVD coatings for high-speed machining application. *Surf. Coat. Technol.* 2004, 177–178, *30*, 800–811.
40. Amini, E.J. A complete analysis for prediction of stresses at the chip-tool interface. *J. Strain Anal.* 1968, 3, 206–213.
41. Astakhov, V.P. *Metal Cutting Mechanics*. CRC Press: Boca Raton, FL, 1998.
42. Green, A.P. A theory of plastic yielding due to the bending of cantilever and fixed-ended beams: Part 1. *J. Mech. Phys. Solids* 1954, 3, 1.
43. Green, A. P. A theory of plastic yielding due to the bending of cantilever and fixed-ended beams: Part 2. *J. Mech. Phys. Solids* 1954, 3, 143.

Index

A

Adaptability
 adaptation process, 9,282, 377, 382, 411
 adaptive coatings, 5, 231, 266, 267, 268, 290, 297, 298, 301, 315, 326, 327, 405, 406, 407
 adaptive materials, 10, 165, 171, 176, 179, 194, 316
 geometrical adaptation, 9, 411, 412
Adhesion, 56, 78, 127, 131, 151, 152, 155, 163, 171, 178, 179, 232, 233, 235, 236, 237, 242, 248, 249, 250, 251, 270, 271, 284, 293, 299, 307, 388, 389, 390, 393, 394, 402,
Adhesive wear, 129, 134, 136, 137, 138, 151, 152, 178, 235, 247, 400
Aerospace alloys, 327
Alumina, 176, 259, 266, 267, 273, 275, 281, 282, 289, 295, 296, 314, 315, 320, 322, 324, 336, 337, 338, 340, 341, 342, 343, 344, 345, 346, 347, 348, 349, 350, 351, 353, 356, 358, 360, 364, 366, 368, 369, 370, 378, 379, 380,
Alumina film, layer, 275, 281, 282, 347, 349, 350, 358, 360, 364, 366, 367
Aluminum oxides, 176, 187, 279, 280, 314, 406
Amorphization, 118, 159, 188, 192, 253, 280, 342, 360
Amorphous films, 187, 378
Amorphous-like structures, 192, 254, 261
Anti-frictional layer, 245, 246, 247
Anti-frictional properties, 65, 66, 67, 72, 247, 267
Auger Electron Spectroscopy (AES), 81, 83, 88, 89, 118, 226, 307, 342, 400

B

Bearings, 60, 62, 63, 64, 65, 66, 69, 74, 75, 76, 77, 78
Belousov, 15, 17, 35
Belousov–Zhabotinski reaction, 15, 17, 35
Benard and Taylor effect, 15, 16, 17, 29, 34, 36, 48, 77
Benard's cells, 15, 16, 17, 29, 34, 36, 48, 77
Bershadsky L. I., 4, 14, 38, 56, 77
Bifurcation, 34, 35
Brake systems, 60, 69
Build-up layer, 142, 144, 153, 163, 168, 179, 182, 255, 281, 443, 449

C

Carno–Clausius inequality, 30, 39, 56
Catalyst, 46, 54, 197, 209, 216, 222, 223, 224, 225, 226
Catastrophic wear, 237, 251, 252, 370
Cemented carbide, 5, 131, 138, 154, 163, 164, 165, 167, 168, 169, 171,172,173,174, 175,176,177, 178, 179, 193, 194,195, 232, 258, 270, 293, 301, 331, 332, 442, 445, 448, 449, 450, 451
Ceramics, 169, 170, 176, 177, 195
Cermet, 169, 170, 177, 178, 193, 194, 195
Chaotic condition, state, 15, 20, 34, 35,
Chipping, 142, 171, 173, 174, 175, 436
Chips characteristics, 260
Chromium effect, 336, 349
Chromium oxides, 346
Closed system, 18, 19, 20, 22
Coatings
 AlCrN, 139, 140, 141, 143, 147
 CVD coatings, 231, 232, 258, 291, 294
 diamond like (DLC), 405, 406, 409
 duplex, 231, 238, 240, 242, 243, 244, 245, 246
 equilibrium, 261, 271, 275, 279
 filtered (FAD), 236, 237, 269
 multi-layered, 298, 329
 nano-composite, 265, 293, 294, 295
 nano-crystalline, 269, 274, 276, 277, 278, 279
 nano-laminated, 366
 non-equilibrium, 252, 253, 265, 268, 269, 359, 360, 362, 364, 365, 366, 368, 373, 374
 oxidation-resistant, 335, 367
 quaternary, 297, 335, 351
 self-adaptive, 238, 266, 298, 299, 301, 369, 371
 self-lubricating, 231, 244, 245, 295
 smart, 245, 246, 257
 super-hard, 261, 263
 synergistically alloyed, 306, 307, 315, 323, 358, 364, 365, 367, 368, 377
 TiN arc PVD 121, 125, 129, 130, 131, 132, 133, 231, 233, 237
 TiAlN, 132, 135, 245, 258, 259, 261, 265, 411
 TiN filtered arc PVD (FAD), 134,
 wear-resistant, 10, 142, 245, 258
Coefficient of friction, 6, 8, 15, 36, 41, 45, 47, 48, 49, 50, 51, 52, 53, 60, 62, 66, 67, 68, 69, 129, 134, 179, 198, 199, 235, 236, 298, 369, 386
Cubic boron nitride (CBN), 176
Current collection, 197
Cutting, 3, 5, 7, 9, 121, 126, 151, 152, 153, 163, 266, 298, 367, 386, 392, 411
Cutting/stamping tools, 3, 5, 6, 7, 11, 87, 121, 122,123,124, 151, 152, 163, 167, 266, 386, 411
CVD coating, 231, 232, 258, 291, 294

D

Defects, lattice imperfections, 7, 116, 118, 234, 254, 257, 259
Diamond, 176
Diffusion, mass transfer, 25, 32, 35, 36, 40, 231, 232, 240, 254, 350
Dissipative heterogeneity, 4, 14
Dissipative structures, 3, 6, 15, 16, 17, 30, 33, 34, 36, 37, 41, 45, 46, 198, 200, 376
DLC coatings, 405, 406, 409
Dry friction, 69, 72, 74, 79
Dry machining, 298, 301, 306, 309, 310, 311, 312, 313, 314, 378, 381
Duhem F., 31
Duplex coatings, 231, 238, 240, 242, 243, 244, 245, 246

E

Electric contacts, 198, 200, 206, 222, 223
Electric current, 197, 198, 199, 200, 201, 202, 203, 210, 223
End mill cutting, 135, 178, 194, 196, 297, 298, 301, 369, 449
Energy dissipation, 5, 7, 11, 14, 27, 39, 40, 41, 47, 56, 129
Energy loss fine structure spectroscopy (EELFS), 83, 92, 119, 154, 159
Entropy
  entropy production, 14, 16, 22, 24, 25, 26, 27, 28, 29, 33, 35, 36, 38, 41, 42, 43, 44, 45
  entropy production negative, 16
  entropy reduction, 37, 41
Equilibrium coatings, 261, 271, 275, 279
Equilibrium far from, 15, 16, 32

F

FAD pulsed technique, 263, 268, 269, 270, 280
Fatigue, 60, 64, 65, 66, 67, 68
Fatigue wear, 134, 135, 138, 160
Filtered coatings (FAD), 236, 237, 269
Friction
  friction coefficient, 6, 8, 15, 36, 41, 45, 47, 48, 49, 50, 51, 52, 53, 60, 62, 66, 67, 68, 69, 129, 134, 179, 198, 199, 235, 236, 298, 369, 386
  friction tests, 128, 134
  passivation processes at, 6, 8
Functionally-graded tool materials, 169, 176, 194

G

Geometrical adaptability, 411, 448,
Gibbs J., 31
Graphite-based material, 207, 210, 212

H

H/E ratio, 129, 130, 140, 141
Heat conductivity, 35, 47, 48, 49
Heavily Loaded Tribo-Systems (HLTS), 3, 4, 5, 6
Hexagonal boron nitride, 190
High performance machining, 10, 231, 238, 258, 259, 289, 301
High resolution electron energy losses spectroscopy (HREELS), 97, 100, 101, 102, 260, 400
High-speed machining, 10, 231, 238, 258, 259, 289, 301
Hot hardness, 127, 232, 233, 258
HREELS, 97, 100, 101, 102, 260, 400
HSS high-speed tool steels, 167, 168, 178, 179, 180

I

Intermetallics, 336, 337, 338, 340, 343, 344, 345, 347, 348
Ion mixing, 245, 246, 249, 252, 253, 254, 255
Ion modification (ion implantation), 247, 248, 249, 250, 253, 254, 290
Ion nitriding, 239, 240, 244, 246, 254, 256
Irreversible thermodynamics, 3, 4, 13, 14, 15, 38, 54

K

Klamecki, B. E., 4, 14
Klimantovich, Y.L., 36,
Kostetsky, B. 4, 38, 39

L

Lapunov A.M., 29, 30, 33, 286
Lattice imperfections, 7, 116, 118, 234, 254, 257, 259
Le Chatelier, H., 28, 29, 32, 33
Le Chatelier-Brown principle, 51, 52
Linear processes, 25, 35, 36
Linear thermodynamic area, 25, 35, 49
Liquid lubrication, 60, 65, 66, 72, 193,
Lubricants
  cutting fluids, 382, 383, 384, 386
  cutting oils, 383
  gaseous coolants, 383
  plastic lubricants, 384
  synthetic cutting fluids, 384
  water-based cutting fluids, 383
Lubrication
  lubricating current effect, 199, 200
  lubricious films, 7, 153
  lubricious layers, 171
  lubricious oxides, 168, 267, 405
  lubricious polymer-based films, 382
  lubricious tribo-films, 7, 153, 258, 266
  minimum quantity lubrication, 392
  mixed lubrication, 62, 66
  perfluorpolyether (PFPE) lubricating films, 393, 394, 395
Lapunov's function (theorem), 29, 30, 33, 286

M

Microhardness Dissipation Parameter (MDP), 133, 134, 135
Multi-layered adaptive/smart coating, 245, 246, 257, 298, 329

N

Nano-crystalline structure, 269, 274, 276, 277, 278, 279
Nanoindentation, 129, 130, 132, 138, 139, 140, 142
Nanoindentation testing, 129, 130, 132, 138, 139, 140, 142
Nano-laminates, 366
Nanostructured coatings
  nano-composite, 265, 293, 294, 295
  nano-crystalline, 269, 274, 276, 277, 278, 279
  nano-laminated, 366
  nano-multilayered coatings, 315, 368
Nano-tribology, 4
Nature-mimetics, 10
Non-equilibrium state, condition
  non-equilibrium coatings, 252, 253, 265, 268, 269, 359, 360, 362, 364, 365, 366, 368, 373, 374
  non-equilibrium phase transformations, 3, 34, 42, 66
  non-equilibrium thermodynamics, 55
  non-equilibrium unsteady state, 79
Non-linear equations, 34
Non-linear processes, 77

O

Open system, 25, 34, 36, 38, 56
Open thermodynamic system, 13, 38,
Ordered state, 34, 35,
Orthogonal cutting, 412, 420, 427
Oxidation resistance, 176, 247, 280, 336, 342, 343, 348, 349, 350, 353, 355, 356, 358
Oxidation
  oxide films, 247, 258, 268, 279, 281, 283, 307, 308, 309, 315, 322, 338
  oxide layer, 60, 171, 275, 315, 320, 348, 350, 353
  oxide tribo-film, 285, 307, 320, 376
  oxides, 56, 69, 71, 201, 215, 232, 242, 249, 251, 298, 299, 336, 343

P

Palmquist toughness, 235, 236, 237, 240
Perfluorpolyether (PFPE), 393, 394, 395, 396, 397, 399, 400
Perfluorpolyether (PFPE) lubricating films, 393, 398
Phase transformation, 34, 35, 42, 44
Phase transition, 7, 66
Post running-in stage of wear, 64, 259
Prigogine I., 3, 6, 11, 14, 26, 34, 79, 285, 368
Prigogine theorem, 16, 26, 27, 33
PVD coatings, 121, 125, 129, 130, 131, 132, 133, 231, 232, 233, 237
PVD nitride coatings, 121, 125, 129, 130, 131, 132, 133, 231, 232, 233, 237

Q

Quaternary coatings, 368, 369

S

Scratch tests, 139, 140, 141, 142, 248
Secondary Ions Mass Spectroscopy (SIMS), 107, 108, 109, 110, 179, 240
Secondary structures, 4, 6, 7, 40, 41, 74
Seizure, 62, 63, 66, 235, 245, 247, 259, 275, 309, 314, 367
Self-organization
  self-adaptive coatings, 238, 266, 298, 299, 301, 369, 371
  self-adaptive hard coatings, 238, 266, 298, 299, 301, 369, 371
  self-healing, 4, 10
  self-lubricating coatings, 244
  self–organization, 3, 4, 5, 6,7,8, 10, 11, 13, 14,15,16,17,33, 34, 35, 36, 40, 41, 43, 44, 45, 46, 47, 48, 49, 50, 51, 53, 54, 59, 62, 74, 75, 77, 78, 153, 157, 163, 169, 170, 179, 182, 197, 198, 209, 233, 237, 266, 267, 284, 334, 359, 362, 364, 365, 366, 368, 369, 370, 403, 412,
  self-organization phenomena, 3, 4, 5, 6,7,8, 10, 11, 13, 14,15,16,17,33, 34, 35, 36, 40, 41, 43, 44, 45, 46, 47, 48, 49, 50, 51, 53, 54, 59, 62, 74, 75, 77, 78, 153, 157, 163, 169, 170, 179, 182, 197, 198, 209, 233, 237, 266, 267, 284, 334, 359, 362, 364, 365, 366, 368, 369, 370, 403, 412
  self-protection, 4, 10,
SIMS, 107, 108, 109, 110, 179, 240
Sliding electric contacts, 197, 198, 222
Solid lubricants, 385, 408
Stamping, 3, 151, 152, 153, 155, 168, 170
State
  equilibrium, 14, 15, 16, 17, 19, 20, 30
  non-equilibrium, 14, 15, 19, 20
Structural adaptation, 10
Super-hard coatings, 261, 263
Surface engineering
  CVD, 231, 232, 258, 291, 294
  FAD pulsed technique, 263, 268, 269, 270, 280
  ion mixing, 245, 246, 249, 252, 253, 254, 255
  ion modification , 247, 248, 249, 250, 253, 254, 290
  ion implantation), 247, 248, 249, 250, 253, 254, 290
  ion nitriding, 239, 240, 244, 246, 254, 256
  PVD, 121, 125, 129, 130, 131, 132, 133, 231, 232, 233, 237
Surface finish, 232
Surface-active materials (SAM), 393
Surface structures
  dissipative structures, 3, 6, 15, 16, 17, 30, 33, 34, 36, 37, 41, 45, 46, 198, 200, 376
  lubricating films, 7, 153
  lubricious tribo-films, 7, 153, 258, 266

secondary structures, 4, 6, 7, 40, 41, 74
steady structures, 14, 16, 25, 26, 33, 40
tribo-ceramics, 7, 267
tribo-films, 7, 153, 258, 266, 285, 307, 320, 376
tribo-oxides, 285, 307, 320, 376
Synergistic alloying, 306, 307, 315, 323, 358, 364, 365, 367, 368, 377

T

Ternary alloys, 336, 337, 342
Ternary coatings, 315, 316
Thermal barrier properties, 7, 238, 261
Thermo-diffusion, 25, 35
Thermodynamic
instability, 35, 36
stability, 35, 37
Thermodynamics
equilibrium, 14, 28
irreversible processes, 3, 4, 13, 14, 15, 38, 54
non-equilibrium, 14, 15, 38
second law of, 14, 20, 24
Threshold of external friction, 127, 128
Tool/workpiece interactions
Tribo-ceramic films, 7, 267
Tribo-films, 7, 153, 258, 266, 285, 307, 320, 376
Tribological compatibility, 4, 5, 6, 8, 60, 65, 72, 73, 127, 129
Tribology, 4, 5, 9, 29, 77, 127, 165, 181, 291, 333, 380, 394
Tribo-oxidation, 285, 307, 320, 376
Tribo-oxides, 285, 307, 320, 376
Tribo-system, 3, 4, 5, 6, 7, 8, 9, 10, 11, 38, 47, 59, 60, 121, 126, 158, 160, 195, 200, 246, 252, 327, 369, 376, 402, 411
Trigger effect, 285, 289, 367
Tungsten oxides, 316, 322, 377, 407

W

Waspalloy, 331, 332
Wear stage
catastrophic wear, 238, 251
post running-in period, 64, 259, 261
running-in stage, 6, 7, 8, 14, 42, 43, 56, 60, 127, 129, 155, 157, 164, 177, 179, 218, 233, 237, 244, 257, 297, 298, 368, 371, 402, 411, 412, 443, 448
steady (stationary state), 25, 49
Wear rate, 14, 15, 16, 26, 27, 37

X

X-ray photoelectron spectroscopy (XPS), 83, 84,85, 252, 308

Z

Zhabotinski A.M., 15, 17, 35

OTHER RELATED TITLES OF INTEREST INCLUDE:

Handbook of Lubrication and Tribology: Volume 1 Application and Maintenance, Second Edition
George Totten, Portland State University
ISBN 084932095X

Handbook of Micro/Nano Tribology, Second Edition
Bharat Bhushan, Ohio State University
ISBN 0849384028

Fundamentals of Machining and Machine Tools, Third Edition
Geoffrey Boothroyd, Boothroyd Dewhurst, Inc.
Winston A. Knight, University of Rhode Island
ISBN 1574446592

MicroMechatronics
Kenji Uchino, Penn State University
Jayne Giniewicz, Indiana University of Pennsylvania
ISBN 0824741099

Surface Engineering of Metals: Principles, Equipment, Technologies
Tadeusz Burakowski, Institute of Precision Mechanics
Tadeusz Wierzchon, Warsaw University of Technology
ISBN 0849382254